Springer Series in

MATERIALS SCIENCE 147

Springer Series in

MATERIALS SCIENCE

Editors: R. Hull C. Jagadish R.M. Osgood, Jr. J. Parisi Z. Wang

The Springer Series in Materials Science covers the complete spectrum of materials physics, including fundamental principles, physical properties, materials theory and design. Recognizing the increasing importance of materials science in future device technologies, the book titles in this series reflect the state-of-the-art in understanding and controlling the structure and properties of all important classes of materials.

Please view available titles in *Springer Series in Materials Science* on series homepage http://www.springer.com/series/856

D.B. Sirdeshmukh
L. Sirdeshmukh
K.G. Subhadra

Atomistic Properties of Solids

With 506 Figures

Springer

Professor Dinker B. Sirdeshmukh
Lalitha Sirdeshmukh
Kakatiya University
Santoshnagar Colony 23A, 500059 Hyderabad, India
E-mail: dsirdeshmukh@yahoo.com, lsirdeshmukh@yahoo.com

K.G. Subhadra
Kakatiya University, Manasarovar Heights- Phase II
832/9, 500009 Secunderabad, India
E-mail: kgsubhadra@yahoo.com

Series Editors:

Professor Robert Hull
University of Virginia
Dept. of Materials Science and Engineering
Thornton Hall
Charlottesville, VA 22903-2442, USA

Professor Jürgen Parisi
Universität Oldenburg, Fachbereich Physik
Abt. Energie- und Halbleiterforschung
Carl-von-Ossietzky-Straße 9–11
26129 Oldenburg, Germany

Professor Chennupati Jagadish
Australian National University
Research School of Physics and Engineering
J4-22, Carver Building
Canberra ACT 0200, Australia

Dr. Zhiming Wang
University of Arkansas
Department of Physics
835 W. Dicknson St.
Fayetteville, AR 72701, USA

Professor R. M. Osgood, Jr.
Microelectronics Science Laboratory
Department of Electrical Engineering
Columbia University
Seeley W. Mudd Building
New York, NY 10027, USA

Springer Series in Materials Science ISSN 0933-033X
ISBN 978-3-642-19970-7 e-ISBN 978-3-642-19971-4
DOI 10.1007/978-3-642-19971-4
Springer Heidelberg Dordrecht London New York

Library of Congress Control Number: 2011935154

Cover design: eStudio Calamar Steinen

Printed on acid-free paper

Springer is part of Springer Science+Business Media (www.springer.com)

Preface

The authors have been engaged in offering courses on solid state physics for more than three decades at different levels. These courses were based on material culled from a large number of sources. There are several textbooks on solid state physics, each with its strengths and weaknesses. It is observed that these books differ in the relative importance given to the experiment and theory; the experiment generally gets a raw deal. Further, the books also differ in the degree of detail in the treatment of different topics. Also, some new topics deserve inclusion in textbooks.

This book attempts to bring together comprehensive information on essential aspects of solid state physics. The present day expanse of solid state physics is vast. It was therefore decided to limit this book to the atomistic properties, i.e., properties that depend upon factors like the crystal structure, interatomic forces and atomic displacements caused by stimuli like temperature, stress or electric fields. Other equally important topics like electrical, electronic and magnetic properties that could not be included in this book may form another volume.

Comprehensiveness is the main feature of this book. Each chapter is, in a way, an exhaustive essay on the topic. Another feature is that this book gives equal importance to experiment and theory. Experimental methods are described in detail and experimental data are quoted, so that readers can get a feel of magnitude and accuracy. In discussing theories, a logical approach is adopted to lead the reader from the simpler to the more elaborate theories. This book lays emphasis on the role of symmetry in crystal properties. Neumann's principle is discussed with a large number of examples. The effect of symmetry on 2nd, 3rd and 4th rank tensor properties is worked out.

Crystal growth, the tensor nature of physical properties, mechanical properties of crystals and crystal structure determination are topics of basic importance. In this book, full chapters are dedicated to these topics; in fact, it opens with a chapter on crystal growth. Similarly, some topics that are not usually discussed in solid state textbooks have been included. Some of them are: zone refinement, higher order elasticity, local modes in lattice vibrations, the Rietveld method of structure determination, photoelasticity, non-linear optical properties and methods of observation of dislocations. Chapter 16 introduces readers to the areas which have gained importance in recent times like nanocrystals, polycrystals, thin films, liquid crystals and quasicrystals.

Textbooks are generally addressed to readers of a particular level. The authors of this book, however, did not keep any specific level in mind. Instead, this book is addressed to a multilevel readership. Thus, it may cater to the requirements of courses like B.Sc. (Hon.), M.Sc. (or M.S.), the 5-year integrated M.Sc. courses offered by some Indian universities and, to some extent, even introductory research courses. By perusing the detailed contents, a particular reader may decide which part is the most relevant to his needs.

We are thankful to Academic Press (Elsevier), Akadmiai Kiado, American Physical Society, Bell and Co. (University of Reading Archives), Benjamin-Cummins, Cambridge University Press, Dover Publications, Elsevier Publications, General Electric Research Laboratory, Indian Academy of Sciences, Institute of Physics, International Union of Crystallography, Methuen Publishing (Routledge), National Institute of Science Communication and Information Resources, Nature Publishing Group, New Age International, North-Holland (Elsevier), Oxford University Press, Pergamon Press (Elsevier), Prentice Hall, Prentice Hall-India, Royal Society, Saunders College, South Asian Publishers, Springer, Vakils-Feffer-Simmons, VCH (Wiley-VCH), W.B. Saunders, Wiley (Wiley-VCH), World Scientific and several individual authors for permission to use figures/data from books, journals and research papers published by them.

Prof. K.G. Bansigir and Prof. K.G. Prasad are acknowledged for interest and encouragement. Thanks are due to Prof. M.J. Joshi and Prof. K. Kishan Rao for sharing some of their results on crystal growth. We are also grateful to Prof. C.S. Sunandana (University of Hyderabad), Prof. B.J. Rao (Tata Institute of Fundamental Research, Bombay), Dr. K. Ravikumar (Indian Institute of Chemical Technology, Hyderabad), Dr. Deepak Sirdeshmukh (North Carolina State University) and Dr. Harish Dixit (Jawaharlal Nehru Centre for advanced Research, Bangalore) for providing rare books and papers from their libraries.

Finally, we offer our grateful thanks to Dr. Claus E. Ascheron, Executive Editor, Springer, for encouragement and advice throughout the processing of this book. Colleagues at the Springer Editorial office are acknowledged for their cooperation.

Hyderabad, India
January 2011

Dinker B. Sirdeshmukh
Lalitha Sirdeshmukh
K.G. Subhadra

Contents

Chapter 1
Introduction

1.1 General

Crystallography and crystal properties were mostly studied by mineralogists in the later part of the nineteenth century and the earlier part of the twentieth century. Solid state physics developed its own identity in the 1930s and 1940s. Today, solid state physics is an almost independent branch of science. Among solids, there are two types. In the first type, the internal arrangement of atoms has regularity and symmetry. These are called crystalline solids or, simply, crystals. In the other type, there is no regularity in the internal arrangement of atoms. These are called non-crystalline or amorphous solids. We are interested in the physical properties of crystalline solids. This book is about the atomistic properties of solids, i.e. properties determined by factors like atomic positions (crystal structure), interatomic forces and atomic displacements. Properties of solids depending on other factors like the excited electron states, band structure and magnetism are not included in this volume. This chapter provides a brief introduction to the main aspects of the book. This preview will help readers (particularly those without prior exposure to solid state physics) in understanding and appreciating the detailed treatment that follows.

1.2 Growth of Crystals

There are three main processes by which crystals can be grown. Firstly, crystals are formed by the deposition of the vapour of a substance on a cooled solid surface; this is vapour growth. Secondly, crystals are formed when the solution of a substance in a liquid becomes supersaturated. The reduction of the solvent by evaporation results in the precipitation of the substance in the solid form; this is solution growth. The third method is to melt a solid and allow the molten liquid to cool through its melting point in a controlled fashion; this is melt growth. The common feature in all these methods is that crystal growth takes place by the deposition of the constituent atoms

D.B. Sirdeshmukh et al., *Atomistic Properties of Solids*, Springer Series in Materials Science 147, DOI 10.1007/978-3-642-19971-4_1,

Fig. 1.1 Some crystals grown from solution

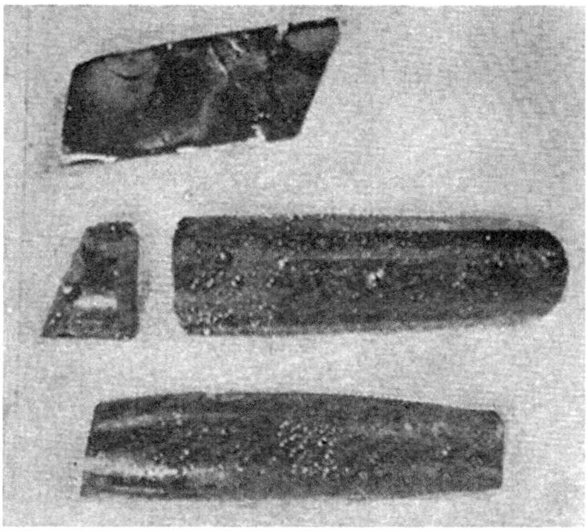

Fig. 1.2 Tellurium crystals grown from melt

around a nucleus. All these methods are used for growing crystals in a laboratory. Some laboratory-grown crystals are shown in Figs. 1.1 and 1.2.

Impurities have considerable influence on the magnitude of a physical property. Hence, it is desirable to have highly pure crystals. A method called zone refinement helps to purify a solid. The method is based on the principle that certain impurities move from one end of a solid ingot to the other end by melting the ingot section by section from one end to the other. This section-wise melting is carried out by the use of ring heaters of narrow width.

It is interesting to note that crystals are being formed in nature all the while by these processes. Ice crystals in the atmosphere are an example of vapour growth in nature. Similarly, the formation of rock salt on sea shores is an example of solution growth in nature. The formation of kidney stones in the human body is another example of solution growth in nature. Deep inside the earth, substances melt and when the molten liquid reaches cooler parts, mineral crystals are formed.

Laboratory-grown crystals are generally of cm-dimensions. Crystals of very small size 10^{-4} cm in diameter (whiskers) and massive crystals a few kg in weight have also been grown; the latter are crystals of industrial applications like Rochelle salt and potassium dihydrogen phosphate.

1.3 Symmetry in Crystals

In a crystal, the atoms are located on an infinite array of points such that the environment of each atom is identical to that of any other equivalent atom. The atoms are arranged in regular rows in three directions; these directions are the crystal axes. The intersection of these rows divides the crystal space into parallelepipeds called unit cells. A unit cell is described by six parameters – three sides a, b, c and three angles α, β and γ. Seven unit cells are permitted; these are the seven crystal systems: cubic, tetragonal, orthorhombic, trigonal, hexagonal, monoclinic and triclinic. The triclinic system with $a \neq b \neq c$ and $\alpha \neq \beta \neq \gamma$ is the least symmetric, while the cubic system with $a = b = c$ and $\alpha = \beta = \gamma = 90°$ is the most symmetric.

Further, the atomic arrangements satisfy certain symmetry conditions. The symmetry is described in terms of symmetry elements. A symmetry element is said to exist if the relevant symmetry operation leaves the crystal invariant (indistinguishable in appearance). The symmetry elements are: planes of symmetry, rotation axes, centre of inversion and rotation–inversion axes. Only 32 combinations of these symmetry elements are permitted; these are the 32 point groups or crystal classes.

The above-mentioned symmetry operations are called non-translational operations as, in each of them, at least one point remains unaffected. In addition, there are translational symmetry operations. These are glide planes and screw axes. In a glide plane, the lattice is reflected in a plane and then translated along an axis. Similarly, in a screw axis, the lattice is rotated about an axis and then translated along the same axis. There are 230 combinations of all these symmetry elements (non-translational and translational). Each combination is called a space group. The crystal system, the point group and space group of some crystals are given in Table 1.1. The symmetry of internal atomic arrangement manifests in the symmetry of the external form. Thus, sodium chloride crystals appear as cubes, sodium bromate crystals as triangular pyramids and quartz as hexagonal prisms with six-sided pyramids at the ends.

Table 1.1 Symmetry of some crystals

Crystal	Crystal system	Point group	Space group
NaCl	Cubic	$m3m$	$F\frac{4}{m}\bar{3}\frac{2}{m}$
Zinc blende	Cubic	$\bar{4}3m$	$F\,43m$
Te	Trigonal	32	$P3_121$
$CaWO_4$	Tetragonal	$4/m$	$I4_1/a$
Rutile (TiO_2)	Tetragonal	$4/mmm$	$P4_2/mmm$
Graphite	Hexagonal	$6/mmm$	$P\frac{6_3}{m}\frac{2}{m}\frac{2}{c}$
ZnO	Hexagonal	$6mm$	$P6_3mc$

1.4 Structure of Crystals

Some information about the structure of crystals could be obtained from a study of their external forms. In this way, mineralogists and crystal growers were able to detect the presence of pure symmetry elements like rotation axes and planes of symmetry. Physical properties like optical activity and piezoelectricity indicated the absence of centre of symmetry. With these tools, one could identify the crystal system and crystal class. Further, studying the form and interfacial angles of a number of specimens of a given crystal, it was even possible to estimate the ratios of the unit cell parameters ($a : b : c$).

However, the real clue to the inner structure of crystals was provided only after the discovery of X-ray diffraction by crystals by Laue in 1912. The crystal lattice acts like a grating and diffracts X-rays. The pattern of these diffraction maxima, generally recorded on a photographic film as a set of lines or spots, is related to the structure of the crystal. Thus, the symmetry of the so-called Laue photograph reveals the presence of rotation axes and symmetry planes whereas a rotating crystal photograph provides a way to estimate the unit cell parameters.

Vital information about the structure is contained in the intensities of the diffraction maxima. It is interesting to note that even absent reflections provide important information. The centring of faces (Bravais lattice) and the presence of glide planes and screw axes prohibit the occurrence of families of reflections. This phenomenon is systematized, i.e. the Miller indices of absent reflections and the related symmetry element have been tabulated. Thus by noting the absent reflections, the symmetry elements can be inferred and hence the space group can be identified.

The intensities of the recorded reflections are related to the atomic positions. Elaborate analytical methods are available to deduce the atomic coordinates from the experimentally measured intensities. Two structures, one simple and the other complex, determined by X-ray diffraction are shown in Fig. 1.3.

1.5 Interatomic Forces

Why can't atoms in a crystal move about freely and what holds them together? This is a fundamental question. Two well-separated atoms do not exert any force on each other and act independently. However when they come close, say within a few Angstroms, they, sort of, see each other's electrons and exert forces on each other. These forces depend on the charge distribution of the atoms. Thus, we have the following types of interactions or bonds: ionic bond, covalent bond, metallic bond, van der Waals bond and hydrogen bond.

Let us consider the ionic bond. This comes into play between two neighbouring atoms having incomplete outermost electronic shells. For example, in NaCl the Na atom has a single s electron and the Cl atom has five p electrons. The transfer of

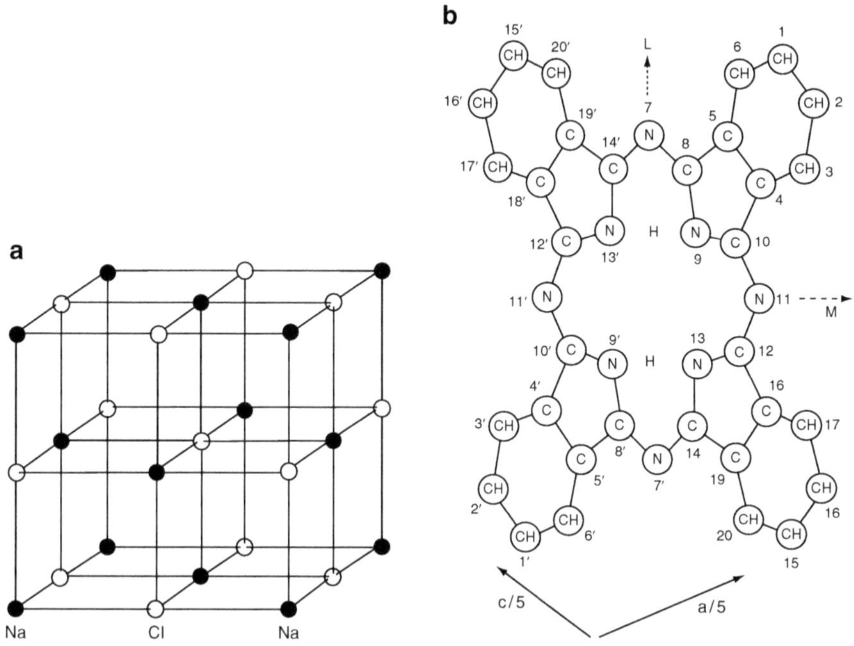

Fig. 1.3 Crystal structure of (**a**) sodium chloride and (**b**) pthalocyanine

the *s* electron from Na to Cl makes the outer shell of both atoms complete, thus increasing the stability of both. In this process, the atoms become Na^+ and Cl^- ions and exert Coulomb force on each other. On the other hand, if two atoms are both short of electrons, they may share each others' electrons such that each has a closed shell structure. This sharing results in an attractive interaction which has a quantum mechanical origin; this is the covalent bond. The bond between the carbon atoms in diamond is a classic example.

In the metallic bond, which occurs in metals, the valence electrons are loosely attached to the atomic cores. They get easily detached; these free electrons form an electron gas and are free to move about. The atomic cores are embedded in this electron gas. In some solids, like solid inert gases, the atoms are neutral with closed outer electron shells. There is no scope either for electron transfer or for electron sharing or for the electrons to get detached. In such atoms, although the total charge of an atom (nuclear charge + electron charge) is zero, at any given instant, there are dipoles and quadrupoles. These instantaneous entities interact and exert a force of attraction. This is called the van der Waals bond. The van der Waals forces are weak. Also they are universal, i.e. they occur in all solids though they may be masked by more dominant forces. Finally, in hydrogen-containing solids, the hydrogen atom provides attraction between neighbouring electronegative atoms; this is the hydrogen bond, e.g. O–H$\cdots$O, O–H$\cdots$S.

When the atoms are so close that their electron clouds overlap, a repulsive force comes into play; this again is a universal quantum mechanical interaction. The atoms in a crystal assume equilibrium positions such that the repulsive force equals the attractive force due to the interactions discussed above. The atoms do work against these forces in moving from infinity to their respective positions; this is the cohesive energy of the solid.

1.6 Tensor Nature of Crystal Properties

Crystal properties are, in general, anisotropic, i.e. direction-dependent. Hence, to describe a property fully, a single coefficient is not enough; the number of coefficients depends upon the property and also upon the symmetry of the crystal. The magnitude of a property in any particular direction in the crystal can be calculated from the several coefficients.

Another way of describing the anisotropic nature of a physical property is to express the property as a tensor. A tensor is defined by the way the coefficients are transformed when the coordinate axes are rotated. For example, a second-rank tensor is defined by the transformation law

$$P'_{ij} = \alpha_{ik}\alpha_{jl}P_{kl}, \tag{1.1}$$

where P'_{ij} and P_{kl} are the coefficients after and before the rotation of the axes; α_{ik}, α_{jl} are the cosines of the angles between the original axes and the axes after rotation. The laws of transformation are more elaborate for tensors of higher ranks. Some examples are given in Table 1.2.

1.7 Atomistic Nature of Crystal Properties

Several properties of crystals are atomistic in nature. Atomic positions, atomic displacements and interatomic forces play an important role in these properties. Mechanical, thermal and dielectric properties are examples of atomistic properties. We shall discuss some of these properties.

Table 1.2 Some crystal property tensors

Rank of tensor	Transformation law	Physical property	Maximum number of coefficients
0 (scalar)	Invariant	Density, specific heat	1
1 (vector)	$P'_i = \alpha_{ij}P_j$	Pyroelectric coefficient	3
2	$P'_{ij} = \alpha_{ik}\alpha_{jl}P_{kl}$	Dielectric constant, thermal expansion	6
3	$P'_{ijk} = \alpha_{il}\alpha_{jm}\alpha_{kn}P_{lmn}$	Piezoelectric constants	18
4	$P'_{ijkl} = \alpha_{im}\alpha_{jn}\alpha_{ko}\alpha_{lp}P_{mnop}$	Elastic constants	21

The atomic positions are entirely different in some crystals with identical chemical composition. This is called allotropism. Thus diamond and graphite are both made of carbon atoms but the positions of carbon atoms are very different resulting in difference in physical properties. The same is the case in calçite and aragonite (both having formula $CaCO_3$) and rutile and anatase (both having formula TiO_2). In some crystals, atomic positions change significantly at some definite temperature causing a change in structure. This is called polymorphism and the temperature is called the transition temperature. For example, CsCl undergoes a phase transition at a high temperature to the NaCl structure.

The application of heat causes displacement of atoms. These displacements are resisted by restoring forces in the lattice resulting in thermal vibrations. These vibrations are quantized and are called phonons. They are responsible for the specific heat of crystals and also for the reduction in X-ray diffraction intensities as the temperature increases. The thermal vibrations are anharmonic. This anharmonicity causes a crystal to expand as the temperature is increased. Also, the process of thermal conduction in insulators is due to the anharmonicity of thermal vibrations. At high temperatures, the amplitude of thermal vibration becomes large and when it becomes comparable with the interatomic distance, the lattice becomes unstable and finally it melts.

A mechanical stress also causes atomic displacement resulting in lattice strain. This leads to the property of crystal elasticity. The elastic constants are fourth-rank tensors. When an electric field is applied to an insulator, the ions are displaced causing an electric polarization. This is the origin of the dielectric property. The behaviour of some ferroelectric materials is sensitive to subtle displacements of atoms (e.g. H in KDP, O in $BaTiO_3$).

Interatomic forces have a profound effect on crystal properties. Stronger the forces, larger are properties like melting point, Debye temperature and hardness, and smaller are properties like thermal expansion and compressibility. The covalent bond is the strongest and the van der Waals bond the weakest. This makes diamond the hardest and graphite the softest of crystals.

1.8 Defects in Crystals

We defined a crystal as a three-dimensional array of atoms arranged such that the environment of a given atom is identical to that of another similar atom. Such a crystal is an "ideal" or "perfect" crystal. Most theories of crystal properties assume a perfect crystal.

As we study various properties of crystals, we are confronted with results that cannot be explained by the perfect-crystal model. We have to assume that a crystal departs from the ideal description and that, indeed, there are "defects" in it.

Anything that makes the crystal deviate from the ideal-crystal criterion constitutes a defect. Thus an atom missing from its site (a vacancy) or an atom which has wandered to a wrong site (an interstitial) are defects; these are called point defects.

These two defects are very important as they are necessary for the thermodynamic equilibrium of the crystal. Their concentration increases exponentially with temperature.

There are also line defects called dislocations in which the environment of atoms all along a line is different from the environment of atoms in any other parallel line. Dislocations are of three types: edge, screw and mixed. There are several experimental methods for the observation of dislocations.

Defects have a profound effect on crystal properties. For example, in ionic crystals vacancies increase the ionic conductivity by several orders. Similarly, dislocations reduce the mechanical strength of crystals by several orders.

1.9 Theories

To account for the various properties and phenomena, a number of theories were proposed. Some of them are comprehensive in the sense that they cover a wide range of observations; we shall consider some of them.

Born developed the theory of cohesion for ionic crystals. For a crystal like NaCl, he calculated the Coulomb energy due to attraction between an ion and the ions in the rest of the lattice and for stability he assumed a repulsive force due to overlap of electron clouds. Thus he could compute the cohesive energy of alkali halides. In due course, this simple model was found useful in providing estimates of infrared absorption frequencies, elastic constants and defect-formation energies of ionic crystals.

A theory to account for the low temperature behaviour of specific heats was proposed by Einstein in which he assumed that all atoms in a solid vibrate with one single frequency. The theory was improved by Debye by assuming that the vibrating atoms have a spectrum of frequencies with an upper limit. Born developed a theory of lattice vibrations in which he took into account the structure of the crystal and the nature of the interatomic forces. The vibration frequencies calculated from the theoretical models have been verified with the experimental technique of neutron inelastic scattering. Though the initial objective of these theories was to explain the temperature variation of specific heats, eventually they accounted for other phenomena like infrared absorption, impurity-induced optical absorption and certain aspects of dielectric behaviour. With the inclusion of the concept of anharmonicity, the theory of lattice vibrations could account for other phenomena like thermal expansion and thermal conductivity. New concepts like Debye temperature, phonons, Brillouin zones and dispersion relations were introduced.

The explanation of the various experimental observations on ferroelectric crystals required theories with different approaches. The first such theory was the dipole theory which assumed the presence of permanent electric dipoles which could orient in an electric field. The theory is similar to the Langevin–Weiss theory of ferromagnetism. The local field theory did not assume permanent dipoles and yet could account for the experimentally observed "polarization catastrophe". Devonshire proposed a phenomenological theory based on thermodynamics. This theory accounts for the ferroelectric behaviour in the vicinity of the "paraelectric–ferroelectric transition". The order parameter in the theory is the electric polarization. A major development in ferroelectric theories is Cochran's theory based on lattice dynamics. Cochran introduced the idea of "soft modes", i.e. lattice vibration modes whose frequency assumes a low value in the vicinity of the transition temperature. These soft modes cause large increase in dielectric constant at the transition temperature.

1.10 Other Crystalline Forms

The above discussion relates to "regular" crystals. There are other crystal forms which differ from the "regular" crystals in one way or the other. Thus, there are "quasicrystals" which deviate from laws of crystal symmetry and are found in localized regions. "Nanocrystals" are extremely small in size (1–100 nm). They are crystallographically similar to the bulk material but differ in physical properties. There are still other forms like thin films, polycrystalline aggregates and liquid crystals.

1.11 Practical Applications of Solids

The emphasis in this book is on physics of phenomena in solid state physics and not so much on applications. Yet for completeness, we shall mention some applications of solids in science and in day-to-day life. Consistent with the theme of this book, only applications based on atomistic properties are mentioned.

Alkali halide (NaCl, CsBr) prisms are used in infrared transmission. Crystals like calcite and tourmaline are used as optical polarization devices. Calcite, graphite and LiF are used as monochromators in X-ray diffraction. Many minerals are used as precious stones (gems) because of their beautiful colours and shapes.

Metals with very low thermal expansion are used as length standards. Combinations of metals with different expansion are used as temperature controlling devices. Solids with low thermal conductivity are useful as containers in special situations.

Very hard materials like tungsten carbide are used as cutting tools whereas very soft materials like graphite are used as solid lubricants.

Ferroelectrics have large dielectric constant which is exploited in fabrication of capacitors. An important application of ferroelectrics is as memory devices; this is based on the property of hysteresis.

Solid state devices are popular as radiation detectors. NaI (Tl), Ge and Si are used in X-ray detection. Hg(Cd-Te) alloy has a special application in infrared detection.

A very important application of solids is as laser hosts. $CaWO_4$, CaF_2, garnets and ruby are commonly used in solid state lasers. Doped LiF is specially used as a tuned laser. The ability to generate harmonics shown by some nonlinear optical crystals like KDP makes them useful in communication.

Piezoelectric crystals like quartz and PZT have a number of applications in electronics and ultrasonics.

Liquid crystals have many applications. The most popular is as liquid crystal display in many modern appliances like mobile phones and TV.

Chapter 2
Crystal Growth

2.1 Introduction

It was seen in Chap. 1 that solid state physics is all about the behaviour of crystals. According to Pfann [2.1], the single crystal is the sine qua non of the experiment. It is appropriate that a study of solid state physics should begin with a study of the experimental methods and the theoretical models of crystal growth. A detailed discussion of the former is given by Gilman [2.2], Brice [2.3] and Tarjan and Matrai [2.4] and of the latter by Ohara and Reid [2.5] and Sangwal [2.6].

Crystals can be grown by melting of a substance followed by cooling, by precipitation from solution or by condensation of a vapour. The basic principle common to all these methods is that a nucleus is first formed and it grows into a single crystal when ions or molecules get deposited on it. The general approach is to see that this deposition is slow and multiple nucleation is minimised.

In this chapter, several methods of crystal growth are described. The more commonly used methods – solution growth and melt growth – are discussed in some detail. A few other methods which are useful in special situations are discussed briefly. Theoretical aspects are touched upon for completeness.

2.2 Crystal Growth from Solution

2.2.1 Normal Solution Growth

2.2.1.1 Principle

By normal solution growth, we mean crystal growth from solutions at room temperature (or moderately elevated temperatures) and atmospheric pressure. The principle of crystal growth from solution is based on the phenomena of solubility,

D.B. Sirdeshmukh et al., *Atomistic Properties of Solids*, Springer Series in Materials Science 147, DOI 10.1007/978-3-642-19971-4_2,

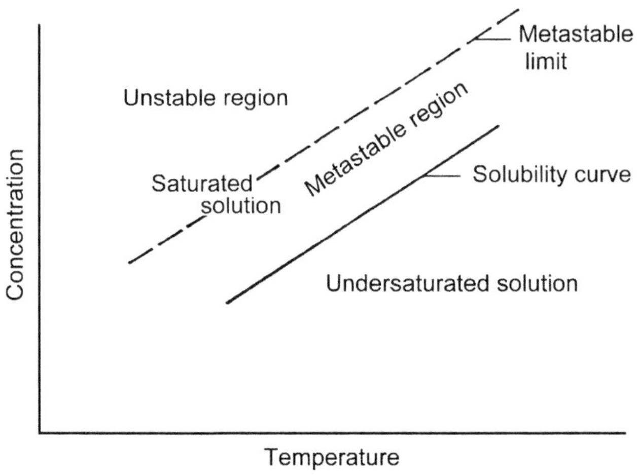

Fig. 2.1 A typical solubility curve

supersaturation and nucleation. The solvent is generally water but organic solvents are also used. At a given temperature, only a limited amount of the substance (solute) dissolves in a solvent. This amount is called the solubility of the substance in that solvent at a particular temperature. The solution containing this limiting amount is called a saturated solution; the solvent cannot accept an amount of the solute exceeding this amount. The solubility is a function of temperature. The solubility generally increases with increasing temperature. A typical solubility curve is shown in Fig. 2.1. If a saturated solution is prepared at a certain temperature and its temperature is lowered, then it contains more solute than that permitted by its solubility at the lowered temperature. The same thing happens if the temperature remains unaltered but the solvent is allowed to evaporate. In either situation, the solution becomes "supersaturated" (Fig. 2.1). This is a metastable stage and at the slightest disturbance, the solution rejects the excess salt and returns to the saturated state. If this precipitation is slow and takes place around preferred centres (nuclei, seed), a single crystal grows.

The molecules or ions in the solution may congregate to form a growth nucleus or growth embryo which provides a starting point of growth. This is called auto- or self-nucleation. This is possible when such formation results in the reduction of the free energy of the system. These nuclei are formed and they redissolve in the solvent unless their radius r is larger than a critical value r^* given by

$$r^* = -(2\gamma_{sl}/\Delta G_V), \tag{2.1}$$

where γ_{sl} is the solid–liquid interfacial energy and ΔG_V the change in free energy. Apart from such self-nucleation, nucleation is also facilitated by impurities, solid foreign particles like dust, cracks and crevices in the container parts in contact with the solution and even local fluctuations in supersaturation within the solution.

Controlled single crystal growth is possible by deliberately introducing a "seed" into the saturated solution before inducing supersaturation. The best seed, of course, is a small crystal of the substance to be grown. If that is not available, a small crystal of another substance with the same structure and having a matching lattice constant can be used, e.g. KBr seed in KCl solution. If even this is not available, any crystal sample like mica may be used as seed in initial experiments and the crystal grown on it could be used in subsequent experiments.

Other factors which control the growth of a crystal are the pH of the solution, stirring of the solution during crystal growth and the deliberate addition of impurities (called additives).

2.2.1.2 Experimental Details

In Table 2.1, the solubilities of some inorganic substances are given at select temperatures. It can be seen that in most substances, the solubility increases with temperature. In some substances, e.g. NaCl, the increase in solubility is very little. There are some substances, e.g. Na_2SO_4, for which the solubility decreases with increasing temperature at elevated temperatures. As mentioned earlier, the addition of some impurities aids the growth of some crystals; some additives are listed in Table 2.2.

It is desirable to have "seeds" for growing large crystals. If a saturated solution is prepared and left to evaporate, several small crystals are found at the bottom of the container. These are filtered out and used as seeds in subsequent experiments.

Table 2.1 Solubility (g salt/100 g H_2O) in water at various temperatures of some crystalline materials

Material	Temperature [°C]						
	20	30	40	50	60	70	80
$AgNO_3$	222	300	376	455	525	–	669
$BaSO_4$	2.4×10^{-4}	2.85×10^{-4}	–	–	–	–	–
$CuSO_4$ (anhydr.)	20.7	25.0	28.5	33.3	40.0	–	55.0
KBr	65.2	70.6	75.5	80.2	85.5	90.0	95.0
KCl	34.0	37.0	40.0	42.6	45.5	48.3	51.1
KI	144	152	160	168	176	184	192
$KMnO_4$	6.4	9.0	12.56	16.89	22.2	–	–
$MgSO_4{\cdot}7H_2O$	26.2	29	31.3	–	–	–	–
NH_4Br	75.5	83.2	91.1	99.2	107.8	116.8	126
NH_4Cl	37.2	41.4	45.8	50.4	55.2	60.2	65.6
NH_4I	172.3	181.4	190.5	199.6	208.9	218.7	228.8
NaCl	36.0	36.3	36.6	37.0	37.3	37.8	38.4
$NaNO_3$	88	96	104	114	124	–	148
$PbCl_2$	0.99	1.20	1.45	1.70	1.98	–	2.62
$Pb(NO_3)_2$	56.5	66	75	85	95	–	115
$KAl(SO_4)_2 \cdot 12H_2O$	5.9	8.39	11.7	17.0	24.75	40.0	71.0

Table 2.2 Additives which aid growth of some crystals

Material	Additives
$AgBrO_3$	Fe^{3+}, Al, Zr, W^{6+}, Th
$Ba(NO_3)_2$	(10% alcohol solution) Mg, Te^{4+}
CsCl	La, Ce^{3+}, Nb^{3+} (all of slight effect)
$KBrO_3$	Pb, Th, Te^{4+}, V^{5+}
KCl	Pb, Bi, Sn^{2+}, Ti, Zr, Th, Cd, Hg, Fe
KI	Pb, Ti, Sn, Bi, Fe
KNO_3	Pb, Th, Bi
K_2SO_4	UO_2^{2+}, VO_2^{+}, Cd, Mn, Fe, Ce, Cu, Al, Mg, Bi
LiCl. H_2O	Cd, Mn^{2+}, Sn^{2+}, Sn^{4+}, Co, Ni, Fe^{3+}, Ti, Cr, Th
NaCl	Pb, Mn^{2+}, Bi, Sn^{2+}, Ti, Cd, Fe, Hg
$NaNO_2$	Ca
$(NH_4)_2C_2O_4 \cdot H_2O$	Cu, Ca, Fe, Th
NH_4Cl	Zr, Cd, Mn, Fe^{2+}, Cu, Co, Ni, Fe^{3+}, Cr
NH_4F	(In alcohol saturated with NH_3) Ca
$NH_4H_2PO_4$	Fe^{3+}, Cr, Al, Sn
NH_4IO_3	Cd, Ca
RbCl	Pb, Sn, Zr, Ti

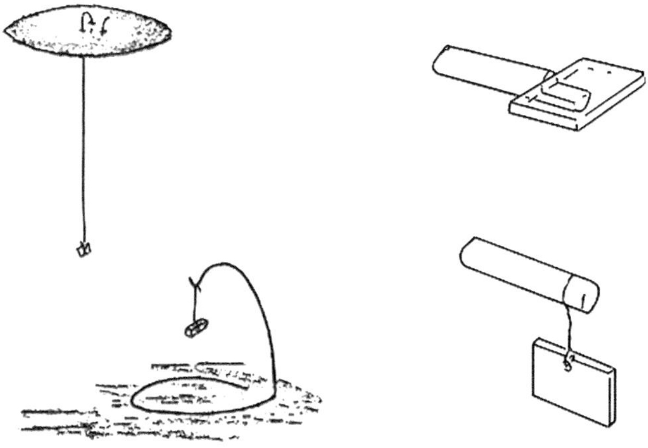

Fig. 2.2 Methods of introducing a seed

Different methods of suspending a seed crystal in a growth vessel are shown in Fig. 2.2.

Figure 2.3 shows the simplest growth set-up recommended by Holden and Singer [2.7] with which even amateurs can grow crystals. The saturated solution is taken in a glass jar and the seed is suspended in it with a string. To permit slow evaporation, the jar is covered with filter paper or a lid with perforations. In the set-up shown in Fig. 2.4, there is provision for stirring the solution, for temperature control and for maintaining the saturation of the solution. The three tanks (7, 1 and 5) are maintained at different temperatures T_1, T_2 and T_3 by separate thermostats. T_1

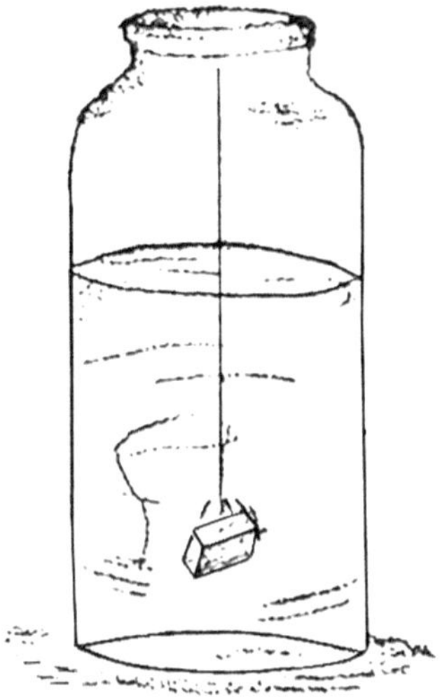

Fig. 2.3 A simple arrangement for solution growth

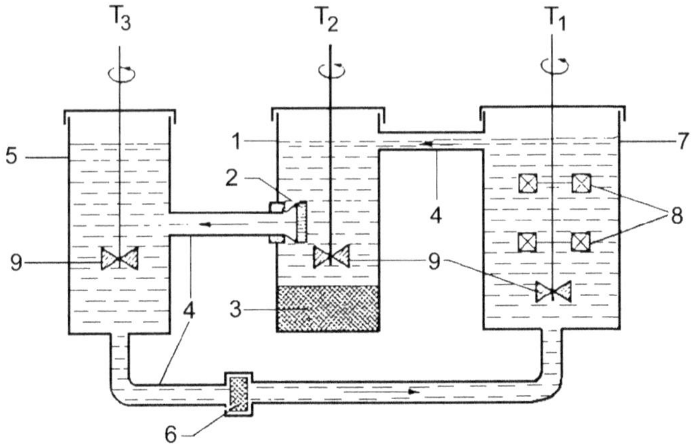

Fig. 2.4 Solution growth set-up with circulation: 1 – solution make-up tank, 2 – glass filter, 3 – polycrystalline material, 4 – heated connecting tube, 5 – holding tank, 6 – membrane pump, 7 – growth tank, 8 – seed crystals, 9 – mixer

is at the desired temperature of growth (say 45°C). T_2 and T_3 are at slightly higher temperatures (45.5 and 46°C). As the crystals grow on the seeds, the solution in their vicinity is depleted of the solute and this diluted solution rises to the top of the growth tank. It flows into tank 1 which has an amount of the solute at its base. The diluted solution which has flown into it dissolves the solute and its saturation is restored. The solution now flows into the holding tank 5 from which it is circulated back to the bottom of the growth tank 1. Figure 2.5 shows a still more sophisticated crystal growth system [2.8] in which all the growth parameters are under better control. Several crystals grown by the solution method are shown in Fig. 2.6.

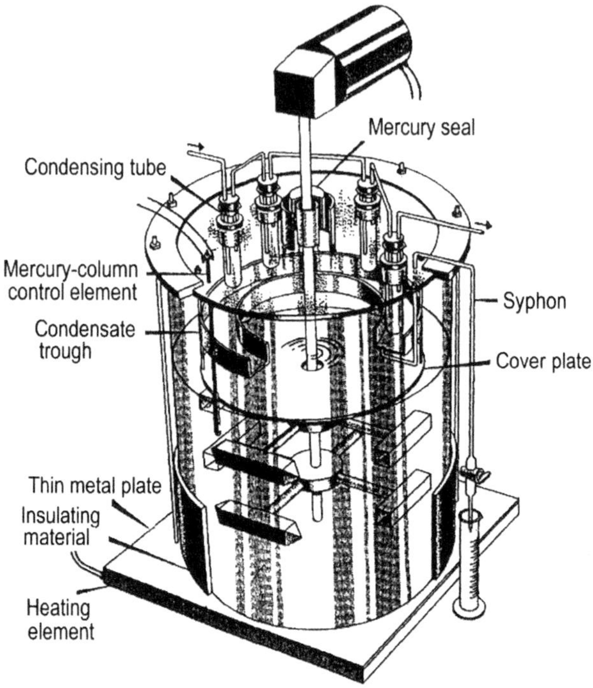

Fig. 2.5 A fully controlled solution growth set-up

2.2.1.3 Growth Forms

To look at the faceted forms of solution-grown crystals is an aesthetic experience. Different crystals grow in different forms. In fact crystals of the same substance may grow with different forms when grown in different conditions.

A crystal grows with a certain growth rate. If this growth rate was the same for all faces (i.e. in all directions), the crystal would grow as a sphere. But the different forms that are seen are the result of different faces having different growth rates. These growth rates depend not only on the orientation of the face but also on other growth conditions like relative supersaturation, temperature, impurities.

In Fig. 2.7, the cross section of a growing crystal is shown; 1, 2 and 3 are three adjacent faces. The normals to these faces meet at some point O within the crystal. In unit time, the faces are displaced to position "a". Let v_1, v_2 and v_3 be the growth rates normal to the three faces. If $v_1/v_2 = k_1$ and $v_3/v_2 = k_2$ (k_1 and k_2 being constants), the crystal shape at any stage of growth will be the same. But if $v_1 > k_1v_2$ and $v_3 > k_2v_2$, the size of plane 2 increases relative to the other faces (position b). On the other hand, if $v_1 < k_1v_2$ and $v_3 < k_2v_2$, the size of plane 2 decreases (position c). Thus, if a crystal face grows at a higher rate than that given by the above conditions, its size decreases and the face finally disappears allowing the other faces to grow.

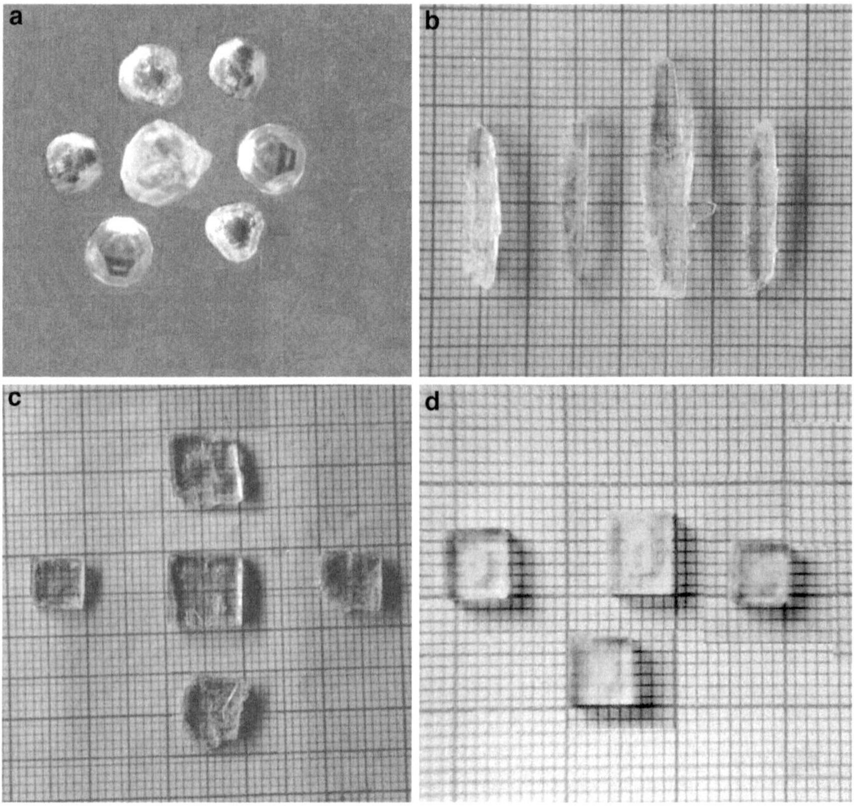

Fig. 2.6 Some solution-grown crystals: (**a**) Alum, (**b**) ADP, (**c**) potassium ferrocyanide, (**d**) NH_4Cl

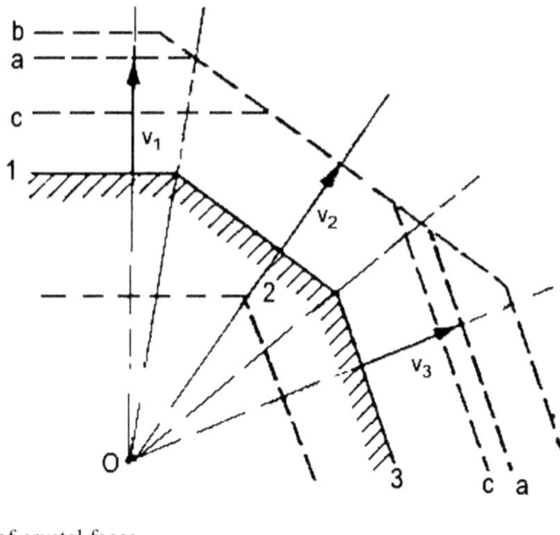

Fig. 2.7 Growth of crystal faces

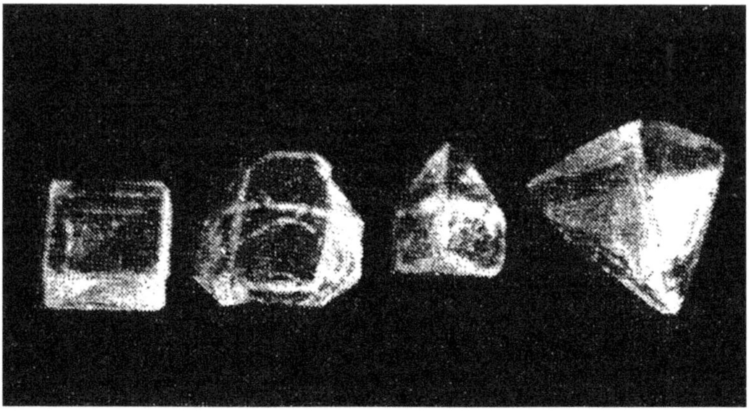

Fig. 2.8 Change in habit of $NaClO_3$ crystals from solutions containing increasing concentration of borax (from *left* to *right*)

Some impurities deliberately added to the solution have an influence on the growth form. Figure 2.8 shows sodium chlorate ($NaClO_3$) crystals grown with and without borax as impurity. While the crystal grown from pure aqueous solution is bounded by the cubic faces, that grown from the solution with borax as impurity is bounded by tetrahedral faces. The effect of impurities on the growth forms of several crystals is summarised in Table 2.3. The numbers given in column 2 of this table and elsewhere in this chapter to indicate planes and faces are Miller indices. This system of indexing is explained in Chap. 3.

The impurity atoms (or ions or molecules) are adsorbed on the surface of the crystal. Thus, in NaCl, $CdCl_2$ gets adsorbed on the (111) planes and the $Na_3Fe(CN)_6 \cdot H_2O$ complex ion on the (100) planes (Fig. 2.9). In Fig. 2.9a, region A shows the normal NaCl lattice. Region B shows the top layers. Here Cd^{2+} replaces Na^+ and forms a two-dimensional $CdCl_2$ lattice.

The degree of adsorption depends on the atomic structure of the plane, the sites allowed for adsorption and the binding energy for adsorption. Thus the degree of adsorption differs from plane to plane. The adsorbed impurity may have different effects on different faces; it may hinder the growth of one face and promote the growth of another. As a result, the relative growth rates of different faces in an impurity-containing solution may differ from those in a pure solution.

2.2.2 *Flux Growth*

Unlike conventional growth in which either water or organic liquids are used as solvents, flux growth employs molten metals or fused inorganic salts as solvents. Knowledge of the equilibrium phase diagram between the various participants in flux growth is helpful in planning a flux growth experiment.

Table 2.3 Habit changes due to impurities in some crystals grown from aqueous solution

Substance	Impurity	Crystal polyhedron faces
KCl	–	100, hexahedron
	Urea	411
	Bromobenzene, Aniline, phenol	Hexoctahedron
	Pb^{2+}	111, 100
	Hg^{2+}	100, 111
KBr	–	100, hexahedron
	Pb^{2+}	111, 100
	Ti^{4+}	100, 111, 110
CsCl	–	100, 110, 111
	La^{3+}, Ce^{3+}, Gd^{3+}	100
NaCl	–	100, hexahedron
	Urea	100, 111, 110
	$HgCl_2$	100, 110
	Alanine	111
	Glycine	110
$Pb(NO_3)_2$	–	111
	Methylene blue	100
$NaClO_3$	–	Cube
	Na_2SO_4	Tetrahedron
$MgSO_4 \cdot 7H_2O$	–	Rhombic prismatic
	$Na_2B_4O_7 \cdot 10H_2O$	Tetrahedron
$KAl(SO_4)_2 \cdot 7H_2O$	–	001, 111, 011, etc.
	KOH, Na_2CO_3	001

Flux growth is carried out at high temperatures (1,000–1,500°C) at which both the solvent and the solute are in liquid state. On cooling, the crystal is formed and sometimes the solvent too is solidified; the latter is removed by dissolving it in another medium called the leaching solution. A few examples of crystal growth by the flux method will be considered which will bring out the problems involved in this method.

2.2.2.1 GaP

A mixture of phosphorous and gallium is taken in an evaporated silica tube, the amount of Ga being far in excess of that needed to form stoichiometric GaP. Ga is a liquid at 30°C and when the (Ga + P) mixture is heated to 400–500°C, GaP is formed and it precipitates in the liquid Ga. On heating to ~1,150°C, the GaP particles dissolve in the liquid Ga to form a saturated solution. Subsequent slow cooling leads to formation of GaP crystals which are extricated by decanting off the excess liquid Ga. Some GaP crystals grown in this way are shown in Fig. 2.10.

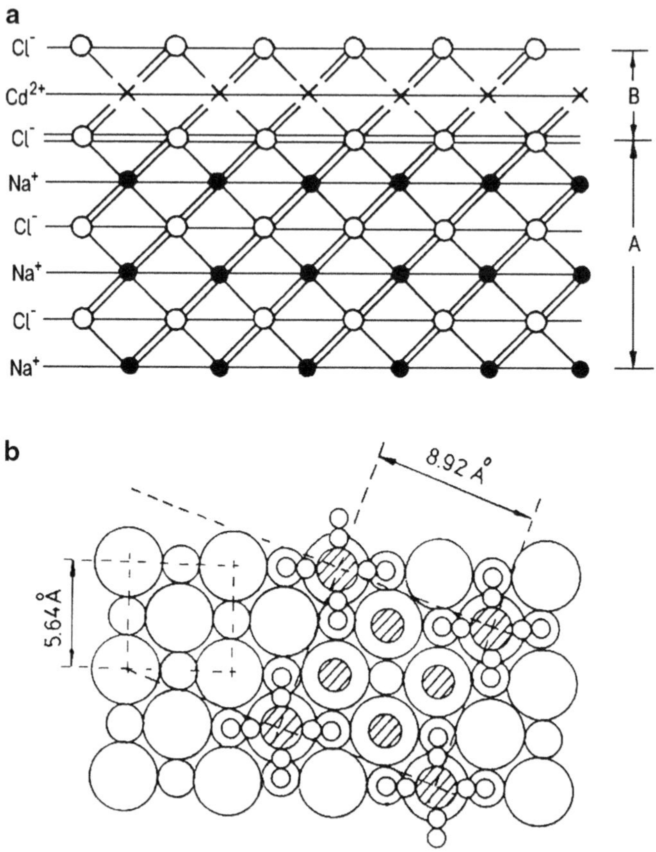

Fig. 2.9 (**a**) $CdCl_2$ adsorbed on NaCl planes and (**b**) $Na_3Fe(CN)_6$. H_2I adsorbed on NaCl planes

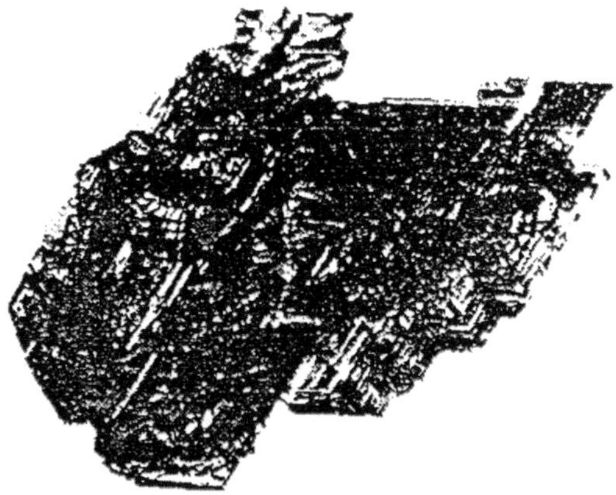

Fig. 2.10 GaP crystal grown from molten Ga

Fig. 2.11 $BaTiO_3$ crystals grown from KF flux

2.2.2.2 $BaTiO_3$

Trials with different solvents like fused BaF_2, $BaCl_2$ and KF proved that KF is most satisfactory. It does not attack the platinum crucible, crystallised $BaTiO_3$ sinks in KF and, at the end, the residual KF can be easily leached away.

$BaTiO_3$ is held in molten KF at ~1,150–1,200°C for about 8 h in a platinum crucible in a furnace. At the end of the soaking period, the furnace is cooled at the rate of 20–50°C/h to a temperature of 900–1,000°C. $BaTiO_3$ crystals are formed at the bottom of the crucible. Liquid KF is decanted off and the $BaTiO_3$ crystals are allowed to cool to room temperature. Some residual KF which has solidified around the $BaTiO_3$ crystals is washed away with hot water. Some $BaTiO_3$ crystals grown in this way by Laudisse [2.9] are shown in Fig. 2.11.

2.2.2.3 Yttrium Iron Garnet

Molten PbO is used as the solvent for growing yttrium iron garnet (YIG) crystals. The starting material is a mixture of Fe_2O_3 and Y_2O_3. After some initial trials and failures, it was found that best results are obtained when the ratio Fe_2O_3:Y_2O_3 is substantially greater than 5:3. The resulting YIG crystals were removed by leaching out the flux with hot HNO_3. Molten BaO–B_2O_3 is also a satisfactory flux for YIG. In fact good large crystals of YIG have been grown by seeding in a solution of YIG in a BaO–B_2O_3 flux. An arrangement for the seeded flux growth of YIG and the resulting crystals are shown in Fig. 2.12.

The advantages of the flux method are that the required equipment is simple, a choice of solvents is available and the growth rate is fast. On the other hand, the disadvantages are that impurity control is difficult and the operating temperatures are high. Details of the flux growth of a few other important crystals are given in Table 2.4.

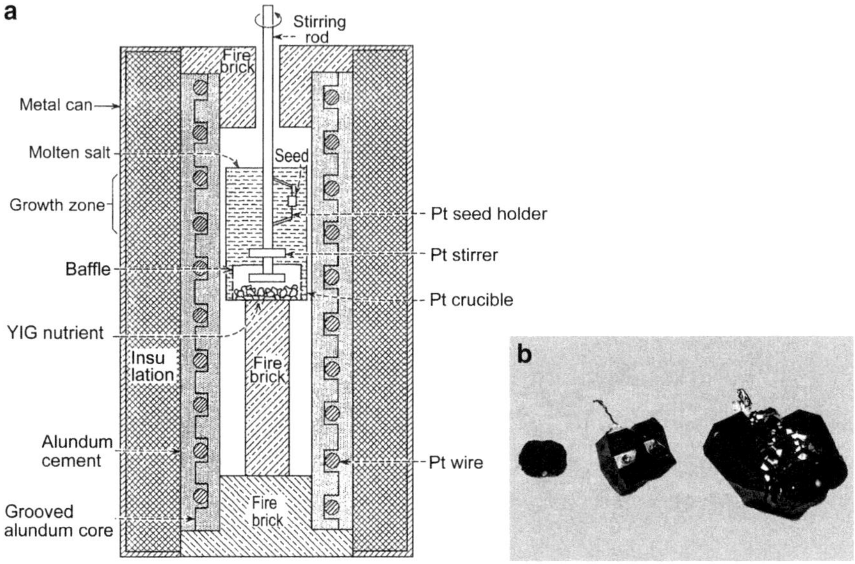

Fig. 2.12 (**a**) Set-up for growth of YIG from flux. (**b**) YIG crystals grown from flux

2.2.3 Hydrothermal Growth

Certain substances are not amenable to crystal growth by conventional methods; α-quartz is an example. This is the low temperature form of quartz which is stable below 575°C. Growth from normal solution growth method is ruled out as no solvent is known in which SiO_2 has a reasonable solubility. Melt and vapour growth are also ruled out. In such circumstances, the hydrothermal growth method is useful.

The hydrothermal crystal growth method is based on two phenomena. Firstly, when water boils, the vapour has its own pressure. This vapour pressure increases with temperature and the increased pressure increases the boiling point so that, at these high pressures we have water much above its normal boiling point. The vapour pressure depends also upon a parameter "c" called the "charge" or "degree of filling" which is the volume of the water in the enclosure expressed as a percentage of the total volume of the enclosure. The $P - T$ curves for water at different charge values are shown in Fig. 2.13 [2.10]. The second phenomenon is that substances insoluble in a solvent at normal temperatures and pressures become soluble when the solvent is at a high pressure and temperature. Thus, SiO_2 which is otherwise insoluble in water develops solubility at higher temperatures. The solubility–temperature curves for quartz at different pressures are shown in Fig. 2.14.

Thus, the requirements in this method are an arrangement for heating a solvent–nutrient mixture in an enclosure which can withstand the high vapour pressure and the usual arrangement for crystal growth from solution with seeds. The growth vessel (called autoclave) is an enclosure with a well-fitting closure. It holds the quantity of water necessary to obtain the desired pressure at the chosen growth

Table 2.4 Details of flux growth of some crystals [2.2]

Crystal	Formula	Flux and temperature range of cooling	Cooling rate	Typical charge [mol%]	Crystals leached from flux with
Potassium orthoniobate	$KNbO_3$	KF or KCl ~1,100 to 800°C	~1–10°C/h	50% K_2CrO_3, 50% Nb_2O_3 amount of flux not given	Hot water
Lead orthozirconate	$PbZrO_3$	PbF_2 1,250 to 1,000°C	50°C/h	50% $PbZrO_3$, 50% PbF_2	Not chemically separable
Yttrium aluminium garnet	$Y_3Al_5O_{12}$	$PbO–B_2O_3$ 1,250 to 950°C	1°C/h	32% Y_2O_3, 5.3% Al_2O_3, 18.5% B_2O_3 and 73% PbO	Hot HNO_3
Nickel ferrite	$NiFe_2O_4$	$Na_2B_4O_7 \cdot 10H_2O$ 1,330 to 1,250°C	2°C/h	32.1% NiO, 32.1% Fe_2O_3, 35.8% $Na_2B_4O_7 \cdot 10H_2O$	Hot HNO_3
Zinc sulphide (β)	ZnS	ZnF_2 1,000 to 800°C	1–5°C/h	21% ZnS, 79% ZnF_2	Hot conc. NH_4OH
Zinc oxide	ZnO	PbF_2 1,150 to 800°C	1–10°C/h	30% ZnO, 70% PbF_2	Not chemically separable

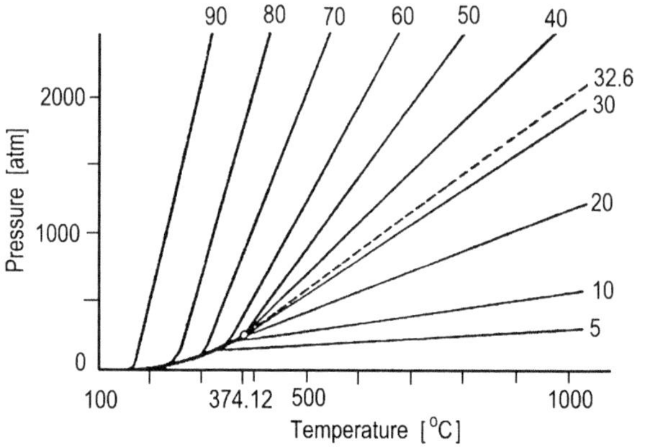

Fig. 2.13 Temperature dependence of vapour pressure of water for various charges

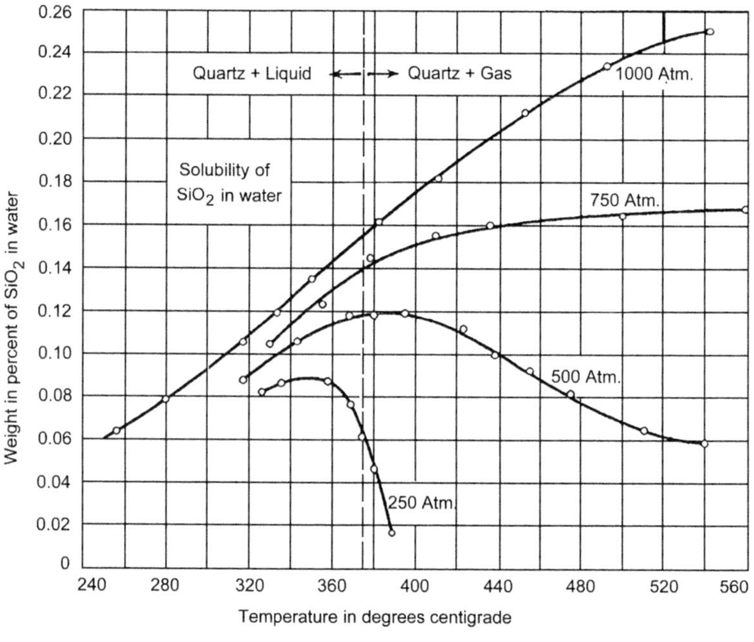

Fig. 2.14 Solubility of quartz at different temperatures and pressures

temperature. Once the charge "*c*" and the temperature (T) are chosen, the vapour pressure (P) is known from the P–T–c curves. The autoclave is designed to withstand this pressure (generally, thousands of atmospheres). Autoclave designs are shown in Fig. 2.15. These vessels are generally made of low-carbon steels and, if the closure (sealing) is proper, they can withstand pressures of several thousand atmospheres at temperatures of 300–500°C. The autoclave containing the solvent and the nutrient is then placed in a two-zone furnace (Fig. 2.16) with a vertical

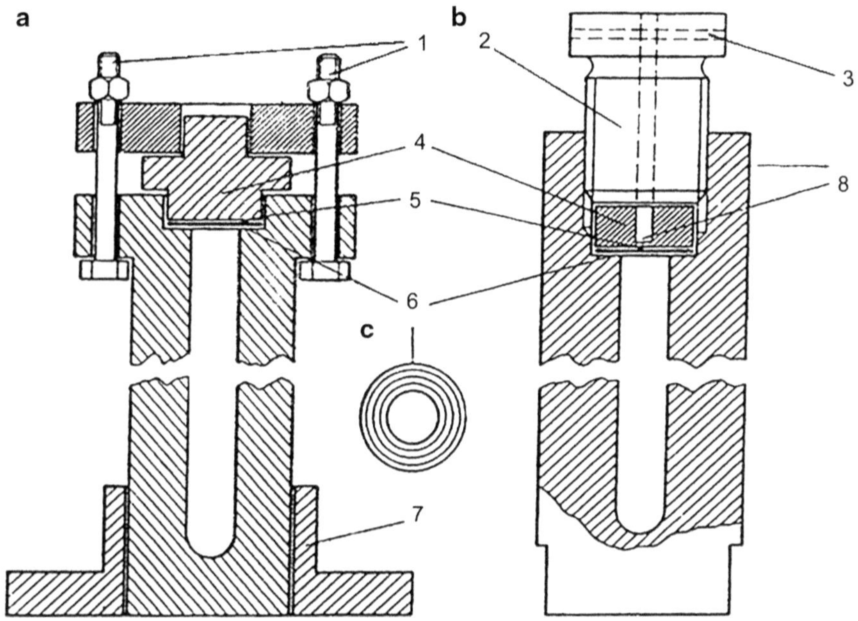

Fig. 2.15 Autoclave types: (**a**) closing with the aid of six to eight connecting screws 1; (**b**) closing with a single screw 2; it is fixed by a steel rod put through a transverse boring 3; (**c**) the closing surface viewed from above; 1 – connecting bolt, 2 – fixing bolt, 3 – transverse boring, 4 – closing steel disc, 5 – sealing silver plate, 6 – closing surface, 7 – support fixed to the table to prevent overturning, 8 – safety diaphragm

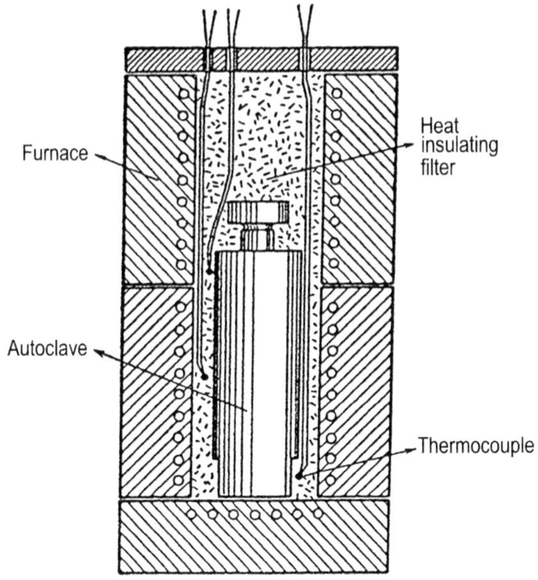

Fig. 2.16 An autoclave in a furnace

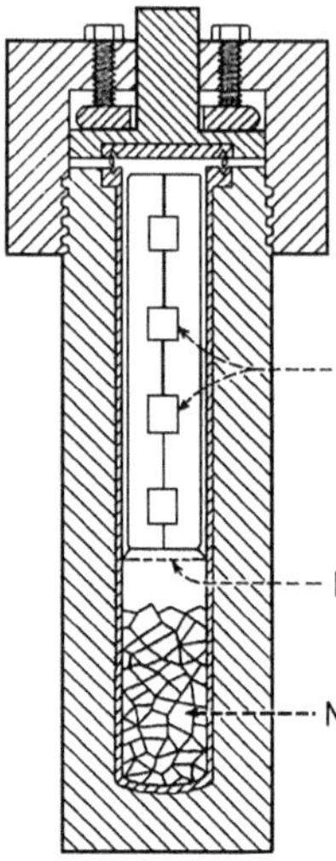

Fig. 2.17 The arrangement inside an autoclave

Table 2.5 Operating conditions in hydrothermal growth of quartz

Operating parameter	Value
Dissolving temperature	400°C
Crystallising temperature	380°C
Temperature differential	40°C
Degree of fill (c)	80%
Baffle opening	5%
Pressure	21,000 psi

temperature gradient (low temperature in upper zone and elevated temperature in the lower zone). The arrangement for crystal growth inside the autoclave is shown in Fig. 2.17.

The necessary quantities of the nutrient and the solvent are taken in the autoclave and the temperature of the furnace is raised to the desired value. The quartz dissolves and the saturated quartz solution from the lower warmer zone rises up to the cooler zone where it becomes supersaturated. The seed plates start growing reducing the supersaturation of the solution. Due to a convection current maintained by the temperature gradient, the depleted solution moves down and fresh saturated solution moves up. This cycle of solution and deposition maintains the growth of the seeds. These convection currents can sometimes become so strong that they may drive solid nutrient particles upward. This is prevented by the baffle shown in Fig. 2.17. Typical operating conditions in the growth of quartz are given in Table 2.5.

Single crystals of quartz, sapphire and $AlPO_4$ grown by the hydrothermal method are shown in Fig. 2.18. Some other crystals grown by this method are tourmaline, magnetite, ferrites, garnets, ZnO and CdO.

The great advantage of the method is that it facilitates the growth of crystals which are not amenable to other methods. Further, the resulting crystals are faceted and the growth rates are reasonably fast (0.65 cm/day). The disadvantage is that

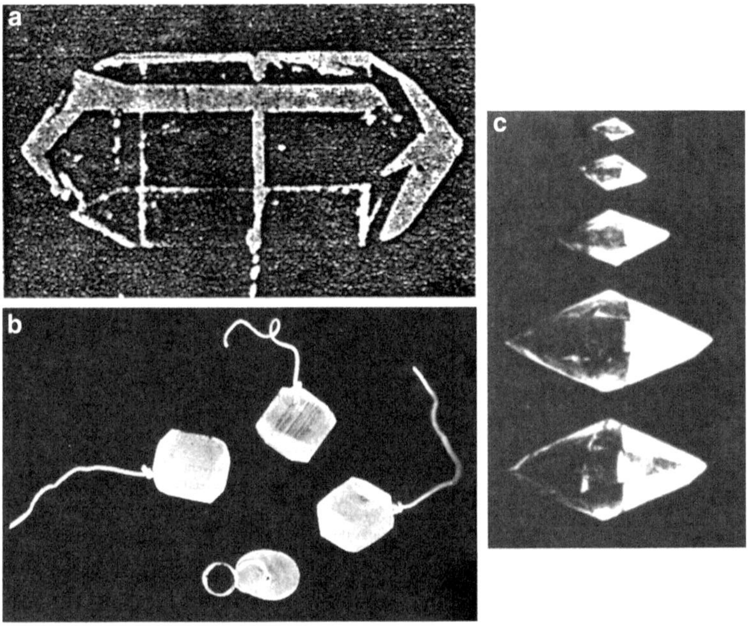

Fig. 2.18 Crystals grown by the hydrothermal method: (**a**) quartz, (**b**) saphire, (**c**) $AlPO_4$

growth takes place in a closed enclosure and the success or failure of the growth is known only at the end.

2.3 Crystal Growth from Melt

2.3.1 General

Crystal growth from melt involves melting of a substance and subsequent cooling. As the melt cools, the atoms/ions deposit around a deliberately introduced seed. While this is the basic process, the actual crystal growth is carried out by following a variety of procedures. These procedures and the materials required for growth from melt are discussed here.

2.3.2 Conventional Melt Growth Methods

2.3.2.1 The Czochralski and Kyropoulos Methods

The set-up used in the Czochralski method is shown in Fig. 2.19. The substance to be crystallised is taken in a crucible. The crucible is then kept in a furnace. The furnace provides a large constant-temperature zone such that the entire material in the furnace melts. A seed is mounted at the lower end of a tubular vertical shaft.

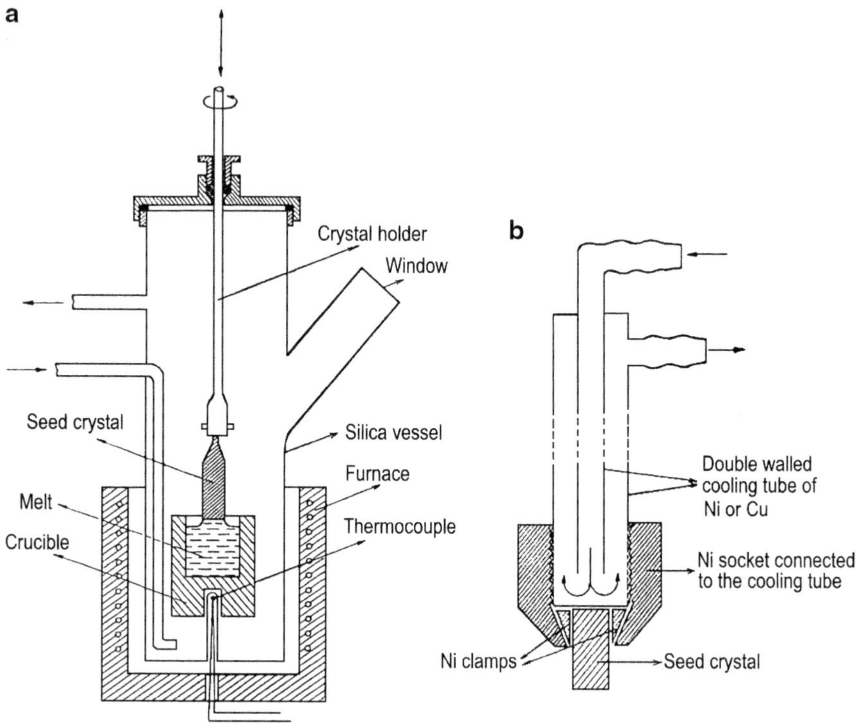

Fig. 2.19 (**a**) A Czochralski set-up; (**b**) details of seed-holder

This tube has ports for circulating water. The seed-holder is slowly lowered till the tip of the seed melts. If the seed is now raised, a part of the melt rises with it because of surface tension but it soon cools and solidifies as the seed crystal in contact with the water-cooled end of the holder acts as a heat sink. The raising of the seed is now continued. The slow pulling of the seed and the cooling by the seed-holder causes the growth of a large single crystal. For a better (and more uniform) size, the seed-holder is rotated. The growth process is continued till all the material in the crucible is lifted out. Crystals grown by this method are generally 2–5 cm in diameter and 5–10 cm in length. A crystal grown by this method is shown in Fig. 2.20.

The Kyropoulos method (Fig. 2.21) is a modification of the Czochralski method. Here, the furnace is a gradient furnace and the crucible is kept in the region with a temperature gradient. After the initial contact of the seed with the melt, the temperature is slightly lowered so that the molten material at the top of the melt solidifies in continuation of and around the lower tip of the seed. But below that, the material is still molten. The furnace temperature is gradually lowered. More and more of the molten material solidifies. The growth of the crystal proceeds from the top of the crucible downward until the whole material is used up. The growth process stops here and as the cooling of the furnace continues, the crystal gets annealed. Crystals grown in this way have larger lateral dimensions of 4–12 cm (Fig. 2.22).

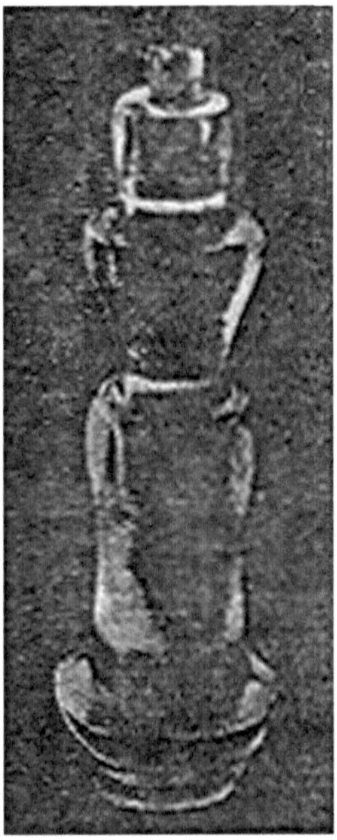

Fig. 2.20 NaCl crystal grown by Czochralski method

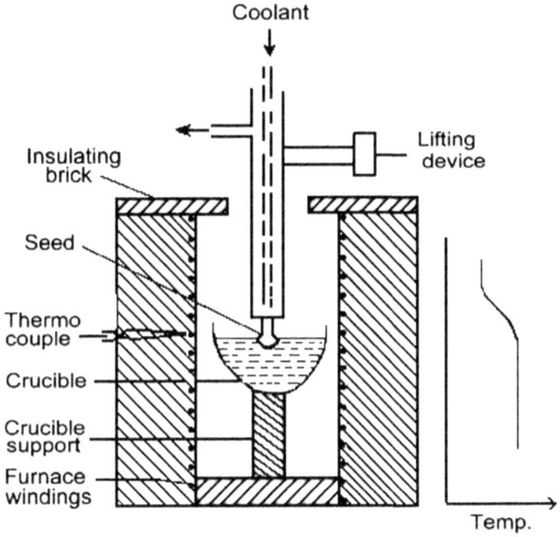

Fig. 2.21 A Kyropoulos set-up

Fig. 2.22 NaCl crystal grown by Kyropoulos method

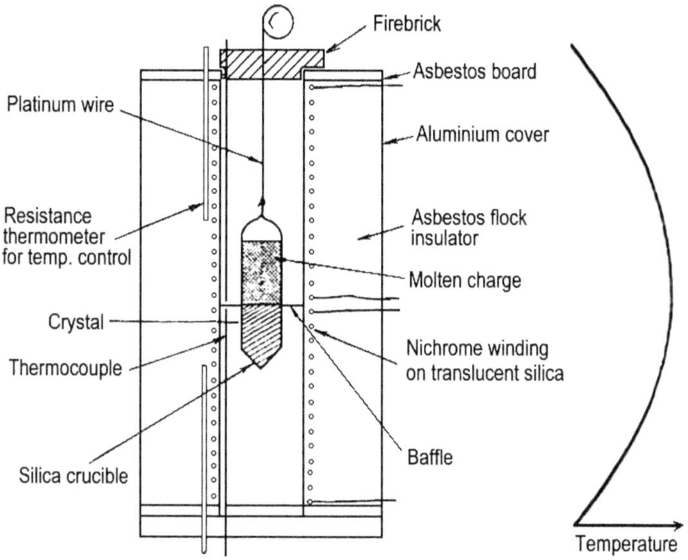

Fig. 2.23 A Bridgman set-up

2.3.2.2 Bridgman and Stockbarger Methods

In the Bridgman method, the polycrystalline charge is taken in a tube. The evacuated and sealed tube is suspended in a vertical gradient furnace (Fig. 2.23). Depending on the melting point of the material, a fused quartz or pyrex tube is used. The gradient is such that the middle part of the furnace is at a higher temperature. When held in the middle region, the complete charge melts. With the help of a lowering mechanism, the tube is gradually lowered so that it enters the lower temperature region. The molten material at the tip of the tube is the first to solidify and this solidified part

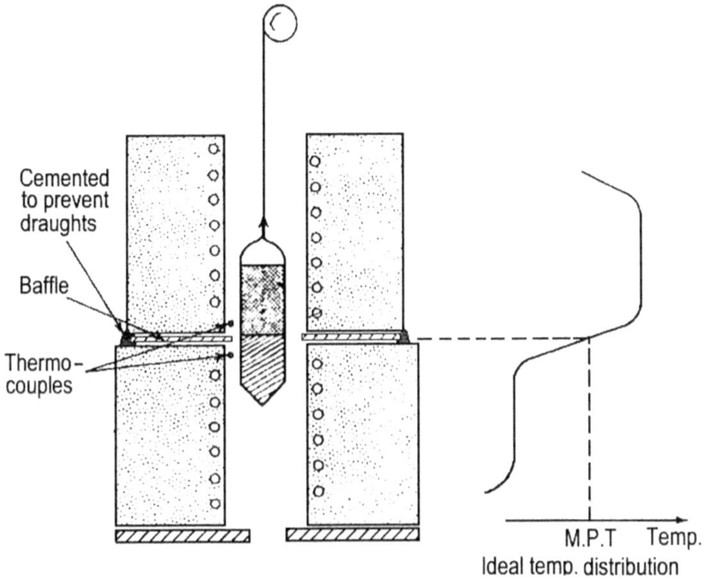

Fig. 2.24 A Stockbarger set-up

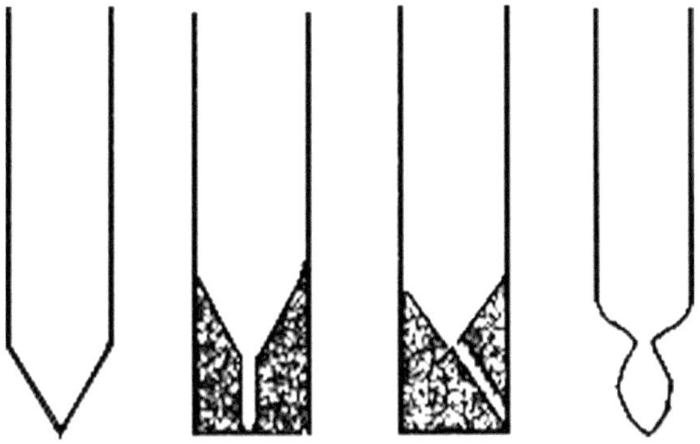

Fig. 2.25 Containers for Bridgman growth

acts as a seed for further growth. As the tube is lowered further, more and more of the melt solidifies as a crystal in continuation with the seed.

A variant of the Bridgman method is the Stockbarger method. The main difference is the use of two furnaces which provide an abrupt change in temperature (Fig. 2.24).

If the mass that solidifies first at the tip of the tube is polycrystalline, the resulting ingot too will be polycrystalline. On the other hand, if the first solidified mass is a single crystallite, the whole ingot will be a single crystal with the same orientation as that of the seed crystal formed at the tip. To enhance chances of single crystal formation, tubes of special shapes like those shown in Fig. 2.25 are employed.

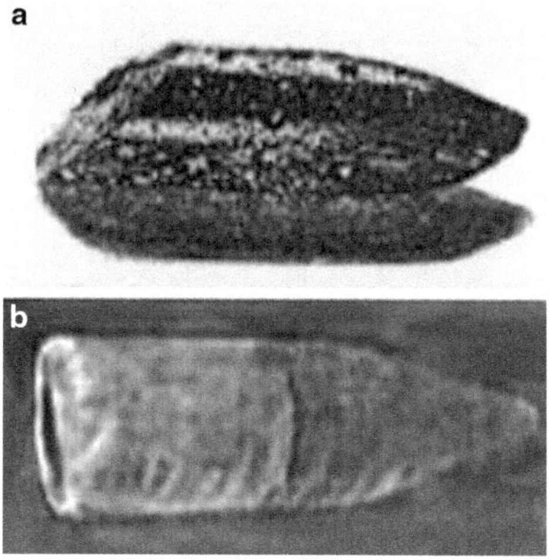

Fig. 2.26 Crystals grown by Bridgman method: (**a**) PbTe and (**b**) anthracene

The Bridgman method is generally used to grow single crystals of materials having not very high melting points like Sb, Bi, Cd, In, Pb, Se, Te and PbTe and low melting point compounds like anthracene. Some crystals grown by the Bridgman method are shown in Fig. 2.26.

2.3.3 *Verneuil Method*

In this method developed by Verneuil in 1902, a seed crystal mounted on an upright pedestal is heated in such a way that only the top of the crystal is heated to near melting temperature. The material to be grown is taken in the form of a fine powder which is continuously sprinkled on the molten top of the seed crystal. As the powder particles touch the molten top of the seed crystal, they melt and then freeze by losing heat to the seed crystal below, which acts as a heat sink. As the powder material solidifies at the top, the seed crystal is slowly lowered. Thus, a Verneuil set-up should have an arrangement to drop the powder, a gas-flow to ensure even distribution of the powder, a heating arrangement which provides high temperature in a narrow region (top of the seed crystal) and a lowering arrangement.

In initial attempts an oxy-hydrogen flame was used for heating. Later, oxy-acetylene and oxy-carbon monoxide flames were also used. To obtain a pointed flame, burners of different designs are used. A tricone burner is shown in Fig. 2.27 in which three concentric tubes are used. The inner tube carries hydrogen and the outer two carry oxygen. The gas flame heating is now superseded by R.F. induction heating. A typical set-up by Bauer and Field [2.11] employing R.F. heating is shown in Fig. 2.28 along with the vibrator-dropper.

Another heating technique is the plasma torch where streams of ionised gas pass through high currents. There are different plasma torches like D.C. plasma torch, A.C. plasma torch and inductively coupled plasma torch. In the last type (Fig. 2.29)

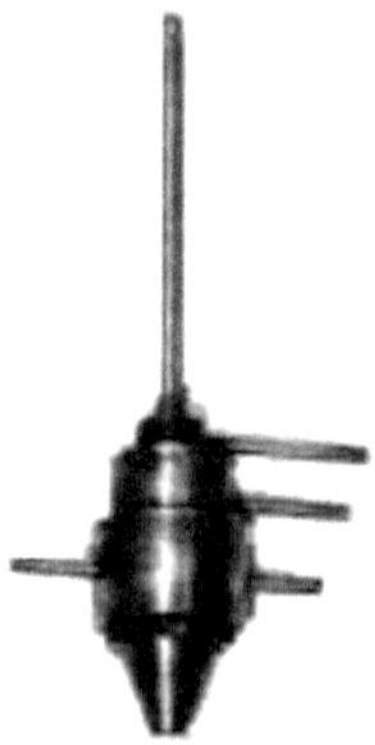

Fig. 2.27 A tricone burner

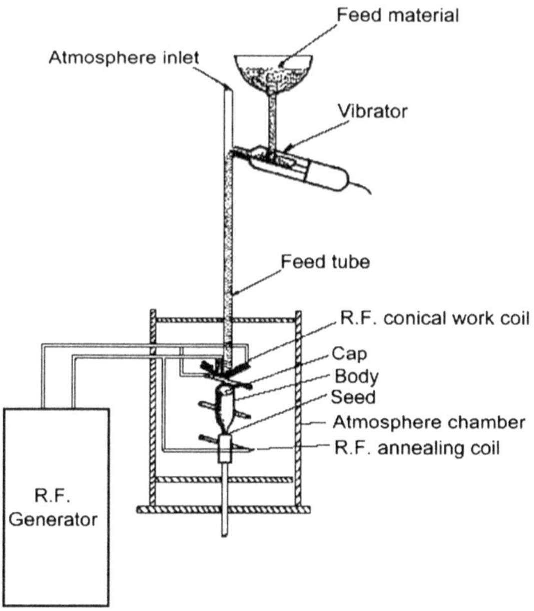

Fig. 2.28 A Verneuil set-up

there are three concentric quartz tubes with the induction coil just below the exit of the middle tube. The outer orifice is the inlet for the swirling gas. The torch is started by heating a graphite rod in the R.F. field till gaseous breakdown occurs; the hot probe is now withdrawn. Energisation of the plasma is done by the R.F. field.

Yet another heating technique is the arc-image method. The principle is shown in Fig. 2.30. The heat source is a high temperature electric arc. Parabolic reflectors are employed to collect and focus the radiation from the arc. Absorption of the focused light by an object can result in heating it to ~3,000°C.

Fig. 2.29 Inductively coupled plasma torch

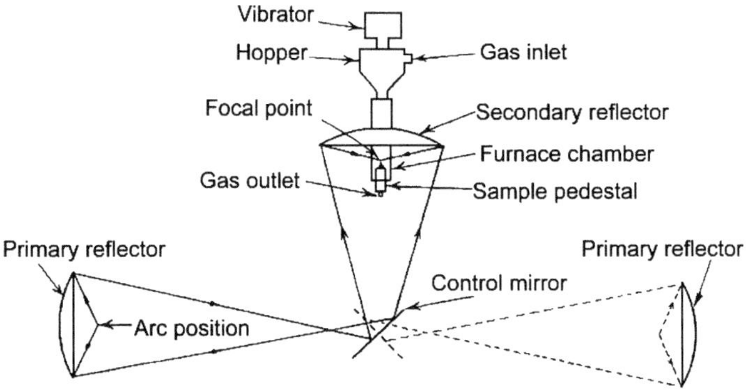

Fig. 2.30 Arc-image crystal growth system

The main advantage in this method is that it does not require a container. Thus this method is useful for growing crystals of materials which react with crucibles and tubes used in other melt growth techniques. At the same time, it has the disadvantage that crystals grown by this method contain impurities from the gases used and mechanical strains. Crystals of ferrites, spinels, garnets, rutile and gems like sapphire and ruby have been grown by the Verneuil method.

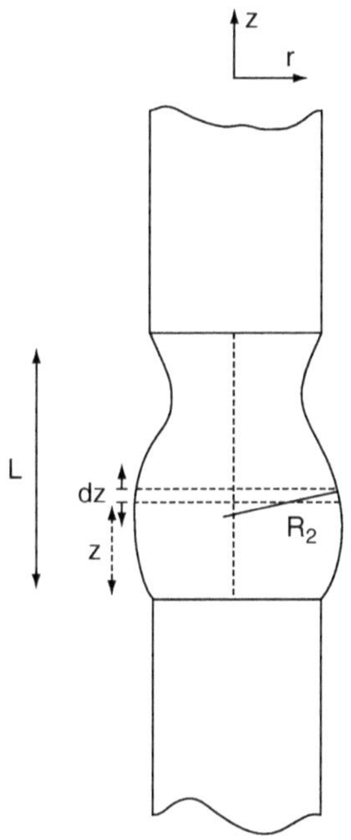

Fig. 2.31 Shape of liquid zone

2.3.4 *Floating Zone Melting Method*

The floating zone melting technique developed by Keck and Golay in 1953 opened up a new way for growing single crystals of very high melting point (>1,500°C) elements, alloys and semiconductors. This is another crucibleless method. As the term implies, the technique involves melting a narrow zone of the charge at a time and then moving the zone in a controlled manner. A polycrystalline rod of the material is held vertically and a zone of short length is heated to melting. When melting occurs as shown in Fig. 2.31, it is as if a floating molten zone is held between two rods. This molten zone is supported by the surface tension of the liquid. The mechanical stability of the floating zone is determined by the radius of the rods, the length of the zone, the surface tension of the melt, the density of the liquid and the shape of the liquid–solid interface.

Let us consider the factors that affect the maximum length and radius of the molten zone. Assuming cylindrical symmetry the outward hydrostatic pressure acting on an element of width dz at height z is $\rho_L\, g(L - z)$, ρ_L being the density of the molten liquid and L the length of the zone. There is also an inward pressure due to surface tension; this is $\gamma(R_1^{-1} + R_2^{-1})$ where γ is the surface tension and R_1 and R_2 are the radii of curvature in the plane of the figure and normal to it. It may be noted that R_1 and R_2 are related to how r varies with z. It can be shown that

$$R_1 = [1 + (\mathrm{d}z/\mathrm{d}r)^2]^{1/2}/(\mathrm{d}^2z/\mathrm{d}r^2) \tag{2.2}$$

and

$$R_2 = r[1 + (\mathrm{d}z/\mathrm{d}r)^2]^{1/2}/(\mathrm{d}z/\mathrm{d}r). \quad (2.3)$$

For stability these two pressures should balance. Thus, we have

$$\frac{(\mathrm{d}^2z/\mathrm{d}r^2)}{[1 + (\mathrm{d}z/\mathrm{d}r)^2]^{1/2}} + \frac{(\mathrm{d}z/\mathrm{d}r)}{r[1 + (\mathrm{d}z/\mathrm{d}r)^2]^{1/2}} = \frac{(L - z)\rho_\mathrm{L} g}{\gamma}. \quad (2.4)$$

From (2.4), the maximum zone length L_{max} and the maximum radius r_{max} consistent with stability are obtained as

$$L_{\mathrm{max}} = 2.84(\gamma/\rho_\mathrm{L} g)^{1/2}, \quad (2.5)$$

and

$$r_{\mathrm{max}} = 0.65(2\gamma/\rho_\mathrm{L} g)^{1/2}. \quad (2.6)$$

A typical experimental arrangement for the floating zone method is shown in Fig. 2.32. The starting rod is held at its end in chucks. For heating, several options are available like the conventional metal wire heating, r.f. induction, electron bombardment or radiation heating. For crystal growth to occur, the molten material has to be slowly moved out of the zone so that it cools and crystallises and a fresh part of the rod melts. This movement can be achieved either by moving the heater or the rods-zone assembly. The direction of travel also appears to matter. Thus,

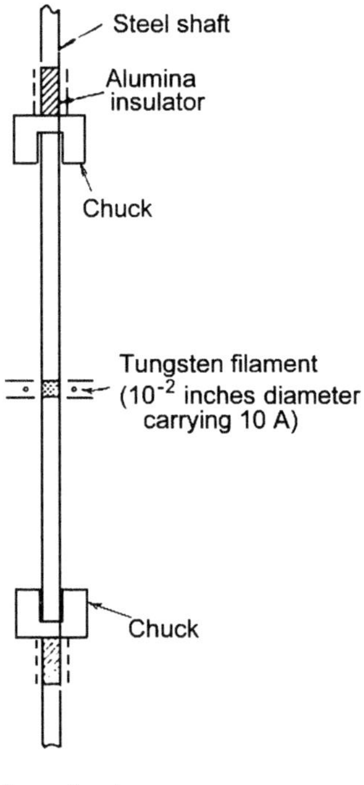

Fig. 2.32 Floating zone crystal growth set-up

it is observed that for growth of silicon crystals, upward movement of the zone is favourable. For metals, both upward and downward movements are equally suitable. If a seed is not used, the resulting single crystal will have random orientation. It is possible to have crystal growth with desired orientation by using an oriented single crystal seed as the upper rod.

A great advantage in this method is the high purity of the end product. The absence of a crucible and the possibility of providing vacuum, both contribute to purity. Further, if the starting material is pure, controlled amounts of a desired impurity (dopant) can be added. Ni, Gd, Fe, Ti, Pt, Rh, V, Mo and Ni–Fe alloys are some materials of which single crystals have been grown by this method.

2.3.5 *Materials and Accessories Used in Melt Growth*

The methods described in the preceding sections are high temperature experiments requiring equipment and accessories which can withstand these high temperatures. Some of these aspects will be discussed in this section.

2.3.5.1 Furnaces

Although flames, electron bombardment and arc-imaging are occasionally employed, the common method of heating is an electrically heated furnace. The furnace is often made of alumina or refractory cement; fused quartz tubes are also employed. The heating elements are wires of nichrome, Zr, Pt or (Pt-10% Rh) alloy for temperatures up to 1,800°C. Wires are generally preferred for making gradient furnaces as it is possible to vary the number of turns in the different parts of the furnace. For higher temperatures, rods of graphite, SiC, Ta or W are used.

2.3.5.2 Container Material

Crucibles should be made of materials which do not react with the charge, have a high shock resistance and low coefficients of thermal expansion and thermal conductivity. Pt, Pt-10% Rh alloy, graphite, alumina, beryllia and silicon are commonly used materials for crucibles and boats. For low melting point charges (M.P. < 500°C), pyrexware can be employed.

2.3.5.3 Temperature Measurement

The most commonly employed tool for measurement of temperature is the thermocouple. It occupies very little space, follows temperature changes rapidly and its positioning can be easily manoeuvred. The thermocouples in common use are (temperature range is given in parenthesis): copper–constantan (600°C), chrome–nickel–constantan (800°C), iron–constantan (900°C), Pt–Pt (10% Rh) (1,500°C) and tungsten–(tungsten/rhenium) (2,000°C).

Resistance thermometers made of Ni and Pt are useful up to 300 and 1,000°C respectively. The equation used to estimate temperature from a platinum resistance thermometer is

$$(R - R_0)/R_0 = At + Bt^2, \tag{2.7}$$

where $R$and R_0 are the values of the resistance at temperature t and 0°C respectively. A and B are constants with values 3.978×10^{-3}/degree and 5.84×10^{-7}/degree^2 respectively. The resistance thermometers occupy more space than the thermocouples.

Another instrument for measurement of high temperatures is the optical pyrometer. In this, the visible radiation from a hot body (furnace, crucible, growing crystal) is focused on the incandescent filament of a tungsten lamp. When observed, the images of both the filament and the hot body are seen, one darker than the other. The current in the filament is adjusted until its image merges into that of the hot body (i.e. the contrast between them vanishes). This happens when the temperature of the filament equals that of the hot body. The filament temperature is previously calibrated. The advantage of this method is that it is a no-contact temperature probe.

2.3.5.4 Temperature Control

As important as the production and measurement of temperature is the control of temperature. The details of several electronic temperature controllers are to be found in books on electronics and instrumentation. As an example, a temperature controlling system based on the use of a resistance is shown in Fig. 2.33. The unit is set for a desired temperature. Any change in the temperature changes the resistance in the circuit. This is restored by changing the current in the furnace windings, i.e. by changing the temperature back to the set temperature.

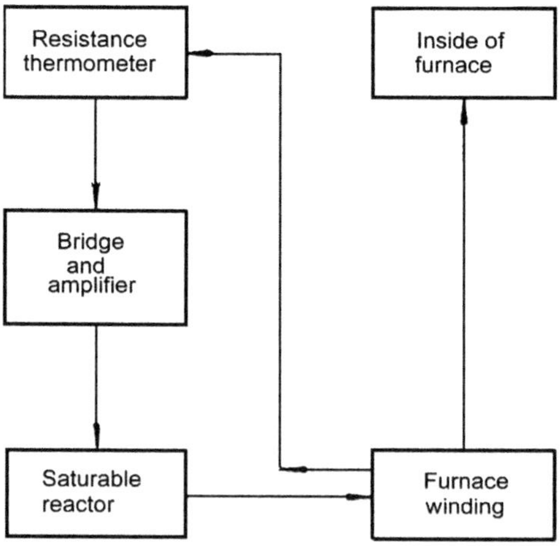

Fig. 2.33 Resistance temperature control system

2.3.5.5 Atmosphere

A clean atmosphere inside the crystal growth unit is desirable to ensure the growth of a pure crystal. Vacuum of the order of 10^{-5} torr is possible with routinely available vacuum machines. For handling charges which are reactive, vacuum of the order of 10^{-11} torr is required. In some situations, after evacuation of air, H_2, A or He gases are let in.

2.4 Crystal Growth from Vapour

2.4.1 General

The method of crystal growth from vapour consists in producing the vapour of a substance in coexistence with some unevaporated solid nutrient, i.e. the vapour is saturated. The saturated vapour is then transported from a region at temperature t_1 to a region at temperature $t_2 (t_1 > t_2)$ resulting in supersaturation. If the supersaturation is controlled properly, crystalline nuclei are formed which allow further development of a good single crystal. For single crystal growth, the vapour pressure should be between 10^{-3} and 1 atm; otherwise there is a tendency to form dendrites. To facilitate crystal growth, the system should be in equilibrium with a small temperature difference between the nutrient and the growing crystal surface. The parameters to be controlled are the supersaturation, the growth temperature, the rate of vapour transport and the flow of any entraining gas.

Vapour growth is carried out in two ways: direct condensation and chemical transport. In the first method, as the term indicates, the vapour of the material to be grown directly condenses on the walls of the growth vessel or a substrate. In the second method, the material forms a compound in the vapour state which is then transported to another region where it dissociates and then deposits on a substrate. To bring out more details of the vapour growth method, some specific examples are now discussed.

2.4.2 Vapour Growth by Condensation

2.4.2.1 Static Container Method

The set-up for the growth of CdO crystals is shown in Fig. 2.34a. CdO powder is compressed into small spheres and placed in a cylindrical alumina crucible which is placed in a furnace with a precalibrated temperature gradient. The lower part of the crucible is at ~1,000°C. At high temperature, the CdO charge sublimes according to the reaction

$$2\ \mathrm{CdO\ (s)} = 2\ \mathrm{Cd\ (g)} + \mathrm{O_2\ (g)}. \qquad (2.8)$$

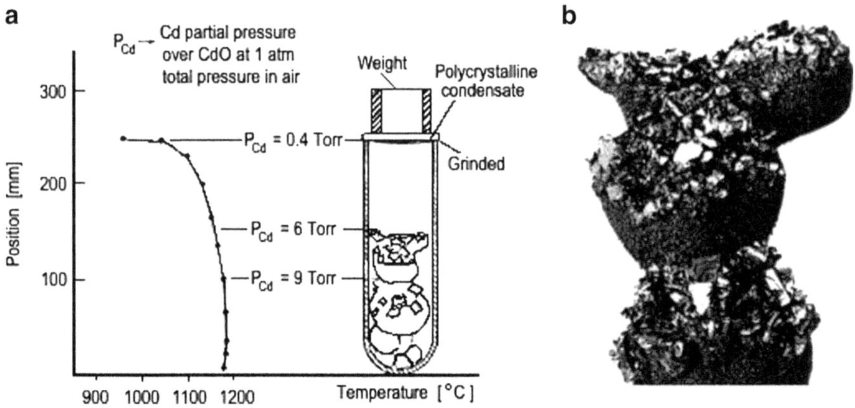

Fig. 2.34 (**a**) Vapour growth static container method (SCM). (**b**) CdO crystals grown by SCM

The vapour pressures of Cd and O_2 vary along the height of the crucible as shown in the figure. The Cd and O_2 vapours rise into the upper cooler regions; here, they combine to form CdO and condense on the unsublimed surface of the charge. The crystals are 60–150 mm^3 in size (~3–6 mm linear dimensions) with the (100) faces showing up (Fig. 2.34b).

2.4.2.2 Moving Container Method

In this method (Fig. 2.35), the charge (CdS, for example) is taken in an evacuated quartz tube (Fig. 2.36) with a conical upper end. The tube is suspended in the furnace, very much as in the Bridgman method. The furnace is heated and the tube is positioned at the middle where the temperature is ~1,200°C. At temperatures above 1,000°C, CdS dissociates according to the equation

$$2\,\mathrm{CdS} \leftrightarrow 2\,\mathrm{Cd} + \mathrm{S}_2. \tag{2.9}$$

As the tube is pulled upward, its upper end is the first to enter the lower temperature region. The vapours condense and form a crystal seed at the top, which grows into a single crystal as the tube moves up. A large CdS crystal grown in this way is shown in Fig. 2.37.

2.4.3 Vapour Growth by Chemical Transport

As an example of crystal growth by chemical transport, we shall consider the growth of thin germanium layers. Growth of Ge by conduction of Ge vapour directly on a substrate is difficult as Ge evaporates at temperatures which are too high to retain

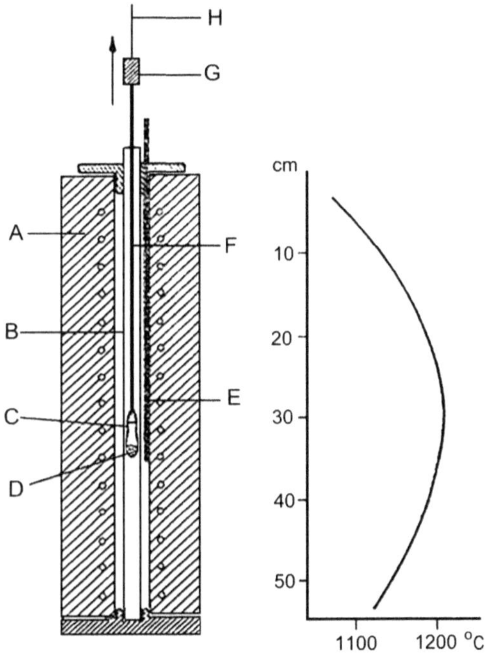

Fig. 2.35 Vapour growth by moving container method (MCM): A – furnace, B – quartz tube lining the furnace, C – ampoule, D – polycrystalline CdS, E – Pt–PtRh thermocouple, F – quartz tube suspending the ampoule, G – counterweight, H – stranded Kanthal or tungsten wire

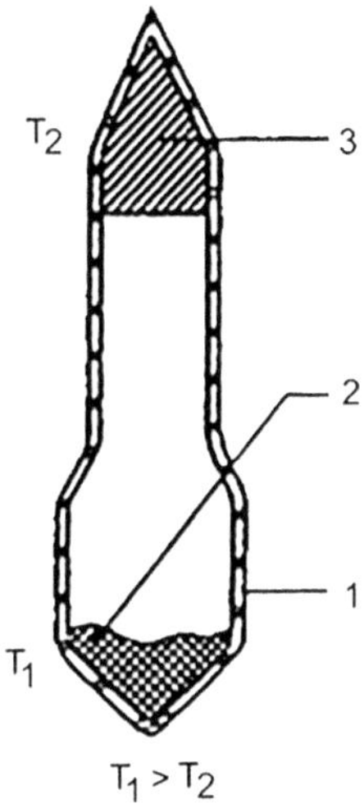

Fig. 2.36 Details of ampoule used in MCM: 1 – closed glass or quartz ampoule, 2 – polycrystalline material, 3 – single crystal

stability of the substrate. The chemical transport method facilitates the growth of Ge at much lower temperatures (400–600°C) by using the reaction

$$2\ \mathrm{Ge} + 2\mathrm{I_2} \xrightarrow{600^\circ\mathrm{C}} 2\ \mathrm{GeI_2} \xrightarrow{400^\circ\mathrm{C}} \mathrm{GeI_4} + \mathrm{Ge}. \qquad (2.10)$$

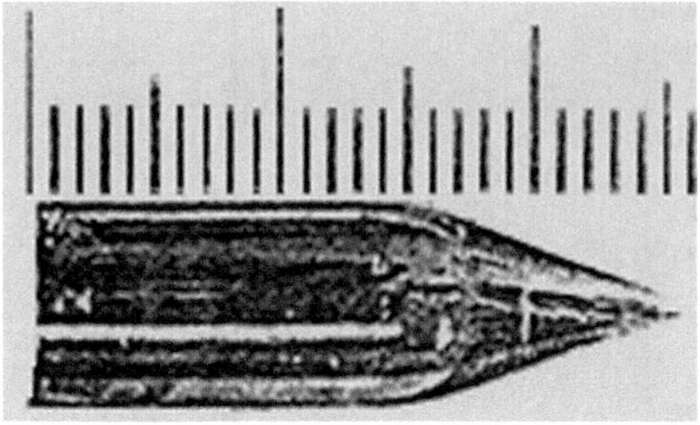

Fig. 2.37 CdS crystal grown by MCM

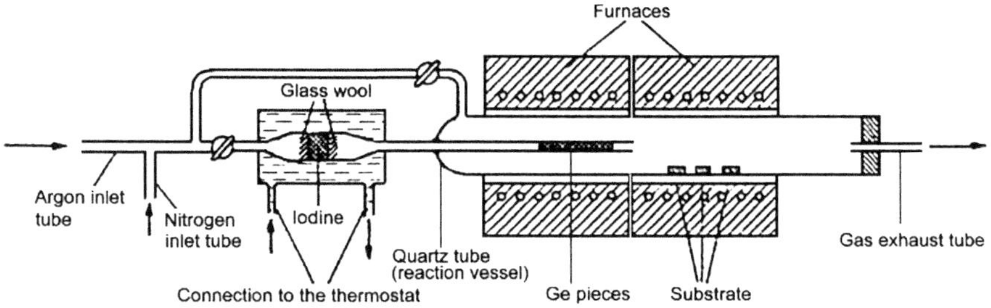

Fig. 2.38 Chemical transport system set-up for growth of Ge

The necessary set-up is shown in Fig. 2.38. Iodine powder is kept in a pyrex glass tube which is in a thermostat that maintains a temperature of ~70°C. The iodine sublimes and the vapour is transported by a stream of argon gas to the quartz reaction vessel which contains some polycrystalline Ge. This region is in the furnace which is at 600°C. Here the I_2 vapour reacts with Ge to form GeI_2. The argon stream transports the GeI_2 vapour to the right where the furnace is at 400°C. The GeI_2 disassociates and forms free Ge which deposits on the substrate placed at the bottom of the vessel. The crystalline Ge layers are about 10 μm thick. The orientation of the layer is dependent on the crystallography of the substrate; generally, Ge, GaAs or GaP crystals are used as substrate for growth of Ge.

This method is employed when crystals with large surface areas are required as in semiconductor device technology.

2.5 Epitaxial Growth

The oriented growth of a thin layer of a crystal on another crystal (substrate) is called epitaxy. The deposition of the epitaxial layer may be from any phase – solution, melt or vapour; we shall consider an example of each type.

Fig. 2.39 Epitaxial growth of NaBr from solution on NaCl

Fig. 2.40 Epitaxial growth of KBr from melt on mica

2.5.1 Epitaxial Growth from Solution

We shall consider, as an example, the growth of NaBr on NaCl substrate. A drop of 20% aqueous solution of NaBr is placed on a freshly cleaved surface of NaCl. On evaporation, tiny oriented NaBr crystals are formed (Fig. 2.39). The edges of the square or rectangular NaBr crystals are parallel to the ⟨100⟩ edges of the substrate NaCl crystal.

2.5.2 Epitaxial Growth from Melt

As an example, we shall consider the growth of KBr crystals from melt using mica as substrate. A small quantity of KBr powder is spread on a thin mica sheet. The mica sheet is heated from below on a gas flame till the KBr melts. When the molten KBr is allowed to cool, KBr crystallites are formed (Fig. 2.40). The crystallites are triangular and the orientation is such that the (111) planes of KBr grow on the (100) plane of mica.

Fig. 2.41 Epitaxial growth of KBr from vapour on mica

2.5.3 *Epitaxial Growth from Vapour*

Here, again, we shall consider growth of KBr on mica. A little KBr powder is introduced in the crevice of a partially cleaved mica sheet. On heating, the KBr particles melt and, on further heating, KBr vapour is formed. The upper mica sheet is torn off when the smoke-like KBr vapour condenses on the lower mica layer. It can be seen under the microscope (Fig. 2.41) that in this case also the epitaxial KBr is in the form of triangles.

These are oversimplified examples of epitaxial growth. Although epitaxial growth is possible from any phase, generally growth from vapour is preferred. Epitaxial growth is very important in fabrication of semiconductor devices. Hence, elaborate equipment is designed for epitaxial growth from vapour with control on vacuum, temperature and thickness. A further advanced technique is molecular beam epitaxy (MBE). The details of MBE are to be found in specialised texts.

2.6 Crystal Growth by Gel Method

The gel method of crystal growth is a variant of the solution method. Here, the substance to be crystallised is precipitated in a reaction between two solutions. This reaction takes place in a gel. Typically, the pores in a gel have an effective diameter of 50–100 Å and the pores are separated by a solid film of 2×10^{-5} cm thickness. Diffusing through the gel, the two reactants reach the critical saturation at some places which results in the formation of a few nuclei. Further supply of the nutrients to the nuclei helps them to grow into crystals. The gel framework acts like a three-dimensional matrix or a container in which the crystal nuclei are delicately held.

Certain crystals grow well in certain gels. Hence a choice of gel is desirable. The following gels are commonly used:

Silica gel: This is prepared by mixing aqueous solution of sodium silicate or sodium metasilicate with an acid (1–4 N of a mineral acid or an organic acid).

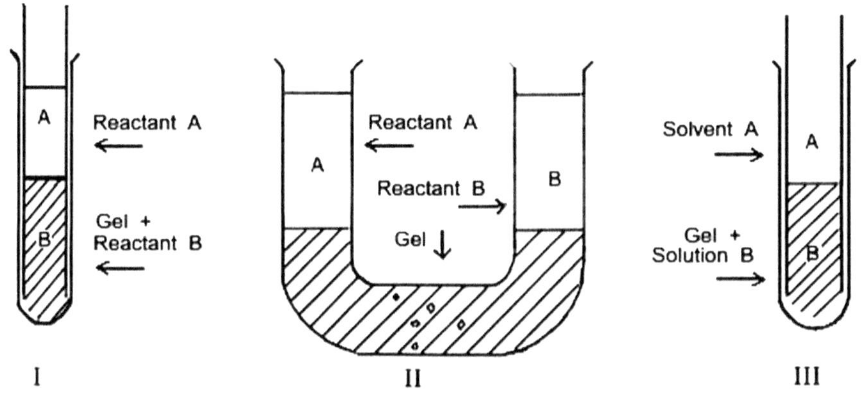

Fig. 2.42 Methods of crystal growth from gel

Table 2.6 Reactions for growth of some crystals by gel method

Crystal to be grown	Reaction
Calcium tartarate tetrahydrate	$C_4H_6O_6 + CaCl_2 = CaC_4H_4O_6 \cdot 4H_2O + 2HCl$
$CaC_4H_4O_6 \cdot 4H_2O$	(Tartaric acid)
Potash alum	$Al_2(SO_4)_3 + K_2SO_4 = 2KAl(SO_4)_2 \cdot 12H_2O$
$KAl(SO_4)_2 \cdot 12H_2O$	
Calcite	$CaCl_2 + Na_2CO_3 = CaCO_3 + 2NaCl$
$CaCO_3$	
$BaSO_4$	$BaCl_2 + Na_2SO_4 = BaSO_4 + 2NaCl$
PbI_2	$Pb(NO_3)_2 + 2KI = PbI_2 + 2KNO_3$
$PbMoO_4$	$Pb(NO_3)_2 + (NH_4)_2MoO_4 = PbMoO_4 + 2NH_4NO_3$

Gelatin: It is prepared by dissolving 5–6 g of gelatin powder in water at a temperature of 50°C and cooling it to room temperature. Addition of 1 ml of formaldehyde improves the gel.

Agar: This is prepared by dissolving a few grams of agar powder in water, boiling the solution and then cooling it to room temperature.

In the commonly employed version of the method (method I), one of the reactants is taken in the form of solution and is dispersed in the gel. The other reactant taken as solution is kept above the set gel and diffuses slowly into the gel. The slow reaction results in nucleation and subsequent growth. In a variation (method II), both reactants are taken in the form of solution and are allowed to diffuse into the set gel from two sides. In another modification (method III), the substance to be crystallised is dissolved in solvent A (generally water) and the solution is dispersed in the gel. Another solvent B is now allowed to diffuse into the gel. This solvent is so chosen that the solubility of the substance is less in B than in A. Then as B diffuses in the gel, due to the reduced solubility the substance precipitates resulting in nucleation and growth.

The experimental arrangements for methods I and II are shown in Fig. 2.42. For these, pyrex test tubes and U-tubes with diameter 1–4 cm and 20–30 cm in height are suitable. The arrangement for method III is also shown in Fig. 2.42.

Table 2.7 Growth of some crystals by method III

Crystal	Aqueous solution in gel	Solvent above gel
NaCl	NaCl	HCl or CH_3CH_2OH
KH_2PO_4	KH_2PO_4	CH_3CH_2OH
$(NH_4)H_2PO_4$	$(NH_4)H_2PO_4$	NH_4Cl

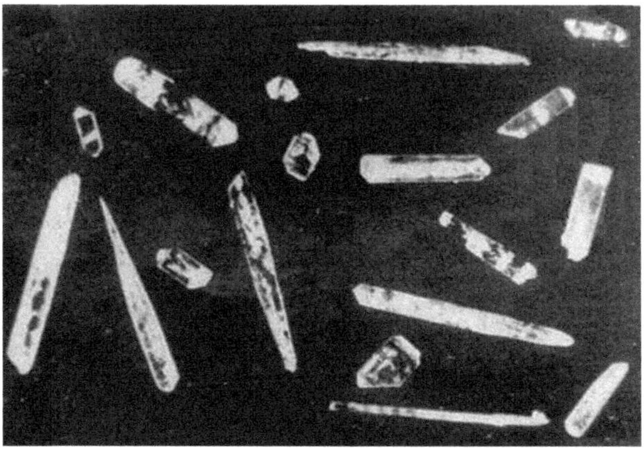

Fig. 2.43 Some crystals grown from gel

The reactions for growth of some crystals by methods I and II are given in Table 2.6. Similarly, the reactants for the growth of some crystals by method III are given in Table 2.7. Some crystals grown by the gel method are shown in Fig. 2.43.

The growth of crystals by the gel method depends on the concentration of the reactant solutions and the pH of the gel. Gel growth is one of the simpler techniques of crystal growth and can be tried in any laboratory. Though the method is simple, it is a slow process and it takes several days, if not weeks, for a crystal to grow. The resulting crystals are small in size, typical linear dimensions being a few mm. The crystals have to be removed carefully from the gel. The crystals grown by the gel method are good in quality with a low dislocation density.

2.7 Miscellaneous Methods

Some of the more commonly used methods of crystal growth have been discussed in earlier sections. A few other methods used in special situations are briefly described here.

2.7.1 Crystal Growth by Electrolysis

Metal crystals can be grown by electrolysis of solutions or molten salts containing metals as cations. Thus, iron crystallites are grown by the electrolysis of acidified

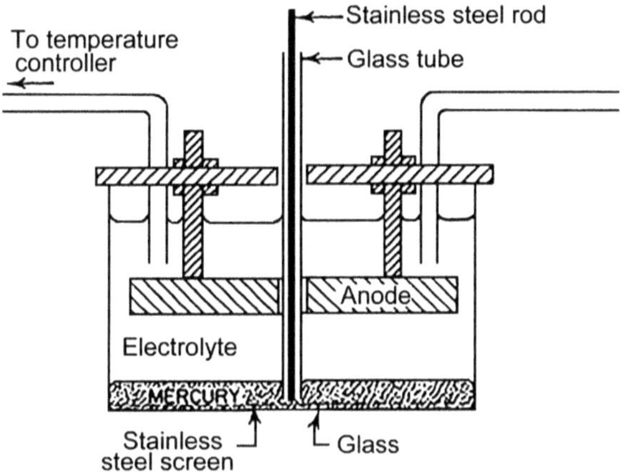

Fig. 2.44 Cell for electrodeposition of iron crystals

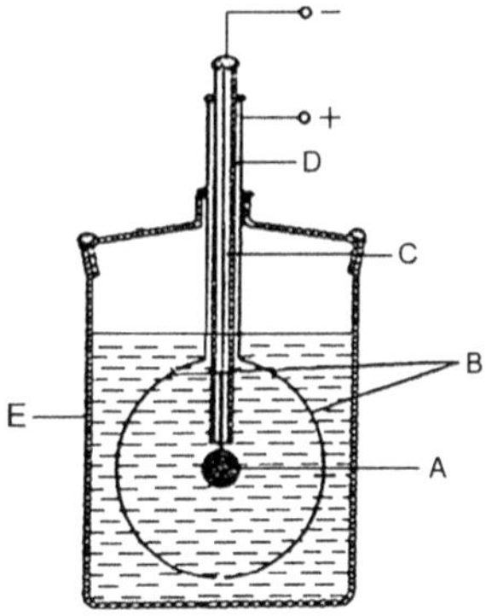

Fig. 2.45 Electrolysis tank for growth of Ag crystals: A – silver seed as cathode, B – silver shells as anode, C – silver wire to hold seed, D – porcelain tube, E – cylindrical electrolysis tank

ferrous sulphate solution. The cell used for electrolysis is shown in Fig. 2.44. The pool of mercury acts as the cathode. Copper crystals can be grown by the electrolysis of CuCl or CuBr solutions. In these cases, the electrolysis is done at room temperature. Single crystals of silver can be grown by electrolysis of molten AgCl or AgBr or AgI. Obviously, electrolysis is carried out at elevated temperatures by keeping the electrolysis cell in a furnace. Further, a microcrystal of Ag is used as a seed. The electrodes are two hemispheric shells made of silver and the suspended seed crystal is at the centre of these shells (Fig. 2.45). This configuration focuses the electric field onto the seed crystal. A faceted silver crystal is shown in Fig. 2.46.

2.7.2 *Recrystallisation*

Recrystallisation is a process in which individual single grains in a polycrystalline sample are made to grow at the expense of neighbouring smaller grains. The recrystallisation process involves alternately deforming and annealing the sample at elevated temperatures.

Fig. 2.46 Electrolytically grown Ag crystal

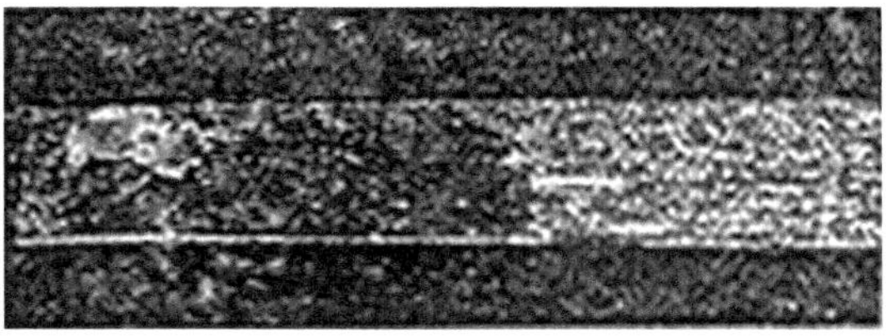

Fig. 2.47 Recrystallised Al crystal (the bright region on the *right* is the recrystallised metal)

As an example, let us consider the recrystallisation of Al. A polycrystalline Al strip reveals single crystalline grains when observed under a microscope. The average grain size can be measured. Next, the strip is deformed by rolling to 50–70%. After deformation, it is slowly moved through a furnace at 600°C. An increase in grain size is observed. By repeating this process of alternate deformation and annealing, single crystals 1–2 cm in length can be grown. A partially recrystallised Al sample is shown in Fig. 2.47. Depending on the sample, the deformation can be carried out by stretching (a wire) or compressing (a powder) in a die.

2.7.3 Crystal Growth of Diamond

Diamond cannot be grown by any of the methods described so far. It can be grown by heating graphite under high pressure. Metals such as Mn, Fe, Ni and Pt act as catalysts in the high pressure–high temperature conversion of graphite into diamond.

The equipment to provide the desired temperature and pressure is shown in Fig. 2.48; it was designed by Hall [2.12] and is known as the "Belt". Two tapered pistons compress a sample held in a short tapered cylinder. These are made of cemented tungsten carbide. The sample itself is held in pyrophylite and is heated by electrical resistance. The apparatus can provide pressures up to 100 kbar and temperatures up to 3,000°C. The reaction capsule is shown in Fig. 2.49. A central graphite cylinder is capped at each end by the catalyst metal. When the assembly is heated to 2,000–2,500°C at pressures of 60–80 kbar, the graphite forms an alloy

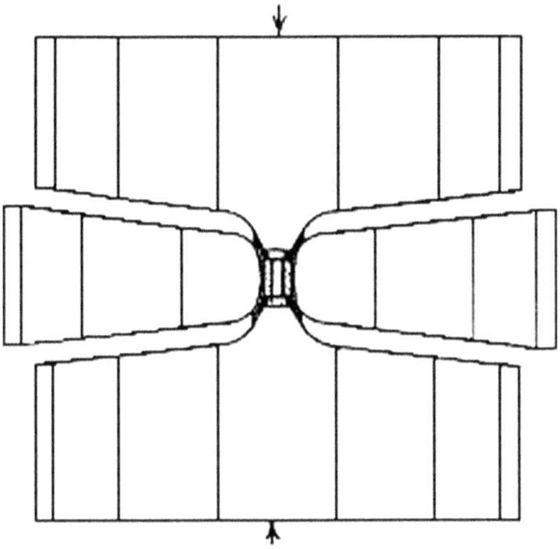

Fig. 2.48 High-pressure high-temperature cell for growing diamond

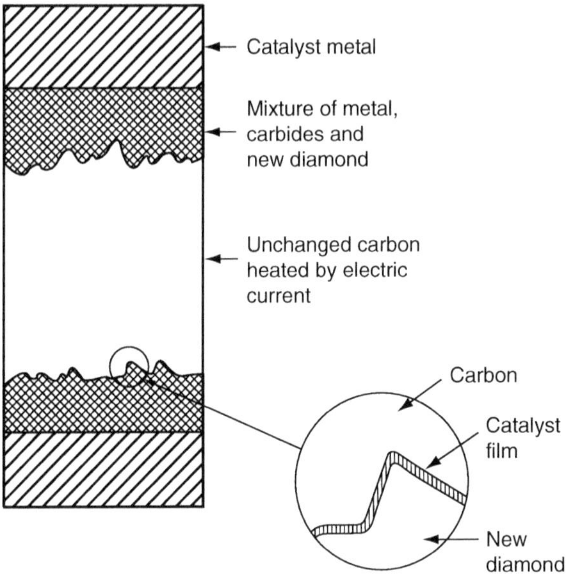

Fig. 2.49 Reaction cell for diamond growth

with the metal. The catalytic act of the metal now comes into play and a thin layer of graphite gets converted into diamond. As time passes, the process progresses leaving the entire mass as a mixture of diamond crystals and the catalyst alloy. A diamond crystal grown in this way is shown in Fig. 2.50 [2.13].

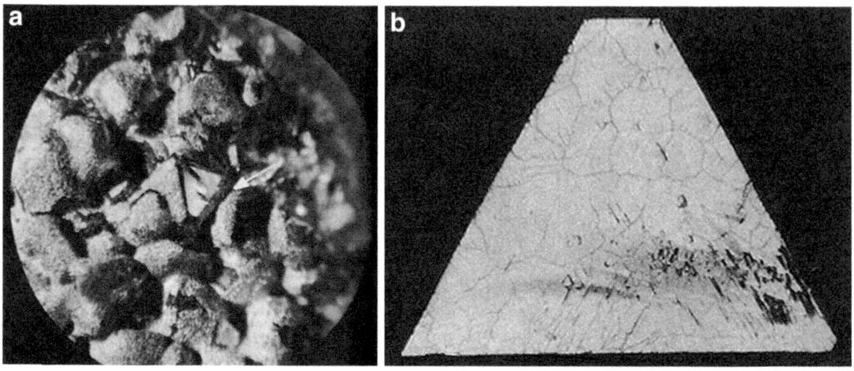

Fig. 2.50 (**a**) Diamond crystals on graphite pellet. (**b**) Enlarged view of a diamond crystal

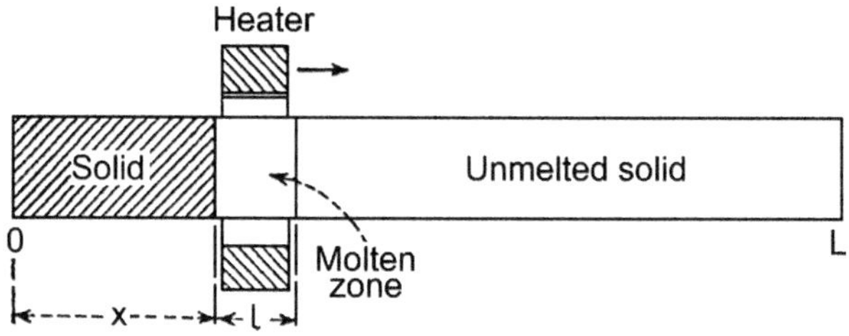

Fig. 2.51 Molten zone in a solid rod

2.8 Zone Refinement

2.8.1 Principles

Pfann [2.1] raises a thoughtful question: "Of what use is a detailed investigation of some substance which contains unspecified impurities or imperfections that later are shown to have a critical bearing on its behaviour?" Methods of crystal growth and experimental techniques of property measurement have advanced so much that it has become necessary to have crystals of high purity and perfection. The aspects of perfection will be taken up in a later chapter. Here we shall consider the purification of materials. The need for purity is all the more in the case of semiconductor materials as unknown impurities can affect device performance. On the other hand, device performance can be improved and controlled by the deliberate addition of chosen impurities. Zone refinement introduced by Pfann [2.14] is a powerful method to attain high purity in materials. The theory and practice of zone refinement is discussed in detail by Pfann [2.1, 2.15]. Here, we shall have a brief and simplified discussion of the basic principles.

Let us consider a bar of a solid (Fig. 2.51) of which a narrow zone is heated till it melts. It is assumed that there is equilibrium between the solid and the liquid (molten) phases. The solid and its molten liquid are regarded as two solvent phases,

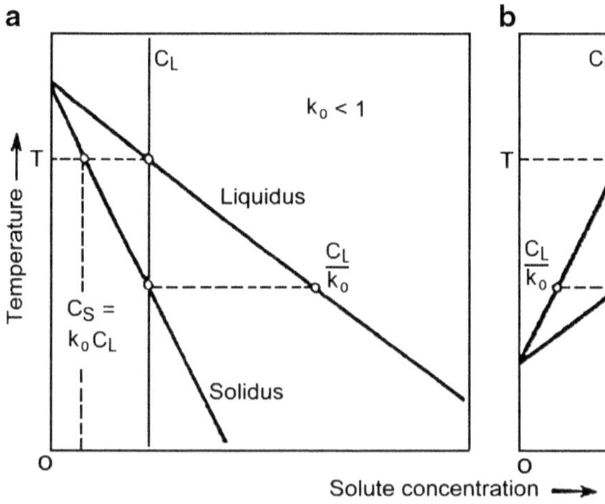

Fig. 2.52 Constitutional diagrams for (**a**) $k_0 < 1$ and (**b**) $k_0 > 1$

in which an impurity acts as a solute, with equilibrium concentrations C_S and C_L respectively. We define an equilibrium distribution coefficient or segregation coefficient k_0 as

$$k_0 = C_S/C_L. \tag{2.11}$$

Typical constitutional diagrams for an impurity and a solvent are shown in Fig. 2.52. It is seen that some impurities (Fig. 2.52a) lower the melting point of the solvent. In this case $C_S < C_L$ and $k_0 < 1$. On the other hand, there are impurities (Fig. 2.52b) which raise the melting point; here, $C_S > C_L$ and $k_0 > 1$.

Let us assume that the solid bar in Fig. 2.51 contains an impurity with $k_0 < 1$, i.e. $C_S < C_L$. If the molten zone moves to the right, the melt freezes, and in so doing, it rejects the excess amount of impurities (remember, $C_S < C_L$), into the newly formed molten zone through the liquid–solid interface. Thus as the zone continues to move, it carries with it more and more of the impurity. The end of the bar which is the last to solidify is rich in impurity and is cut off. For an impurity with $k_0 > 1$, the reverse happens, i.e. the liquid will reject the impurity into the solid. Thus the freezing interface rejects certain impurities and accepts others. The k_0 values of some common impurities are given in Table 2.8. It is seen that in most cases $k_0 < 1$ but there are impurities like boron for which $k_0 > 1$.

We shall now consider the distribution coefficient in greater detail. The definition given in (2.11) applies when the freezing proceeds slowly, i.e. at low velocities of the liquid–solid interface. At high interface velocities, the advancing solid rejects the solute (impurity) more rapidly than it can diffuse into the bulk of the liquid and hence a layer enriched in the solute develops ahead of the interface. The C_S/C_L ratio is now dependent on the concentration of the solute in the layer rather than in the bulk of the liquid. The distribution coefficient is no longer k_0 but has an effective

Table 2.8 Segregation coefficients (k_0) for Ge and Si

Impurity	k_0 (Ge)	k_0 (Si)
P	0.12	0.04
As	0.04	0.07
Sb	0.003	0.002
B	20	0.68
Al	0.1	0.0016
Ga	0.1	0.004
In	0.001	0.0003

value k given by

$$k = k_0/[k_0 + (1 - k_0)\exp(-\Omega)]. \tag{2.12}$$

with

$$\Omega = f\,\delta/D. \tag{2.13}$$

Here, f is the growth rate of the solid, δ the thickness of the liquid film in which transport is only by diffusion and D the diffusivity of the liquid. The growth parameter Ω is a dimensionless quantity which may be regarded as a normalised growth velocity. It can be seen that for small values of Ω, $k \rightarrow k_0$ and for large values of Ω, $k \rightarrow 1$; k thus lies between k_0 and 1.

Once a zone passes across the length of the bar, the solute concentration in the solid is no longer uniform but becomes a function of the distance x along the length of the bar. The concentration $C_S(x)$ after a single pass of the zone is

$$C_S(x) = C_0[1 - (1 - k)\exp(-k\,x/l)]. \tag{2.14}$$

Here, C_0 is the initial concentration of the solute at $x = 0$ and l the zone width. This equation does not apply to the last section of width l of the bar. The solute concentration for a single pass according to (2.14) is shown in Fig. 2.53 for different values of k. For further purification, the traverse of the zone across the length of the bar should be repeated. However, the time required can be considerably saved by having a train of closely spaced heaters (Fig. 2.54). A single traverse of a system having n heaters is equivalent to n traverses with a single heater. After a number of passes, the solute concentration reaches an ultimate distribution. This ultimate distribution is given by

$$C_S(x) = A\exp(Bx), \tag{2.15}$$

where A and B are constants obtainable from

$$k = Bl/\exp(Bl) - 1, \tag{2.16}$$

$$A = C_0 BL[\exp(BL) - 1]. \tag{2.17}$$

The ultimate distribution attainable for several values of k for an ingot of length $L = 10l$ is shown in Fig. 2.55. As an exercise, let us estimate the number of passes required to achieve the ultimate distribution. If $k = 0.1$, $L = 10$ and $l = 1$,

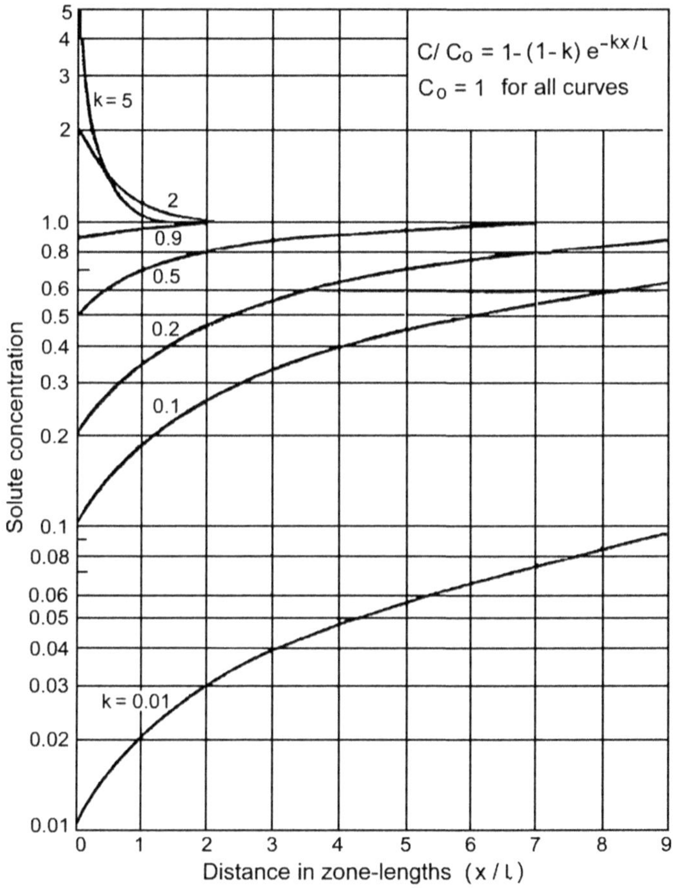

Fig. 2.53 Solute concentration versus distance curves for single-pass zone melting for various values of the distribution coefficient k

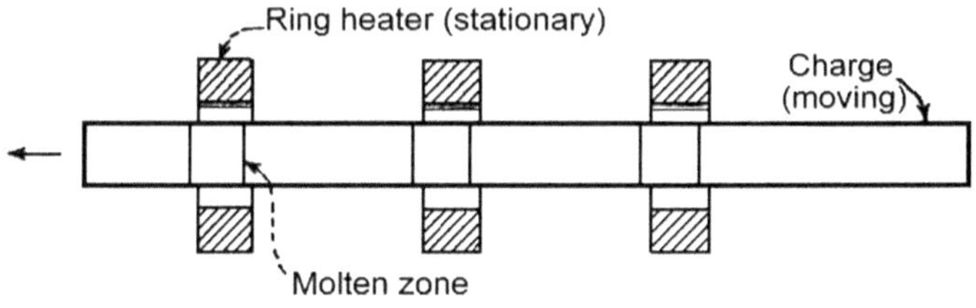

Fig. 2.54 Schematic of multiple-pass zone melting

the ultimate value of C_S at $x = 0$ is 10^{-14} C_0. Since the concentration at $x = 0$ decreases by a factor k in each pass, at least 14 passes are necessary to reach the ultimate distribution. It should be emphasised that for given values of k, l and L, it is not possible to purify a rod beyond the ultimate distribution.

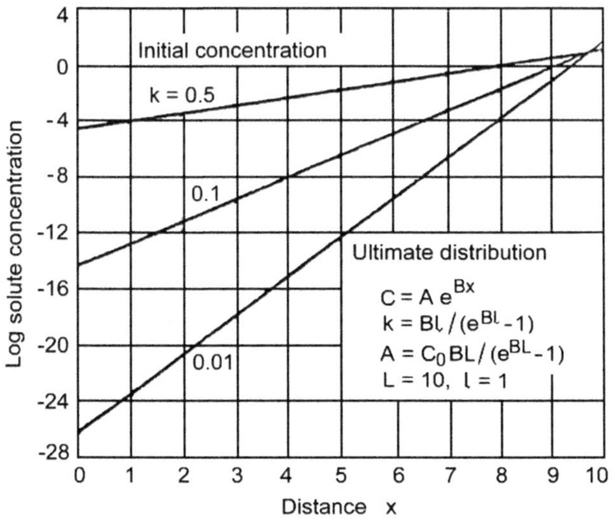

Fig. 2.55 Ultimate distribution in zone melting

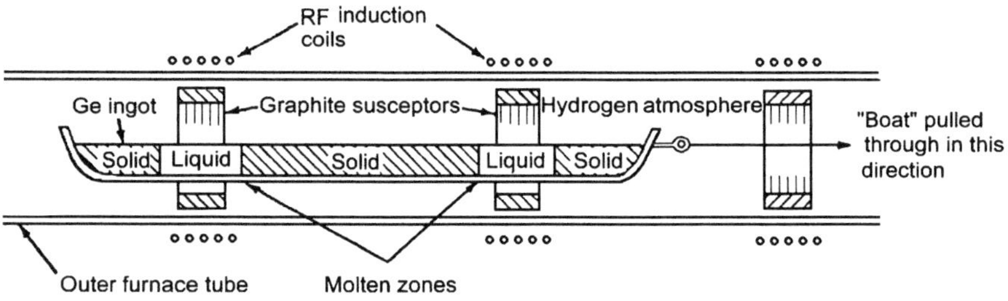

Fig. 2.56 Zone refining set-up

2.8.2 Experimental Details

Figure 2.56 shows a typical arrangement for zone refinement. The starting material is taken in a silica boat coated inside with graphite. The boat can slide over polished graphite rails. The boat is enclosed in a transparent silica tube. The tube is closed at both ends but allows inlet and outlet ports for flushing the tube with a 3:1 mixture of N_2 and H_2 gases. The R.F. induction coils operate at 500 kHz. In the figure three zones are shown. The movement of the zones is effected by pulling the boat (as shown in the figure) or by pulling the induction heaters in the opposite direction.

In historic experiments, Pfann and Oslen [2.16] zone-refined Ge using a set-up with six heaters. They succeeded in reducing the concentration of impurities like Cu, Ni, Fe and the group II, III, IV, VI elements to less than 1 in 10^{10} atoms of Ge; ingots of 0.5–1 kg in weight were produced. Ge being important industrially, the

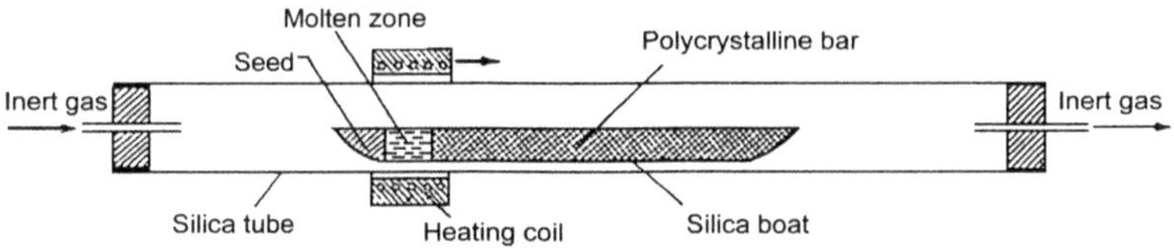

Fig. 2.57 Crystal growth by zone melting

reduction of various elemental impurities has been systematically investigated and the depletion of all solid impurities "detectable on the solid mass spectrograph" to a level of 1 in 10^7 atoms has been achieved.

Zone refinement has been used for the purification of materials like Si, other elements and III–V compounds and alkali halides. Zone refinement equipment is now commercially available.

2.8.3 Applications of Zone Refinement

Two important applications of zone refinement deserve mention. The first is that if the zone velocity is sufficiently slow, the zone refinement set-up can as well be used for simultaneous purification and single crystal growth. In fact by introducing an oriented single crystal seed at the starting end of the boat (Fig. 2.57), a single crystal with desired orientation can be grown. Single crystals of Sn, InSb, KCl and several other materials have been grown by this method.

The other application is the use of zone refinement in the formation of *p–n* junctions. Suppose a charge of Ge contains both boron (which is an acceptor) and phosphorous (which is a donor). The k_0 values of B and P are 20 and 0.12. If the Ge charge is now subjected to zone refinement, B and P are driven to opposite ends and there will be a section where adjacent regions are B-rich and P-rich, i.e. that section is a *p–n* junction. Such *p–n* junctions in Ge observed accidentally by Pfann and Scaff [2.17] are shown in Fig. 2.58. This technique was subsequently systematised and it is possible to grow not only *p–n* junctions but also *n–p–n* junctions by zone refinement.

2.9 Crystal Growth Theory

2.9.1 General

As mentioned in Sect. 2.1 in this chapter, the emphasis is on the experimental aspects of crystal growth. Accordingly, a large number of experimental methods have been

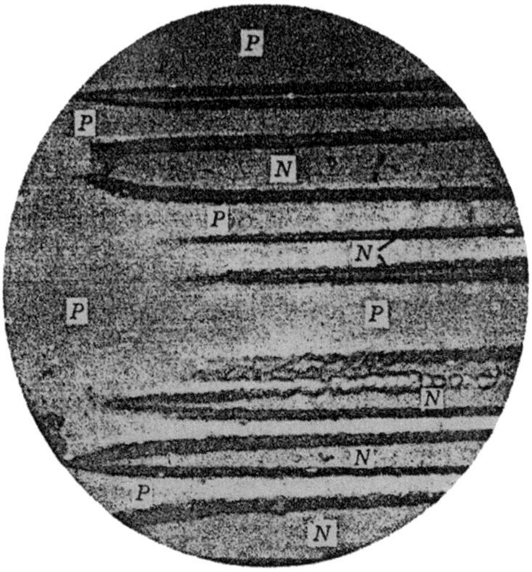

Fig. 2.58 *p–n* junction formation during zone refinement

described. For completeness, theoretical aspects of crystal growth, particularly the growth mechanisms, will be briefly and qualitatively discussed.

In crystal growth, we start with a homogeneous phase (phase I) which may be a vapour, a solution or melt. The vapour is saturated at pressure P_0 and becomes supersaturated when the pressure changes to P; the coefficient of supersaturation is defined by $S = P/P_0$ and the relative supersaturation is defined by $\sigma = S + 1$. Similarly, in a solution if the saturation concentration C_0 changes to $C, S = C/C_0$ and $\sigma = S + 1$. A melt which is at temperature T_{m} may be supercooled to T; the corresponding parameters are: coefficient of supercooling ($= T/T_{\mathrm{m}}$) and the coefficient of relative supercooling [$= (T–T_{\mathrm{m}})/T$].

Supersaturation (and also supercooling) is a metastable state and in this condition, crystallisation commences in the form of growth nuclei; this is phase II. In this process of phase change (I → II), the free energy changes by ΔG_V for unit volume. ΔG_V for the three states is given by

$$\begin{aligned} \Delta G_V &= -aRT \ln(P/P_0) = -aRT \ln S \quad \text{(Vapour)}, \\ &= -aRT \ln(C/C_0) = -aRT \ln S \quad \text{(Solution)}, \\ &= \Delta H_{\mathrm{m}}(T - T_{\mathrm{m}})/T_{\mathrm{m}} \quad \text{(Melt)}. \end{aligned} \tag{2.18}$$

Here a is the number of ions formed from one molecule, R the gas constant and ΔH_{m} the enthalpy of melting. The decrease in the total free energy (ΔG) has a contribution due to the creation of a new surface. Denoting the radius of the nucleus by r and the solid–liquid interfacial surface energy by γ_{SL}, we get

$$\Delta G = (4/3)\pi r^3 \Delta G_V + 4\pi r^2 \gamma_{SL}. \quad (2.19)$$

The minimum size r^* of a stable nucleus is obtained by differentiation of (2.19) with respect to r. r^* is given by

$$r^* = -2\gamma_{SL}/\Delta G_V, \quad (2.20)$$

and the critical free energy ΔG^* is given by

$$\Delta G^* = 16\pi \gamma_{SL}^3/3(\Delta G_V)^2. \quad (2.21)$$

Nuclei with radii less than r^* are formed but they redissolve whereas nuclei with $r \geq r^*$ continue to grow. The rate of increase in the number of nuclei (J) is given by

$$J = J_0 \exp(-\Delta G^*/kT). \quad (2.22)$$

But this is not to be mistaken as the growth rate of the crystal as such. For this, we will have to consider the frequency (or probability) with which nuclei deposit themselves on the crystal surface. We shall consider the growth mechanisms and growth rates of crystals from vapours, solutions and melt separately.

2.9.2 Growth from Vapour and Solution

Figure 2.59 shows the surface of a microcrystal and various sites on it where a nucleus may deposit itself. These sites are not energetically equivalent; site 5 (kink in a step) is energetically most favourable. A nucleus depositing on the surface

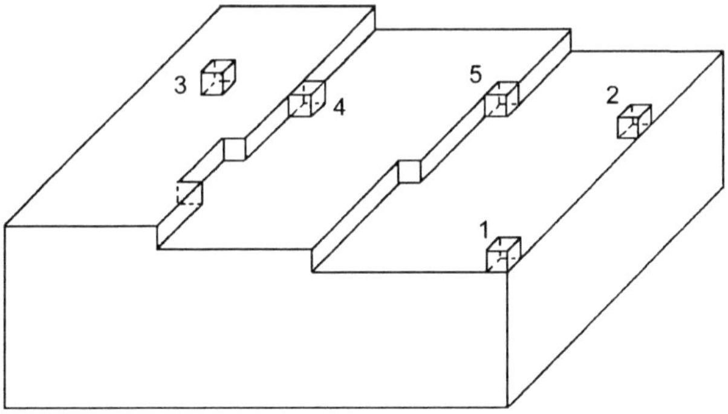

Fig. 2.59 Various positions of surface atoms in a crystal: 1 – atom at a corner, 2 –atom at an edge, 3 – atom at a terrace, 4 – atom at a ledge and 5 – atom at a kink site

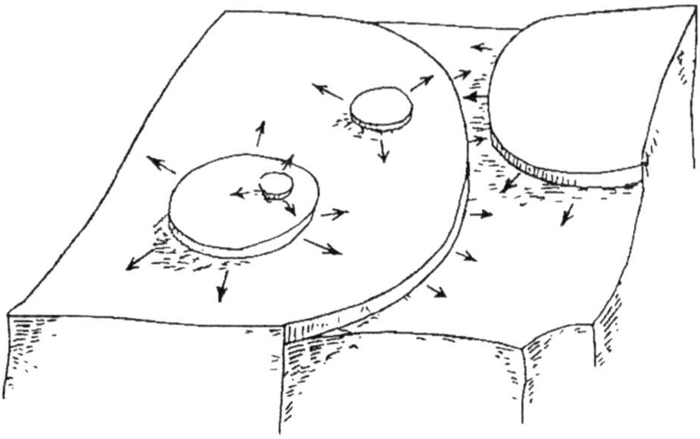

Fig. 2.60 Birth-and-spread model

loses one degree of freedom (normal to the surface) and effectively becomes a two-dimensional nucleus. It can freely slide over the surface. When it reaches one of the sites (particularly the kink), it integrates into the crystal lattice and thereafter it grows. This growth takes place according to two mechanisms – the birth-and-spread mechanism (BSM) and the screw dislocation mechanism.

2.9.2.1 Birth-and-Spread Mechanism

We shall discuss only the physical picture. A nucleus deposits at one of the sites on the surface and starts growing, forming an island on the surface (Fig. 2.60). This island grows laterally and may occupy the whole surface. In the meantime, another growth island may form on the previously formed island and this in turn may start growing. Thus, the growth takes place by the birth and spread of the two-dimensional island; the crystal grows layer by layer. The growth rate f of the crystal according to this mechanism is given by

$$f = C_1(S-1)^{2/3}(\ln S)^{1/6}\exp[-C_2/3(T^2\ln S)], \tag{2.23}$$

where C_1 and C_2 are constants. According to this equation, the growth is negligible at very low values of S and is measurable at supersaturation greater than about 25% following the functional dependence given in the equation. Another implication of this model is that the growing surface of the crystal has a stepped structure.

2.9.2.2 Spiral Growth Mechanism (SGM)

This is popularly known as the BCF model after Burton, Cabrera and Frank [2.18] who proposed it. The intersection of a screw dislocation with the surface of a

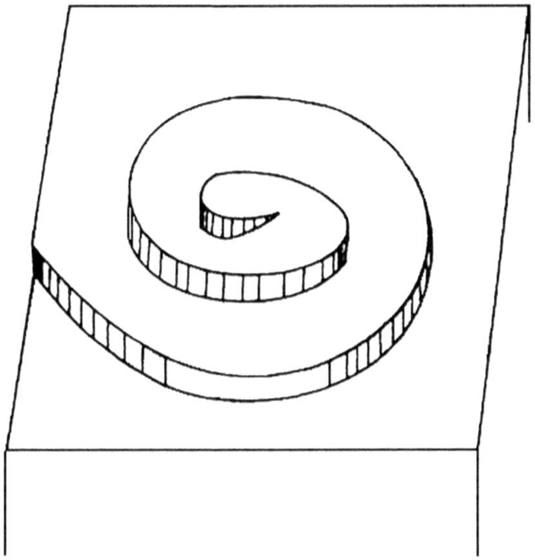

Fig. 2.61 Development of growth spiral (BCF model)

growing crystal creates a step which provides a favourable site (kink) for deposition of nuclei. Because of the lattice environment around a screw dislocation, the original step creates new steps in the course of its growth. This ensures continuous growth of the surface in the form of a spiral (Fig. 2.61). The growth rate f is given by

$$f = C_3[(S-1)^2/C_4]\tanh[C_4/(S-1)], \tag{2.24}$$

where C_3 and C_4 are constants. Equation (2.24) predicts measurable growth even at low values of supersaturation in contrast to (2.23). For low supersaturation $(S-1) \ll C_4$ and $f \approx C_3(S-1)^2/C_4$, i.e. the growth rate varies parabolically with $(S-1)$. On the other hand, for $(S-1) \gg C_4$, i.e. high supersaturation, $f = C_3(S-1)$, i.e. the growth rate varies linearly with $(S-1)$. In the intermediate range of S, the dependence of f on S is given by (2.24). Another implication of this model is that the surface grows like a spiral ramp.

These ideas and results apply to solution growth also with minor modifications.

2.9.3 Growth from Melt

As mentioned in Sect. 2.9.1, (2.22) merely gives the rate of increase J in the number of nuclei and the rate of crystal growth involves other factors. In melt growth, the growth rate f involves several parameters like the enthalpy of melting ΔH_m, the melting point T_m, the supercooling temperature T, the diffusion coefficient D, the coefficient of thermal conductivity in the solid phase K_S, the molar volume V, the diameter of the molecules in the melt λ, the solid–liquid surface energy γ_{SL}

Table 2.9 Growth rate f for melt growth

Type of interface	Growth rate f	Dependence on ΔT
Rough sharp interface	$\frac{D\Delta H_{\mathrm{m}}\,\Delta T}{\lambda RTT_{\mathrm{m}}}$	$f \propto \Delta T$
Stepped interface	$\frac{K_{\mathrm{S}}\Delta T}{a\,\Delta H_{\mathrm{m}}}$	$f \propto \Delta T$
Interface with screw dislocations	$\frac{D\Delta H_{\mathrm{m}}^2\,\Delta T^2 V_{\mathrm{m}}}{4\pi\gamma_{\mathrm{SL}}TT_{\mathrm{m}}^2 R}$	$f \propto (\Delta T)^2$

and (if there are steps) the step height a. In addition, the growth rate depends on the nature of the interface. The growth rates for different conditions are given in Table 2.9. Apart from the different expressions, the qualitative difference with regard to dependence on the supercooling ΔT needs to be noted.

A phenomenological expression for the growth rate can be obtained by considering the heat flow across an interface. If A is the area of cross section, K_{S} and K_{L} the coefficients of thermal conductivity of the solid and the melt, G_{S} and G_{L} the temperature gradients in the solid and the melt, ΔH_{m} the enthalpy of melting and f the linear rate of crystal growth, then, the heat flow equation is

$$K_{\mathrm{S}}AG_{\mathrm{S}} = K_{\mathrm{L}}AG_{\mathrm{L}} + Af\Delta H_{\mathrm{m}}. \tag{2.25}$$

The growth rate will have a maximum value f_{max} if $G_{\mathrm{L}} = 0$. In such a case

$$f_{\mathrm{max}} = K_{\mathrm{S}}G_{\mathrm{S}}/\Delta H_{\mathrm{m}}. \tag{2.26}$$

In a typical case (tin, for example), $K_{\mathrm{S}} = 0.14\,\mathrm{cal/(cm°C)}$, $\Delta H_{\mathrm{m}} = 102\,\mathrm{cal/cm^3}$ and assuming $G_{\mathrm{S}} = 20°\mathrm{C/cm}$, f_{max} calculated from (2.26) turns out to be $0.027\,\mathrm{cm/s}$. In deriving (2.25), the heat radiated to the surroundings has been ignored. If it is included, (2.26) will be modified.

2.9.4 Experimental Evidence

Growth mechanisms in specific crystals have been studied by two methods: growth rate studies (GRS) and surface feature studies (SFS). In the GRS method, growth rates are carefully measured at different supersaturation and the results are fitted to equations pertaining to different mechanisms. Examples are shown in Fig. 2.62. In the SFS method, the surface of the growing or as-grown crystal is observed under an appropriate microscope (optical, phase contrast, interference or SEM) to see whether the surface shows a stepped structure or a growth spiral (Fig. 2.63). Results of such studies, collected from literature, are summarised in Table 2.10.

We may conclude that in vapour growth and solution growth, different crystals grow either by the BSM or by SGM. There are also cases where the same crystal

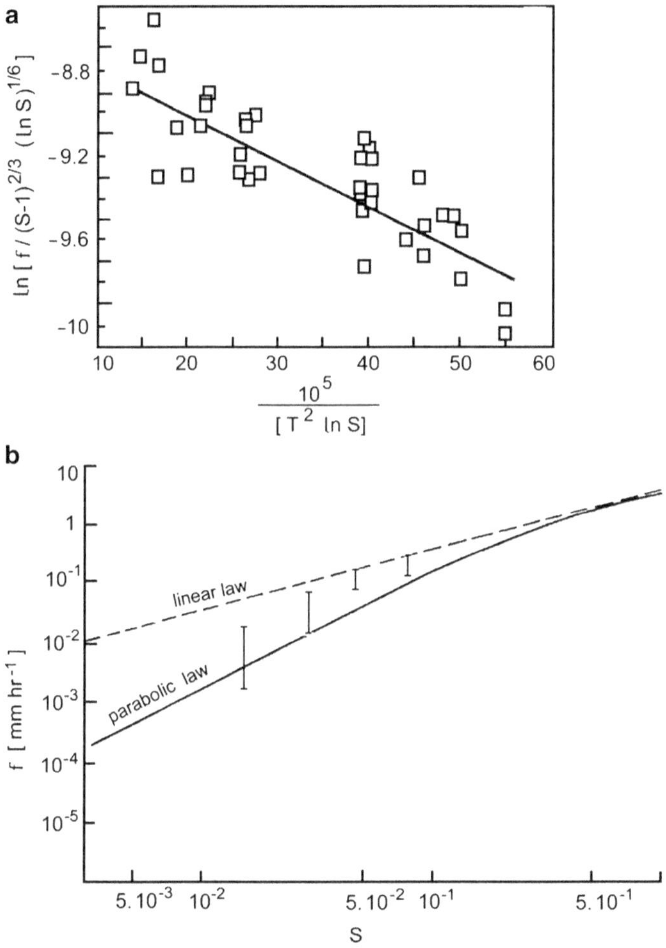

Fig. 2.62 (**a**) Plot of data on supersaturation and growth rate for $NaBrO_3$ according to (2.23). (**b**) Plot of data on supersaturation and growth rate for I_2 according to (2.24)

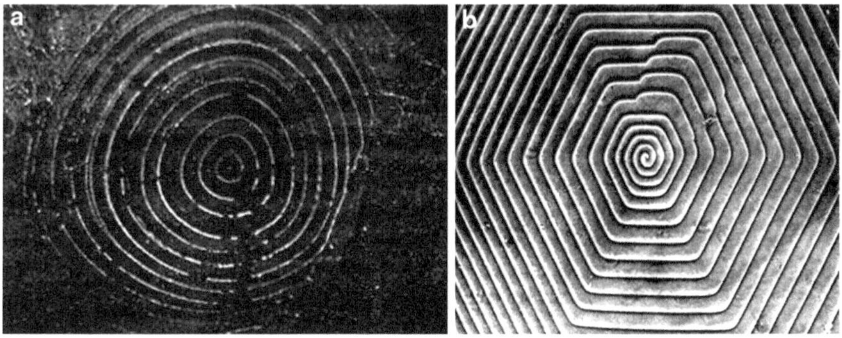

Fig. 2.63 (**a**) Concentric steps on $NaBrO_3$. (**b**) Growth spiral on SiC crystal

Table 2.10 Experimental studies of growth mechanisms

Growth method	Crystal	Experimental method	Conclusion
Vapour			
	Hg	GRS	SGM
	I_2	GRS	SGM
	P_4	GRS	SGM
Solution			
	K-Alum	GRS	SGM
		SFS	SGM
	$NaClO_3$	GRS	SGM at low *S*
		GRS	BSM at high *S*
	$NaBrO_3$	GRS	BSM
		SFS	BSM
	NaCl	SFS	SGM
	KCl	SFS	SGM
		SFS	BSM
	KBr	GRS	SGM at low *S*
			BSM at high *S*
	$Ba(NO_3)_2$	SFS	SGM
Melt			
	Glycerol	GRS	SGM
	Tin	GRS	SGM
	Phosphorous	GRS	SGM

GRS growth rate studies, SFS surface feature studies, SGM spiral growth mechanism, BSM birth-and-spread mechanism

grows by one mechanism at low supersaturations and by the other at higher supersaturations. Melt grown crystals appear to favour the SGM.

Problems

1. The solubility of Na_2SO_4 in water is given below at different temperatures. Draw the temperature–solubility plot and discuss the strategy to be adopted to grow crystals of Na_2SO_4 at (a) 35°C and (b) 50°C.

Temp. [°C]	Solubility [g/100 g H_2O]	Temp. [°C]	Solubility [g/100 g H_2O]
25	27	38	48
27	34	44	47
30	42	60	45
33	49	75	43

2. The observed rates of crystal growth (f_{obs}) at different supersaturation (σ) of an aqueous solution of KCl are given below. Show that these data fit the BCF growth equation.

σ	f_{obs} [μm/h]	σ	f_{obs} [μm/h]	σ	f_{obs} [μm/h]
0.15	0.02	0.79	0.53	14.9	14.75
0.20	0.04	0.89	0.64	29.8	30.0
0.30	0.09	0.94	0.70	50.1	49.5
0.40	0.15	2.5	2.4	89	88.2
0.59	0.33	7.0	7.0	100	99

3. Using the heat flow equation, calculate the rate of growth of Mg crystal growing under a temperature gradient of 50°C/cm ($\Delta H = 153\,\mathrm{cal/cm}$, $K = 0.41\,\mathrm{cal\,cm^2/s°C}$).
4. In a multipass zone refinement equation, an impurity with segregation coefficient $k = 0.2$ in a metal is reduced to 10^{-16} of its starting value after a number of passes. Estimate the minimum number of passes required to reach this purity.
5. What method/s should be chosen to grow (a) a faceted crystal, (b) a long crystal, (c) a very thin crystal and (d) a crystal of a material that decomposes on heating?

References

2.1. W.G. Pfann, Solid State Phys. **4**, 423 (1957)
2.2. J.J. Gilman, *The Art and Science of Growing Crystals* (John Wiley and Sons, New York, 1963)
2.3. J.C. Brice, *Growth of Crystals from Melt* (North Holland Pub. Co., Amsterdam, 1965)
2.4. I. Tarjan, M. Matrai, *Laboratory Manual on Crystal Growth* (Akademiai Kiado, Budapest, 1972)
2.5. M. Ohara, R.C. Reid, *Modelling Crystal Growth Rates from Solutions* (Prentice-Hall, New Jersey, 1973)
2.6. K. Sangwal, *Etching of Crystals* (North Holland Pub. Co., Amsterdam, 1987)
2.7. A. Holden, P. Singer, *Crystals and Crystal growing* (Vakils, Feffer and Simons, Bombay, 1965)
2.8. P.H. Egli, R. Johnson, in *The Art and Science of Growing Crystals*, ed. by J.J. Gilman (John Wiley and Sons, New York, 1963), P. 194
2.9. R.A. Laudisse, in *The Art and Science of Growing Crystals*, ed. by J.J. Gilman (John Wiley and Sons, New York, 1963), P. 252
2.10. G.C. Kennedy, Am. J. Sci. **248**, 540 (1950)
2.11. W.H. Bauer, W.G. Field, in *The Art and Science of Growing Crystals*, ed. by J.J. Gilman (John Wiley and Sons, New York, 1963), P. 398
2.12. H.T. Hall, Rev. Sci. Instr. **31**, 125 (1960)
2.13. R.H. Wenterf Jr., in *The Art and Science of Growing Crystals*, ed. by J.J. Gilman (John Wiley and Sons, New York, 1963), P. 176
2.14. W.G. Pfann, Trans. Am. Inst. Mining Met. Eng. **194**, 747 (1952)
2.15. W.G. Pfann, *Zone Melting* (John Wiley and Sons, New York, 1958)
2.16. W.G. Pfann, K.M. Oslen, Phys. Rev. **89**, 322 (1953)
2.17. W.G. Pfann, J.H. Scaff, Trans. Am. Inst. Mining Met. Eng. **185**, 389 (1949)
2.18. W.K. Burton, N. Cabrera, F.C. Frank, Philos. Trans. Roy. Soc. **A243**, 299 (1951)

Chapter 3
Crystallography

3.1 Introduction

Crystallography is the science of the geometry and symmetry of the external forms as well as the internal structure of crystals.

The first crystals that became available for such studies were, of course, the mineral crystals. The initial observations were made by mineralogists. Looking at the external forms of crystals, a number of simple but fundamental laws were noted.

Thus, it was noted that crystals were bound by perfectly plane faces (Fig. 3.1). These faces meet at straight edges which, in turn, intersect at sharp corners. In fact, the number of faces (F), edges (E) and corners (C) follow the simple relation $F + C = E + 2$.

Making measurements of interfacial angles Steno observed in the seventeenth century that in different sections of the same crystal, the angle between two chosen faces is always the same. He thus arrived at the law of constancy of interfacial angles. A complementary law is the law of rational ratios of intercepts stated by Hauy in the eighteenth century according to which the intercepts of a plane on crystal axes (generally three converging edges of a crystal) are rational; this further resulted in a notation for labelling crystal faces (the Miller indices).

Hauy also found that when a rhombohedral calcite crystal was cleaved, the resulting fragments were, again, rhombohedral. He repeatedly cleaved successive fragments and found that the minutest fragment was still a rhombohedron. Hauy propounded the view that "continued cleavage would ultimately lead to a smallest unit... by a repetition of which the whole crystal is built up". Hauy had almost arrived at the idea of a unit cell. Continuing such external observations, crystallographers could classify crystals into seven crystal systems and 32 point groups.

All these results were based on observations of external forms. On the other hand, the appearance of plane faces on mineral crystals and solution-grown crystals intuitively pointed at the existence of an internal regular arrangement. Independently of the morphological approach, an abstract mathematical study of the symmetry of arrays of points in space was undertaken by Sohncke, Frankenheim, Barlow,

D.B. Sirdeshmukh et al., *Atomistic Properties of Solids*, Springer Series in Materials Science 147, DOI 10.1007/978-3-642-19971-4_3,

Fig. 3.1 Natural crystals of quartz showing the occurrence of plane faces

Schoenflies and Fedorov in the late nineteenth century. They pointed out that such arrays possess not only the symmetry elements shown by external forms of crystals but also translational symmetry elements which result in their classification into 14 space lattices and 230 space groups. These mathematical crystallographers were able to anticipate the existence of internal crystal structure which was experimentally revealed only after the discovery of X-ray diffraction by Laue in 1912.

In this chapter we develop the concepts of symmetry operations, symmetry elements, unit cells, crystal systems, point groups, Bravais lattices and space groups. The elegant concept of the reciprocal lattice is also introduced. Treatments of these topics in different degrees of detail are given by Kaelble [3.1], Levy [3.2], Blakemore [3.3], Phillips [3.4], Ladd and Palmer [3.5] and Sirotin and Shaskolskaya [3.6].

3.2 External Form and Habit of Crystals

A striking feature of crystals, as mentioned earlier, is the appearance of a series of smooth plane surfaces called "faces" which appear to have some regularity of arrangement. These crystal faces are the result of a natural process. Hence a crystal may be defined as a solid bound by naturally formed plane faces. The regularity of arrangement of the faces imparts to the crystal a certain form of symmetry. More often, actual crystals deviate from a perfect polyhedral form. Equivalent faces of individual crystals may vary in size, or may even be completely missing due to growth conditions. When the conditions for growth of a particular face are more favourable than for other faces, the former grows at the expense of the latter. This results in crystals of a given chemical compound showing different forms and variations in habit. The habits of calcite ($CaCO_3$) in its different forms are shown in Fig. 3.2. It is well known that addition of small amounts of impurities during growth affects the habit of the crystals formed. For example, sodium chlorate crystallises as

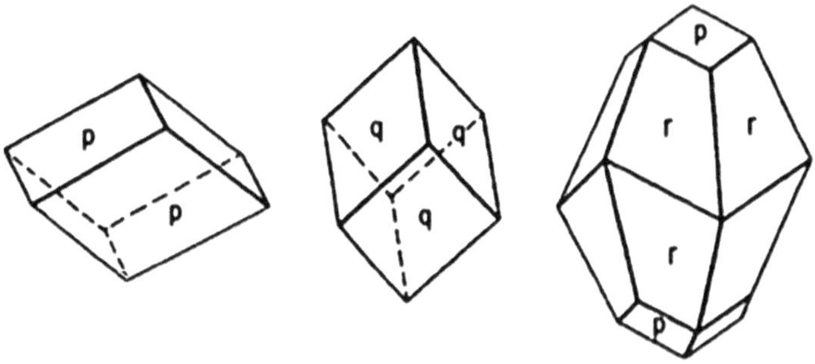

Fig. 3.2 Habit variations in $CaCO_3$ equivalent faces are labelled with the *same letter*

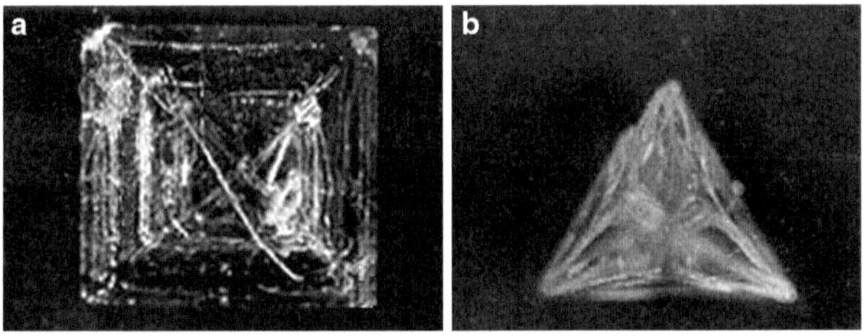

Fig. 3.3 Habit of $NaClO_3$ grown from (**a**) pure solution and (**b**) solution containing borax

cubes from solution when pure water is used. Addition of trace amounts of impurity like borax changes its habit completely to the tetrahedral form (Fig. 3.3). The internal structure in a given species exhibiting different habits, however, remains the same.

3.3 Lattice and Unit Cell

In an ideally perfect crystal, there is an infinite three-dimensional arrangement of atoms in a periodic array. We may imagine points in space about which these atoms are located. These points in space are called lattice points and the collection of such lattice points defines a space lattice. Thus, a lattice is an infinite array of physically equivalent points repeated regularly in three-dimensional space. The environment of a given point is identical to that of every other equivalent point in the lattice. The lattice points may be generated by a translation operation.

For simplicity consider a two-dimensional lattice (Fig. 3.4) defined by two fundamental vectors **a** and **b**. A point P may be located by the vector $2\mathbf{a} + \mathbf{b}$ and the point Q by $\mathbf{a} + 2\mathbf{b}$. The other points in the lattice may be located by suitable combinations of **a** and **b**. The parallelogram defined by the vectors **a** and **b** forms

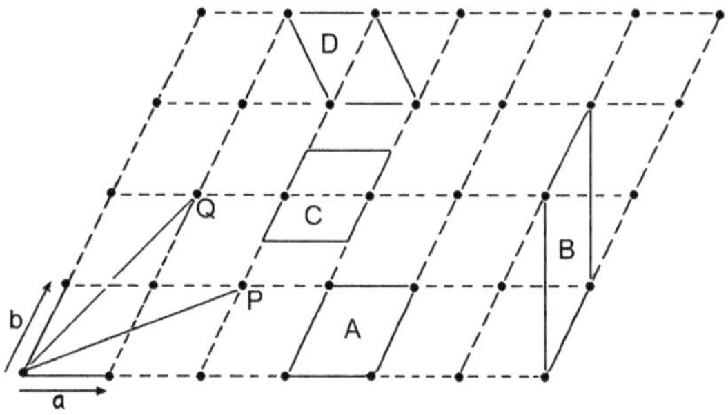

Fig. 3.4 A two-dimensional lattice showing different ways of choosing primitive translations

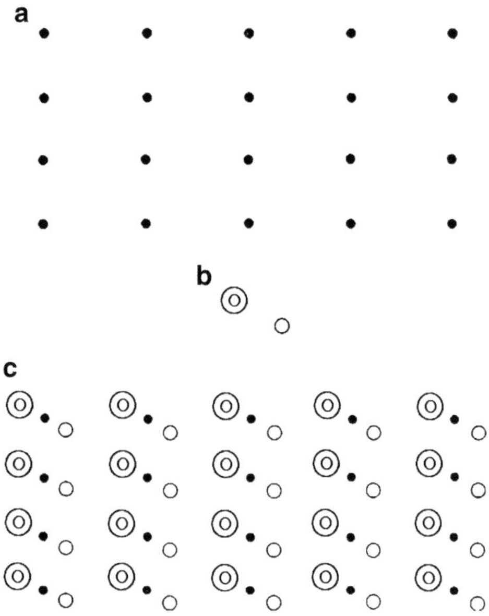

Fig. 3.5 (**a**) Space lattice; (**b**) the basis containing two different atoms and (**c**) crystal structure resulting by placing the basis at every lattice point

a "unit cell". The lattice vectors **a**, **b** are called primitive translations if they form the smallest cell, which when translated parallel to itself along the two directions generates the lattice. The choice of primitive translation vectors to define a particular lattice is not unique. Some ways of choosing primitive translations are shown in Fig. 3.4. A unit cell is called a primitive cell if it has lattice points only at its corners.

A lattice is a mathematical abstraction. A crystal structure is formed when an identical group of atoms is attached to each lattice point. The group of atoms is called the basis. The basis may consist of a single atom or a group of atoms (Fig. 3.5).

For a three-dimensional lattice, the translation operation **T** is given by

$$\mathbf{T} = n_1\mathbf{a} + n_2\mathbf{b} + n_3\mathbf{c}, \tag{3.1}$$

where **a**, **b**, **c** are three non-coplanar fundamental vectors and n_1, n_2 and n_3 are arbitrary integers. When the translation operation is applied to any point **r**, another lattice point **r**′ results given by

$$\mathbf{r}' = \mathbf{r} + \mathbf{T} = \mathbf{r} + n_1\mathbf{a} + n_2\mathbf{b} + n_3\mathbf{c}, \tag{3.2}$$

where **r**′ is identical in all respects to the original point **r**. We conclude that **r**′ is indistinguishable from **r**. The vectors **a**, **b** and **c** are called primitive translations or unit translations or unit vectors if they form the smallest cell that serves as the building block for the structure. For a three-dimensional lattice, the parallelepiped constructed with the three unit vectors is the unit cell. When a lattice is used to describe a crystal, **a**, **b** and **c** are called the crystal axes. The crystal axes are not always orthogonal (Fig. 3.6). The volume of a unit cell may be calculated from the relation

$$V = abcN, \tag{3.3}$$

where

$$N^2 = 1 - \cos^2\alpha - \cos^2\beta - \cos^2\gamma + 2\cos\alpha\cos\beta\cos\gamma. \tag{3.4}$$

Here α is the angle between **b** and **c**, β that between **c** and **a** and γ between **a** and **b** (Fig. 3.6).

Sometimes it is useful to define a lattice by a set of unit vectors that generates the complete lattice if some or all of the coefficients n_1, n_2 and n_3 in (3.2) are non-integral. Such a set of unit vectors is called a non-primitive set. The non-primitive unit vectors **a**, **b**, **c** are preferred over the primitive unit vectors **a**′, **b**′, **c**′ for the sake of simplicity and convenience of mutually orthogonal axes. A familiar example in which non-primitive unit vectors are useful is the face-centred cubic lattice. In Fig. 3.7, the conventional, non-primitive unit cell of the face-centred cubic lattice is outlined by the full lines. The primitive cell (not usually used) is outlined by the dashed lines. The primitive cell has lattice points at all eight corners. Since each lattice point belongs to adjoining eight cells, the primitive unit cell is said to have

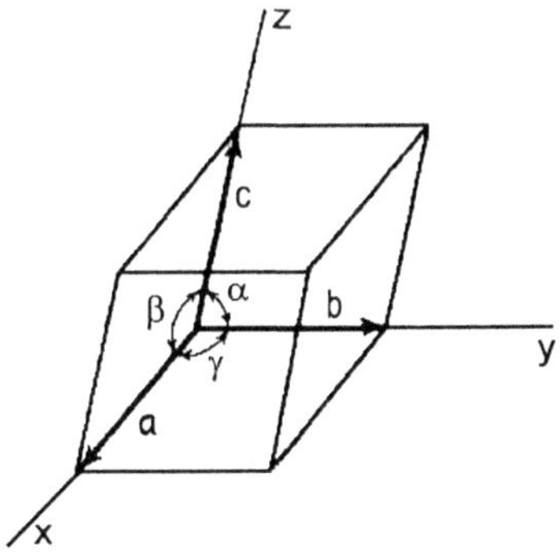

Fig. 3.6 The crystal axes and the elementary parallelepiped

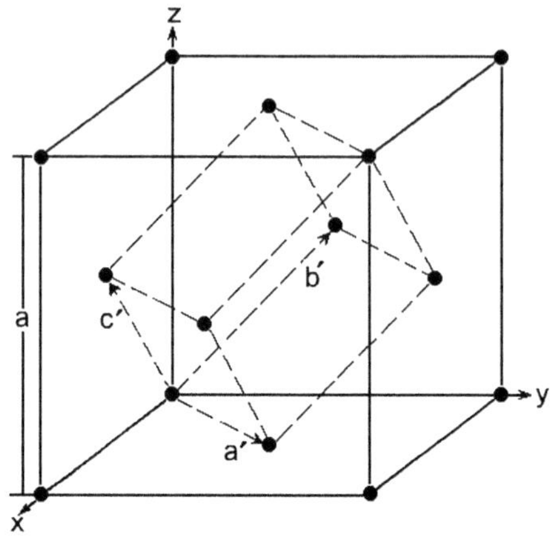

Fig. 3.7 The primitive unit cell (enclosed by *dashed lines*) and the non-primitive unit cell (*full lines*) of the face centred cube

only one lattice point. Non-primitive unit cells are characterised by more than one lattice point. Thus the face-centred cubic lattice has four lattice points.

3.4 Symmetry Operations and Symmetry Elements

A symmetry operation is an operation such as rotation about a fixed axis performed on an object or pattern which brings it into coincidence with itself, i.e. the object or pattern remains indistinguishable from its initial position after the operation. While the act of rotation is a symmetry operation, the axis about which such an operation is possible is a symmetry element. In crystals, in addition to lattice translation operation, there are other symmetry operations which leave the crystal invariant. We shall consider various types of symmetry elements that occur in crystals.

3.4.1 Centre of Symmetry

A prominent feature of crystals is the occurrence of similar faces in parallel pairs on opposite sides of the crystal, i.e. the feature which occurs a given distance away from the exact centre in one direction will also occur the same distance in the opposite direction. Expressed in terms of three mutually perpendicular axes x, y, z, with the origin at the centre, for any feature at x_1, y_1, z_1 there will be a similar feature at $-x_1$, $-y_1$, $-z_1$. The crystal is then said to have a centre of symmetry. If only some of the faces occur in parallel pairs, while others have no similar face parallel to them, the crystal does not possess a centre of symmetry.

3.4.2 Planes of Symmetry

A crystal may have a plane that cuts it into two parts which are mirror images of each other. In fact, a crystal may have several such planes. The crystal is then said to possess plane(s) of symmetry. If in the rectangular coordinate system, the plane of symmetry is the yz plane, to every feature at x_1, y_1, z_1, there should occur a similar feature at $-x_1, y_1, z_1$.

In addition to three planes of symmetry parallel to its faces, a cube has six diagonal planes of symmetry. A rectangular parallelepiped does not have such diagonal planes. Though the rectangular parallelepiped can be geometrically divided into two similar wedges, they are not mirror reflections of each other in the plane. The only crystallographic planes of symmetry in this case are parallel to the three pairs of faces.

3.4.3 Axes of Symmetry

We say that a crystal possesses an n-fold axis of rotation if the crystal remains invariant on rotation through an angle $(2\pi/n)$ about this axis. The value of n determines the degree of the axis. If $n = 1$, the crystal must be rotated through 360° before it can assume a congruent position. Such an axis is called an identity axis and every crystal possesses any number of such axes. $n = 2$ denotes a rotation through 180° and the axis has twofold symmetry. Such an axis is termed a diad axis. $n = 3$ denotes a rotation through 120°. The axis having a threefold symmetry is termed a triad axis. Similarly, $n = 4$ denoting a rotation through 90° is termed a tetrad axis and $n = 6$ denoting a rotation through 60° is termed a hexad axis.

We may show that one-, two-, three-, four- and sixfold axes are the only permissible axes of symmetry in crystals. A degree of symmetry higher than six is not allowed and a pentad axis is not a symmetry axis in crystals. Consider a row P of atoms in a crystal separated by a repeat distance a (Fig. 3.8). Let us assume that an angle ϕ is a rotation allowed by the symmetry of this lattice. An anticlockwise rotation ϕ about atom B would move atom A to position A′ and a clockwise rotation

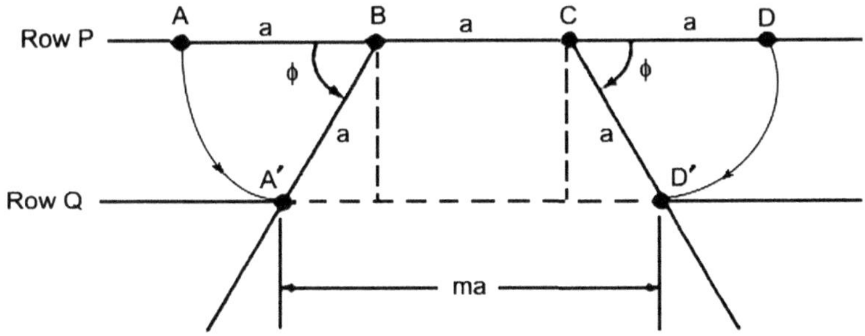

Fig. 3.8 Rotational symmetry in crystal lattices

Table 3.1 Possible rotation axes in crystals

M	$\cos\phi$	ϕ	$n = 2\pi/\phi$
−2	−1	π	2
−1	$-\frac{1}{2}$	$\frac{2\pi}{3}$	3
0	0	$\frac{\pi}{2}$	4
1	$\frac{1}{2}$	$\frac{\pi}{3}$	6
2	1	$0(2\pi)$	1

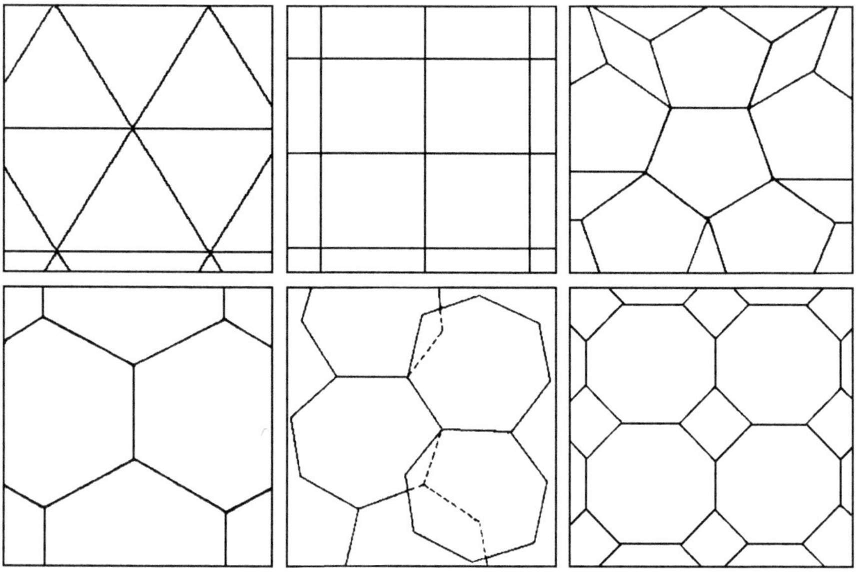

Fig. 3.9 Attempts at arranging regular polygons in a close-packed array

ϕ about atom C would move atom D to position D′. The atoms A′ and D′ must lie on another lattice row Q and the distance between them must be ma where m is an integer.

From Fig. 3.8 it may be seen that

$$ma = a + (2a\cos\phi). \tag{3.5}$$

Hence

$$\cos\phi = \frac{(m-1)}{2} = \frac{M}{2}, \tag{3.6}$$

where $M = (m-1)$ is another integer. Since $-1 \leq \cos\phi \leq 1$, the only permitted values of M are 0, ±1 and ±2. The rotation symmetries for these values of M are given in Table 3.1. It may be seen that only rotation axes of degree 1, 2, 3, 4 and 6 are possible. Rotation axes of any other degree are not possible. We may also demonstrate this by considering a two-dimensional case (Fig. 3.9) where it is seen that it is possible to fill an area with regular triangles, squares and hexagons in a close-packed array, whereas polygons with other rotational symmetry leave empty spaces or overlap with each other.

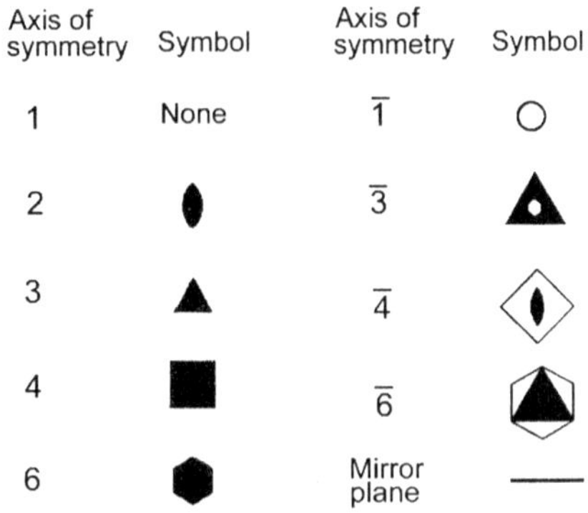

Fig. 3.10 Symbolic representation of symmetry elements

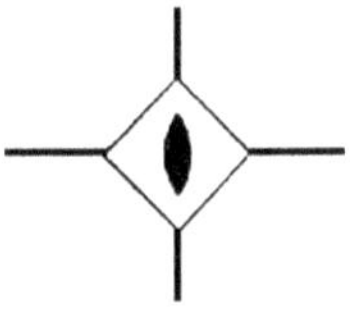

Fig. 3.11 A combination of symmetry elements

3.4.4 Inversion Axis

If a crystal occupies a congruent position after undergoing a rotation about an axis followed by inversion through the centre, the symmetry element representing it is called a rotation–inversion axis or simply an inversion axis and is denoted as $\bar{n}$. The operation of rotation–inversion is a combined operation involving symmetry with respect to an axis and a point. Thus a crystal may have a fourfold axis of rotation–inversion ($\bar{4}$) even though it may not have either a fourfold axis or a centre of inversion as independent symmetry elements.

The symmetry elements may be represented symbolically as in Fig. 3.10. A twofold axis of rotation–inversion is equivalent to a mirror plane. $\bar{1}$ is equivalent to a centre of symmetry. The centre of symmetry is also known as inversion centre. A combination of symmetry elements is shown in Fig. 3.11; the open square indicates inversion axis $\bar{4}$, the solid almond indicates a twofold axis coincident with it and the thick straight lines indicate the two mirror planes.

A cube has a centre of symmetry (or an inversion centre), nine planes of symmetry (three parallel to the cube faces and six diagonal planes), three tetrad axes, four triad axes and six diad axes. These symmetry elements are shown in Fig. 3.12.

The octahedron and the rhombic dodecahedron (Fig. 3.13), though quite different in shape from the cube, possess the same group of symmetry elements as the cube. A rhombohedron, on the other hand, shows considerably less symmetry, viz., centre of symmetry, one triad axis, three diad axes and three planes of symmetry.

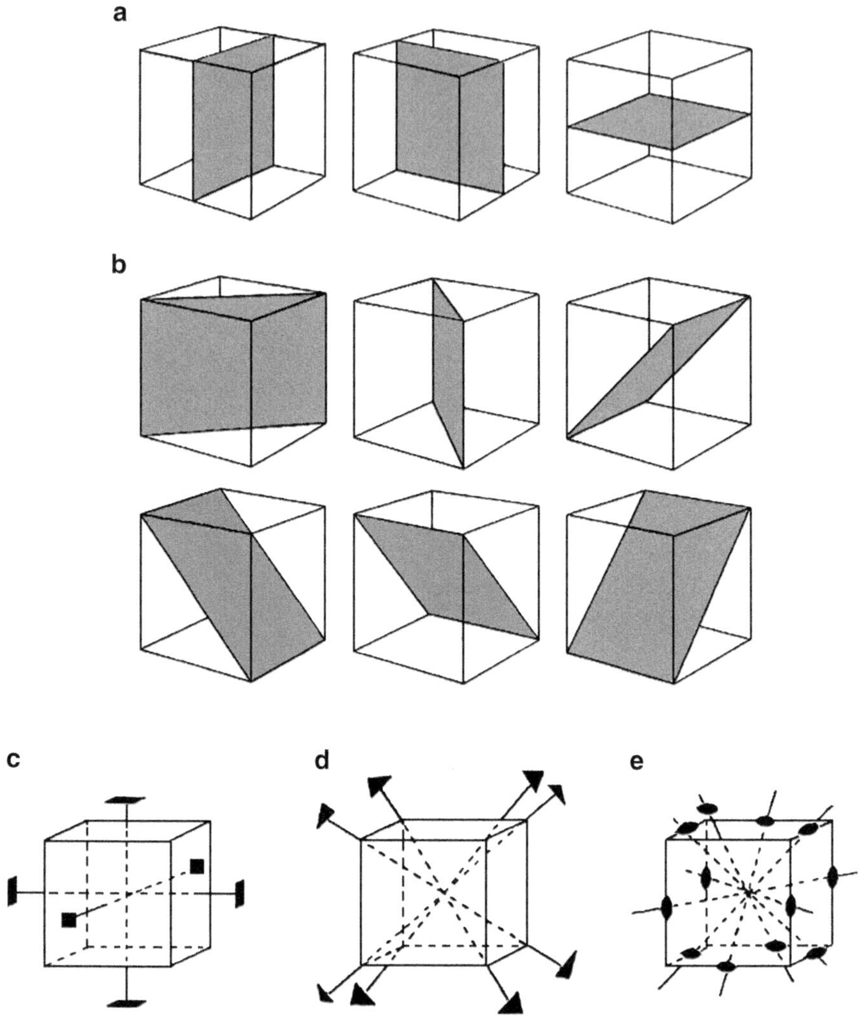

Fig. 3.12 Various symmetry elements of a cube: (**a**) the three planes of symmetry parallel to cube faces, (**b**) six diagonal planes of symmetry, (**c**) three tetrad axes, (**d**) four triad axes and (**e**) six diad axes. The inversion centre is not shown in the figure

3.5 Crystal Systems

The unit cell and hence the crystal structure are characterised by three axes a, b, c and three interaxial angles, α, β, γ. All crystals may be classified into seven crystal systems based on the relations between the lengths of their axes and their orientation. These are the triclinic, monoclinic, orthorhombic, tetragonal, cubic, trigonal and hexagonal systems. The seven crystal systems and their interaxial

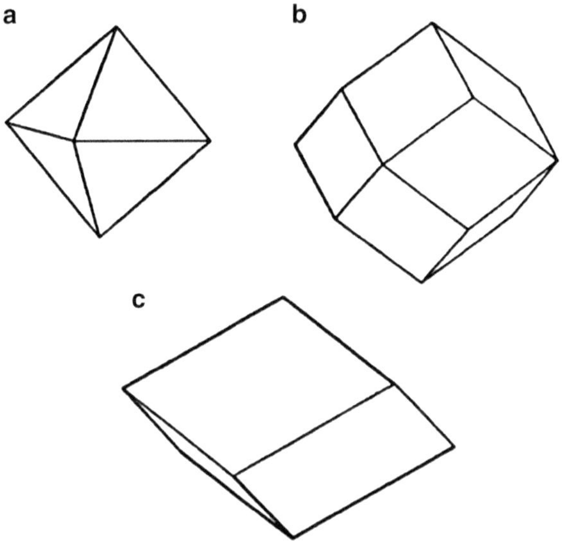

Fig. 3.13 (**a**) The octahedron; (**b**) the rhombic dodecahedron and (**c**) a rhombohedron

Table 3.2 The seven crystal systems

System	Restrictions on cell dimensions and angles	Characteristic axes of symmetry
Triclinic	$a \neq b \neq c$ $\alpha \neq \beta \neq \gamma$	No axes of symmetry
Monoclinic	$a \neq b \neq c$ $\alpha = \gamma = 90° \neq \beta$	One diad axis and no axes of higher degree
Orthorhombic	$a \neq b \neq c$ $\alpha = \beta = \gamma = 90°$	Three diad axes
Tetragonal	$a = b \neq c$ $\alpha = \beta = \gamma = 90°$	One tetrad axis
Trigonal	$a = b = c$ $\alpha = \beta = \gamma < 120°, \neq 90°$	One triad axis
Hexagonal	$a = b \neq c$ $\alpha = \beta = 90°, \gamma = 120°$	One hexad axis
Cubic	$a = b = c$ $\alpha = \beta = \gamma = 90°$	Four triad axes

relationships are given in Table 3.2. The characteristic axes of symmetry associated with each crystal system are also given in the table.

3.6 Point Groups

The external symmetry exhibited by crystals can therefore be described by a combination of symmetry elements discussed in Sect. 3.4. A collection of symmetry elements forms a group. The total number of symmetry groups which can be constructed from all possible kinds of crystallographic symmetry elements is limited. A combination of symmetry elements may result in additional symmetry elements. If a symmetry group consists of one plane of symmetry and one axis of rotation (whether diad, triad, tetrad or hexad), the axis must necessarily be normal to the plane, as otherwise, on reflection about the plane the axis would give rise to a second axis. If a crystal possesses another axis of symmetry inclined to the first, then additional rotation axes may result. For example, if a crystal possesses a second triad inclined to the first there must be three such inclined axes for congruence on rotation through 120°. Hence, if there is more than one triad, there must be four. Likewise, a threefold axis cannot have a single mirror plane parallel to it because every feature of the object including its mirror plane must be repeated three times in a complete revolution around the axis.

Some symmetry elements are inherent in others. For example, a twofold axis is always inherent in a fourfold axis. Some symmetry elements may be equivalent to others. $\bar{6}$ is equivalent to a threefold axis and a mirror plane normal to it.

The combined operation on an object that leaves at least one point in the object unmoved during the operations as well as keeping distances between all points unchanged is called a point group. Every symmetry operation must be consistent with every other symmetry operation in the point group, i.e. while certain combinations of symmetry operations can make a point group, others cannot.

This interdependence of symmetry operations in a point group and the equivalence of some symmetry elements limit the number of crystallographic point groups to 32, distributed among the seven crystal systems. All the point groups in a crystal system possess the characteristic symmetry of the system. Thus a crystal in the trigonal system may possess only a triad axis, or a triad axis and a centre of symmetry, or a combination of a triad axis with diad axes or symmetry planes, or with both. The point groups are also referred to as the crystal classes.

It is important to have notations and symbols for distinguishing these 32 point groups without ambiguity. Two such notations are in common use. The first is the Schoenflies [3.7] notation and is described by the following symbols and their combinations:

C_n = groups having a single rotation axis of degree n,
C_{nh} = groups having an n-fold rotation axis with a horizontal symmetry plane normal to it,
C_{nv} = groups having an n-fold rotation axis with vertical planes of symmetry,
D_n = groups possessing an n-fold principal axis and n twofold axes normal to it,
d = diagonal symmetry planes as in D_{2d},
s = a mirror plane when the normal axis is of degree 1,
S_n = groups having an n-fold rotary reflection axis which combines a rotation about an axis with a reflection across a mirror plane normal to the axis,
T = tetrahedral groups having the four threefold and the three twofold axes of a regular tetrahedron,

O= octahedral groups possessing three fourfold axes, four threefold axes, and six twofold axes distributed as in a cube,

i = an inversion.

The second notation is based on the symbols proposed by Hermann [3.8, 3.9] and Mauguin [3.10] in which the point group symbol represents at a glance all the symmetry elements of the group. The numbers 1, 2, 3, 4 and 6 are used to represent the rotation axes, and $\bar{3}, \bar{4}$ and $\bar{6}$ for inversion axes in the Hermann–Mauguin notation. A symmetry centre is represented by the symbol $\bar{1}$ and a plane of symmetry by m (mirror). $\bar{2}$ is equivalent to m. In representing a point group, the symmetry of the principal axis is given first followed by a symmetry plane normal to it. The secondary axes are next indicated followed by any additional mirror planes. If we denote a principal axis of any degree by the general symbol X, we have the following combinations:

R a rotation axis alone
$\overline{R}$ an inversion axis alone
R/m a rotation axis normal to a plane of symmetry
Rm a rotation axis and a plane of symmetry parallel to it
$\bar{R}m$ an inversion axis and a plane of symmetry parallel to it
$R2$ rotation axis with a diad axis normal to it
R/mm rotation axis normal to a plane of symmetry and another plane of symmetry normal to the first.

For example, the symbol $\frac{2}{m}$ (or $2/m$) implies a twofold axis with a plane of symmetry perpendicular to it. This is a monoclinic point group and is the equivalent of C_{2h} in the Schoenflies notation. Similarly $4/mmm$ has a symmetry plane normal to a fourfold axis and two additional sets of mirror planes parallel to the principal axis. While writing such symbols it is not necessary to indicate more symmetry elements than the minimum number required to completely define the symmetry of the group. For example, the symbol 62 defines the point group which has a sixfold axis and six twofold axes normal to it. The addition of a twofold axis normal to a sixfold axis immediately creates a second twofold axis 90° from the first one through the action of sixfold symmetry operation of the principal axis. Further, this action immediately introduces four more twofold axes in the plane of the original two, each of the six axes making angles 30° with each other. The group $\bar{6}m2$ has a principal axis of sixfold inversion, three mirror planes parallel to it and three twofold axes normal to the principal axis. It is also possible to abbreviate symbols which completely define the symmetry. Thus, $\frac{2}{m}\frac{2}{m}\frac{2}{m} = mmm$ since three planes of symmetry automatically give rise to twofold axes normal to each plane.

Special convention must be observed for the cubic system. All the five crystal classes in the cubic system have four triad axes which automatically introduce three diad axes at right angles. The triad axes are along the cube diagonals and are therefore denoted in the symbol of each class by the figure 3 appearing in a position other than the first such as 23. Further, since a cube may have planes of symmetry parallel to the cube faces and/or diagonal planes, the symbol m may precede 3 in the

Table 3.3 The 32 point groups

Crystal system	Point group symbol	
	Schoenflies	Hermann–Mauguin
Triclinic	C_1	1
	$S_2 = C_i$	$\bar{1}$
Monoclinic	C_2	2
	$C_{1h} = C_h$	$m = \bar{2}$
	C_{2h}	$2/m$
Orthorhombic	$V = D_2$	222
	C_{2v}	$2mm = mm$
	$V_h = D_{2h}$	$\frac{2}{m}\frac{2}{m}\frac{2}{m} = mmm = 2/mm$
Tetragonal	S_4	$\bar{4}$
	C_4	4
	$V_d = D_{2d}$	$\bar{4}2m$
	D_4	42
	C_{4h}	$4/m$
	C_{4v}	$4mm$
	D_{4h}	$\frac{4}{m}\frac{2}{m}\frac{2}{m} = 4/mmm$
Trigonal	C_3	3
	C_{3i}	$\bar{3}$
	D_3	32
	C_{3v}	$3m$
	D_{3d}	$\bar{3}\frac{2}{m} = \bar{3}m$
Hexagonal	C_{3h}	$\bar{6} = 3/m$
	D_{3h}	$\bar{6}m2 = 3/mm$
	C_6	6
	D_6	62 (or 622)
	C_{6h}	$6/m$
	C_{6v}	$6mm$
	D_{6h}	$\frac{6}{m}\frac{2}{m}\frac{2}{m} = 6/mmm$
Cubic	T	23
	O	$432 = 43$
	T_h	$2/m3 = m3$
	T_d	$\bar{4}3m$
	O_h	$4/m3m = m3m$

former type (e.g. $m3$) or m may be placed after 3 in the latter type (e.g. $\bar{4}3m$). The 32 point groups in both the Schoenflies and Hermann–Mauguin notation are given in Table 3.3.

We may also represent point groups by plane diagrams which are stereographic projections of the symmetry elements, and of the equivalent general points, in each point group. These are projections of point groups on a plane normal to the principal axis of the group, which is thought to be located at the centre of a large circle. Points shown as solid circles are above the plane of projection and the points shown as open circles refer to points below the plane. Each type of point is brought into coincidence with other points of its type by rotation about the principal axis or reflection about a vertical plane of symmetry. The action of an inversion axis,

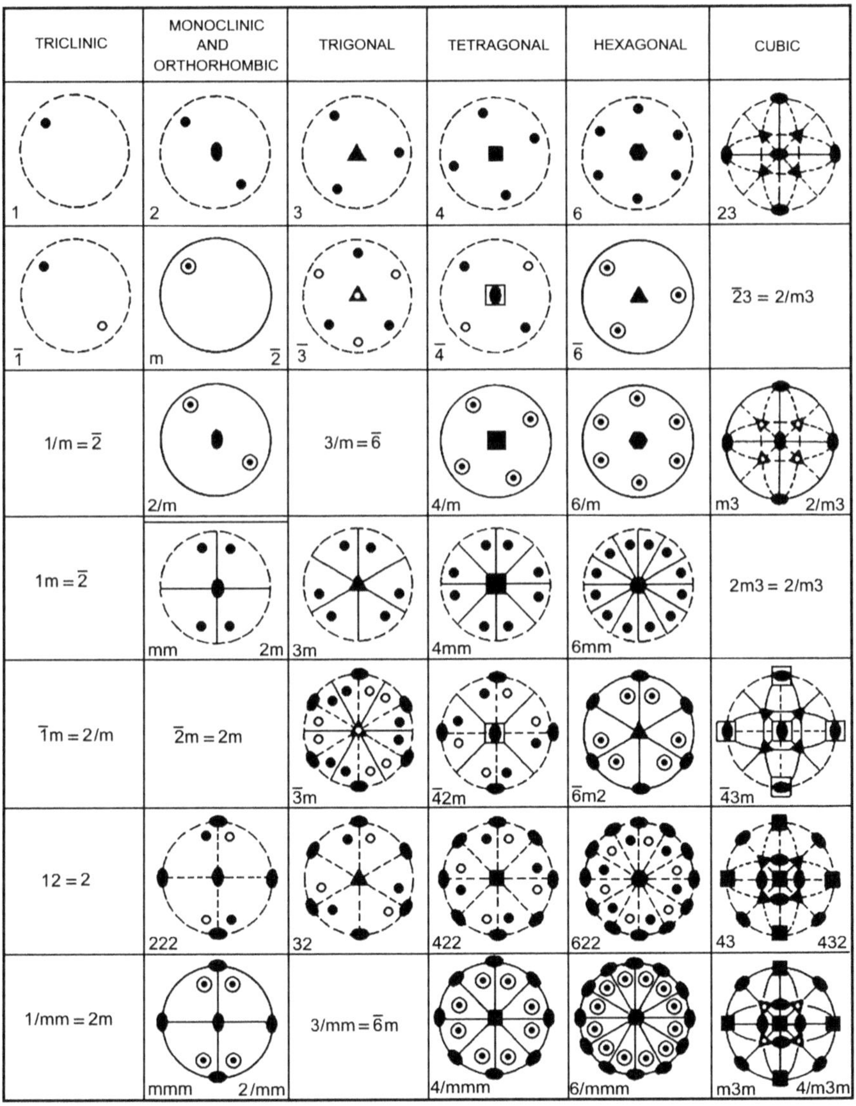

Fig. 3.14 Stereograms of the 32 crystallographic point groups

a horizontal reflection plane or a horizontal twofold axis brings a solid point into coincidence with open points. The vertical mirror planes are indicated by solid radial lines on the stereograms. A horizontal mirror plane is indicated by a solid circle rather than the dotted stereogram circle. The stereograms of the 32 point groups are dealt with in detail by Phillips [3.4] and are shown in Fig. 3.14.

3.7 Bravais Lattices

Lattices can now be classified on the basis of their symmetry. We shall first consider the simpler hypothetical two-dimensional lattices before discussing real three-dimensional lattices.

3.7.1 Two-Dimensional Bravais Lattices

The most general kind of two-dimensional lattice is the oblique lattice in which there is no special relationship between the translation vectors and the angle between them. Such a lattice has only twofold rotation axes and inversion symmetry. Oblique lattices in general do not have reflection symmetry. The lattice may possess rotation axes of higher order (i.e. fourfold and sixfold axes) and reflection symmetry if the translation vectors are related in some special way, or when the angle between them has a special value. These restrictions result in five two-dimensional lattices.

To demonstrate this, let us consider an oblique lattice with fundamental translation vectors **a** and **b**, ϕ being the angle between them. Let the a-axis lie along the x-axis of a Cartesian coordinate system (Fig. 3.15). If reflection symmetry is imposed on this lattice, we wish to find out if any relationship results between **a** and **b**. From Fig. 3.15, we have

$$\mathbf{a} = a_x\mathbf{i}, \tag{3.7}$$

$$\mathbf{b} = b_x\mathbf{i} + b_y\mathbf{j}. \tag{3.8}$$

Let us reflect the lattice about the x-axis. This operation gives us a new set of vectors $\mathbf{a}'$ and $\mathbf{b}'$ where

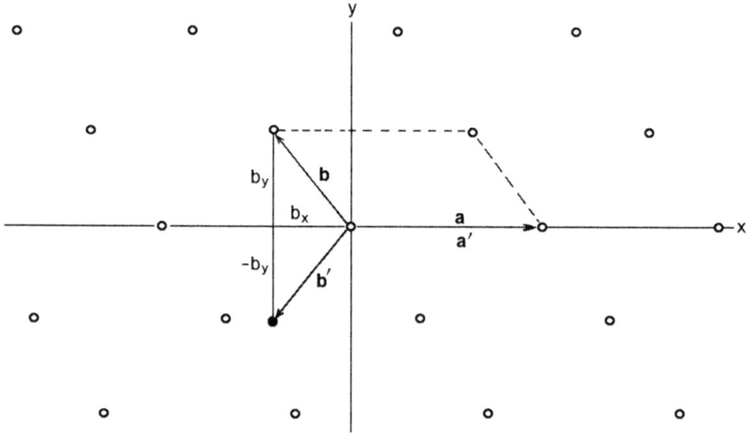

Fig. 3.15 Reflection symmetry imposed on an oblique lattice resulting in a rectangular lattice

$$\mathbf{a}' = a_x\mathbf{i}, \tag{3.9}$$

$$\mathbf{b}' = b_x\mathbf{i} - b_y\mathbf{j}. \tag{3.10}$$

Since $\mathbf{a}' = \mathbf{a}$, $\mathbf{a}'$ is a translation vector, but $\mathbf{b}'$ is not necessarily a translation vector. It will be a translational vector only if it satisfies the condition for translational symmetry, i.e. $\mathbf{b}' = n_1\mathbf{a} + n_2\mathbf{b}$. For the x-axis to be a true line of reflection symmetry, $\mathbf{a}'$ and $\mathbf{b}'$ must be translational vectors. Therefore,

$$\mathbf{b}' = n_1\mathbf{a} + n_2\mathbf{b} = n_1 a_x\mathbf{i} + n_2(b_x\mathbf{i} + b_y\mathbf{j}). \tag{3.11}$$

Equating coefficients of **i** and **j** in (3.10) and (3.11) we get

$$b_x = n_1 a_x + n_2 b_x \quad \text{and} \quad -b_y = n_2 b_y \quad \text{or} \quad n_2 = -1. \tag{3.12}$$

It follows that

$$b_x = n_1 a_x/2 \tag{3.13}$$

and b_y is arbitrary. If $n_1 = 0$, $b_x = 0$ and $\mathbf{a}\cdot\mathbf{b} = 0$. Therefore, $\phi = 90°$ and we have the primitive rectangular lattice. Such a lattice has a twofold symmetry with two perpendicular sets of mirror lines. If $n_1 = 1$, $b_x = a_x/2$ and $2b_x = a_x$. Since $\mathbf{a}\cdot(2\mathbf{b} - \mathbf{a}) = 0$ for this case, the rectangular translation vectors would be **a**, (2**b** – **a**) and we have the centred rectangular lattice with non-primitive translation vectors. Though it is possible to describe the unit cell in this case with primitive translation vectors **a**, **b** with the angle $\phi = \cos^{-1}(a/2b)$, it is convenient to use the non-primitive unit cell which is rectangular. The centred-rectangular lattice has the twofold rotation symmetry associated with the primitive oblique unit cell and also the reflection symmetries associated with the non-primitive rectangular cell.

Another special lattice type, the square lattice results when the lattice has a fourfold rotational symmetry. In this case $a = b$ and $\phi = 90°$ Yet another lattice, the hexagonal lattice results when $a = b$ and $\phi = 60°$ In this case the primitive unit cell is a rhombus with $\phi = 120°$ or $60°$ Three such primitive cells make up the hexagonal unit cell. The lattice has a sixfold rotational symmetry and six reflection lines at 30° with respect to each other.

These five lattice types – oblique, primitive, rectangular, centred-rectangular, square and hexagonal – are known as the five Bravais lattices in two dimensions (Fig. 3.16).

3.7.2 *Three-Dimensional Bravais Lattices*

In three dimensions, the most general lattice is the triclinic for which the primitive unit cell is a parallelepiped with translation vectors **a**, **b** and **c** of different lengths and angles α, β and γ unequal and different from 90°. Thirteen more lattices result when special relationships between sides and angles are considered. In 1848 Bravais

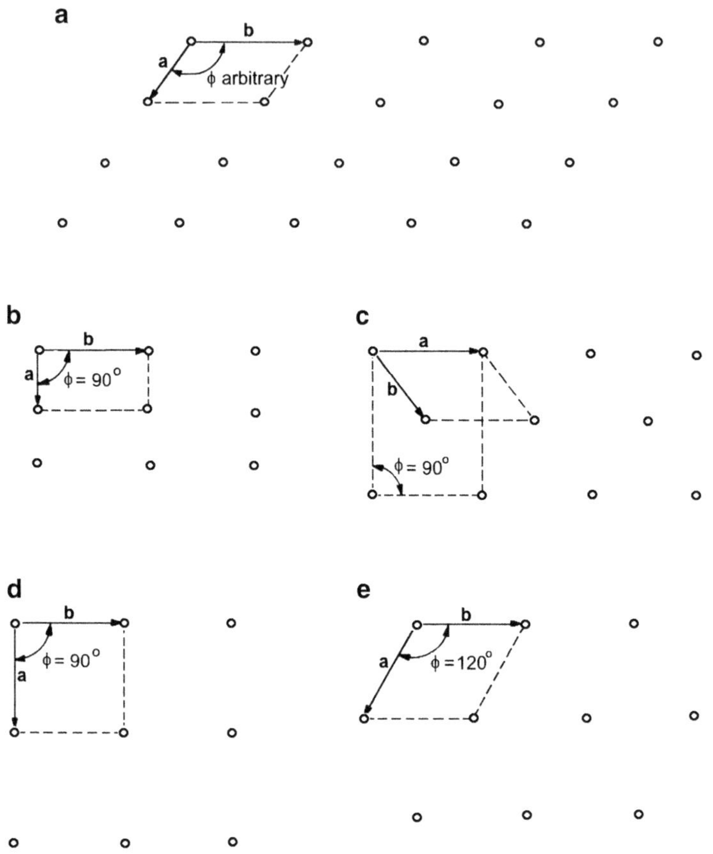

Fig. 3.16 The five Bravais lattices in two dimensions: (**a**) oblique, (**b**) rectangular, (**c**) centred rectangular, (**d**) square and (**e**) hexagonal

demonstrated that because of the symmetry of a crystal associated with rotation, reflection, inversion or a combination of these, there can be only 14 possible ways of arranging points in space which will have translational periodicity. The environment of any one point in the lattice is identical with that of any other point in the lattice. These point lattices are called the Bravais lattices. For some of these lattices, non-primitive cells are chosen as they demonstrate the symmetry more clearly. The primitive cell with lattice points at the corners is denoted by P. The body-centred lattice (I) has a lattice point at the intersection of body diagonals in addition to those at the corners. A face-centred lattice (F) has atoms at face centres as well as at the corners. If only the (001) faces are centred, we have the base-centred or C lattice. Centring the (100) or (010) faces results in A or B lattice. In the trigonal system, the shape of the unit cell is a rhombohedron. Body centring or face centring this unit cell does not produce a new type of arrangement. Thus the only arrangement possible is a P lattice, but it is conventionally given the symbol R and is described

Table 3.4 The seven crystal systems and the associated Bravais lattices

System	Number of lattices in the system	Bravais lattices in the system	Symmetry of lattice
Triclinic	1	*P*	$\bar{1}$
Monoclinic	2	*P*	2/*m*
		C	
Orthorhombic	4	*P*	*mmm*
		C	
		I	
		F	
Tetragonal	2	*P*	4/*mmm*
		I	
Trigonal	1	*R*	$\bar{3}m$
Hexagonal	1	*P*	6/*mmm*
Cubic		*P*	*m*3*m*
		I	
		F	

as the trigonal space lattice. The 14 Bravais lattices associated with the seven crystal systems are given in Table 3.4 and are illustrated in Fig. 3.17.

3.8 Crystal Planes, Directions and Miller Indices

A crystal usually grows with plane faces. The lattice points in a space lattice can be arranged in an infinite number of ways in parallel equidistant sheets or planes. The atomic density on these planes is different. The densely packed sheets are far apart, while others are closer together. Hence several planes oriented in different directions are possible in a crystal. Some planes that may exist in a crystal lattice are shown in Fig. 3.18.

3.8.1 Directions

In discussing crystals it is necessary to identify the various faces and directions without ambiguity. To do this we choose the crystal axes (a, b, c) as the reference axes (x, y, z). We consider a two-dimensional lattice network for simplicity (Fig. 3.19a). The directions are indicated by the coordinates of the first whole numbered point (x, y) they pass. For the direction OM, the coordinates would be [2,1]. For ON it is [1,1]. This may be extended in three dimensions in terms of the reference axes a, b, c. Figure 3.19b shows some directions in a crystal lattice. The point P is at $(1, 0, 0)$. Since "a" represents unit repeat distance in the direction OP, the direction OP is denoted as [100]. Similarly, direction OQ is indicated as [110], OR as [010] and OS as [111]. The points T and U are at (1/2, 1/2, 1) and (1, 0,

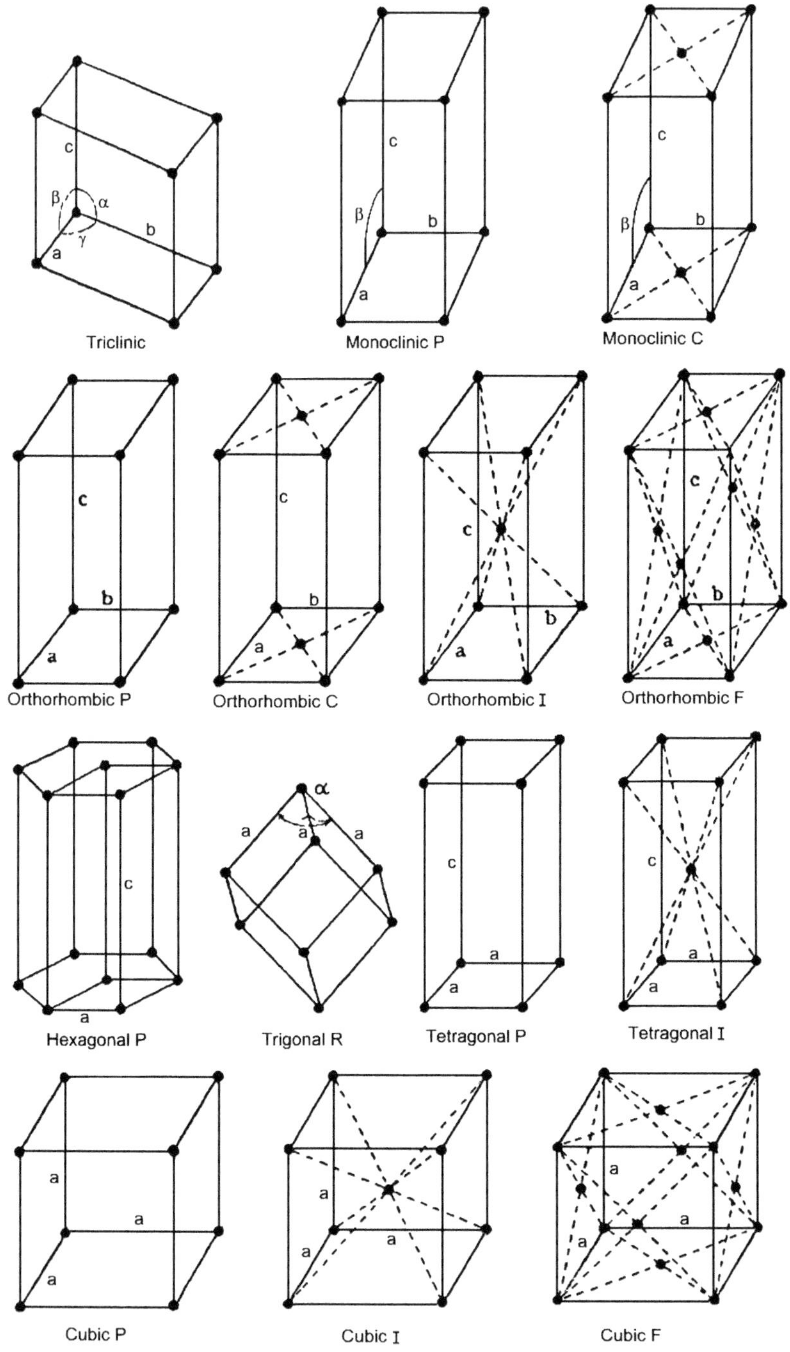

Fig. 3.17 The 14 Bravais lattices in three dimensions

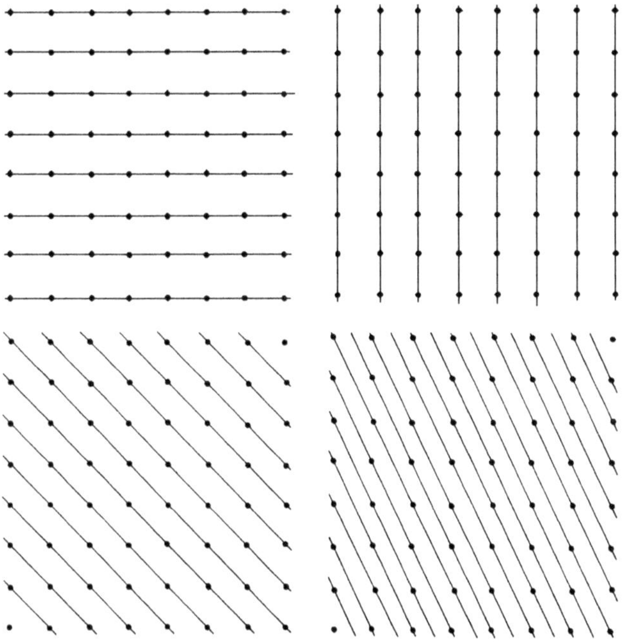

Fig. 3.18 Some lattice planes in two dimensions

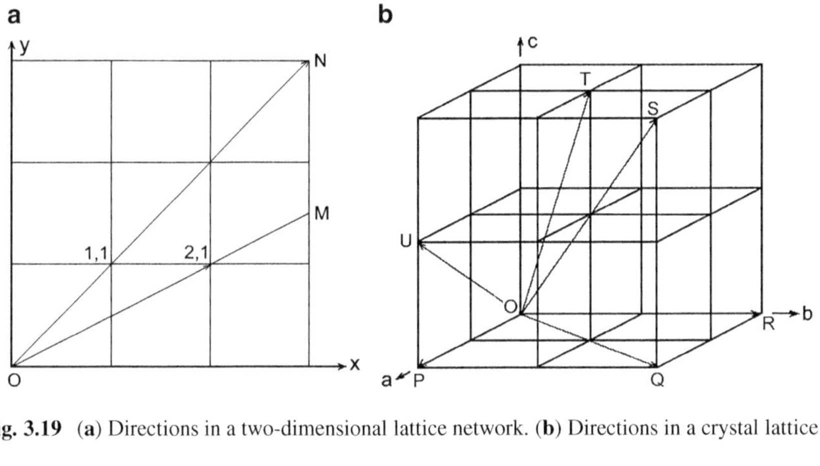

Fig. 3.19 (**a**) Directions in a two-dimensional lattice network. (**b**) Directions in a crystal lattice

1/2) respectively. Hence directions OT and OU are represented by [112] and [201], respectively, in terms of whole numbers. By convention, integers in square brackets [*uvw*] are used to indicate a direction. Negative directions are indicated with a bar above the number. A lattice vector connects a lattice point to the origin and is given as $\mathbf{r} = u\mathbf{a}+v\mathbf{b}+w\mathbf{c}$. By choosing different values of the integers u, v, w it is possible to find all lattice points. A family of directions are denoted as $\langle uvw \rangle$. For example, the family of $\langle 100 \rangle$ directions are [100], [010], [001], $[\bar{1}00]$, $[0\bar{1}0]$ and $[00\bar{1}]$.

3.8.2 Miller Indices

As mentioned earlier, several planes are possible in a crystal. To identify these planes, a set of axes which bears a close relationship to the symmetry of the crystal is chosen. These axes are not necessarily orthogonal. After the choice of axes are made, a plane intersecting all three axes is chosen as the unit plane or parametral plane; a, b, c are the intercepts made by the plane on the crystal axes. It is now possible to express the intercepts of all other planes as ma, nb and pc where m, n and p are integers. That this is so follows from the law of rational indices stated by Hauy according to which, the external symmetric form exhibited by crystals require that only those faces which bear a rational relationship to one another are possible. The three numbers m, n and p might be used as indices to denote a given face. Sometimes the index would have the value infinity which is troublesome in mathematical calculations. It is therefore more convenient to use numbers proportional to the reciprocals m, n and p as indices viz., $h \propto 1/m$; $k \propto 1/n$ and $l \propto 1/p$. The indices h, k and l are expressed as integers without a common divisor. These are known as Miller indices of the face or plane and written as (hkl).

The crystal axes being axes of reference may extend in positive as well as in negative directions from the origin. If the plane intersects an axis on its negative side, the Miller index is negative and denoted by a bar placed above the index. For example (hkl) and $(\bar{h}\bar{k}\bar{l})$ denote two parallel faces on opposite sides of the crystal. The indices (hkl) denote a set of planes (hkl), $(\bar{h}kl)$, $(h\bar{k}l)$, $(hk\bar{l})$, $(h\bar{k}\bar{l})$, $(\bar{h}k\bar{l})$, $(\bar{h}\bar{k}l)$, $(\bar{h}\bar{k}\bar{l})$ and are said to belong to a form; the form is denoted by a set of indices in curly brackets as $\{hkl\}$.

The Miller indices of a plane may be determined as follows:

1. Find the intercepts of the plane on the axes a, b, c in terms of the lattice constants.
2. Take the reciprocals of these numbers and then reduce to the smallest three integers h, k, l. The indices of the plane are then (hkl).

For the plane whose intercepts are $(1, 2, 3)$, the reciprocals are 1, 1/2 and 1/3 and the Miller indices are (632). For a plane parallel to one of the crystal axes, the intercept on that axis is ∞ and the corresponding Miller index is zero. The Miller indices of some important planes in a cubic crystal are indicated in Fig. 3.20.

In hexagonal crystals, the unit cell is a right prism based on a 60° rhombus. The vertical prism edge is chosen as the direction of the z-axis. The horizontal edges of the prism are normal to z direction. Two of these edges can be chosen as the x and y axes. The question arises as to which of the three possible directions should be selected as x direction and which as y. Since all three directions are equivalent, all three directions are used. Thus hexagonal crystals are described in terms of

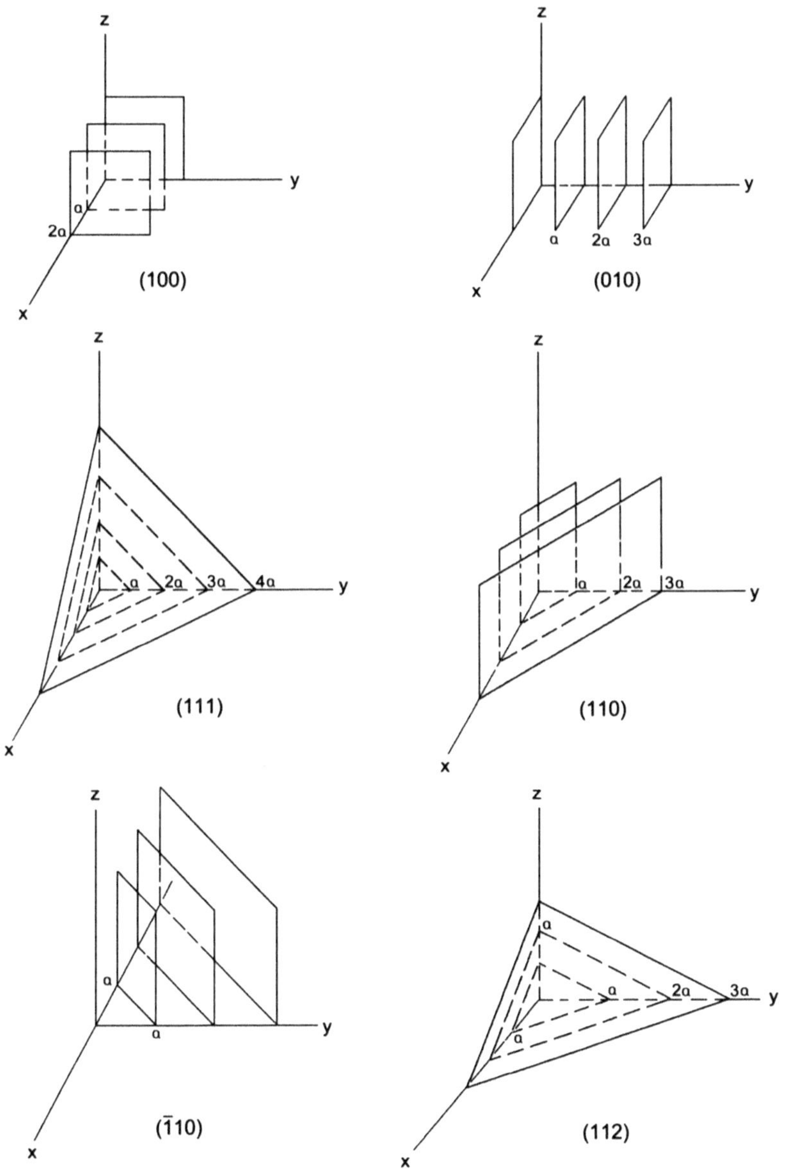

Fig. 3.20 The Miller indices for some planes in cubic crystals

four crystallographic axes; three horizontal axes x, y and u at an angle of 120° to each other and normal to the vertical z axis (Fig. 3.21). In describing the planes of hexagonal crystals, Bravais adopted the Miller symbols ($hkil$) referring to the axes in the order x, y, u, z. These indices are therefore known as the Bravais–Miller indices. It may be shown that $h + k = -i$ so that $h + k + i = 0$. Sometimes the hexagonal indices are found in the form ($hk{\cdot}l$) when the third index is said to be suppressed.

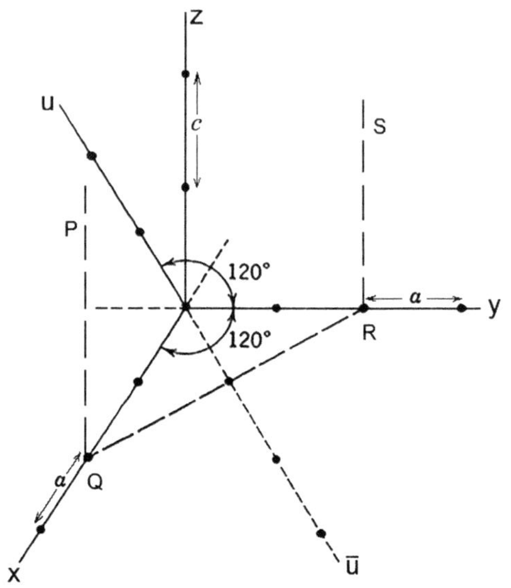

Fig. 3.21 Miller–Bravais indices (*hkil*). The plane shown is $(11\bar{2}0)$

3.8.3 Equation for the Plane (hkl)

For simplicity, we shall derive the equation of the plane (*hkl*) for the cubic case. Consider the plane PQR in Fig. 3.22 making intercepts OP, OQ and OR on the x, y,z axes. From analytical geometry, the equation for this plane is

$$\frac{x}{\mathrm{OP}} + \frac{y}{\mathrm{OQ}} + \frac{z}{\mathrm{OR}} = 1. \tag{3.14}$$

For the cubic case, OP $= pa$, OQ $= qa$ and OR $= ra$ where a is the repeat distance along each of the axes. On substitution (3.14) becomes

$$\frac{x}{pa} + \frac{y}{qa} + \frac{z}{ra} = 1 \tag{3.15}$$

or,

$$xqr + ypr + zpq = apqr. \tag{3.16}$$

The Miller indices of the plane are $h \propto 1/p$; $k \propto 1/q$ and $l \propto 1/r$. Clearing fractions by multiplying by pqr gives $h \propto qr$; $k \propto pr$ and $l \propto pq$. Hence

$$h = qr/t; k = pr/t \quad \text{and} \quad l = pq/t, \tag{3.17}$$

where t is their common divisor. Combining (3.16) and (3.17) we get

$$hx + ky + lz = apqr/t. \tag{3.18}$$

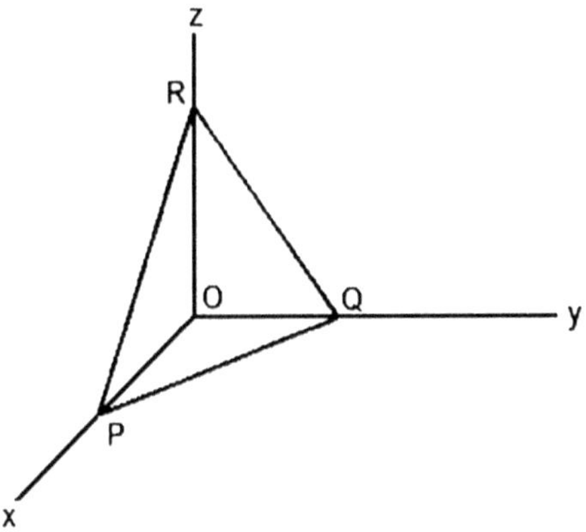

Fig. 3.22 A lattice plane in a crystal

The equation for the plane (*hkl*) is then

$$hx + ky + lz - Da = 0, \tag{3.19}$$

where $D = pqr/t$. The equation for the plane (*hkl*) passing through the origin is

$$hx + ky + lz = 0. \tag{3.20}$$

For the orthorhombic system, the equation for the plane (*hkl*) is

$$\frac{hx}{a} + \frac{ky}{b} + \frac{lz}{c} - D = 0, \tag{3.21}$$

where a, b and c are unit lengths along the axes.

3.8.4 *Interplanar Spacings*

In a space lattice, each Miller index triplet represents a series of parallel equidistant planes containing the lattice points. The perpendicular distance between successive planes of a *hkl* series is denoted as d_{hkl}. We shall develop the expression for d_{hkl} for the cubic lattice first and then indicate the expressions for d_{hkl} for systems of lower symmetry. The equation to the plane with Miller indices (*hkl*) in the cubic lattice is

$$hx + ky + lz - Da = 0.$$

Its distance from the origin is

$$\frac{Da}{(h^2 + k^2 + l^2)^{1/2}}.$$

Let us now consider a parallel plane passing through the lattice point $P_1(x_1y_1z_1)$. Its equation is

$$hx + ky + lz - (hx_1 + ky_1 + lz_1) = 0. \tag{3.22}$$

Its distance from the origin is

$$\frac{hx_1 + ky_1 + lz_1}{(h^2 + k^2 + l^2)^{1/2}}.$$

Similarly, the distance from the origin of another parallel plane passing through the lattice point $P_2(x_2y_2z_2)$ is

$$\frac{hx_2 + ky_2 + lz_2}{(h^2 + k^2 + l^2)^{1/2}}.$$

The distance between the two parallel planes passing through points P_1 and P_2 is therefore,

$$\frac{h(x_1 - x_2) + k(y_1 - y_2) + l(z_1 - z_2)}{(h^2 + k^2 + l^2)^{1/2}}.$$

For a cubic lattice, the unit distance along all three axes is the same, equal to a. The coordinates of P_1 and P_2 can thus be expressed as $x_1 = r_1a$; $y_1 = s_1a$; $z_1 = t_1a$ and $x_2 = r_2a$; $y_2 = s_2a$; $z_2 = t_2a$, where r_1, s_1, t_1 and r_2, s_2, t_2 are integers. The distance between the parallel planes through P_1 and P_2 is now given as

$$\frac{a\{h(r_1 - r_2) + k(s_1 - s_2) + l(t_1 - t_2)\}}{(h^2 + k^2 + l^2)^{1/2}}.$$

Since h, k and l are always integral, the coefficient of a must be integral, and the minimum possible separation for the two planes is given when this coefficient has the smallest possible value, unity. The two planes are coincident when the coefficient is zero. For any set of three integers h, k and l we can always choose three other integers r, s and t such that the equation

$$hr + ks + lt = 1 \tag{3.23}$$

is satisfied. Thus for the cubic lattice, we get

$$d_{hkl} = \frac{a}{(h^2 + k^2 + l^2)^{1/2}}. \tag{3.24}$$

For the tetragonal lattice, the expression for d_{hkl} is obtained as

$$d_{hkl} = \frac{a}{[h^2 + k^2 + (la/c)^2]^{1/2}}. \tag{3.25}$$

Table 3.5 Formulas for calculating interplanar spacings d_{hkl}

System	Axial translations and angles	d_{hkl}
Cubic	$a = b = c$ $\alpha = \beta = \gamma = 90^\circ$	$a(h^2 + k^2 + l^2)^{-1/2}$
Tetragonal	$a = b \neq c$ $\alpha = \beta = \gamma = 90^\circ$	$[(h^2/a^2) + (k^2/a^2) + (l^2/c^2)]^{-1/2}$
Orthorhombic	$a \neq b \neq c$ $\alpha = \beta = \gamma = 90^\circ$	$[(h^2/a^2) + (k^2/b^2) + (l^2/c^2)]^{-1/2}$
Hexagonal	$a = b \neq c$ $\alpha = \beta = 90^\circ$, $\gamma = 120^\circ$	$[(4/3a^2)(h^2 + k^2 + hk) + (l^2/c^2)]^{-1/2}$
Trigonal	$a = b = c$ $\alpha = \beta = \gamma < 120^\circ$, $\neq 90^\circ$	$a\left[\frac{(h^2 + k^2 + l^2)\sin^2\alpha + 2(hk + hl + kl)(\cos^2\alpha - \cos\alpha)}{1 + 2\cos^3\alpha - 3\cos^2\alpha}\right]^{-1/2}$
Monoclinic	$a \neq b \neq c$ $\alpha = \gamma = 90^\circ \neq \beta$	$\left[\frac{(h^2/a^2) + (l^2/c^2) - (2hl/ac)\cos\beta}{\sin^2\beta} + \frac{k^2}{b^2}\right]^{-1/2}$
Triclinic	$a \neq b \neq c$ $\alpha \neq \beta \neq \gamma$	$V[S_{11}h^2 + S_{22}k^2 + S_{33}l^2$ $+ 2S_{12}hk + 2S_{23}kl + 2S_{13}hl]^{-1/2}$

The expressions for the interplanar spacings for crystals of lower symmetry may be obtained from similar reasoning and are given in Table 3.5. Expressions for the interplanar angles may also be obtained from simple methods and are given in Table 3.6. In Tables 3.5 and 3.6,

$$S_{11} = b^2c^2\sin^2\alpha;\ S_{12} = abc^2(\cos\alpha\cos\beta - \cos\gamma);$$
$$S_{22} = a^2c^2\sin^2\beta;\ S_{23} = a^2bc(\cos\beta\cos\gamma - \cos\alpha);$$
$$S_{33} = a^2b^2\sin^2\gamma S_{13} = ab^2c(\cos\gamma\cos\alpha - \cos\beta);$$

and V is the volume of the unit cell.

3.9 Space Groups

In Sect. 3.6, the point group symmetry elements were enumerated and it was pointed out that each combination of symmetry elements constitutes a point group; there are 32 point groups.

But the crystal lattice is infinite in extent and is, in fact generated by translating the Bravais unit cell parallel to itself along the three crystallographic axes. This infinite nature of the lattice bestows on it two new types of translational symmetry elements, the screw axes and glide planes.

Table 3.6 Formulas for calculating angle ϕ between planes with Miller indices $h_1 k_1 l_1$ and $h_2 k_2 l_2$ and spacings d_1 and d_2; V is the volume of the unit cell

System	$\cos\phi$
Cubic	$\dfrac{h_1h_2+k_1k_2+l_1l_2}{\sqrt{(h_1^2+k_1^2+l_1^2)(h_2^2+k_2^2+l_2^2)}}$
Tetragonal	$\dfrac{\dfrac{h_1h_2+k_1k_2}{a^2}+\dfrac{l_1l_2}{c^2}}{\sqrt{\left(\dfrac{h_1^2+k_1^2}{a^2}+\dfrac{l_1^2}{c^2}\right)\left(\dfrac{h_2^2+k_2^2}{a^2}+\dfrac{l_2^2}{c^2}\right)}}$
Orthorhombic	$\dfrac{\dfrac{h_1h_2}{a^2}+\dfrac{k_1k_2}{b^2}+\dfrac{l_1l_2}{c^2}}{\sqrt{\left(\dfrac{h_1^2}{a^2}+\dfrac{k_1^2}{b^2}+\dfrac{l_1^2}{c^2}\right)\left(\dfrac{h_2^2}{a^2}+\dfrac{k_2^2}{b^2}+\dfrac{l_2^2}{c^2}\right)}}$
Hexagonal	$\dfrac{h_1h_2+k_1k_2+\frac{1}{2}(h_1k_2+h_2k_1)+\dfrac{3a^2}{4c^2}l_1l_2}{\sqrt{\left(h_1^2+k_1^2+h_1k_1+\dfrac{3a^2}{4c^2}l_1^2\right)\left(h_2^2+k_2^2+h_2k_2+\dfrac{3a^2}{4c^2}l_2^2\right)}}$
Trigonal	$\dfrac{a^4d_1d_2}{V^2}\left[\sin^2\alpha(h_1h_2+k_1k_2+l_1l_2)+(\cos^2\alpha-\cos\alpha)(k_1l_2+k_2l_1+l_1h_2+l_2h_1+h_1k_2+h_2k_1)\right]$
Monoclinic	$\dfrac{d_1d_2}{\sin^2\beta}\left[\dfrac{h_1h_2}{a^2}+\dfrac{k_1k_2\sin^2\beta}{b^2}+\dfrac{l_1l_2}{c^2}-\dfrac{(l_1h_2+l_2h_1)\cos\beta}{ac}\right]$
Triclinic	$\dfrac{d_1d_2}{V^2}[S_{11}h_1h_2+S_{22}k_1k_2+S_{33}l_1l_2+S_{23}(k_1l_2+k_2l_1)+S_{13}(l_1h_2+l_2h_1)+S_{12}(h_1k_2+h_2k_1)]$

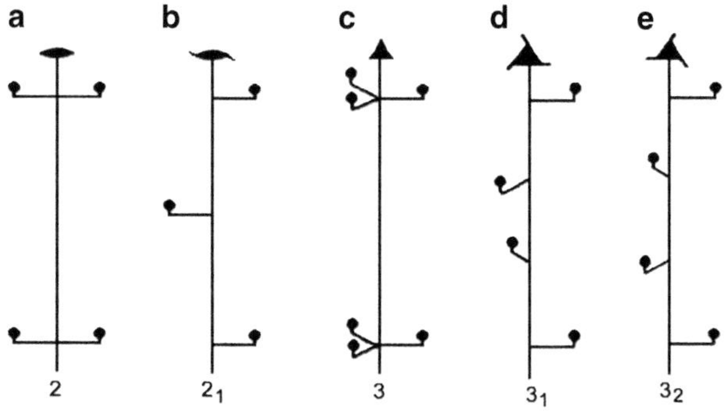

Fig. 3.23 The operation of 2, 2_1, 3, 3_1 and 3_2 axes

3.9.1 Screw Axes

Suppose the external symmetry of the crystal reveals a vertical diad axis, then we may suppose that some structural unit is arranged in pairs about this direction (Fig. 3.23a). But another kind of twofold regularity is possible by a rotation through

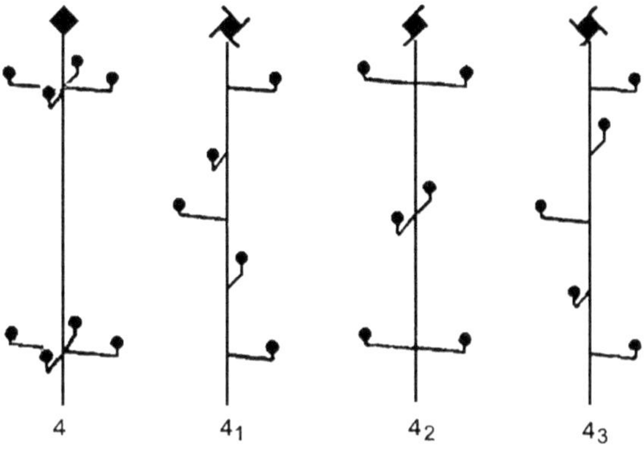

Fig. 3.24 The operation of 4, 4_1, 4_2 and 4_3 axes

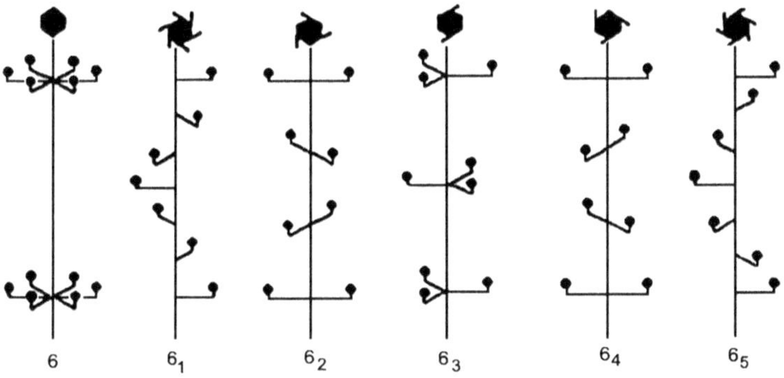

Fig. 3.25 The operation of 6, 6_1, 6_2, 6_3, 6_4 and 6_5 axes

180° about the diad axis and a translation parallel to the axis (Fig. 3.23b). Such an axis is termed a screw diad axis, and is denoted 2_1. The screw character is indicated by the tails affixed to the symbol corresponding to the rotation axis. If a structure shows threefold symmetry, the axis may be a true rotation axis 3 (Fig. 3.23c) or a screw triad axis where every rotation through 120° is combined with translation parallel to the axis by one-third the repeat distance. In this case two kinds of screw triads are possible; one corresponding to an anticlockwise rotation (3_1) (Fig. 3.23d) and the other (3_2) (Fig. 3.23e) to a clockwise rotation with the same translation. It can be seen from Fig. 3.23 that the two screw triads are enantiomorphs of each other.

For fourfold rotational symmetry, there are four possibilities (4, 4_1, 4_2 and 4_3) shown in Fig. 3.24. For sixfold symmetry we have six possibilities (6, 6_1, 6_2, 6_3, 6_4 and 6_5) shown in Fig. 3.25.

3.9.2 *Glide Planes*

Just as a rotation axis may represent a screw axis in the same direction in the internal structure, external planes of symmetry may also represent planes of symmetry

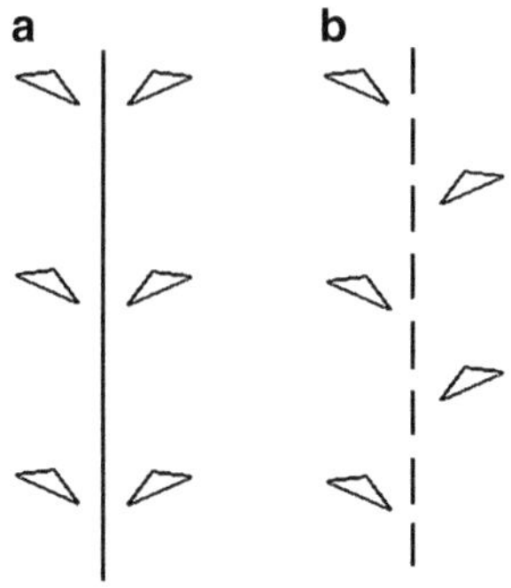

Fig. 3.26 The illustration of (**a**) mirror reflection and (**b**) glide reflection

parallel to this direction, in which the structural unit undergoes a translation after reflection about the plane. Thus if we have a true reflection plane m and a scalene triangle representing a structural unit then we have Fig. 3.26a. In Fig. 3.26b, the reflection is combined with the translation in the plane of the diagram. Such a plane is called a glide reflection plane. The difference between these two types of internal arrangement would not be appreciable externally. When external symmetry indicates more than one plane of symmetry, any or all may correspond to a glide reflection internally.

We shall now list out all the possible symmetry elements (non-translational and translational) that space lattices can have (Table 3.7). A space group is a combination of these symmetry elements. These symmetry elements are arranged on the space lattice. The type of Bravais lattice and the location of the point group symmetry elements and the translational symmetry elements over the space lattice constitute a space group. These symmetry elements taken together determine the positions of equivalent points within the unit cell.

Let us try to understand the space group symbol. The first letter (in capitals) indicates the type of Bravais lattice (P – primitive, I – body-centred, F – face-centred, A – side-centred, additional translation $\frac{b}{2} + \frac{c}{2}$, C – side-centred, additional translation $\frac{a}{2} + \frac{b}{2}$). The next three positions indicate symmetry elements along the a, b, c axes. Some symmetry elements, though present, are not indicated in the symbol as they arise due to the result of the other indicated symmetry elements. Some examples are given below.

$P2_1/c$: Primitive, monoclinic unit cell; c-glide plane $\perp b$;2_1 axis $||b$; centrosymmetric

$Ibca$: Body-centred, orthorhombic unit cell; b-glide plane $\perp a$; c-glide plane $\perp b$; a-glide plane $\perp c$; centrosymmetric

$P4_12_12$: Primitive, tetragonal unit cell; 4_1 axis $||c$; 2_1 axis $||a$ and b; twofold axes at 45° to a and b; non-centrosymmetric

$F\bar{4}3c$: Face-centred, cubic unit cell; $\bar{4}$ axes $||a$, b and c; threefold axes $||\langle 111\rangle$; c-glide planes $\perp\langle 110\rangle$; non-centrosymmetric

We shall now consider a few space groups to gain familiarity with the representation of space groups.

Table 3.7 Space group symmetry elements

Type of symmetry element	Written symbol	Graphical symbol	
Bravais lattice	*P, F, I, C, A*		
Centre of symmetry	$\bar{1}$	○	
		Perpendicular to paper	In plane of paper
Mirror plane	*m*	———	┐ ⌝\
Glide planes	*a* *b* *c*	--------	↰ ↴
		Glide in plane of paper	Arrow shows glide direction
		.... Glide out of plane of paper	
	n	—·—·—·—	⬋
Rotation axes	2	⬮	⟶
	3	▲	
	4	◆	
	6	⬢	
Screw axes	2_1	⬮	——⇀
	3_1, 3_2	▲ ▲	
	4_1, 4_2, 4_3	◆ ◆ ◆	
	6_1, 6_2, 6_3, 6_4, 6_5	⬢ ⬢ ⬢ ⬢ ⬢	
Inversion axes	$\bar{3}$	▲	
	$\bar{4}$	◆	
	$\bar{6}$	⬢	

3.9.3 *Space Group P*1

This is a primitive triclinic lattice *P*. Only an identity axis is associated with each point of the lattice. Hence the symbol is *P*1. A portion of the pattern based on this space group is shown in Fig. 3.27. A repeating unit of the structure is represented by a scalene triangle. A + symbol indicates the unit is at some height above the base of the unit cell.

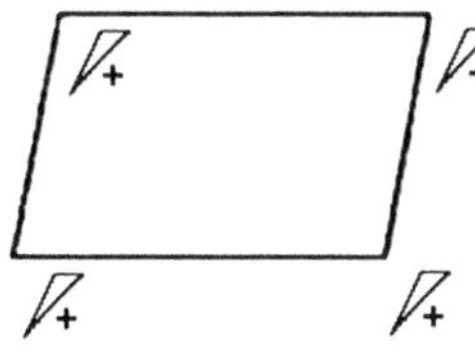

Fig. 3.27 Arrangement of repeating units for space group $P1$

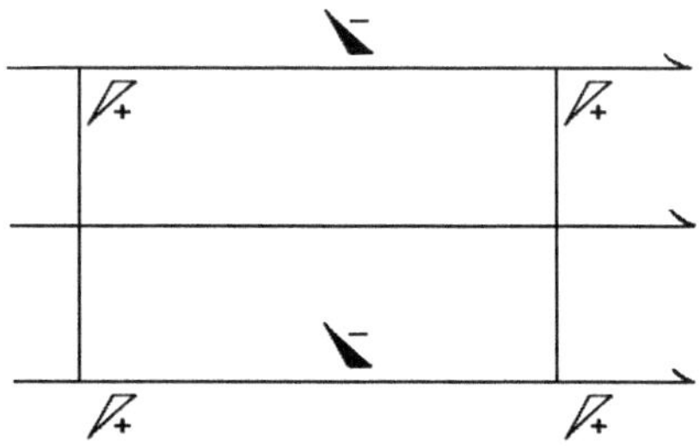

Fig. 3.28 The symmetry elements and the repeating units for the space group $P2_1$

3.9.4 *Space Group* $P2_1$

This monoclinic primitive lattice is shown in Fig. 3.28. The horizontal screw diad axes in the y direction are drawn as single-based arrows. Further screw diad axes arise (not indicated in diagram), in addition to those along the edges of the unit cell. A shaded triangle with a – symbol indicates a structural unit at an equal distance below the plane of the diagram.

3.9.5 *Space Group* $I4$

This is a body-centred tetragonal space group (Fig. 3.29). Apart from the tetrad axes, the translation 1/2 1/2 1/2 gives rise to 4_2 axes parallel to the rotation tetrad axes. Also due to the combination of the other symmetry elements, diad axes also arise parallel to the tetrad axes. In the figure $\frac{1}{2}+$ indicates a structural unit at a height + above the base of the unit cell raised to a height $\frac{1}{2}+$.

3.9.6 *Space Group Pnma*

This is an orthorhombic space group with a P lattice. The symmetry elements and the repeating structural units are shown in Fig. 3.30. These include 2_1 screw axes and centres of inversion. The symmetry operation about these symmetry elements generates a pattern of eight symmetrically equivalent points. If the glide plane n is along the axes b or c, the space group is called *Pmnb* or *Pcmn* but they are not new and independent space groups.

Several examples of space groups are discussed by Phillips [3.4] and Ladd and Palmer [3.5]. The space groups of a few crystals are given in Table 3.8. Schoenflies

Fig. 3.29 The symmetry elements and the repeating units for the space group $I\bar{4}$

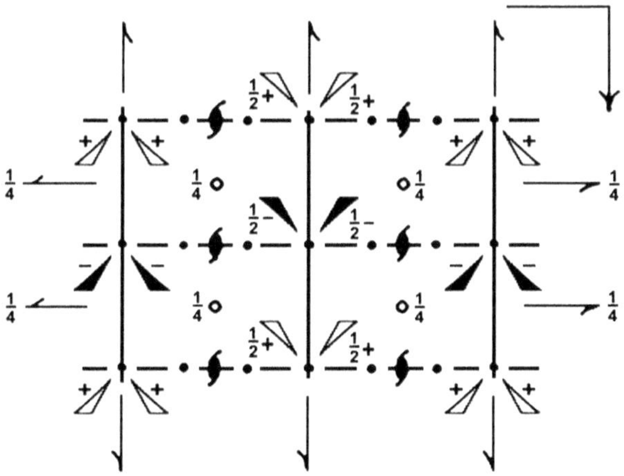

Fig. 3.30 The symmetry elements and the repeating units for the space group *Pnma*

[3.7], Barlow [3.11] and Fedorov [3.12] showed that the total number of permissible combinations of the symmetry elements, i.e. the number of space groups is 230. These space groups are given in the International Tables for X-ray Crystallography [3.13]. A complete list is also given by Kaelble [3.1].

3.10 Packing in Crystals

In a solid, the atoms are arranged in a manner depending on the interatomic bonds. Though we have so far assumed the crystal structure as an array of lattice points separated from each other, this is not really so. We assume the atoms or molecules to be hard spheres arranged to touch each other. The most favourable

Table 3.8 Space groups of some crystals

Crystal	Crystal system	Space group
NaCl	Cubic	$Fm3m$
CsCl	Cubic	$Pm3m$
Diamond (C)	Cubic	$Fd3m$
Pyrite (FeS_2)	Cubic	$Pa3$
$NaClO_3$	Cubic	$P2_13$
$CaWO_4$	Tetragonal	$I4_1/a$
$BaTiO_3$ (<120°C)	Tetragonal	$P4mm$
KDP	Tetragonal	$I\bar{4}2d$
Graphite	Hexagonal	$P6_3mmc$
Calcite ($CaCO_3$)	Trigonal	$R\bar{3}c$
L-arginine dihydrate $(^{+}H_2N)_2CNH\ (CH_2)_3CH(NH_2)COO^{-}$	Orthorhombic	$P2_12_12_1$
$HfSi_2$	Orthorhombic	$Cmcm$
Oxalic acid dehydrate $(COOH)_2 \cdot 2H_2O$	Monoclinic	$P2_1/n$
Naphthalene ($C_{10}H_8$)	Monoclinic	$P2_1/c$
Copper sulphate pentahydrate	Triclinic	$P\bar{1}$

arrangement is that of close packing where the constituent atoms, ions or molecules are surrounded by as many neighbours as possible. The coordination number is defined as the number of equidistant nearest neighbours that these constituents have in a given structure. Greater the coordination number, the more closely packed will the structure be. For the simple cubic (sc) structure, there is one atom at each of the eight corners of the unit cell. Each atom is surrounded by six equidistant neighbours, the distance between nearest neighbours being "a". The coordination number in this case is 6. The body-centred cubic (bcc) lattice has eight atoms at the corners of the unit cell and one at the centre. The central atom is surrounded by eight equidistant neighbours and the nearest neighbour distance is $a\sqrt{3}/2$. The face-centred cubic (fcc) unit cell has eight atoms at the corners and one each at the centre of six faces. The number of nearest neighbours is 12 and the nearest neighbour distance is $a/\sqrt{2}$. We note that every lattice point at the corner of a unit cell is shared by eight adjacent unit cells. Hence the contribution of each lattice point to the unit cell will be 1/8. The simple cubic lattice therefore has one lattice point per cell. The lattice point at the centre of a bcc lattice belongs to one cell only and the effective number of lattice points for bcc is 2. A lattice point at the face centre is shared by two adjacent cells and contributes one-half to each cell. Thus in a fcc unit cell, the number of lattice points per cell would be 4.

A close-packed structure has the maximum number of nearest neighbours. A sphere can be surrounded by at the most 12 other spheres of the same size. The arrangement of the 12 spheres around a central sphere can be made in two different ways resulting in two close-packed structures – the face-centred cubic and the hexagonal close-packed structures (Fig. 3.31).

Consider a central sphere surrounded by six spheres in a close-packed plane. Three spheres each may be arranged in the plane above and below this middle plane at positions x or positions y giving the total of 12 spheres around the central sphere. If the spheres sit at the same position, x (or$y$), in both the planes above and

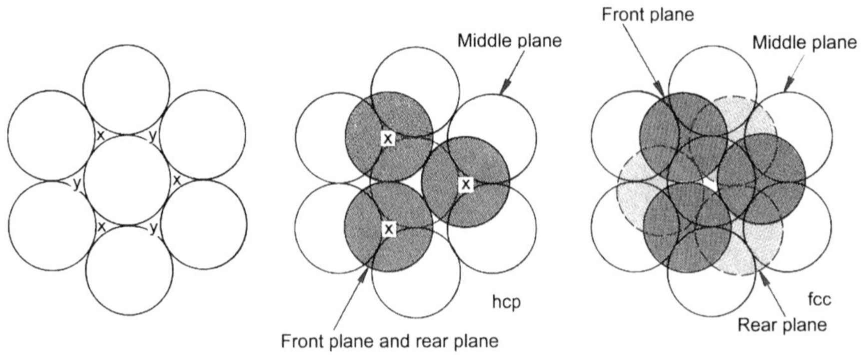

Fig. 3.31 Origin of two possible close-packed configurations

Table 3.9 Characteristics of cubic and hcp lattices

	Simple cubic (sc)	Body-centred cubic (bcc)	Face-centred cubic (fcc)	Hexagonal close-packed (hcp)
Volume of unit cell	a^3	a^3	a^3	$3\sqrt{2}a^3$
Lattice points per unit cell	1	2	4	6
Lattice points per unit volume	$1/a^3$	$2/a^3$	$4/a^3$	$\sqrt{2}/a^3$
Coordination number	6	8	12	12
Nearest neighbour distance	a	$a\sqrt{3}/2$	$a/\sqrt{2}$	$a/2$
Packing fraction	0.524	0.68	0.74	0.74

below, the structure is hexagonal close-packed. If however, the spheres in the planes above and below are in different positions, then the structure is face-centred cubic. The space lattice for the hexagonal close-packed structure is simple hexagonal with the basis consisting of two atoms, one at 000 and the other at 2/3,1/3,1/2. The coordination number for both structures is 12. Physical properties like thermal expansion, electrical and thermal conductivity measured along the a and c axis are different for hcp solids and are said to be anisotropic. Seventy percent of metals crystallise in one of these two structures. The characteristics of the three types of cubic lattices and the hcp lattice are summarised in Table 3.9.

The packing factor (fraction) is the ratio of the volume of the atoms occupying the unit cell to the volume of the unit cell. We shall now calculate the atomic packing factor (APF) for some simple structures.

3.10.1 Calculation of Atomic Packing Factor

3.10.1.1 Simple Cubic (sc)

The simplest structure is the simple cubic structure shown in Fig. 3.32. The unit cell has one lattice point at each of its eight corners. The coordination number for the

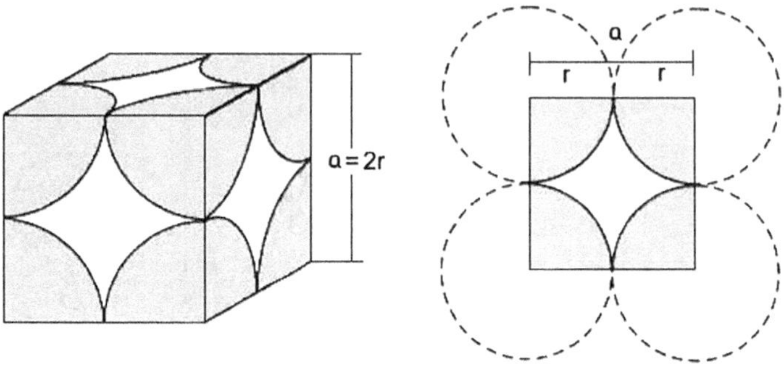

Fig. 3.32 The simple cubic (sc) structure

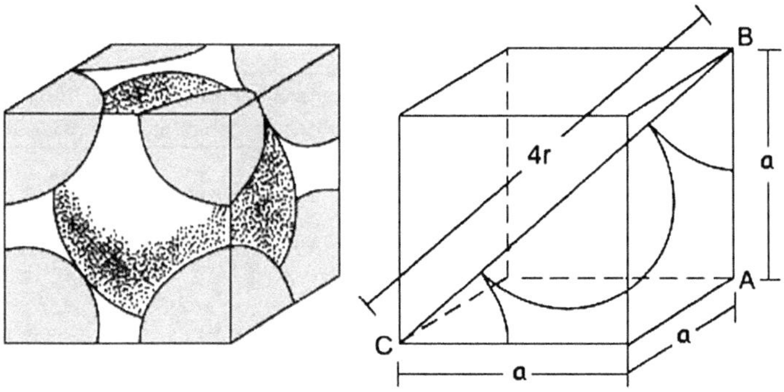

Fig. 3.33 The body-centred cubic (bcc) structure

simple cubic lattice is 6. The equivalent number of atoms per unit cell is 1 as each atom at the corner is shared by eight unit cells. Since the atoms are assumed to touch each other, the lattice constant $a = 2r$, where r is the radius of the atom.

$$\text{The atomic packing factor} = \frac{\text{volume occupied by the atoms in the unit cell}}{\text{volume of the unit cell}}$$

$$= \frac{4}{3}\frac{\pi r^3}{a^3} = \frac{4}{3}\frac{\pi r^3}{(2r)^3} = \frac{\pi}{6} = 0.524.$$

3.10.1.2 Body-Centred Cubic (bcc)

The body-centred cubic structure is shown in Fig. 3.33. The unit cell has an atom at each of the eight corners and an atom at the body centre. The coordination number for the bcc structure is 8 and the number of atoms per unit cell is 2. Each corner atom touches the atom at the body centre along the body diagonal. From Fig. 3.33 we have

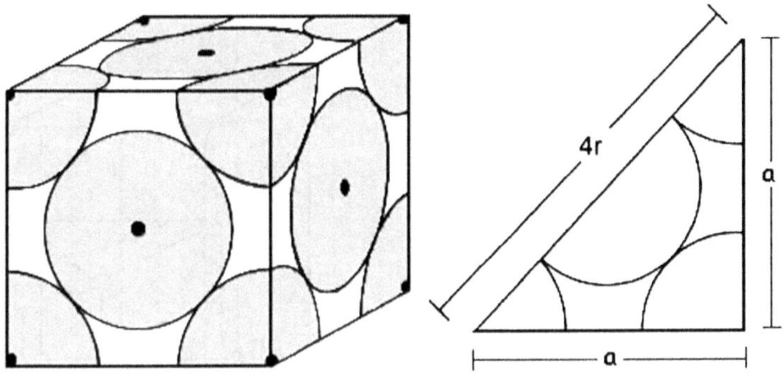

Fig. 3.34 The face-centred cubic (fcc) structure

$$(\mathrm{BC})^2 = (\mathrm{AC})^2 + (\mathrm{AB})^2.$$

$$(4r)^2 = \left(\sqrt{2}a\right)^2 + a^2 \text{ or } a = \frac{4}{\sqrt{3}}r.$$

$$\text{The atomic packing factor} = \frac{2 \times \frac{4}{3}\pi r^3}{a^3} = \frac{\sqrt{3}}{8}\pi = 0.68.$$

3.10.1.3 Face-Centred Cubic (fcc)

The face-centred cubic structure is shown in Fig. 3.34. There are eight atoms at the eight corners of the cube and six atoms at the centres of the six cube faces. The atoms touch each other along the face diagonal. Each atom has six nearest neighbours at a distance half the face diagonal. Hence the coordination number is 12. Since each of the atoms at the face centres are shared by two adjoining unit cells, the number of atoms in the unit cell is 4. From Fig. 3.34, we have

$$(4r)^2 = a^2 + a^2 \quad \text{or} \quad a = 2\sqrt{2}r,$$

$$\text{The atomic packing, factor} = \frac{4 \times \frac{4}{3}\pi r^3}{a^3} = \frac{\sqrt{2}}{6}\pi = 0.74.$$

3.10.1.4 Hexagonal Close-Packed Structure

The ideal hexagonal close-packed (hcp) structure is shown in Fig. 3.35. The top and bottom planes of the unit cell have hexagonal shape with one atom at each corner of the hexagon, one atom each at the centre of the hexagonal faces and three more atoms in the plane midway between the hexagonal faces forming an equilateral triangle. Each atom touches three atoms in the layer below its plane, six atoms in its own plane and three atoms in the layer above. Hence the coordination number for this structure is 12.

The atoms at the corners in the top and bottom layers are shared by six surrounding hexagonal cells and the atom at the centre is shared by two adjacent

Fig. 3.35 The ideal hexagonal close-packed (hcp) structure

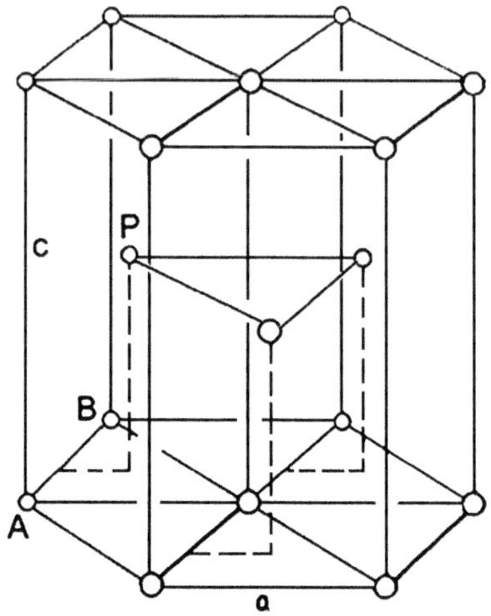

cells. The three atoms in the mid-plane are within the body of the cell. Thus the equivalent number of atoms per unit cell is 6.

Let "a" be the edge of the unit cell and "c" its height. Since the atoms touch each other along the edge of the hexagon, $a = 2r$, where r is the radius of the atom.

The volume occupied by all the atoms in the unit cell $= 6 \times \frac{4}{3}\pi r^3 = 8\pi r^3$.

The volume of the unit cell is = area of the basal plane × height of the unit cell.

$$\text{Area of the basal plane} = 6 \times \text{ area of each equilateral triangle}$$

$$= 6 \times \frac{a}{2} \times \frac{\sqrt{3}}{2}a = \frac{3\sqrt{3}}{2}a^2.$$

Hence volume of the unit cell $= \frac{3\sqrt{3}}{2}a^2 \times c$.

$$\text{The atomic packing factor} = \frac{8\pi r^3}{(3\sqrt{3}/2)a^2c} = \frac{2\pi}{3\sqrt{3}(c/a)}.$$

To calculate the c/a ratio for the ideal hcp structure we note that the three atoms in the mid-plane are at a height $c/2$ from the centroid of the alternate triangles at the top or base of the hexagonal cell. From Fig. 3.36, we have PQ $= c/2$, AS = PS = BS $= a$, QS = (2/3)RS and RS = BS cos 30° $= a\sqrt{3}/2$.

Hence

$$\text{QS} = \frac{2}{3}a\frac{\sqrt{3}}{2} = \frac{a}{\sqrt{3}}.$$

We know from ΔPQS that PS^2 = QS^2 + PQ^2. Substituting for PS, QS and PQ we get

$$a^2 = (c/2)^2 + a^2/3,$$

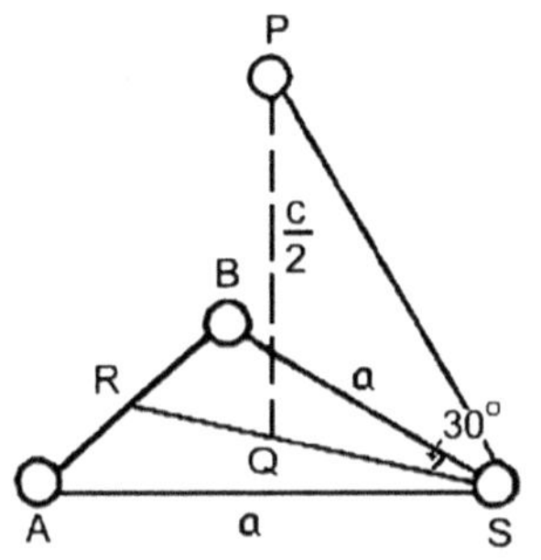

Fig. 3.36 Base of the hcp unit cell and position P of an atom in the mid-plane (ΔABS represents one-sixth of the base)

which gives

$$\frac{c}{a} = \sqrt{\frac{8}{3}} = 1.633.$$

$$\text{The atomic packing factor} = \frac{2\pi}{3\sqrt{3}(c/a)} = \frac{2\pi}{3\sqrt{3}}\sqrt{\frac{3}{8}} = \frac{\pi}{3\sqrt{2}} = 0.74.$$

3.11 Some Commonly Occurring Crystal Structures

We shall here discuss some representative structures of crystals which have either ionic or covalent bonds or a combination of both. These are the NaCl, CsCl, diamond and zinc blende structures. The structure which is energetically most favourable is determined by the anion–cation radius ratio and must be such as to avoid anion–anion or cation–cation contacts.

3.11.1 Sodium Chloride (Rock-Salt Structure)

The sodium chloride (NaCl) structure is shown in Fig. 3.37. The space lattice is face-centred cubic. The sodium and chloride ions are arranged on face-centred cubic sites of two lattices; the two interpenetrating fcc lattices being displaced from each other by one-half of the cube diagonal resulting in a simple cubic configuration. There are four molecules of NaCl in a unit cell, with the ions situated at

$$\text{Na}: \frac{1}{2}\frac{1}{2}\frac{1}{2};\ 00\frac{1}{2};\ 0\frac{1}{2}0;\ \frac{1}{2}00$$

$$\text{Cl}:\ 000;\ \frac{1}{2}\frac{1}{2}0;\ \frac{1}{2}0\frac{1}{2};\ 0\frac{1}{2}\frac{1}{2}$$

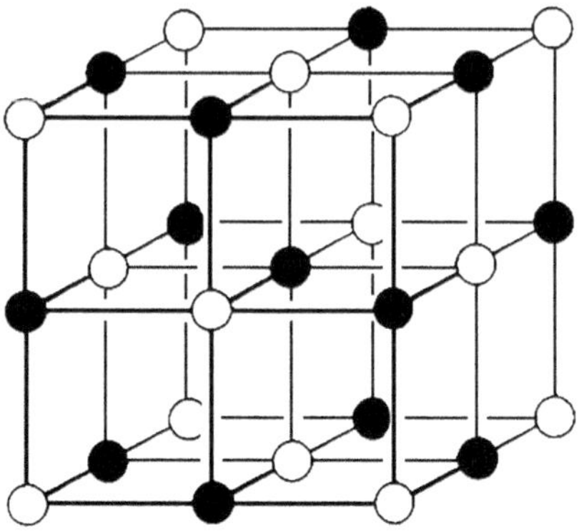

Fig. 3.37 The sodium chloride (rock salt) structure

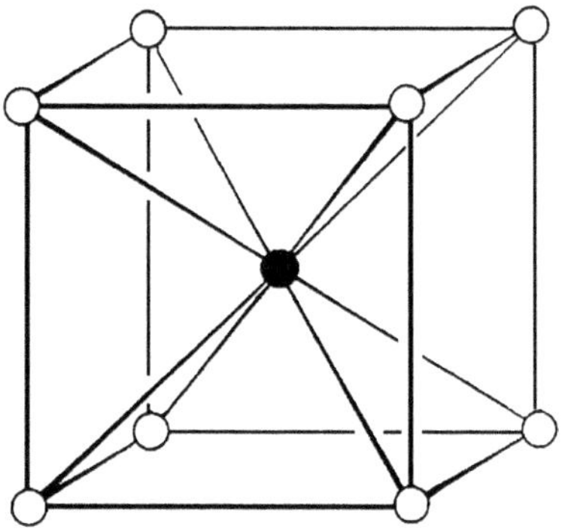

Fig. 3.38 The caesium chloride (CsCl) structure

In the sodium chloride structure each ion is surrounded by six nearest neighbours of the opposite ion type (characteristic of simple cubic coordination) and 12 next-nearest neighbours of the same ion type. Some crystals which crystallise with this structure are the halides of sodium potassium and rubidium, the silver halides, lead chalcogenides, etc.

3.11.2 Caesium Chloride Structure

The caesium chloride (CsCl) structure is shown in Fig. 3.38. This structure consists of two interpenetrating simple cubic lattices of Cs^+ and Cl^- ions. Each ion is at the centre of a cube of ions of the opposite kind. The number of nearest neighbours or the coordination number is thus 8. There is one molecule of CsCl in a unit cell with the ions at the corners 000 and body-centred positions $\frac{1}{2}\frac{1}{2}\frac{1}{2}$. Some crystals crystallising with this structure are the CsCl, CsBr, CsI, TlBr, TlI, NH_4Cl, NH_4Br, CuZn (β-brass)1, AgMg, LiHg, etc.

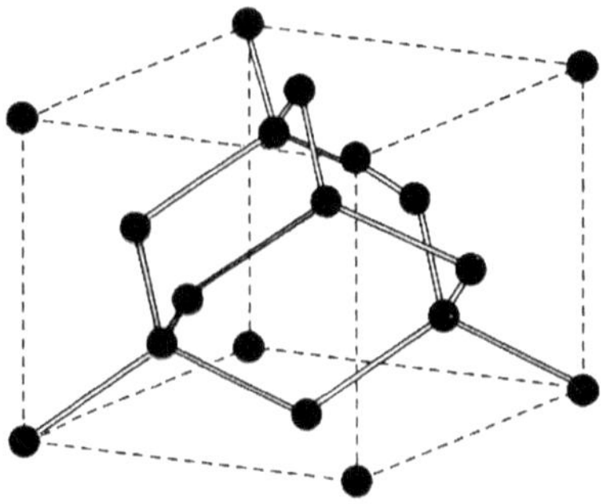

Fig. 3.39 The diamond structure

3.11.3 *Diamond* (C)

Carbon crystallises in two forms as diamond and graphite. Diamond is cubic while graphite has the hexagonal structure. The diamond space lattice is fcc. Each lattice point is associated with a basis of two identical atoms at 000 and $\frac{1}{4}\frac{1}{4}\frac{1}{4}$. The diamond lattice can be considered to be made up of two interpenetrating fcc lattices of carbon displaced from each other by one-quarter of the body diagonal. Each atom has four nearest neighbours and 12 next-nearest neighbours. The unit cell has eight atoms. The diamond structure is shown in Fig. 3.39.

The diamond lattice is loosely packed with a packing factor of 34%. Silicon and germanium crystallise in this structure. The diamond structure has the tetrahedral bond arrangement as a result of directional covalent bonding.

3.11.4 *Cubic Zinc Sulphide or Zinc Blende Structure* (ZnS)

Zinc sulphide (ZnS) crystallises in the cubic (zinc blende) or the hexagonal (wurtzite) structures. The zinc blende structure is identical to the diamond structure except that the atoms on the two interpenetrating fcc sublattice are different; Zn atoms are placed on one fcc lattice and S atoms on the other fcc lattice. The conventional cell of this structure is a cube shown in Fig. 3.40. There are four molecules of ZnS per conventional cell. The coordinates of atoms are

$$Zn : 000;\ \frac{1}{2}\frac{1}{2}0;\ \frac{1}{2}0\frac{1}{2};\ 0\frac{1}{2}\frac{1}{2}$$

$$\mathrm{S} : \frac{1}{4}\frac{1}{4}\frac{1}{4};\ \frac{1}{4}\frac{3}{4}\frac{3}{4};\ \frac{3}{4}\frac{1}{4}\frac{3}{4};\ \frac{3}{4}\frac{3}{4}\frac{1}{4}$$

Each atom has four equally distant atoms of the opposite kind arranged at the four corners of a regular tetrahedron. Some compounds which crystallise with this structure are CuCl, ZnS, ZnSe, ZnTe, InAs, InSb, SiC, CdS, CdSe, CdTe, etc.

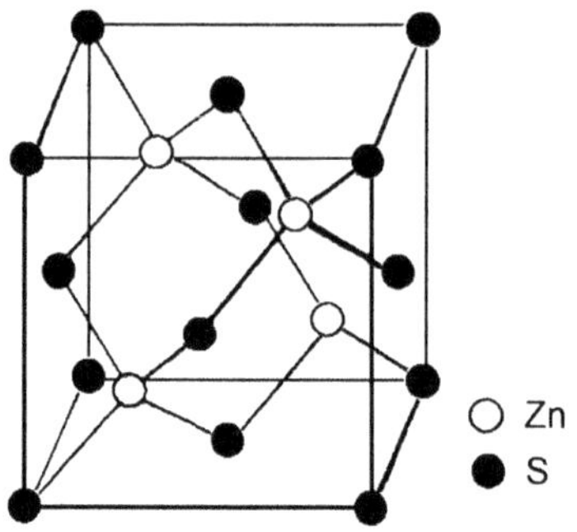

Fig. 3.40 The zinc blende structure

3.12 Reciprocal Lattice

The concept of the reciprocal lattice was proposed by Ewald in 1921 to facilitate the understanding of the phenomenon of X-ray diffraction in crystals. For each direct lattice, a corresponding reciprocal lattice may be constructed. It has the same symmetry as that of the direct lattice.

3.12.1 Definition and Properties

The reciprocal lattice is a set of imaginary points constructed in such a way that the direction of a vector from one point to another coincides with the direction of a normal to the real lattice planes. The separation of the reciprocal lattice points is equal to the reciprocal of the real interplanar distance.

We may construct a reciprocal lattice corresponding to a direct crystal lattice in the following manner:

1. To every plane in the direct lattice, construct a normal from an arbitrary origin.
2. Then limit the length of each normal so that it equals the reciprocal of the interplanar spacing of the planes.
3. Place a point at the end of each limited normal.

The collection of such points represents

(a) A collection of the slopes of the direct lattice planes in the form of the directions of the normals
(b) A collection of interplanar spacings of the direct lattice in the form of reciprocal spacings.

Thus, to each plane (*hkl*) of the crystal, we draw a normal of length

$$\sigma_{hkl} = 1/d_{hkl}, \tag{3.26}$$

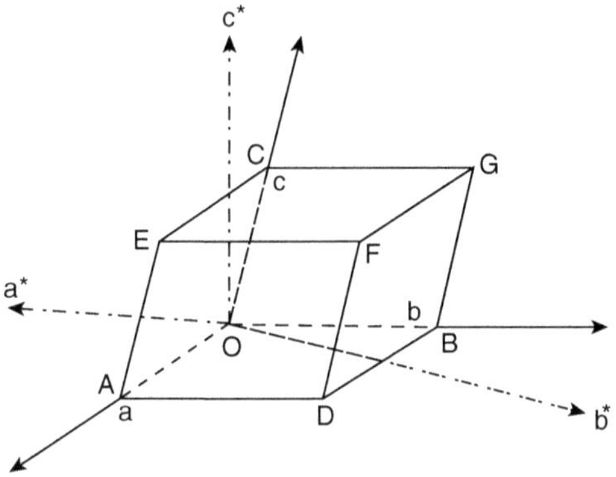

Fig. 3.41 Direct lattice vectors (**a**, **b**, **c**) and reciprocal lattice vectors (**a***, **b***,**c***)

where d_{hkl} is the interplanar distance of the *hkl* planes. The sum total of points at the end of these normals forms a lattice array, which constitutes a new lattice, the reciprocal lattice. Equation (3.26) can be written in vector form as

$$\mathbf{r}^*_{hkl} = \sigma_{hkl}\,\mathbf{n} = (1/d_{hkl})\mathbf{n}, \tag{3.27}$$

where **n** is a unit vector in the direction of the normal to the (*hkl*) plane and $\mathbf{r}^*$ represents the reciprocal lattice vector. Therefore

$$\mathbf{r}^*_{001} = (1/d_{001})\mathbf{n} = \mathbf{c}^* \tag{3.28}$$

and

$$|\mathbf{c}^*| = (1/d_{001}). \tag{3.29}$$

The volume V of the unit cell in the direct lattice (Fig. 3.41) is the area of the parallelogram with sides a and b, times the altitude d_{001}.

Hence

$$V = \text{area} \quad \text{AOBD} \cdot d_{001} = ab \sin\gamma \cdot d_{001}. \tag{3.30}$$

This gives

$$\frac{1}{d_{001}} = \frac{ab \sin\gamma}{V} = \frac{ab \sin\gamma}{\mathbf{a}\cdot\mathbf{b}\times\mathbf{c}} = |\mathbf{c}^*| \tag{3.31}$$

using (3.29). It follows that

$$\mathbf{c}^* = \frac{\mathbf{a}\times\mathbf{b}}{\mathbf{a}\cdot\mathbf{b}\times\mathbf{c}}. \tag{3.32}$$

Here a, b, c are the primitive translations in the direct lattice and α, β,γ the interaxial angles. By analogy, similar expressions can be written for $\mathbf{r}^*_{100}$ and $\mathbf{r}^*_{010}$. These vectors ($\mathbf{a}^*, \mathbf{b}^*, \mathbf{c}^*$) are chosen as the three reciprocal axes. Thus

$$\mathbf{a}^* = \mathbf{r}^*_{100} = \frac{\mathbf{b}\times\mathbf{c}}{\mathbf{a}\cdot\mathbf{b}\times\mathbf{c}},$$

$$\mathbf{b}^* = \mathbf{r}^*_{010} = \frac{\mathbf{c} \times \mathbf{a}}{\mathbf{a} \cdot \mathbf{b} \times \mathbf{c}}$$

$$\mathbf{c}^* = \mathbf{r}^*_{001} = \frac{\mathbf{a} \times \mathbf{b}}{\mathbf{a} \cdot \mathbf{b} \times \mathbf{c}} \tag{3.33}$$

From (3.33) it may be seen that $\mathbf{a}^*$ is normal to the *bc* plane, $\mathbf{b}^*$ to the *ac* plane and $\mathbf{c}^*$ to the *ab* plane. Thus $\mathbf{a}^*$, $\mathbf{b}^*$, $\mathbf{c}^*$ have the property that

$$\mathbf{a}^* \cdot \mathbf{a} = \mathbf{b}^* \cdot \mathbf{b} = \mathbf{c}^* \cdot \mathbf{c} = 1$$

and

$$\mathbf{a}^* \cdot \mathbf{b} = \mathbf{b}^* \cdot \mathbf{c} = \mathbf{c}^* \cdot \mathbf{a} = \mathbf{a} \cdot \mathbf{b}^* = \mathbf{b} \cdot \mathbf{c}^* = \mathbf{c} \cdot \mathbf{a}^* = 0. \tag{3.34}$$

If a lattice is constructed using the reciprocal lattice vectors defined above, it follows that the successive points in the $\mathbf{a}^*$ direction represent successive submultiples h of the spacing of (100) planes; in the $\mathbf{b}^*$ direction, successive submultiples of k of the spacing of (010) planes and, in the $\mathbf{c}^*$ direction, submultiples l of the (001) planes. The reciprocal lattice vector $\mathbf{r}^*$ is written as

$$\mathbf{r}^* = h\mathbf{a}^* + k\mathbf{b}^* + l\mathbf{c}^*. \tag{3.35}$$

Hence every crystal structure has two lattices associated with it, the crystal lattice and the reciprocal lattice. The two lattices are related by the definitions (3.33). Vectors in the crystal lattice have the dimensions of [length]; vectors in the reciprocal lattice have the dimensions of $[\text{length}]^{-1}$. When the crystal is rotated, we rotate both the crystal lattice and the associated reciprocal lattice. The reciprocal lattice is of particular use in that sets of planes are replaced by sets of points.

The reciprocal of the reciprocal lattice is the direct lattice. The volumes of the unit cells in direct and reciprocal space are inverse. Since

$$\mathbf{a}^* = \frac{\mathbf{b} \times \mathbf{c}}{V} \quad \text{and } \mathbf{a} = \frac{\mathbf{b}^* \times \mathbf{c}^*}{V^*}$$

we have

$$\mathbf{a} \cdot \mathbf{a}^* = \frac{(\mathbf{b}^* \times \mathbf{c}^*) \cdot (\mathbf{b} \times \mathbf{c})}{VV^*}$$

It can be shown that $(\mathbf{b}^* \times \mathbf{c}^*) \cdot (\mathbf{b} \times \mathbf{c}) = 1$. It therefore follows that $VV^* = 1$. A monoclinic direct lattice and the corresponding reciprocal lattice are shown in Fig. 3.42.

We can show that (a) the vector $\mathbf{r}^*_{hkl}$ to the point *hkl* in the reciprocal lattice R is normal to the planes whose Miller indices are (*hkl*) in the direct lattice L and (b) that $|\mathbf{r}^*_{hkl}|$ is the reciprocal of the spacing d_{hkl} of the lattice L.

In the direct lattice L, let ABC be the plane of the set (*hkl*) that passes nearest to the origin O (Fig. 3.43). Then the intercepts OA, OB, OC on the a, b, c axes are a/h, b/k and c/l, respectively. We have to show that the vector r^*_{hkl} is normal to

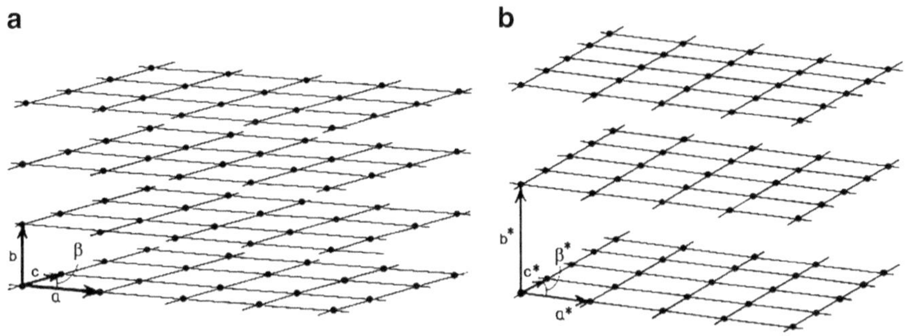

Fig. 3.42 (**a**) A monoclinic direct lattice and (**b**) the corresponding reciprocal lattice

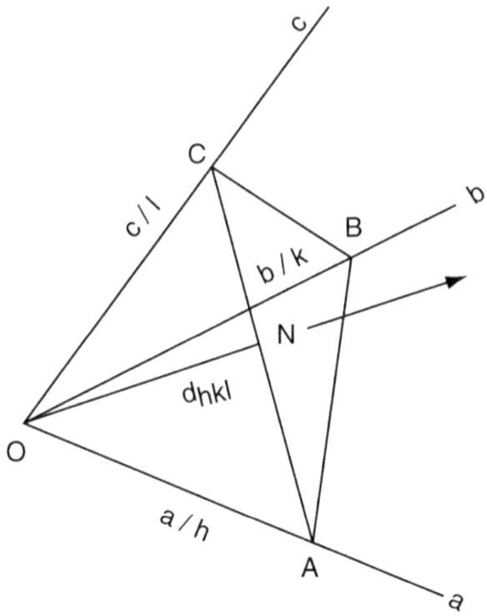

Fig. 3.43 The *hkl* plane

the plane ABC. To do this it is sufficient to show that it is perpendicular to any two of the vectors representing the sides of the triangle ABC. From Fig. 3.43, it may be seen that

$$\mathbf{r}^*_{hkl} \cdot \mathbf{AB} = (h\mathbf{a}^* + k\mathbf{b}^* + l\mathbf{c}^*).\left(\frac{\mathbf{b}}{k} - \frac{\mathbf{a}}{h}\right) = \mathbf{b}^* \cdot \mathbf{b} - \mathbf{a}^* \cdot \mathbf{a} = 0. \tag{3.36}$$

Since the scalar product of two vectors vanishes, they must be mutually perpendicular. In a similar manner, it can be shown that r^*_{hkl} is perpendicular to BC and CA. Hence r^*_{hkl} is perpendicular to the ABC plane with Miller indices (*hkl*).

We now show that $|r^*_{hkl}| = 1/d_{hkl}$. From Fig. 3.43, it may be seen that the length of the normal onto the ABC plane from the origin is the spacing d_{hkl} of the *hkl* planes in the direct lattice. If $\mathbf{n}$ is a unit vector in the direction of the normal, the length of ON $= \mathbf{OA} \cdot \mathbf{n}$ and

$$\mathbf{n} = \frac{\mathbf{r}^*_{hkl}}{|\mathbf{r}^*_{hkl}|} \tag{3.37}$$

since r^*_{hkl} is in the direction of the normal ON. Thus

$$\begin{aligned} \mathrm{ON} = d_{hkl} &= \mathbf{n} \cdot \mathbf{OA} = \mathbf{n} \cdot \frac{\mathbf{a}}{h} \\ &= \frac{\mathbf{r}^*_{hkl}}{|\mathbf{r}^*_{hkl}|} \cdot \frac{\mathbf{a}}{h} = \frac{(h\mathbf{a}^* + k\mathbf{b}^* + l\mathbf{c}^*)}{|\mathbf{r}^*_{hkl}|} \cdot \frac{\mathbf{a}}{h} \\ &= \frac{1}{|\mathbf{r}^*_{hkl}|}. \end{aligned} \tag{3.38}$$

Usually the length of the reciprocal vector is expressed as

$$|\mathbf{r}^*_{hkl}| = k \frac{1}{d_{hkl}}, \tag{3.39}$$

where k is chosen to be 1 or λ in X-ray diffraction and 2π in solid state physics, for convenience.

We shall now find the reciprocal lattices to some cubic lattices.

3.12.2 Reciprocal Lattice to a Simple Cubic (sc) Lattice

The primitive translation vectors of a simple cubic lattice may be taken as

$$\mathbf{a} = a\mathbf{x}; \mathbf{b} = b\mathbf{y}; \mathbf{c} = c\mathbf{z}, \tag{3.40}$$

where $\mathbf{x}$, $\mathbf{y}$, $\mathbf{z}$ are orthogonal unit vectors along the a, b, c axes. The volume of the primitive cell is

$$V = \mathbf{a} \cdot \mathbf{b} \times \mathbf{c} = a^3.$$

Using (3.31) the primitive translation vectors of the reciprocal lattice are

$$\mathbf{a}^* = \frac{\mathbf{b} \times \mathbf{c}}{\mathbf{a} \cdot \mathbf{b} \times \mathbf{c}} = \frac{\mathbf{x}}{a}; \ \mathbf{b}^* = \frac{\mathbf{c} \times \mathbf{a}}{\mathbf{a} \cdot \mathbf{b} \times \mathbf{c}} = \frac{\mathbf{y}}{a}; \ \mathbf{c}^* = \frac{\mathbf{a} \times \mathbf{b}}{\mathbf{a} \cdot \mathbf{b} \times \mathbf{c}} = \frac{\mathbf{z}}{a}.$$

Thus the reciprocal lattice of a simple cubic lattice is another simple cubic lattice of lattice constant $(1/a)$.

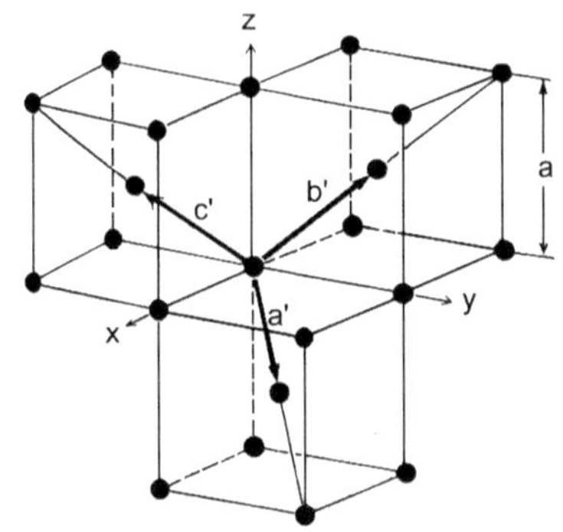

Fig. 3.44 Primitive translation vectors of the bcc lattice

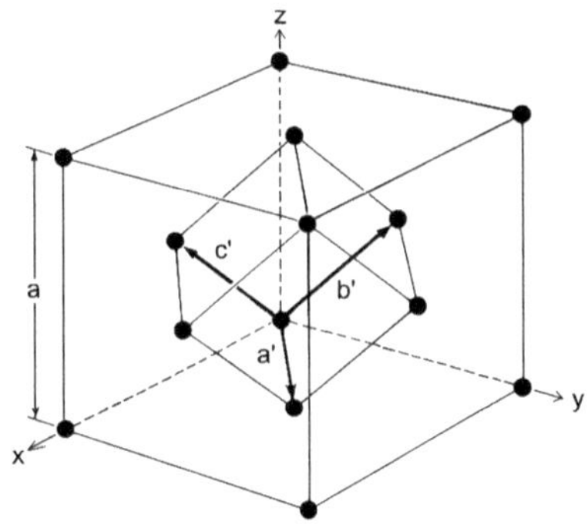

Fig. 3.45 Primitive translation vectors of the fcc lattice

3.12.3 Reciprocal Lattice to a Body-Centred Cubic (bcc) Lattice

The primitive translation vectors of the bcc lattice shown in Fig. 3.44 are

$$\mathbf{a}' = \frac{1}{2}a(\mathbf{x}+\mathbf{y}-\mathbf{z});\mathbf{b}' = \frac{1}{2}a(-\mathbf{x}+\mathbf{y}+\mathbf{z});\mathbf{c}' = \frac{1}{2}a(\mathbf{x}-\mathbf{y}+\mathbf{z}), \tag{3.41}$$

where a is the side of the conventional cube and $\mathbf{x}$, $\mathbf{y}$, $\mathbf{z}$ are orthogonal unit vectors along the cube edges. The volume of the primitive cell is

$$V = \mathbf{a}' \cdot \mathbf{b}' \times \mathbf{c}' = \frac{1}{2}a^3.$$

Using (3.33) and (3.41) the primitive translation vectors of the reciprocal lattice are given by

$$\mathbf{a}^* = \frac{1}{a}(\mathbf{x}+\mathbf{y});\mathbf{b}^* = \frac{1}{a}(\mathbf{y}+\mathbf{z});\mathbf{c}^* = \frac{1}{a}(\mathbf{x}+\mathbf{z}). \tag{3.42}$$

These are just the primitive translation vectors of a fcc lattice (Fig. 3.45). Thus the reciprocal lattice of the bcc lattice is the fcc lattice.

3.12.4 Reciprocal Lattice to a Face-Centred Cubic (fcc) Lattice

The primitive translation vectors of a fcc lattice are shown in Fig. 3.45 and are given by

$$\mathbf{a}' = \frac{a}{2}(\mathbf{x}+\mathbf{y});\ \mathbf{b}' = \frac{a}{2}(\mathbf{y}+\mathbf{z});\ \mathbf{c}' = \frac{a}{2}(\mathbf{z}+\mathbf{x}). \quad (3.43)$$

The volume of the primitive cell is

$$V = \mathbf{a}' \cdot \mathbf{b}' \times \mathbf{c}' = \frac{1}{4}a^3.$$

The primitive translation vectors of the lattice reciprocal to fcc lattice are

$$\mathbf{a}^* = \frac{1}{a}(\mathbf{x}+\mathbf{y}-\mathbf{z});\ \mathbf{b}^* = \frac{1}{a}(-\mathbf{x}+\mathbf{y}+\mathbf{z});\ \mathbf{c}^* = \frac{1}{a}(\mathbf{x}-\mathbf{y}+\mathbf{z}) \quad (3.44)$$

These are just the primitive translation vectors of a bcc lattice. Thus the reciprocal of the fcc lattice is the bcc lattice.

3.12.5 Use of Reciprocal Lattices in Crystal Geometry

Many problems that arise in the geometry of space lattices are simplified by the reciprocal lattice. Some of these are as follows.

3.12.5.1 Calculation of Lattice Spacings d_{hkl}

From (3.36) we have

$$d_{hkl} = \frac{1}{\left|\mathbf{r}^*_{hkl}\right|}$$

which gives

$$\left|\mathbf{r}^*_{hkl}\right|^2 = \frac{1}{(d_{hkl})^2}$$

Now

$$\begin{aligned}|\mathbf{r}^*_{hkl}|^2 &= |ha * + kb^* + lc^*|^2 \\ &= h^2a^{*2} + k^2b^{*2} + l^2c^{*2} + 2hl\mathbf{a}^* \cdot \mathbf{b}^* + 2kl\mathbf{b}^* \cdot \mathbf{c}^* + 2hl\mathbf{a}^* \cdot \mathbf{b}^* \\ &= h^2a^{*2} + k^2b^{*2} + l^2c^{*2} + 2hla^*b^* \cos\gamma^* + 2klb^*c^* \cos\alpha^* \\ &\quad + 2hla^*b^* \cos\beta^*. \end{aligned} \quad (3.45)$$

Hence d_{hkl} can be calculated.

3.12.5.2 Calculation of the Angle Between Planes

The angle ϕ between two sets of planes $(h_1k_1l_1)$ and $(h_2k_2l_2)$ is equal to the angle between the normals to the planes and, hence, to the angle between the two reciprocal lattice vectors $\mathbf{r}^*_{h_1k_1l_1}$ and $\mathbf{r}^*_{h_2k_2l_2}$ or between $\mathbf{r}^*_1$ and $\mathbf{r}^*_2$. Now,

$$\mathbf{r}^*_1 \cdot \mathbf{r}^*_2 = r^*_1 r^*_2 \cos\phi = \cos\phi / d_1 d_2$$

where d_1 and d_2 are the corresponding spacings. Hence

$$\cos\phi = [(h_1\mathbf{a}^* + k_1\mathbf{b}^* + l_1\mathbf{c}^*)] \cdot [(h_2\mathbf{a}^* + k_2\mathbf{b}^* + l_2\mathbf{c}^*)] d_1 d_2 \tag{3.46}$$

or

$$\cos\phi = [h_1^2 a^{*2} + k_1 b^{*2} + l_1 c^{*2} + (h_1k_2 + h_2k_1)a^*b^* \cos\gamma + (k_1l_2 + k_2l_1)b^*c^* \cos\alpha^* + (l_1h_2 + l_2h_1)c^*a^* \cos\beta^*] d_1 d_2. \tag{3.47}$$

Problems

1. Calculate the angles between the normals to the following pairs of planes in a cubic crystal: (a) (100), (010); (b) (100), (210); (c) (121), (111)
2. Calculate the packing fraction for the diamond structure.
3. Draw the figure of a tetrahedron and enumerate its symmetry elements.
4. Determine the parameters of the reciprocal cell of the unit cell with parameters $a = 5\text{Å}$, $b = 10\text{Å}$, $c = 15\text{Å}$.
5. What are (a) the crystal system, (b) point group and (c) symmetry elements of the space group $P2_12_12_1$?
6. What is the space group of the crystal for which the equivalent points are as shown in the figure given below?

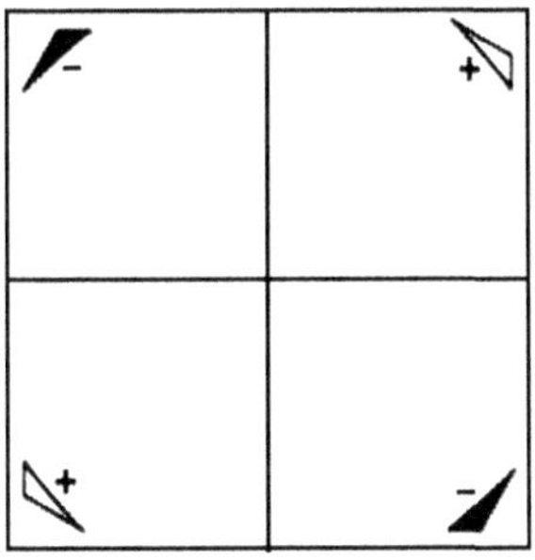

References

3.1. E.F. Kaelble, *Handbook of X-rays* (McGraw Hill, New York, 1967)
3.2. R.A. Levy, *Principles of Solid State Physics* (Academic Press, New York, 1968)
3.3. J.S. Blakemore, *Solid State Physics* (W.B. Saunders, London, 1969)
3.4. F.C. Phillips, *An Introduction to Crystallography* (Wiley, New York, 1971)
3.5. M.F.C. Ladd, R.A. Palmer, *Structure Determination by X-ray Crystallography* (Plenum Press, New York, 1973)
3.6. Yu.I. Sirotin, M.P. Shaskolskaya, *Fundamentals of Crystal Physics* (Mir Publishers, Moscow, 1982)
3.7. A. Schoenflies, *Krystallsysteme und Krystallstruktur* (Leipzig, 1891)
3.8. C. Hermann, Z. Kristallogr. **68**, 257 (1928)
3.9. C. Hermann, Z. Kristallogr. **76**, 559 (1931)
3.10. C. Mauguin, Z. Kristallogr. **76**, 542 (1931)
3.11. W. Barlow, Z. Kristallogr. **23**, 1 (1894)
3.12. E. Fedorov, Z. Kristallogr. **24,** 209 (1895)
3.13. *International Tables for X-ray Crystallography*, Vol. 1 (Kynoch Press, Birmingham, 1952)

Chapter 4
Diffraction of Radiation by Crystals (Principles and Experimental Methods)

4.1 Introduction

What is the nature of X-rays? Do crystals have a regular internal structure which is responsible for their regular external forms? These questions vexed crystallographers and physicists for many years. Laue's discovery of X-ray diffraction in 1912 answered both the questions: X-rays are short wavelength electromagnetic waves and crystals have a regular internal structure; further, the wavelength of X-rays and the interatomic spacing in crystals are of the same order. Besides answering these questions, the discovery of X-ray diffraction opened up a new field of crystal structure determination.

The concept of matter waves proposed by De Broglie in 1923 suggested that if the particle waves have wavelengths of the same order as spacings in crystals, they too would undergo diffraction by crystals. This led to the discovery of diffraction of electrons and neutrons by crystals.

For convenience, the chapter is divided into parts A, B and C. In part A, the principles of X-ray diffraction are discussed in detail. Some of the important experimental techniques are described. The factors which contribute to the intensity of diffracted rays are discussed and the expression for the integrated intensity is formulated; this expression forms the basis for the determination of crystal structure. Electron and neutron diffraction are discussed in parts B and C, respectively.

Fairly detailed treatments of X-ray diffraction are given by James [4.1], Kaelble [4.2], Klug and Alexander [4.3], Warren [4.4] and Chatterjee [4.5]. Electron diffraction is discussed by Sproull [4.6], Bragg [4.7] and Kaelble [4.2]. The subject of neutron diffraction is discussed by Shull and Wollan [4.8], Bacon [4.9] and Dachs [4.10].

D.B. Sirdeshmukh et al., *Atomistic Properties of Solids*, Springer Series in Materials Science 147, DOI 10.1007/978-3-642-19971-4_4,

Part A: X-ray Diffraction

4.2 Discovery of X-ray Diffraction

X-rays were discovered by Roentgen in 1895. Roentgen himself noted many of their properties. But there was no clue as to the nature of X-rays. Roentgen ended his discovery paper with the conjecture that X-rays were perhaps longitudinal electromagnetic waves. The nature of X-rays remained shrouded for many years.

Assuming that the external regular form and cleavage in crystals were manifestations of an internal regularity, crystallographers had estimated from a knowledge of density and molecular weight that lattice spacings were of the order of 10^{-8} cm. From the optical properties of X-rays, Sommerfeld estimated that, whatever their nature, X-rays have wavelengths of about 10^{-9} cm. From these two facts, it occurred to Max von Laue, by sheer intuition, that crystals may serve as gratings for X-rays and display diffraction. The diffraction of X-rays from a crystal was successfully observed by Laue (Box 4.1) with the assistance of his colleagues Friedrich and Knipping in 1912 [4.11].

The experimental set-up used to observe X-ray diffraction is shown in Fig. 4.1. T was an X-ray tube developed by Roentgen himself. The X-rays were collimated by passing them successively through openings B_1 in the lead screen S, B_2 in the lead box K and B_3 and B_4 in lead blocks. The collimated X-rays fell on a copper sulphate crystal C mounted on the goniometer G. Photographic plates P_1–P_4 were placed in different positions and at different distances from the crystal. The direct beam was absorbed in the exit tube R.

The result was stunning; the photographs (Fig. 4.2) showed the presence of spots besides the one produced by the direct rays. Clearly, these spots were created by X-rays travelling in various directions from the crystal. To quote Amoros et al. [4.12] "Laue's intuition was confirmed ... (this) was experimental proof of the periodic nature of X-rays".

Box 4.1

Max Von Laue, a German physicist discovered X-ray diffraction in 1912. Laue's discovery confirmed the electromagnetic nature of X-rays and also the regular arrangement of atoms in a crystal. The discovery was the beginning of a new discipline of X-ray crystallography. Laue was awarded the Nobel Prize in physics in 1914.

Max Von Laue (1879–1960)

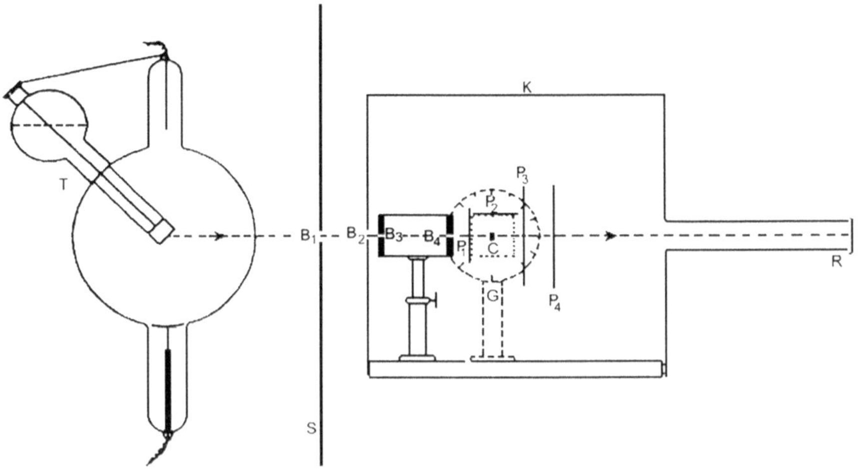

Fig. 4.1 Experimental set-up used by Friedrich et al. (1912)

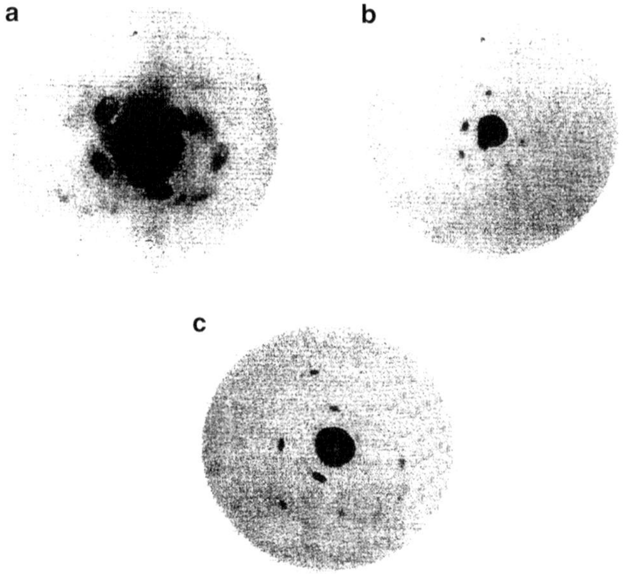

Fig. 4.2 (**a**) First X-ray diffraction photograph of copper sulphate crystal, (**b**) and (**c**) photographs of the same crystal with improved collimation and orientation

4.3 Bragg's Law

Soon after the discovery of X-ray diffraction, Laue [4.13] undertook a theoretical interpretation of the X-ray diffraction pattern treating diffraction as a result of the scattering of electromagnetic waves by the electrons in the atoms. Another

Box 4.2

Sir William Lawrence Bragg, a British physicist is known for his contribution in the field of X-ray diffraction and the law named after him. W. L. Bragg was awarded the Nobel Prize for Physics in 1915. W.L. Bragg is the youngest scientist to get the Nobel Prize at the age of 25 years

W.L. Bragg (1890–1971)

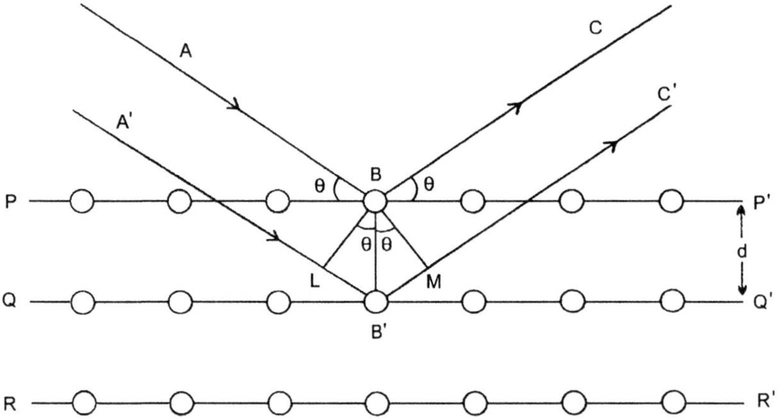

Fig. 4.3 Reflection of X-rays by lattice planes

interpretation was provided by Bragg [4.14] who assumed that the X-rays are "reflected" at atomic planes. Laue's treatment is more rigorous and leads to the same result as Bragg's. We shall take up Bragg's treatment first since it is simpler and easier to grasp.

4.3.1 Bragg's Derivation

Bragg (Box 4.2) derived the law for diffraction of X-rays by crystals by assuming that the X-ray waves are reflected at the crystal planes and applying optical laws for the interference of the reflected rays.

In Fig. 4.3, PP′, QQ′ and RR′ represent consecutive parallel atomic planes in a crystal with spacing d between successive planes. An X-ray wave of wavelength λ proceeding along AB is incident on the plane PP′ at a glancing angle θ. This ray is reflected at B and proceeds along BC. Note that in X-ray optics, we consider the glancing angle and not "the angle made with the normal" as in general optics. Again, let another parallel ray A′B′ incident on the next plane be reflected and proceed along B′C′. Let BL and BM be normals from B on A′B′ and B′C′, respectively. The path difference between the two sets of rays is (B′L + B′M) which equals $2d \sin\theta$. These parallel rays interfere constructively if

$$2d \sin\theta = n\lambda, \tag{4.1}$$

where n is an integer. We say that the rays have been diffracted at an angle θ. Equation (4.1) is known as Bragg's law and θ is called the Bragg angle. It may be noted that for θ to have realistic values d should be of the same order as but not smaller than, $\lambda/2$.

If the Miller indices of the planes are h, k, l, (4.1) becomes

$$\frac{2a \sin\theta}{(h^2 + k^2 + l^2)^{1/2}} = \lambda \tag{4.2}$$

for a cubic crystal with lattice constant a. The relation for crystals with lower symmetry will be correspondingly more complicated. The integer n in (4.1) is now included in the indices h, k, l.

4.3.2 *Laue's Derivation*

Laue did not start with the assumption that X-rays are reflected. Instead, he assumed that they were scattered by the atoms in the crystal.

For simplicity, let us consider atoms along a linear lattice PP′ with spacing a (Fig. 4.4). Let AB and A′B′ be two parallel rays with wavelength λ incident on the

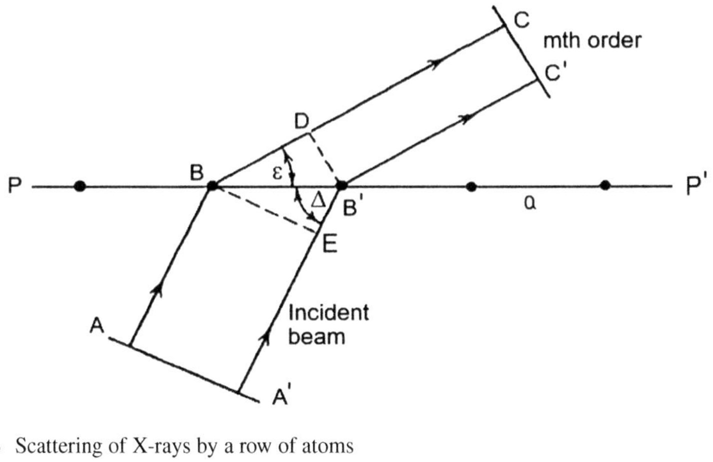

Fig. 4.4 Scattering of X-rays by a row of atoms

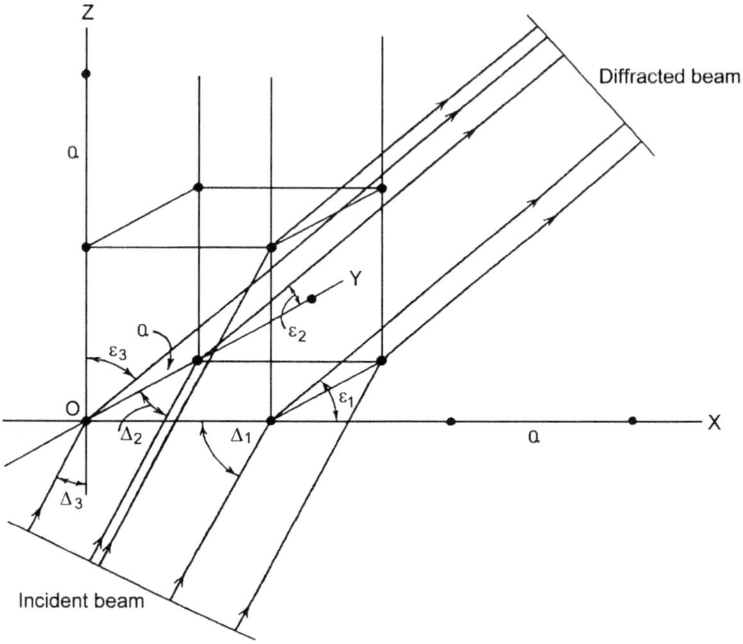

Fig. 4.5 Scattering of X-rays by a simple cubic lattice

linear lattice at angle Δ. The X-rays will be scattered in all directions by the atoms at B and B$'$. Out of these scattered rays, let us consider the rays scattered along BC and B$'$C$'$ making angle ε with the lattice. Let BE and B$'$D be normals from B and B$'$ on A$'$B$'$ and BC, respectively. The path difference between the rays ABC and A$'$B$'$C$'$ is a (cos ε–cos Δ). These rays will interfere constructively and result in a diffraction maximum if

$$a(\cos\varepsilon - \cos\Delta) = m\lambda, \tag{4.3}$$

where m is an integer.

We shall now extend this result to a three-dimensional case. For simplicity, we shall consider a simple cubic lattice with lattice constant a (Fig. 4.5). We may treat this lattice as a set of three linear lattices along the X, Y and Z directions with O as the origin. We shall draw a set of parallel rays incident on this lattice. These rays make angles Δ_1, Δ_2 and Δ_3 with the three axes. Out of the rays spherically scattered by the atoms, we shall consider a set of parallel rays scattered in a chosen direction shown in the figure which makes angles ε_1, ε_2 and ε_3 with the axes.

We may treat the rays scattered by atomic rows in any one direction and apply the diffraction condition as in (4.3). Repeating this procedure successively for all three directions, we get

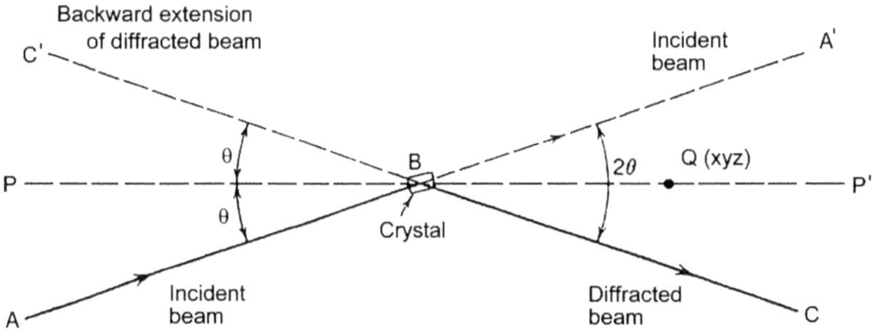

Fig. 4.6 Relation between incident ray, diffracted ray and the lattice plane defined by Laue equations

$$
\begin{aligned}
a(\cos\Delta_1 - \cos\varepsilon_1) &= p\lambda,\\
a(\cos\Delta_2 - \cos\varepsilon_2) &= q\lambda,\\
a(\cos\Delta_3 - \cos\varepsilon_3) &= r\lambda,
\end{aligned}
\tag{4.4}
$$

where p, q and r are integers. The three equations (4.4) are known as Laue's equations. Each set of equations represents diffraction due to the three non-coplanar rows of atoms in the lattice. Note that $\cos\Delta_1, \cos\Delta_2$ and $\cos\Delta_3$ are the direction cosines of the incident rays while $\cos\varepsilon_1, \cos\varepsilon_2$ and $\cos\varepsilon_3$ are the direction cosines of the diffracted ray.

To understand the significance of these equations, let us redraw the rays. In Fig. 4.6, let AB be the incident ray and BC the diffracted ray. Let us extend the incident ray to A′ and the diffracted ray to C′. Let us denote the angle of deviation A′BC by 2θ. Recalling the properties of direction cosines, we have

$$
\begin{aligned}
\cos 2\theta &= \cos\varepsilon_1\cos\Delta_1 + \cos\varepsilon_2\cos\Delta_2 + \cos\varepsilon_3\cos\Delta_3,\\
\cos^2\Delta_1 &+ \cos^2\Delta_2 + \cos^2\Delta_3 = 1,\\
\cos^2\varepsilon_1 &+ \cos^2\varepsilon_2 + \cos^2\varepsilon_3 = 1.
\end{aligned}
\tag{4.5}
$$

Now squaring both sides of (4.4) and adding up, we get

$$
\begin{aligned}
&a^2[(\cos^2\Delta_1 + \cos^2\Delta_2 + \cos^2\Delta_3) + (\cos^2\varepsilon_1 + \cos^2\varepsilon_2 + \cos^2\varepsilon_3)\\
&-2(\cos\Delta_1\cos\varepsilon_1 + \cos\Delta_2\cos\varepsilon_2 + \cos\Delta_3\cos\varepsilon_3)] = \lambda^2(p^2+q^2+r^2).
\end{aligned}
\tag{4.6}
$$

Substituting from (4.5), we get

$$2a^2(1-\cos 2\theta) = \lambda^2(p^2+q^2+r^2). \tag{4.7}$$

But

$$1-\cos 2\theta = 2\sin^2\theta. \tag{4.8}$$

Hence, we have

$$4a^2 \sin^2 \theta = \lambda^2(p^2 + q^2 + r^2). \tag{4.9}$$

If the numbers p, q, r have a common factor n so that $p = np_0, q = nq_0$ and $r = nr_0$, (4.9) may be written as

$$\frac{2a \sin \theta}{(p_0{}^2 + q_0{}^2 + r_0{}^2)^{1/2}} = n\lambda. \tag{4.10}$$

Equation (4.10) relates the wavelength λ with the lattice constant a, half the angle of deviation θ and the numbers p_0, q_0 and r_0. We shall presently see the significance of these numbers.

In Fig. 4.6, let PP′ passing through B be the trace of a plane which bisects the angle of deviation, i.e. ∠A′BP′ = ∠CBP′ = θ. Now, consider a point Q(xyz) in this plane PP′. Let QB make an angle δ with BA′ and BC. Then, from the properties of direction cosines, we have

$$\begin{aligned}\cos\delta &= \frac{x}{\mathrm{BQ}}\cos\Delta_1 + \frac{y}{\mathrm{BQ}}\cos\Delta_2 + \frac{z}{\mathrm{BQ}}\cos\Delta_3 \\ &= \frac{x}{\mathrm{BQ}}\cos\varepsilon_1 + \frac{y}{\mathrm{BQ}}\cos\varepsilon_2 + \frac{z}{\mathrm{BQ}}\cos\varepsilon_3.\end{aligned} \tag{4.11}$$

By rearrangement,

$$x(\cos\Delta_1 - \cos\varepsilon_1) + y(\cos\Delta_2 - \cos\varepsilon_2) + z(\cos\Delta_3 - \cos\varepsilon_3) = 0. \tag{4.12}$$

Substituting from (4.4), we get

$$\frac{p\lambda}{a}x + \frac{q\lambda}{a}y + \frac{r\lambda}{a}z = 0$$

or

$$p_0x + q_0y + r_0z = 0. \tag{4.13}$$

Equation (4.13) is an equation to a lattice plane with Miller indices p_0, q_0 and r_0. Thus, the plane that bisects the angle between the undeviated incident ray and the diffracted ray (and thus makes equal angles) is a lattice plane with Miller indices (p_0, q_0, r_0). The spacing of this set of planes, $d_{p_0q_0r_0}$, is $a/(p_0{}^2 + q_0{}^2 + r_0{}^2)^{1/2}$. We now see that (4.10) is identical with Bragg's law (4.2). Thus, Laue's equations lead to Bragg's law. There is no assumption of reflection. On the other hand, it emerges in a logical way that the plane which makes equal angles with the incident and diffracted ray happens to be a lattice plane. Having noted that the numbers p_0, q_0, r_0 are Miller indices, using conventional notation for Miller indices we may rewrite (4.4) as

$$\begin{aligned}a(\cos\Delta_1 - \cos\varepsilon_1) &= h\lambda, \\ a(\cos\Delta_2 - \cos\varepsilon_2) &= k\lambda, \\ a(\cos\Delta_3 - \cos\varepsilon_3) &= l\lambda.\end{aligned} \tag{4.14}$$

Further, for a lower symmetry crystal like, say, an orthorhombic crystal, (4.14) becomes

$$\begin{aligned} a(\cos\Delta_1 - \cos\varepsilon_1) &= h\lambda, \\ b(\cos\Delta_2 - \cos\varepsilon_2) &= k\lambda, \\ c(\cos\Delta_3 - \cos\varepsilon_3) &= l\lambda, \end{aligned} \tag{4.15}$$

where a, b, c are the lattice parameters.

4.3.3 *Refraction Correction for Bragg's law*

In Bragg's derivation of (4.1), it was assumed that the X-rays travel through the crystal without undergoing any refraction. Actually, X-rays do undergo refraction. The paths of the X-rays allowing for refraction are shown in Fig. 4.7. The observed Bragg angle is denoted by θ_{obs} and that at the inner plane by θ_{calc}; θ_{calc} is calculated from the Bragg equation $\lambda = 2d \sin\theta_{\text{calc}}$. The refractive index n based on Snell's law is $n = \cos\theta_{\text{obs}}/\cos\theta_{\text{calc}}$. It is found that n is slightly less than unity. If we define $\delta = (1\text{–}n)$, the experimental value of δ is of the order of $10^{-6}\text{–}10^{-5}$.

From the geometry of the rays in Fig. 4.7 and with some approximations, Darwin [4.15] showed that

$$(\theta_{\text{obs}} - \theta_{\text{calc}}) = \delta \sec\theta_{\text{obs}}\,\text{cosec}\theta_{\text{obs}} \tag{4.16}$$

and derived the modified Bragg's law

$$\lambda = 2d\left(1 - \frac{\delta}{\sin^2\theta_{\text{obs}}}\right). \tag{4.17}$$

From (4.17), we can obtain an approximate correction for the lattice constant a as

$$a_{\text{corr}} = a_{\text{uncorr}}\left(1 - \frac{\delta}{\sin^2\theta_{\text{obs}}}\right). \tag{4.18}$$

The uncorrected and corrected values of the lattice constants for some metals [4.2] are given in Table 4.1. It can be seen that the correction is very small.

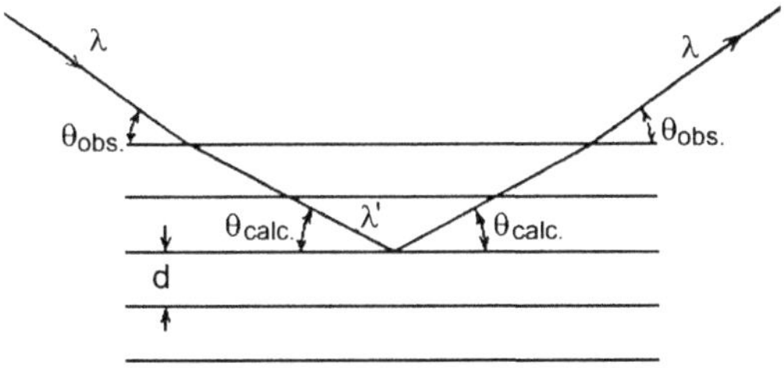

Fig. 4.7 Refraction of X-rays

Table 4.1 Refraction correction to lattice constant a

Substance	a[Å] Uncorrected	Corrected
Al	4.04944	4.04948
Pb	4.95052	4.95066
W	3.16494	3.16510

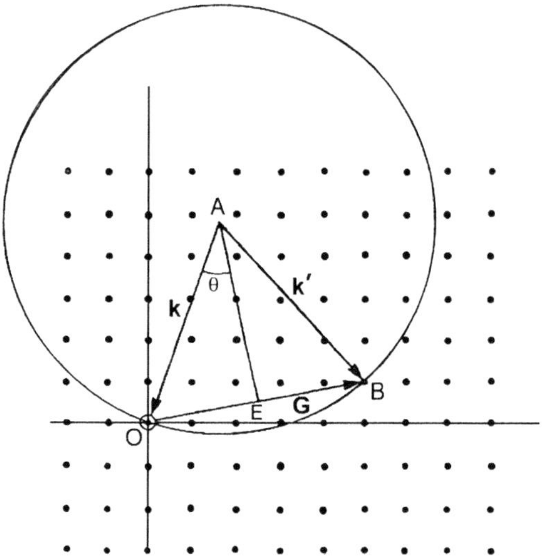

Fig. 4.8 Incident and diffracted X-rays in reciprocal lattice

4.3.4 Bragg's Law in Reciprocal Lattice

A part of a reciprocal lattice is shown in Fig. 4.8. To start with consider a point A (which need not be a reciprocal lattice point). Let us join A to O, the origin of the reciprocal lattice. Let AO be the direction of an incident ray and let AO be equal to $1/\lambda$ where λ is the wavelength. Now construct a sphere with A as centre and AO as radius; the circle shown in the figure is the intersection of the sphere with the plane of the paper. Let B be a reciprocal lattice point lying on the circle. Finally, let us draw the normal AE from A on OB.

By definition of a reciprocal lattice point, B represents a set of parallel lattice planes normal to OB and having spacing d given by OB $= 1/d$; let their Miller indices be h, k, l. The normal AE is one such plane and the angle OAE is the glancing angle θ.

By the nature of the construction, OE $=$ EB and $\angle$BAE $= \angle$OAE $= \theta$. We have

$$\mathrm{OE} = \left(\frac{1}{2}\right)\left(\frac{1}{d}\right). \tag{4.19}$$

Also

$$\mathrm{OE} = \mathrm{AO}\sin\theta = (1/\lambda)\sin\theta. \tag{4.20}$$

Equating (4.19) and (4.20), we have

$$\lambda = 2d \sin \theta, \tag{4.21}$$

which is Bragg's law. We conclude that the Bragg condition is satisfied for a given wavelength λ provided the surface of a sphere of radius $(1/\lambda)$ drawn about a point passes through a reciprocal lattice point representing a set of planes with spacing d. This construction is called the Ewald construction and the sphere is called the Ewald sphere. It may be noted that if $|\mathbf{OA}| < (1/2)(1/a)$, the sphere would not pass through any reciprocal lattice point; thus diffraction is not possible if $\lambda > 2a$.

The Ewald diagram enables us to express Bragg's law in vector form. Let us denote the incident ray vector AO (Fig. 4.8) by $\mathbf{k}$, the reciprocal lattice vector OB by $\mathbf{G}$ and the vector AB by $\mathbf{k}'$. Since $\mathbf{k}'$ makes angle θ with AE, i.e. to the planes represented by B, $\mathbf{k}'$ is the scattered ray. From the diagram it can be seen that $|\mathbf{AB}| = |\mathbf{AO}|$, i.e.

$$k^2 = k'^2 = (\mathbf{k} + \mathbf{G})^2. \tag{4.22}$$

Hence,

$$2\mathbf{k} \cdot \mathbf{G} + G^2 = 0. \tag{4.23}$$

This is the vector form of Bragg's law. Further, it can be seen that

$$\mathbf{k}' = \mathbf{k} + \mathbf{G} \quad \text{or} \quad \mathbf{k}' - \mathbf{k} = \mathbf{G}. \tag{4.24}$$

Thus, scattering only changes the direction of $\mathbf{k}$ and the scattered wave differs from the incident wave by a reciprocal lattice vector $\mathbf{G}$.

4.4 Experimental Aspects

We shall first consider general matters like generation, collimation, monochromatisation and detection of X-rays. We shall then discuss the principles of some of the techniques used in X-ray diffraction. Details are found in the sources referred in Sect. 4.1.

4.4.1 General

X-rays are generated when high-speed electrons collide with a metallic target. The production takes place in an X-ray tube (Fig. 4.9). This is an evacuated and sealed lead glass tube with a tungsten filament at one end and a metal target at the other end. Electrons are emitted by the tungsten filament heated by a current. A high DC voltage (20–60 kV) from a rectifier–transformer combination is applied between the filament and the target so that the electrons are accelerated and travel towards

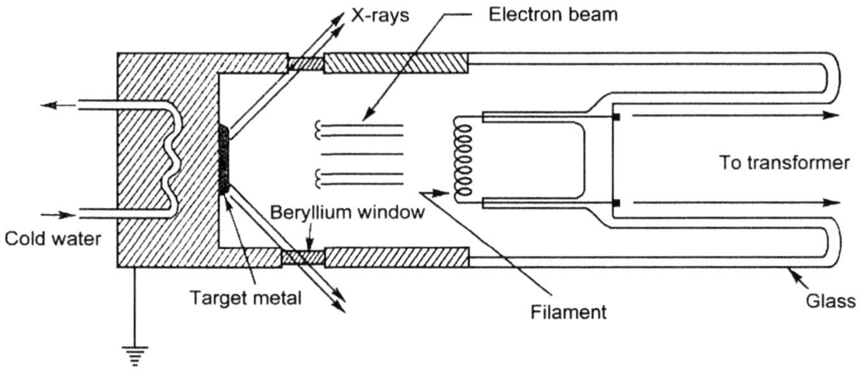

Fig. 4.9 An X-ray tube

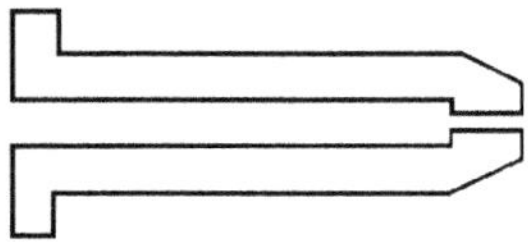

Fig. 4.10 A collimator

the target. The X-rays are produced at the point (focal spot) where the electrons strike the target. Only a small part of the electron energy is converted into X-rays and a larger part is transferred to the target as heat. Hence the target is cooled by circulating water. The X-rays emerge out of the tube through beryllium windows sealed on to the tube.

It has been mentioned earlier that the refractive index of X-rays for materials is only negligibly different from unity. Therefore, conventional optical methods cannot be employed to focus them or to collimate them. X-rays emerge from an X-ray tube in the form of a divergent beam. The only way to collimate them is to allow the unwanted part of the beam to be absorbed by heavy absorbers like brass or lead. In practice, the rays from the tube pass through a long tube (Fig. 4.10) made of brass with the first end having a bore of 2–3 mm and the second end having a bore of 0.5–1 mm.

Figure 4.11 shows the spectrum of X-rays emitted by a target. It consists of some sharp peaks (characteristic X-rays) superposed over a background (continuous X-rays). When X-rays pass through a material, they are absorbed. The change in intensity due to absorption is given by

$$I = I_0 e^{-(\mu/\rho)\rho t}, \tag{4.25}$$

where I_0 is the initial intensity, I the intensity after passing through a slab of thickness t, ρ the density and μ the absorption coefficient; (μ/ρ) is called the mass absorption coefficient. The mass absorption coefficient depends on the material

Fig. 4.11 Spectrum of X-rays from a Cu target

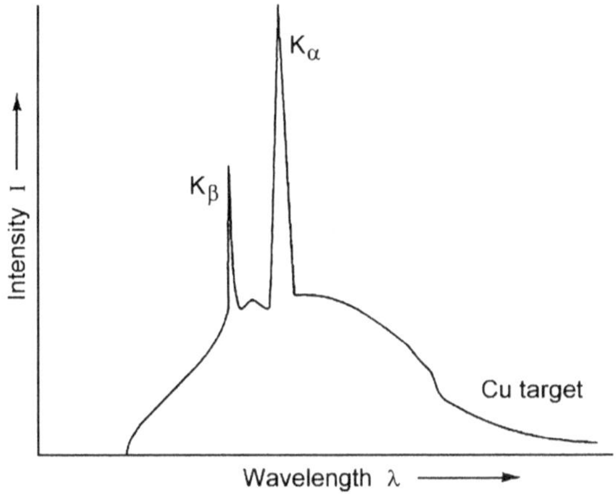

Fig. 4.12 Absorption spectrum

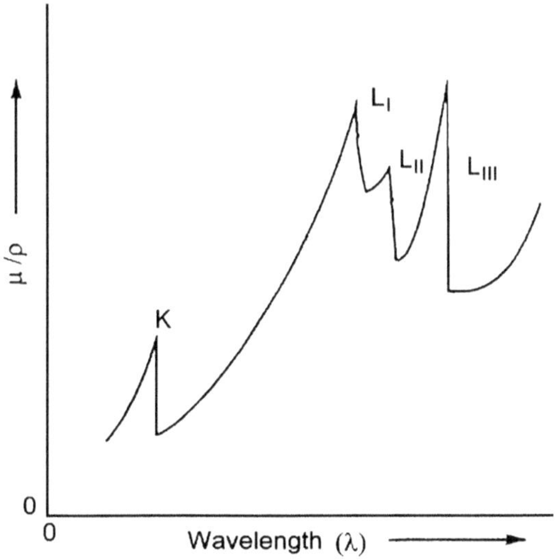

and the wavelength. Typical variation of the mass absorption coefficient with wavelength is shown in Fig. 4.12. The absorption increases with the wavelength and suddenly drops in the vicinity of a characteristic peak. These discontinuities are called absorption edges. This phenomenon of absorption is used to choose a desired wavelength (monochromatisation) from an otherwise polychromatic beam. The principle is shown in Fig. 4.13 where the absorption curve of Ni is superposed on the X-ray spectrum from a copper target. The K-edge of Ni occurs just before the $K\alpha$ peak of Cu. Hence the use of a Ni filter (a thin disc of Ni) placed in front of the beryllium window results in eliminating most of the wavelengths on the low wavelength side of the $K\alpha$ peak. If a second filter (Co) is used (Fig. 4.14), it eliminates the radiation on the higher wavelength side of the $K\alpha$ peak. Such a double filter is called a Ross filter. In recent times, for more efficient

Fig. 4.13 Action of Ni filter

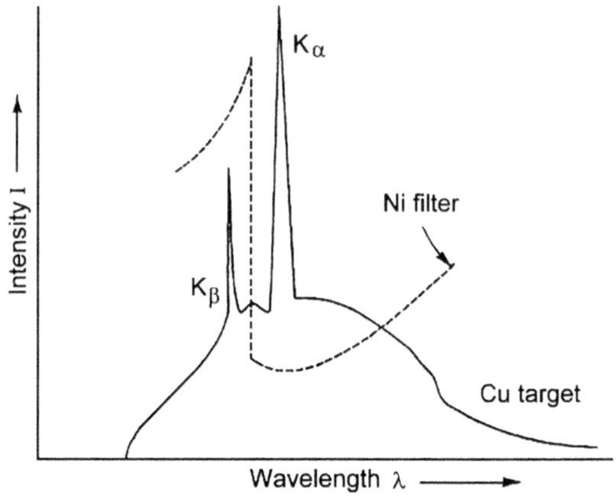

Fig. 4.14 Ross filter

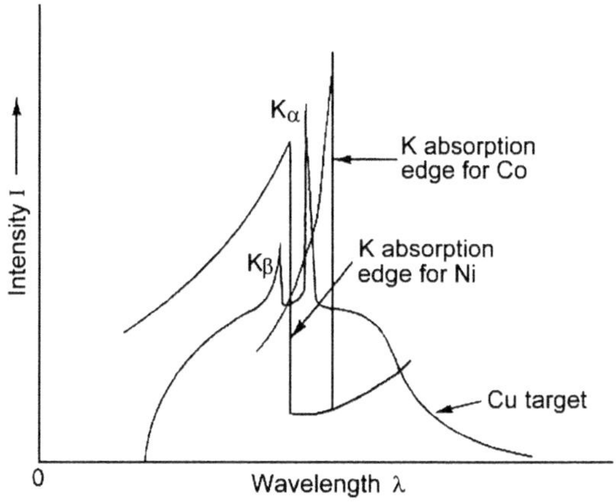

monochromatisation, crystal monochromators are used. A crystal diffracts one of the characteristic wavelengths and this diffracted beam, which is essentially monochromatic, is used as the incident beam for other experiments. NaCl, LiF, quartz and graphite are some crystals used as monochromators.

In many techniques, the diffraction patterns are recorded on photographic films. If the intensities of the reflections are required, the film has to be passed through a microphotometer. In recent times, detectors like the Geiger–Müller counter, the proportional counter, the scintillation counter and solid state detectors like Ge and Si are used. The advantage is that they provide a direct measure of the intensities.

4.4.2 Laue Method

The Laue method of recording a diffraction pattern is shown in Fig. 4.15. Unfiltered (polychromatic) X-rays pass through a collimator and fall on a single crystal. The

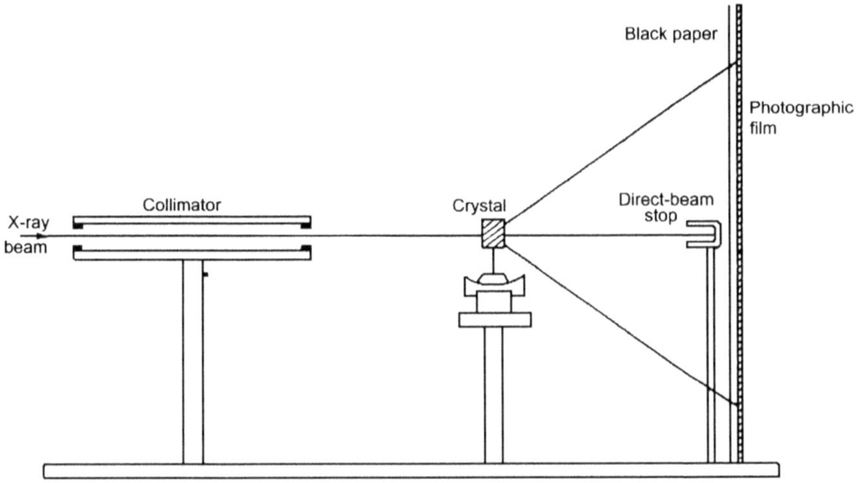

Fig. 4.15 Laue camera

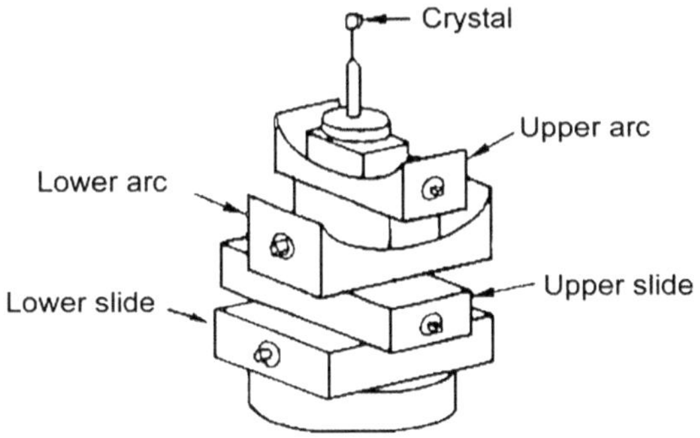

Fig. 4.16 A goniometer head

diffracted rays are received on a photographic film; the direct beam is absorbed by a lead stop. The crystal is mounted on a gadget called goniometer head. This device is an essential part of all single crystal techniques. The goniometer head (Fig. 4.16) has provision for linear displacements (lower slide and upper slide) in two mutually perpendicular directions in the horizontal plane. Also, there is a provision for angular displacements (lower arc and upper arc) in two perpendicular directions. Using these facilities, the crystal can be oriented (by looking at it through a telescope) such that the desired crystal axis coincides with the axis of the holder.

As a typical example, the Laue photograph of a quartz crystal taken about the trigonal axis is shown in Fig. 4.17. It consists of spots arranged symmetrically. The threefold symmetry is clearly seen.

Different lattice planes choose appropriate wavelengths from the polychromatic beam of X-rays and diffract them at different angles such that the Bragg equation

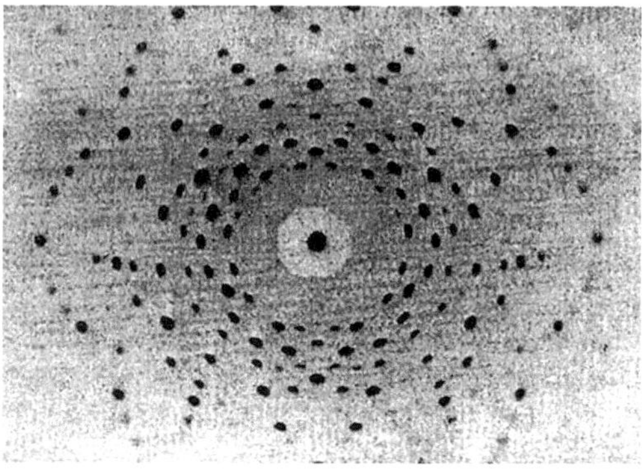

Fig. 4.17 Transmission Laue photograph of quartz

is fulfilled by each combination of d value, wavelength and Bragg angle. Thus, for different spots recorded on the film, we have

$$\begin{aligned} 2d_1 \sin\theta_1 &= \lambda_1, \\ 2d_2 \sin\theta_2 &= \lambda_2, \\ 2d_3 \sin\theta_3 &= \lambda_3, \text{ etc.} \end{aligned} \tag{4.26}$$

Photographs with spots arranged symmetrically are obtained only when the crystal is properly oriented with one of the principal axes of the crystal being normal to the X-ray beam. Thus, the Laue method is used to orient crystals for further experiments.

4.4.3 Debye–Scherrer Method

The principle of the Debye–Scherrer method is shown in Fig. 4.18. A beam of filtered X-rays enters the cylindrical camera through the collimator C. The sample S is a thin cylindrical rod made out of the powdered material. The rod is rotated about its own axis (which is also the camera axis). The direct beam leaves the camera through the exit tube E. The diffracted rays are received on the photographic film which rests against the cylindrical surface of the camera. Two such rays (1, 1 and 2, 2) are shown in the diagram. A typical Debye–Scherrer photograph is shown in Fig. 4.19. The pattern consists of concentric rings; these are called powder lines.

Let us try to understand the origin of these rings. The monochromatic rays are diffracted by different lattice planes which make appropriate angles with the incident direction to fulfill Bragg's law. Thus, two planes d_1 and d_2 make angles θ_1 and θ_2 such that they satisfy the Bragg relations

$$\begin{aligned} 2d_1 \sin\theta_1 &= \lambda, \\ 2d_2 \sin\theta_2 &= \lambda, \end{aligned} \tag{4.27}$$

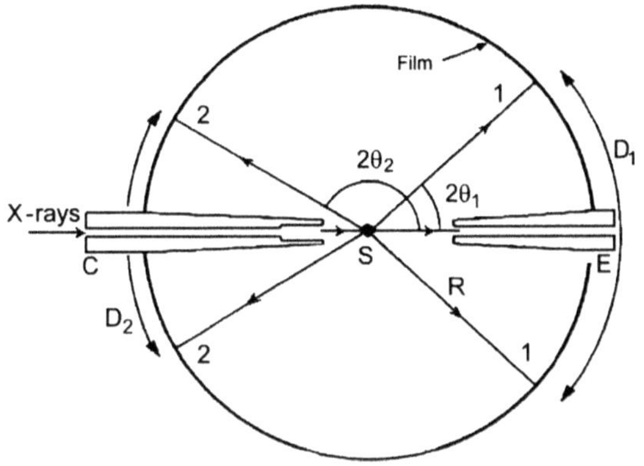

Fig. 4.18 Debye–Scherrer camera

Fig. 4.19 Debye–Scherrer pattern of aluminium

and so on. Each of these equations is satisfied by several crystallites in the sample so that, instead of a single diffracted ray, a cone of rays travels towards the film (Fig. 4.20). To ensure that sufficient number of crystallites are oriented to diffract at a given angle, the sample is rotated. Let the planes d_1 and d_2 relate to the rays 1, 1 and 2, 2 shown in Fig. 4.18. Then if R is the radius of the camera and D_1 and D_2 are the diameters of the rings corresponding to the two sets of rays, the Bragg angles are given by

$$\theta_1 = \left(\frac{D_1}{4\pi R}\right) 180°$$

and

$$(90 - \theta_2) = \left(\frac{D_2}{4\pi R}\right) 180°. \tag{4.28}$$

The region for which $\theta < 45°$ is called the transmission (or front reflection) region whereas the region with $\theta > 45°$ is called the back reflection region.

The Debye–Scherrer method is used (1) to identify unknown materials by comparing their patterns with standard patterns and (2) to calculate the lattice constant.

By differentiation of Bragg's law (4.1), we get

$$\left(\frac{\Delta d}{d}\right) = -\cot\theta \cdot \Delta\theta. \tag{4.29}$$

This shows that the error $\Delta d/d$ resulting from an error $\Delta\theta$ in the measurement of the Bragg angle decreases from the front reflection region to the back reflection

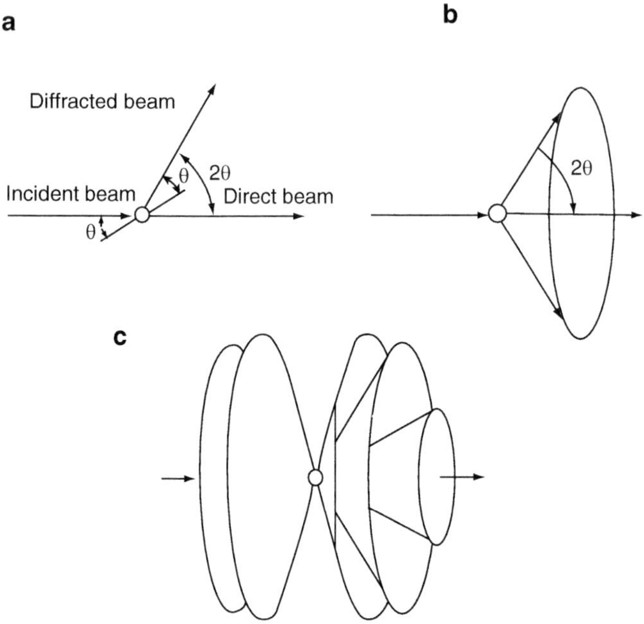

Fig. 4.20 Formation of powder diffraction lines: (**a**) single ray, (**b**) cone of rays from a plane, (**c**) cones of rays from various planes

Table 4.2 Accurate values of the lattice constants (a) of some cubic crystals determined using a Debye–Scherrer camera at room temperature

Substance	a[Å]	Accuracy [Å]	Reference
$Sr(NO_3)_2$	7.7798	±0.0002	Deshpande et al. [4.16]
NH_4Cl	3.8759	±0.0001	Suryanarayana and Sirdeshmukh [4.17]
NH_4Br	4.0600	±0.0001	Suryanarayana and Sirdeshmukh [4.17]
RbCl	6.5941	±0.0001	Srinivas et al. [4.18]
RbBr	6.8892	±0.0002	Srinivas et al. [4.18]

region and finally tends to zero as θ tends to 90°. Hence for accurate determination of lattice constants, reflections with larger values of θ are preferred. Accurately determined values of the lattice constants of some crystals are given in Table 4.2.

4.4.4 Rotating Crystal Method

The arrangement in a rotating crystal camera is shown in Fig. 4.21. Monochromatic X-rays enter through the collimator C into a cylindrical camera. Compared to a Debye–Scherrer camera, the rotating crystal camera has a larger height. The sample in the form of a single crystal S is mounted with one of its principal axes coinciding

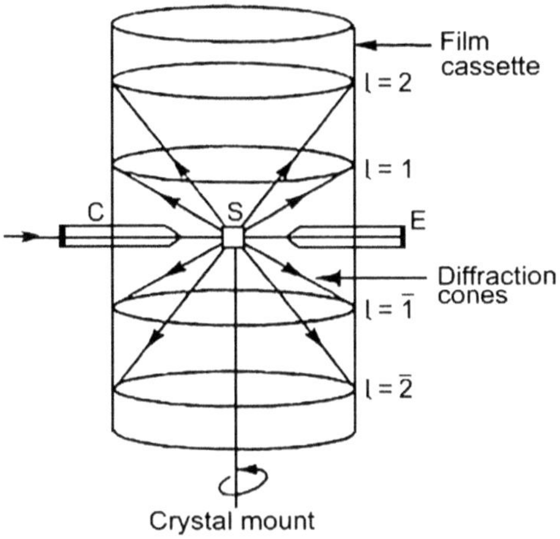

Fig. 4.21 Rotating crystal camera

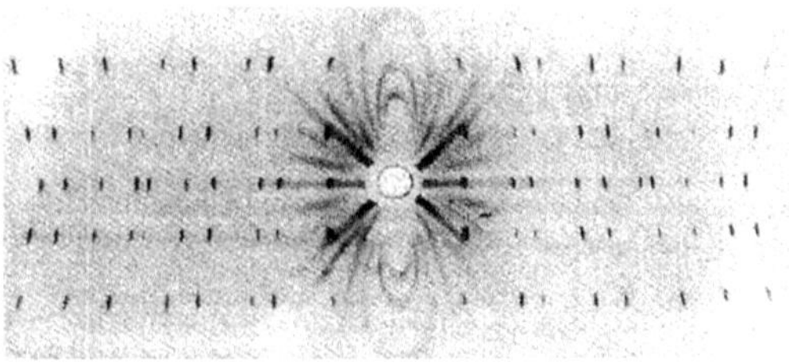

Fig. 4.22 Rotation pattern of quartz about c-axis

with the axis of the camera. There is provision to rotate the crystal about its axis. The direct beam is absorbed by the lead glass in the exit tube E. The diffracted rays proceed along cones and produce a pattern on the photographic film which consists of a number of spots arranged in lines; these lines are called layer lines. A typical X-ray diffraction photograph obtained with a rotating crystal camera is shown in Fig. 4.22.

Let us consider the geometry of diffraction (Fig. 4.23). The incident ray makes angle ε_3 with the c-axis of the crystal and the diffracted ray makes angle Δ_3. Then, according to the third equation of (4.15),

$$\cos\Delta_3 - \cos\varepsilon_3 = l\lambda / c. \tag{4.30}$$

But $\varepsilon_3 = 90°$. Hence

$$\cos\Delta_3 = l\lambda / c. \tag{4.31}$$

Thus, various lattice planes having different values for the indices h and k but a common l index will diffract at the same angle Δ_3 and their reflections will lie on a line. If the height of the first layer line is D_1 and the camera radius is R, then (4.31) becomes

$$c = l\lambda\left[(R^2 + D_1^2)^{1/2} / D_1\right]. \tag{4.32}$$

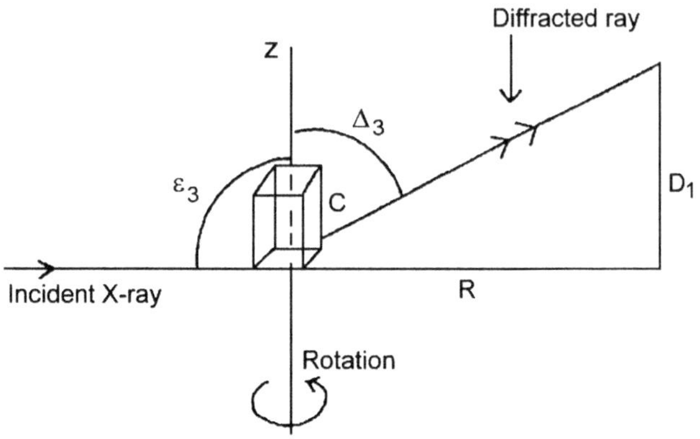

Fig. 4.23 Geometry of ray in a rotation camera

Note that (4.32) enables us to estimate the lattice parameter merely by measuring the height of the first layer line. If the crystal is oriented with the a and b axes successively along the camera axis, the corresponding patterns will facilitate the determination of the a and b lattice parameters. Thus, this method provides a preliminary estimate of the lattice parameters.

It may also be noted that layer lines are perfectly straight when the crystal axis is exactly along the camera axis. If there is a misset the layer lines become slanted; correct orientation can be achieved by tilting the crystal such that the slant disappears. Thus, this method is also useful in correctly orienting crystals.

A modification of the rotation camera is the oscillation camera. It differs from the rotation camera essentially in the angular range covered. Whereas in a rotation camera the crystal rotates through 360°, in the oscillation camera the angular range is much smaller; generally, the ranges are 5°, 10° or 15°. The oscillation is achieved by coupling the shaft on which the crystal is mounted to eccentric cams.

4.4.5 Weissenberg Method

In the rotating crystal photograph shown in Fig. 4.22 the diffraction spots in a given layer line are well separated. Similar photographs with crystals with low symmetry and large lattice parameter show overcrowding and overlapping of spots. This can be seen in the rotation crystal photograph of a monoclinic crystal (Fig. 4.24). It will be seen later that, for a realistic determination of the crystal structure, a large number of diffraction spots need to be recorded and their intensities measured. Overlapping of reflections has, hence, to be avoided. This is achieved ingenuously in the Weissenberg method.

The principle of the Weissenberg camera is shown in Fig. 4.25. Figure 4.25a shows the geometry and Fig. 4.25b gives an inside view. The Weissenberg camera is

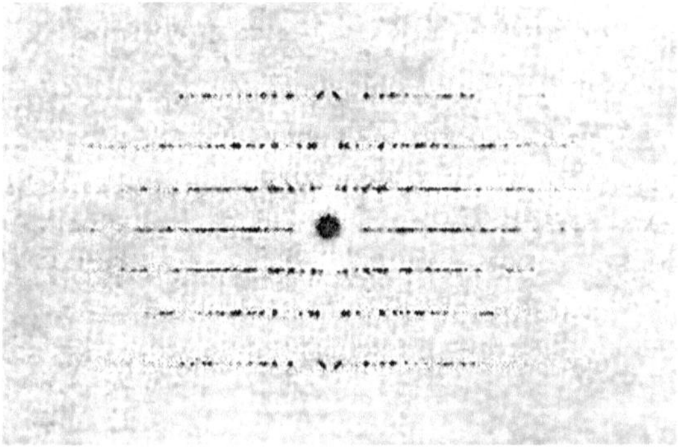

Fig. 4.24 Example of a rotating crystal photograph showing crowding and overlapping of reflections

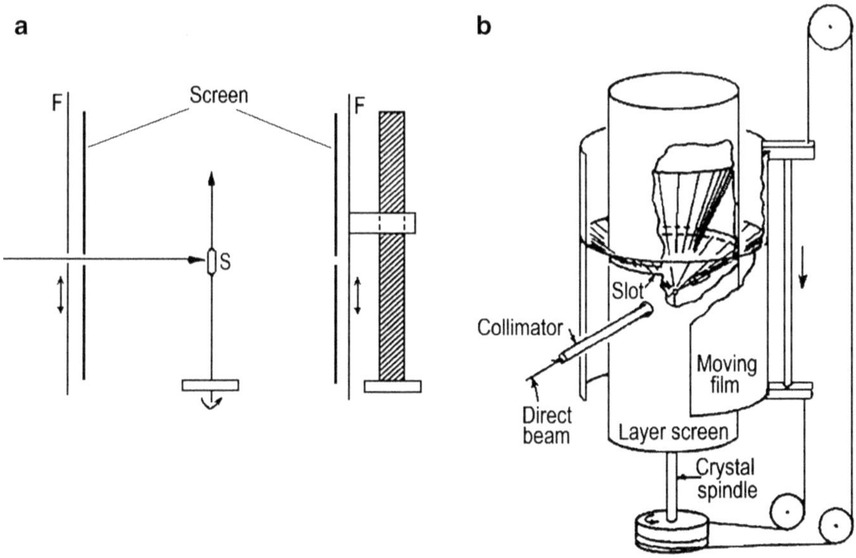

Fig. 4.25 Weissenberg camera: (**a**) schematic, (**b**) inside view

essentially a modified cylindrical rotating crystal camera. The modification consists firstly in introducing a metallic screen with a slot in front of the film such that the slot allows reflection only in one layer line to reach the film and, secondly, in giving an oscillatory motion to the film in a direction parallel to the cylinder axis. In this way, the reflections which would have been recorded in a single layer line in a static film camera are now spread over a curve thus avoiding overlapping. The diffraction pattern of a crystal obtained with a Weissenberg camera is shown in Fig. 4.26. The Weissenberg technique is immensely useful in providing intensities of a large number of reflections essential in structure determination.

Fig. 4.26 Example of a Weissenberg photograph

4.4.6 *Precession Method*

The design of Buerger's precession camera is shown in Fig. 4.27. The crystal oscillates about two mutually perpendicular axes both normal to the incident beam. A flat film is used (unlike in the rotation and Weissenberg cameras). The geometrical relation between the crystal and the film can be seen in the figure. The design and the setting procedures are a little intricate but with one crystal setting, all the lattice parameters can be determined. Further, the precession camera records the reciprocal lattice without any distortion. Another advantage is that the photographic pattern is simple and indexing is obvious and unmistakable. A typical precession photograph is shown in Fig. 4.28.

4.4.7 *X-ray Diffractometer*

4.4.7.1 Powder Diffractometer

This technique facilitates direct recording of the complete diffraction pattern, i.e. the Bragg angles as well as integrated intensities. A simplified diagram of an X-ray

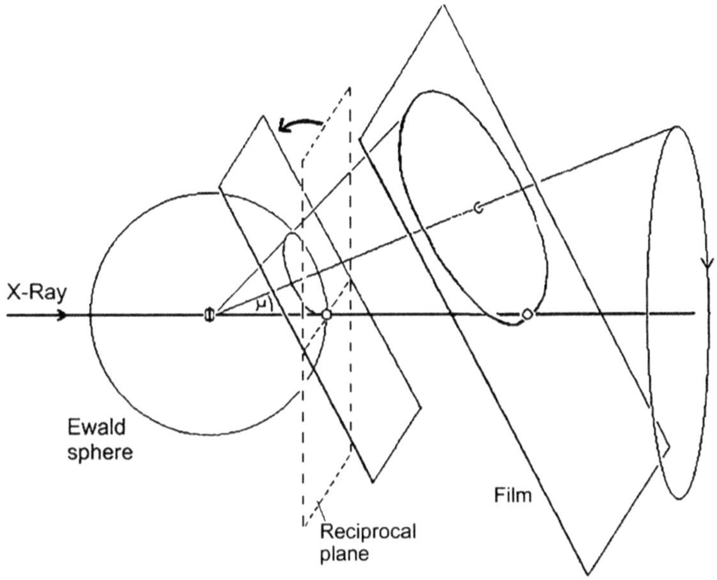

Fig. 4.27 Diagram of a precession camera

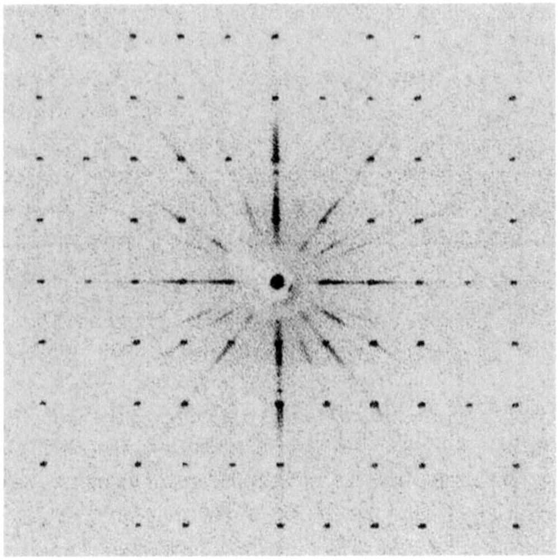

Fig. 4.28 Example of a precession photograph

powder diffractometer is shown in Fig. 4.29. A divergent monochromatic X-ray beam originating at A falls on a flat-faced powder sample S. The sample rotates about an axis at S normal to the plane of the paper at a rotational speed ω. The diffracted rays are focussed at D where they are received into a detector. The detector rotates at a speed 2ω so that it is always in position to receive the diffracted rays.

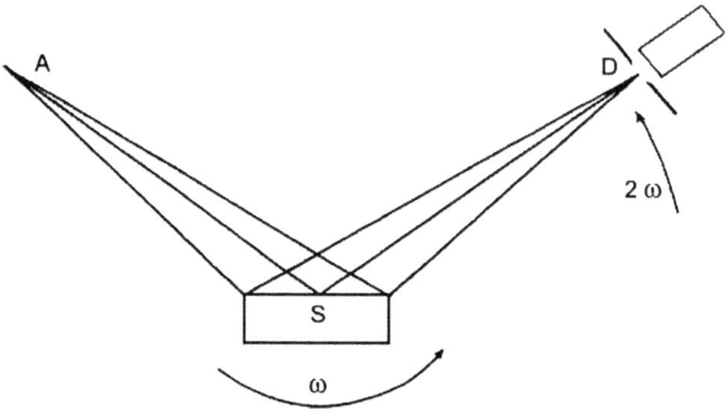

Fig. 4.29 Principle of X-ray diffractometer

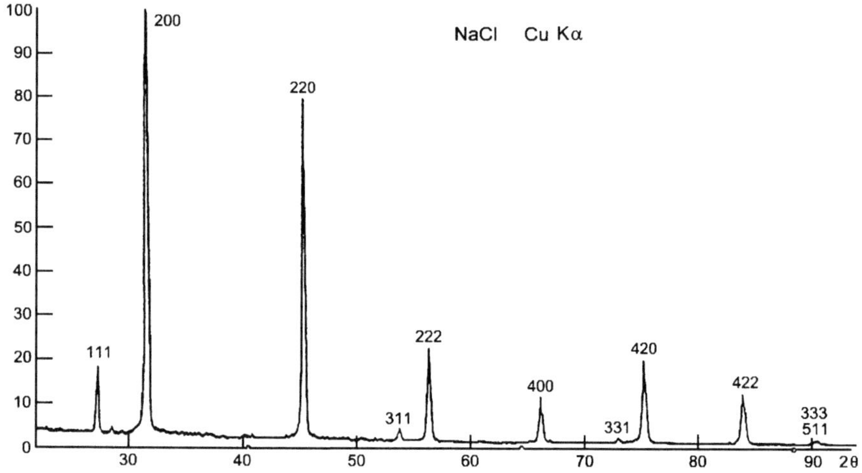

Fig. 4.30 Diffraction record of NaCl powder sample with Cu radiation

The detector could be one of the counters mentioned in Sect. 4.4.1. The pulses in the detector are amplified and fed to a chart recorder. The diffractometer record for a powder sample of NaCl is shown in Fig. 4.30 and that of a reflection recorded at slow speed in Fig. 4.31.

Powder diffractometer traces are useful in identification of unknown samples and in lattice constant determination. X-ray powder diffractometers with different degrees of sophistication are commercially available.

4.4.7.2 Single Crystal Diffractometer

The modern instrument to record intensities of single crystal diffraction spots is the four circle diffractometer shown in Fig. 4.32. The crystal is mounted on a

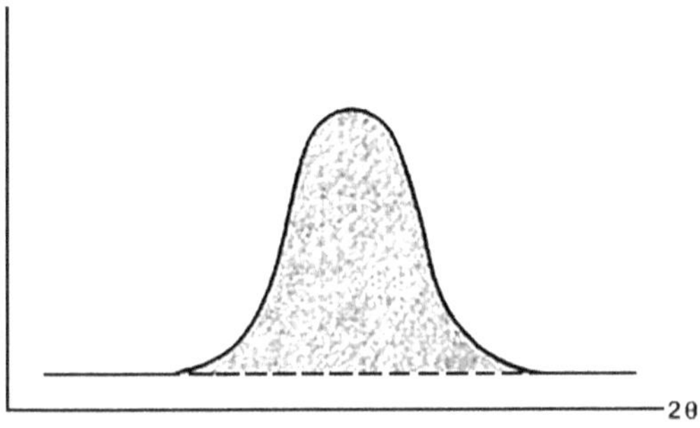

Fig. 4.31 Diffraction curve of a single peak

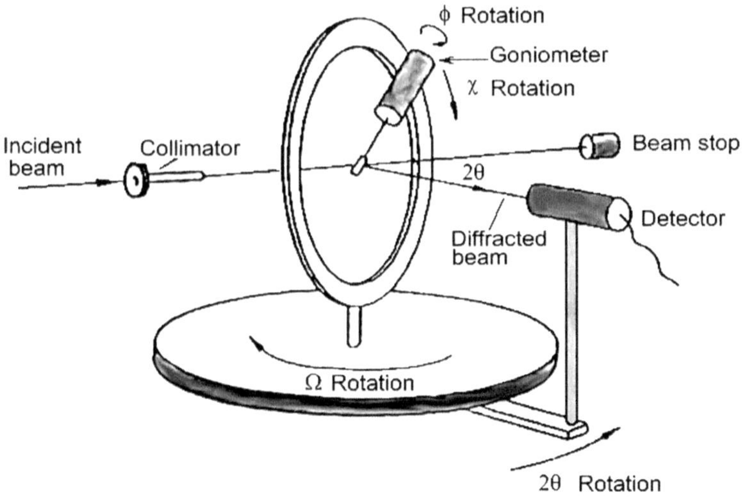

Fig. 4.32 Four circle diffractometer; the counter rotates about the 2θ axis in one plane, and the crystal is oriented about the three axes ϕ, χ and Ω so that the incident and reflected beams lie in the horizontal (2θ) plane

goniometer and set so that one of the crystal axes is parallel to the goniometer axis. Thereafter, the crystal can be rotated about the goniometer axis (ϕ rotation). The goniometer as a whole can be rotated on a circular frame (χ rotation) and the circular frame can be rotated about a vertical axis (Ω rotation). Once the angles ϕ, ψ and Ω are adjusted to bring a plane in position to diffract, the detector which rotates in a horizontal plane (2θ rotation) is placed in the proper position to receive the diffracted beam. Once the setting is adjusted for any one reflection, the settings for other reflections can be easily decided. In practice, all the four angular adjustments (ϕ, χ, Ω and 2θ) are done automatically by a computer without any intervention by the crystallographer.

4.5 Intensity of Diffracted X-rays

The intensity of diffracted X-rays depends on a large number of factors, some not depending on the structure of the crystal and some intimately related to it. Further, it depends critically on the degree of perfection of the crystal. We shall consider these various factors individually and, finally, consider the expression for the intensity.

4.5.1 Absorption Factor

As mentioned earlier, X-rays are absorbed as they pass through the crystal which results in reduction of intensity. This reduction is denoted by the absorption factor A. The absorption factor is different for different techniques. However, in every case, it depends on the absorption coefficient μ of the sample, its thickness t and the angle of diffraction θ. Values of A for different values of μ, t and θ are given in tables for each technique (Peiser et al. [4.19]). In case of a flat-faced sample, the absorption factor is independent of θ.

4.5.2 Multiplicity Factor

Let us consider a cubic crystal (Fig. 4.33) mounted such that one of its axes is parallel to the z-axis. Let us consider a beam of X-rays incident on the (110) plane and diffracted at an angle θ. This plane will diffract twice in the same direction in a full rotation about the z-axis. Further, another equivalent plane $(1\bar{1}0)$ also diffracts in the same direction twice in a rotation. Thus, effectively, the (110) plane diffracts four times in one rotation of the crystal. The number of times a set of equivalent planes diffract in the same direction and thereby contribute to the intensity is called its multiplicity factor (J). This factor depends on (1) the experimental technique, (2) the point group and (3) the Miller indices of the plane. For instance, for a cubic crystal and in powder diffraction, $J = 6$, 12 and 8 for the (100), (110) and (111) planes, respectively.

The multiplicity factors for different techniques and for different sets of planes in crystals of different crystal classes are given in standard books on X-ray diffraction. Klug and Alexander [4.3] have given multiplicity factors specifically for the powder technique.

4.5.3 Lorentz-Polarization Factor

This factor combines two effects, the polarized nature of the X-ray beam and the velocity of rotation of the sample.

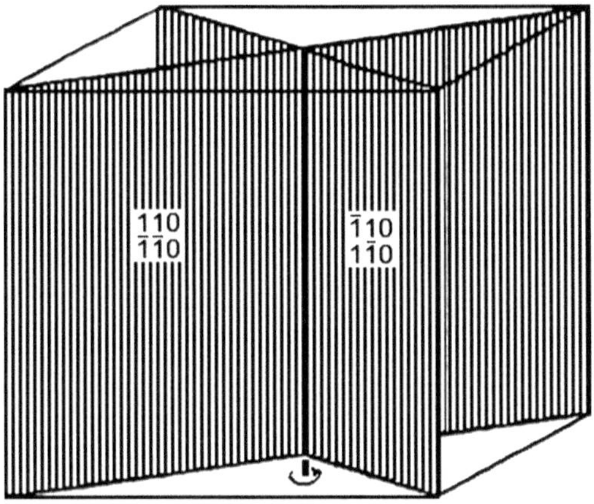

Fig. 4.33 Multiplicity of 110 plane

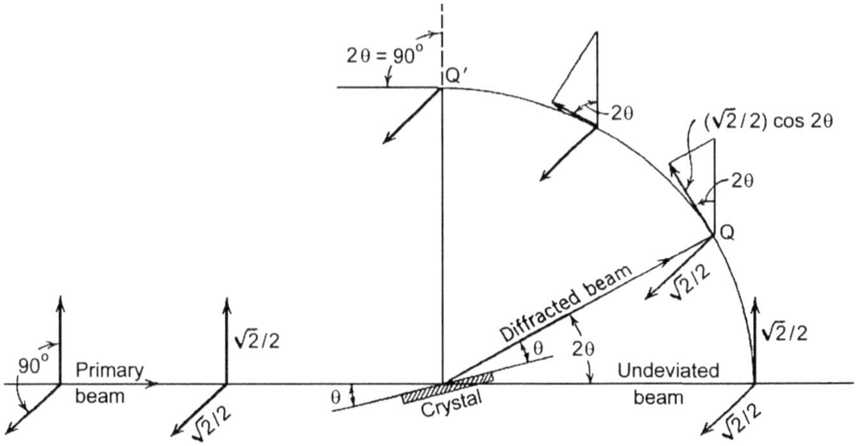

Fig. 4.34 Horizontal and vertical components of a diffracted X-ray beam at different angles of diffraction

Let us consider the effect of polarization. The X-ray beam from an X-ray tube is unpolarized. After diffraction, the beam gets polarized; the degree of polarization varies with the angle of diffraction.

Consider a beam incident on a crystal and diffracted at an angle θ (Fig. 4.34). In this unpolarized incident beam, the electric vector vibrates in all directions. Any one of these randomly oriented vectors may be assumed to be a unit vector and may be resolved into two equal perpendicular vectors of magnitude $\sqrt{2}/2$. The electrons in the crystal perform forced vibrations parallel to these vectors in the plane perpendicular to the incident beam. At the point Q′, with $2\theta = 90°$, only the horizontal component $\sqrt{2}/2$ is effective. On the other hand, at the point Q along the diffraction direction, the horizontal component is $\sqrt{2}/2$ but the effective vertical component is $(\sqrt{2}/2)\cos 2\theta$. At this point the total contribution to intensity is $(1/2)(1 + \cos^2 2\theta)$. It is interesting to note that this expression for the polarization factor was derived by Thomson [4.20] much before the discovery of X-ray diffraction in his theory of the scattering of free electrons by matter. This "polarization factor" is common to all methods of X-ray diffraction. In recent experimental techniques, particularly in diffractometers, a crystal-diffracted monochromatic beam is used as the incident beam. In this case, the polarization factor is

$$\left[\frac{1 + \cos^2 2\theta_1 \cos^2 2\theta_2}{1 + \cos^2 2\theta_1}\right],$$

where θ_1 is the diffraction angle of the monochromator crystal and θ_2 is the diffraction angle of the sample crystal.

Even a supposedly monochromatised beam will have a wavelength spread. Further, even a collimated beam will not be strictly parallel but will have a degree of divergence. In addition, in most techniques, the sample is rotated. These three factors determine the "opportunity" of a plane to diffract. Due to these factors, the intensity will be proportional to a factor which depends on the angle of diffraction. The derivation of this factor for the Laue case was done by Lorentz in a class room lecture. This Lorentz factor depends on the experimental technique.

It is the practice to combine these two factors and call them the Lorentz-polarization (L.P.) factor. The L.P. factors for some techniques are given in Table 4.3. Their values for different θ values are given in books on X-ray diffraction. Lorentz-polarization factors for the oscillation, Weissenberg and precession cameras are more complicated; they are tabulated in Kaelble [4.2].

4.5.4 Structure Factor

The structure factor F_{hkl} is the resultant of the amplitudes of electromagnetic waves scattered by the several atoms in the unit cell of a crystal expressed as a fraction of the amplitude scattered by a single free electron. As the name indicates, F_{hkl} contains information about the structure of the crystal. We shall carry out the derivation in stages.

Table 4.3 The Lorentz polarization factor (L.P.) for some techniques

Technique	L.P. factor
Laue	$\left[\frac{1}{\sin^2\theta}\right]\left[\frac{1}{2}(1 + \cos^2 2\theta)\right]$
Powder	$[\mathrm{cos\,ec}2\theta\ \mathrm{cos\,ec}\,\theta]\left[\frac{1}{2}(1 + \cos^2 2\theta)\right]$
Rotating crystal	$\left[\frac{1}{\sin\theta\cos\theta}\right]\left[\frac{1}{2}(1 + \cos^2 2\theta)\right]$

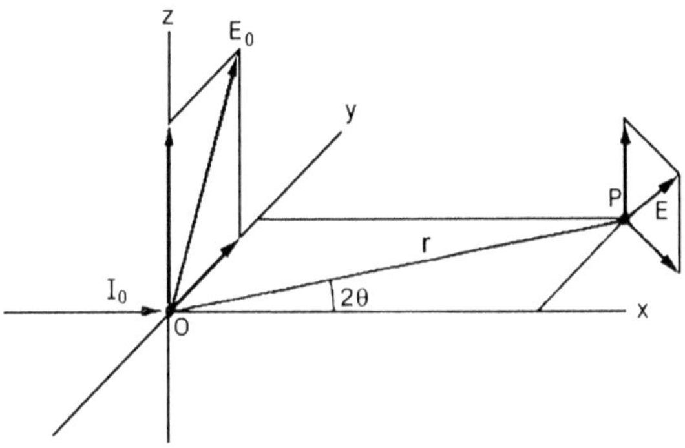

Fig. 4.35 Scattering of an X-ray beam by a single free electron

4.5.4.1 Scattering by a Single Free Electron

Let us consider a beam of X-rays incident on an electron with charge e and mass m located at the point O (Fig. 4.35). Let its amplitude and intensity be E_0 and I_0, respectively. Under the effect of the X-ray wave, the electron is set into forced vibrations and becomes a secondary source of electromagnetic waves. Let us consider the scattered wave at point P at a distance r from O. Further, OP makes an angle 2θ with the incident beam direction. It was shown by Thomson that the amplitude of the scattered wave E_e is given by

$$E_\mathrm{e} = \left(\frac{e^2}{mc^2 r}\right) E_0. \tag{4.33}$$

Since the intensity of an em wave is $(c/8\pi)$ times the square of the amplitude, the intensity I_e of the scattered wave at P is

$$I_\mathrm{e} = \left(\frac{e^2}{mc^2 r}\right)^2 I_0. \tag{4.34}$$

The polarization effect discussed in Sect. 4.5.3 is not included in (4.34). The numerical value of (e^4/m^2c^4) is $7.94 \times 10^{-26}\ \mathrm{cm}^2$ in cgs units.

4.5.4.2 Scattering by an Atom

Consider an atom with centre B, lying in the lattice plane PP′ (Fig. 4.36). AB is an X-ray wave incident at angle θ on the lattice plane and BC is the wave scattered at the same angle by an electron supposed to be located at B. The radius of the atom is R. Now let us consider another electron located at B′ a distance x below B. A wave A′B′ parallel to AB is scattered by the electron at B′ along B′C′ which is parallel to BC. The path difference between the two sets of rays ABC and A′B′C′ is $2x \sin\theta$. So far, the situation is similar to that in the derivation of Bragg's law.

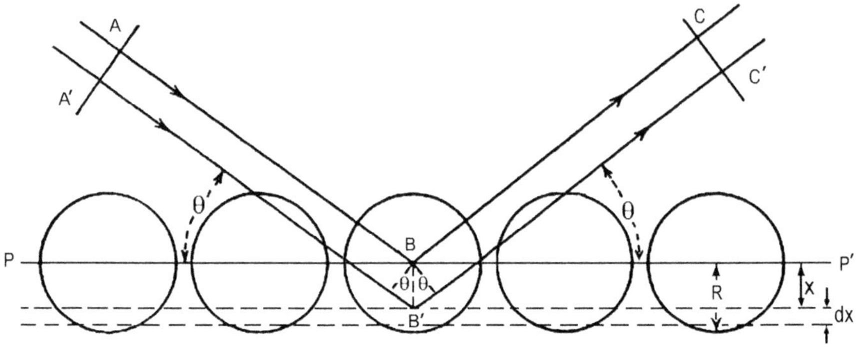

Fig. 4.36 X-rays scattered by different electrons in an atom

But the electron at B′ is not in a lattice plane; hence, we cannot equate the path difference $2x \sin\theta$ to $n\lambda$. Instead, as in conventional optics, we shall add up the amplitude contribution of the wave scattered by the electron at B′, which is in phase with the wave scattered by the electron supposed to be at B. This contribution [4.6] is

$$E_x = E_\mathrm{e} \cos\left(\frac{4\pi x}{\lambda} \sin\theta\right), \tag{4.35}$$

where E_e is the amplitude of the wave scattered by a single electron. At this stage, we have to introduce the important point that in the quantum mechanical picture an electron cannot be said to be definitely located at a point like B′; instead there is only a probability $P(x)\,\mathrm{d}x$ of its existence in the region between x and $\mathrm{d}x$. Hence, (4.35) becomes

$$E_x \mathrm{d}x = P(x) E_\mathrm{e} \cos\left(\frac{4\pi x}{\lambda} \sin\theta\right) \mathrm{d}x. \tag{4.36}$$

The resultant amplitude E_a due to the X-rays scattered by the entire charge (Ze) distributed over the atom is

$$E_\mathrm{a} = \int_{-R}^{R} E_x \mathrm{d}x = ZE_\mathrm{e} \int_{-R}^{R} P_x \cos\left(\frac{4\pi x}{\lambda} \sin\theta\right) \mathrm{d}x, \tag{4.37}$$

where Z is the atomic number. The ratio $(E_\mathrm{a}/E_\mathrm{e})$ is called the atomic scattering factor f which is now given by

$$f = Z \int_{-R}^{R} P_x \cos\left(\frac{4\pi x}{\lambda} \sin\theta\right) \mathrm{d}x. \tag{4.38}$$

Also,

$$E_a = fE_e = f E_0 e^2 / mc^2 r. \tag{4.39}$$

Thus, the atomic scattering factor is the ratio of the amplitude at any point of the radiation scattered by all the elements of electronic charge distributed over an atom to the amplitude of the radiation scattered by a free electron.

To know the value of f, we should know the probability function $P(x)$. According to quantum mechanics, the total electronic charge ze is distributed over the atom and $P(x)$ is the charge density at x. There are different models for the charge distribution. A comparative description of these models is given by James [4.1]. In Fig. 4.37 the atomic scattering curves for some atoms are shown based on three models. In Fig. 4.38, the f-curves for a large number of atoms are shown

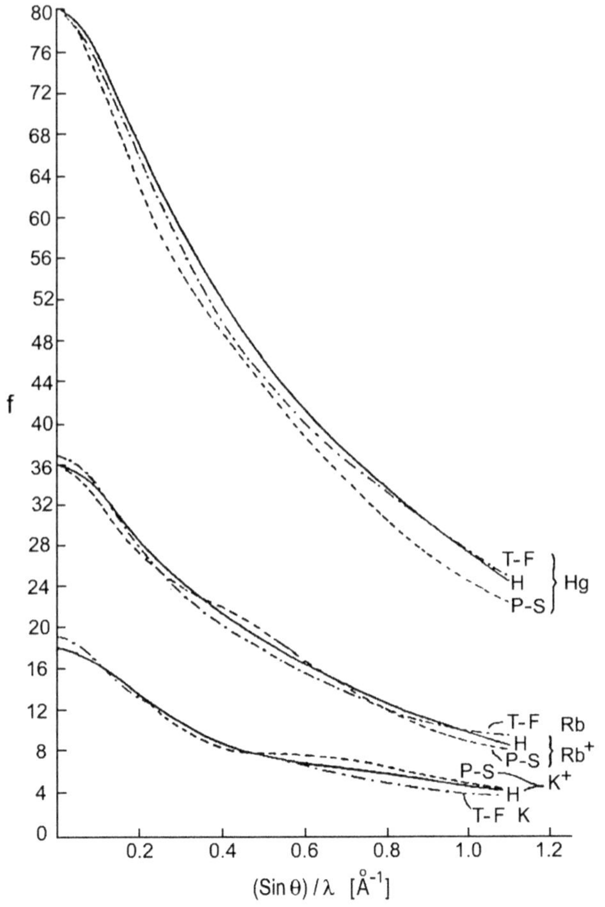

Fig. 4.37 Atomic scattering factors of Hg, Rb and K calculated from Hartree (H), Pauling–Sherman (P–S) and Thomas–Fermi (T–F) models

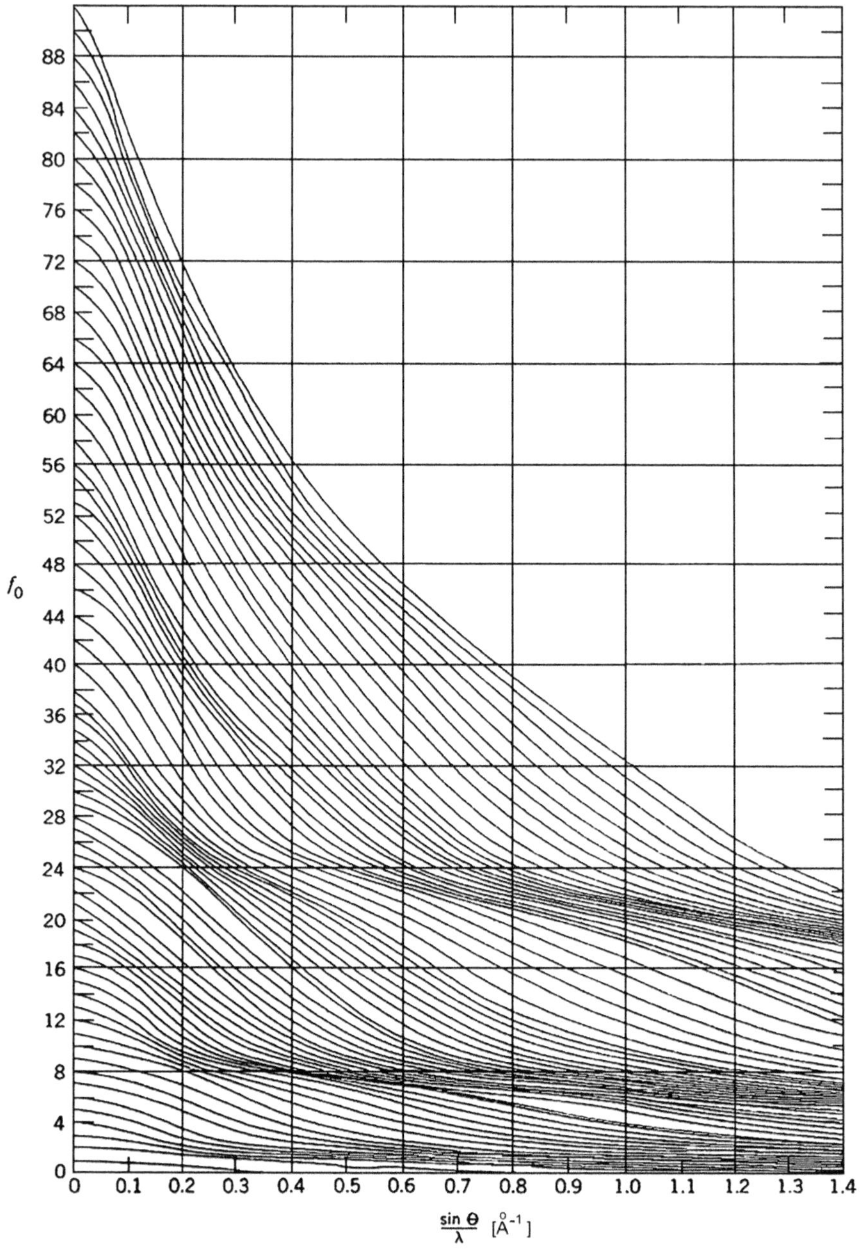

Fig. 4.38 Variation of atomic scattering factors of various atoms with (sin θ/λ); calculations based on Pauling–Sherman model

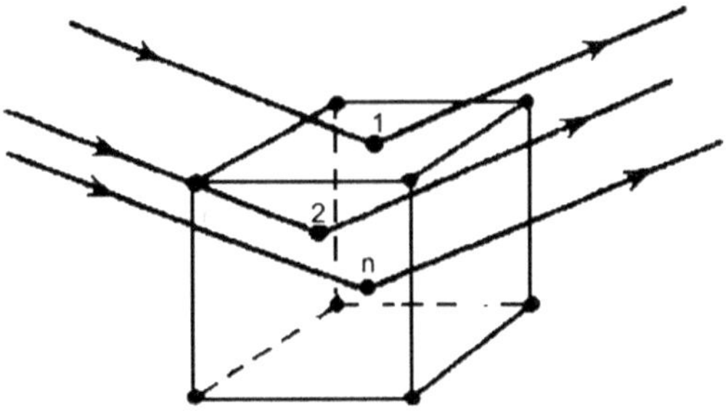

Fig. 4.39 X-rays scattered by different atoms in a unit cell

based on a single model, the Pauling–Sherman model [4.3]. From these curves, we note the following features:

1. f varies with $(\sin\theta/\lambda)$. The variation is steep at low values of $\sin\theta/\lambda$ and becomes slower at larger values of $\sin\theta/\lambda$ At $\sin\theta/\lambda = 0$, f becomes equal to Z.
2. The f curves are higher for high-Z atoms.
3. The values of f depend upon the assumed model of electron distribution.

Calculated values of the atomic scattering factor for neutral atoms and ions corresponding to different models are available in the form of tables in the original publications and also in books on X-ray diffraction.

4.5.4.3 Scattering from a Unit Cell

Consider a unit cell (Fig. 4.39). The position coordinates of the n th atom may be represented by x_n, y_n, z_n. These are the actual coordinates expressed as fractions of the lattice parameters. Let us consider the rays diffracted at the *hkl* plane at angle θ. Each atom scatters with its respective amplitude f_n. However, the scattered waves from all the atoms are not in phase since the atoms are located at different positions in the unit cell. Thus, the scattered wave from the n th atom has a phase $\phi_n = 2\pi(hx_n + ky_n + lz_n)$. The resultant of waves from all atoms in the unit cell is

$$F_{hkl} = f_1 e^{i\phi_1} + f_2 e^{i\phi_2} + f_3 e^{i\phi_3} + \cdots + f_N e^{i\phi_N} = \sum_{n=1}^{N} f_n e^{i\phi_n}, \qquad (4.40)$$

where N is the total number of atoms in the unit cell. F_{hkl} is called the structure factor. It is a function of the atomic scattering factors, the position coordinates of the atoms in the unit cell and the Miller indices of the diffracting plane. The structure factor F_{hkl} is the resultant of the waves scattered by individual atoms located at different positions in the unit cell. Recalling how the atomic scattering factor was defined, it may be stated that the structure factor represents the combined effect of scattering of X-rays by the entire electron charge in the unit cell.

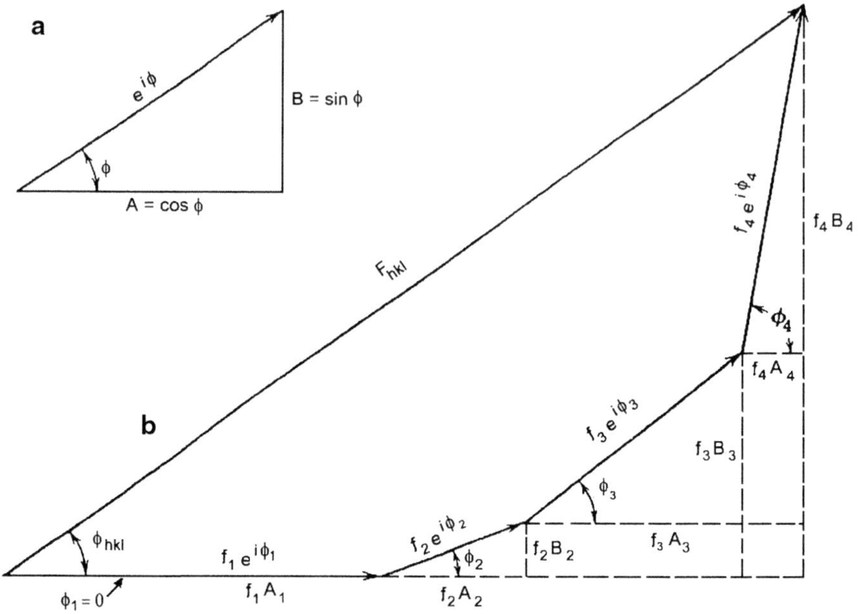

Fig. 4.40 Vector diagram of the complex structure factor

In general, the structure factor F_{hkl} is a complex quantity having an amplitude $|F_{hkl}|$ and a phase ϕ_{hkl}. Let us represent the various terms in (4.40) on an Argand diagram (Fig. 4.40). Since $e^{i\phi} = \cos\phi + i\sin\phi$, each term in (4.40) splits into components $f_n \cos\phi_n = f_n A_n$ and $f_n \sin\phi_n = f_n B_n$. From the diagram, it follows that

$$|F_{hkl}| = [(f_1A_1 + f_2A_2 + \cdots + f_N A_N)^2 + (f_1B_1 + f_2B_2 + \cdots + f_N B_N)^2]^{1/2}$$

$$= \left\{\left[\left(\sum_{n=1}^{N} f_n A_n\right)^2\right] + \left[\left(\sum_{n=1}^{N} f_n B_n\right)^2\right]\right\}^{1/2} \quad (4.41)$$

or

$$|F_{hkl}| = \left\{\left[\sum_{n=1}^{N} f_n \cos 2\pi(hx_n + ky_n + lz_n)\right]^2 + \left[\sum_{n=1}^{N} f_n \sin 2\pi(hx_n + ky_n + lz_n)\right]^2\right\}^{1/2}. \quad (4.42)$$

It also follows that

$$\phi_{hkl} = \tan^{-1}\frac{\sum_{n=1}^{N} f_n B_n}{\sum_{n} f_n A_n} = \tan^{-1}\frac{\sum_{n=1}^{N} f_n \sin 2\pi(hx_n + ky_n + lz_n)}{\sum_{n=1}^{N} f_n \cos 2\pi(hx_n + ky_n + lz_n)}. \tag{4.43}$$

It may be noted that to evaluate ϕ_{hkl} we should know the position coordinates of the atoms.

As an example, we shall calculate the structure factors for the NaCl structure. There are four molecular units in the unit cell for this structure with the atoms located at:

Na (0,0,0), (1/2,1/2,0), (1/2,0,1/2), (0,1/2,1/2)

Cl (1/2,1/2,1/2), (0,0,1/2), (0,1/2,0), (1/2,0,0)

Substituting these coordinates in (4.42), we get

$$\begin{aligned}
|F_{hkl}|^2 = &\left\{ f_{\mathrm{Na}}\left[\cos 2\pi(0) + \cos 2\pi\left(\frac{h+k}{2}\right) + \cos 2\pi\left(\frac{h+l}{2}\right) + \cos 2\pi\left(\frac{k+l}{2}\right)\right]\right.\\
&\left. + f_{\mathrm{Cl}}\left[\cos 2\pi\left(\tfrac{h+k+l}{2}\right) + \cos 2\pi\left(\frac{l}{2}\right) + \cos 2\pi\left(\frac{k}{2}\right) + \cos 2\pi\left(\frac{h}{2}\right)\right]\right\}^2\\
&+\left\{ f_{\mathrm{Na}}\left[\sin 2\pi(0) + \sin 2\pi\left(\frac{h+k}{2}\right) + \sin 2\pi\left(\frac{h+l}{2}\right) + \sin 2\pi\left(\frac{k+l}{2}\right)\right]\right.\\
&\left. + f_{\mathrm{Cl}}\left[\sin 2\pi\left(\frac{h+k+l}{2}\right) + \sin 2\pi\left(\frac{l}{2}\right) + \sin 2\pi\left(\frac{k}{2}\right) + \sin 2\pi\left(\frac{h}{2}\right)\right]\right\}^2.
\end{aligned} \tag{4.44}$$

Substituting specific values of h,k,l in (4.44), we get

$$\begin{aligned}
|F_{hkl}| &= 4(f_{\mathrm{Na}} + f_{\mathrm{Cl}}) \quad \text{for } h, k, l, \text{ all even,}\\
&= 4(f_{\mathrm{Na}} - f_{\mathrm{Cl}}) \quad \text{for } h, k, l, \text{ all odd,}\\
&= 0 \qquad \text{for } h, k, l, \text{ mixed.}
\end{aligned} \tag{4.45}$$

The structure factors for a few simple structures are given in Table 4.4.

4.5.5 The Temperature Factor

In the above treatment, it was assumed that the atoms are at rest. But at all finite temperatures, the atoms and the associated electrons are in a state of vibration. These lattice thermal vibrations reduce the scattering power of the atoms. This problem was examined by Debye [4.21] soon after he developed the theory of

Table 4.4 Structure factor $|F|$ for some simple structures

S. no.	Structure	Coordinates	Indices	$\|F\|$
1.	fcc	$000;\ \frac{1}{2}\frac{1}{2}0;\ \frac{1}{2}0\frac{1}{2};\ 0\frac{1}{2}\frac{1}{2};$	h, k, l all odd or all even	$4f$
2.	hcp	$000; \frac{1}{3}\frac{2}{3}\frac{1}{2}$	$h + 2k = 3n, l$ even	$2f$
			$h + 2k = 3n \pm 1, l$ odd	$\sqrt{3}f$
			$h + 2k = 3n \pm 1, l$ even	f
3.	NaCl	Na: 000; fcc sites	h,k,l all even	$4f_{Na} + 4f_{Cl}$
		Cl: $\frac{1}{2}\frac{1}{2}\frac{1}{2}$; fcc sites	h,k,l all odd	$4f_{Na} - 4f_{Cl}$
4.	CsCl	Cs: 000	$h + k + l$ even	$f_{Cs} + f_{Cl}$
		Cl: $\frac{1}{2}\frac{1}{2}\frac{1}{2}$	$h + k + l$ odd	$f_{Cs} - f_{Cl}$
5.	ZnS	Zn:000; fcc sites	$h + k + l = 4n$	$4f_{Zn} + 4f_S$
		S: $\frac{1}{4}\frac{1}{4}\frac{1}{4}$; fcc sites	$h + k + l = 4n + 2$	$4f_{Zn} - 4f_S$
			$h + k + l = 4n + 1$	$4(f_{Zn}^2 + f_S^2)^{1/2}$
6.	CaF_2	Ca: 000; fcc sites	$h + k + l = 4n + 1$	$4f_{Ca}$
		F: $\frac{1}{4}\frac{1}{4}\frac{1}{4}$; fcc sites	$h + k + l = 4n$	$4f_{Ca} + 8f_F$
		F: $\frac{3}{4}\frac{3}{4}\frac{3}{4}$; fcc sites	$h + k + l = 4n + 2$	$4f_{Ca} - 4f_F$

specific heat of solids. According to Debye's theory, modified by Waller [4.22], the atomic scattering factor f of an atom is modified to $f\mathrm{e}^{-M}$ or $f\mathrm{e}^{-B(\sin\theta/\lambda)^2}$. B is called the Debye–Waller factor and is given by

$$B = \frac{6h^2}{m_a k_B \theta_D}\left[\frac{\phi(x)}{x} + \frac{1}{4}\right] \text{ where } \phi(x) = \frac{1}{x}\int\limits_0^x \frac{\xi \mathrm{d}\xi}{\mathrm{e}^{\xi} - 1}. \tag{4.46}$$

Here m_a is the mass of the vibrating atom, θ_D the Debye temperature and $x = \theta_D/T$. The values of the expression in the brackets for different values of x are available in literature.

Equation (4.40) is now modified to

$$F_{hkl}^T = \sum_{n=1}^{N} f_n \mathrm{e}^{-B(\sin\theta/\lambda)^2} \mathrm{e}^{2\pi \mathrm{i}(hx_n + ky_n + lz_n)}, \tag{4.47}$$

where F_{hkl}^T is the structure factor at temperature T. As a consequence, the observed diffraction intensities get reduced by a factor e^{-2M} or $\mathrm{e}^{-2B(\sin\theta/\lambda)^2}$ which is called the temperature factor.

4.5.6 Integrated Intensity

In an ideal situation, we should expect the X-ray reflection to be extremely sharp with a unique value for the intensity. However, due to various physical reasons, what is observed is a peak with some width. Thus, the intensity has a peak value

Fig. 4.41 A mosaic crystal

at some value of θ and then it falls off on either side of the peak becoming zero at some angle $\theta \pm \Delta\theta$. The total intensity reflected over this range is called the integrated intensity (I_{hkl}). I_{hkl} is, of course, proportional to the factors J and $L.P.$ The relation between I_{hkl} and the structure factor F_{hkl} is not straightforward. The theory connecting I_{hkl} with F_{hkl} has been worked out by Darwin [4.23] and Ewald [4.24, 4.25]. It is discussed in detail by James [4.1] and Warren [4.4]. We shall only quote the results.

The theory which assumes the crystal to be perfect and also transparent is called the dynamical theory. According to this theory

$$I_{hkl} = \left(\frac{8}{3\pi}\right)\left(\frac{I_0}{\omega}\right)\left(\frac{e^2}{mc^2}\right)(N\lambda^2)\,(J\,L.P.)\left|F^T_{hkl}\right|. \tag{4.48}$$

On the other hand, the kinematic theory assumes an imperfect (mosaic) and absorbing crystal. A mosaic crystal (Fig. 4.41) consists of minute crystallites which are slightly misaligned with respect to one another. For such a crystal, the integrated intensity is given by

$$\begin{aligned} I_{hkl} &= \left(\frac{1}{2\mu}\right)\left(\frac{I_0}{\omega}\right)\left(\frac{e^2}{mc^2}\right)^2(N^2\lambda^3)\,(J\,L.P.)|F^T_{hkl}|^2 \qquad (4.49)\\ &= S_1(JL.P.)|F^T_{hkl}|^2, \end{aligned}$$

where the constant S_1 is called the scale factor. In (4.48) and (4.49), μ is the linear absorption coefficient, I_0 the intensity of the incident beam, ω the angular velocity of the crystal and N is the number of atoms in unit volume of the crystal. The other symbols have been defined in Sect. 4.5.4.

Some differences between (4.48) and (4.49) may be noted. The absorption coefficient appears only in (4.49). The term (e^2/mc^2) appears as it is in (4.48), but in the square form in (4.49). Again, the powers of N and λ are different in the two cases. But, by far, the most important difference is in the dependence of the intensity on the structure factor in the two theories; in (4.48), I is proportional to $|F|$ whereas in (4.49) it is proportional to $|F|^2$. The intensity is much more for an imperfect crystal than for a perfect crystal.

Part B: Electron Diffraction

4.6 Principle of Electron Diffraction

In 1912, De Broglie proposed the concept of "matter waves" according to which particles behave like waves in certain situations. The De Broglie waves associated with particles have a wavelength λ given by

$$\lambda = h/mv, \tag{4.50}$$

where h is Planck's constant, m the mass of the particle and v its velocity.

Let us apply this idea to an electron. If the electron is accelerated through a voltage V, its kinetic energy is

$$\frac{1}{2}m_0v^2 = \frac{Ve}{300}. \tag{4.51}$$

m_0 now is the rest mass of the electron and e its charge. Substituting (4.51) in (4.50), we get

$$\lambda = \frac{h}{(m_0e)^{1/2}}\left(\frac{150}{V}\right)^{1/2} = \left(\frac{150h^2}{m_0eV}\right)^{1/2}. \tag{4.52}$$

On substitution of the values of h, m_0 and e, (4.52) becomes

$$\lambda = \left(\frac{150}{V}\right)^{1/2} \tag{4.53}$$

with λ in Å. If higher voltages are applied to accelerate the electron, the relativistic expression has to be used in (4.51). The wavelength is now given by

$$\lambda = \left(\frac{150h^2}{m_0eV}\right)^{1/2}\left(1 + \frac{eV}{1{,}200m_0c^2}\right)^{-1}. \tag{4.54}$$

Equations (4.53) and (4.54) give values of 1, 0.1227 and 0.0856Å for λ at voltages of 150 V, 10 kV and 20 kV, respectively. These wavelengths are of the same order as lattice spacings in crystals and Bragg diffraction of these waves is, in principle, possible. Attempts were undertaken for the observation of diffraction of electrons by crystals.

4.7 Experiments

4.7.1 Davisson and Germer's Experiment

The diffraction of electrons was discovered by Davisson and Germer [4.26] and independently by Thomson [4.27]. The experimental set-up used by Davisson and

Box 4.3

Clinton Joseph Davisson, an American physicist, is known for his discovery of electron diffraction. Davisson along with Lester Germer obtained the electron diffraction pattern of a crystal. The experiment confirmed the de Broglie hypothesis and established the wave-particle duality of electrons. Davisson was awarded the Nobel Prize for physics in 1937.

C.J. Davisson (1881–1958)

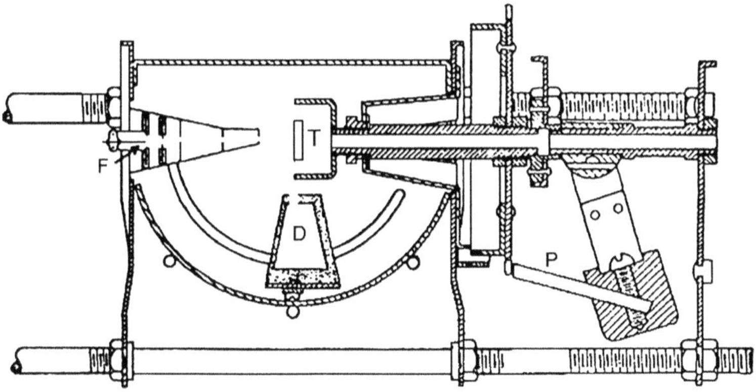

Fig. 4.42 Davisson and Germer's set-up to observe electron diffraction

Germer (see Box 4.3) is shown in Fig. 4.42. Electrons from a filament F accelerated by voltages in the range 65–600 V were used. The target T was an oriented nickel single crystal with the (111) face exposed to the electron beam. The diffracted beam was collected by the Faraday detector D. By rotating the crystal and moving the detector on a circular arm, different orders of the (111) reflection were recorded (Fig. 4.43).

The experiment clearly demonstrated the diffraction of electron waves. However, the method was of limited use as the wavelengths produced at low voltages had low penetration into the crystal surface which led to weak intensities.

4.7.2 Thomson's Experiment

Thomson [4.27] (Box 4.4) discovered electron diffraction independently of Davisson and Germer. His experimental set-up is shown in Fig. 4.44. The electrons were produced in discharge tube A operated at high voltages of 10–60 kV. After passing

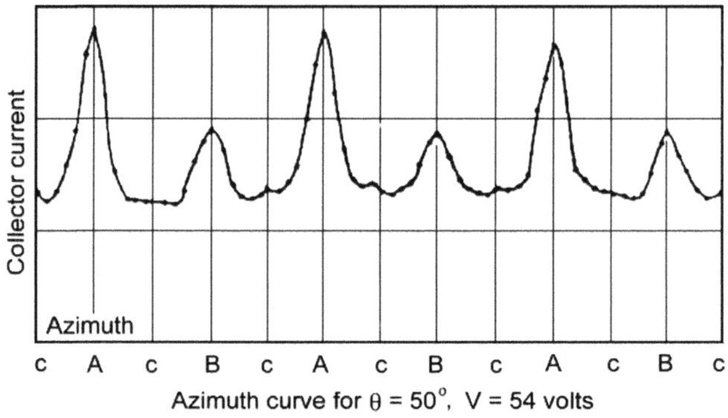

Fig. 4.43 Electron diffraction reflections of different orders from the (111) face of Ni crystal

Box 4.4

Sir George Paget Thomson, an English physicist, independently discovered electron diffraction. He shared the Nobel Prize in Physics with Davisson in 1937.

G.P. Thomson (1892–1975)

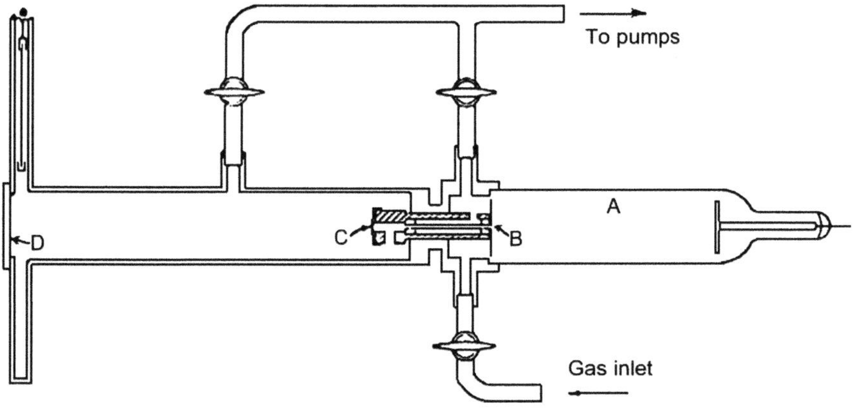

Fig. 4.44 Thomson's apparatus to observe electron diffraction

through a narrow tube B, they were diverted towards the sample C which was a gold foil. The entire chamber was evacuated to prevent the electrons from being absorbed by air. The diffracted beam could be observed on a fluorescent screen D or recorded on a photographic film. The diffraction pattern (Fig. 4.45) revealed circular rings.

Fig. 4.45 Electron diffraction pattern from a gold foil

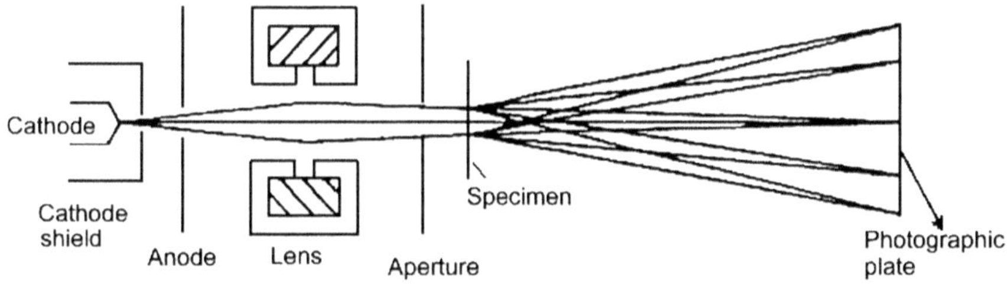

Fig. 4.46 Schematic of a modern electron diffraction set-up

The discovery of electron diffraction is of great importance because it proved beyond doubt the existence of De Broglie waves.

4.7.3 *Refinements*

Electron diffraction equipment (Fig. 4.46) is supplied as an attachment to electron microscopes. The source of electrons is a hot-filament electron gun. A high DC voltage is applied to an evacuated chamber. There is facility to introduce and remove a sample. The most important improvement is the manipulation (collimation and focussing) of the electron beam by electrostatic and magnetic fields called "lenses". Due to the short De Broglie wavelength, the entire diffraction pattern is confined within an angular range of $\sim 10^\circ$. The resolution is enhanced by having a long specimen-to-screen distance (~ 50 cm).

4.8 Scattering of Electron Waves

Let us revert back to the scattering of X-rays by an atom. If we denote the amplitude of the incident wave by A' and that of the wave scattered by the atom by A'', then from (4.39), we have

$$A''/A'(\text{X}-\text{rays}) = (1/r) \times 2.82 \times 10^{-13} f. \tag{4.55}$$

While X-rays are scattered only by electrons, the electrons are scattered by the electrons in the atom as well as by the nucleus. Hence both f (the atomic scattering factor for X-rays) and the atomic number Z will be involved in the scattering of electrons. Born showed that the ratio A''/A' in the case of electron waves is

$$A''/A'(\text{electrons}) = (1/r) \times 2.82 \times 10^{-10} (Z - f) \left[\frac{\lambda}{\sin \theta} \times 10^{8} \right]^{2}. \tag{4.56}$$

There are two important differences between (4.55) and (4.56). Firstly, the term f in (4.55) is replaced by $(Z-f)$ in (4.56). Secondly, the numerical coefficient 10^{-13} in (4.55) is replaced by 10^{-10} – an enhancement by a factor 10^3; the scattered electron wave is stronger than the X-ray wave by this factor. To emphasise this aspect, the ratios A''/A' for X-rays and electrons are compared in Fig. 4.47 as a function of $(\sin \theta/\lambda)$. It may be noted that to bring the two ratios on the same scale, the values of A''/A' for electrons have been reduced by 10^4.

4.9 Intensity of Diffracted Electrons

The intensity $(I_{hkl})_{\text{electron}}$ of a beam of electrons diffracted by a crystal is given by

$$(I_{hkl})_{\text{electrons}} = \frac{S_2 J \, |F_{hkl}|^2_{\text{electron}}}{\sin^2 \theta}. \tag{4.57}$$

Here S_2 is the scale factor. The polarization factor $(1 + \cos^2 \theta)/2$ which occurs in X-rays is absent. Further $|F_{hkl}|_{\text{electron}}$ is a function of the atomic scattering factor for electrons. As discussed in the preceding section, this is much larger than its X-ray counterpart. Hence the intensity of the diffracted beam is very high so that a tiny speck of a crystal yields an intense diffraction pattern in an exposure of a few seconds. On the other hand, electrons are absorbed easily so that they can penetrate only a few surface layers ($\sim$ 3,000Å thick) of the crystal. As a result, the electron diffraction patterns from single crystals reflect their two-dimensional structure.

Part C: Neutron Diffraction

4.10 Principle of Neutron Diffraction

Neutrons are generally emitted with high velocities. This fact, together with the large mass of the neutron, makes the De Broglie wavelength very short. It was pointed out by Elsasser [4.28] that thermal neutrons pertaining to 20°C have energy

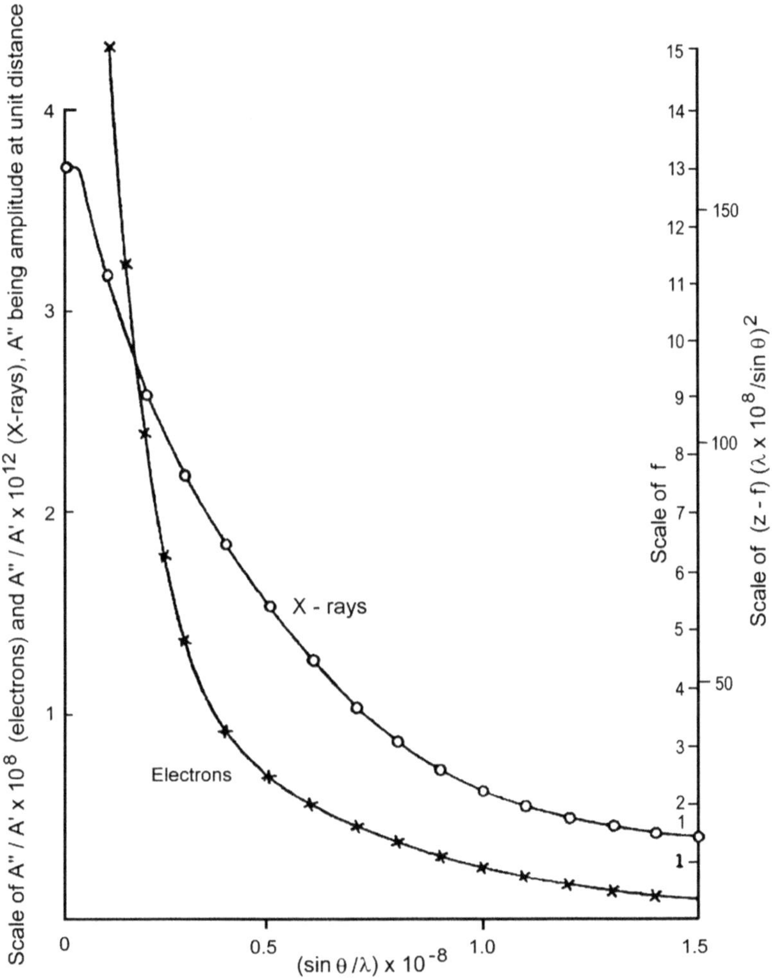

Fig. 4.47 Atomic scattering factors for X-rays and electrons for aluminium

$\sim$0.025 eV and velocity $\sim$ 2.2 km/s. Their De Broglie wavelength is $\sim$ 1.8Å which is comparable with lattice spacings. Such waves should be amenable to diffraction by crystals.

4.11 Experiments

4.11.1 Experiment by Halbanjun and Preiswerk

Following Elssasser's suggestion, an attempt to observe the diffraction of neutrons by crystals was made by Halbanjun and Preiswerk [4.29]. Their set-up is shown in

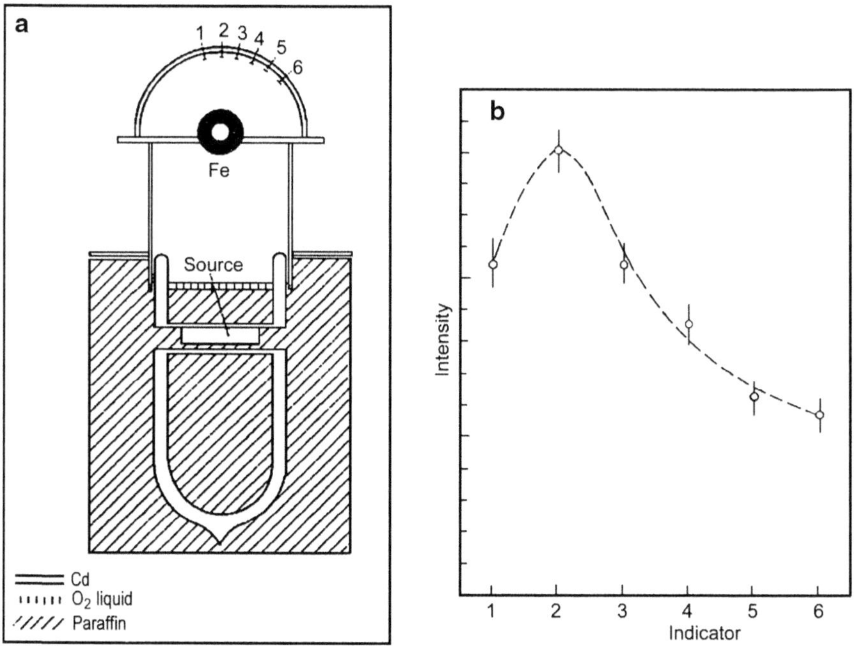

Fig. 4.48 (**a**) Set-up to observe diffraction of neutrons, (**b**) angular intensity of scattered neutrons

Fig. 4.48a. The neutrons were from a Ra–Be source. A polycrystalline sample of iron was used. Dysprosium detectors were arranged around the sample at intervals of 13°. The angular distribution of the intensity of scattered neutrons (Fig. 4.48b) agreed approximately with that calculated by Elsasser assuming that the sample diffracted neutrons.

4.11.2 Experiment by Mitchell and Powers

Mitchell and Powers [4.30] conducted a more direct and convincing experiment on neutron diffraction almost simultaneously and independently of Halbanjun and Preiswerk. Their set-up is shown in Fig. 4.49. The Ra–Be source was cladded with cadmium; the neutrons had a De Broglie wavelength of 1.6Å. Several MgO single crystals were arranged in a ring. MgO has a lattice constant of $\sim$ 4Å and the (111) plane would diffract the De Broglie waves at a precalculated Bragg angle of 22°. The MgO crystals were so arranged that rays diffracted by them would converge at a point. An ionisation chamber with BF_3 was placed at this point to detect the neutrons. Absorbers were placed such that they stopped all neutrons except those travelling towards the MgO crystals. The detector recorded the neutrons scattered by the MgO crystals clearly demonstrating the diffraction effect.

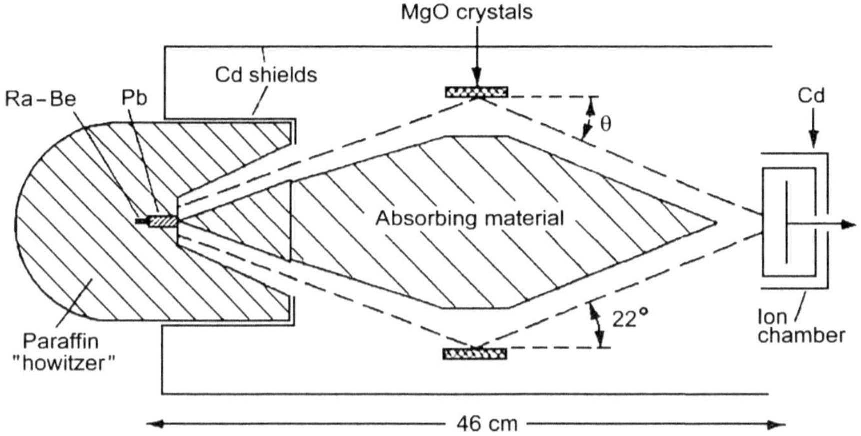

Fig. 4.49 Set-up of Mitchell and Powers (1936) to observe neutron diffraction

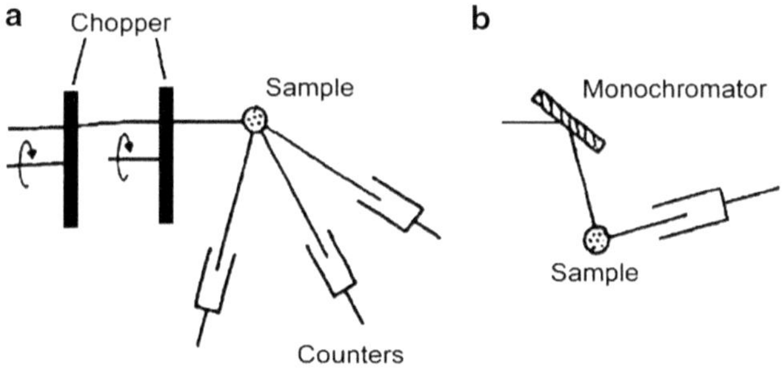

Fig. 4.50 Methods of monochromatisation of neutrons: (**a**) chopper, (**b**) crystal diffraction

4.11.3 Refinements

The weak Ra–Be neutron sources are no longer used. Instead the modern source of neutrons is a nuclear reactor. A nuclear rector provides an intense neutron beam. At the same time reactor-generated neutrons are highly polychromatic and monochromatisation becomes important.

Monochromatisation of neutrons can be affected by passing polychromatic neutrons through two rotating discs made of an absorbing material and having "holes", i.e. windows of non-absorbing materials. These discs are placed a fixed distance apart. Out of the neutrons passing through the first disc only those neutrons will pass through a hole in the second which travel the distance between the discs in the time taken by the hole to be in line of the beam; this time depends on the speed of rotation. A more commonly adopted method of monochromatisation is to allow the neutron beam to be diffracted by a crystal. The diffracted beam will contain neutrons of a given wavelength. This beam can then be used to study diffraction from the sample. These two methods are shown in Fig. 4.50.

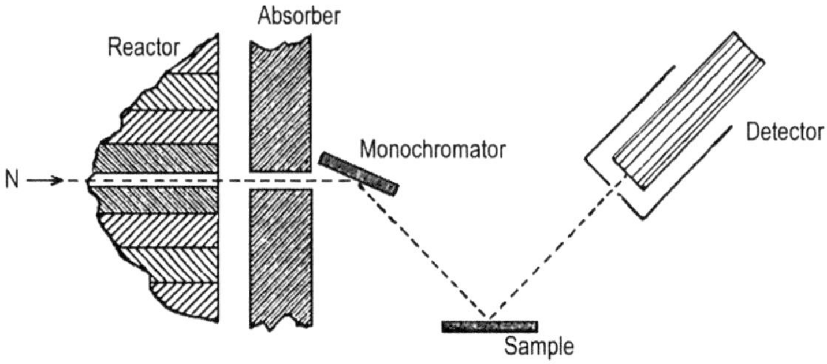

Fig. 4.51 Configuration of main parts of a neutron diffractometer

Table 4.5 Neutron scattering amplitudes $b(10^{-12}$ cm)

Nucleus	*b*	Nucleus	*b*	Nucleus	*B*
H	−0.378	Al	0.35	Cu	0.75
D	0.65	K	0.35	Zn	0.59
Li	−0.18	Ca	0.49	Zr	0.62
Be	0.774	Ti	−0.38	Nb	0.69
C	0.661	V	−0.05	Ag	0.61
N	0.940	Mn	−0.36	I	0.52
O	0.58	Fe	0.96	Hf	0.88
Na	0.351	Co	0.28	Pb	0.96
Mg	0.54	Ni	1.03		

The configuration of the various parts in a neutron diffractometer is shown in Fig. 4.51. The beam from a reactor, after collimation by slits in an absorbing material and after monochromatisation reaches the sample. The beam diffracted by the sample is received in a BF_3 counter. Compared to an X-ray diffractometer and an electron diffractometer, a neutron diffractometer is massive and occupies more space.

4.12 Nuclear Scattering of Neutrons

Neutrons are scattered by atoms by the mechanism of neutron–nucleus interaction. Neutron scattering is expressed in terms of the neutron scattering amplitude (or scattering length); the units of b are 10^{-12} cm. We shall not go into the theory of neutron scattering; instead, we shall quote some results. The neutron scattering amplitudes of several elements were experimentally determined by Fermi and Marshall [4.31] and Shull and Wollan [4.32]. The values for some nuclei are given in Table 4.5 [4.32] and also shown in Figs. 4.52 and 4.53. We shall discuss some features in these values.

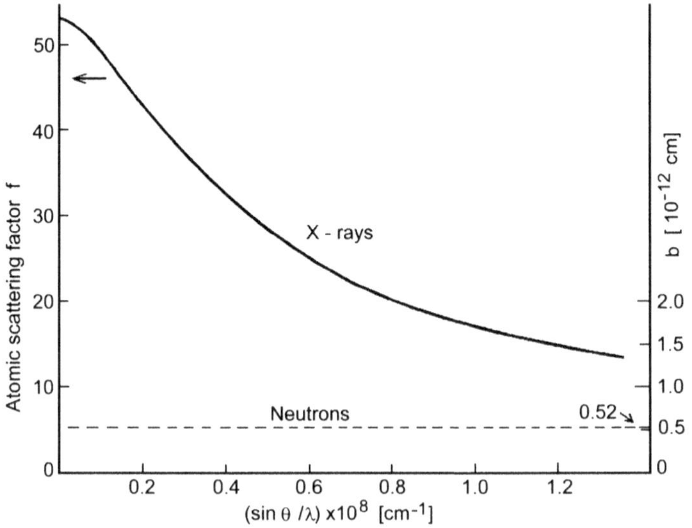

Fig. 4.52 X-ray atomic scattering factor f and neutron neutron scattering amplitude b of iodine

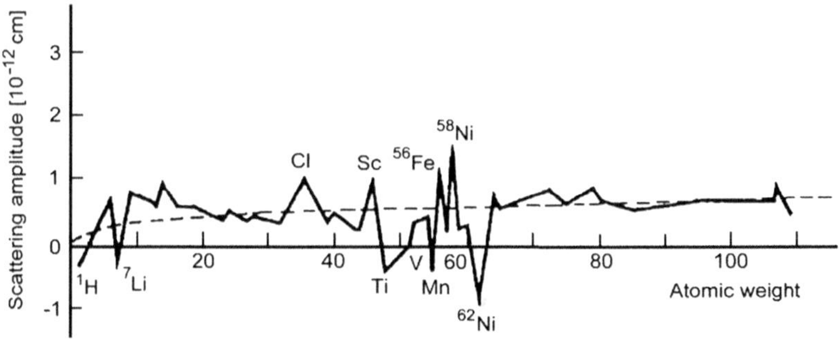

Fig. 4.53 Neutron scattering amplitudes of atoms as a function of atomic weight

1. The neutron scattering amplitude of iodine is shown in Fig. 4.52 along with the atomic scattering factor for X-rays. The main difference between the two is that whereas the X-ray atomic scattering factor varies with $(\sin\theta/\lambda)$, the neutron scattering amplitude is isotropic.
2. The neutron scattering amplitudes of several elements are plotted in Fig. 4.53 as a function of the atomic weight. The X-ray atomic scattering increases progressively from lighter to heavier atoms. In fact, over the periodic table, f increases by a factor of nearly 100. On the other hand, the neutron scattering amplitudes do not show any systematic variation. The variation in the values is within a factor of 2. The average value is 0.6×10^{-12} cm.
3. For some elements, the amplitudes are negative, i.e. these elements scatter neutrons with a phase change of 180°.

4. The amplitude for H is -0.38×10^{-12} cm. This is comparable with the amplitudes of several other heavier nuclei, e.g. 0.66 for C, 0.49 for Ca, 0.59 for Zn. This is very important when we recall that the atomic scattering factor for X-rays is the least for H making H a poor X-ray scatterer.
5. Some neighbouring elements in the periodic table like Fe, Co and Ni, Mn have nearly equal X-ray atomic scattering factors making it difficult to differentiate them in the structure of their alloys. In contrast, the neutron scattering amplitudes of these elements are sufficiently different to permit them to be distinguished.

4.13 Intensity of Diffracted Neutrons

The expression for the intensity I_{hkl} of a diffracted neutron beam is

$$(I_{hkl})_{\text{neutron}} = \frac{S_3 A J |F_{hkl}|^2_{\text{neutron}}}{\sin\theta \sin 2\theta}, \tag{4.58}$$

where S_3 is the scale factor, A the absorption factor and J the multiplicity factor. The polarization factor $(1 + \cos^2 2\theta)/2$ is absent as in the case for electron diffraction. The structure factor $(F_{hkl})_{\text{neutron}}$ is given by

$$(F_{hkl})_{\text{neutron}} = \sum_n b_n e^{2\pi i(hx_n + ky_n + lz_n)}, \tag{4.59}$$

where the b_n's are neutron scattering amplitudes.

4.14 Magnetic Scattering of Neutrons

Besides the neutron–nucleus scattering, another mechanism of scattering of neutrons was pointed out by Bloch [4.33]. This is the scattering due to the interaction between the magnetic moment of the neutron and the magnetic moment of some atoms with non-zero spin. This scattering shows up when the spins of these atoms take up ordered orientations as in ferro-, antiferro and ferromagnetic materials. This aspect makes neutrons an important tool in exploring "magnetic structures".

Problems

1. In a Laue photograph of NaCl, the spot due to diffraction of wavelength $\lambda = 0.416$Å by the (151) plane occurs at Bragg angle 11.05°. X-rays of another wavelength are diffracted by the (131) plane at Bragg angle 17.26°. What is the wavelength?

2. In the Debye–Scherrer photograph of a cubic crystal [Sr $(NO_3)_2$], the (200) plane diffracts at 17.20°. At what angle will the (222) plane diffract?
3. The Cu $K\alpha$ radiation consists of two wavelengths $K\alpha_1 = 1.5405$ Å and $K\alpha_2 =$ 1.5433 Å. The reflections due to these wavelengths overlap at low Bragg angles but are well separated at high Bragg angles. Explain this phenomenon.
4. Graphite is hexagonal with four atoms per cell in positions 000; $\frac{1}{3}\frac{2}{3}0$; $00\frac{1}{2}$; $\frac{2}{3}\frac{1}{3}\frac{1}{2}$. Show that the structure factor is given by

$$F = 4f\cos^2\pi(h+2k)/3 \quad \text{for } l \text{ even}$$

$$F = i2f\sin 2\pi(h+2k)/3 \quad \text{for } l \text{ odd}$$

5. Assuming the NaCl crystal to be imperfect, calculate the quantity $(I_{hkl}\omega/I_0)$ for the 200 reflection given; $N = 5.58 \times 10^{21}$, $\lambda = 0.711$Å, $F_{200} = 87.5$, $L.P. =$ 0.25 and $\mu = 17.7$.

References

4.1. R.W. James, *The Optical Principles of the Diffraction of X-rays* (G. Bell, London, 1962)
4.2. E.F. Kaelble, *Handbook of X-rays* (McGraw Hill, New York, 1967)
4.3. H.P. Klug, L.E. Alexander, *X-ray Diffraction Procedures* (Wiley, New York, 1967)
4.4. B.E. Warren, *X-ray Diffraction* (Dover, New York, 1990)
4.5. S.K. Chatterjee, *X-ray Diffraction* (Prentice Hall, New Delhi, 1999)
4.6. W.T. Sproull, *X-rays in Practice* (McGraw-Hill, New York, 1946)
4.7. W.L. Bragg, *The Crystalline State* (G. Bell, London, 1966)
4.8. C.G. Shull, E.O. Wollan, Solid State Phys. **2**, 137 (1956)
4.9. G.E. Bacon, *X-ray and Neutron Diffraction* (Pergamon, New York, 1966)
4.10. H. Dachs, *Neutron Diffraction* (Springer-Verlag, Berlin, 1978)
4.11. W. Friedrich, P. Knipping, M. Laue, in *Proceedings of the Bavarian Academy of Sciences*, 1912, p. 303
4.12. J.L. Amoros, M.J. Buerger, M.C. Amoros, *The Laue Method* (Academic Press, New York, 1975)
4.13. M. Laue, Proc. Bavarian Acad. Sci., 368 (1912)
4.14. W.L. Bragg, Proc. Cambridge Phil. Soc. **17**, 43 (1913)
4.15. C.G. Darwin, Philos. Mag. **27**, 315 (1914)
4.16. V.T. Deshpande, D.B. Sirdeshmukh, V.M. Mudholker, Acta Cryst. **12**, 257 (1959)
4.17. K. Suryanarayana, D.B. Sirdeshmukh, in *Proc. National Seminar on Cryst.* Calcutta, 1981
4.18. K. Srinivas, M. Ateequddin, D.B. Sirdeshmukh, Pramana **28**, 281 (1987)
4.19. H.S. Peiser, H.P. Rooksby, A.J.C. Wilson, *X-ray Diffraction by Polycrystalline Materials* (Institute of Physics, London, 1955)
4.20. J.J. Thomson, *Conduction of Electricity Through Gases* (Cambridge University Press, London, 1903)
4.21. P. Debye, Ann. Phys. **43**, 49 (1914)
4.22. I. Waller, Z. Phys. **17**, 398 (1923)
4.23. C.G. Darwin, Phil. Mag. **27**, 315, 675 (1914)
4.24. P.P. Ewald, Ann. Phys. **49**, 1, 117 (1916)
4.25. P.P. Ewald, Ann. Phys. **54**, 519 (1917)
4.26. C.J. Davisson, L.G. Germer, Phys. Rev. **30**, 705 (1927)

4.27. G.P. Thomson, Proc. Roy. Soc. (Lond.) **117**, 630 (1928)
4.28. W.M. Elsasser, Comp. Rend. **202**, 1029 (1936)
4.29. M. Halbanjun, P. Preiswerk, Comp. Rend. **203**, 73 (1936)
4.30. D.P. Mitchell, P.N. Powers, Phys. Rev. **50**, 486 (1936)
4.31. E. Fermi, L. Marshall, Phys. Rev. **71**, 666 (1947)
4.32. C.G. Shull, E.O. Wollan, Phys. Rev. **81**, 527 (1951)
4.33. F. Bloch, Phys. Rev. **50**, 259 (1936)

Chapter 5
Principles of Crystal Structure Determination

5.1 Introduction

The principles and experimental methods of diffraction of radiation by crystals were discussed in the preceding chapter. In this chapter, we shall discuss the application of these principles to the determination of crystal structure. By "structure" we mean (1) the shape of the unit cell, i.e. the crystal system, (2) the size of the unit cell, i.e. the unit cell parameters, (3) the contents of the unit cell, i.e. the number of formula units per unit cell, (4) the symmetry of the internal atomic arrangement, i.e. the space group and (5) the atomic coordinates.

Indicative of the importance of this subject is the vastness of literature in the form of reference books, text books, reviews and research publications. We shall cite only some select sources: Lipson and Cochran [5.1], Peiser et al [5.2], Buerger [5.3], Kaelble [5.4], Klug and Alexander [5.5], Carpenter [5.6], Woolfson [5.7], Ladd and Palmer [5.8] and Massa [5.9].

5.2 The Unit Cell

5.2.1 Crystal System

In principle, it is possible to deduce the crystal system from a powder diffraction photograph from the fact that the $\sin^2\theta$ values show existence of ratios among themselves which are characteristic of the crystal system.

For the cubic system, we have

$$\sin^2\theta = A(h^2 + k^2 + l^2) \quad \text{and} \quad A = \lambda^2/4a^2. \tag{5.1}$$

Here θ is the Bragg angle, λ the wavelength, a the lattice constant and h, k, l the Miller indices. The factor $(h^2 + k^2 + l^2)$ is a sum of three whole numbers. By

D.B. Sirdeshmukh et al., *Atomistic Properties of Solids*, Springer Series in Materials Science 147, DOI 10.1007/978-3-642-19971-4_5,

considering possible combination of values for h, k, l it can be verified that the $\sin^2\theta$ values will be in the ratio 1:2:3:4:5:6:8:9, etc. The values 7, 15, 23 will be absent as they cannot be expressed as a sum of squares of three numbers. Similarly, for the tetragonal system we have

$$\sin^2\theta = A(h^2 + k^2) + Bl^2 \tag{5.2}$$

with $A = \lambda^2/4a^2$ and $B = \lambda^2/4c^2$. Here a and c are the two lattice constants. For the (h,k,0) type of reflections, the $\sin^2\theta$ values will be in the ratio 1:2:4:5:8:9:10:13:16:17, etc. Note the presence of the numbers 1, 2, 4, 8, 16 in these ratios. These numbers increase by a factor of 2; this is characteristic of the tetragonal system. Again, for the hexagonal system, we have

$$\sin^2\theta = A(h^2 + hk + k^2) + Bl^2 \tag{5.3}$$

with $A = \lambda^2/3a^2$ and $B = \lambda^2/4c^2$. Here a and c are the lattice parameters of the hexagonal crystals. In this case, for (hk0) type of reflections, the $\sin^2\theta$ values will be in the ratio 1:3:4:7:9:12:13:16:19:21, etc. The ratio 3 between the first two and between several other pairs is characteristic of the hexagonal system. Thus by an inspection of the $\sin^2\theta$ values, the cubic, tetragonal and hexagonal systems can be identified. The low symmetry systems can also be identified by a close examination of the $\sin^2\theta$ values but the procedure is understandably more complicated [5.10–5.12].

However, single crystal methods are preferred for the determination of crystal system. As discussed in the preceding chapter, if the crystal is properly oriented, the rotation photograph shows layer lines, the height of which is related to the unit cell parameter along the axis of rotation; a set of rotation photographs taken with different crystal axes will indicate the crystal system. The Weissenberg and precession cameras provide the information even more easily.

5.2.2 Unit Cell Parameters

Equations (5.1)–(5.3) contain the principle of the determination of unit cell parameters; the expression for $\sin^2\theta$ includes all the unit cell parameters of a crystal. Once the values of $\sin^2\theta$ are known from measurements and the various reflections are indexed, the unit cell parameters can be calculated by solving one or more equations of the above types. The minimum number of reflections needed for determining the cell parameters is 1 for cubic, 2 for tetragonal, 3 for orthorhombic, 4 for monoclinic and 6 for the triclinic systems.

For accurate determination, the following conditions are desirable: (1) large number of reflections, (2) larger camera radius for accurate determination of θ, (3) reflections at high Bragg angles and (4) methods for correction of systematic errors. For these reasons, the powder technique is preferred to single crystal techniques in

Table 5.1 Axial ratios for some crystals

Crystal	$a:b:c$		Ref.
	Optical measurements	X-ray diffraction	
$CuSO_4 \cdot 5H_2O$	0.5715:1:0.5575	0.571:1:0.558	[5.16]
Fichtelite ($C_{18}H_{22}$)	1.433:1:1.756	1.435:1:1.758	[5.17]
K_2MnO_4	0.7570:1:0.568	0.741:1:0.570	[5.18]

accurate determination of unit cell parameters. Detailed treatment of this topic can be found in [5.2, 5.4, 5.5, 5.13].

5.2.3 *Optical Measurements*

Long before the discovery of X-ray diffraction, mineralogists developed the procedure for determining interfacial angles and axial lengths from measurements on well-faceted minerals. These optical measurements were extended to crystals grown from solutions. While angles between the principal faces could be directly determined, the axial lengths could be obtained only as ratios $a{:}b{:}c$; by convention b is taken as unity. Thousands of crystals were examined in this way by Mitscherlich in 1831 (quoted by Mellor [5.14] and Groth [5.15]). Some results are given in Table 5.1; subsequent results from X-ray diffraction are included for comparison. The accuracy of the optical measurements deserves appreciation.

5.2.4 *Formula Units per Unit Cell*

From the unit cell parameters, the volume V of the unit cell is obtained ($V = a^3$, a^2c, abc, etc). When this is combined with the molecular weight M and the density ρ, the number of formula units (N) in the unit cell is obtained from the relation

$$N = \frac{\rho\,(\mathrm{g\,cm^{-3}})\; V\,(\mathrm{A^3})}{1.660\, M\,(\mathrm{awu})}. \qquad (5.4)$$

The density is measured by conventional methods using the specific gravity bottle or the flotation method. Occasionally, particularly when the sample is tiny, the centrifuge is used. When the density of the liquid (or liquid mixture) matches the density of the crystal sample, it remains static in the centrifuge [5.19]. Values of N for some crystals are given in Table 5.2. The calculated values are close to, but not exactly equal to, whole numbers. These deviations could be due to errors in V or ρ. Since N has necessarily to be a whole number, its calculated value is rounded off to the nearest whole number. Values of 2 or 4 are frequently observed. However, other values like 1, 3, 6, 8, 10, 16 and 32 have also been obtained [5.21].

Table 5.2 Number of formula units (N) per unit cell

Crystal	Volume V [Å^3]	Densityρ ρ [g cm^{-3}]	Molecular weight M [awu]	N Calculated	True	Ref.
N-methyl acetamide	453.6	1.02	73.095	3.81	4	[5.4]
Napthalene	359.2	1.152	128.2	1.94	2	[5.8]
FeS_2	158.1	4.87	120.0	3.87	4	[5.8]
Dimethyl thallium bromide $Tl(CH_3)_2Br$	275.3	3.790	314.35	2.02	2	[5.20]

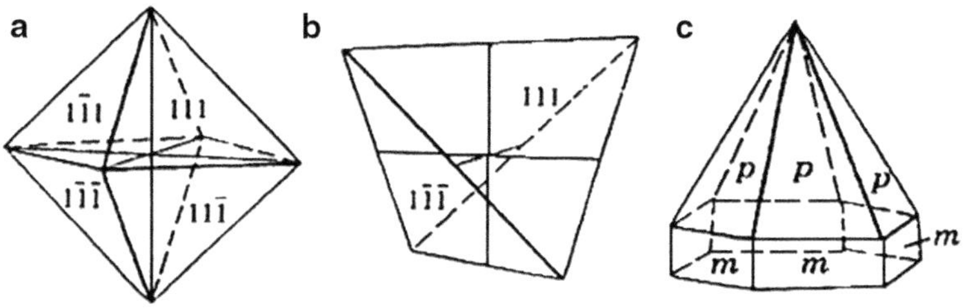

Fig. 5.1 External forms of crystals: (**a**) holohedral (alum), (**b**) hemihedral (zinc blende), (**c**) hemimorphic (zincite)

5.3 Space Group Determination

The space group represents the total symmetry of the atomic arrangement in the crystal. This symmetry consists of point symmetry elements (mirror planes, rotation axes, centre of symmetry, rotation–inversion axes) as well as translational symmetry elements (centring of faces, glide planes, screw axes). We shall now discuss the strategies for identifying the different symmetry elements.

5.3.1 Point Symmetry Elements

Much information can be obtained from the external forms of well-faceted crystals. Mirror planes and rotation axes can be identified. Some crystal forms are shown in Fig. 5.1. An alum crystal (Fig. 5.1a) shows all the eight faces of an octahedron. This is called the holohedral form. When a crystal shows only four of these faces as in (b), it is called the hemihedral form; this is the form of a ZnS (zinc blende) crystal. In another form (c), half the planes of the holohedral form are grouped at one end of an axis of symmetry; this happens to be the form of zincite. It is called the hemimorphic form. The two latter forms lack a centre of symmetry.

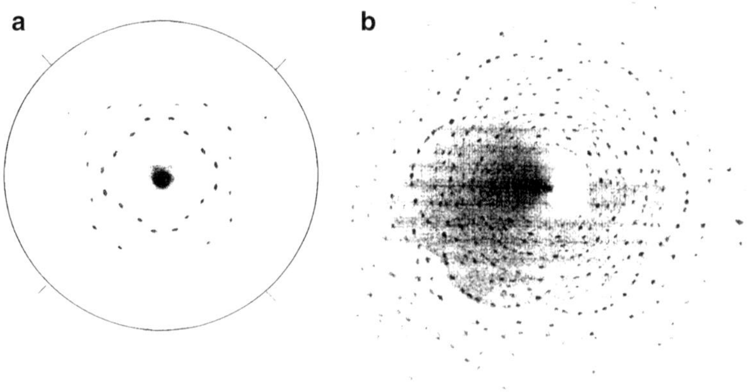

Fig. 5.2 Laue photographs: (**a**) zinc blende, incident radiation parallel to (100), (**b**) beryl, incident radiation parallel to (001)

Information about the planes and axes of symmetry can also be obtained from X-ray Laue photographs. Laue photographs of zinc blende and beryl are shown in Fig. 5.2. The zinc blende photograph shows a fourfold rotation axis and mirror planes whereas the beryl photograph shows mirror planes and a sixfold rotation axis. A serious limitation of the Laue method (and any X-ray method for that matter) is that it suppresses the lack of inversion symmetry, i.e. it shows the presence of a centre of symmetry whether it exists in the crystal or not.

The centre of symmetry is an important symmetry element. In fact the crystalline world can be divided in two groups, centrosymmetric and noncentrosymmetric. To determine whether a centre of symmetry exists or not, recourse is often taken to optical and electrical properties. Thus, optical activity and second harmonic generation are sure tests of lack of centre of symmetry. Similarly, the presence of pyro- and piezoelectricity indicates lack of centre of symmetry.

5.3.2 *Translational Symmetry Elements*

The methods discussed in the preceding section fail to give any information regarding the translational features like centring of lattices, glide planes and screw axes. For this, we have to turn to the effect of these symmetry features on the intensities of X-ray reflections. We shall show that specific symmetry features reduce the intensity of specific reflections to zero. This effect is called systematic absence or space group extinction. The same effect may be expressed by indicating which class of reflections are not affected by symmetry; these conditions are then called limiting conditions. We shall now consider some examples.

5.3.2.1 Centring of (001) Faces

In this case, for every atom at x_n, y_n, z_n, there is an identical atom at $x_n + 1/2$, $y_n + 1/2$, z_n. The structure factor (4.40) may now be written as

$$\begin{aligned} F_{hkl} &= \sum_{n=1}^{N} f_n \mathrm{e}^{2\pi i(hx_n+ky_n+lz_n)} \\ &= \sum_{n=1}^{N/2} \left[f_n \mathrm{e}^{2\pi i(hx_n+ky_n+lz_n)} + f_n \mathrm{e}^{2\pi i\{h(x_n+\frac{1}{2})+k(y_n+\frac{1}{2})+lz_n\}} \right] \\ &= \sum_{n=1}^{N/2} \left[f_n \mathrm{e}^{2\pi i(hx_n+ky_n+lz_n)} \{1 + \mathrm{e}^{2\pi i(h+k)/2}\} \right] \end{aligned} \tag{5.5}$$

The last factor in (5.5) has only two possible values, 2 for $(h + k)$ even and 0 for $(h + k)$ odd. Thus whatever the values of x_n, y_n, z_n, the reflections with $(h + k)$ odd are absent. Conversely, if reflections with $(h + k)$ odd are systematically absent, it is inferred that the (001) faces are centred. Similarly, if the (100) faces are centred, the $(k + l)$ odd reflections will be absent; also, if (010) faces are centred, the reflections with $(l + h)$ odd will be absent.

5.3.2.2 Face-Centred Cubic Lattice

In this lattice, for every atom at x_n, y_n, z_n, there are three other identical atoms at $x_n + 1/2$, $y_n + 1/2$, z_n; x_n, $y_n + 1/2$, $z_n + 1/2$ and $x_n + 1/2$, y_n, $z_n + 1/2$. Extending the calculation in the earlier case, it can be shown that the structure factor equals $4f$ for h, k, l all even or all odd and 0 for mixed values of h, k, l. Thus, the presence of reflections with all even or all odd h, k, l indicates that the lattice is face centred.

5.3.2.3 Glide Plane *c*

Let us consider a glide plane passing through the origin and normal to the b-axis with a translation of 1/2 along the z-direction. In this case, for every atom at x_n, y_n, z_n there is an identical atom at $x_n, \bar{y}_n, z_n + \frac{1}{2}$. Taking pairs of such atoms, the structure factor may be written as

$$\begin{aligned} F_{hkl} &= \sum_{n=1}^{N} f_n \mathrm{e}^{2\pi i(hx_n+ky_n+lz_n)} \\ &= \sum_{n=1}^{N/2} \left[f_n \mathrm{e}^{2\pi i(hx_n+ky_n+lz_n)} + f_n \mathrm{e}^{2\pi i\{hx_n-ky_n+l(z_n+\frac{1}{2})\}} \right]. \end{aligned} \tag{5.6}$$

For $k = 0$, (5.6) becomes

$$F_{hkl} = \sum_{n=1}^{N/2} \left[f_n e^{2\pi i(hx_n + lz_n)} \{1 + e^{2\pi i \frac{l}{2}}\} \right]. \tag{5.7}$$

It can be seen that F_{h0l} will be zero for l odd. Thus, the systematic absence of $h0l$ reflections with l odd indicates the presence of a glide plane normal to the b-axis with a glide in the c-direction.

5.3.2.4 Screw Axis

To complete the discussion, let us consider the effect of a twofold screw axis parallel to the y-axis with a translation of 1/2. Now, for every atom at x_n, y_n, z_n, there is an identical atom at $\bar{x}_n$, $y_n + \frac{1}{2}$, $\bar{z}_n$. Taking pairs of such atoms, the structure factor may be written as

$$\begin{aligned} F_{hkl} &= \sum_{n=1}^{N} f_n e^{2\pi i(hx_n + ky_n + lz_n)} \\ &= \sum_{n=1}^{N/2} \left[f_n e^{2\pi i(hx_n + ky_n + lz_n)} + f_n e^{2\pi i\{-hx_n + k(y_n + \frac{1}{2}) - lz_n\}} \right]. \end{aligned} \tag{5.8}$$

For $h = l = 0$, (5.8) becomes

$$F_{hkl} = \sum_{n=1}^{N/2} \left[f_n e^{2\pi i(ky_n)} \{1 + e^{2\pi i \frac{k}{2}}\} \right]. \tag{5.9}$$

Again the last factor becomes zero for k odd. Hence, systematic absence of reflections $(0k0)$ with k odd indicates the presence of a twofold screw axis along the y-direction.

In this manner, the effect of various translational symmetry elements with all possible orientations has been worked out leading to rules of systematic absences (or systematic nonextinctions). These rules are given in Table 5.3 (Raman and Katz [5.22]).

5.3.3 *Space Group Absences*

The symmetry elements associated with each space group are known. From Table 5.3, we can put together the rules of systematic extinctions (or nonextinctions) for a space group. These provide the guidelines for determining that particular space group. Niggli [5.23] was the first to tabulate the systematic absence rules for various

Table 5.3 Symmetry interpretations of extinctions

Class of reflection	Symmetry element	Condition for nonextinction (n= an integer)
hkl	Body centred lattice	$h+k+l=2n$
	C-centred lattice	$h+k=2n$
	B-centred lattice	$h+l=2n$
	A-centred lattice	$k+l=2n$
	Face centred lattice	h,k,l all odd or all even
	Rhombohedral lattice indexed on hexagonal reference system	$-h+k+l=3n$
	Hexagonal lattice indexed on rhombohedral reference system	$h+k+l=3n$
$0kl$	(100) glide plane, component $b/2$	$k=2n$
	(100) glide plane, component $c/2$	$l=2n$
	(100) glide plane, component $b/2+c/2$	$k+l=2n$
	(100) glide plane, component $b/4+c/4$	$k+l=4n$
$h0l$	(010) glide plane, component $a/2$	$h=2n$
	(010) glide plane, component $c/2$	$l=2n$
	(010) glide plane, component $a/2+c/2$	$h+l=2n$
	(010) glide plane, component $a/4+c/4$	$h+l=4n$
$hk0$	(001) glide plane, component $a/2$	$h=2n$
	(001) glide plane, component $b/2$	$k=2n$
	(001) glide plane, component $a/2+b/2$	$h+k=2n$
	(001) glide plane, component $a/4+b/4$	$h+k=4n$
hhl	(110) glide plane, component c/2	l = 2n
	(110) glide plane, component $a/2+b/2$	$h=2n$
	(110) glide plane, component $a/4+b/4+c/4$	$h+l=2n$
	(110) glide plane, component $a/2+b/4+c/4$	$2h+l=4n$
$h00$	[100] screw axis, component $a/2$	$h=2n$
	[100] screw axis, component $a/4$	$h=4n$
$0k0$	[010] screw axis, component $b/2$	$k=2n$
	[010] screw axis, component $b/4$	$k=4n$
$00l$	[001] screw axis, component $c/2$	$l=2n$
	[001] screw axis, component $c/3$	$l=3n$
	[001] screw axis, component $c/4$	$l=4n$
	[001] screw axis, component $c/6$	$l=6n$
$hh0$	[110] screw axis, component $a/4+b/4$	$h=2n$

space groups. Eventually, these rules have become available in the International Tables for X-ray Crystallography [5.24]. As an example, the limiting conditions for monoclinic space groups are given in Table 5.4.

To get a feel of how these rules operate, we shall consider two typical cases. In Table 5.5, the indices of reflections observed in the diffraction photograph of a monoclinic crystal (which we shall call crystal I) are given. It is noted that there is no restriction on hkl type of reflections. On the other hand, $h0l$ reflections all have l even and $0k0$ reflections all have k even. By reference to Table 5.4, we conclude that

Table 5.4 Limiting conditions for monoclinic space groups

Conditions limiting possible X-ray reflections	Possible space groups
hkl: none $h0l$: none $0k0$: none	$P2$, Pm, $P2/m$
hkl: none $h0l$: none $0k0$: $k = 2n$	$P2_1$, $P2_1/m$
hkl: none $h0l : l = 2n$ $0k0$: none	Pc, $P2/c$
hkl: none $h0l : l = 2n$ $0k0 : k = 2n$	$P2_1/c$
hkl: $h + k = 2n$ $h0l : h = 2n$ $0k0 : k = 2n$	$C2$, Cm, $C2/m$
hkl: $h + k = 2n$ $h0l : l = 2n, h = 2n$ $0k0 : k = 2n$	Cc, $C2/c$

Table 5.5 Some reflection data for monoclinic crystal (I)

hkl	hkl	hkl	hkl	hkl
100	202	110	113	040
200	204	310	311	060
300	402	111	322	080
400	502	122	020	

Table 5.6 Some reflection data for monoclinic crystal (II)

hkl	hkl	hkl	hkl	hkl
200	400	110	114	020
201	401	111	310	040
202	402	112	311	060
203	600	113	510	080

the space group is $P2_1/c$. The reflections occurring in another monoclinic crystal (II) are given in Table 5.6. The reflections satisfy the following rules:

$$hkl : h + k = 2n$$

$$h0l : h = 2n$$

$$0k0 : k = 2n.$$

By reference to Table 5.4, we see that three space groups are possible: $C2$, Cm, $C2/m$. These space groups have the same translational symmetry elements and hence the same systematic nonextinctions; they differ in the point symmetry elements. It is not possible to uniquely decide the space group in this case from systematic absences alone. Such ambiguity occurs in several other cases. We shall revert to this problem of ambiguity in space groups in Sect. 5.5.

5.4 Atomic Coordinates

5.4.1 Preliminaries

5.4.1.1 Terminology of Atomic Coordinates

Atomic coordinates are expressed not as the actual distances along the crystal axes but as fractions of the unit cell parameters. Thus, if the actual coordinates of an atom are X_n, Y_n, Z_n and if the unit cell parameters are a, b, c, then for convenience, the coordinates are expressed as x_n, y_n, z_n given by

$$x_n = \frac{X_n}{a}, \quad y_n = \frac{Y_n}{b}, \quad z_n = \frac{Z_n}{c}. \tag{5.10}$$

As can be seen, these are dimensionless. In the early days of crystallography, the atomic coordinates were sometimes expressed as angles: $\theta_1 = (360X_n/a)^\circ$, $\theta_2 = (360Y_n/b)^\circ$, $\theta_3 = (360Z_n/c)^\circ$ [5.25, 5.26]. But the system of dimensionless coordinates (5.10) is most commonly employed.

Atomic coordinates are classified as general and special. General coordinates are specified by three coordinates, e.g. x, y, z; $\bar{x}, \bar{y}, \bar{z}$, etc. No point symmetry element is present at any of these positions. On the other hand, some atoms lie on symmetry elements. These are called special positions. They are specified by fewer non-zero coordinates, e.g. 0, 0, 0; 0, 1/2, 1/2, etc.

Another aspect of atomic coordinates is the equipoint. A number of positions of a given atomic species may be related by a symmetry element. The collection of such points is called an equipoint. Associated with an equipoint is the symmetry factor S. The symmetry factor represents the resultant of waves scattered by the atoms in the equipoint. If there are m atoms in an equipoint,

$$S = e^{i\phi_1} + e^{i\phi_2} + \cdots + e^{i\phi_m} = \sum_{j=1}^{m} e^{i\phi_j} = \sum_{j=1}^{m} e^{2\pi i(hx_j + ky_j + lz_j)}. \tag{5.11}$$

The structure factor F_{hkl} may be expressed in terms of the symmetry factor S_j as

$$F_{hkl} = \sum_j f_j S_j. \tag{5.12}$$

5.4.1.2 Measurement of Intensities

While space group is determined from absent reflections, the determination of atomic positions, as will be seen later, is dependent entirely on the intensities of recorded reflections. The measurement of intensities was briefly discussed in the preceding chapter. In view of the importance of intensities in structure determination, we shall undertake a more detailed discussion.

In the early work on X-ray crystallography, relative intensities were estimated visually in a qualitative manner using such terms as vs (very strong), s (strong), ms (medium strong), w (weak), vw (very weak), etc. [5.16]. In the 1940s, intensities were still estimated visually, but with the improvement that they were compared with calibrated scales. These scales were prepared by recording a direct beam attenuated by a metallic absorber or a crystal-diffracted beam on a strip of film. By changing the time of exposure, a series of spots were obtained. The scale is extended by exposing not a single film but a pack of films. In this way, the intensities ranging from 1 to 1,000 could be recorded. The intensity of a given reflection was estimated by holding the X-ray diffraction photograph and the calibrated strip in juxtaposition. The human eye is a fairly reliable judge of these intensities. While judging intensities in the range 1–1,000 is routinely done, Sime and Abrahams [5.27] made the rare claim of estimating weakest and strongest reflections in the intensity ratio 1:8,000 and even 1:39,000. The next stage in estimating intensities was to put the film through a densitometer.

By the 1960s, several counter detectors (Geiger Muller, proportional, scintillation, solid state) came into use. Of late, the charge-coupled device detectors are also in use. However, the photographic method remained popular till the 1980s until the advent of the computer controlled automatic four-circle diffractometer.

5.4.1.3 Relative and Absolute Structure Factors

The structure amplitudes $|F_{hkl}|$ obtained from the equation

$$I_{hkl} = S|F_{hkl}|_R^2 \tag{5.13}$$

are called relative structure amplitudes. Here I_{hkl} is the measured intensity corrected for absorption, multiplicity and Lorentz-polarization effect; S is the scale factor. The effect of temperature on intensities was discussed in Chap. 4. In (5.13), the temperature effect is included within the structure factor. If it is separated out, we get the absolute structure amplitude $|F_{hkl}|_A$. Assuming a single common Debye–Waller factor B for all the atoms, (5.13) gets modified to

$$I_{hkl} = S\mathrm{e}^{-2B(\sin\theta/\lambda)^2}|F_{hkl}|_A^2. \tag{5.14}$$

Now we may write $|F_{hkl}|_A^2$ as

$$|F_{hkl}|_A^2 = F_{hkl} \cdot F_{hkl}^*, \tag{5.15}$$

where F^* is the complex conjugate of F. Equation (5.15) maybe expanded as

$$|F_{hkl}|_A^2 = \sum_n f_n e^{2\pi i(hx_n+ky_n+lz_n)} \sum_m f_m e^{-2\pi i(hx_m+ky_m+lz_m)}$$
$$= \sum_n f_n^2 + \sum_{n\neq m} f_n f_m e^{2\pi i\{h(x_n-x_m)+k(y_n-y_m)+l(z_n-z_m)\}}\,. \quad (5.16)$$

If we take the average values, the second term on the rhs becomes zero since there will be as many negative values as the positive values. Thus, combining (5.14) and (5.16), we get

$$\langle I_{hkl}\rangle = S e^{-2B\langle(\sin\theta/\lambda)^2\rangle}\langle|F_{hkl}|_A^2\rangle = S e^{-2B\langle(\sin\theta/\lambda)^2\rangle}\sum_n f_n^2$$

or

$$\log\frac{\langle I_{hkl}\rangle}{\sum f_n^2} = \log S - 2B\langle(\sin\theta/\lambda)^2\rangle\,. \quad (5.17)$$

In (5.17), $\langle I_{hkl}\rangle$ is the average of the intensities of reflections in a narrow range of $(\sin\theta/\lambda)$ with average value $\langle(\sin\theta/\lambda)^2\rangle$. In practice, a few ranges of $(\sin\theta/\lambda)$ are chosen, the intensities averaged and $\sum f_n^2$ calculated. A plot of log $[\langle I_{hkl}\rangle/\sum f_n^2]$ and $\langle(\sin\theta/\lambda)^2\rangle$ is then plotted. The plot is linear with log S as intercept and –$2B$ as slope. Such a plot is shown in Fig. 5.3. Once the values of S and B are known, we can revert to (5.14) and get the experimental values of $|F_{hkl}|_A$. This method was suggested by Wilson [5.28] and a plot like Fig. 5.3 is called a Wilson plot.

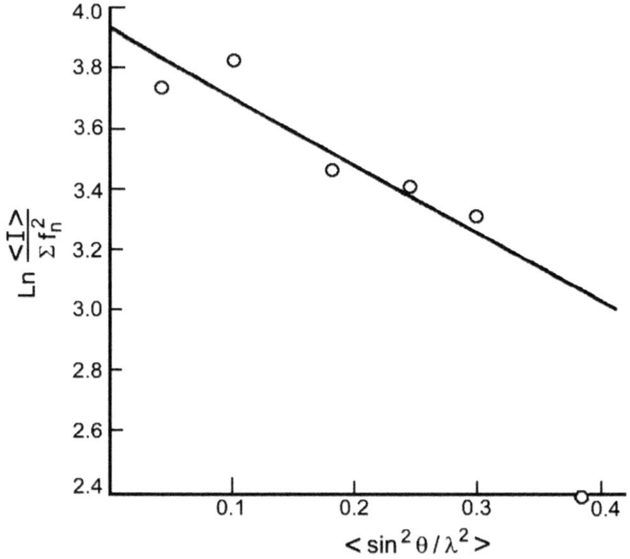

Fig. 5.3 Wilson plot for metatolidine dihydrochloride

5.4.2 Trial and Error Method

Equation (5.14) represents the relation between the intensity of a reflection and the structure amplitude. The structure amplitude is a function of the coordinates. In fact,

$$|F_{hkl}|^2 = \left[\sum_n f_n \cos 2\pi(hx_n + ky_n + lz_n)\right]^2 + \left[\sum_n f_n \sin 2\pi(hx_n + ky_n + lz_n)\right]^2. \tag{5.18}$$

In principle, then, it should be possible to determine the coordinates from the intensities. But we cannot solve (5.18) for the atomic coordinates, i.e. we cannot express the atomic coordinates as functions of the intensities.

In the trial-and-error method, what is done is to assume a set of coordinates for the several atoms, consistent with space group symmetry and construct the structure amplitudes from (5.18). The next step is to calculate the intensities from (5.14) and to compare them with observed intensities. If there is reasonable agreement between the calculated and observed intensities, the assumed structure is correct. If not, another set of coordinates is assumed and the process is repeated. In fact, the process is repeated until we hit upon the right structure.

Buerger [5.3] has illustrated the use of this method by solving the structures of some simple crystals like NaCl, FeS_2 and valentine. We shall consider the case of NaCl. To start with, we shall assume a structure (I) in which the atomic coordinates are: Na at (0, 0, 0), (1/2, 0, 1/2), (0, 1/2, 1/2), (1/2, 1/2, 0) and Cl at (1/4, 1/4, 1/4), (1/4, 3/4, 3/4), (3/4, 1/4, 3/4), (3/4, 3/4, 1/4). The structure factors for some reflections are calculated and, finally, the intensities. These intensities are given in Table 5.7 along with observed intensities. Strong differences are observed between the calculated and observed intensities for the (200), (420), (600) and (640) reflections. We then assume another structure (II) with atomic coordinates: Na at (0, 0, 0), (1/2, 0, 1/2), (0, 1/2, 1/2), (1/2, 1/2, 0) and Cl at (1/2, 1/2, 1/2), (1/2, 0, 0), (0, 1/2, 0), (0, 0, 1/2). The calculated intensities with these coordinates are also given in Table 5.7. The calculated and observed relative intensities are now in agreement. Thus, structure I is wrong while structure II is correct. These two structures are now known as the sphalerite and NaCl structures.

However, the journey from a wrong set of coordinates to a right one is neither short nor easy. This method is cumbersome and uncertain and can be applied only to the simplest of structures. It may be noted that for each atom, there are three position coordinates and, in rigorous work, six thermal parameters to be determined, besides the scale factor. Thus, if there are N atoms, the number of unknowns to be determined is $9N + 1$. In inorganic compound crystals there are 20–30 atoms to be handled. This number goes up to even 100 or more in the case of biological crystals. The trial-and-error method starts with the assumption of atomic coordinates. What we need is a method which does not assume coordinates but rather leads to them in a systematic and logical manner. Such a method is the Fourier series method.

Table 5.7 Calculated ($I_{calc.}$) and observed ($I_{obs.}$) intensities of some NaCl reflections

Reflection	Intensity		
	$I_{calc.}$		$I_{obs.}$
	Structure I	Structure II	
200	1,300	44,000	43,000
220	20,000	20,000	20,500
400	8,500	8,500	7,900
420	550	13,000	11,500
440	4,300	4,300	3,800
600	420	4,400	3,900
620	9,600	9,600	8,800
640	3,500	29,000	26,000

5.4.3 Principle of Fourier Synthesis

The structure factor F_{hkl} is defined as

$$F_{hkl} = \sum_{n=1}^{N} f_n e^{2\pi i(hx_n + ky_n + lz_n)}. \tag{5.19}$$

Here n is the number of an atom in the unit cell and N the total number of atoms. As stated in Chap. 4, the significance of the structure factor is that it represents the resultant of the waves scattered by all the electrons in the unit cell. In writing (5.19), we have treated the electrons as located at the atom centres. Instead, let us treat the electron charge to be continuously distributed over the unit cell space. We shall introduce ρ_{xyz} as the density of electron charge at the point x, y, z. The electron charge in a volume element $\mathrm{d}x\mathrm{d}y\mathrm{d}z$ is $\rho_{xyz}\mathrm{d}x\mathrm{d}y\mathrm{d}z$. We shall add up the waves scattered by all such charge density elements. Then, in the spirit in which we defined F_{hkl} in (5.19), we shall now write

$$F_{hkl} = \int_{x=0}^{1}\int_{y=0}^{1}\int_{z=0}^{1} V\rho_{xyz} e^{2\pi i(hx+ky+lz)}\mathrm{d}x\mathrm{d}y\mathrm{d}z, \tag{5.20}$$

where V is the volume of the unit cell.

The unit cell repeats itself along three directions with the periodicity of the lattice. The charge density function, which is now a property of the unit cell, also repeats itself with the same periodicity. Thus ρ_{xyz} is a periodic function which, like any other periodic function, can be expressed as a Fourier series. The Fourier series for the charge density is

$$\rho_{xyz} = \sum_{h'}\sum_{k'}\sum_{l'=-\infty}^{\infty} C_{h'k'l'} e^{2\pi i(h'x+k'y+l'z)}, \tag{5.21}$$

where $C_{h'k'l'}$ is a coefficient and h', k', l' are three whole number indices; the similarity of these indices with the Miller indices will become clear soon. Substituting (5.21) in (5.20), we get

$$F_{hkl} = \int_0^1 \int_0^1 \int_0^1 \sum\sum\sum C_{h'k'l'} e^{2\pi i(h'x+k'y+l'z)} e^{2\pi i(hx+ky+lz)} (V\mathrm{d}x\mathrm{d}y\mathrm{d}z). \tag{5.22}$$

It is the property of these integrals that they reduce to zero for all values of h', k', l' except when $h' = -h$, $k' = -k$ and $l' = -l$ for which they become equal to 1. Thus, (5.22) becomes

$$F_{hkl} = VC_{\bar{h}\bar{k}\bar{l}} = VC_{hkl} \tag{5.23}$$

since the planes hkl and $\bar{h}\bar{k}\bar{l}$ are equivalent. Thus, the coefficients $C_{h'k'l'}$ are $(1/V)$ times the structure factor F_{hkl}. We may now write (5.21) as

$$\rho_{xyz} = \frac{1}{V} \sum_h \sum_k \sum_{l=-\infty}^{\infty} F_{hkl} e^{-2\pi i(hx+ky+lz)}. \tag{5.24}$$

Equation (5.24) holds the key to the problem of structure determination. We may start with a crystal with unknown structure. If we have the information about the structure factors, we can use (5.24) to calculate the electron density at various points in the unit cell. We may then find that ρ_{xyz} is low at most places but has large values at some select places; obviously, these are the locations of atoms.

5.4.3.1 The Phase Problem

We have seen in Chap. 4 that the structure F_{hkl} is a complex quantity given by

$$F_{hkl} = |F_{hkl}| e^{i\phi_{hkl}}, \tag{5.25}$$

where $|F_{hkl}|$ is the amplitude and ϕ_{hkl} is the phase given by

$$\phi_{hkl} = \tan^{-1}\left[\frac{\sum f_n \sin 2\pi(hx_n + ky_n + lz_n)}{\sum f_n \cos 2\pi(hx_n + ky_n + lz_n)}\right]. \tag{5.26}$$

While the amplitude $|F_{hkl}|$ is obtained directly from the measured intensity, ϕ_{hkl} is not measurable and is a function of atomic coordinates. Thus to know x_n, y_n, z_n, we need ρ_{xyz}, but to know ρ_{xyz} we need ϕ_{hkl} which is a function of x_n, y_n, z_n. This vicious circle is a stumbling block in structure determination; it is called the phase problem. Fortunately, with ingenuity, methods have been developed to solve the phase problem. We shall consider some of these methods and then continue further discussion of Fourier synthesis.

5.4.3.2 Centrosymmetric Structures

The phase angle ϕ_{hkl} can take any value in the range 0–2π in noncentrosymmetric structures. Let us consider what happens in centrosymmetric structures. Here, for every atom at (x_n, y_n, z_n,), there is an identical atom at ($-x_n$, $-y_n$, $-z_n$,). Then ϕ_{hkl} becomes

$$\begin{aligned}\phi_{hkl} &= \tan^{-1}\left[\frac{\sum\limits_{n=1}^{N} f_n \sin 2\pi(hx_n + ky_n + lz_n)}{\sum\limits_{n=1}^{N} f_n \cos 2\pi(hx_n + ky_n + lz_n)}\right] \\ &= \tan^{-1}\left[\frac{\sum\limits_{n=1}^{N/2} f_n \sin 2\pi(hx_n + ky_n + lz_n) + \sum\limits_{n=1}^{N/2} f_n \sin\{-2\pi(hx_n + ky_n + lz_n)\}}{\sum\limits_{n=1}^{N/2} f_n \cos 2\pi(hx_n + ky_n + lz_n) + \sum\limits_{n=1}^{N/2} f_n \cos\{-2\pi(hx_n + ky_n + lz_n)\}}\right] \\ &= \tan^{-1} 0.\end{aligned} \tag{5.27}$$

Hence, ϕ_{hkl} is 0 or π and the real part of $e^{i\phi_{hkl}}$ is $+1$ or -1. Thus in centrosymmetric structures, the problem of phase determination is reduced to assigning $+$ or $-$ signs to the structure amplitudes $|F_{hkl}|$. We shall now consider some methods of determining the phases.

5.4.3.3 Patterson Synthesis

Patterson [5.29] introduced a procedure which avoids the phases altogether. Let us consider two points located at (x, y, z) and ($x+u$, $y+v$, $z+w$); obviously, u, v and w are the components of the vector joining the two points. The electron charge density at these points may be denoted by ρ_{xyz} and $\rho_{x+u,y+v,z+w}$, respectively. Patterson introduced a function P defined by

$$P_{uvw} = V \int\limits_{x=0}^{1} \int\limits_{y=0}^{1} \int\limits_{z=0}^{1} \rho_{xyz}\rho_{x+u,y+v,z+w}\,\mathrm{d}x\mathrm{d}y\mathrm{d}z. \tag{5.28}$$

Substituting for ρ from (5.24), we get

$$P_{uvw} = \frac{1}{V} \int\limits_{x=0}^{1} \int\limits_{y=0}^{1} \int\limits_{z=0}^{1} \sum_{h}\sum_{k}\sum_{l}\sum_{h'}\sum_{k'}\sum_{l'=-\infty}^{\infty} F_{hkl} e^{-2\pi i(hx+ky+lz)} \times F_{h'k'l'} e^{-2\pi i[h'(x+u)+k'(y+v)+l'(z+w)]}\mathrm{d}x\mathrm{d}y\mathrm{d}z. \tag{5.29}$$

The integrals vanish for all values of h', k', l' except for $h' = -h, k' = -k, l' = -l$. Thus, the integration signs and the second summation disappear and we get

$$\begin{aligned} P_{uvw} &= \frac{1}{V}\sum_h\sum_k\sum_{l=-\infty}^{\infty} F_{hkl}F_{\bar{h}\bar{k}\bar{l}}e^{2\pi i(hu+kv+lw)} \\ &= \frac{1}{V}\sum_h\sum_k\sum_{l=-\infty}^{\infty} |F_{hkl}|e^{i\phi_{hkl}} \times |F_{\bar{h}\bar{k}\bar{l}}|e^{i\phi_{\bar{h}\bar{k}\bar{l}}}e^{2\pi i(hu+kv+lw)} \\ &= \frac{1}{V}\sum_h\sum_k\sum_{l=-\infty}^{\infty} |F_{hkl}|\ |F_{\bar{h}\bar{k}\bar{l}}|e^{2\pi i(hu+kv+lw)}. \end{aligned} \tag{5.30}$$

Since $|F_{hkl}| = |F_{\bar{h}\bar{k}\bar{l}}|$, we finally have

$$P_{uvw} = \frac{1}{V}\sum_h\sum_k\sum_{l=-\infty}^{\infty} |F_{hkl}|^2\, e^{2\pi i(hu+kv+lw)}. \tag{5.31}$$

The function P_{uvw} is called the Patterson function. This function contains only the structure amplitude and the phases are no longer present. Since the structure amplitudes $|F_{hkl}|$ are directly determined from the measured intensities, it is possible to evaluate P_{uvw} at various points in the unit cell; u, v, w can take all values from 0 to 1 in the unit cell.

Let us note some properties of the Patterson function. Firstly, by expressing the exponential term in (5.31) in terms of sine and cosine and writing the terms for a pair of positive and negative hkl values, it can be shown that (5.31) reduces to

$$P_{uvw} = \frac{1}{V}\sum_h\sum_k\sum_{l=-\infty}^{\infty} |F_{hkl}|^2 \cos 2\pi(hu + kv + lw). \tag{5.32}$$

Thus, we see that P_{uvw} is always a real quantity. Secondly, $P_{000} = \frac{1}{V}\sum\sum\sum|F_{hkl}|^2$ which is proportional to $\sum Z_n^2$ where Z_n is the number of electrons in atom n. Thirdly, by substituting positive and negative values of uvw in (5.31) we see that $P_{uvw} = P_{-u,-v,-w}$, i.e. within the unit cell the function P_{uvw} has a centre of symmetry.

Let us now consider the most important aspect viz., the physical significance of the Patterson function. Let Fig. 5.4 be the projection of a structure on the xy plane. Let OP be a vector with components u, v. In the figure, the circles denote atoms. Let O′P′, O″P'' and O‴P‴ represent three vectors equal and parallel to OP. Consider O′P′. Here, both ends are away from the atoms; consequently the charge densities $\rho_{O'}$ and $\rho_{P'}$ are zero. The Patterson function for this pair of atoms is zero. Next consider O″P″. Here P″ coincides with an atom while O″ is away from an atom. The charge density $\rho_{P''}$ has a finite value while $\rho_{O''}$ is zero. In this case also the Patterson function has a zero value. Finally, let us consider O‴P‴. Here both ends

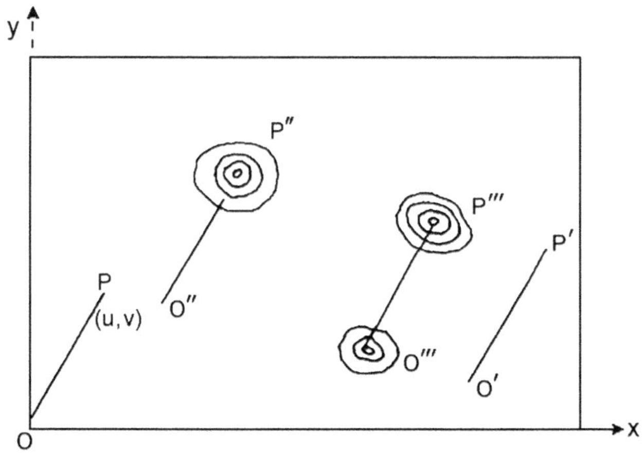

Fig. 5.4 Physical significance of the Patterson function

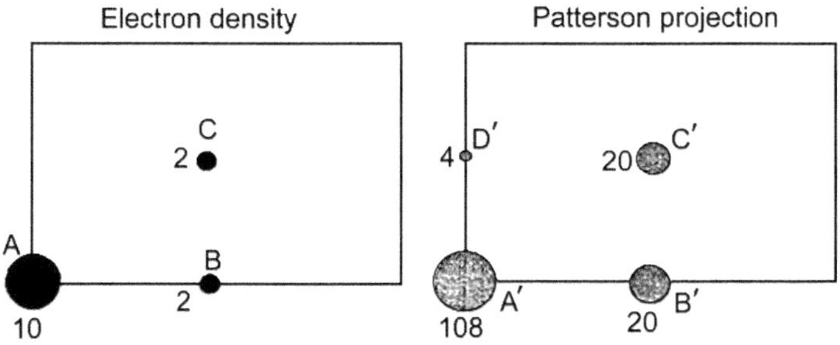

Fig. 5.5 Electron density (*lhs*) and Patterson (*rhs*) for a unit cell containing atoms at A, B, C

of the vector terminate at atomic sites where the charge densities $\rho_{O'''}$ and $\rho_{P'''}$ both have finite values. Thus, P_{uv} will have a large value. Hence if the function P_{uv} is large at a point with coordinates uv, the vector OP represents an interatomic vector. The argument can be extended to three dimensions. This point is further illustrated in Fig. 5.5 which shows the projection of a structure with three atoms A, B, C in the unit cell. Their Z values which are a measure of the charge density are also shown in the lhs figure. The Patterson function for this structure is shown on the rhs. There are Patterson peaks at A′, B′, C′ and D′ with relative values of $P_{A'} = \rho_A^2 + \rho_B^2 + \rho_C^2$, $P_{B'} = \rho_A\rho_B$, $P_{C'} = \rho_A\rho_C$ and $P_{D'} = \rho_B\rho_C$.

As an example, the projection of the Patterson function for cysteinyl-glycine sodium iodide on the (010) plane is shown in Fig. 5.6b along with the atomic positions (Fig. 5.6a).

The Patterson synthesis gives only interatomic vectors; it does not, by itself, give the atomic coordinates. The retrieval of atomic coordinates from the Patterson map involves a long process. Yet, to quote Lipson and Cochran [5.1], "the Patterson method has been used to give information that could not be obtained in any other way".

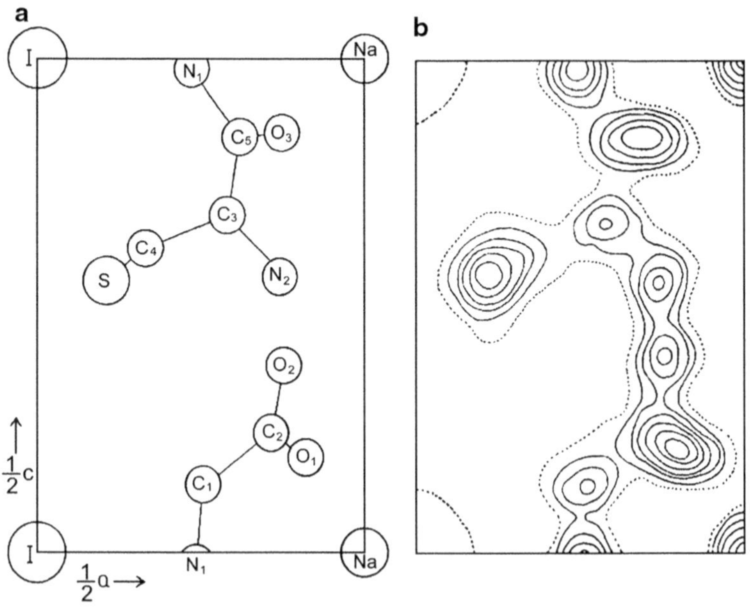

Fig. 5.6 Atomic positions (**a**) and Patterson projection (**b**) for cysteinyl-glycine sodium iodide

5.4.3.4 Isomorphous Replacement Method

This method was proposed by Cork [5.30]. As the name indicates, the method can be applied if the crystal under study has an isomorphous crystal, the two differing by one atom.

Suppose a crystal MABCD has an isomorph NABCD. We shall denote the residue ABCD by R and the two crystals by MR and NR. We shall express the structure factor for the *hkl* reflection as

$$^{\mathrm{MR}}F_{hkl} = {}^{\mathrm{M}}F_{hkl} + {}^{\mathrm{R}}F_{hkl} \tag{5.33}$$

and

$$^{\mathrm{NR}}F_{hkl} = {}^{\mathrm{N}}F_{hkl} + {}^{\mathrm{R}}F_{hkl}. \tag{5.34}$$

The difference between the two may be written as

$$^{\mathrm{MR}}F_{hkl} - {}^{\mathrm{NR}}F_{hkl} = {}^{\mathrm{M}}F_{hkl} - {}^{\mathrm{N}}F_{hkl}. \tag{5.35}$$

Since the two atoms are similarly located, they have the same equipoint and the same symmetry factor S_{hkl}. Then (5.35) can be written as

$$^{\mathrm{MR}}F_{hkl} - {}^{\mathrm{NR}}F_{hkl} = \left({}^{\mathrm{M}}f - {}^{\mathrm{N}}f\right)\ S_{hkl}, \tag{5.36}$$

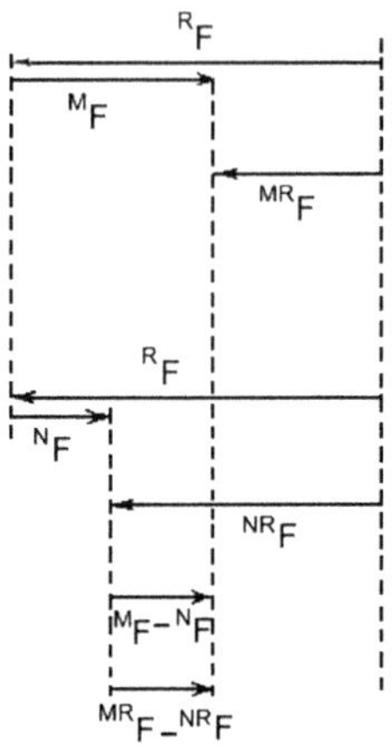

Fig. 5.7 Geometrical interpretations of (5.35) under the condition that ${}^{M}F$–${}^{N}F$ is positive. Here ${}^{R}F$ is negative

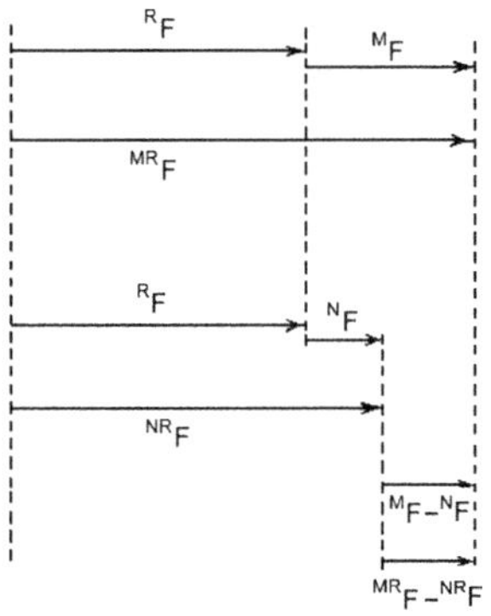

Fig. 5.8 Geometrical interpretations of (5.35) under the condition that ${}^{M}F$–${}^{N}F$ is positive. Here ${}^{R}F$ is positive

where f is the atomic scattering factor. If the atomic number Z of the M atom is greater than that of the N atom, then $({}^{M}f - {}^{N}f)$ is positive and the phase of the lhs is determined by S_{hkl}.

The interpretation of (5.35) is not very obvious. Let us represent the various terms on Figs. 5.7 and 5.8 which are drawn with the assumption that $({}^{M}F - {}^{N}F)$ is positive. The diagrams show that we get this result with ${}^{R}F$ positive or negative. The correct signs are obtained by trying both alternatives. This point is further clarified by an examination of experimental data on chlorine–camphor and bromine–camphor [5.31] given in Table 5.8. Different signs are tried for ${}^{BrR}F$ and ${}^{ClR}F$ and the best choice is decided by comparison with $({}^{Br}F - {}^{Cl}F)$.

5.4.3.5 Heavy Atom Method

The presence of a heavy atom helps in deciding the phases of reflections. The heavy atom may be present in the original crystal (for example, K and P in KDP) or may be chemically introduced in the crystal (for example Pt in pthalocyanine).

Table 5.8 Determination of phases for chlorine–camphor and bromine–camphor

$h0l$	$\lvert {}^{\mathrm{BrR}}F \rvert$	$\lvert {}^{\mathrm{ClR}}F \rvert$	${}^{\mathrm{BrR}}F\_{}^{\mathrm{ClR}}F$				Computed ${}^{\mathrm{Br}}F\_{}^{\mathrm{Cl}}F$	Deduced signs	
			(++)	(+−)	(−+)	(−−)		${}^{\mathrm{BrR}}F$	${}^{\mathrm{ClR}}F$
001	45	36	+9	+81	−81	−9	+18	+	+
100	13	10	+3	+23	−23	−3	−3	−	−
$10\bar{1}$	15	35	−20	+40	−40	+20	+27	−	−
103	32	34	−2	+66	−66	+2	+1		
300	7	< 4	> +3	< +11	> −11	< −3	+7	+	
$20\bar{3}$	19	11	+8	+30	−30	−8	+25	+	−
$30\bar{1}$	27	10	+17	+37	−37	−17	−18	−	−
301	40	19	+21	+59	−59	−21	+24	+	+
004	22	12	+10	+34	−34	−10	−14	−	−
$00\bar{7}$	11	7	+4	+18	−18	−4	−6	−	
606	9	7	+2	+16	−16	−2	−6	−	
$80\bar{1}$	9	7	+2	+16	−16	−2	+5	+	+
800	8	4	+4	+12	−12	−4	+4	+	+

To understand the principle of the method, let us denote the contributions of the heavy atom (H) and the residue (R) to a reflection by ${}^{\mathrm{H}}F_{hkl}$ and ${}^{\mathrm{R}}F_{hkl}$, respectively. Then,

$$F_{hkl} = {}^{\mathrm{H}}F_{hkl} + {}^{\mathrm{R}}F_{hkl}. \tag{5.37}$$

If ${}^{\mathrm{H}}S_{hkl}$ is the symmetry factor for the heavy atom and ${}^{\mathrm{H}}f$ its atomic scattering factor, (5.37) may be written as

$$F_{hkl} = {}^{\mathrm{H}}f\ {}^{\mathrm{H}}S_{hkl} + {}^{\mathrm{R}}F_{hkl}. \tag{5.38}$$

For simplicity, consider a crystal with a centre of symmetry and with a heavy atom whose position is known from other information. The first term on the right-hand side of (5.38) can be computed. For a centrosymmetric crystal, the terms in (5.38) are real and can take either positive or negative values. The maximum value of the second term is ${}^{\mathrm{R}}[\sum f_n]_{hkl}$. Whenever the heavy atom contribution is greater than this maximum value of the contribution of the residue, the sign of F_{hkl} can be taken as that of the heavy atom term.

The structure determination of Pt-pthalocyanine by Robertson and Woodward [5.32] is considered to be the first application of the heavy atom method. In 302 reflections of type ($h0l$), where the Pt atoms were making a large positive contribution, the sign of the reflections was taken to be positive and the structure was completely solved.

5.4.3.6 Method of Inequalities

A new general approach to tackle the problem was initiated by Harker and Kasper [5.33]. The method is in the form of inequalities between the structure factors and structure amplitudes of certain reflections. The derivation of these inequalities is based upon relations known in mathematics as Cauchy's and Schwarz's inequalities. Further it makes use of (5.13), (5.20), (5.24) and (5.25). This derivation is lengthy and we shall only quote the results. Further, the derivation assumes that the charge density is positive at all points. The inequalities are expressed in terms of unitary structure factors U_{hkl} defined by

$$U_{hkl} = F_{hkl}/Z, \tag{5.39}$$

where Z is the total number of electrons ($Z = \sum_{n=1}^{N} Z_n$).

Harker and Kasper [5.33] showed that for a crystal with a centre of inversion,

$$U_{hkl}^2 \leq \frac{1}{2} + \frac{1}{2} U_{2h,2k,2l}. \tag{5.40}$$

This means that if U_{hkl}^2 is greater than 1/2, then $U_{2h,2k,2l}$ is positive. Similarly, if $|U_{2h,2k,2l}|$ is 1/2 and U_{hkl}^2 is larger than 1/4, then also $U_{2h,2k,2l}$ is positive.

In the case of a crystal containing a twofold rotation axis in the y-direction, Harker and Kasper showed that

$$|U_{hkl}|^2 \leq \frac{1}{2} + \frac{1}{2} U_{2h,0,2l}. \tag{5.41}$$

The implication of (5.41) is similar to that of (5.40) but here all $|U_{hkl}|^2$ with constant h, l but various values of k can be used to determine the sign of $U_{2h,0,2l}$. Other symmetry elements lead to corresponding inequalities. Several inequalities derived by Harker and Kasper are given in Table 5.9. While these inequalities are useful in general, they are particularly useful when the methods of isomorphous replacement and heavy atom methods cannot be used.

The method of inequalities initiated by Harker and Kasper has become a trendsetter. Many more inequalities were derived by Gilles [5.34], Karle and Hauptman [5.35] (Box 5.1), Sayre [5.36] and Cochran [5.37].

5.4.3.7 Anomalous Scattering Method

Generally the intensities of reflections from the *hkl* and $\bar{h}\bar{k}\bar{l}$ planes are equal; this is known as Friedel's law. If we consider the effect of dispersion, the atomic scattering factor is not f_0 but $f_0 + f' + if''$; the additional terms are called the dispersion correction. Even with this modification, Friedel's law still holds. However, it fails if the k-absorption edge of the scattering edge is close to the wavelength λ of the radiation used. In this case, the intensities of *hkl* and $\bar{h}\bar{k}\bar{l}$ are unequal which, in turn, means that the structure amplitude $|F_{hkl}| \neq |F_{\bar{h}\bar{k}\bar{l}}|$.

Table 5.9 Harker–Kasper inequalities

Symmetry element $\|\|_c$	Inequality
1	$\|U_{hkl}\|^2 \leq 1$
$\bar{1}$	$\|U_{hkl}\|^2 \leq \frac{1}{2} + \frac{1}{2}U_{2h\,2k\,2l}$
2	$\|U_{hkl}\|^2 \leq \frac{1}{2} + \frac{1}{2}U_{2h\,2k\,0}$
2_1	$\|U_{hkl}\|^2 \leq \frac{1}{2} + \frac{1}{2}(-1)^l\,U_{2h\,2k\,0}$
$\bar{2} = m$	$\|U_{hkl}\|^2 \leq \frac{1}{2} + \frac{1}{2}U_{0\,2k\,0}$
a	$\|U_{hkl}\|^2 \leq \frac{1}{2} + \frac{1}{2}(-1)^h\,U_{0\,0\,2l}$
3	$\|U_{hkl}\|^2 \leq \frac{1}{3} + \frac{2}{3}\|U_{(h-k)(h+2k)0}\|\cos 2\pi\phi_{(h-k)\,(h+2k)\,0}$
$3_1, 3_2$	$\|U_{hkl}\|^2 \leq \frac{1}{3} + \frac{2}{3}\|U_{(h-k)(h+2k)0}\|\cos 2\pi(\phi_{(h-k)\,(h+2k)\,0} + \frac{1}{3}l)$
$\bar{3} = 3 + \bar{1}$	$\|U_{hkl}\|^2 \leq \frac{1}{6} + \frac{1}{6}U_{2h\,2k\,2l} + \frac{1}{3}U_{h\,k\,2l} + \frac{1}{3}U_{(h-k)\,(h+2k)\,0}$
4	$\|U_{hkl}\|^2 \leq \frac{1}{4} + \frac{1}{4}U_{2h\,2k\,0} + \frac{1}{2}U_{(h-k)\,(h+k)\,0}$
$4_1, 4_3$	$\|U_{hkl}\|^2 \leq \frac{1}{4} + \frac{1}{4}(-1)^l\,U_{2h\,2k\,0} + \frac{1}{2}(\cos 2\pi\frac{1}{4}l)U_{(h-k)\,(h+k)\,0}$
4_2	$\|U_{hkl}\|^2 \leq \frac{1}{4} + \frac{1}{4}U_{2h\,2k\,0} + \frac{1}{2}(-1)^l\,U_{(h-k)\,(h+k)\,0}$
$\bar{4}$	$\|U_{hkl}\|^2 \leq \frac{1}{4} + \frac{1}{4}U_{2h\,2k\,0} + \frac{1}{2}\|U_{(h-k)(h+k)2l}\|\cos 2\pi\phi_{(h-k)\,(h+k)\,2l}$
6	$\|U_{hkl}\|^2 \leq \frac{1}{6} + \frac{1}{6}U_{2h\,2k\,0} + \frac{1}{3}U_{(h-k)\,(h+2k)\,0} + \frac{1}{3}U_{hk0}$
$6_1, 6_5$	$\|U_{hkl}\|^2 \leq \frac{1}{6} + \frac{1}{6}(-1)^l\,U_{2h\,2k\,0} + \frac{1}{3}(\cos 2\pi\frac{1}{3}l)U_{(h-k)\,(h+2k)\,0}$ $+\frac{1}{3}(\cos 2\pi\frac{1}{6}l)U_{hk0}$
$6_2, 6_4$	$\|U_{hkl}\|^2 \leq \frac{1}{6} + \frac{1}{6}U_{2h\,2k\,0} + \frac{1}{3}(\cos 2\pi\frac{1}{3}l)U_{(h-k)\,(h+2k)\,0}$ $+\frac{1}{3}(\cos 2\pi\frac{1}{3}l)U_{hk0}$
6_3	$\|U_{hkl}\|^2 \leq \frac{1}{6} + \frac{1}{6}(-1)^l\,U_{2h\,2k\,0} + \frac{1}{3}U_{(h-k)\,(h+2k)\,0} + \frac{1}{3}(-1)^l\,U_{hk0}$
$\bar{6} = 3/m$	$\|U_{hkl}\|^2 \leq \frac{1}{6} + \frac{1}{6}(-1)^l\,U_{0\,0\,2l} + \frac{1}{3}\|U_{(h-k)\,(h+2k)\,0}\|\cos 2\pi\phi_{(h-k)\,(h+2k)\,0}$ $+\frac{1}{3}\|U_{(h-k)\,(h+2k)\,2l}\|\cos 2\pi\phi_{(h-k)\,(h+2k)\,2l}$

Ramachandran and Raman [5.38] showed that this effect can be used to determine the phase ϕ_{hkl} of a reflection. The phase ϕ_{hkl} is given by

$$\phi_{hkl} = \frac{\pi}{2} + \phi_A - \theta \tag{5.42}$$

Here ϕ_A is the contribution of the anomalous scatterer to the total phase; this can be calculated from knowledge of the coordinates of the anomalous scatterers. The angle θ is given by

$$\theta = \Delta F/2F'', \tag{5.43}$$

where

$$\Delta F = \left[|F_{hkl}|^2 - |F_{\bar{h}\bar{k}\bar{l}}|^2\right]^{1/2}.$$

F'', again, is the contribution of the anomalous scatterer to the structure factor.

Ramachandran and Raman [5.38] used this method to determine the phases of reflections in ephedrine hydrochloride. This effect is prominent in halogens, sulphur, C, N and O when Cu $K\alpha$ radiation is used. It is therefore very useful in the study

Box 5.1

Herbert A. Hauptman (born in 1917) developed a mathematical method for the determination of the three-dimensional molecular structure of proteins and enzymes. Hauptman was awarded the Nobel Prize for chemistry in 1985.

H.A. Hauptman

Jerome Karle

Jerome Karle, an American chemist (born in 1918) shared the Nobel Prize for chemistry in 1985 with H.A. Hauptman for their achievements in the development of direct methods for the determination of crystal structures.

Box 5.2

Dorothy Crowfoot Hodgkin an English chemist is known for her work in the field of X-ray crystallographic studies of biomolecules. She was awarded the Nobel Prize for chemistry in 1964 for determining the structures of Penicillin, vitamin B_{12} *and insulin.*

D.C. Hodgkin (1910–1994)

of biological crystals. It may be mentioned that this method was used by Dorothy Hodgkin [5.39] (Box 5.2) in her study of the structure of vitamin B_{12}.

5.4.4 Fourier Synthesis in Practice

Once the intensities are measured, we can determine the structure amplitude $|F_{hkl}|$. The phase ϕ_{hkl} of a reflection can be determined by one of the methods discussed in Sect. 5.4.3. With these two determinations, we have full knowledge of the structure factor F_{hkl} and the stage is set for the evaluation of the electron density ρ_{xyz} using the relation

$$\rho_{xyz} = \frac{1}{V} \sum_h \sum_k \sum_{l=-\infty}^{\infty} F_{hkl} \mathrm{e}^{-2\pi \mathrm{i}(hx+ky+lz)}. \tag{5.24}$$

There is a complex term in (5.24) and, in fact, another in F_{hkl} because of the phase. But the charge density is a real quantity and so we shall express (5.24) as

$$\rho_{xyz} = \left(\frac{1}{V}\right) F_{000} + \frac{2}{V}\sum_{h=0}^{\infty}\sum_{k=-\infty}^{\infty}\sum_{l=-\infty}^{\infty}\big[|F_{hkl}|\cos\phi_{hkl}\cos 2\pi(hx+ky+lz)$$
$$+|F_{hkl}|\sin\phi_{hkl}\sin 2\pi(hx+ky+lz)\big]. \qquad (5.44)$$

Here $F(000)$ corresponds to $h = k = l = 0$ and is equal to the total number of electrons in the unit cell. Further we are confining the actual summation to half the complete reciprocal lattice and hence have included the factor 2 before the summation.

Let us note some practical aspects of carrying out the calculation of ρ_{xyz} from (5.44). Firstly, a large number of intensities have to be used. This is a necessary requirement. According to the theory of Fourier series, the number of terms should be infinite (as indicated by the limits of h, k, l). In reality, this is not possible but efforts are made to use as many reflections as experimentally possible by using the techniques discussed in Chap. 4. Secondly, ρ_{xyz} has to be calculated point by point. For this, the unit cell edges are divided into intervals. In early work on structure determination, the interval used to be 1/20. For better resolution, smaller intervals are desirable but that increases the number of sample points as may be seen from Table 5.10. This makes the evaluation of ρ in three dimensions, a formidable exercise. Thirdly, because of this division of the unit cell into coordinate points, there is no dimension available to indicate the electron density [5.3]. For all these reasons, three-dimensional Fourier synthesis is carried out only rarely and that too in the final stages of structure determination.

For limited purposes, a one-dimensional Fourier synthesis is useful. If the synthesis is to be done along a cell edge, only $(h00)$ type of reflections are used and the relevant form of (5.24) is

$$\rho_{x00} = \frac{1}{a}\sum_{h=-\infty}^{\infty} F_{h00}\mathrm{e}^{-2\pi i hx}. \qquad (5.45)$$

It is clear that such a synthesis will give the variation of the charge density along the cell edge. Such one-dimensional evaluation of the charge density can be done in other directions too by choosing reflections with appropriate indices. As an example, the charge density variation along the 111 direction in NaCl is shown in Fig. 5.9. The sodium and chlorine peaks are clearly seen.

Table 5.10 Number of sample points in three-dimensional Fourier series calculation

Interval as fraction of cell edge	Number of sample points
1/50	$50^3 = 125{,}000$
1/60	$60^3 = 216{,}000$
1/100	$100^3 = 1{,}000{,}000$
1/120	$120^3 = 1{,}728{,}000$

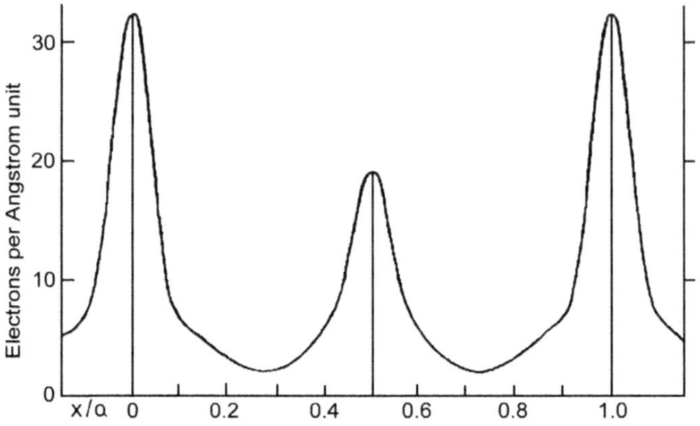

Fig. 5.9 Electron density along 111 line in NaCl; *large* and *small peaks* represent chlorine and sodium ions

Table 5.11 Number of sample points in two-dimensional Fourier series calculation

Interval as fraction of cell edge	Number of sample points
1/50	$50^2 = 2{,}500$
1/60	$60^2 = 3{,}600$
1/100	$100^2 = 10{,}000$
1/120	$120^2 = 14{,}400$

Two-dimensional Fourier synthesis is the most popular. For such synthesis, reflections with one index equal to zero are used. Thus, $(hk0)$ type of reflections will yield electron density distribution in the xy plane. The charge density equation takes the form

$$\rho_{xy} = \frac{1}{S} \sum_h \sum_{k=-\infty}^{\infty} F_{hk0} e^{-2\pi i(hx+ky)}, \tag{5.46}$$

where S is the area of the projection plane. The great advantage of two-dimensional Fourier synthesis over three-dimensional Fourier synthesis can be appreciated if we consider the number of sample points in this case (Table 5.11) for the same intervals. When the charge density is calculated, its value is marked at various points of the coordinate net and the points of equal ρ are joined; these are the contours (Fig. 5.10). One can easily note regions of high electron density; atoms are located at the centre of gravity of these high ρ regions. By counting the total charge in each region and knowing the chemical constitution, the exact atom at a location is identified. It is clear that from such an electron density map, the x, y coordinates of the atoms can be estimated. Similar synthesis with $(0kl)$ and $(h0l)$ reflections will yield the y, z and z, x coordinates. Figure 5.11 shows the projections of electron density on three planes of diopside; the corresponding atomic positions are also shown.

The two-dimensional synthesis discussed earlier gives the electron density in a rational plane. It is also possible to obtain the electron density distribution in a

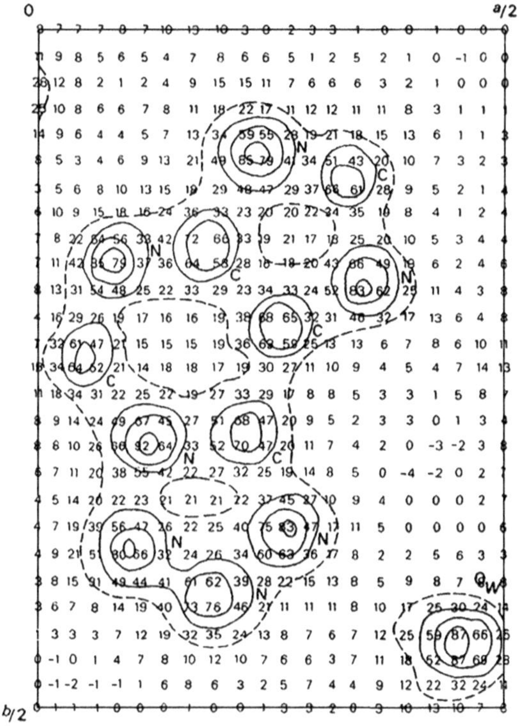

Fig. 5.10 Two-dimensional electron density map (ρ_{xy}) for $C_5H_3N_7 \cdot H_2O$; peak marked O_W represents the oxygen atom of water molecule

plane parallel to the rational plane (110) at a distance, say, z_1. The electron density equation is

$$\rho_{xyz_1} = \frac{1}{V} \sum_h \sum_k \sum_{l=-\infty}^{\infty} F_{hkl} \mathrm{e}^{-2\pi \mathrm{i}(hx+ky+lz_1)}. \tag{5.47}$$

Since z_1 is constant, (5.47) may be written as

$$\begin{aligned} \rho_{xyz_1} &= \frac{1}{V} \sum_h \sum_k \sum_{l=-\infty}^{\infty} F_{hkl} \mathrm{e}^{-2\pi \mathrm{i}(hx+ky)} \mathrm{e}^{-2\pi \mathrm{i} lz_1} \\ &= \frac{1}{V} \sum_h \sum_k \left\{ \sum_{l=-\infty}^{\infty} F_{hkl} \mathrm{e}^{-2\pi \mathrm{i} lz_1} \right\} \mathrm{e}^{-2\pi \mathrm{i}(hx+ky)}. \end{aligned} \tag{5.48}$$

Introducing $Q_{hk} = \left\{ \sum_{l=-\infty}^{\infty} F_{hkl} \mathrm{e}^{-2\pi \mathrm{i} lz_1} \right\}$, (5.48) may be written as

$$\rho_{xyz_1} = \frac{1}{V} \sum_h \sum_k Q_{hk} \mathrm{e}^{-2\pi \mathrm{i}(hx+ky)}. \tag{5.49}$$

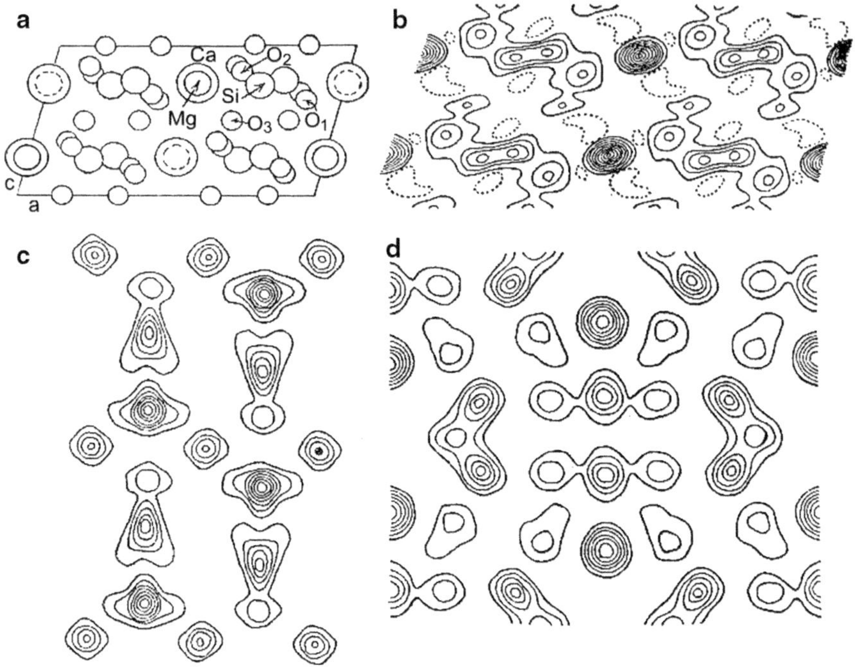

Fig. 5.11 (**a**) Atomic positions and (**b**), (**c**) and (**d**) electron density projections on 010, 100 001 planes, respectively, of diopside

Such electron density distributions may be carried out for different values of z, e.g. z_1, z_2, etc. If these distributions are drawn on transparent sheets of perspex or acetate and several of them are stacked at relevant heights and viewed in diffuse light, what is seen is a two-dimensional projection of the full three-dimensional distribution of electron density. In Fig. 5.12, the electron density distribution in the xz plane (seen along the b-axis) is shown for euphenyl iodoacetate. This was obtained by stacking 17 sections. The molecular model deduced from this electron density map is also shown.

The Fourier synthesis method for the determination of crystal structure was proposed by W.H. Bragg [5.40] (see Box 5.3). But it was actually used by W.L. Bragg [5.25] to verify a previously determined structure of diopside. The first application of the method to determine an unknown structure is credited to Beevers and Lipson [5.16] in their study of $CuSO_4 \cdot 5H_2O$.

5.4.5 *Refinement*

Electron density maps obtained by Fourier synthesis reveal maxima in charge density from which atomic coordinates can be read off. Using these coordinates

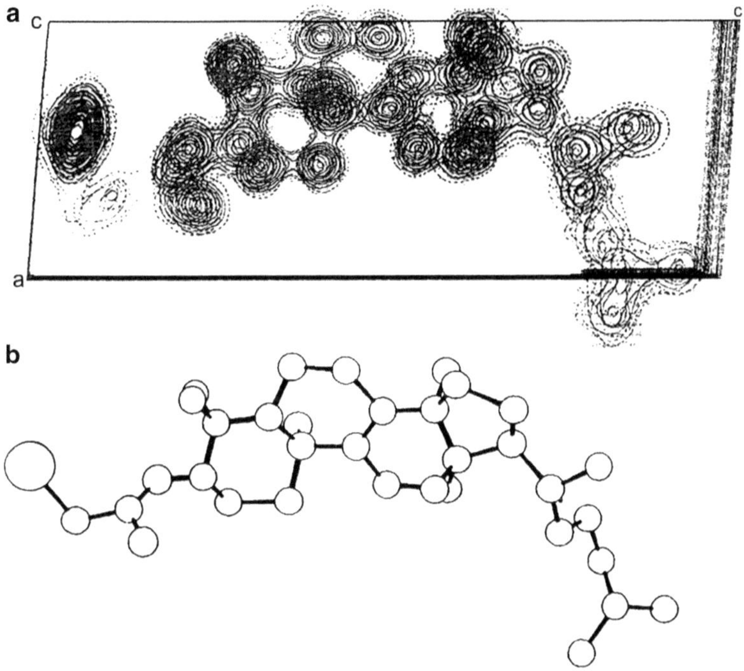

Fig. 5.12 (**a**) Three-dimensional electron density map and (**b**) molecular model for euphenyl iodoacetate

Box 5.3

Sir William Henry Bragg was a British scientist. He introduced the method of Fourier synthesis and laid the foundation for structure determination. W.H. Bragg was awarded the Nobel Prize for physics in 1915.

W.H. Bragg (1862–1942)

in (5.18), the structure amplitude can be calculated; we shall call this $|F_C|$ to distinguish it from $|F_O|$ the structure amplitudes obtained from observed intensities. If the coordinates are correct $|F_C|$ should be equal to $|F_O|$. However, this does not happen and the two sets differ due to various errors. Some errors are (1) errors in measured intensities, (2) errors in absorption correction, (3) errors in extinction correction and (4) approximation in the temperature factor. These errors cause errors in the atomic coordinates. What is the difference between the set of coordinates obtained from electron density maps and the set of true coordinates? This difference is expressed in terms of reliability index R (R-factor) defined by

$$R = \frac{\sum [|F_O| - |F_C|]}{\sum |F_O|}. \tag{5.50}$$

A few other definitions also exist but we shall use the one given by (5.50). According to Wilson [5.41], values of $R > 0.828$ for a centrosymmetric structure indicate that the structure is wrong; the corresponding value for a noncentrosymmetric structure is 0.586. Efforts are made to reduce R by incorporating two improvements. Firstly, shifts are introduced in the values of the atomic coordinates. Secondly, a more elaborate temperature factor is included. In place of the single common B-factor, individual B factors are included for each atom. Further, instead of isotropic B-factors, anisotropic B-factors are employed having the following form:

$$M = \frac{1}{4}(h^2a^{*2}B_{11} + k^2b^{*2}B_{22} + l^2c^{*2}B_{33} + 2hka^*b^*B_{12} + 2klb^*c^*B_{23} + 2lhc^*a^*B_{31}), \tag{5.51}$$

where a^*, b^*, c^* are the reciprocal cell parameters. After incorporating these changes $|F_C|$'s and R are recalculated. A value of $R < 0.5$ indicates that the structure (i.e. the set of atomic coordinates) is close to (but not exactly the same as) the true set.

At this stage, the process of refinement begins. Refinement is the process of successive calculation of electron densities, atomic positions, $|F_C|$'s and R so that the value of R is as low as 0.05. A further reduction may not be possible due to residual random errors.

Refinement is a long and laborious process. It is possible to carry out refinement in a systematic manner by using the method of least squares. Let x_i be the value of an atomic coordinate estimated by an initial Fourier synthesis and let e_i be the error in it, i.e. the true value is $(x_i + e_i)$. Then

$$|F_C| = f(x_i) \quad \text{and} \quad |F_O| = f(x_i + e_i). \tag{5.52}$$

Here $f(x_i)$ means a function of x_i. Obviously $|F_C| = |F_O|$ when $e_i \to 0$. Since the e_i's are small, $|F_O|$ can be expressed as a Taylor series:

$$|F_O| = |F_C| + \left[\sum_{i=1}^{N} \frac{\partial F_C}{\partial x_i} e_i\right] + \frac{1}{2}\left[\sum_{i=1}^{N} \left(\frac{\partial F_C}{\partial x_i} e_i\right)^2\right] + \cdots. \tag{5.53}$$

We shall consider only the first term in e_i and ignore the rest. Then

$$\Delta F = |F_O| - |F_C| = \sum_{i=1}^{N} \frac{\partial F_C}{\partial x_i} e_i. \tag{5.54}$$

Let us multiply both sides by $\dfrac{\partial F_C}{\partial x_j}$. Then

$$\Delta F^q \frac{\partial F_C^q}{\partial x_j} = \sum_{i=1}^{N} \frac{\partial F_C^q}{\partial x_j} \frac{\partial F_C^q}{\partial x_i} e_i . \tag{5.55}$$

Here q represents h,k,l. If there are Q reflections, there will be Q such equations. Adding up all such equations, we have

$$\underbrace{\sum_{q=1}^{Q} \Delta F^q \frac{\partial F_C^q}{\partial x_j}}_{Q \text{ terms}} = \underbrace{\sum_{i=1}^{N}}_{N \text{ terms}} \left[\left(\underbrace{\sum_{q=1}^{Q} \frac{\partial F_C^q}{\partial x_i} \frac{\partial F_C^q}{\partial x_j}}_{Q \text{ terms}} \right) \right] e_i . \tag{5.56}$$

In these equations, j is fixed. Equation (5.56) may be rewritten as

$$K_j = \sum_{i=1}^{N} a_{ij} e_i \tag{5.57}$$

with $K_j = \sum_{q=1}^{Q} \Delta F^q \dfrac{\partial F_C^q}{\partial x_j}$ and $a_{ij} = \sum_{q=1}^{Q} \dfrac{\partial F_C^q}{\partial x_i} \dfrac{\partial F_C^q}{\partial x_j}$.

There will be N such equations one for each j with j taking values $1, 2, 3 \ldots N$. Equation (5.57) can be written in matrix form as

$$[K_j] = [a_{ij}]\ [e_i] . \tag{5.58}$$

We shall multiply (5.58) by $[b_{ij}]$ where $[b_{ij}]$ is the inverse of $[a_{ij}]$. Then

$$[K_j]\ [b_{ij}] = [a_{ij}]\ [b_{ij}]\ [e_i] = [e_i] . \tag{5.59}$$

Thus, we can evaluate e_i and apply the necessary correction to the coordinates x_i. Finally, after this correction, R is again calculated. To get the best values for the atomic coordinates, the procedure of refinement is repeated 10–12 times.

5.5 Ambiguous Space Groups

In Sect. 5.3, it was stated that space groups can be determined from systematic absences. Some space groups can be determined with certainty as the relevant rules for systematic absences are unique for them. But there are several cases where the rules of systematic absences are common to two or more space groups. In such cases different strategies have to be adopted to identify the correct space group. Some examples are discussed below.

5.5.1 Methods Based on Physical Properties

Crystals grown from solution generally have well-developed facets which reveal the symmetry of the faces. Sometimes this provides crucial information. In the determination of the structure of $CuSO_4 \cdot 5H_2O$ (which is triclinic), Beevers and Lipson [5.16] made use of an earlier observation by Tutton [5.42] that the crystals of $CuSO_4{\cdot}5H_2O$ show well-developed (001) parallel faces. This indicates that there is a centre of symmetry. The space group was thus taken to be $P\bar{1}$. Again, in the structure determination of Ag_3AsS_3 (proustite), Harker [5.43] observed that reflections of the type (0*kl*) with *l* odd were systematically absent which indicated that the space group is either $R3c$ or $R\bar{3}c$. Harker observed that the crystals of proustite have a hemihedral growth habit which indicates a lack of centre of symmetry. The space group was thus taken to be $R3c$ and not $R\bar{3}c$.

Pyroelectricity, piezoelectricity, optical activity and second harmonic generation are properties associated with the absence of a centre of symmetry. If two crystals differ in being either centrosymmetric or noncentrosymmetric, they can be distinguished by looking for these properties. Thus, in the case of bishydroxydurylmethane (Chaudhari and Hargreaves [5.44]) observed that (*hkl*) reflections with $(h + k)$ odd are absent. This condition is common to space groups $C2/m$, $C2$ or Cm. The presence of pyroelectricity eliminated the possibility of space groups $C2/m$ and Cm leaving $C2$ as the correct space group. The limitation of these properties is that they are often very feeble for measurement.

5.5.2 Methods Based on Diffraction Effects

We shall now consider some methods based on diffraction effects which are helpful in resolving ambiguity in space group.

The determination of a Patterson map sometimes reveals symmetry (or lack of symmetry) which is crucial to the problem. We shall cite the case of calcium uranate (Ca_3UO_6) studied by Rietveld [5.45]. In the X-ray pattern, $0k0$ reflections with k odd were systematically absent which meant that the space group is either $P2_1$ or $P2_1/m$. A three-dimensional Patterson synthesis showed that the U–Ca peaks were elongated in the y-direction indicating that they were lying outside the xz plane. This eliminates the possibility of xz being a mirror plane and leaves $P2_1$ as the only possible space group.

Another approach is the statistical distribution of intensities. Let $<I>$ be the average intensity of a group of reflections in a range of $\sin\theta$ values. If I is the intensity of a particular reflection, we shall define the ratio $I/<I>$ as the quantity z. The number of reflections in the group having intensity ratio $\leq z$ is denoted by $N(z)$. Howells et al. [5.46] showed that the relationship between $N(z)$ and z is entirely different for centrosymmetric and noncentrosymmetric structures. Thus,

$$N(z) = 1 - \mathrm{e}^{-z} \text{ for centrosymmetric structures}$$

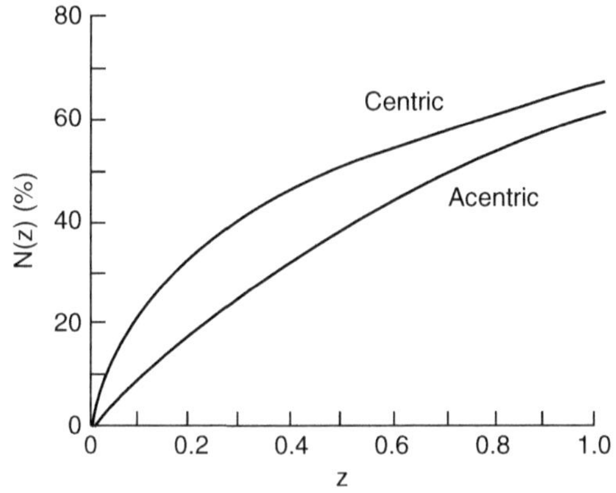

Fig. 5.13 Theoretical $N(z)$ versus z curves for a centric and an acentric crystal

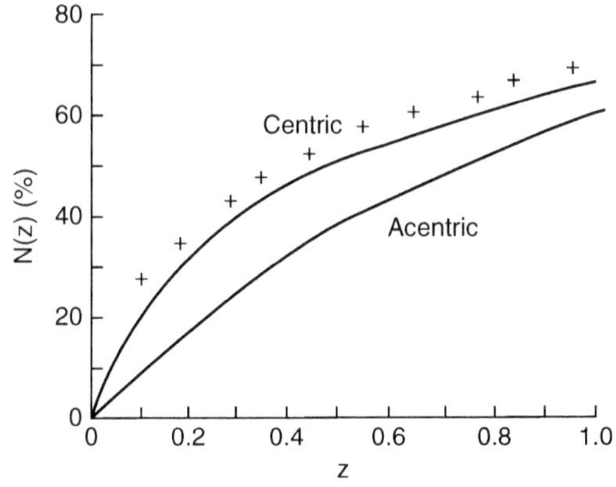

Fig. 5.14 $N(z)$ versus z plot of experimental points for anthrahydroquinone dibenzoate (+)

and

$$N(z) = 1 - \text{erf}(z/2)^{1/2} \text{ for noncentrosymmetric structures.} \tag{5.60}$$

Here 'erf' stands for error function. The $N(z)$ versus (z) curves for the two cases are shown in Fig. 5.13. For the triclinic crystal $C_{28}H_{18}O_4$ it had to be decided whether the space group is $P1$ or $P\bar{1}$. The $N(z)$ values from observed intensities are shown in Fig. 5.14 (Iball and Mackay, [5.47]). It is clear that the structure is centrosymmetric and the space group is $P\bar{1}$. In a monoclinic crystal metatolidine dihydrochloride, X-ray data [5.48] could not decide between space groups Pm, $P2$ or P/m. The plot between $N(z)$ and (z) (Fig. 5.15) showed that the structure is noncentrosymmetric and, so, the space group can only be $P2$.

If all these methods do not help, the last resort is to assume atomic positions conforming to each of the ambiguous space groups and refine each structure to the lowest R value; the structure with the lowest R is taken as acceptable. As an example we shall discuss the structure determination of α-K_2SO_4 which is hexagonal at 630°C. The powder pattern gave 21 reflections with measurable intensities. Systematic absence of hhl reflections with $l = 2n + 1$ indicated that

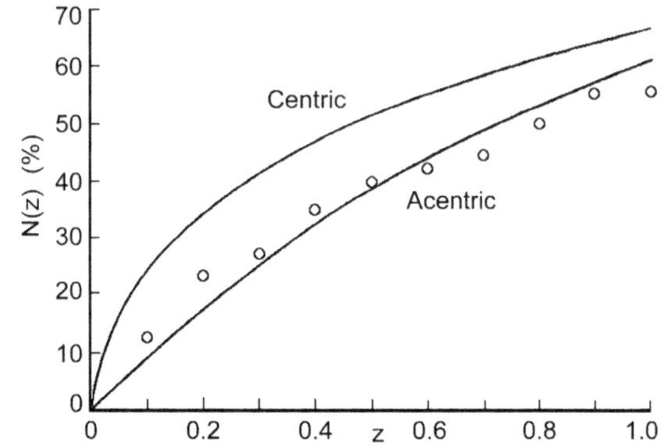

Fig. 5.15 $N(z)$ versus z plot of experimental points for metatolidine dihydrochloride

the space group could be $P31c$, $P\bar{3}1c$, $P6_3mc$, $P\bar{6}2c$ or $P6_3/mmc$. The structure assuming each of these space groups was refined and the minimum R was found for $P6_3/mmc$ [5.49].

5.6 Hydrogen Atom Positions

The hydrogen atom has the least scattering power for X-rays (smallest f). As a result, in a structure containing hydrogen atoms, the contribution of the hydrogen atoms to the structure factor is very small compared to that of the rest of the atoms and the electron density map derived from the structure factors does not reveal the presence of hydrogen atoms. We shall discuss two methods by which hydrogen positions can be determined: X-ray difference synthesis and neutron diffraction.

5.6.1 X-ray Difference Fourier Synthesis

Let ρ_O be the electron density obtained from Fourier synthesis of observed structure factors F_O and ρ_C that obtained from calculated structure factors F_C. For simplicity, let us consider a two-dimensional synthesis using ($hk0$) reflections. Then

$$\rho_O = \frac{1}{S}\sum_{h}\sum_{k=-\infty}^{\infty} F_O(hk0)e^{-2\pi i(hx+ky)} \tag{5.61}$$

and

$$\rho_C = \frac{1}{S}\sum_{h}\sum_{k=\infty}^{-\infty} F_C(hk0)e^{-2\pi i(hx+ky)}. \tag{5.62}$$

The difference $D = \rho_O - \rho_C$ is given by

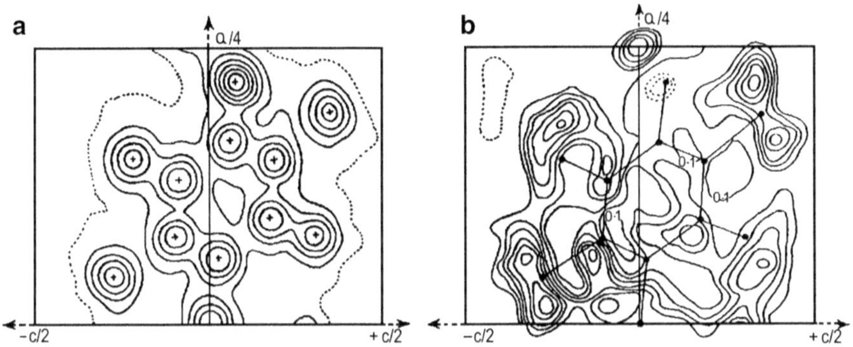

Fig. 5.16 Projections on the (010) plane of (**a**) electron density and (**b**) (F_O–F_C) synthesis for bishydroxydurylmethane; extra features in (**b**) relate to hydrogen atoms

$$D = \Delta\rho = \rho_O - \rho_C = \frac{1}{S}\sum_h \sum_{k=\infty}^{-\infty} [F_O(hk0) - F_C(hk0)]\, e^{-2\pi i(hx+ky)}. \quad (5.63)$$

The evaluation of D as a function of x, y is called difference Fourier synthesis. The method was introduced by Cochran [5.50]. It may be noted that difference Fourier synthesis is a point-by-point subtraction of the electron density obtained from observed structure factors and the true electron density. A D-map has the following properties:

1. If the calculated structure is the true structure, the D-map will be a flat featureless map.
2. If an atom is incorrectly placed in the ρ_C map, it shows up as a negative feature in the D-map.
3. Any series termination error in the ρ_C map is eliminated in the D-map.
4. Light atoms such as hydrogen which are not seen in the ρ_C map are seen in the D-map. This is the most important application of difference synthesis. However, due to the inherently feeble scattering by the H atoms, the positions of H atoms cannot be determined as accurately as those of heavier atoms.

As an example of the use of difference Fourier synthesis, the ρ and D maps for bishydroxydurylmethane obtained by Chaudhuri and Hargreaves [5.44] are shown in Fig. 5.16; hydrogen atoms not seen in Fig. 5.16a are seen in Fig. 5.16b.

5.6.2 Neutron Diffraction

We have noted in Chap. 4 and again in Sect. 5.6.1 that the limitation in X-ray diffraction in locating hydrogen atoms is due to the severe difference in the atomic factors for light and heavy atoms. It is precisely in this respect that neutron diffraction has an advantage over X-ray diffraction. The neutron scattering

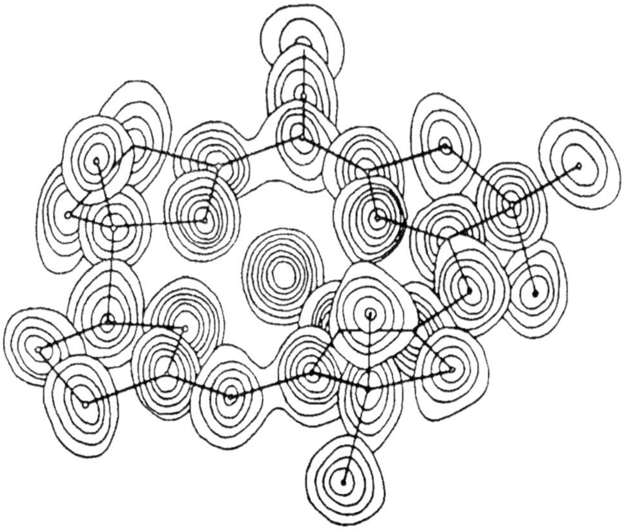

Fig. 5.17 Electron density map for nickel corrin derivative (vitamin B_{12})

amplitudes for light and heavy atoms do not differ severely; they are comparable. Hence in neutron diffraction, the contribution of H atoms to the structure factor is comparable to the contribution of the other nonlight atoms. As a result, in the Fourier synthesis of neutron structure factors the hydrogen atoms show up as naturally as the other atoms. The diffraction procedures are the same in X-ray and neutron diffraction with the difference that in the former, we use atomic scattering factors and in the latter, the neutron scattering amplitude. Also, the Fourier synthesis of neutron structure factors leads to neutron scattering density in place of the electron density obtained from X-ray Fourier synthesis.

As an example, the electron density map for the corrin derivative (vitamin B_{12}) obtained by Dunitz and Mayer [5.51] from X-ray data is shown in Fig. 5.17. This may be compared with the neutron scattering density map (Fig. 5.18) for the same compound obtained by Moore and Willis [5.52]. Hydrogen atoms not seen in Fig. 5.17 are clearly seen in Fig. 5.18.

Finally, we shall discuss the results of X-ray diffraction study (Sime and Abrahams [5.27]) and neutron diffraction study (Bacon and Curry [5.53]) of 4, 4′-dichloro-diphenyl sulphone. The electron density map is shown in Fig. 5.19; this shows atoms other than hydrogen. The neutron scattering density map is shown in Fig. 5.20a. As expected, it shows the atoms in Fig. 5.19 and also the hydrogen atoms (broken contours). Bacon and Curry also carried out a difference Fourier synthesis (F_{CH}–F_C) where F_{CH} and F_C were the neutron structure factors with and without the H atoms. The projection (Fig. 5.20b) now shows only the hydrogen atoms. This is a beautiful example that shows the power of neutron diffraction as well as that of difference synthesis.

Fig. 5.18 Neutron scattering density map for corrin nucleus (vitamin B_{12}); *dotted contours* represent hydrogen atoms

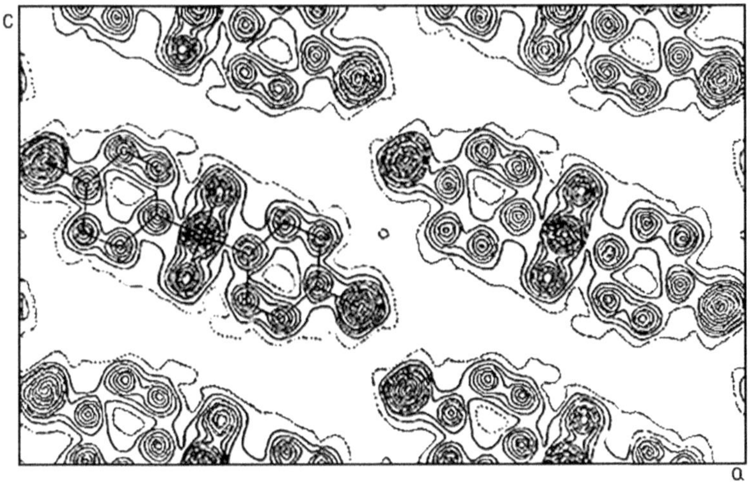

Fig. 5.19 Electron density projection along *b*-axis for 4, 4′dichloro-diphenyl sulphone

5.7 Rietveld Method

In diffraction patterns, considerable information is lost because of overlapping of reflections. This is particularly so in powder patterns. On the other hand, powder pattern is the sole permissible technique when single crystals are not available. A method to obtain structural parameters from powder patterns was introduced by Rietveld [5.54, 5.55] which does not use integrated intensities but, instead, makes use of intensities at every point on the profile of a reflection.

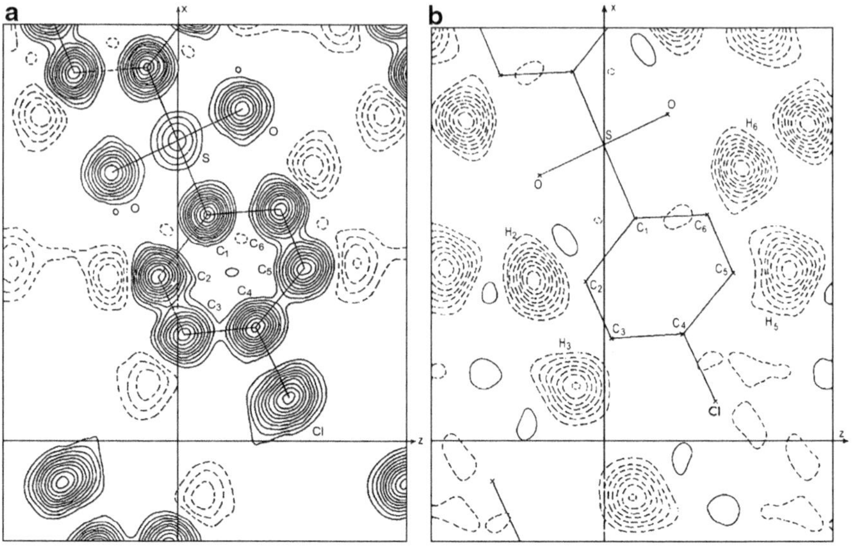

Fig. 5.20 (**a**) Neutron scattering density projection along b-axis for the crystal in Fig. 5.19, (**b**) the same obtained by $(F_{\mathrm{CH}}-F_{\mathrm{C}})$ difference synthesis

Let us assume that the line profile is Gaussian. Let θ_k be the Bragg angle of a reflection k and θ_i be the angle at a particular point on the profile. The intensity y_i at θ_i is given by

$$y_i = tS_k^2 j_k L_k \frac{2\sqrt{\log 2}}{H_k\sqrt{\pi}} \exp[-4\log 2\{(2\theta_i - 2\theta_k)/H_k\}^2]. \tag{5.64}$$

Here t = step width of the detector, j_k = multiplicity, L_k = Lorentz factor, H_k = full width at half maximum and

$$S_k^2 = F_k^2 + J_k^2.$$

The angular dependence of H_k is given by

$$H_k^2 = U\ \tan 2\theta_k + V\ \tan\theta_k + W. \tag{5.65}$$

The constants U, V, W are obtained by manually fitting the measured value of H_k for a few peaks to (5.65). If neutron diffraction is used $S_k^2 = F_k^2 + J_k^2$, where F_k is the structure factor calculated with nuclear scattering amplitudes and J_k is the structure factor calculated with magnetic amplitudes (Halpern and Johnson [5.56]). On the other hand, if X-ray diffraction is used, $S_k^2 = F_k^2$ where F_k is the conventional structure factor. In X-ray diffraction, the line shape is necessarily Gaussian. If another line shape is used (5.64) is modified. Equation (5.64) may be

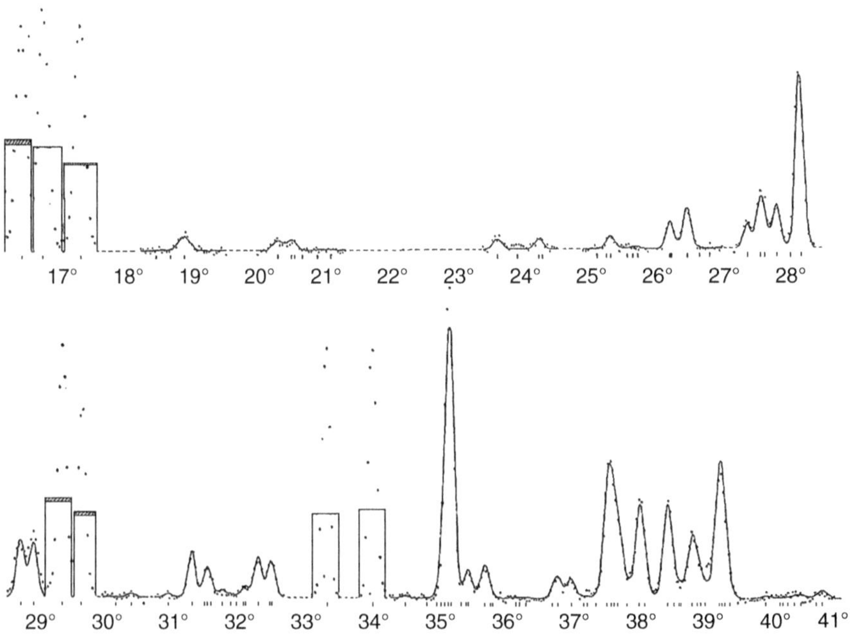

Fig. 5.21 Neutron powder diagram of WO_3 (intensity versus 2θ). *Solid lines* indicates Rietveld calculated profile and the *dots* represent measured intensities

conveniently written as

$$y_i = I_k \exp[-\{b_k(2\theta_i - 2\theta_k)^2\}], \tag{5.66}$$

where $b_k = 4 \log 2/H_k^2$ and I_k represents the pre-exponential factor in (5.64).

Computer programmes are available [5.54, 5.57] for least squares refinement based on (5.64, 5.66). The programmes offer a choice of line profiles. The output is atomic coordinates, Debye–Waller factors and intensities calculated therefrom. The recorded (continuous line) and calculated (dots) neutron diffraction pattern for WO_3 is shown in Fig. 5.21.

5.8 Flow Sheet of Structure Analysis

The real process of structure determination starts after the preliminary X-ray measurements have been made. These consist of recording the diffraction pattern, measurement of Bragg angles, identification of crystal class, indexing of the reflections, determination of unit cell dimensions, number of formula units in a unit cell, measurement of intensities and determination of space group. The final stage viz. determination of atomic positions is the most important part and involves the Fourier methods.

Table 5.12 Flow sheet of structure determination of cobyric acid

B factors	Process	Atoms	R
	$F^2(hkl)$, $F^2(h\bar{k}\bar{l})$		
	↓		
	P_{xyz}		
	↓		
	Co		
	↓		
	$\|F\|$ and phase angles		
B: Co, 2.7, all other atoms, 3.0	ρ1 set 1	67 atoms	28.5
	ρ2 set 2	67 atoms	23.2
	least squares		
B: individual	set 3		19.7
	$\Delta\rho$3		
	set 4		16.4
	$\Delta\rho$4		
	set 5	78 atoms +52 H atoms	16.0
	$\Delta\rho$5 and ρ5		
	set 6		15.7
B: Co, anisotropic, other atoms isotropic	5 rounds of least squares		
	set 11		
	set 12		13.5
	least squares		
B: all atoms anisotropic	set 13		13.3
	ρ6 and $\Delta\rho$6		
	set 14		12.7
	ρ7 and $\Delta\rho$7		12.05

To get a feel of how these methods lead us from an unknown structure to a well-determined structure, we shall consider the structure determination by Venkatesan et al. [5.58] of cobyric acid ($C_{46}H_{66}O_9N_{11}Co{\cdot}11H_2O$) which is related to the vitamin B_{12} structure. It is a monoclinic crystal. More than 6,000 reflections were recorded with a Weissenberg camera and the intensities were estimated visually. The progress of structure determination is shown in Table 5.12. In the flow sheet, P_{xyz} represents Patterson synthesis, ρ_{xyz} Fourier synthesis of F_{hkl} and $\Delta\rho$ difference synthesis.

To start with, the intensities were used to plot Patterson maps from which the position of the cobalt atom could be determined. Next, the phases of reflections were determined using the anomalous scattering and heavy atom methods. The structure factors F_{hkl} (structure amplitudes $|F_{hkl}|$ and phase angles ϕ_{hkl}) were then subjected to Fourier synthesis. The initial Fourier synthesis (ρ) revealed positions of some atoms. The Fourier synthesis and difference synthesis ($\Delta\rho$) were repeated revealing more atoms. The structure was progressively improved by (1) inclusion of additional atoms and their positions and (2) inclusion of individual anisotropic temperature factors. In all 130 atoms were located. This meant determination of $(130 \times 9 + 1)$ parameters (three position coordinates and six thermal parameters for each atom + a scale factor) by 11 Fourier syntheses and 5 difference syntheses; in the process the R-factor got reduced from 28.5 to 12.05%.

5.9 An Update

It may be emphasised that what we have discussed so far are the principles of structure determination. To start with crystal structure analysis was initiated in physics and mineralogy laboratories but very soon it spread to chemistry, biology and medical laboratories. Here, the interest was in 'large crystals', i.e. crystals containing as many as a few hundred atoms in the unit cell; this meant determining many more unknown parameters. A number of developments have taken place which has made modern structure determination a sophisticated, highly automatised technique.

On the experimental side, a great leap was achieved with the design of the four-circle single crystal diffractometer. With modest, preliminary manoeuvres by the investigator, this instrument proceeds to record intensities of thousands of reflections automatically without any further intervention by the investigator. The discovery of the charged coupled device area detector and the availability of intense X-ray beam from a synchrotron have made the study of soft biological crystals possible. On the theoretical side, the direct methods and the method of inequalities in several forms enable the study of structures which were not amenable to heavy atom and isomorphous-replacement methods.

But the real problem was the calculations involved in the various types of synthesis and refinements. Different computing devices were introduced to do these calculations like the Beevers–Lipson strips, mechanical and electrical calculators and finally the computer. Powerful dedicated computers are now employed in crystal structure analysis. They are equipped with software for every stage of structure analysis like indexing, crystal system identification, unit cell parameter determination, L.P and absorption correction for intensities, space group identification, phase determination, trial structures, refinement and even graphics. Some of the programmes are:

Programme	Function
POWDERX	Unit cell parameters from powder data
SAINT	Unit cell parameters from single crystal data
SADABS – 2002	Absorption
SMART	Intensity data collection
SHELXS	To solve structure
SHELXL	Refinement
Busing and Levy [5.59]	Anisotropic thermal parameters
Busing et al. [5.60]	Refinement
Sheldrik [5.61]	Refinement

Many more programmes are listed by Massa [5.9]. Mention may also be made of the books on crystallographic computing by Ahmed et al. [5.62], Sayre [5.63] and Hall and Ashida [5.64].

In the period, 1930–1950, complete structure determination of a crystal with 10–20 atoms in the unit cell used to take a year or even more. In the initial days of computerisation (1960–1980), the time taken was reduced to a few months if not less. Today, it is not unusual for a crystallographer to complete structure determination of a moderately complex crystal in a few days. Very large crystals like proteins, however, still involve lot of time and labour.

Problems

1. Derive the rules of systematic absences for a structure with (a) a 100 glide plane with translation component $b/2$ and (b) a [100] screw diad axis with translation component $a/2$.
2. Derive the extinction conditions for the space group $P2_12_12_1$. An orthorhombic crystal has the following reflections in its diffraction photograph.

hkl	*hkl*	*hkl*	*hkl*
111	011	110	020
112	021	120	040
212	012	310	060
312	101	200	002
322	203	400	004
332	303	600	006

Show that this pattern conforms to space group $P2_12_12_1$.
3. An available computer can determine a maximum of 200 variables. How many atoms can be refined with (a) a single mean temperature factor, (b) with individual isotropic temperature factors and (c) with individual anisotropic temperature factors?
4. Atoms A and B of a cubic crystal are located at A: 000 and B: $1/2, 1/2, 1/2$. Show that the crystal diffracts only reflections of the type $(h + k + l)$ all even or all odd.
5. Using Wilson's method, estimate the scale factor and the mean B-factor for a crystal with data given below.

$\log \sum_i f_i^2(hkl) / < \lvert F_0(hkl)^2 \rvert >$	$< (\sin\theta)^2/\lambda^2 >$
4	0.10
5.6	0.20
6.5	0.30
7.9	0.40
9.4	0.50

References

5.1. Lipson, W. Cochran, *Determination of Crystal Structures* (G. Bell and Sons, London, 1953).
5.2. H.S. Peiser, H.P Rooksby, A.J.C. Wilson, *X-ray Diffraction by Polycrystalline Materials* (Institute of Physics, London, 1955)
5.3. M.J. Buerger, *Crystal Structure Analysis* (John Wiley and Sons, New York, 1960)
5.4. E.F. Kaelble, *Handbook of X-rays* (McGraw-Hill Company, New York, 1967)
5.5. H.P. Klug, L.E. Alexander, *X-ray Diffraction Procedures* (John Wiley and Sons, New York, 1967)
5.6. G.B. Carpenter, *Principles of Crystal Structure Determination* (W.A. Benjamin, New York, 1969)
5.7. M.M. Woolfson, *An Introduction to X-ray Crystallography* (Vikas Publications, New Delhi, 1980)
5.8. M.F.C. Ladd, R.A. Palmer, *Structure Determination by X-ray crystallography* (Plenum Press, New York, 1977)
5.9. W. Massa, *Crystal Structure Determination* (Springer, Berlin, 2000)
5.10. R. Hesse, Acta Cryst. **1**, 200 (1948)
5.11. H. Lipson, Acta Cryst. **2**, 43 (1949)
5.12. T. Ito, *X-ray Studies on Polymorphism* (Maruzen, Tokyo, 1950)
5.13. D.B. Sirdeshmukh, L. Sirdeshmukh, K.G. Subhadra, *Micro- and Macro-Properties of Solids* (Springer, Berlin, 2001)
5.14. J.W. Mellor, *A Comprehensive Treatise on Inorganic and Theoretical Chemistry*, Vol. 12 (Orient Longman's, London)
5.15. P. Groth, *Chemische Krystallographie*, 5 vols. (Engelmeann, Leipzig, 1906–1914)
5.16. C.A. Beevers, H. Lipson, Proc. R. Soc. **A146**, 570 (1934)
5.17. D. Crowfoot, J. Chem. Soc. 1241 (1938)
5.18. F.H. Herbstein, Acta Cryst. **13**, 357 (1960)
5.19. J.D. Bernal, D. Crowfoot, Nature 1**34**, 809 (1934)
5.20. H.M. Powell, D.M. Crowfoot, Z. Kristallogr. Kristallgeom. **87**, 370 (1934)
5.21. R.W.G. Wyckoff, *Crystal Structures*, Vol. 1 (Interscience, New York, 1965)
5.22. S. Raman, J.L. Katz, in *Handbook of X-rays,* ed. by E.F. Kaelble (McGraw-Hill, New York, 1967)
5.23. P. Niggli, *Geometrische Kristallographic des Discontinuums* (Gobruder Born Traeger, Leipzig, 1919)
5.24. *International Tables for Crystallography*, Vol. I, International Union of Crystallography (Kynoch Press, Bermingham, 1982)
5.25. W.L. Bragg, Proc. R. Soc. (Lond.) **A123**, 537 (1929)
5.26. J.M. Robertson, J. Chem. Soc. 1195 (1936)
5.27. J.G. Sime, S.C. Abrahams, Acta Cryst. **13**, 1 (1960)
5.28. A.J.C. Wilson. Nature, **150**, 152 (1942)
5.29. A.L. Patterson, Phys. Rev. **46**, 372 (1934)
5.30. J.M. Cork, Philos. Mag. **4**, 688 (1927)
5.31. H. Wiebenga, C.J. Crom, Recl. Trav. Chim. **65**, 663 (1946)
5.32. J.M. Robertson, I. Woodward, J. Chem. Soc. 36 (1940)
5.33. D. Harker, J.S. Kasper, Acta Cryst. **1**, 70 (1948)
5.34. J. Gillis, Acta Cryst. **1**, 76 (1948)
5.35. J. Karle, H. Hauptman, Acta Cryst. **3**, 181 (1950)
5.36. D. Sayre, Acta Cryst. **5**, 60 (1952)
5.37. W. Cochran, Acta Cryst. **5**, 65 (1952)
5.38. G.N. Ramachandran, S. Raman, Curr. Sci. **25**, 348 (1956)
5.39. D. Hodgkin, Nobel Lect. (Chemistry) (1965)
5.40. W.H. Bragg, Phil. Trans. R. Soc. **210**, 253 (1915)
5.41. A.J.C. Wilson. Acta Cryst. **3**, 397 (1950)

5.42. A. Tutton, *Crystallography and Practical Crystal Measurement* (Macmillan, London, 1922)
5.43. D. Harker, J. Chem. Phys. **4**, 381 (1936)
5.44. B. Chaudhuri, A. Hargreaves, Acta Cryst. **9**, 793 (1956)
5.45. H.M. Rietveld, Acta Cryst. **20**, 508 (1966)
5.46. E.R. Howells, D.C. Phillips, D. Rogers, Acta Cryst. **3**, 210 (1950)
5.47. J. Iball, K.J. Mackay, Acta Cryst. **15**, 148 (1962)
5.48. F. Fowweather, A. Hargreaves, Acta Cryst. **3**, 81 (1950)
5.49. A.J. Berg, F. Tuinstra, Acta Cryst. **B34**, 3177 (1978)
5.50. W. Cochran, Acta Cryst. **4**, 81 (1951)
5.51. J.D. Dunitz, E.E. Meyer, quoted by Dorothy Crowfoot Hodgkin, Les Prix Nobel, p. 157, 1965.
5.52. F.M. Moore, B.T.M. Willis, Nature, **214**, 129 (1967)
5.53. G.E. Bacon, Curry, Acta Cryst. **13**, 10 (1960)
5.54. H.M. Rietveld, Acta Cryst. **22**, 151 (1967)
5.55. H.M. Rietveld, J. Appl. Cryst. **2**, 65 (1969)
5.56. O. Halpern, M.H. Johnson, Phys. Rev. **55**, 898 (1939)
5.57. D.B. Wiles, R.A. Young, J. Appl. Cryst. **14**, 149 (1981)
5.58. K. Venkatesan, D. Dale, D. Crowfoot-Hodgkin, C.E. Nockolds, F.H. Moore, B.H. O'Connor, Proc. R. Soc. **A 323**, 455 (1971)
5.59. W.R. Busing, H.A. Levy, Acta Cryst. **11**, 450 (1958)
5.60. W.R. Busing, K.O. Martin, H.A. Levy, ORFLS (Oakridge National Laboratory, Oakridge, 1962)
5.61. G.M. Sheldrick, Acta Cryst. **A64**, 112 (2008)
5.62. F.R. Ahmed, K. Huml, B. Sedlacek, *Crystallographic Computing Techniques* (Munskgaard, Copenhagen, 1976)
5.63. D. Sayre, *Computational Crystallography* (Clarendon Press, Oxford, 1982)
5.64. S.R. Hall, T. Ashida, *Methods and Applications of Crystallographic Computing* (Clarendon Press, Oxford, 1984)

Chapter 6
Cohesion

6.1 Introduction

The constituents (atoms/ions/molecules) of gases and liquids are free to move about in the body of the medium. In contrast, the entities in a solid are fixed in positions determined by its crystal structure. What, then, holds a crystal together? In fluids, interatomic distances are of the order of 30 Å while in solids they are much shorter – of the order of 2–4 Å. When atoms, ions or molecules come so close to one another, they exert forces upon each other. These forces depend on the outer electronic structure of the atomic entities. These interatomic forces or bonds are:

1. Ionic
2. Covalent
3. Partially ionic or partially covalent
4. Metallic
5. Van der Waals
6. Hydrogen bond

We discuss the origin of these bonds. Because of the interatomic forces, the crystal has a cohesive energy. The calculation of the cohesive energy of crystals of different bond types is discussed. The relation between the type of bond and the physical properties is also discussed.

The calculation of cohesive energy of ionic crystals is discussed in detail by Tosi [6.1], Kittel [6.2] and Dekker [6.3]. Kubo and Nagamiya [6.4] and Aschroft and Mermin [6.5] discuss the calculation of cohesive energy of crystals of other bond types also.

D.B. Sirdeshmukh et al., *Atomistic Properties of Solids*, Springer Series in Materials Science 147, DOI 10.1007/978-3-642-19971-4_6,

6.2 Types of Bonds

6.2.1 Ionic Bond

Atoms with a closed outer-shell electronic structure are stable. Thus, Ne with $2p^6$ electrons in its outer shell and Ar with $3p^6$ electrons in its outer shell are stable. Atoms which do not have a closed-shell structure tend to assume a closed-shell structure either by giving up an electron or by attracting an electron.

As an example let us consider a sodium and a chlorine atom. As shown in Fig. 6.1a their outer-shell structures are $1s^2(2s^22p^6)3s^1$ and $1s^2(2s^22p^6)(3s^23p^5)$. Thus, the sodium atom can have closed-shell structure if it can lose an electron. Similarly, the chlorine atom can have a closed-shell structure if it can have an additional electron. If the atoms come close, an electron transfer is possible between the two atoms (Fig. 6.1b). For such an electron transfer to take place, one atom (like Na) should have a low ionisation energy and the other (like Cl) should have a high electron affinity. The electronic structure after the electron transfer is shown in Fig. 6.1c; this is a stable structure with a positively charged sodium ion and a negatively charged chlorine ion (Fig. 6.1d). If r is the distance between the ions, a

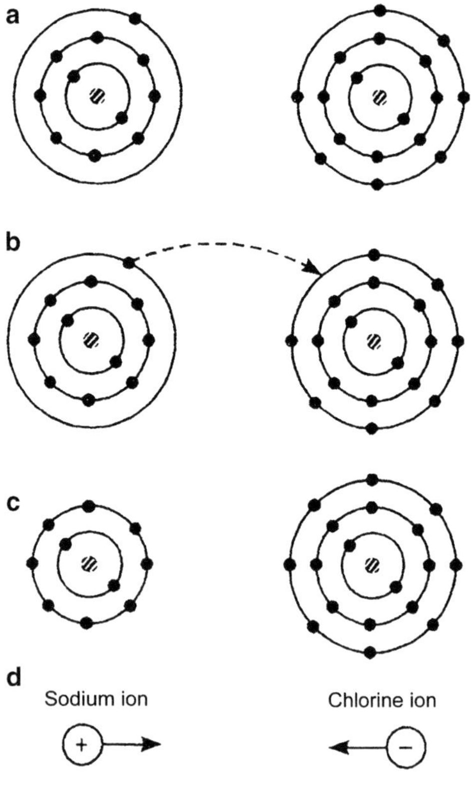

Fig. 6.1 Formation of ionic bond between Na and Cl

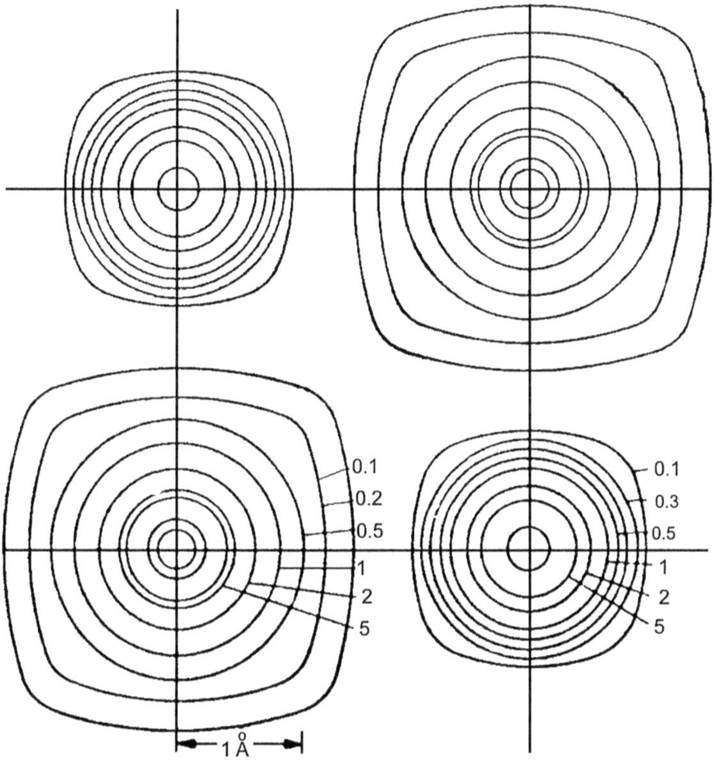

Fig. 6.2 Map of electron density (in electrons/$Å^3$) of NaCl

Coulomb attractive force – e^2/r^2 and an attractive energy – e^2/r develops between them. That such an electron transfer really takes place is seen in the electron charge distribution map (Fig. 6.2) for NaCl obtained from X-ray diffraction intensities. Numerical integration of the electron densities yields net charges of $+e$ and $-e$ for the Na^+ and Cl^- ions. Another point to be noted is that the charge density is near-zero midway between the two ions.

Sodium chloride is a classic example of an ionic crystal. All other alkali halides, silver halides, alkaline earth and transition metal oxides and alkaline earth fluorides are other examples of ionically bonded crystals.

6.2.2 Covalent Bond

Another way by which atoms with incomplete outer shells can assume a closed-shell structure is by sharing electrons with other atoms. As an example, let us consider two chlorine atoms. The electronic configuration of a Cl atom (Fig. 6.3a) with

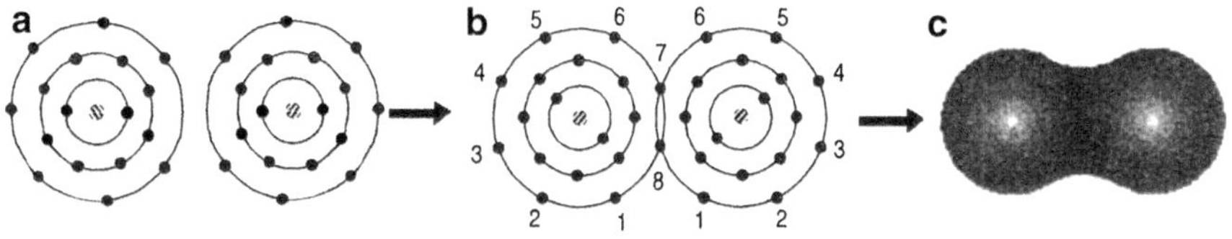

Fig. 6.3 Formation of covalent bond in chlorine molecule

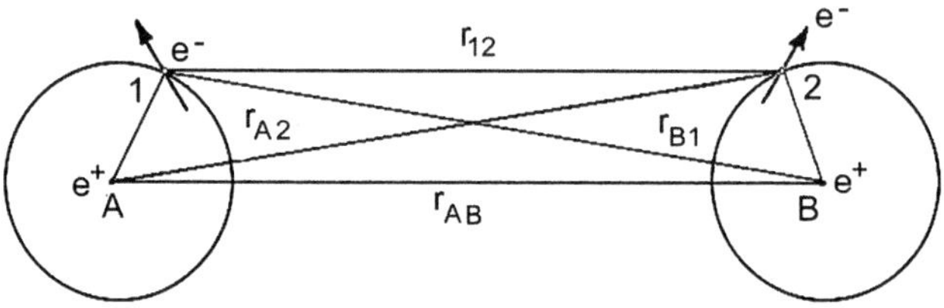

Fig. 6.4 The hydrogen molecule

17 electrons is $1s^2(2s^22p^6)(3s^23p^5)$. If two chlorine atoms are sufficiently close, they may share an electron such that each has six electrons in the *p*-state (Fig. 6.3b). This is the formation of a covalent bond between the two Cl atoms which results in a Cl_2 molecule (Fig. 6.3c).

How does sharing of electrons cause cohesion? To understand this qualitatively, we shall consider the hydrogen molecule H_2. But before considering the H_2 molecule, let us consider the hydrogen atom with a proton and an electron at a distance r from the proton. The Schrodinger equation for this system is:

$$\nabla^2\psi + \frac{8\pi^2 m}{h^2}\left(E + \frac{e^2}{r}\right)\psi = 0, \tag{6.1}$$

where m and e are the electron mass and charge, E the energy and ψ the wave function. When solved, the equation leads to energy eigenvalues E_n and wave functions ψ_n.

The hydrogen molecule is shown in Fig. 6.4. There are two protons A and B and the associated electrons 1 and 2. The distances between these particles are $r_{A1}, r_{B2}, r_{A2}, r_{B1}, r_{12}$ and r_{AB}. If the two hydrogen atoms are far apart and there is no interaction between them, the combined Schrodinger equation is:

$$\nabla^2\psi + \frac{8\pi^2 m}{h^2}\left(E + \frac{e^2}{r_{A1}} + \frac{e^2}{r_{B2}}\right)\psi = 0. \tag{6.2}$$

The solutions of (6.2) are ψ_a and ψ_b; these are similar to the solutions of (6.1).

On the other hand, if the two hydrogen atoms are so close that they share the two electrons, the Schrodinger equation becomes

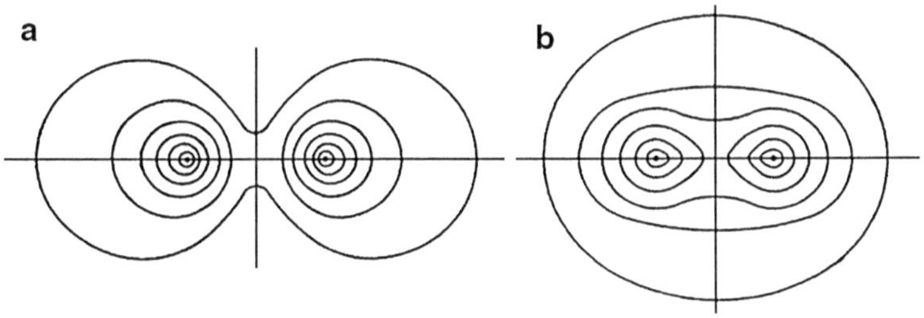

Fig. 6.5 Electron charge densities pertaining to the solutions (**a**) ψ_1 (attraction) and (**b**) ψ_2 (repulsion) for the H_2 molecule

$$\nabla^2\psi + \frac{8\pi^2 m}{h^2}\left(E + \frac{e^2}{r_{A1}} + \frac{e^2}{r_{B2}} - \frac{e^2}{r_{AB}} - \frac{e^2}{r_{12}}\right)\psi = 0. \tag{6.3}$$

It is difficult to solve (6.3) rigorously. But as an approximation, we assume solutions ψ_1 and ψ_2 which are linear combinations of ψ_a and ψ_b. These are

$$\psi_1 = \psi_a + \psi_b \quad \text{and} \quad \psi_2 = \psi_a - \psi_b. \tag{6.4}$$

ψ_1 and ψ_2 are called symmetric and antisymmetric solutions. The electron charge densities ρ calculated from these solutions are shown in Fig. 6.5. It is seen that the effect of the first solution is to cause attraction and that of the second to cause repulsion.

This simple model can be extended to atoms with several electrons. If the outer shells of the participating atoms are incomplete, both the attraction and repulsion are present. If on the other hand, the outer shells are complete, i.e. there is no sharing of electrons but only an overlapping of the shells, only the repulsion effect is present. Thus, the repulsion effect is universal. Covalent bonds are formed when atoms share 1, 2 or more electrons but the bond is stable (saturated) only when two electrons with opposite spins are shared.

The simple hydrogen molecule model can be extended to solids. The atomic number of carbon is 6 and its electronic configuration is $1s^2(2s^2 2p^2)$; it needs four electrons to have a stable structure. When carbon atoms come together in the solid state, each carbon atom forms a covalent bond with its four neighbours, sharing one electron with each of its neighbours. This is shown in Fig. 6.6. This is the situation in diamond which is a classic example of covalent bonding in solid state. Other examples of covalent solids are Ge, Si, grey tin and SiC. The characteristics of the covalent bond are that it is very strong, it is directional and there is an accumulation of charge midway between the bonding atoms. Such an accumulation of charge is seen in the electron charge density map for diamond obtained from X-ray diffraction intensities (Fig. 6.7).

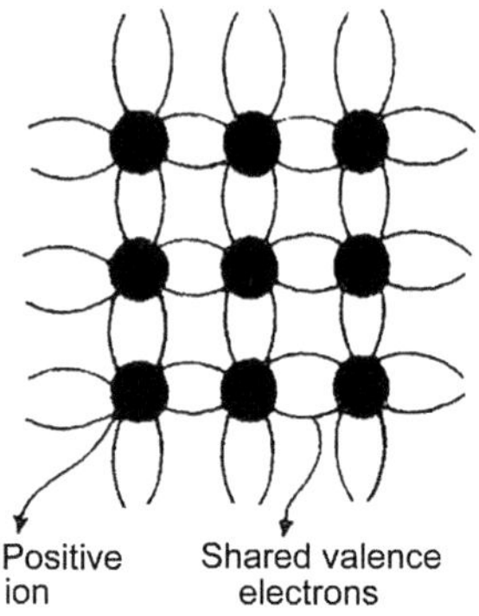

Fig. 6.6 Sharing of electrons among carbon atoms in diamond

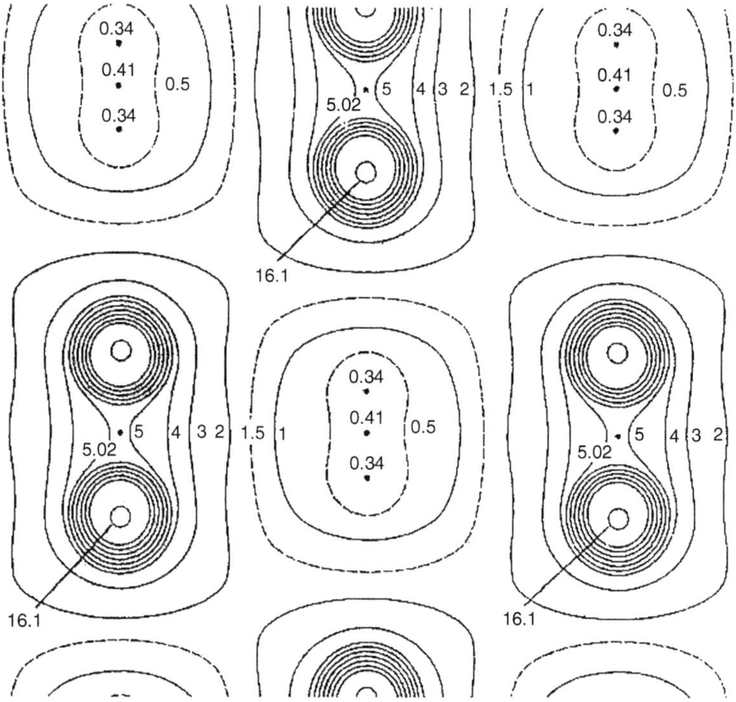

Fig. 6.7 Electron charge distribution on a plane of the cubic cell of diamond. The numbers along the constant density curves are electrons/Å^3. Electron density is high at points midway between the atoms

6.2.3 *Partially Ionic (Partially Covalent) Bond*

In several crystals, the charge transfer is not complete or the shared charge is less than an electron. Such crystals are called partially ionic or partially covalent. There

Table 6.1 Values of the effective ionic charge *s* for some covalent, partially covalent and ionic crystals

Covalent crystals		Partially covalent crystals		Ionic crystals	
Crystal	*s*	Crystal	*s*	Crystal	*s*
C	0	GaP	0.58	LiF	0.87
Ge	0	GaAs	0.51	NaF	0.93
Si	0	GaSb	0.33	NaCl	0.74
		InP	0.68	KCl	0.80
		InAs	0.56	RbCl	0.84
		InSb	0.42	CsCl	0.84

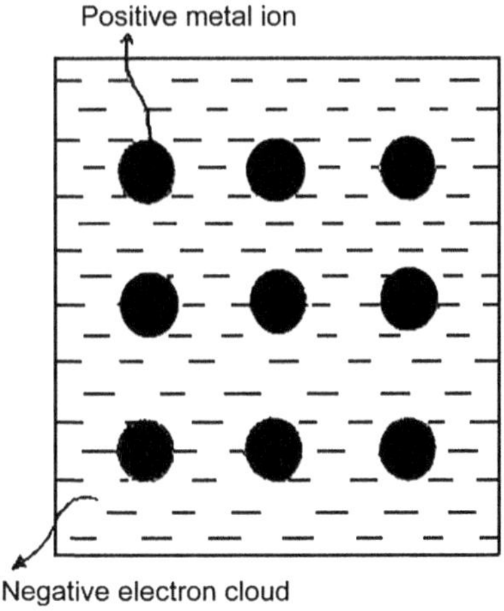

Fig. 6.8 Electron charge in metals

is X-ray evidence showing that only a fraction of an electron charge accumulates in between neighbouring atoms. Szigeti (details in Chap. 11) developed a theory of ionic crystals in which he introduced a parameter "*s*" (since called the Szigeti charge or effective ionic charge) which takes values ranging from 0 (for fully covalent crystals) to 1 (for fully ionic crystals). The partially ionic crystals have intermediate values for *s* (Table 6.1).

6.2.4 *Metallic Bond*

In metals, the valence electrons are loosely attached to the ion cores so that in the solid state they get easily detached and move freely in the body of the crystal. The situation in a metal (Fig. 6.8) may be described as an array of ion cores in a nearly uniform cloud of electron gas. The cohesion is provided by the interaction between the electron gas and the positively charged ion cores.

That the ion cores retain charge equal to the atomic number minus the charge of the valence electrons is seen in the electron charge distribution map for Al (Fig. 6.9)

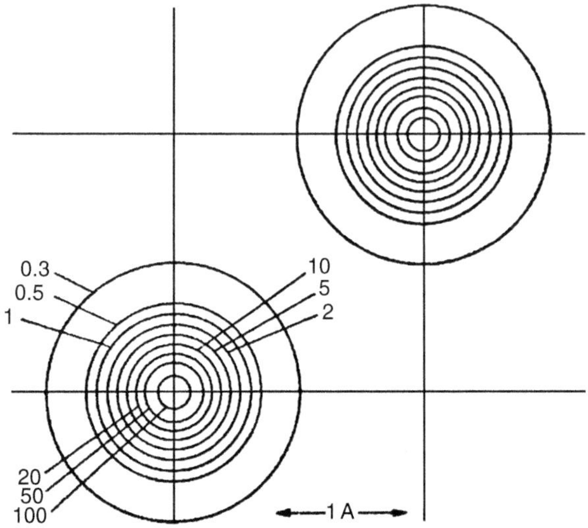

Fig. 6.9 Electron density map for *xyo* plane in aluminium

obtained from X-ray diffraction intensities. The cores have the charge of the closed shell and the valence-electron-cloud charge has an average density of 0.21 el./Å^3.

The alkali elements, transition metal elements, rare earth element and the alkaline earth elements are examples of solids where cohesion is provided by the metallic bond.

6.2.5 *Van der Waals Bond*

Let us consider two atoms both having a closed outer-shell structure. When they come close, there is no possibility of electron transfer (ionic bond) or electron sharing (covalent bond). Yet there are crystalline solids like the inert gas solids which consist of atoms with closed outer-shell structure. Surely, some attraction holds them together. Let us consider what is the source of cohesion in such solids.

Let us consider a charge e_i located at a point A having coordinates x_i, y_i, z_i. Let us consider another point B having coordinates x, y, z. The potential ϕ_i at B due to the charge at A is

$$\phi_i = \frac{e_i}{r_i} \tag{6.5}$$

where $r_i (= \sqrt{(x-x_i)^2 + (y-y_i)^2 + (z-z_i)^2})$ is the distance of B from A.

If instead of a single electron, there is a cluster of electrons around A, like the electrons of an atom, the total potential ϕ at B is

$$\phi = \sum \phi_i = \sum \frac{e_i}{r_i} = \sum \frac{e_i}{\sqrt{(x-x_i)^2 + (y-y_i)^2 + (z-z_i)^2}}. \tag{6.6}$$

Remembering that $x \gg x_i$, $y \gg y_i$ and $z \gg z_i$ and introducing $r = \sqrt{x^2 + y^2 + z^2}$, we can express ϕ as the series

$$\phi = \left[\frac{1}{r}\sum e_i\right] - \left[\frac{\partial}{\partial x}\left(\frac{1}{r}\right)\sum e_i x_i + \frac{\partial}{\partial y}\left(\frac{1}{r}\right)\sum e_i y_i + \frac{\partial}{\partial z}\left(\frac{1}{r}\right)\sum e_i z_i\right].$$
$$-\left[\text{higher order terms such as} \sum e_i x_i^2, \text{ etc.}\right]. \quad (6.7)$$

Let us consider the significance of the terms in the three brackets in (6.7). The term in the first bracket is simply the Coulomb term. If the cluster of electrons at A is the electrons associated with a neutral atom, then this term becomes zero since $\Sigma e_i = 0$. The second term contains quantities like $\Sigma e_i x_i$, etc. Each of these is the component of the dipole moment which is a vector. These vectors may have different orientations and their average may be zero. However, at a given instant each component has a non-zero value. Similarly, the terms in the third bracket like $\Sigma e_i x_i^2$ are the elements of a quadrupole moment. If an atom was located at B, the cluster of electrons associated with it would also be equivalent to instantaneous dipole moments and quadrupole moments. Slater [6.6] has pointed out that the interaction between the dipole moments of the two charge distributions results in an attractive energy ($-C/r^6$) and the interaction between the dipole moment of one atom with the quadrupole moment of the other results in an attractive energy ($-D/r^8$).

Thus, irrespective of whether two atoms have a closed or incomplete outer-shell structure, there results an attractive energy ($-C/r^6$) due to the dipole–dipole interaction and ($-D/r^8$) due to the dipole–quadrupole interaction. This is a universal effect present in all crystals (ionic, covalent, metallic). The contribution of this van der Waals energy is a small percentage of the total energy in ionic, covalent and metallic crystals. However in crystals with closed outer-shell atoms, where other interactions are absent, the van der Waals energy is the sole source of cohesion. The inert gas solids and some organic solids are examples of van der Waals bonding.

6.2.6 Hydrogen Bond

Hydrogen has a single electron. So, it can be expected to form a covalent bond with another atom, particularly a highly electronegative atom like O, F, Cl, etc. However, it is found that the H atom forms a bond with a second atom thus forming units like O–H. . . O, O–H. . . N, etc. Such a bond is called a hydrogen bond. In such hydrogen-linked units, the hydrogen is not equidistant from the two atoms but is nearer to the O on the left thus forming a dipole moment. The H bond is a weak bond. Ice crystals, KHF_2 and potassium dihydrogen phosphate (KH_2PO_4) are examples of H-bonded crystals.

6.3 Cohesive Energy

6.3.1 General

Cohesive energy is defined as the energy required to separate the constituents (atoms/ions/molecules) from their positions in the crystal to an infinite distance. Numerically, it is also equal to the energy which is released when ions (or atoms) are brought from infinite distance to appropriate positions in the crystal.

Cohesive energy is the result of the attractive and repulsive forces between the constituents of the crystal. The dependence of the interatomic forces and the resulting energy upon the interatomic distance is shown in Fig. 6.10. As the atoms come close, the attractive (F_A) and repulsive (F_R) forces (along with their respective

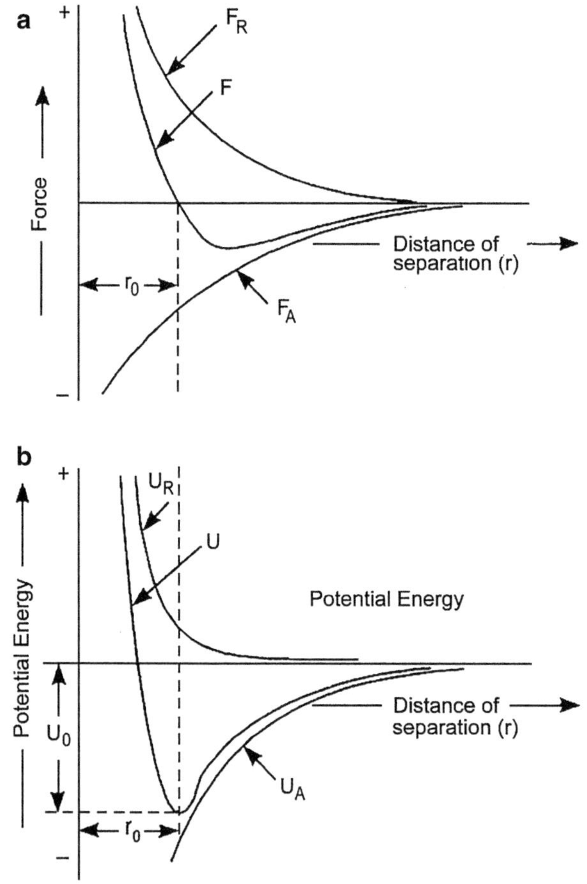

Fig. 6.10 Variation along the interatomic distance of (**a**) interatomic forces and (**b**) potential energy

signs) increase. At a value $r = r_0$ called the equilibrium distance, the total force (F) is zero. However, the variation of the attractive (U_A) and repulsive (U_R) energies is such that the total energy U attains a minimum value ($-U_0$) at $r = r_0$. At these values of $r(= r_0)$ and $U(= U_0)$, the system is in equilibrium and most stable.

It can be seen from Fig. 6.10 that the cohesive energy, which is the minimum energy of the system, is a negative quantity. In fact, this negative sign is the condition for stability of the crystal [6.5, 6.7].

It is confusing to find that the cohesive energy values are given with positive sign by some authors. These positive values are to be taken as absolute values $|U_0|$ [6.4]; the true value of U_0 is to be always taken as negative.

Cohesive energies are expressed in units of kcal/mol or eV/molecule (23.05 kcal/mol = 1 eV/molecule).

6.3.2 *Ionic Crystals*

6.3.2.1 Coulomb Energy and Madelung Constant

In 1910, Born made the first calculation of cohesive energy of ionic crystals. The Coulomb energy between two ions with charges e and $-e$ and separation r is $-e^2/r$. But as the Coulomb force is a long range force, a given ion interacts with not just the nearest neighbours but with more distant neighbours too.

Let us consider a linear NaCl lattice which is a chain of ions with alternate signs (Fig. 6.11), the separation of each ion from its neighbour being r_0. Let us consider the positive ion at 0 as the reference ion. The Coulomb energy of the reference ion due to several of its neighbours is

$$\begin{aligned} U_{\text{Coul}} &= -\frac{2e^2}{r_0} + \frac{2e^2}{2r_0} - \frac{2e^2}{3r_0} - \frac{2e^2}{4r_0} + \ldots \\ &= -\frac{2e^2}{r_0}\left[1 - \frac{1}{2} + \frac{1}{3} - \frac{1}{4} + \ldots\right] \\ &= -\frac{\alpha_M e^2}{r_0} \end{aligned} \tag{6.8}$$

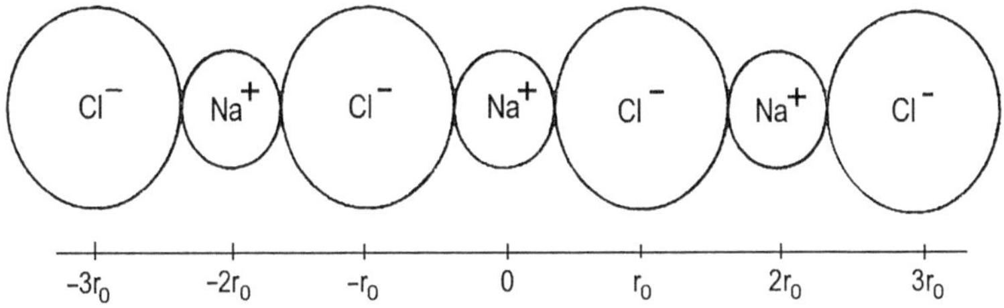

Fig. 6.11 A hypothetical linear NaCl lattice

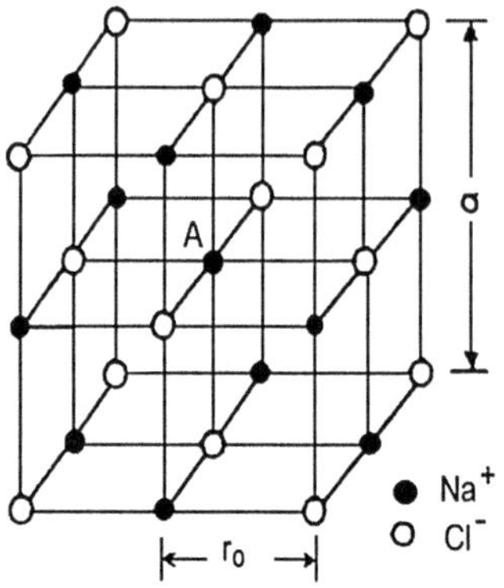

Fig. 6.12 The three-dimensional NaCl lattice

Table 6.2 Madelung constant a_M for some ionic structures

Structure	a_M
NaCl	1.74756
CsCl	1.76267
CaF_2	5.03805
ZnS (zinc blende)	1.63805
ZnS (wurtzite)	1.641
TiO_2 (rutile)	19.077

where the constant α_M is called the Madelung constant. The series $2[1 - \frac{1}{2} + \frac{1}{3} - \frac{1}{4} + \ldots]$ adds up to $2\log 2\,(= 1.38)$.

Let us now consider the NaCl lattice in three dimensions (Fig. 6.12). The distance between the reference ion at A and its nearest neighbour is r_0 and the lattice constant $a_0 = 2r_0$. This reference ion Na^+ has as its neighbours 6 Cl^- ions at distance r_0, 12 Na^+ ions at distance $\sqrt{2}r_0$, 8 Cl^- ions at distance $\sqrt{3}r_0$, again 6 Na^+ ions at distance $\sqrt{4}r_0$, and so on. The total Coulomb energy U_{Coul} of the Na^+ ion at A due to all the neighbouring ions is

$$U_{\mathrm{Coul}} = -\frac{e^2}{r_0}\left[6 - \frac{12}{\sqrt{2}} + \frac{8}{\sqrt{3}} - \frac{6}{\sqrt{4}} + \ldots\right] \tag{6.9}$$
$$= -\frac{\alpha_M e^2}{r_0}.$$

The Madelung constant α_M of the NaCl lattice represented by the series in the bracket in (6.9) has the value 1.747565. The Madelung constants for some common ionic structures are given in Table 6.2.

It may be noted that the Madelung constant is a lattice sum. It is dependent on the structure but not upon the interatomic distance. The first calculation of the Madelung constant was done by Madelung [6.8]. The terms in the series are alternately +ve and –ve and hence the convergence of the series is very slow. Methods to calculate the Madelung constant with better convergence have been proposed by Ewald [6.9]) and Evjen [6.10].

Equation (6.9) relates to ionic structures containing singly charged ions. For divalent ionic crystals like MgO, the Coulomb energy is $-\alpha_M(4e^2)/r_0$. In general if the ions have charges $\pm ze$, the Coulomb energy is $\alpha_M(ze)^2/r_0$. If the two ions have dissimilar charges, z is the highest common factor of the two charges. Thus for CaF_2, z is 1 and for TiO_2, z is 2. Equation (6.9) relates to a pair of ions. For a mole containing $2N_A$ ions, the Coulomb energy is $-N_A\,\alpha_M\,(ze)^2/r_0$.

6.3.2.2 Repulsive Energy

The Coulomb force brings the oppositely charged ions close to each other. When they are close their electron clouds start to overlap (or penetrate). Such overlapping is against Pauli's exclusion principle and a repulsion comes into play. A reference to this quantum phenomenon has been made in Sect. 6.2.3. In the initial stages of ionic theory, the repulsive energy was represented by an inverse power term B/r^n where B is a constant and n is called the repulsive index or exponent. The index is of the order of 10 but its exact value differs from crystal to crystal and can be calculated from crystal properties. The total cohesive energy U is

$$U = N_A\left[-\frac{\alpha_M e^2}{r} + \frac{B}{r^n}\right]. \tag{6.10}$$

6.3.2.3 Equilibrium Cohesive Energy

We have seen that at equilibrium the cohesive energy is minimum. Hence,

$$dU/dr = 0. \tag{6.11}$$

Applying this condition to (6.10), we have

$$\frac{dU}{dr} = N_A\left[\frac{\alpha_M e^2}{r^2} - \frac{Bn}{r^{n+2}}\right] = 0 \tag{6.12}$$

which leads to

$$\frac{B}{r^n} = \frac{\alpha_M e^2}{n}. \tag{6.13}$$

Substituting in (6.10), we have, for the value of the equilibrium cohesive energy U_0;

$$U_0 = -\frac{N_A\alpha_M e^2}{r_0}\left(1 - \frac{1}{n}\right). \tag{6.14}$$

6.3.2.4 Evaluation of Repulsion Index

Let us see how the repulsion index n can be estimated from the compressibility ψ. By definition,

$$\psi = -\frac{1}{V}\frac{\mathrm{d}V}{\mathrm{d}P}. \tag{6.15}$$

From thermodynamics we have

$$\mathrm{d}U = -P\mathrm{d}V \quad \text{and} \quad \frac{\mathrm{d}P}{\mathrm{d}V} = -\frac{\mathrm{d}^2U}{\mathrm{d}V^2}. \tag{6.16}$$

Thus,

$$\frac{1}{\psi} = V\frac{\mathrm{d}^2U}{\mathrm{d}V^2}. \tag{6.17}$$

The volume $V = N_{\mathrm{A}}Cr^3$ where C is a constant for a given structure; for the NaCl structure $C = 2$ and

$$V = 2N_{\mathrm{A}}r^3. \tag{6.18}$$

From (6.17) and (6.18) and remembering that $\mathrm{d}U/\mathrm{d}r = 0$ at equilibrium, we get

$$\frac{1}{\psi} = \frac{1}{18N_{\mathrm{A}}r_0}\left(\frac{\mathrm{d}^2U}{\mathrm{d}r^2}\right)_{r=r_0}. \tag{6.19}$$

Applying (6.19) to (6.10), we get

$$n = 1 + \frac{18r_0^4}{\psi\alpha_{\mathrm{M}}e^2}. \tag{6.20}$$

Slater [6.6] calculated the values of n from (6.20) for some alkali halides using experimental values of r_0 and ψ and used these n values to calculate the cohesive energy U_0 from (6.14). These n and U_0 values are given in Table 6.3.

Table 6.3 Repulsive index n and the cohesive energy U_0 of some alkali halides

Crystal	$r_0[10^{-8}$ cm]	$\psi[10^{-12}$ cm^2/dyn]	n	U_0 [kcal/mol] Calc. from (6.14)	Expt. from (6.21)
LiF	2.014	1.53	5.8	−238	−240
LiCl	2.570	3.41	6.75	−191	−199
LiBr	2.751	4.31	6.95	−180	−188
NaCl	2.820	4.20	7.66	−179	−183
NaBr	2.989	5.08	7.97	−169	−175
KF	2.674	3.31	7.90	−189	−190
KCl	3.147	5.62	8.75	−163	−165
RbBr	3.445	7.93	8.82	−149	−154
RbI	3.671	9.61	9.37	−141	−145

Table 6.4 Contributions of various terms to the cohesive energy U_0 of some alkali halides

Interaction	Contribution [eV/molecule]		
	LiF	NaCl	RbI
Coulomb	−12.4	−8.85	−6.79
Repulsive	+1.9	+1.02	+0.67
Van der Waals	−0.20	−0.125	−0.175
Zero-point	+0.17	+0.075	+0.031
Total U_0	−9.53	−7.90	−6.26

6.3.2.5 Improved Calculations

Some improvements in the estimation of cohesive energy of ionic crystals have been made by Born and Mayer [6.11].

Firstly, in the original Born theory, the repulsion energy due to overlap of electron clouds was empirically represented by the term B/r^n. Rigorous quantum mechanics leads to a term $B'e^{-r/\rho}$ where B' is another constant in place of B and ρ is a constant (sometimes called the hardness parameter). While this form of repulsive energy is theoretically superior, it does not change the results substantially.

Secondly, it was pointed out in Sect. 6.2.4 that the dipole–dipole and dipole–quadrupole interactions between the charge distributions of respective atoms/ions lead to attractive energy terms of the form $-C/r^6$ and $-D/r^8$. These "van der Waals" terms have to be included in the cohesive energy.

Finally, a correction has to be added for the zero-point energy of atomic oscillations. From quantum mechanics, the energy of an oscillator in its lowest state is $1/2\,h\nu$. Combining this with the Debye model, the energy of an ion pair is $(9/4)\,h\nu_D$ where ν_D is the Debye frequency.

The contributions of the various terms to the cohesive energy of a few alkali halides are given in Table 6.4. It can be seen that, typically, the repulsive contribution is about 15% and the van der Waals contribution is about 2% of the total.

6.3.2.6 Experimental Values

There is no way of determining the cohesive energies through a single laboratory experiment. However, an "experimental" value can be obtained by developing a formula involving several experimentally determined quantities. Such a formula was developed by Born [6.12] by imagining the breaking up of, say, the NaCl crystal into its constituents and recreating it through several processes. Born's processes were presented in the form of a diagram by Haber [6.13] and the method is known as the Born–Haber cycle.

The Born–Haber cycle for a crystal like NaCl is shown in Fig. 6.13. In reaction (i), the NaCl crystal is decomposed into the constituent ions with energy U_0. In reaction (ii), the ions are converted into neutral atoms; this process requires energy equal to the ionisation energy I of the Na atom less the electron affinity A of the

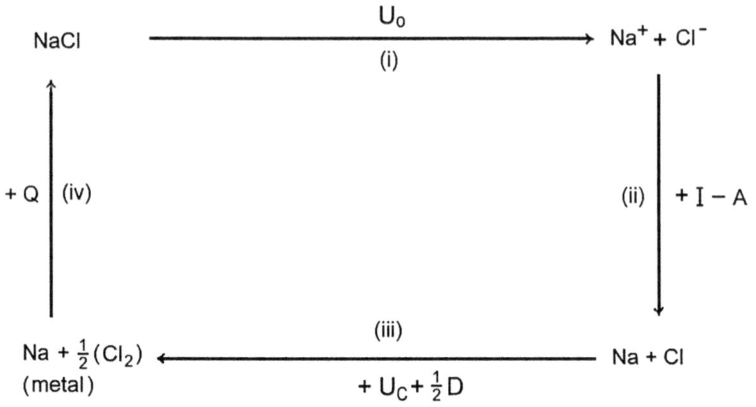

Fig. 6.13 The Born–Haber cycle for NaCl

Cl atom. In reaction (iii), the neutral atoms are converted into their standard state (Na metal and Cl_2 gas) requiring energy U_C which is the sublimation energy of the Na metal and half the dissociation energy (1/2) D of Cl_2 gas molecule. Finally in process (iv), the solid metal atoms and the Cl_2 molecules are united to form NaCl crystal by using the heat of formation Q. Obviously, the sum of these energy terms should be zero, i.e.

$$U_0 + I - A + U_C + \tfrac{1}{2}D + Q = 0$$

or

$$U_0 = -(I - A + U_C + \tfrac{1}{2}D + Q). \tag{6.21}$$

Thus, U_0 can be calculated with all the quantities on the r.h.s of (6.21) which are experimentally determined. For instance, for NaCl, $Q = 98.6$, $U_C = 26.0$, $D = 57.9$, $I = 119$ and $A = 83.1$ (all in kcal/mol) which lead to a value of $U_0 = -183$ kcal/mol. The cohesive energies of several alkali halides from the Born–Haber cycle are given in Table 6.3. These "experimental" values agree fairly with those calculated from the theoretical expression (6.14).

6.3.3 *Van der Waals Crystals*

The cohesive energy of van der Waals crystals may be represented by

$$U = -\frac{C}{r^6} - \frac{D}{r^8} + Be^{-r/\rho}. \tag{6.22}$$

A simpler form is

$$U = -\frac{C}{r^6} + \frac{A}{r^m} \tag{6.23}$$

Table 6.5 Cohesive energy of inert gas solids [6.4]

Crystal	r_0 [Å]	U_0 [kcal/mol]	
		Calculated (6.22)	Experiment
A	4.077	−1.724	−1.850
Ne	3.310	−0.493	−0.447

where C and D are lattice sums and A a constant. The index m can be determined from the compressibility of the crystal following the same procedure as followed in the case of ionic crystals but it is generally assumed to be 12; then (6.23) is called the 6–12 potential. The cohesive energies of A and Ne with $m = 12$ are given in Table 6.5.

6.3.4 Metals and Covalent Crystals

The calculation of cohesive energy of metals and covalent crystals requires quantum mechanical methods which are necessarily complicated; we shall limit ourselves to empirical methods.

Slater [6.6] suggested that the cohesive energy of a metal may be empirically represented by the Morse function

$$U = L\{e^{[-2a(r-r_0)]} - 2e^{[-a(r-r_0)]}\}. \tag{6.24}$$

Here L is the heat of evaporation, a the Morse parameter, r the interatomic distance and r_0 its equilibrium value. In support of (6.24), Slater derived the compressibility and Gruneisen constant and found that the calculated values agree with the experimental ones.

Another form in use for the cohesive energy of metals is

$$U = A\left(\frac{V_0}{V}\right) + B\left(\frac{V_0}{V}\right)^{2/3} - C\left(\frac{V_0}{V}\right)^{1/3}. \tag{6.25}$$

where A, B and C are constants, V the volume and V_0 its equilibrium value.

6.3.5 Hydrogen-Bonded Crystals

Here, again, the method to calculate cohesive energy is difficult. Attempts are limited to calculating the electrostatic part of the cohesive energy. As an example we shall consider the case of ice. The calculation is done by assuming the charge distribution. Campbell [6.14] assumed a charge of $0.361e$ located at a point 0.96 Å from one of the oxygens and the two O–O lines and an equal negative charge of –0.361e at 0.054 Å from each oxygen on the same two O–O lines. He also assumed the distance between adjacent oxygens as 2.72 Å. The electrostatic energy is calculated by the Ewald method. The calculated energy is –5.6 kcal/mol which is to be compared with the experimental value of –11.36 kcal/mol.

6.4 Bonding and Crystal Properties

Crystal properties are influenced by bonding type in two ways – directly through the electronic process and indirectly through the strength of the bond.

Some properties are governed by the charge distribution. Thus, metals are good electrical conductors whereas ionic and van der Waals crystals are insulators. Similarly optical properties like the transparency and absorption vary across bond types.

Some properties depend upon the strength of the interatomic bonding. Thus, mechanical properties like bulk moduli and hardness, melting points, Debye temperatures have large values for high bonding strength and lower values for lower bond strength. On the other hand, thermal expansion is low for high bond strength and large for weaker bonding. The bonding is very strong in covalent crystals, moderately strong in ionic and metallic crystals and weak in van der Waals crystals. The mechanical and thermal properties mentioned earlier show a gradation as we pass from covalent to ionic and metallic to van der Waals crystals. These trends in properties of differently bonded crystals are summarised in Table 6.6.

6.5 Structures with Complex Bonding

In Sect. 6.2, we discussed different types of bonding. In that connection, we gave examples of crystals in which the bonding is predominantly of one type. Thus, the bonding is predominantly ionic in NaCl, covalent in diamond, metallic in sodium and van der Waals in inert gas solids. We shall now consider some crystals (Table 6.7) in which there is an interplay of more than one type of bond.

6.5.1 Graphite

Graphite is hexagonal (Fig. 6.14). The carbon atoms form a honeycomb array in layers perpendicular to the c-axis. Three out of four outer electrons from each carbon atom are used in forming covalent bonds within the layer; the fourth electron is free. The C–C distance within a layer is 1.42 Å. The weak van der Waals bonding between the layers is reflected in a large interlayer distance of 3.4 Å.

The strong intra-layer bonding in the a-direction and the weak inter-layer bonding in the c-direction make graphite highly anisotropic. For instance, the

Table 6.6 Bond types and crystal properties

Aspect	Covalent bond	Ionic bond	Metallic bond	Van der Waals bond
Mechanism	Sharing of electron	Transfer of electrons between two dissimilar atoms	Interaction between ion cores and free electrons	Interaction between dipoles and quadrupoles formed by fluctuation in charge distributions
Typical crystals	Diamond, Si, Ge, SiC	Alkali halides, silver halides	Alkali metals, noble metals	H_2, Cl_2, inert gas solids
Electron charge distribution revealed by X-rays	Accumulation of charge in the region between the atoms	Electron charge around nuclei diminished or enhanced due to electron transfer and near-zero charge between ions	Ion cores deprived of valence electrons and uniform charge density between ion cores	
Electrical conduction	Insulators or semiconductors	Insulators at low temperature; slight conductivity at high temperatures due to defects	Good conductors	Insulators
Optical properties	Strongly absorbing for photon energies above intrinsic edge; transparent for larger wavelengths	Restrahlen absorption in IR; otherwise transparent	Opaque and highly reflecting	Transparent in far UV
Binding energy	Moderate to high	Moderate to high	Low	Very low
Hardness	Very hard	Moderately hard	Moderately hard	Soft
Melting point	High melting point	Moderately high melting point	Moderately high melting point	Low melting point
Debye temperature	Very high	Moderately high	Moderately high	Low
Thermal expansion	Low expansion	Moderate	Moderate	Large expansion

thermal expansion in the two principal directions is: $\alpha_a = 1.3 \times 10^{-6}/°C$ and $\alpha_c = 17.2 \times 10^{-6}/°C$ [6.15]. More importantly, the layers slide easily over each other which makes graphite a solid lubricant.

6.5.2 *Tellurium*

The structure of tellurium is trigonal but it can also be considered as hexagonal (Fig. 6.15). If we look down the structure along the *c*-axis, the tellurium atoms

Table 6.7 Crystals with complex bonding

Crystal	Bonding
Graphite	Covalent, van der Waals
Tellurium	Covalent, van der Waals, metallic
Cadmium iodide	Ionic, van der Waals
Calcite	Covalent, ionic
Ice	Covalent, hydrogen bond
Potassium dihydrogen phosphate	Covalent, ionic, hydrogen bond

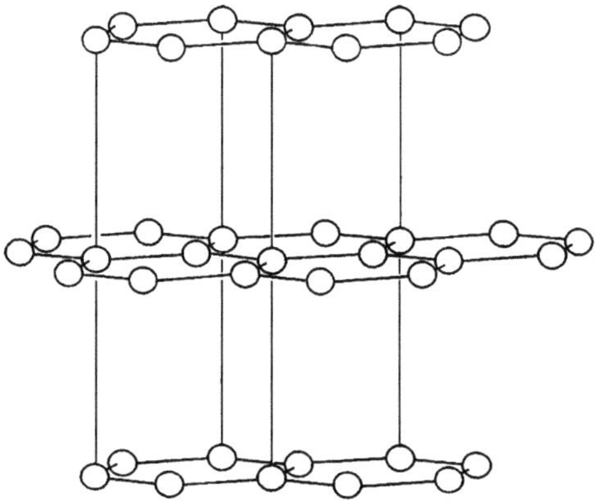

Fig. 6.14 Structure of graphite

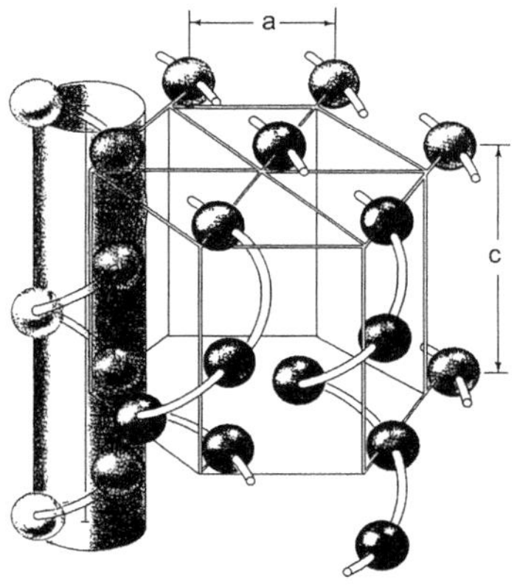

Fig. 6.15 Structure of tellurium

appear to be located along a spiral chain. The bonding between the atoms in the chain is covalent but adjacent chains are held together by a combination of van der Waals and metallic bond. The crystal shows a linear chain-like behaviour in some properties viz. specific heats (Chap. 10). The bonding is much stronger in the c-direction than in the a-direction. This causes a strong anisotropy in physical properties; for instance, the thermal expansion coefficients are $\alpha_a = 38 \times 10^{-6}/°C$ and $\alpha_c = -2.5 \times 10^{-6}/°C$ [6.16].

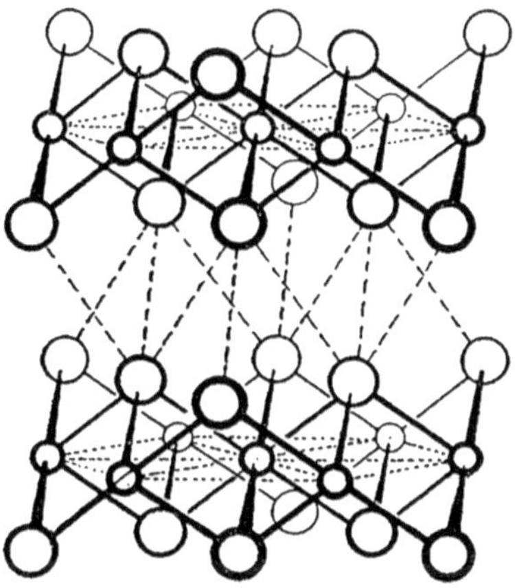

Fig. 6.16 Two layers of CdI_2 structure: *small circles*, Cd; *large circles*, I

6.5.3 Cadmium Iodide

Cadmium iodide (CdI_2) has a hexagonal structure (Fig. 6.16). The Cd-to-iodine distance is ~3 Å and the iodine-to-iodine distance is 4.2 Å. These are roughly equal to the sums of the ionic radii; as such the structure may be considered as a packing of iodine ions bonded to the Cd ions. The Cd ions form a layer between the two layers of iodine ions. These iodine–Cd–iodine sandwiches are held by weak van der Waals forces. This combination of weak van der Waals forces and relatively strong ionic forces results in a layer structure; the layers cleave easily making the substance flaky in nature.

6.5.4 Calcite

Calcite ($CaCO_3$) is trigonal but can be considered hexagonal. Its structure shown in Fig. 6.17 consists of Ca^{++} ions and $(CO_3)^-$ ions. The $(CO_3)^-$ ion is a triangular ion with its plane perpendicular to the c-axis. The C–O length is much shorter than the sum of the ionic radii; the C–O bond is thus a covalent bond. The electrostatic force between the Ca^{++} ion and the $(CO_3)^-$ ion acts in a direction making an angle with the a-axis; its component along the a-axis further strengthens the forces in the plane perpendicular to c. The other component of the electrostatic force acts along the c-direction. Thus, the binding in the a-direction is stronger than that in the c-direction; the thermal expansion in the a-direction is much less than that in the c-direction.

6.5.5 Ice

The structure of ice is similar to the hexagonal wurtzite structure. As seen in Fig. 6.18, there are four hydrogen bonds to each oxygen at tetrahedral angles. The

Fig. 6.17 Structure of calcite

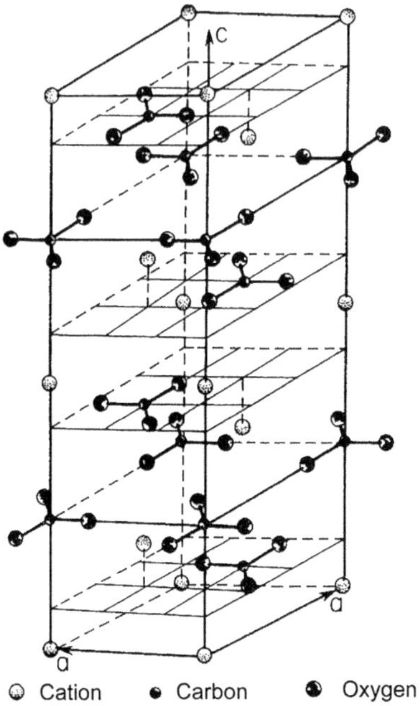

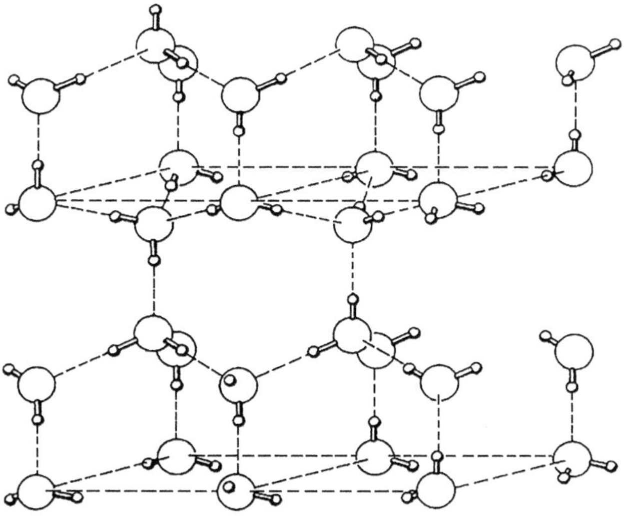

Fig. 6.18 Structure of ice: large circles are oxygen, *small circles* are protons

O–O distance is 2.76 Å. Each proton is 1.01 Å from the oxygen of its molecule and 1.75 Å from the oxygen of the neighbouring molecule. These distances are typical of covalent bonding within the molecule and hydrogen bonding between molecules.

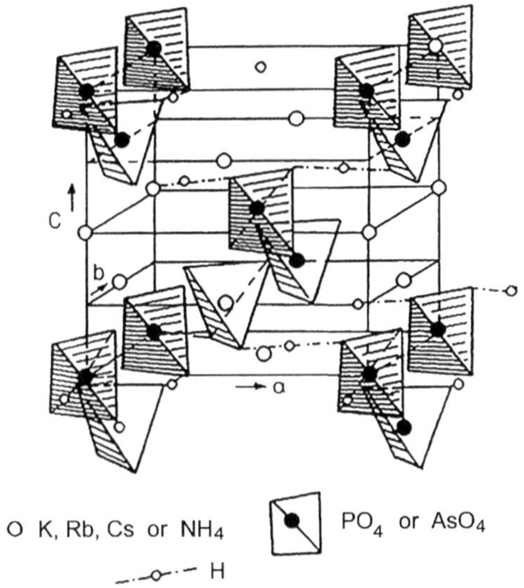

Fig. 6.19 Structure of KDP-type crystals

6.5.6 Potassium Dihydrogen Phosphate

The phosphates and arsenates of K, Rb and NH_4 form a tetragonal structure (Fig. 6.19). The structure (KDP) is made up of (PO_4) tetrahedra and K^+ ions. The oxygens of the tetrahedra are linked with oxygens of other tetrahedra through O–H. . . O bonds. These hydrogen bonds are more or less parallel to the basal planes and strengthen the structure in the a-direction. This makes the thermal expansion and compressibility less in the a-direction than in the c-direction. Another result of the presence of the hydrogen bonds is to contribute to the ferroelectricity in KDP.

The internal bonding in the PO_4 and AsO_4 groups is covalent. The piezoelectric constants of the arsenates in this group of crystals are larger than those in the phosphates. Adhav [6.17] attributed these relative values to a weaker covalent bonding in (AsO_4) than in (PO_4).

The structure can also be considered to be made up of K^+ and $(H_2PO_4)^-$ ions. Hartman [6.18] showed that the growth habit of these crystals is caused by the ionic interaction between these ions. Sirdeshmukh and Rao [6.19] pointed out that the mean hardness of the crystals in this family is influenced by the ionic nature of these crystals.

Problems

1. Estimate the Born repulsion index for LiF for which $r_0 = 2.014$ Å and $\psi = 1.53 \times 10^{-12}\,\mathrm{cm^2/dyn}$.
2. Estimate the cohesive energy of NaBr for which $r_0 = 2.989$ Å and the Born repulsion index is 8.

3. Derive the Madelung constant for the CsCl structure.
4. Derive expressions for the cohesive energy of NaCl using the term $B'\mathrm{e}^{-r/\rho}$ for the repulsion.
5. Discuss the nature of bonding in (1) Solid N_2, (2) GaAs, (3) grey tin and (4) NH_4F.
6. Suppose NaCl crystallises in the usual NaCl structure and also in the ZnS (zinc blende) structure. Assume that in both forms $r_0 = 2.82\,\text{Å}$ and $\psi = 8\times10^{-12}\,\text{cm}^2/\text{dyn}$. Calculate the cohesive energies for the two forms and discuss the significance of the relative values.

References

6.1. M. Tosi, Solid State Phys. **16**, 1 (1964).
6.2. C. Kittel, *Introduction to Solid State Physics* (John Wiley & Sons, New York, 1996).
6.3. A.J. Dekker, *Solid State Physics* (Macmillan Press, London, 1981).
6.4. R. Kubo, T. Nagamiya, *Solid State Physics* (McGraw-Hill, New York, 1968).
6.5. N.W. Ashcroft, N.D. Mermin, *Solid State Physics* (Holt, Rinchart and Winston, Philadelphia, 1976).
6.6. J.C. Slater, *Introduction to Chemical Physics* (McGraw Hill, New York, 1939).
6.7. F.C. Brown, *The Physics of Solids* (W.A. Benjamin, New York, 1966).
6.8. E. Madelung, Phys. Z. **19**, 524 (1918).
6.9. P.P. Ewald, Ann. Phys. **64**, 253 (1921).
6.10. H.M. Evjen, Phys. Rev. **39**, 675 (1932).
6.11. M. Born, J.E. Mayer, Z. Phys. **75**, 1 (1932).
6.12. M. Born, Verh D. D. Phys. Ges. **21**, 679 (1919).
6.13. E.S. Campbell, J. Chem. Phys. **20**, 1411 (1952).
6.14. F. Haber, Verh. D. D. Phys. Ges. **21**, 750 (1919).
6.15. J. Pierrey, C. R. Acad. Sci. Paris 223, 501 (1946).
6.16. V.T. Deshpande, R. Pawar, Physica **31**, 671 (1965).
6.17. R.S. Adhav, J. Appl. Phys. **46**, 2808 (1975).
6.18. P. Hartman, Acta Cryst. **9**, 721 (1956).
6.19. D.B. Sirdeshmukh, K.K. Rao, Indian J. Pure Appl. Phys. **16**, 860 (1978).

Chapter 7
Tensor Nature of Physical Properties

7.1 Introduction

A striking feature of physical properties of crystals is their dependence on direction or anisotropy. Some properties (like the density or specific heat) are independent of direction or isotropic in all crystals. They are called scalar properties. There are a few other properties like thermal conductivity or dielectric constant which are isotropic in only cubic crystals but anisotropic in non-cubic crystals. On the other hand, elastic properties are anisotropic even in cubic crystals. This anisotropy of crystal properties arises out of their tensor nature.

In this chapter, we first develop some essential properties of tensors and matrices and then consider some tensor properties. We consider the second-rank tensor properties in detail; the concepts and methods developed here will be useful in handling higher rank tensor properties in later chapters.

The tensor nature of crystal properties is treated well by Nye [7.1], Smith [7.2], Bhagavantam [7.3] and Sirotin and Shaskolskaya [7.4].

7.2 Matrices and Tensors

In this section, we shall consider only those aspects of matrices and tensors which are necessary for appreciating the main theme of this chapter, namely, the tensor nature of crystal properties.

7.2.1 The Transformation Matrix

Let us consider the transformation of the components of a vector due to the rotation of axes. Consider a vector **p** referred to coordinate axes $Ox_1x_2x_3$ (Fig. 7.1). Let its

D.B. Sirdeshmukh et al., *Atomistic Properties of Solids*, Springer Series in Materials Science 147, DOI 10.1007/978-3-642-19971-4_7,

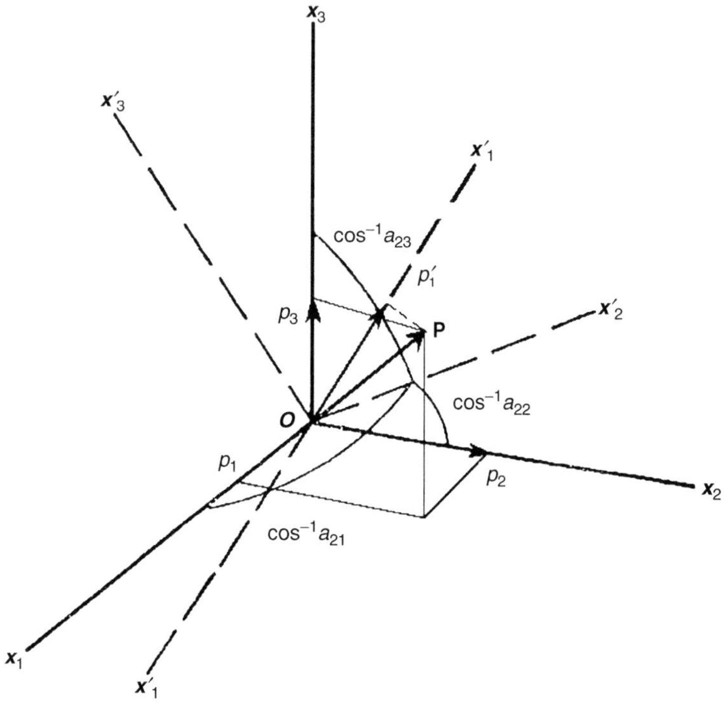

Fig. 7.1 Rotation of coordinate axes and transformation of vector components

components along the three coordinate axes be p_1, p_2 and p_3. Let us consider a set of axes $Ox'_1x'_2x'_3$ which are rotated with reference to the old system $Ox_1x_2x_3$. Let the components of the vector **p** along the new coordinate axes be p'_1, p'_2 and p'_3. Now, p'_3 is the sum of the components of p_1, p_2 and p_3 resolved along the Ox'_1 axis. Thus, we have

$$p'_1 = p_1 \cos \widehat{x'_1 x_1} + p_1 \cos \widehat{x'_1 x_2} + p_1 \cos \widehat{x'_1 x_3}\,. \tag{7.1}$$

Similarly,

$$p'_2 = p_1 \cos \widehat{x'_2 x_1} + p_1 \cos \widehat{x'_2 x_2} + p_1 \cos \widehat{x'_2 x_3}, \tag{7.2}$$

and

$$p'_3 = p_1 \cos \widehat{x'_3 x_1} + p_2 \cos \widehat{x'_3 x_2} + p_3 \cos \widehat{x'_3 x_3}\,. \tag{7.3}$$

Denoting the cosines by a_{11}, a_{12}, etc. we can write (7.1)–(7.3) as

$$p'_i = \sum_{j=1}^{3} a_{ij}\, p_j\,. \tag{7.4}$$

This is the transformation law for the old components in terms of the new components. The first subscript in the a's refers to the "new" axes and the second to the "old". Similarly, we can express the old components in terms of the new components as:

$$p_i = \sum_{j=1}^{3} a_{ji} p'_j. \tag{7.5}$$

Note that the index j comes in succession in (7.4) whereas it is separated in (7.5). Equation (7.4) is the transformation law for the components of a vector and the matrix

$$\left(a_{ij}\right) = \begin{pmatrix} a_{11} & a_{12} & a_{13} \\ a_{21} & a_{22} & a_{23} \\ a_{31} & a_{32} & a_{33} \end{pmatrix}, \tag{7.6}$$

is called the transformation matrix. In general, $a_{ij} \neq a_{ji}$.

For convenience we may adopt a shorter notation wherein (7.4) and (7.5) may be written as

$$p'_i = a_{ij} p_j \quad (i, j = 1, 2, 3), \tag{7.7}$$

and

$$p_i = a_{ji} p'_j. \quad (i, j = 1, 2, 3). \tag{7.8}$$

This is called the dummy suffix notation which may be stated as: when a suffix occurs twice in the same term it is to be automatically understood that there is summation of that term with respect to that suffix. The suffix j in (7.7) and (7.8) is called the dummy suffix and occurs in pairs in each term. Summation with respect to j is understood.

7.2.2 Representation of a Symmetry Operation by a Matrix

Let us consider the point P in Fig. 7.1. Its coordinates are x_1, x_2 and x_3 with respect to axes $Ox_1x_2x_3$. These are also the components of the vector OP. If x'_1, x'_2 and x'_3 are the coordinates of P with respect to the rotated axes $Ox'_1x'_2x'_3$, then

$$\begin{aligned} x'_1 &= a_{11}x_1 + a_{12}x_2 + a_{13}x_3 \\ x'_2 &= a_{21}x_1 + a_{22}x_2 + a_{23}x_3 \\ x'_3 &= a_{31}x_1 + a_{32}x_2 + a_{33}x_3 \end{aligned} \tag{7.9}$$

or

$$x'_i = a_{ij} x_j. \tag{7.10}$$

Similarly, the old coordinates can be expressed in terms of the new coordinates as:

$$x_i = a_{ji} x'_j. \tag{7.11}$$

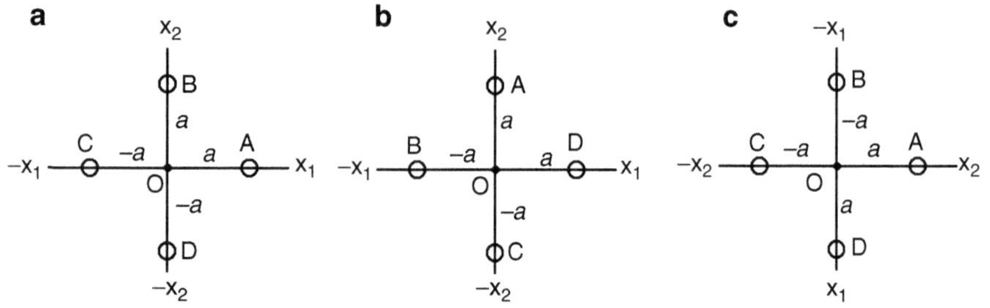

Fig. 7.2 Fourfold rotation: (**a**) original configuration, (**b**) configuration after rotation of atoms, (**c**) configuration after rotation of coordinate axes (rotation axis is normal to the plane of the figure)

Comparing (7.10) and (7.11) with (7.4) and (7.5), we see that the transformation matrix for the coordinates of a point on rotation of axes is the same as the transformation matrix for the components of a vector given by (7.6).

Let us consider symmetry operations. A symmetry operation involves the shifting of every point in the lattice to some other position. In other words, the coordinates of the given point change. We shall show that this change of coordinates can be brought about either by actually shifting the points to new positions or by a rotation of the coordinate axes. Let us consider four atoms A, B, C, D in the Ox_1x_2 plane as shown in Fig. 7.2a. Their coordinates are: A $(a, 0)$; B $(0, a)$; C$(-a, 0)$; D$(0, -a)$. Let the atomic positions be rotated by 90° in anticlockwise direction. The new coordinates of the atoms (Fig. 7.2b) are: A$(0, a)$; B$(-a, 0)$; C$(0, -a)$; D$(a, 0)$. Instead of rotating the atomic positions, let us rotate the coordinate axes by 90° in the clockwise direction. The new axes are shown in Fig. 7.2c. The atomic coordinates now are: A$(0, a)$; B$(-a, 0)$; C$(0, -a)$; D$(a, 0)$. Thus, whether the atoms are shifted to new positions or whether the coordinate axes are rotated, the new coordinates are the same. Hence the symmetry operation of a fourfold rotation about the Ox_3 axis is equivalent to the fourfold rotation of coordinate axes about Ox_3. The transformation matrix for such a rotation of axes is $\begin{pmatrix} 0 & 1 & 0 \\ -1 & 0 & 0 \\ 0 & 0 & 1 \end{pmatrix}$.

The transformation matrices associated with various symmetry operations and combinations of symmetry operations are given in Table 7.1. These transformation matrices will be useful when we take up the effect of symmetry on crystal properties.

7.2.3 *Definition and Transformation Laws of Tensors*

We have seen how the components of a vector transform on rotation of axes (7.4). There are other quantities which transform like the products of components of two or more vectors, e.g.

$$T'_{ij} = a_{ik}a_{jl}T_{kl}, \tag{7.12}$$

Table 7.1 Transformation matrices associated with individual and combinations of symmetry operations

Symmetry operation	Orientation of symmetry element	Transformation matrix
$\bar{1}$	Centre of inversion at origin	$\begin{pmatrix} -1 & 0 & 0 \\ 0 & -1 & 0 \\ 0 & 0 & -1 \end{pmatrix}$
2	Twofold rotation axis coincident with Ox_1 axis	$\begin{pmatrix} 1 & 0 & 0 \\ 0 & -1 & 0 \\ 0 & 0 & -1 \end{pmatrix}$
3	Threefold axis coincident with Ox_1 axis	$\begin{pmatrix} 1 & 0 & 0 \\ 0 & -1/2 & \sqrt{3}/2 \\ 0 & -\sqrt{3}2 & -1/2 \end{pmatrix}$
	Threefold axis along <111>	$\begin{pmatrix} 0 & 0 & 1 \\ 1 & 0 & 0 \\ 0 & 1 & 0 \end{pmatrix}$
	Threefold axis along <11$\bar{1}$>	$\begin{pmatrix} 0 & 1 & 0 \\ 0 & 0 & -1 \\ -1 & 0 & 0 \end{pmatrix}$
4	Fourfold axis coincident with Ox_1 axis	$\begin{pmatrix} 1 & 0 & 0 \\ 0 & 0 & 1 \\ 0 & -1 & 0 \end{pmatrix}$
	Fourfold axis coincident with Ox_3 axis	$\begin{pmatrix} 0 & 1 & 0 \\ -1 & 0 & 0 \\ 0 & 0 & 1 \end{pmatrix}$
6	Sixfold axis coincident with Ox_3 axis	$\begin{pmatrix} 1/2 & \sqrt{3}/2 & 0 \\ -\sqrt{3}/2 & 1/2 & 0 \\ 0 & 0 & 1 \end{pmatrix}$
m	Mirror plane lying in x_2x_3 plane	$\begin{pmatrix} -1 & 0 & 0 \\ 0 & 1 & 0 \\ 0 & 0 & 1 \end{pmatrix}$
$2m$	Twofold axis coincident with Ox_1 axis and mirror plane lying in x_2x_3 plane	$\begin{pmatrix} 1 & 0 & 0 \\ 0 & -1 & 0 \\ 0 & 0 & 1 \end{pmatrix}$

Table 7.2 Transformation laws for tensors

Rank of tensor [r]	Transformation law: New in terms of old	Old in terms of new
0 (Scalar)	$\phi' = \phi$	$\phi = \phi'$
1 (Vector)	$p'_i = a_{ij} p_j$	$p_i = a_{ji} p'_j$
2	$T'_{ij} = a_{ik} a_{jl} T_{kl}$	$T_{ij} = a_{ki} a_{lj} T'_{kl}$
3	$T'_{ijk} = a_{il} a_{jm} a_{kn} T_{lmn}$	$T_{ijk} = a_{li} a_{mj} a_{nk} T'_{lmn}$
4	$T'_{ijkl} = a_{im} a_{jn} a_{ko} a_{lp} T_{mnop}$	$T_{ijkl} = a_{mi} a_{nj} a_{ok} a_{pl} T'_{mnop}$
5	$T'_{ijklm} = a_{in} a_{jo} a_{kp} a_{lq} a_{mr} T_{nopqr}$	$T_{ijklm} = a_{ni} a_{oj} a_{pk} a_{ql} a_{rm} T'_{nopqr}$
6	$T'_{ijklmn} = a_{io} a_{jp} a_{kq} a_{lr} a_{ms} a_{nt} T_{opqrst}$	$T_{ijklmn} = a_{oi} a_{pj} a_{qk} a_{rl} a_{sm} a_{tn} T'_{opqrst}$

and

$$T'_{ijk} = a_{il} a_{jm} a_{kn} T_{lmn}. \tag{7.13}$$

It may be noted that each of the indices takes values 1, 2 and 3. Such quantities are called tensors. For each tensor type, there is a transformation law. This transformation law determines the rank r of the tensor. The number of quantities in the tensor (or components of the tensor) is 3^r. The transformation laws [7.1] for tensors of various ranks are given in Table 7.2.

7.3 Second-Rank Tensor Properties

7.3.1 Physical Properties Relating Two Vectors

Let us consider two vectors P and Q with components (p_1, p_2, p_3) and (q_1, q_2, q_3), respectively, with respect to the $Ox_1x_2x_3$ axes. Let each component of one be linearly related to all the components of the other through relations:

$$\begin{aligned} p_1 &= T_{11}q_1 + T_{12}q_2 + T_{13}q_3 \\ p_2 &= T_{21}q_1 + T_{22}q_2 + T_{23}q_3 \\ p_3 &= T_{31}q_1 + T_{32}q_2 + T_{33}q_3 \end{aligned} \tag{7.14}$$

or

$$p_i = T_{ij} q_j. \tag{7.15}$$

Let us consider the nature of the coefficients T_{ij}. For this purpose, we rotate the axes $Ox'_1x'_2x'_3$. Let the components of the vectors and the coefficients T_{ij} in the new system be p'_i, q'_i and T'_{ij}. We are interested in the transformation of T_{ij} to T'_{ij}. We carry out this transformation in the sequence

$$p' \rightarrow p \rightarrow q \rightarrow q'.$$

Table 7.3 Second-rank tensor properties relating vectors

Applied vector	Resulting vector	Relation	Second-rank tensor property
Electric field (E_i)	Electric current density (j_k)	$E_i = \rho_{ik} j_k$	Electrical resistivity (ρ_{ik})
Electric field (E_j)	Dielectric polarization (P_i)	$P_i = \alpha_{ij} E_j$	Dielectric polarizability (α_{ij})
Magnetic field (H_j)	Intensity of magnetisation (I_i)	$I_i = \mu_0 \chi_{ij} H_j$	Magnetic susceptibility (χ_{ij})
Temperature gradient $-(\partial T/\partial x_j)$	Heat flow density (h_i)	$h_i = -k_{ij}(\partial T/\partial x_j)$	Thermal conductivity (k_{ij})

Then we have,

$$p'_i = a_{ik} p_k = a_{ik} T_{kl} q_l = a_{ik} a_{jl} T_{kl} q'_j. \tag{7.16}$$

Comparing (7.16) with (7.15), we get

$$T'_{ij} = a_{ik} a_{jl} T_{kl}. \tag{7.17}$$

This is the same as the transformation law for a second-rank tensor (Table 7.2). Hence, if a physical property results due to the linear relation between the components of two vectors, we shall infer that it is a second-rank tensor. Some examples of such physical properties of crystals are given in Table 7.3.

We have shown that a physical property that relates two vectors is a second-rank tensor. For completeness, it may be mentioned that a physical property that connects a second-rank tensor to a scalar is also a second-rank tensor.

As mentioned earlier a second-rank tensor has nine components. All second-rank tensor properties given in Table 7.3 are symmetric, i.e. $T_{ij} = T_{ji}$. Further all second-rank tensor properties are centrosymmetric, i.e. if we reverse all p_i's and q_j's, the value of T_{ij} remains unaltered.

7.3.2 *The Representation Quadric*

Let us consider the equation:

$$\sum_{i,j=1}^{3} S_{ij} x_i x_j = 1. \tag{7.18}$$

Here, x_i and x_j are the coordinates of a point and the S_{ij}'s are coefficients; i and j, as usual, have values 1, 2, 3. The expanded form of (7.18) is:

$$\begin{aligned} &S_{11}x_1^2 + S_{12}x_1x_2 + S_{13}x_1x_3 \\ &\quad + S_{21}x_2x_1 + S_{22}x_2^2 + S_{23}x_2x_3 \\ &\quad + S_{31}x_3x_1 + S_{32}x_3x_2 + S_{33}x_3^2 = 1 . \end{aligned} \tag{7.19}$$

Assuming that $S_{ij} = S_{ji}$, we get

$$S_{11}x_1^2 + S_{22}x_2^2 + S_{33}x_3^2 + 2S_{12}x_1x_2 + 2S_{13}x_1x_3 + 2S_{23}x_2x_3 = 1. \quad (7.20)$$

This is the general equation of a second degree surface with its centre at the origin. It is called a quadric.

Let us consider the effect of rotating the coordinate axes from $Ox_1x_2x_3$ to $Ox_1'x_2'x_3'$. From (7.11) we have

$$\begin{aligned} x_i &= a_{ki}x_k' \\ x_j &= a_{lj}x_l'. \end{aligned} \quad (7.21)$$

Substituting in (7.18), we get

$$\sum_{ij} S_{ij}x_ix_j = \sum_{ijkl} S_{ij}a_{ki}a_{lj}x_k'x_l' = 1, \quad (7.22)$$

or,

$$\sum_{kl} S_{kl}'x_k'x_l' = 1. \quad (7.23)$$

This is the equation to the quadric in the new coordinate system. From (7.18), (7.22) and (7.23), we see that

$$S_{kl}' = \sum_{ij} a_{ki}a_{lj}S_{ij}. \quad (7.24)$$

Thus, the coefficients S_{ij} transform like the components of a second-rank tensor. Therefore, the components of a second-rank tensor can be represented by a quadric; the quadric (7.18) and (7.23) is called the representation quadric.

The representation quadric has its centre at the origin. It has three orthogonal principal axes which are, in the general form, not coincident with the coordinate axes. These principal axes can be found by the following procedure. Equate the following determinant to zero:

$$\begin{vmatrix} S_{11}-\lambda & S_{12} & S_{13} \\ S_{21} & S_{22}-\lambda & S_{23} \\ S_{31} & S_{32} & S_{33}-\lambda \end{vmatrix} = 0, \quad (7.25)$$

where λ is a constant. Expansion of the determinant leads to a cubic equation in λ; this is called the secular equation. The three roots of this equation, λ_1, λ_2, λ_3 define the directions of the principal axes of the quadric. Once these directions are known, we can transform the S_{ij}'s to the new coordinate system formed by the principal axes. This reduces (7.18) and (7.20) to

$$S_1x_1^2 + S_2x_2^2 + S_3x_3^2 = 1. \quad (7.26)$$

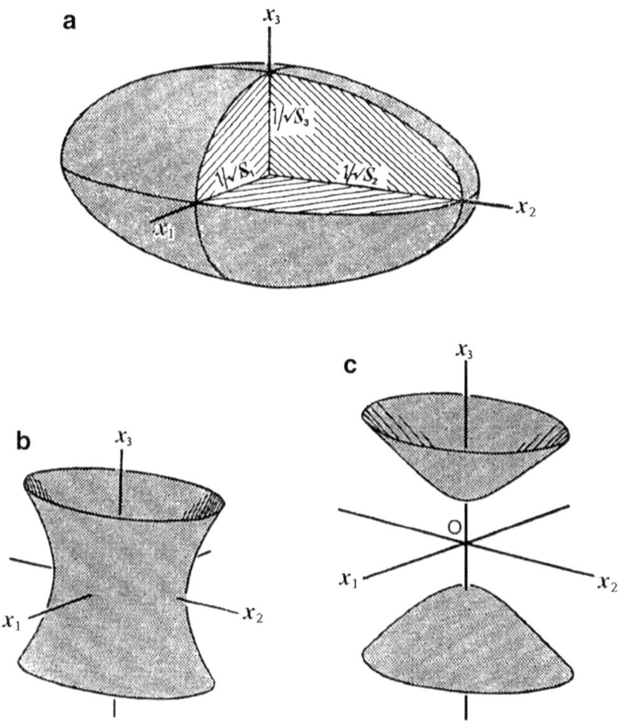

Fig. 7.3 Representation quadrics: (**a**) ellipsoid, (**b**) hyperboloid of one sheet, (**c**) hyperboloid of two sheets

Comparing (7.26) with the equation to an ellipsoid,

$$\frac{x_1^2}{a^2} + \frac{x_2^2}{b^2} + \frac{x_3^2}{c^2} = 1, \tag{7.27}$$

where a, b, c are the lengths of the semi-axes, it can be seen that $1/\sqrt{S_1}, 1/\sqrt{S_2}$ and $1/\sqrt{S_3}$ are the semi-axes of the representation quadric.

Depending on the values of the S_i's (7.26), the quadric may be an ellipsoid or a hyperboloid. If all the S_i's are +ve, the quadric is an ellipsoid (Fig. 7.3a). If two of the S_i's are +ve and one –ve, it is a hyperboloid of one sheet (Fig. 7.3b). On the other hand, if two S_i's are −ve and one +ve, the quadric is a hyperboloid of two sheets (Fig. 7.3c). Finally, if all the S_i's are – ve, the quadric is an imaginary ellipsoid.

We shall note without going into derivation, that the radius vector r of the representation quadric is related to the magnitude of the property S in that direction by

$$S = 1/r^2; \quad r = 1/\sqrt{S}. \tag{7.28}$$

Further, if **p** and **q** are vectors related by $p_i = S_{ij}q_j$, the magnitude S of the property $[S_{ij}]$ in a particular direction is obtained by applying **q** in that direction and measuring $p_{||}/q$ where $p_{||}$ is the component of **p** parallel to **q**.

7.4 Field Tensors (Stress and Strain Tensors)

The tensors discussed in the preceding section are called *matter tensors*. They represent crystal properties and have a definite orientation within a crystal. Also, they are subject to the symmetry of the crystal. On the other hand, there are *field tensors* which do not represent crystal properties. They can have any orientation vis-à-vis the crystal. It is as if they represent an influence applied on to the crystal. We shall consider two such field tensors viz. the stress tensor and the strain tensor.

7.4.1 The Stress Tensor

Let us consider a unit cube within a body (Fig. 7.4) with its edges parallel to the coordinate axes Ox_1, Ox_2, Ox_3. We shall consider the forces exerted by the material outside the cube on the material inside. Let these forces F_1, F_2 and F_3 act on the faces which are the + ve ends of the coordinate axes. Each of these forces can be resolved into components parallel to the coordinate axes. Let σ_{ij} be the component of the force F_j on the face normal to Ox_j acting along Ox_i. Thus, σ_{12} is the force parallel to Ox_1 acting on the face normal to Ox_2. Since the stress is homogeneous, there must be equal and opposite forces acting on the opposite faces. σ_{11}, σ_{22}, σ_{33} are the normal components of the stress and σ_{12}, σ_{23}, σ_{31} are the shear components.

The cube is in static equilibrium. Hence, the moments about the Ox_1 axis must cancel out. Thus, from Fig. 7.5, we get $\sigma_{23} = \sigma_{32}$. Similarly, considering moments

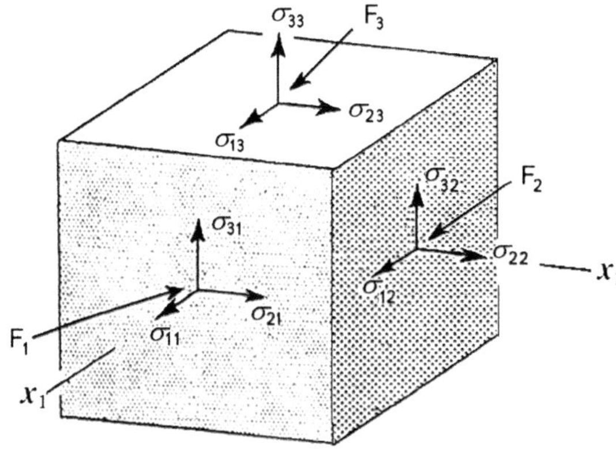

Fig. 7.4 Forces on the faces of a stressed unit cube

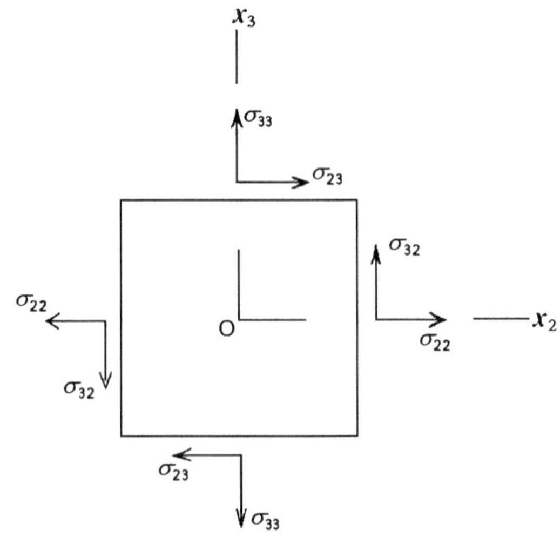

Fig. 7.5 Forces on faces of a unit cube normal to Ox_2 and Ox_3 axes; Ox_1 axis is normal to the plane of the figure

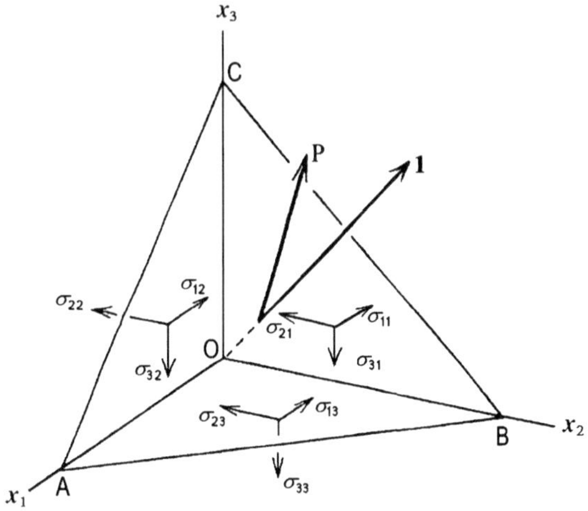

Fig. 7.6 Forces on the faces of the tetrahedron OABC

about the Ox_2 and Ox_3 axes, we can show that $\sigma_{12} = \sigma_{21}$ and $\sigma_{13} = \sigma_{31}$ or, in general,

$$\sigma_{ij} = \sigma_{ji}. \tag{7.29}$$

We shall now prove that the stress components form a second-rank tensor. For this, we shall show that the stress components connect two vector quantities. Consider a small tetrahedron-shaped element of volume OABC in the solid with three of its faces parallel to the coordinate planes (Fig. 7.6). Let a stress **p** act on the face ABC and let **l** be the unit vector normal to ABC. Let p_1, p_2, p_3 be the components of **p** along Ox_1, Ox_2 and Ox_3,respectively. The stresses on the faces of the tetrahedron lying in the coordinate planes are σ_{ij} as shown in Fig. 7.6. Then, resolving the forces parallel to Ox_1, we have

$$p_1(\text{area ABC}) = \sigma_{11}(\text{area BOC}) + \sigma_{12}(\text{area COA}) + \sigma_{13}(\text{area AOB}). \tag{7.30}$$

Considering that BOC, COA and AOB are projections of ABC on the three coordinate planes, we may rewrite (7.30) as

$$p_1(\text{area ABC}) = \sigma_{11}l_1(\text{area ABC}) + \sigma_{12}l_2(\text{area ABC}) + \sigma_{13}l_3(\text{area ABC}), \quad (7.31)$$

where l_1, l_2, l_3 are the components of the vector $\mathbf{l}$. Thus, we have

$$p_1 = \sigma_{11}l_1 + \sigma_{12}l_2 + \sigma_{13}l_3.$$

Similarly,

$$p_2 = \sigma_{21}l_1 + \sigma_{22}l_2 + \sigma_{23}l_3 \quad (7.32)$$

and

$$p_3 = \sigma_{31}l_1 + \sigma_{32}l_2 + \sigma_{33}l_3$$

or

$$p_i = \sum_j \sigma_{ij}l_j. \quad (7.33)$$

Thus, the stress components σ_{ij} connect the vector components p_i and l_j and hence σ_{ij} is a second-rank tensor. Since we have shown earlier that $\sigma_{ij} = \sigma_{ji}$, it is a symmetric tensor denoted, in full, by

$$\left[\sigma_{ij}\right] = \begin{bmatrix} \sigma_{11} & \sigma_{12} & \sigma_{13} \\ \sigma_{12} & \sigma_{22} & \sigma_{23} \\ \sigma_{13} & \sigma_{23} & \sigma_{33} \end{bmatrix}. \quad (7.34)$$

When the stress tensor is referred to its own principal axes, it takes the form

$$\left[\sigma_{ij}\right] = \begin{bmatrix} \sigma_1 & 0 & 0 \\ 0 & \sigma_2 & 0 \\ 0 & 0 & \sigma_3 \end{bmatrix}. \quad (7.35)$$

In these circumstances, the stresses acting on the cube are as shown in Fig. 7.7; it is seen that now there are no shear stresses.

It may be noted that a tetrahedral volume element was chosen only for convenience and the formulation of the stress tensor is independent of the shape of the volume element.

The representation quadric for σ_{ij} is called the stress quadric. Its equation is

$$\sigma_{ij}x_i x_j = 1. \quad (7.36)$$

When referred to principal axes, (7.36) becomes

$$\sigma_1 x_1^2 + \sigma_2 x_2^2 + \sigma_3 x_3^2 = 1. \quad (7.37)$$

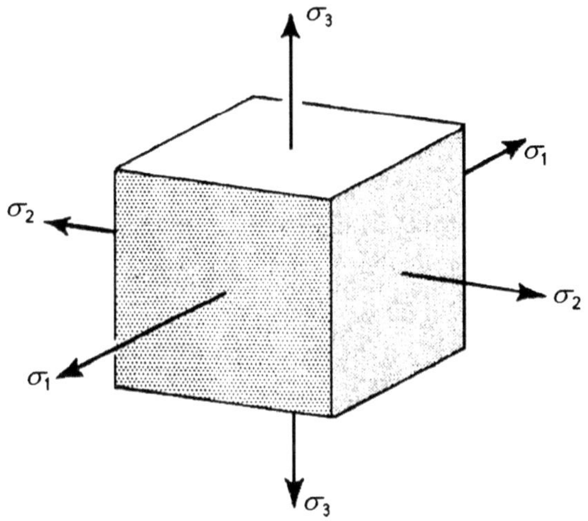

Fig. 7.7 Forces on the faces of a unit cube with its sides parallel to the principal axes

The lengths of the semi-axes are $1/\sqrt{\sigma_1}$, $1/\sqrt{\sigma_2}$ and $1/\sqrt{\sigma_3}$. As in the case of other second-rank tensor quadrics, the stress quadric may be a real or imaginary ellipsoid or hyperboloid.

Some special forms of the stress tensor (referred to the principal axes) are:

1. Uniaxial stress $\begin{bmatrix} \sigma & 0 & 0 \\ 0 & 0 & 0 \\ 0 & 0 & 0 \end{bmatrix}$
2. Biaxial stress $\begin{bmatrix} \sigma_1 & 0 & 0 \\ 0 & \sigma_2 & 0 \\ 0 & 0 & 0 \end{bmatrix}$
3. Pure shear stress $\begin{bmatrix} -\sigma & 0 & 0 \\ 0 & \sigma & 0 \\ 0 & 0 & 0 \end{bmatrix}$
4. Hydrostatic pressure $\begin{bmatrix} -p & 0 & 0 \\ 0 & -p & 0 \\ 0 & 0 & -p \end{bmatrix}$.

7.4.2 *The Strain Tensor*

Under the effect of an external stimulus, a solid is deformed. Let us consider a linear element PQ (Fig. 7.8) parallel to the x-axis. Let the coordinates of P and Q be x and $x + \Delta x$ in the undeformed state. On deformation P is displaced to P′ by an amount

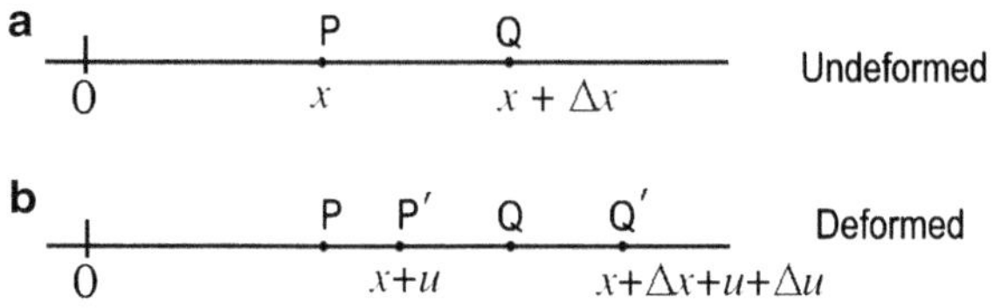

Fig. 7.8 Position of two points along a linear element: (**a**) P, Q before deformation, (**b**) P′, Q′ after deformation

u and Q to Q′ by an amount $u + \Delta u$. $(\Delta u/\Delta x)$ is the deformation per unit length of the undeformed element. The quantity $\mathrm{d}u/\mathrm{d}x$ given by

$$(\mathrm{d}u/\mathrm{d}x) = \underset{\Delta x \to 0}{\mathrm{Lt}} (\Delta u/\Delta x), \tag{7.38}$$

is defined as the linear strain.

Let us now consider the deformation in three dimensions. Again, let us consider two points P and Q having coordinates (x_1, x_2, x_3) and $(x_1 + \Delta x_1, x_2 + \Delta x_2, x_3 + \Delta x_3)$, respectively, in the undeformed state. On deformation, P is displaced to P′ with coordinates $(x_1 + u_1, x_2 + u_2, x_3 + u_3)$. Similarly, Q is displaced to Q′ with coordinates $(x_1 + \Delta x_1 + u_1 + \Delta u_1, x_2 + \Delta x_2 + u_2 + \Delta u_2, x_3 + \Delta x_3 + u_3 + \Delta u_3)$. It is to be noted that each of the quantities Δu_1, Δu_2 and Δu_3 is a function of all the three coordinates of P; these, as we can see, represent the components of the change in the length of PQ. Thus,

$$\begin{aligned}
\Delta u_1 &= (\partial u_1/\partial x_1)\Delta x_1 + (\partial u_1/\partial x_2)\Delta x_2 + (\partial u_1/\partial x_3)\Delta x_3 \\
\Delta u_2 &= (\partial u_2/\partial x_1)\Delta x_1 + (\partial u_2/\partial x_2)\Delta x_2 + (\partial u_2/\partial x_3)\Delta x_3 \\
\Delta u_3 &= (\partial u_3/\partial x_1)\Delta x_1 + (\partial u_3/\partial x_2)\Delta x_2 + (\partial u_3/\partial x_3)\Delta x_3
\end{aligned} \tag{7.39}$$

Let us form an array $[\partial u_i/\partial x_j]$ given by

$$[e_{ij}] = [\partial u_i/\partial x_j] = \begin{bmatrix} \partial u_1/\partial x_1 & \partial u_1/\partial x_2 & \partial u_1/\partial x_3 \\ \partial u_2/\partial x_1 & \partial u_2/\partial x_2 & \partial u_2/\partial x_3 \\ \partial u_3/\partial x_1 & \partial u_3/\partial x_2 & \partial u_3/\partial x_3 \end{bmatrix}. \tag{7.40}$$

The components in the array $[e_{ij}]$ connect the components of the vector Δu_i with the components of the vector Δx_i. Hence $[e_{ij}]$ is a second-rank tensor. This tensor can be expressed as the sum of a symmetric and an antisymmetric tensor as follows:

$$\begin{aligned}
[e_{ij}] = [\partial u_i/\partial x_j] &= \frac{1}{2}\left[\frac{\partial u_i}{\partial x_j} + \frac{\partial u_j}{\partial x_i}\right] + \frac{1}{2}\left[\frac{\partial u_i}{\partial x_j} - \frac{\partial u_j}{\partial x_i}\right] \\
&= [\varepsilon_{ij}] + [\omega_{ij}].
\end{aligned} \tag{7.41}$$

It can be shown that ω_{ij} represents a bodily rotation. We deal with situations where bodily rotation is absent; hence, ω_{ij} can be ignored.

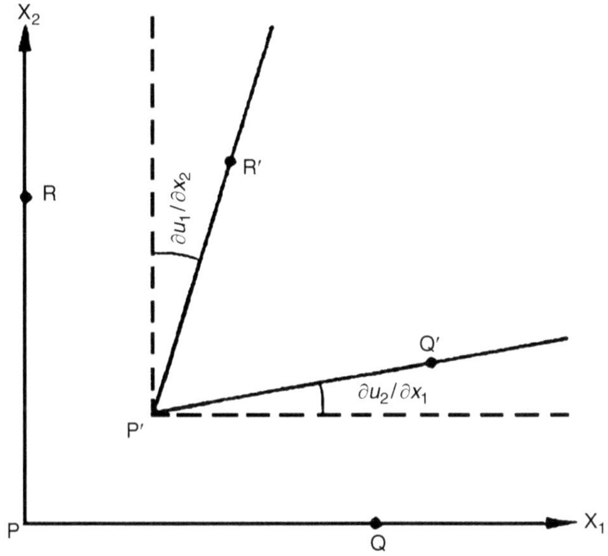

Fig. 7.9 Position of three points in the x_1x_2 plane: (**a**) P, Q, R before shearing, (**b**) P′, Q′, R′ after shearing

Let us now examine the significance of the various components ε_{ij}. If we consider a line element along Ox_1, the increase in length of an element Δx_1 is just $\varepsilon_{11}\,\Delta x_1$; hence, ε_{11} is the linear strain along Ox_1. Similarly, ε_{22} and ε_{33} represent the linear strains along Ox_2 and Ox_3, respectively. In order to understand the meaning of the cross elements ε_{ij} $(i \neq j)$, let us consider the elements PQ parallel to Ox_1 (Fig. 7.9) and PR parallel to Ox_2. On deformation, P, Q and R are displaced to P′, Q′ and R′. PQ $= \Delta x_1$, PR $= \Delta x_2$ and the angle between them is $\pi/2$. On deformation the angle Q′P′R′ is less than the corresponding angle in undeformed state QPR by an amount

$$(\partial u_1/\partial x_2 + \partial u_2/\partial x_1) = 2\varepsilon_{12}. \tag{7.42}$$

This is the shear strain normal to Ox_3. Similarly, ε_{23} and ε_{31} represent the shear strains normal to Ox_1 and Ox_2, respectively. The strain tensor in full form is

$$\left[\varepsilon_{ij}\right] = \begin{bmatrix} \varepsilon_{11} & \varepsilon_{12} & \varepsilon_{13} \\ \varepsilon_{21} & \varepsilon_{22} & \varepsilon_{23} \\ \varepsilon_{31} & \varepsilon_{32} & \varepsilon_{33} \end{bmatrix}. \tag{7.43}$$

From the definition of ε_{ij}, i and j can be interchanged, i.e. $\varepsilon_{ij} = \varepsilon_{ji}$. Hence $[\varepsilon_{ij}]$ is a symmetric second-rank tensor. Its diagonal components represent linear strain and the non-diagonal components represent shear strain. When referred to its principal axes $[\varepsilon_{ij}]$ becomes

$$[\varepsilon_{ij}] = \begin{bmatrix} \varepsilon_1 & 0 & 0 \\ 0 & \varepsilon_2 & 0 \\ 0 & 0 & \varepsilon_3 \end{bmatrix}. \tag{7.44}$$

It may be noted that on deformation, the edges of a unit cube become $(1+\varepsilon_1)$, $(1+\varepsilon_2)$ and $(1+\varepsilon_3)$. Hence the volume strain Δ is

$$\Delta = (1+\varepsilon_1)(1+\varepsilon_2)(1+\varepsilon_3) - 1 \approx \varepsilon_1 + \varepsilon_2 + \varepsilon_3. \tag{7.45}$$

The strain quadric is given by

$$\sum_{ij} \varepsilon_{ij} x_i x_j = 1, \tag{7.46}$$

and, when referred to principal axes, by

$$\varepsilon_1 x_1^2 + \varepsilon_2 x_2^2 + \varepsilon_3 x_3^2 = 1. \tag{7.47}$$

The strain ε in any arbitrary direction is given by

$$\varepsilon = \sum_{i,j=1}^{3} \varepsilon_{ij} l_i l_j, \tag{7.48}$$

or

$$\varepsilon = \varepsilon_1 l_1^2 + \varepsilon_2 l_2^2 + \varepsilon_3 l_3^2, \tag{7.49}$$

when referred to principal axes. The strain is said to be plain strain when one principal strain is zero, i.e.

$$[\varepsilon_{ij}] = \begin{bmatrix} \varepsilon_1 & 0 & 0 \\ 0 & \varepsilon_2 & 0 \\ 0 & 0 & 0 \end{bmatrix}. \tag{7.50}$$

It is said to be a pure shear strain about Ox_3 when $\varepsilon_1 = -\varepsilon_2 = \varepsilon$ and $\varepsilon_3 = 0$, i.e.

$$[\varepsilon_{ij}] = \begin{bmatrix} \varepsilon & 0 & 0 \\ 0 & -\varepsilon & 0 \\ 0 & 0 & 0 \end{bmatrix}. \tag{7.51}$$

7.5 Crystal Properties Which Are Tensors of Higher Rank

In the preceding sections, the discussion was confined to second-rank tensors. This is because a large number of crystal properties happen to be second-rank tensors. However, there are other properties which are tensors of higher ranks. Some of these properties are:

1. The piezoelectric constant tensor $[d_{ijk}]$ is of third rank. It is defined by the equation:
$$P_i = d_{ijk}\sigma_{jk}, \tag{7.52}$$
where P_i are the components of the electric polarization vector **P** and σ_{jk} are components of the stress tensor $[\sigma_{jk}]$.
2. The elastic stiffnesses C_{ijkl} form a fourth rank tensor $[C_{ijkl}]$ defined by the equation
$$\sigma_{ij} = C_{ijkl}\varepsilon_{kl}, \tag{7.53}$$
where σ_{ij} are components of the stress tensor $[\sigma_{ij}]$ and ε_{kl} the components of the strain tensor $[\varepsilon_{kl}]$.
3. Similarly, the photoelastic constants π_{ijkl} form a fourth rank tensor $[\pi_{ijkl}]$ defined by the equation
$$\Delta B_{ij} = \pi_{ijkl}\sigma_{kl}, \tag{7.54}$$
where B_{ij} is a component of the relative dielectric impermeability tensor and σ_{kl} are as defined in 1.
4. The third order elastic constants C_{ijklmn} form a sixth rank tensor $[C_{ijklmn}]$ defined by
$$\phi = \tfrac{1}{2}C_{ijkl}\varepsilon_{ij}\varepsilon_{kl} + C_{ijklmn}\varepsilon_{ij}\varepsilon_{kl}\varepsilon_{mn}, \tag{7.55}$$
where ϕ is the strain energy, C_{ijkl}, the second-order elastic stiffnesses and the ε_{ij}'s the strain components.

7.6 Effect of Crystal Symmetry on Crystal Properties

7.6.1 *Principles*

There is an intimate relationship between the symmetry of a crystal and the symmetry displayed by its properties. The principle governing this relationship is known after Neumann. Neumann stated in 1885 that "the symmetry elements of any crystal property must include the symmetry elements of the point group of the crystal". It is to be noted that what are involved are the point group symmetry elements and not the translation symmetry elements. In view of its importance, some alternate statements have also been proposed [7.3], [7.5]. These are as follows:

1. Every physical property of a crystal must possess at least the symmetry of the point group of the crystal.

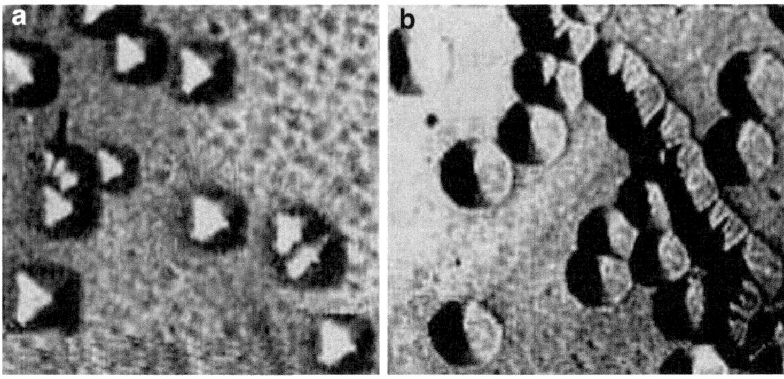

Fig. 7.11 Etch pits on (100) face of NaCl etched with ethyl alcohol and different concentrations of $CdCl_2$ (**a**) 1×10^{-5} mg/cc; (**b**) 4×10^{-5} mg/cc

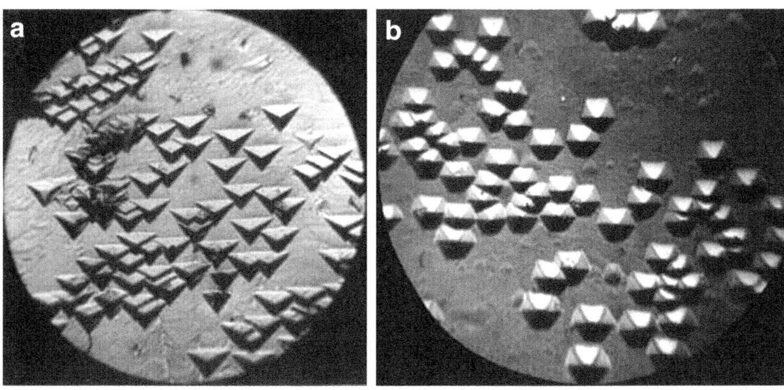

Fig. 7.12 Etch pits on (111) face of $NaBrO_3$ etched with a mixture of acetic acid and formic acid in the ratio 4:1 containing different concentrations of cupric nitrate (**a**) 3 mg/cc, (**b**) 1 mg/cc

Figures 7.11 and 7.12 show a very interesting effect. The exact shape of the etch pits on a face of a crystal, while being consistent with the crystal symmetry, varies with the composition of the etchant. Fig. 7.11 shows etch pits on the 100 face of NaCl obtained with etchants with different compositions. While the shape is square (fourfold axis) in Fig. 7.11a, it is octagonal (eightfold axis) in Fig. 7.11b. Similarly, with variation in the etchant, the triangular etch pits with threefold axis (Fig. 7.12a) on the 111 face of $NaBrO_3$ change to hexagonal shape with sixfold axis (Fig. 7.12b). These are examples which substantiate the statement that a property may show a higher, but not lower, symmetry than that required by crystal symmetry.

7.6.2.2 X-ray Diffraction Photographs

Although X-ray diffraction is a microscopic process, the X-ray diffraction pattern is determined by the point group symmetry and not space group symmetry; the latter

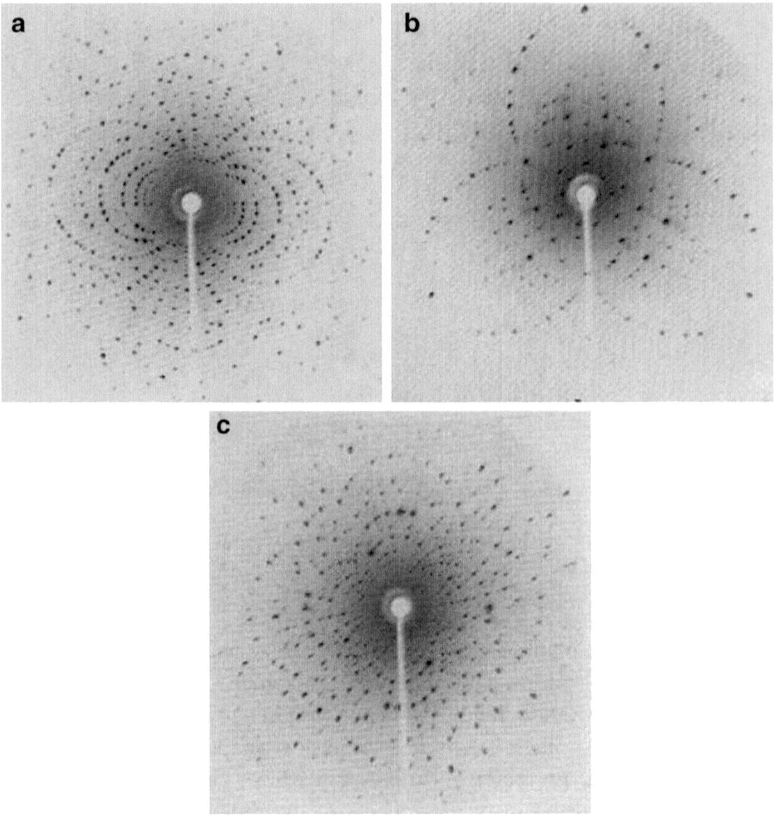

Fig. 7.13 X-ray Laue patterns of (**a**) Kaliporite, twofold axis, (**b**) Dolomite, threefold axis, (**c**) Scapolite, fourfold axis

determines only the absent reflections. The Laue X-ray diffraction photographs for some crystals taken with the X-ray beam along specified symmetry directions are shown in Fig. 7.13a b, c. In these cases, the X-ray beam was parallel to twofold, threefold and fourfold rotation axes. The patterns display the same symmetry.

It may be mentioned that the X-ray diffraction process always adds a centre of symmetry to the pattern, whether the crystal has it or not. This is Friedel's law which states that X-ray diffraction effects cannot reveal whether a crystal has or has not a centre of symmetry. For example, from the X-ray pattern, we cannot distinguish between crystals with point groups $\bar{4}3m$ and $m3m$. Thus, the symmetry shown by an X-ray photograph is the point group symmetry plus a centre of symmetry.

7.6.2.3 Anisotropy of Hardness

In the two examples discussed earlier, interesting inferences could be drawn about the effect of symmetry on crystal properties by merely viewing graphic results

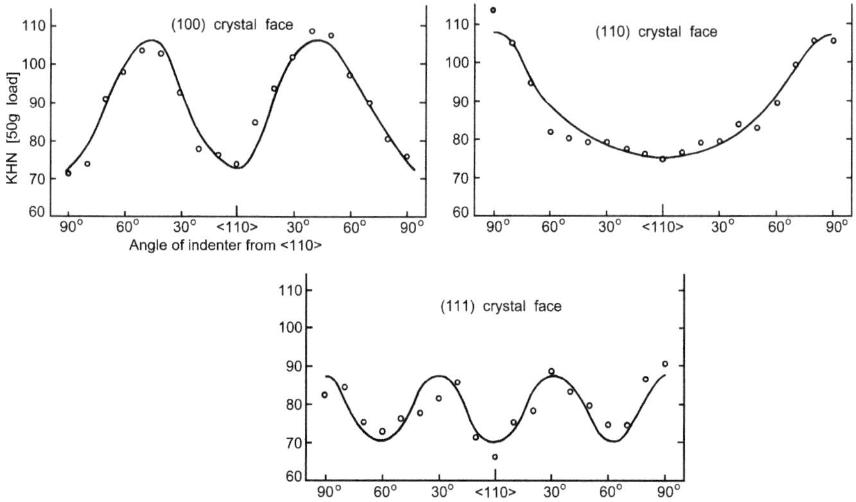

Fig. 7.14 Knoop hardness (KHN) measured with different orientations on the (100), (110) and (111) faces of an iron crystal

(etch pit and X-ray photographs). We shall now consider some results based on measurements of hardness. Hardness is a mechanical property which shows measurable anisotropy, though it has not been formulated as a tensor property.

When a diamond point of definite geometrical shape is pressed on a crystal surface under a load, it creates an impression whose geometry is related to the shape of the diamond. From measurements of the dimensions of the impression (indentation), the hardness is calculated; it is expressed in units of kg/mm^2.

The variation of hardness on the 100, 110 and 111 surfaces of an iron single crystal is shown in Fig. 7.14. These faces have fourfold, twofold and threefold rotational symmetry about directions normal to them. The hardness variation on these faces also shows fourfold, twofold and threefold rotational symmetry consistent with the point group symmetry and also consistent with Neumann's principle.

7.6.3 Effect of Crystal Symmetry on Crystal Properties (Tensor Properties)

In the preceding section, we have considered some simple examples of the effect of crystal symmetry on the physical properties of crystals as a consequence of Neumann's principle. These examples dealt with properties which are not tensors. We shall now deal with properties which are tensors. Here, crystal symmetry has the effect of determining the number of independent constants associated with a property. For simplicity and convenience, we shall consider second-rank tensor properties. We shall adopt two approaches: (1) by considering the shape of the representation quadric and (2) by the use of the transformation matrix.

7.6.3.1 Quadric Surface Method

Let a physical property be represented by the symmetrical second-rank tensor $[T_{ij}]$ given by

$$[T_{ij}] = \begin{bmatrix} T_{11} & T_{12} & T_{13} \\ T_{12} & T_{22} & T_{23} \\ T_{13} & T_{23} & T_{33} \end{bmatrix}. \tag{7.57}$$

In Sect. 7.3.2, we have seen that a symmetrical second-rank tensor can be represented by a quadric surface. For the property $[T_{ij}]$, the equation to the quadric surface is

$$T_{11}x_1^2 + T_{22}x_2^2 + T_{33}x_3^2 + 2T_{12}x_1x_2 + 2T_{13}x_1x_3 + 2T_{23}x_2x_3 = 1. \tag{7.58}$$

A quadric with this equation is an ellipsoid. The cross-section of an ellipsoid normal to any coordinate axis is an ellipse. The symmetry elements associated with an ellipsoid are three diad rotation axes at right angles, three planes of symmetry normal to the diad axes and a centre of inversion. Our method is to seek compatibility between the symmetry elements of the quadric and the symmetry elements of the crystal. The symmetry elements of the crystal systems and the resulting form of $[T_{ij}]$ are given in Table 7.5.

Cubic System

This system has 4 threefold axes along the body diagonals. But the representation quadric which is an ellipsoid cannot have any threefold axes. We invoke Neumann's principle according to which the property can have a higher symmetry. The only way to reconcile the symmetry of the property and that of the quadric is to make the quadric a sphere. The property given by the radius vector has the same value in any direction. The resulting tensor has only one component T_{11} as shown in Table 7.5.

Tetragonal, Hexagonal and Trigonal Systems

The characteristic symmetry elements in these systems are a fourfold axis, a sixfold axis and a threefold axis, respectively, along one of the directions say Ox_3. The only way to associate such an axis with the quadric is to have an ∞-fold axis of rotation along Ox_3. In other words, the quadric becomes a surface of revolution about Ox_3. As a consequence, $T_{ij}(i \neq j) = 0$ and $T_{11} = T_{22} \neq T_{33}$.

Table 7.5 The effect of crystal symmetry on properties represented by symmetrical second-rank tensors

System	Characteristic symmetry	Nature of representation quadric and its orientation	Number of independent coefficients	Tensor referred to axes in the conventional orientation
Cubic	4 threefold axes	Sphere	1	$\begin{bmatrix} T & 0 & 0 \\ 0 & T & 0 \\ 0 & 0 & T \end{bmatrix}$
Tetragonal Hexagonal Trigonal	1 fourfold axis 1 sixfold axis 1 threefold axis	Quadric of revolution about the principal axis $(x_3)(z)$	2	$\begin{bmatrix} T_1 & 0 & 0 \\ 0 & T_1 & 0 \\ 0 & 0 & T_3 \end{bmatrix}$
Orthorhombic	Three mutually perpendicular twofold axes; no axes of higher order	General quadric with axes $(x_1, x_2, x_3)\|\|$ to the diad axes (x, y, z)	3	$\begin{bmatrix} T_1 & 0 & 0 \\ 0 & T_2 & 0 \\ 0 & 0 & T_3 \end{bmatrix}$
Monoclinic	1 twofold axis	General quadric with one axis $(x_3)\|\|$ to the diad axes (x, y, z)	4	$\begin{bmatrix} T_{11} & 0 & T_{13} \\ 0 & T_{22} & 0 \\ T_{13} & 0 & T_{33} \end{bmatrix}$
Triclinic	A centre of symmetry or no symmetry	General quadric. No fixed relation to crystallographic axes	6	$\begin{bmatrix} T_{11} & T_{12} & T_{13} \\ T_{12} & T_{22} & T_{23} \\ T_{13} & T_{23} & T_{33} \end{bmatrix}$

Orthorhombic System

This crystal system has 3 twofold axes along the crystal axes. Thus, the crystal as well as the general quadric has the same symmetry elements. The quadric axes may be oriented to coincide with the crystal axes. Now, $T_{ij}(i \neq j) = 0$ and only T_{11}, T_{22} and T_{33} remain.

Monoclinic System

The characteristic symmetry of this system is a twofold axis. We set the quadric such that one diad axis of the quadric is parallel to the twofold crystal axis. This satisfies the condition that the quadric has the symmetry element of the crystal. There is no other constraint and the ellipsoid may have any orientation. Thus four components are required, three to specify lengths of the semi-principal axes of the quadric and one to indicate the orientation of the quadric with respect to the crystal axes. Hence, there are four non-zero components, T_{11}, T_{22}, T_{33} and T_{13}.

Triclinic System

There are two point groups in this crystal system. One has a centre of symmetry and the other does not have it. Both the point groups have no other symmetry elements.

Thus, the quadric can have any orientation relative to the crystal axes. Hence all the six components are required, three to represent the lengths of the principal axes and three to fix the orientation of the quadric.

By comparing the symmetry elements of the crystal with the symmetry elements of the representation quadric and seeking a reconciliation between the two as required by Neumann's principle, we have arrived at the important result that the number of independent components in a second-rank tensor property varies from one to six depending upon the crystal system.

7.6.3.2 The Matrix Method

We shall now see that the same results can be obtained by using the transformation matrix. The procedure is to apply the symmetry operation associated with the point group of the crystal on the set of equations representing the property and demand that the equations remain invariant. Thus for vectors p_i' and q_j',,we have

$$p_i' = a_{ik} p_k, \tag{7.59}$$

$$q_j' = a_{jl} p_l, \tag{7.60}$$

and for a second-rank tensor,

$$T_{ij}' = a_{ik} a_{jl} T_{kl}. \tag{7.61}$$

where (a_{ik}) is the transformation matrix. For each crystal class, we use the relevant transformation matrix, calculate the T_{ij}'s and apply Neumann's principle.

Cubic System

We have noted that for the cubic system the characteristic symmetry elements are the 4 threefold axes along the body diagonals. Let us consider a threefold rotation about < 111 >. The transformation matrix for this symmetry operation (see Table 7.1) is $\begin{pmatrix} 0 & 0 & 1 \\ 1 & 0 & 0 \\ 0 & 1 & 0 \end{pmatrix}$.

We shall now evaluate T_{11}'. This is given by

$$\begin{aligned} T_{11}' &= a_{11}a_{11}T_{11} + a_{11}a_{12}T_{12} + a_{11}a_{13}T_{13} \\ &+ a_{12}a_{11}T_{12} + a_{12}a_{12}T_{22} + a_{12}a_{13}T_{23} \\ &+ a_{13}a_{11}T_{13} + a_{13}a_{12}T_{23} + a_{13}a_{13}T_{33}. \end{aligned} \tag{7.62}$$

Substituting the values of a_{ij} from the matrix, we get
$T_{11}' = T_{33}$

Similarly, we get

$$T'_{22} = T_{11},\ T'_{33} = T_{22},\ T'_{12} = T_{31},\ T'_{23} = T_{12},\ T'_{31} = T_{23}. \tag{7.63}$$

We shall now perform a threefold rotation about $< 11\bar{1} >$. In this case the transformation matrix is $\begin{pmatrix} 0 & 1 & 0 \\ 0 & 0 & -1 \\ -1 & 0 & 0 \end{pmatrix}$. We get for the T_{ij}'s:

$$\begin{aligned} &T'_{11} = T_{22}, \quad T'_{22} = T_{33}, \quad T'_{33} = T_{11}, \\ &T'_{12} = -T_{23}, \quad T'_{23} = T_{31}, \quad \text{and} \quad T'_{31} = -T_{12}. \end{aligned} \tag{7.64}$$

These two sets of T_{ij}'s can be reconciled with Neumann's principle, if $T_{11} = T_{22} = T_{33}$ and $T_{12} = T_{23} = T_{31} = 0$. Thus, a second-rank tensor property has only one independent component for a cubic crystal and the associated tensor is $\begin{bmatrix} T_{11} & 0 & 0 \\ 0 & T_{11} & 0 \\ 0 & 0 & T_{11} \end{bmatrix}$. This is the same as the result obtained by the quadric method.

Tetragonal System

The characteristic symmetry element for this system is a fourfold rotation axis along, say, the Ox_3axis. The transformation matrix associated with this symmetry operation (Table 7.1) is $\begin{pmatrix} 0 & 1 & 0 \\ -1 & 0 & 0 \\ 0 & 0 & 1 \end{pmatrix}$. Following the same procedure, we get

$$\begin{aligned} &T'_{11} = T_{22}, \quad T'_{22} = T_{11}, \quad T'_{33} = T_{33}, \\ &T'_{12} = -T_{21}, \quad T'_{13} = T_{23} \quad \text{and} \quad T'_{23} = -T_{13}. \end{aligned} \tag{7.65}$$

Invariance as required by Neumann's principle can be ensured by making

$$T_{11} = T_{22} \quad \text{and} \quad T_{12} = T_{23} = T_{13} = 0. \tag{7.66}$$

Thus, there are only two independent components viz. T_{11} and T_{33} and the tensor takes the form $\begin{bmatrix} T_{11} & 0 & 0 \\ 0 & T_{11} & 0 \\ 0 & 0 & T_{33} \end{bmatrix}$. This is the same as the result given in Table 7.5.

The method can be applied to other crystal classes; the results obtained will be same as in Table 7.5.

We have thus seen that crystal symmetry has an important effect on tensor properties. In second-rank tensor properties, the number of independent components is six for the triclinic system. But symmetry reduces this number to four for monoclinic; three for orthorhombic; two for tetragonal, hexagonal and trigonal and only one for the cubic system. While dealing with second-rank tensor properties one could use the quadric method or the matrix method.

Problems

1. Prove that a property that relates a second-rank tensor to a scalar is itself a second-rank tensor.
2. Using matrices show that a twofold rotation about an axis followed by a reflection at a plane normal to the twofold axis is equivalent to inversion.
3. The electrical conductivity tensor $[\sigma_{ik}]$ has the following components referred to axes Ox_1, Ox_2, Ox_3:

$$[\sigma_{ij}] = \begin{bmatrix} 25 & 0 & 0 \\ 0 & 7 & -3\sqrt{3} \\ 0 & 3\sqrt{3} & 13 \end{bmatrix} \quad \text{in units of } 10^7 \text{ohm}^{-1}\text{m}^{-1}.$$

   If the axes are rotated to Ox_1', Ox_2', Ox_3' such that $x_1'\hat{O}x_1 = 0°$, $x_2'\hat{O}x_2 = 30°$, $x_2'\hat{O}x_3 = 60°$ and $x_3'\hat{O}x_3 = 30°$, determine the new components σ_{ij}'.
4. In Problem 3, draw a radius vector OP in the direction whose cosines with respect to old axes are $(0, 1/2, \sqrt{3}/2)$. Determine the length of the radius vector and hence calculate the electrical conductivity in this direction.
5. The deformation of a crystal is defined by the tensor $[e_{ij}]$ as:

$$\left[e_{ij}\right] = \begin{bmatrix} 8 & -1 & -1 \\ 1 & 6 & 0 \\ -5 & 0 & 2 \end{bmatrix} \times 10^{-6}.$$

   Determine the strain tensor $[\varepsilon_{ij}]$ and the rotation tensor $[\omega_{ij}]$.

References

7.1. J.F. Nye, *Physical Properties of Crystals* (Oxford University Press, London, 1957).
7.2. C.S. Smith, Solid State Phys. **6**, 175 (1958).
7.3. S. Bhagavantam, *Crystal Symmetry and Physical Properties* (Academic Press, New York, 1966).
7.4. Yu. I. Sirotin, M.P. Shaskolskaya, *Fundamentals of Crystal Physics* (Mir Publishers, Moscow, 1982).
7.5. J.K. Wadhawan, *Introduction to Ferroic Materials* (Gordon and Breach Publishers, Amsterdam, 2000).

Chapter 8
Mechanical Properties of Solids

8.1 Introduction

As we proceed along, we will see that the elastic properties of solids have twofold importance. Firstly, they indicate the mechanical strength of the solid. Secondly, they are very important in understanding the nature of the interatomic forces and in the analysis of lattice vibrations.

When an elastic solid is stressed, it results in strains. So long as the strains are small, the stress is proportional to the strain. This is known as Hooke's law. We shall first consider the elastic behaviour of an isotropic elastic medium and then discuss the stress–strain relations in crystals. Much of our discussion of the elastic behaviour of solids is within the realm of Hooke's law. At the end, for the sake of completeness, we shall briefly discuss the elastic behaviour beyond Hooke's law.

8.2 Elastic Properties of Isotropic Solids

8.2.1 Definitions

The application of stress creates a strain. According to Hooke's law the ratio of the stress to the strain is a constant; this is called the elastic modulus. But the stress can be applied in three different ways and so there are three moduli; we shall now define them.

Consider a rod-like element in an isotropic elastic medium. Let the element be aligned along the x-axis (Fig. 8.1). Let the length be l in the unstressed state. On the application of a tensile stress σ_x, the length increases by Δl. The longitudinal strain along x is $\varepsilon_x = (\Delta l / l)$.

By Hooke's law

$$\sigma_x = Y\varepsilon_x, \tag{8.1}$$

D.B. Sirdeshmukh et al., *Atomistic Properties of Solids*, Springer Series in Materials Science 147, DOI 10.1007/978-3-642-19971-4_8,

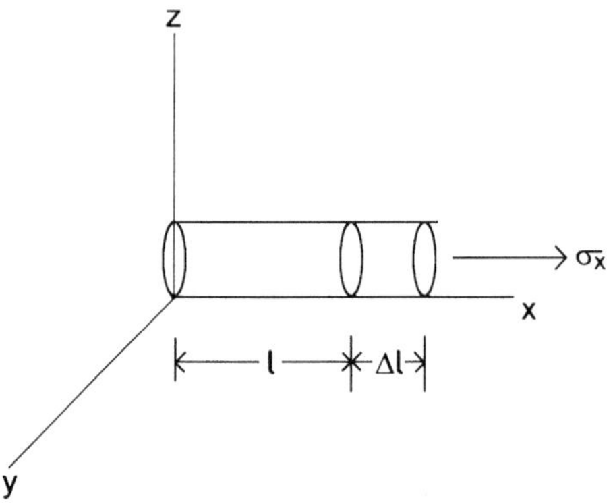

Fig. 8.1 Extension due to tensile stress

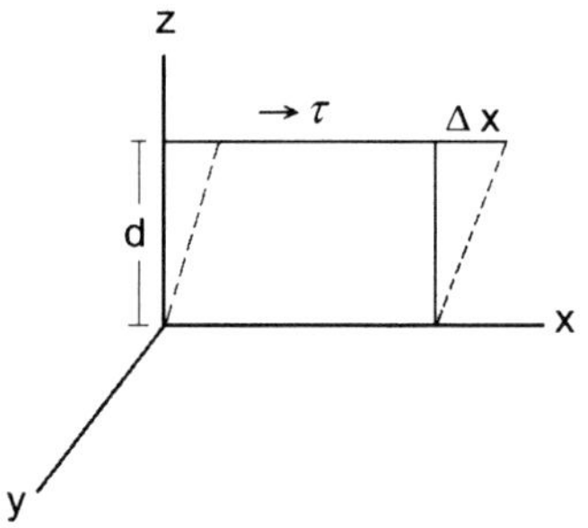

Fig. 8.2 Strain caused by shearing stress

where the constant Y is called Young's modulus. The extension in the x-direction is accompanied by contractions in the y and z directions. Denoting these lateral strains by $-\varepsilon_y$ and $-\varepsilon_z$, we define a property ν as:

$$\nu = -\varepsilon_y/\varepsilon_x = -\varepsilon_z/\varepsilon_x. \tag{8.2}$$

ν is the ratio of the lateral strain to the longitudinal strain and is called the Poisson's ratio; the Poisson's ratio is always a positive quantity.

Let a shear stress τ be applied as shown in Fig. 8.2. Under the effect of this stress, two planes parallel to the xy plane, separated by distance d undergo a relative displacement Δx. The shear strain γ is defined as $(\Delta x/d)$. Again, by Hooke's law,

$$\tau = G\gamma. \tag{8.3}$$

The constant G, which is the ratio of the shearing stress to the shear strain, is called the shear modulus.

Finally, let us consider a cube of volume V in the unstressed medium. If a hydrostatic stress (pressure P) is applied to the cube (Fig. 8.3), its volume is reduced to V–ΔV. The volume strain is (–$\Delta V/V$). Again, by Hooke's law, we have

$$P = B(-\Delta V/V). \tag{8.4}$$

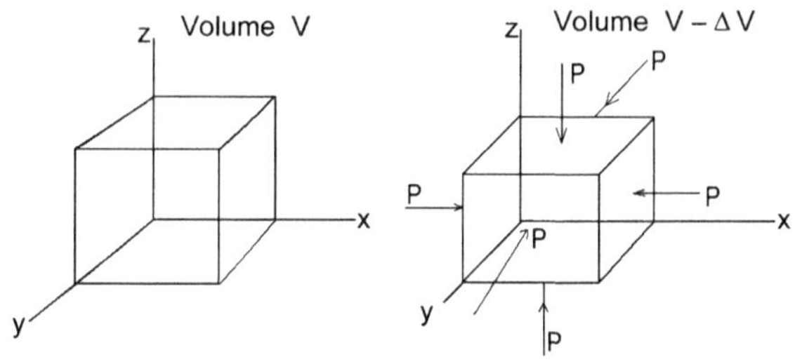

Fig. 8.3 Volume strain caused by hydrostatic stress

The constant B, which is the ratio of the hydrostatic stress to the volume strain, is called the bulk modulus.

8.2.2 Interrelations

It can be shown through elementary derivations that the elastic properties defined earlier are interrelated as follows:

$$Y = 2G(1 + \nu), \quad (8.5a)$$

$$B = Y/[3(1 - 2\nu)], \quad (8.5b)$$

$$G = 3BY/(9B - Y), \quad (8.5c)$$

and

$$\nu = (1/2)(3B - 2G)/(G + 3B). \quad (8.5d)$$

Thus, the elastic behaviour of an isotropic solid is described in terms of the Young's modulus, the shear and bulk moduli and the Poisson's ratio. From (8.5), it can be seen that if any two of the elastic properties are known, the other two can be estimated. It can be seen from (8.5a, b) that the Poisson's ratio has values in the range $0 < \nu < 0.5$. The experimental values of ν for most materials are close to 0.3 but, in any case, are always within the above limits.

8.2.3 Numerical Values

To get an idea about the order of magnitude of the elastic properties, experimental values for some metals are given in Table 8.1 [8.1]. Elastic moduli are expressed in units of 10^{12} dyn/cm^2 or 10^{11}N/m^2.

Table 8.1 Values of Young's modulus Y, shear modulus G, bulk modulus B (all in 10^{12} dyn/cm^2) and the Poisson's ratio ν for some metals

Metal	Y	G	B	ν
Al	0.71	0.26	0.72	0.34
Ag	0.80	0.28	1.01	0.37
Cu	1.23	0.45	1.31	0.34
Fe	2.09	0.82	1.68	0.28
W	3.97	1.53	3.24	0.28

8.3 Compressibility

8.3.1 *Definition*

The compressibility (ψ) is, simply, the reciprocal of the bulk modulus B. Thus, its definition is

$$(\Delta V/V) = -\psi\, P. \tag{8.6}$$

It will be seen later that the compressibility enters into the theories of specific heats, thermal expansion and cohesion. In view of its importance, we shall consider it in some detail.

Depending on the method of measurement, the value of the compressibility may be isothermal (ψ_T) or adiabatic (ψ_S). The two are related as:

$$\psi_\mathrm{T} = \psi_\mathrm{S} + (9\alpha^2 TV/C_\mathrm{P}), \tag{8.7}$$

where α is the coefficient of linear thermal expansion, T the temperature, V the molar volume and C_p the molar specific heat at constant pressure. The correction term is quite small, usually less than 1% of ψ_S.

8.3.2 *Equation of State Parameters*

In thermodynamics, an equation of state relates the parameters P, V and T. At constant T, it is just a volume–pressure relation. In this sense, (8.6) is an equation of state within the limits of Hooke's law, i.e. at moderate pressures and moderate strains. At higher pressures, the volume strain shows a pressure dependence which is generally expressed as a power series in P. Thus, we have

$$(\Delta V/V) = aP + bP^2 + cP^3 + \cdots. \tag{8.8}$$

The number of terms depends on the range of pressure and also on the nature of pressure dependence for a given material. The negative value of the coefficient "a" represents the compressibility (ψ) at zero pressure; the other coefficients b, c, etc. are related to the first and higher order pressure derivatives of the compressibility. Thus, the pressure coefficient $(1/\psi)(\mathrm{d}\psi/\mathrm{d}P)$ is given by $(2b/a)$.

Box 8.1

Percy William Bridgman, an American physicist, is known for the development of high pressure techniques for investigation of properties of matter at high pressure. In particular, he developed the piston displacement method for measurement of compressibility of solids. He was awarded the Nobel Prize for physics in 1946.

P.W. Bridgman (1882–1961)

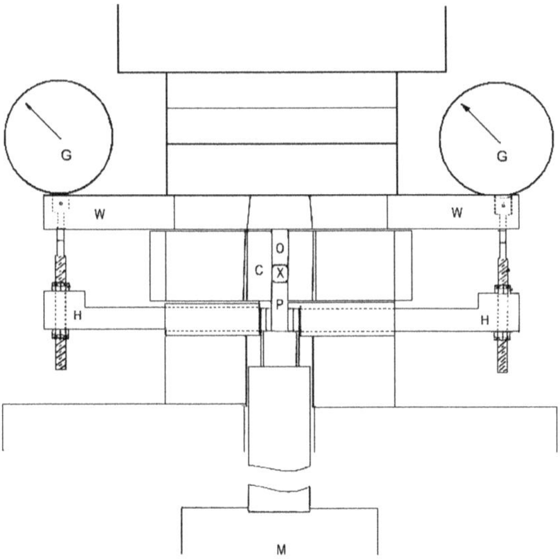

Fig. 8.4 Schematic diagram of the piston displacement set-up

8.3.3 Experimental Measurement

Pioneering experiments for the determination of the compressibility and the equation-of-state parameters were carried out by Bridgman (Box 8.1) using the technique now known as the piston displacement method. A modern set-up based on Bridgman's original design is shown in Fig. 8.4. X is the sample held between the piston P and a fixed tungsten carbide element O. The piston is driven by the hydraulic ram M.

The pressure vessel C is made of tungsten carbide. The relative displacement between the lever arms W and H is measured by the dial gauges G. This is a static method which gives an isothermal value of ψ.

Table 8.2 Values of the parameters a, b (8.8) for some metals over the pressure range 0–45 kbar

Metal	a $[10^{-12}\mathrm{cm}^2/\mathrm{dyn}]$	b $[10^{-24}\mathrm{cm}^2/\mathrm{dyn})^2]$	$\psi = -a$ $[10^{-12}\mathrm{cm}^2/\mathrm{dyn}]$	$(1/\psi)d\psi/dP = 2b/a$ $[10^{-12}\,\mathrm{cm}^2/\mathrm{dyn}]$
Al	−1.27	3.34	1.27	−5.25
Ag	−0.90	2.31	0.90	−5.13
Cu	−0.66	1.09	0.66	−3.30
Fe	−0.58	1.50	0.58	−5.17

8.3.4 Numerical Data on Equation-of-State Parameters

The equation-of-state parameters, the compressibility and its pressure coefficient for some metals are given in Table 8.2 [8.2].

8.4 Single-Crystal Elastic Constants

8.4.1 Definitions

So far, we have considered the elastic properties of an isotropic solid medium. A more elaborate formulation is necessary when we consider an anisotropic crystalline solid. As discussed in Chap. 7, the stress and strain in a single crystal are described by two symmetric second rank tensors $[\sigma_{ij}]$ and $[\varepsilon_{ij}]$ given by

$$\left[\sigma_{ij}\right] = \begin{bmatrix} \sigma_{11} & \sigma_{12} & \sigma_{13} \\ \sigma_{12} & \sigma_{22} & \sigma_{23} \\ \sigma_{13} & \sigma_{23} & \sigma_{33} \end{bmatrix} \tag{8.9}$$

and

$$\left[\varepsilon_{ij}\right] = \begin{bmatrix} \varepsilon_{11} & \varepsilon_{12} & \varepsilon_{13} \\ \varepsilon_{12} & \varepsilon_{22} & \varepsilon_{23} \\ \varepsilon_{13} & \varepsilon_{23} & \varepsilon_{33} \end{bmatrix}. \tag{8.10}$$

We consider only infinitesimal strains. The generalised Hooke's law statement is that each strain component is linearly related to each of the stress components.

Thus,

$$\varepsilon_{ij} = \sum_{kl} S_{ijkl}\sigma_{kl}. \tag{8.11}$$

The coefficients S_{ijkl} are called compliance coefficients or just compliances. Similarly, each stress component is linearly related to all the strain components. Thus,

$$\sigma_{ij} = \sum_{kl} C_{ijkl}\varepsilon_{kl}. \tag{8.12}$$

The coefficients C_{ijkl} are called the stiffness coefficients or, simply, stiffnesses. Together, the S_{ijkl}'s and C_{ijkl}'s are called the second-order elastic constants (SOEC).

Let us now consider the nature of these arrays of coefficients. It may be noted that the stress tensor $[\sigma_{ij}]$ and the strain tensor $[\varepsilon_{ij}]$ are formulated with respect to the coordinate system $Ox_1x_2x_3$. Let the coordinate system be rotated to $Ox_1'x_2'x_3'$. Then $x_i \to x_i'$, $[\sigma_{ij}] \to [\sigma_{ij}']$ and $[\varepsilon_{ij}] \to [\varepsilon_{ij}']$ where the symbol $\to$ reads 'transforms as'. Let us carry out the transformation in the sequence

$$\varepsilon' \to \varepsilon \to \sigma \to \sigma'.$$

Then, following the laws of transformation of tensors, and following the dummy suffix notation (Sect. 7.2.1), we get

$$\begin{aligned} \varepsilon_{ij}' &= a_{ik}a_{jl}\varepsilon_{kl} = a_{ik}a_{jl}S_{klmn}\sigma_{mn} \\ &= a_{ik}a_{jl}S_{klmn}a_{om}a_{pn}\sigma_{op}' = S_{ijop}'\sigma_{op}'. \end{aligned} \tag{8.13}$$

Thus, the coefficient S_{klmn} transforms according to the law

$$S_{klmn} = a_{ik}a_{jl}a_{om}a_{pn}S_{ijop}'. \tag{8.14}$$

This is the transformation law for a fourth-rank tensor (see Table 7.2). A similar transformation equation can be obtained for C_{ijkl}. Thus, the single-crystal elastic constants form a fourth-rank tensor. In general, the number of coefficients in the elastic constant tensor would be 81. However, since the stress and strain tensors are symmetric with only six independent components, the number of independent components in the $[S_{ijkl}]$ and $[C_{ijkl}]$ tensors is 36.

We shall now adopt a notation that enables us to reduce the number of suffixes from 4 to 2. Thus, we change

$$\begin{array}{cccccc} 11 & 22 & 33 & 23,32 & 31,13 & 12,21 \\ \downarrow & \downarrow & \downarrow & \downarrow & \downarrow & \downarrow \\ 1 & 2 & 3 & 4 & 5 & 6 \end{array}$$

and also,

$$\begin{aligned} S_{ijkl} &\to S_{mn} \quad \text{for } m \text{ and } n = 1, 2, 3, \\ 2S_{ijkl} &\to S_{mn} \quad \text{for either } m \text{ or } n = 4, 5, 6 \\ 4S_{ijkl} &\to S_{mn} \quad \text{for both } m \text{ and } n = 4, 5, 6. \end{aligned}$$

With these changes, (8.11) becomes

$$\varepsilon_i = S_{ij}\sigma_j \quad (i, j = 1, 2, \ldots, 6). \tag{8.15}$$

Similarly (8.12) becomes

$$\sigma_i = C_{ij}\varepsilon_j \quad (i,\ j = 1,2,\ldots,6)\,. \tag{8.16}$$

The arrays of these elastic constants are

$$\begin{pmatrix} S_{11} & S_{12} & S_{13} & S_{14} & S_{15} & S_{16} \\ S_{21} & S_{22} & S_{23} & S_{24} & S_{25} & S_{26} \\ S_{31} & S_{32} & S_{33} & S_{34} & S_{35} & S_{36} \\ S_{41} & S_{42} & S_{43} & S_{44} & S_{45} & S_{46} \\ S_{51} & S_{52} & S_{53} & S_{54} & S_{55} & S_{56} \\ S_{61} & S_{62} & S_{63} & S_{64} & S_{65} & S_{66} \end{pmatrix} \quad \text{and} \quad \begin{pmatrix} C_{11} & C_{12} & C_{13} & C_{14} & C_{15} & C_{16} \\ C_{21} & C_{22} & C_{23} & C_{24} & C_{25} & C_{26} \\ C_{31} & C_{32} & C_{33} & C_{34} & C_{35} & C_{36} \\ C_{41} & C_{42} & C_{43} & C_{44} & C_{45} & C_{46} \\ C_{51} & C_{52} & C_{53} & C_{54} & C_{55} & C_{56} \\ C_{61} & C_{62} & C_{63} & C_{64} & C_{65} & C_{66} \end{pmatrix} \tag{8.17}$$

It is important to note that the elastic constants in the two-suffix notation constitute matrices and not tensors. The S_{ij}'s and C_{ij}'s do not transform like tensors; it is only the S_{ijkl}'s and C_{ijkl}'s that do so.

Consider a unit cube in a crystal deformed by an infinitesimal strain $d\varepsilon_i (i = 1, 2, \ldots, 6)$. The strain energy $\mathrm{d}\phi$ is

$$\mathrm{d}\phi = \sigma_i \mathrm{d}\varepsilon_i\,, \tag{8.18}$$

which may be written as

$$\mathrm{d}\phi = C_{ij}\varepsilon_j \mathrm{d}\varepsilon_i\,. \tag{8.19}$$

Differentiating with respect to ε_j, we get

$$\frac{\partial}{\partial\varepsilon_j}\frac{\partial\phi}{\partial\varepsilon_i} = C_{ij}. \tag{8.20}$$

If we had started with strain $\mathrm{d}\varepsilon_j$, we would have obtained

$$\frac{\partial}{\partial\varepsilon_i}\frac{\partial\phi}{\partial\varepsilon_j} = C_{ji}. \tag{8.21}$$

Since the strain energy ϕ is a function of the state of the body, the sequence of differentiation is immaterial. Hence,

$$C_{ij} = C_{ji}. \tag{8.22}$$

Thus, the C_{ij} and S_{ij} matrices are symmetric. This reduces the independent components of the matrices from 36 to 21. The units of the C_{ij}'s are $\mathrm{dyn/cm^2}$ or $\mathrm{N/m^2}$ and those of the S_{ij}'s are $\mathrm{cm^2/dyn}$ or $\mathrm{m^2/N}$. Incidentally, integrating (8.19), we get the expression for the strain energy per unit volume as

$$\phi = \tfrac{1}{2} C_{ij} \varepsilon_i \varepsilon_j = \tfrac{1}{2} S_{ij} \sigma_i \sigma_j . \tag{8.23}$$

8.4.2 *Effect of Symmetry*

We have seen that the full matrix of the elastic constants is

$$\begin{pmatrix} C_{11} & C_{12} & C_{13} & C_{14} & C_{15} & C_{16} \\ & C_{22} & C_{23} & C_{24} & C_{25} & C_{26} \\ & & C_{33} & C_{34} & C_{35} & C_{36} \\ & & & C_{44} & C_{45} & C_{46} \\ & & & & C_{55} & C_{56} \\ & & & & & C_{66} \end{pmatrix},$$

which is a 6×6 symmetric matrix. The 21 independent components are given. The components below the diagonal are given by the relation $C_{ij} = C_{ji}$. This is the elastic constant matrix for the triclinic class. For the other crystal classes, the number of independent components is reduced due to the presence of symmetry elements.

To examine the effect of symmetry on the C_{ij}'s, we shall use the "Direct Inspection Method" [8.3]. Here, we first note how the coordinates change in a symmetry operation (Table 7.1) and how this, in turn, results in a change in the indices of elastic constants in the four-suffix notation (old notation). We then convert these elastic constants to the two-suffix notation (new notation). Finally, we compare the elastic constants before and after the symmetry operation and insist that they remain unaltered following Neumann's principle.

As examples, we shall consider two crystal classes: (1) tetragonal class $\bar{4}$ and (2) cubic classes. Before considering these cases, we shall recall the change from the four-suffix to the two-suffix notation.

Old	11	22	33	23, 32	13, 31	12, 21
New	1	2	3	4	5	6

Tetragonal Class $\bar{4}$: If the operation $\bar{4}$ is performed about the Ox_3 axis, the coordinates change as

$$x_1 \to -x_2, x_2 \to x_1, x_3 \to -x_3$$

Hence the pairs of indices of the elastic constants change as

Old notation		**New notation**	
$11 \to 22$	$23 \to -13$	$1 \to 2$	$4 \to -5$
$22 \to 11$	$13 \to 23$	$2 \to 1$	$5 \to 4$
$33 \to 33$	$12 \to -21$	$3 \to 3$	$6 \to -6$

The C_{ij}'s change as

$$\begin{pmatrix} C_{11} & C_{12} & C_{13} & C_{14} & C_{15} & C_{16} \\ & C_{22} & C_{23} & C_{24} & C_{25} & C_{26} \\ & & C_{33} & C_{34} & C_{35} & C_{36} \\ & & & C_{44} & C_{45} & C_{46} \\ & & & & C_{55} & C_{56} \\ & & & & & C_{66} \end{pmatrix} \rightarrow \begin{pmatrix} C_{22} & C_{21} & C_{23} & -C_{25} & C_{24} & -C_{26} \\ & C_{11} & C_{13} & -C_{15} & C_{14} & -C_{16} \\ & & C_{33} & -C_{35} & C_{34} & -C_{36} \\ & & & C_{55} & -C_{54} & C_{56} \\ & & & & C_{44} & -C_{46} \\ & & & & & C_{66} \end{pmatrix}.$$

Invariance of the elastic constants, as required by Neumann's principle, can be ensured by making $C_{11} = C_{22}$, $C_{44} = C_{55}$, $C_{16} = -C_{26}$, $C_{23} = C_{13}$ and $C_{14} = C_{15} = C_{24} = C_{25} = C_{26} = C_{45} = C_{46} = C_{56} = 0$. Thus, there are only seven surviving independent components and the resulting matrix is

$$\begin{pmatrix} C_{11} & C_{12} & C_{13} & . & . & C_{16} \\ & C_{11} & C_{13} & . & . & -C_{16} \\ & & C_{33} & . & . & . \\ & & & C_{44} & . & . \\ & & & & C_{44} & . \\ & & & & & C_{66} \end{pmatrix}$$

Cubic System: All classes in the cubic system have four triad axes as characteristic symmetry elements. We shall consider threefold rotation about the 111 and $11\bar{1}$ axes. Due to rotation about the 111 axis, the coordinates change as $x_1 \rightarrow x_2$, $x_2 \rightarrow x_3$, $x_3 \rightarrow x_1$ and the indices change as

Old notation		**New notation**	
$11 \rightarrow 22$	$23 \rightarrow 31$	$1 \rightarrow 2$	$4 \rightarrow 5$
$22 \rightarrow 33$	$13 \rightarrow 21$	$2 \rightarrow 3$	$5 \rightarrow 6$
$33 \rightarrow 11$	$12 \rightarrow 23$	$3 \rightarrow 1$	$6 \rightarrow 4$

Consequently, the C_{ij}'s change as

$$\begin{pmatrix} C_{11} & C_{12} & C_{13} & C_{14} & C_{15} & C_{16} \\ & C_{22} & C_{23} & C_{24} & C_{25} & C_{26} \\ & & C_{33} & C_{34} & C_{35} & C_{36} \\ & & & C_{44} & C_{45} & C_{46} \\ & & & & C_{55} & C_{56} \\ & & & & & C_{66} \end{pmatrix} \rightarrow \begin{pmatrix} C_{22} & C_{23} & C_{21} & C_{25} & C_{26} & C_{24} \\ & C_{33} & C_{31} & C_{35} & C_{36} & C_{34} \\ & & C_{11} & C_{15} & C_{16} & C_{14} \\ & & & C_{55} & C_{56} & C_{54} \\ & & & & C_{66} & C_{64} \\ & & & & & C_{44} \end{pmatrix}.$$

Let us now consider a threefold rotation about $11\bar{1}$. The coordinates change as $x_1 \rightarrow -x_3$, $x_2 \rightarrow x_1$, $x_3 \rightarrow -x_2$ and the indices change as

Old notation		New notation	
$11 \to 33$	$23 \to -12$	$1 \to 3$	$4 \to -6$
$22 \to 11$	$13 \to 32$	$2 \to 1$	$5 \to 4$
$33 \to 22$	$12 \to -31$	$3 \to 2$	$6 \to -5$

The C_{ij}'s change as

$$\begin{pmatrix} C_{11} & C_{12} & C_{13} & C_{14} & C_{15} & C_{16} \\ & C_{22} & C_{23} & C_{24} & C_{25} & C_{26} \\ & & C_{33} & C_{34} & C_{35} & C_{36} \\ & & & C_{44} & C_{45} & C_{46} \\ & & & & C_{55} & C_{56} \\ & & & & & C_{66} \end{pmatrix} \to \begin{pmatrix} C_{33} & C_{31} & C_{32} & -C_{36} & C_{34} & -C_{35} \\ & C_{11} & C_{12} & -C_{16} & C_{14} & -C_{15} \\ & & C_{22} & -C_{26} & C_{24} & -C_{25} \\ & & & C_{66} & -C_{64} & C_{65} \\ & & & & C_{44} & -C_{45} \\ & & & & & C_{55} \end{pmatrix}$$

To conform to Neumann's principle, we put $C_{11} = C_{22} = C_{33}$, $C_{12} = C_{13} = C_{23}$, $C_{44} = C_{55} = C_{66}$ and the rest $= 0$. The non-vanishing components are C_{11}, C_{12} and C_{44} and the matrix is

$$\begin{pmatrix} C_{11} & C_{12} & C_{12} & \cdot & \cdot & \cdot \\ & C_{11} & C_{12} & \cdot & \cdot & \cdot \\ & & C_{11} & \cdot & \cdot & \cdot \\ & & & C_{44} & \cdot & \cdot \\ & & & & C_{44} & \cdot \\ & & & & & C_{44} \end{pmatrix}.$$

The procedure can be extended to other crystal systems. The elastic constant matrices for all the seven crystal systems and also for an isotropic solid are given in Table 8.3.

8.4.3 Some Useful Interrelations

8.4.3.1 Conversion of C_{ij}'s into S_{ij}'s

Equations (8.15) and (8.16) can be written in matrix form as

$$(\varepsilon) = (S)\,(\sigma) \tag{8.24}$$

and

$$(\sigma) = (C)\,(\varepsilon), \tag{8.25}$$

where (σ) and (ε) are (6×1) matrices and (C) and (S) are (6×6) matrices. From the algebra of matrices, we have

Table 8.3 Form of the (S_{ij}) and (C_{ij}) matrices [8.4]

KEY TO NOTATION

	·	Zero component
	•	Non-zero component
	•—•	Equal components
	•—o	Components numerically equal, but opposite in sign
For S	⊙	Twice the numerical equal of the heavy dot component to which it is joined
For C	⊙	The numerical equal of the heavy dot component to which it is joined
For S	**x**	$2(S_{11} - S_{12})$
For C	**x**	$(1/2)(C_{11} - C_{12})$

All the matrices are symmetrical about the leading diagonal The number at the right hand foot of a matrix indicates the number of independent components

TRICLINIC
Both classes

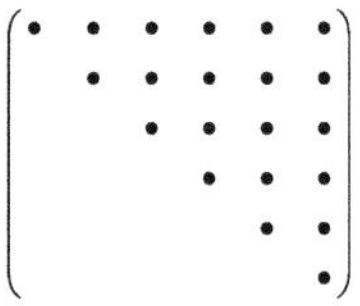
(21)

MONOCLINIC
All classes

Diad ∥ x_2 {standard orientation)

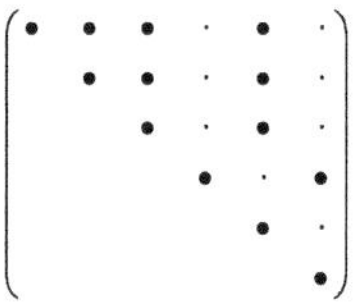
(13)

Diad ∥ x_3

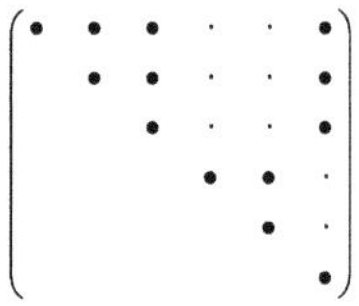
(13)

ORTHORHOMBIC
All classes

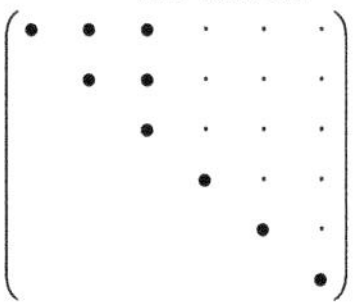
(9)

CUBIC
All classes

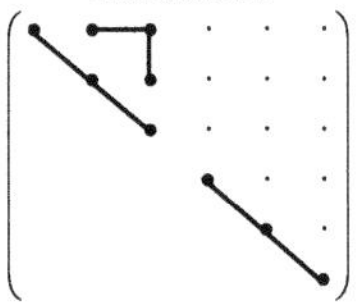
(3)

TETRAGONAL

Classes 4, $\bar{4}$, 4/m

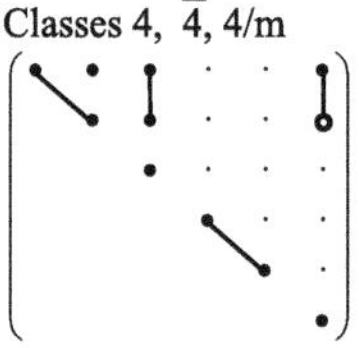
(7)

Classes 4mm, $\bar{4}$2m, 422, 4/mmm

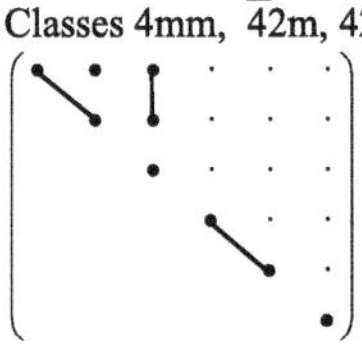
(6)

(*continued*)

Table 8.3 (continued)

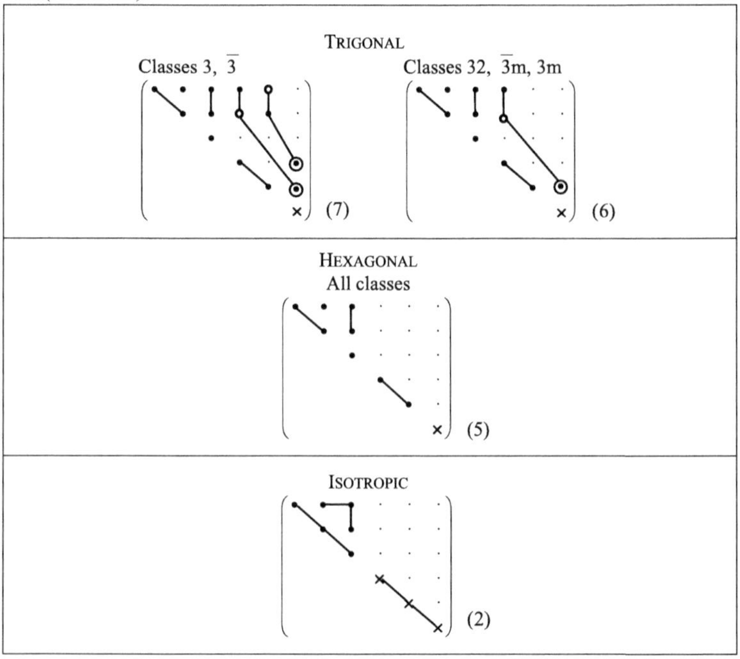

$$(S)\,(C) = (C)\,(S) = \mathrm{I}, \tag{8.26}$$

where I is a unit matrix. Consequently,

$$S_{ij} = (-1)^{i+j}\,\Delta_{ij}^{C}/\Delta C. \tag{8.27}$$

Here Δ_{ij}^{C} is the minor of C_{ij} and ΔC is the determinant of (C). Similarly,

$$C_{ij} = (-1)^{i+j}\,\Delta_{ij}^{S}/\Delta S. \tag{8.28}$$

The explicit relations for some crystal classes are given below.

Cubic System (All Classes)

$$C_{11} = (S_{11} + S_{12})/(S_{11} - S_{12})(S_{11} + 2S_{12}); \quad S_{11} = (C_{11} + C_{12})(C_{11} - C_{12})(C_{11} + 2C_{12})$$

$$C_{12} = -S_{12}/(S_{11} - S_{12})(S_{11} + 2S_{12}); \quad S_{12} = -C_{12}(C_{11} - C_{12})(C_{11} + 2C_{12})$$

$$C_{44} = 1/S_{44}; \quad S_{44} = 1/C_{44}$$

Tetragonal System (Classes 4*mm,* $\bar{4}2m$*,* 422*,* 4/*mmm)*

$$C_{11} + C_{12} = S_{33}/S; \quad C_{11} - C_{12} = 1(S_{11} - S_{12}); \quad C_{13} = -S_{13}/S;$$
$$C_{33} = (S_{11} + S_{12})/S; \quad C_{44} = 1/S_{44}; \quad C_{66} = 1/S_{66}$$

where

$$S = S_{33}(S_{11} + S_{12}) - 2S_{13}^2.$$

Trigonal System (Classes 3*m,* 32*,* $\bar{3}m$*)*

$$C_{11} + C_{12} = S_{33}/S; \quad C_{11} - C_{12} = S_{44}/S'; \quad C_{13} = -S_{13}/S;$$
$$C_{14} = -S_{14}/S'; \quad C_{33} = (S_{11} + S_{12})/S; \quad C_{44} = (S_{11} - S_{12})/S'$$

where

$$S = S_{33}(S_{11} + S_{12}) - 2S_{13}^2 \quad \text{and} \quad S' = S_{44}(S_{11} - S_{12}) - 2S_{14}^2$$

Hexagonal System (All Classes)

$$C_{11} + C_{12} = S_{33}/S; \quad C_{11} - C_{12} = 1/(S_{11} - S_{12}); \quad C_{13} = -S_{13}/S;$$
$$C_{33} = (S_{11} + S_{12})/S; \quad C_{44} = 1/S_{44}$$

where

$$S = S_{33}(S_{11} + S_{12}) - 2S_{13}^2.$$

8.4.3.2 Volume and Linear Compressibility

We have,

$$\varepsilon_{ij} = S_{ijkl}\sigma_{kl}. \tag{8.29}$$

For hydrostatic pressure P, we have

$$\sigma_{kl} = -P\delta_{kl}, \tag{8.30}$$

where δ_{kl} is Kronecker delta. Substituting for σ_{kl}, we get

$$\varepsilon_{ij} = -PS_{ijkl}. \tag{8.31}$$

The volume strain Δ (7.45) is

$$\Delta = \varepsilon_{ii} = -PS_{iikk}. \tag{8.32}$$

The volume compressibility ψ is

$$\psi = -\Delta/P = \sum_{ik} S_{iikk}. \tag{8.33}$$

In the two-suffix notation,

$$\psi = S_{11} + S_{22} + S_{33} + 2(S_{12} + S_{23} + S_{31}). \tag{8.34}$$

For a cubic crystal, (8.34) becomes

$$\psi = 3(S_{11} + 2S_{12}) = 3/(C_{11} + 2C_{12}). \tag{8.35}$$

We define linear compressibility ψ_l as the relative decrease in length of a line in the direction of a unit vector **l** under the application of hydrostatic pressure **P**. It can be shown that

$$\psi_l = S_{ijkk} l_i l_j. \tag{8.36}$$

For a cubic crystal, (8.36) becomes

$$\psi_l = (S_{11} + 2S_{12}). \tag{8.37}$$

Obviously volume compressibility $\psi = 3\psi_l$. The full expressions for ψ_l for other systems are given by Nye [8.4].

8.4.3.3 Elastic Properties from Single-Crystal Elastic Constants

For the elastic properties defined in Sect. 8.2.1, we get for cubic crystals

$$\begin{aligned} B &= 1/[3(S_{11} + 2S_{12})] \\ Y &= 1/S_{11} \\ G &= 1/S_{44} = (C_{11} - C_{12})/2 \\ \nu &= -S_{12}/S_{11} \end{aligned} \tag{8.38}$$

8.4.3.4 Elastic Anisotropy

It can be seen from Table 8.3 that for an isotropic solid

$$(C_{11} - C_{12})/2 = C_{44}. \tag{8.39}$$

Though cubic crystals are isotropic with regard to second-rank tensor properties, they are not isotropic with regard to elastic properties. Hence, the relation (8.39) is not necessarily obeyed. The parameter A given by

$$A = 2C_{44}/(C_{11} - C_{12}) \tag{8.40}$$

is taken as a measure of anisotropy.

8.4.3.5 Stability Conditions

For the mechanical stability of the crystal, the strain energy density must be of a positive definite quadratic form. This condition puts some constraints on the elastic constants. The necessary theory has been discussed by Born and Huang [8.5]. We shall not discuss it as it is beyond our scope but we shall note the results.

Cubic Crystals

$$C_{11}, C_{12}, C_{44} > 0; \quad (C_{11} - C_{12}) > 0.$$

Hexagonal

$$C_{44} > 0;\ C_{11} > |C_{12}|,\ (C_{11} + C_{12})C_{33} > 2C_{13}^2.$$

Tetragonal

$$C_{11}, C_{33}, C_{44}, C_{66} > 0; \quad C_{11} > |C_{12}|; \quad C_{11}C_{66} > C_{16}^2; \quad C_{11}C_{33} > C_{13}^2.$$

All Crystal Classes

Diagonal elements of the elastic constant matrix are all positive.

8.4.3.6 Cauchy Relations

Born and Huang (1954) also discuss the relations between elastic constants resulting due to the presence of a centre of symmetry in the lattice and the existence of central forces of interaction. These relations known as "Cauchy relations" are $C_{23} = C_{44}, C_{31} = C_{55}, C_{12} = C_{66}, C_{14} = C_{56}, C_{25} = C_{64}, C_{36} = C_{45}$. In cubic crystals, this results in $C_{12} = C_{44}$. It may be emphasised that these relations are valid if the assumption of central forces holds. Thus, deviation from Cauchy relations indicates deviation from the assumption of central forces.

8.5 Experimental Determination of Elastic Constants

8.5.1 Principle

Let us consider a plane elastic wave in a crystal [8.6] given by

$$\mathbf{u}(\mathbf{r}, t) = A\mathbf{p}\exp(i\mathbf{k}.\mathbf{r} - i\omega t). \tag{8.41a}$$

Here $\mathbf{u}(\mathbf{r}, t)$ is the displacement at point $\mathbf{r}$ and time t. A is the amplitude, $\mathbf{p}$ the polarization vector (a unit vector in the direction of $\mathbf{u}$), $\mathbf{k}$ the propagation vector and ω the angular frequency. Substituting $(2\pi/\lambda)\mathbf{m}$ for $\mathbf{k}$ where $\mathbf{m}$ is a unit vector in the direction of the wave normal and $(2\pi/\lambda)v$ for ω, where v is the phase velocity, (8.41a) becomes

$$\mathbf{u}(\mathbf{r}, t) = A\mathbf{p}\exp[-(2\pi i/\lambda)(\mathbf{m}.\mathbf{r} - vt)]. \tag{8.41b}$$

The displacement must satisfy the equation of motion

$$C_{ijkl}\frac{\partial^2 u_l}{\partial x_j \partial x_k} = \rho\frac{\partial^2 u_t}{\partial t^2}, \tag{8.42}$$

where u_l and u_t are the parts of u dependent only on r and t, respectively, and ρ is the density of the crystalline medium. Differentiation of (8.41b) with respect to x_j results in multiplication of (8.41b) by $(2\pi i/\lambda)m_{\mathrm{j}}$ and that with respect to t by $(-2\pi i/l)v$. It must be remembered that this has to be done for each component of $\mathbf{p}$. When this is done, we get

$$(L_{ij} - \rho v^2\delta_{ik})p_j = 0, \qquad j = 1, 2, 3, \tag{8.43}$$

where δ_{ik} is Kronecker delta, p_j the direction cosines of the particle displacement direction and L_{ij}'s are functions of the elastic constants C_{ijkl}'s (or C_{ij} in the two-suffix notation) and m_j's, the direction cosines of the wave propagation direction. The solution to the set of three equations represented by (8.43) is obtained from the determinant

Table 8.4 Velocity–C_{ij} relations for the cubic system

Direction of propagation	Direction of polarization	$\rho v^2 =$
(001)	(001)	C_{11}
(110)	(110)	$(C_{11} + C_{12} + 2C_{44})/2$
(110)	(100)	C_{44}
(110)	$(1\bar{1}0)$	$(C_{11}–C_{12})/2$

Table 8.5 Velocity–C_{ij} relations for a hexagonal crystal

Propagation direction	Polarization	$\rho v^2 =$
Parallel to c	Parallel to c	C_{33}
Parallel to c	Normal to c	C_{44}
Parallel to a	Parallel to a	C_{11}
Parallel to a	Normal to c	$C_{66} = (C_{11}–C_{12})$
45° a, c	45° a, c	C_{13}

$$\begin{vmatrix} L_{11} - \rho v^2 & L_{12} & L_{13} \\ L_{12} & L_{22} - \rho v^2 & L_{23} \\ L_{13} & L_{23} & L_{33} - \rho v^2 \end{vmatrix} = 0, \tag{8.44}$$

where L_{ij}'s are functions of the elastic constants. For each chosen direction (m_j's) (8.44) yields three solutions representing three isonormal, mutually perpendicular elastic waves, one longitudinal and two transverse. Equation (8.44) is known as the Christoffel equation.

Thus to determine the elastic constants, we should determine the velocity of sound waves with known propagation and polarization vectors. In actual practice, we transform to a coordinate system with one axis, say Ox_1, parallel to the propagation direction so that the terms in (8.42) involving differentiation with respect to x_2 and x_3 drop out. As examples, the required velocity–elastic constant relations for a couple of crystal systems are given in Tables 8.4 and 8.5.

8.5.2 *Experimental Techniques*

As discussed in the preceding section, the elastic constants are determined from the velocity of sound waves of known polarization propagating in a known direction in the crystal. In most experimental techniques, the velocity determination is made through time-of-flight measurement or through measurement of resonant frequencies of the crystal plates and the known relation between velocity, frequency and crystal thickness. Several methods have been discussed by Huntington [8.7] and Reddy [8.8]. Some representative methods are discussed here.

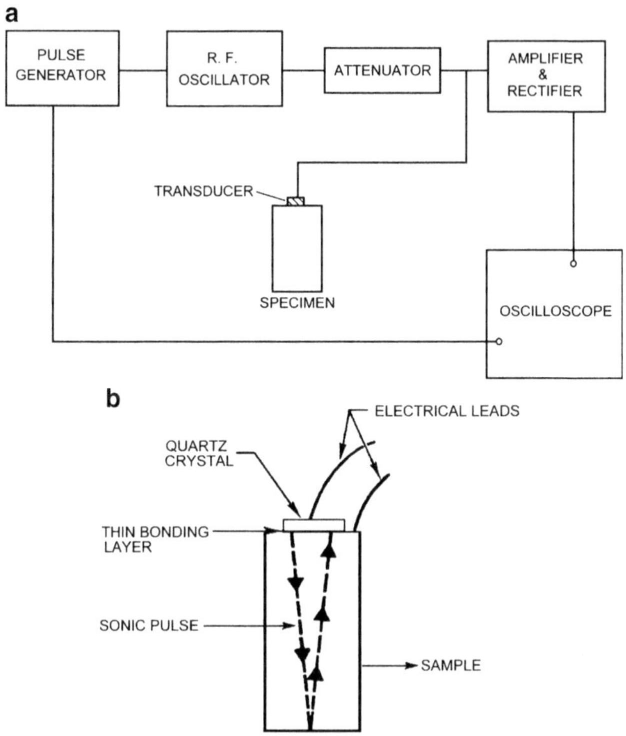

Fig. 8.5 The pulse-echo method: (**a**) block diagram, (**b**) sample holder

8.5.2.1 Pulse Echo Method

In this method (Fig. 8.5a), the time interval between successive echoes is measured using a stable time-marker generator and the velocity is evaluated knowing the acoustic path in the sample. A quartz crystal is cemented to one of two plane parallel faces of the specimen (Fig. 8.5b). A pulse of the order of a microsecond is generated and transmitted through the specimen. On reflection at the opposite face, it returns and when it arrives back at the quartz crystal, it produces an echo signal. Usually, X-cut quartz crystals are used to excite longitudinal waves and Y-cut crystals for transverse waves.

8.5.2.2 Pulse Superposition Method

In this method, the quartz crystal and the sample are attached at the two ends of a buffer rod (Fig. 8.6a). Multiple echoes come as a group at an appreciable time after the initial pulse. The frequency is varied until the condition of constructive interference indicated by a step pattern in the overlapping echoes (Fig. 8.6b).

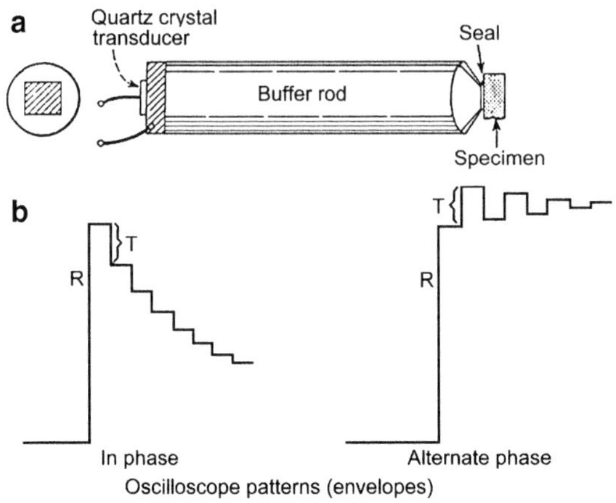

Fig. 8.6 Pulse interferometer: (**a**) sample holder, (**b**) oscilloscope patterns at critical frequencies

Acoustic velocity is obtained from the differences between the resonant frequencies using the relation

$$v = 2tf_n/n, \tag{8.45}$$

where t is the thickness, f_n the resonant frequency and n the number of wavelengths in $2t$.

8.5.2.3 Pulse Comparison Technique

In this technique (Fig. 8.7), two phase-coherent RF pulses are applied in quick succession to the transducer attached to the sample. On transmitting through the sample, each pulse will generate its own wave train of echoes. When the separation of the second pulse from the first is adjusted such that the echoes of the first pulse coincide with the first echo of the second pulse, destructive interference occurs successively at a number of frequencies. From a knowledge of these null frequencies and the thickness of the sample, the ultrasonic velocity can be estimated.

8.5.2.4 Ultrasonic–Optical Method

In this method, a high frequency quartz crystal attached to the side of a transparent cube of material under study serves to excite standing waves (Fig. 8.8a). A beam of monochromatic light transmitted perpendicular to the standing waves produces a diffraction pattern. For anisotropic single crystals, the diffraction pattern consists of closed curves (Fig. 8.8b). From the density of the crystal, the frequency of the quartz crystal and the radii of the diffraction pattern, the elastic constants can be estimated.

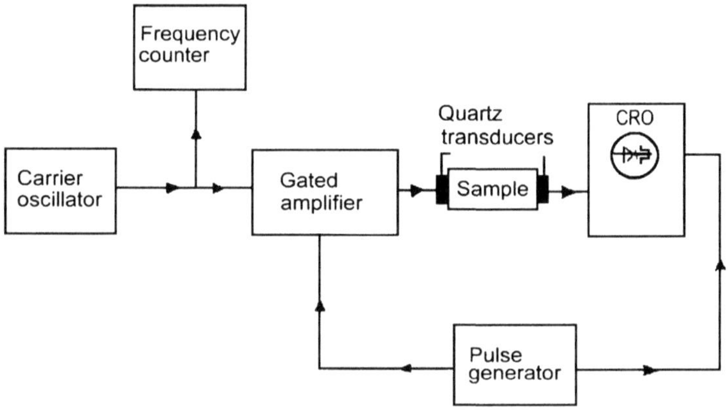

Fig. 8.7 Block diagram for pulse comparison set-up

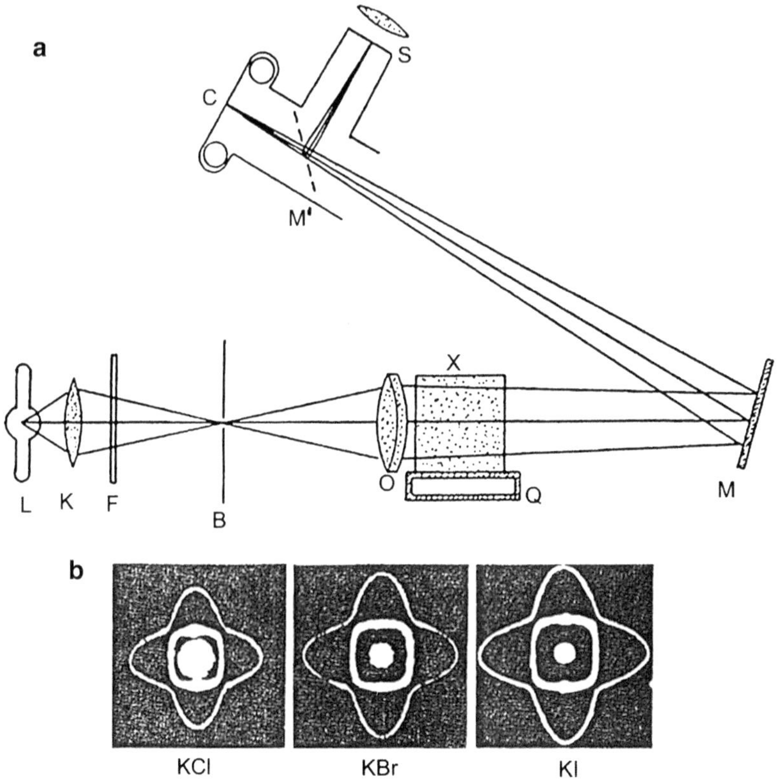

Fig. 8.8 Ultrasonic–optical method. (**a**) Optical arrangement for the diffraction of light by transmission through a crystal vibrating at a high frequency. L light source, K condensing lens, F monochromatising filter, B iris, O objective lens, X crystal, Q transducer, M and M′ mirrors, C camera and S screen; (**b**) elastograms of some alkali halides

Table 8.6 Experimental values of elastic constants (C_{ij}'s in 10^{11} dyn/cm^2 and S_{ij}'s in 10^{-12} cm^2/dyn). Data from [8.7]

Symbol	Material	C_{11}	C_{44}	C_{12}	S_{11}	S_{44}	S_{12}
Cubic Crystals							
Ag	Silver	12.40	4.61	9.34	2.29	2.17	−0.983
C	Diamond	95	43	39	0.138	0.23	−0.04
Li	Lithium	1.48	1.08	1.25	29.5	9.26	−13.5
NaCl	Sodium chloride	4.87	1.26	1.24	2.29	7.94	−0.465
KCl	Potassium chloride	3.98	0.62	0.62	2.62	16.0	−0.35
TlCl	Thallium chloride	4.01	0.76	1.53	3.16	13.2	−0.87
ZnS	Zinc sulphide	9.42	4.36	5.68	1.94	2.29	−0.73
Tetragonal crystals							
Symbol	Material	C_{11}	C_{33}	C_{44}	C_{66}	C_{12}	C_{13}
Sn	Tin	7.35	8.7	2.2	2.26	2.34	2.8
KH_2PO_4	Potassium dihydrogen phosphate	7.4	6.8	1.35	0.63	1.8	2.7
$BaTiO_3$	Barium titanate	27.5	16.5	5.43	11.3	17.9	15.1
Trigonal crystals							
Symbol	Material	C_{11}	C_{33}	C_{44}	C_{12}	C_{13}	C_{14}
Bi	Bismuth	6.28	4.40	1.08	3.50	2.11	−0.42
$CaCO_3$	Calcite	13.74	8.01	3.42	4.40	4.40	−2.03
SiO_2	α -quartz	8.67	10.72	5.79	0.70	1.19	−1.79
Hexagonal crystals							
Symbol	Material	C_{11}	C_{33}	C_{44}	C_{12}	C_{13}	
Cd	Cadmium	12.1	5.13	1.85	4.81	4.42	
H_2O	Ice	1.38	1.50	0.32	0.70	0.58	
Zn	Zinc	16.1	6.10	3.83	3.42	5.01	

8.5.3 Numerical Values of Elastic Constants

In order to get an idea of the relative magnitudes of the elastic constants, data for some high symmetry crystal systems are collected in Table 8.6. Data for several other crystals in these and lower symmetry crystal classes are given in [8.7].

8.6 Theory of Elastic Constants

The theoretical evaluation of elastic constants is quite involved. The general approach is to set up the equations of motion for an atom or ion in the lattice by considering its interactions with the nearest (and, sometimes, also the next nearest) neighbours. Several force constants have to be introduced. These force constants are related, on the one hand with the lattice vibration frequencies and, on the other, with the elastic constants. Thus, if the force constants are known from lattice vibration frequencies, we may use them to estimate the elastic constants or if the force

Table 8.7 Contributions to elastic constants of NaCl type alkali halides (α – Madelung constant; $\delta = r/\rho$; χ–lattice sum = 3.14)

Interaction	C_{11} $[e^2/2r^4]$	C_{12}	C_{44}
Coulomb	$(\alpha/3) - \chi$	$-(5/6)\alpha + (\chi/2)$	$-(\alpha/6) + (\chi/2)$
Repulsion	$(\alpha\delta/3)$	$\alpha/3$	$-\alpha/3$
Total	$(\alpha/3)(\delta + 1) - \chi$	$(\chi - \alpha)/2$	$(\chi - \alpha)/2$

Table 8.8 Theoretical (Th.) and experimental (Exp.) values of C_{ij}'s (10^{11} dyn/cm^2) of some alkali halides

Crystal		C_{11}	C_{12}	C_{44}
LiF	Th.	10.7	4.9	4.9
	Exp.	13.2	6.8	4.2
NaCl	Th.	5.0	1.3	1.3
	Exp.	4.86	1.27	1.28
KCl	Th.	3.9	0.8	0.8
	Exp.	3.98	0.62	0.62

constants are known from the elastic constants, the lattice vibration spectrum can be evaluated.

If the interatomic forces in a given crystal are well defined, it is possible to express the force constants in terms of the interaction forces and then to use them to estimate the elastic constants. This was done by Krishnan and Roy [8.9] for the alkali halides. These crystals are ionic and the relevant interactions are the coulomb electrostatic interaction and repulsion of the overlapping closed shells of electrons. The form of repulsion potential assumed is $A \exp(-r/\rho)$, r is the interionic distance and A and ρ are constants for the crystal. The contributions of these two interactions to the elastic constants are given in Table 8.7. The values of the elastic constants for some alkali halides calculated from the expressions in Table 8.7 are given in Table 8.8 [8.9] along with experimental values. It may be noted that the theoretical expressions for C_{12} and C_{44} are the same because of the assumption of central forces. Experimental values show some difference between C_{12} and C_{44}. This is particularly so in LiF. As mentioned in Sect. 8.4.3, this difference is related to the interaction forces not being purely central in nature.

8.7 Higher Order Elastic Constants (HOEC)

8.7.1 Definitions

We have seen earlier that the strain energy ϕ of a crystal under stress is

$$\phi = (1/2)C_{ijkl}\varepsilon_{ij}\varepsilon_{kl}, \tag{8.46}$$

where the ε's are strains and the C_{ijkl}'s are the SOEC. The C_{ijkl}'s form a fourth-rank tensor. Using a two-suffix notation, they are denoted as C_{ij}. Equation (8.46) holds for strains which obey Hooke's law. For larger strains, Hooke's law does not hold and the expression for ϕ is

$$\phi = (1/2)C_{ijkl}\varepsilon_{ij}\varepsilon_{kl} + (1/6)C_{ijklmn}\varepsilon_{ij}\varepsilon_{kl}\varepsilon_{mn}. \tag{8.47}$$

The elastic constants C_{ijklmn} are called third-order elastic constants (TOEC). They form a sixth rank tensor with 729 components. Due to the symmetry of the strain tensor and thermodynamic conditions, the number of TOEC is reduced to 56, i.e. this is the number of TOEC of a triclinic crystal. Due to the symmetry of the crystal, there can be a further reduction. The number of non-vanishing TOEC for all crystal classes is given by Bhagavantham [8.3]. Thus, the number of TOEC for cubic crystals of class $\bar{4}3m$, 432 and $m3m$ is six and for the classes 23 and $m3$ is 8. Using the two-suffix notation, the TOEC for cubic crystals are C_{111}, C_{112}, C_{123}, C_{144}, C_{166}, C_{456}, C_{113} and C_{155}.

For still larger strains, (8.47) has to be modified to

$$\phi = (1/2)C_{ijkl}\varepsilon_{ij}\varepsilon_{kl} + (1/6)C_{ijklmn}\varepsilon_{ij}\varepsilon_{kl}\varepsilon_{mn} + (1/24)C_{ijklmnop}\varepsilon_{ij}\varepsilon_{kl}\varepsilon_{mn}\varepsilon_{op}. \tag{8.48}$$

Here, $C_{ijklmnop}$ are called the fourth order elastic constants (FOEC); they constitute an eighth rank tensor. In general, an eighth rank tensor has 3^8 components. But the number of components of the FOEC tensor is drastically reduced due to the intrinsic symmetry of the strain tensor and thermodynamic conditions. Thus for class $m3m$, the number of independent FOEC is 11. In contracted notation, these are C_{1111}, C_{1112}, C_{1166}, C_{1122}, C_{1266}, C_{4444}, C_{1123}, C_{1144}, C_{1244}, C_{1456} and C_{4466}.

8.7.2 Experimental Determination

The principle of measurement is to measure the sound velocities in chosen directions in the crystal as a function of pressure. The experimental methods for velocity measurement are the same as for SOEC with the difference that the sample is either under uniaxial stress or hydrostatic stress. The arrangement used by Thurston et al. [8.10] is shown in Fig. 8.9. Since the number of constants to be determined is much larger, it follows that a larger number of velocity measurements are needed.

While the full set of TOEC have been determined for some crystals, in the case of FOEC, only some combinations defined below have been measured.

$$\begin{aligned} \gamma_{11} &= C_{1111} + 4C_{1112} + 2C_{1122} + 2C_{1123} \\ \gamma_{12} &= 2C_{1112} + 2C_{1122} + 5C_{1123} \\ \gamma_{44} &= C_{1144} + 2C_{1166} + 4C_{1244} + 2C_{1266} \end{aligned} \tag{8.49}$$

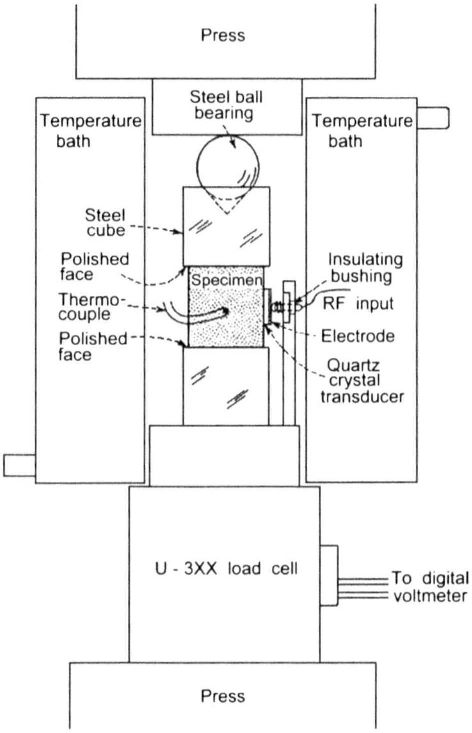

Fig. 8.9 Set-up for measurement of HOEC

Table 8.9 Values of TOEC of some cubic crystals (C_{ijk} in units of 10^{12} dyn/cm^2)

Crystal	C_{111}	C_{112}	C_{123}	C_{144}	C_{166}	C_{456}	C_{113}	C_{155}
Class m3m								
KCl	−7.26	−0.24	0.11	0.23	−0.26	0.16		
NaCl	−8.43	−0.50	0.46	0.29	−0.60	0.26		
LiF	−14.23	−2.64	1.56	0.85	−2.73	0.94		
Class m3								
$KAl(SO_4)_2.12H_2O$	−22.2	−7.1	−13.4	−2.3	−7.44	−2.0	−8.6	−8.02
$NH_4Al(SO_4)_2.12H_2O$	−7.5	−1.1	−1.9	−2.9	−4.94	−0.64	−2.0	−5.62

8.7.3 Numerical Data on HOEC

To get an idea of the magnitudes of the HOEC, values of TOEC for some cubic crystals of class $m3m$ [8.11] and $m3$ [8.12] are given in Table 8.9. As mentioned earlier, the number of TOEC is 6 for some cubic classes and 8 for the others. The determination of FOEC is more difficult; only the values of some combinations γ_{ij} (8.49) are available and these are given in Table 8.10 [8.13].

Table 8.10 Values of γ_{ij} of some alkali halides (γ_{ij} in units of 10^{12} dyn/cm^2)

Crystal	γ_{11}	γ_{12}	γ_{44}
NaCl *structure*			
NaI	92.4	13.1	12.3
KI	81.0	0.45	4.1
RbCl	112.0	−2.0	1.6
RbBr	91.0	−3.0	1.8
RbI	77.0	−0.1	2.1
CsCl *Structure*			
CsCl	73.0	41.0	46.0
CsBr	60.0	36.0	42.0
CsI	49.0	29.0	35.0

8.8 Elastic Behaviour of Polycrystalline Aggregates

We started with the elastic properties of isotropic solids and then took up the elastic behaviour of anisotropic crystals. We shall now consider the elastic behaviour of an isotropic aggregate of crystallites of anisotropic materials. In other words, we shall see how the isotropic elastic properties of an aggregate made of randomly packed crystallites of an anisotropic crystalline material can be estimated from the single-crystal elastic constants of the material.

This problem was studied by Voigt [8.14] who assumed that the aggregate would be as shown in Fig. 8.10a. Here the crystallites are single crystal fibres parallel to the stress axis subject to the same strain but with non-uniform stress. On the other hand, Reuss [8.15] assumed the situation to be as shown in Fig. 8.10b. Here single crystal platelets are arranged in layers perpendicular to the stress axis; the stress is uniform but the strain is non-uniform. Apart from this, it is also assumed that (1) the number of crystallites in the aggregate is large, (2) crystals are oriented randomly, (3) the crystallite size is uniform, (4) elastic properties are homogeneous within a crystallite and in between crystallites and (5) crystallite boundaries do not contribute to elastic properties of the aggregate. With these assumptions, the following expressions were obtained for the isotropic elastic moduli:
Voigt Averages:

$$\begin{aligned} B_V &= (1/3)(L+2M) \\ G_V &= (1/5)(L-M+3N) \\ Y_V &= (L-M+3N)(L+2M)/2L+3M+N) \end{aligned} \tag{8.50}$$

where

$$\begin{aligned} L &= (1/3)(C_{11}+C_{22}+C_{33}) \\ M &= (1/3)(C_{12}+C_{23}+C_{13}) \\ N &= (1/3)(C_{44}+C_{55}+C_{66}) \end{aligned} \tag{8.51}$$

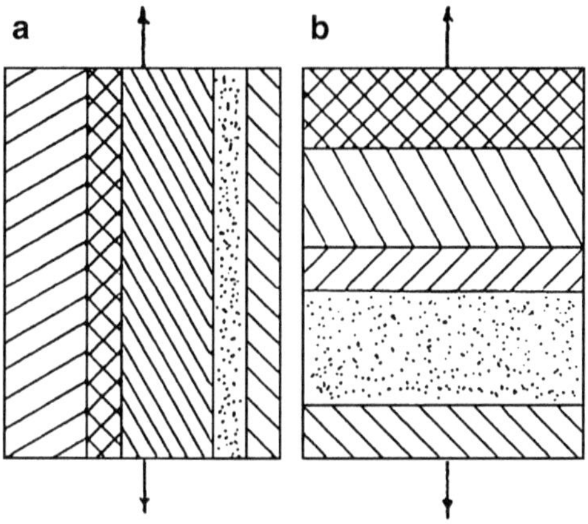

Fig. 8.10 Limiting cases in averaging polycrystalline elastic properties: (**a**) uniform strain in all grains; (**b**) uniform stress in all grains

Reuss' Averages:

$$\begin{aligned} B_R^{-1} &= 3(L' + 2M') \\ G_R^{-1} &= (1/5)(4L' - 4M' + 3N') \\ Y_R^{-1} &= (1/5)(3L' + 2M' + N') \end{aligned} \tag{8.52}$$

where

$$\begin{aligned} L' &= (1/3)(S_{11} + S_{22} + S_{33}) \\ M' &= (1/3)(S_{12} + S_{23} + S_{13}) \\ N' &= (1/3)(S_{44} + S_{55} + S_{66}) \end{aligned} \tag{8.53}$$

For cubic crystals, these expressions lead to

$$\begin{aligned} B_V &= B_R = (1/3)(C_{11} + 2C_{12}) \\ G_V &= (1/5)(C_{11} - C_{12} + 3C_{44}) \\ G_R &= [5C_{44}(C_{11} - C_{12})]/[3(C_{11} - C_{12}) + 4C_{44}] \end{aligned} \tag{8.54}$$

Since the main assumptions made by Voigt and Reuss, respectively, are extremes, we intuitively expect the true values to lie between the Voigt and Reuss values. In fact, using energy density considerations, Hill [8.16] theoretically showed that $B_R \leq B \leq B_V$ and $G_R \leq G \leq G_V$ where the quantities without a suffix are the true values. Further, Hill suggested empirically that the true values may be closer to the arithmetic or geometric mean of the Voigt and Reuss values. These averages are denoted by

Table 8.11 Elastic constants (C_{ij}), elastic anisotropy (A) and polycrystalline shear modulus (G) of some cubic materials (C_{ij}'s and G in units of 10^{12} dyn/cm^2). Data from [8.17]

Material	C_{11}	C_{12}	C_{44}	A	G_V	$G_{\text{VRH-a}}$	$G_{\text{VRH-g}}$	G_R	G (expt)
Cu	1.611	1.199	0.756	3.273	0.546	0.471	0.465	0.396	0.451
Ag	1.222	0.907	0.454	2.883	0.335	0.297	0.295	0.259	0.286
Au	1.929	1.638	0.415	2.852	0.307	0.273	0.271	0.238	0.276
MgO	2.9708	0.9536	1.5613	1.548	1.340	1.310	1.310	1.281	1.292
CaF_2	1.6420	0.4398	0.3370	0.561	0.443	0.426	0.425	0.409	0.409
β-ZnS	1.0462	0.6534	0.4613	2.349	0.355	0.328	0.326	0.300	0.318
ZnSe	0.8096	0.4881	0.4405	2.740	0.329	0.294	0.292	0.260	0.285
CdTe	0.5351	0.3681	0.1994	2.388	0.153	0.141	0.140	0.128	0.139

$$\begin{aligned} G_{\text{VRH-a}} &= (1/2)(G_{\text{V}} + G_{\text{R}}) \\ G_{\text{VRH-g}} &= (G_{\text{V}} G_{\text{R}})^{1/2} \end{aligned} \tag{8.55}$$

Experimental values of the shear modulus for polycrystalline aggregates of some cubic materials are given in Table 8.11 [8.17] along with the average values calculated from single-crystal elastic constants. The difference between the Voigt and Reuss values seems to increase with the elastic anisotropy A (8.40). Most of the experimental values are closer to the Reuss rather than Voigt values. It may be inferred that the elastic behaviour of polycrystalline aggregates conforms more to Reuss uniform stress model than Voigt's uniform strain model.

Problems

1. The equation of state for lithium fluoride is

$$(\Delta V/V_0) = -1.549 \times 10^{-12}\,(\text{cm}^2/\text{dyn}) + 6.53 \times 10^{-24}\,(\text{cm}^2/\text{dyn})^2.$$

Calculate the compressibility ψ and its pressure coefficient $(1/\psi)(\mathrm{d}\psi/\mathrm{d}P)$.

2. Two sets of values have been reported for the elastic constants of the cubic crystal FeS_2 (C_{ij} in units of 10^{11}dyn/cm^2).

	C_{11}	C_{12}	C_{44}
A	35.85	−5.29	10.35
B	38.18	3.10	10.94

Using physical arguments show which set is wrong.

3. The elastic constants of aluminium are $C_{11} = 10.82$, $C_{12} = 6.13$ and $C_{44} = 2.85$ in units of 10^{11} dyn/cm^2. Convert these elastic constants into elastic compliances S_{ij}.

4. Prove that a tetragonal crystal of class 422 has only six non-vanishing elastic constants.
5. Calculate the Voigt and Reuss values of the shear modulus of aluminium, using the values of the single-crystal elastic constants given in problem 3.

References

8.1. K.A. Gschneidner, Jr., Solid State Phys. **16**, 275 (1964)
8.2. S.N. Vaidya, G.C. Kennedy, J. Phys. Chem. Solids **31**, 2329 (1970)
8.3. S. Bhagavantham, *Crystal Symmetry and Physical Properties* (Academic Press, New York, 1966)
8.4. J.F. Nye, *Physical Properties of Crystals* (Oxford University Press, London, 1957)
8.5. M. Born, K. Huang, *Dynamical Theory of Crystal Lattices* (Oxford University Press, London, 1954)
8.6. Y.I. Sirotin, Shaskolskaya, *Fundamentals of Crystal Physics* (Mir Publishers, Moscow, 1982)
8.7. H.B. Huntington, Solid State Phys. **7**, 213 (1958)
8.8. P.J. Reddy, *Crystal Elasticity* (Sri Venkateshwara University, Tirupathi, 1977)
8.9. K.S. Krishnan, S.K. Roy, Proc. Roy. Soc. (London) **A210**, 481 (1952)
8.10. R.N. Thurston, H.J. McSkimmin, P. Andreatch Jr., J. Appl. Phys. **37**, 267 (1966)
8.11. J.R. Drabble, R.E. Strathen. Phys. Soc. **92**, 1090 (1967)
8.12. S. Haussuhl, P. Preu, Acta Cryst. **A34**, 44 (1978)
8.13. D.B. Sirdeshmukh, L. Sirdeshmukh, K.G. Subhadra, *Alkali Halides – A Handbook of Physical Properties* (Springer, Berlin, 2001)
8.14. W. Voigt, Ann. Phys. (Leipz.) **38**, 573 (1889)
8.15. A. Reuss, Z. Angew Math. Phys. **9**, 49 (1929)
8.16. R. Hill, Proc. Phys. Soc. (Lond.) **A65**, 349 (1952)
8.17. H.M. Ledbetter, E.R. Naimon, J. Appl. Phys. **45**, 66 (1974)

Chapter 9
Thermal Properties

9.1 Introduction

In this chapter, we discuss specific heat, thermal expansion and thermal conductivity of solids. Initially, these properties were measured at room temperature. The development of low temperature techniques in the earlier part of the twentieth century led to measurements which revealed many interesting features which, in turn, led to theories involving new concepts.

For each property, we discuss (1) experimental methods of measurement, (2) important experimental results and trends observed in them and (3) the main theories. Finally, the limitations and failures of these theories are pointed out.

General treatments of the theories of thermal properties are given in several text books. For experimental methods and experimental results, reference is made to specific sources.

9.2 Specific Heat

9.2.1 General

Specific heat is the energy required to increase the temperature of a solid by a degree. If energy $\mathrm{d}Q$ is supplied to 1 g of a solid to increase its temperature by $\mathrm{d}T$, $\mathrm{d}Q/\mathrm{d}T$ is the specific heat c per gram. On the other hand, when the energy is supplied to a gram-atom (or a mole) of the material, $\mathrm{d}Q/\mathrm{d}T$ is the specific heat C per gram-atom. If the specific heats of two solids are to be compared, the latter definition is preferable since the same number of atoms is involved. Further, we define two specific heats: the specific heat at constant pressure C_P and the specific heat at constant volume C_V. These are given by

D.B. Sirdeshmukh et al., *Atomistic Properties of Solids*, Springer Series in Materials Science 147, DOI 10.1007/978-3-642-19971-4_9,

$$C_{\mathrm{P}} = \left(\frac{\partial Q}{\partial T}\right)_P, \tag{9.1}$$

and

$$C_{\mathrm{V}} = \left(\frac{\partial Q}{\partial T}\right)_V = \left(\frac{\partial U}{\partial T}\right)_V. \tag{9.2}$$

Here, U is the internal energy. The unit of specific heat is cal/mol-deg; if it is multiplied by 4.186, the unit becomes J/mol-deg. The two specific heats are related by the thermodynamic formula

$$C_{\mathrm{P}} - C_{\mathrm{V}} = \beta^2 VT/\psi, \tag{9.3}$$

where β is the volume coefficient of thermal expansion, V the molar volume and ψ the compressibility. The difference is generally very small at low and moderate temperatures but becomes as large as 10–20% at high temperatures. C_{p} is the experimental value of specific heat. C_{V} is the theoretically calculated value or the experimental value corrected with formula (9.3).

It may be mentioned that specific heat of a solid is a scalar quantity unlike many other physical properties which are tensors.

9.2.2 Experimental Methods

The experimental method of measurement of specific heat of solids is based on the principle involved in (9.1), i.e. to supply energy $\mathrm{d}Q$ and measure the change in temperature $\mathrm{d}T$.

The experimental set-up used by Nernst in much of his pioneering work is shown in Fig. 9.1. Here A is the liquid bath; the coolant is used to provide the chosen starting temperature. B is the specimen. C is a suitable gas which cools B to the temperature of A; this is called the exchange gas. D is an electrical heater which also acts as a resistance thermometer. When the specimen has reached the desired low temperature, the exchange gas is driven out. Energy is supplied to the specimen by the electrical heater at a known rate and the rate of rise of temperature is noted.

Accurate measurement of specific heat needs several precautions, corrections and controls. Thus, leakages have to be plugged, radiation losses to be minimised and temperatures to be precisely measured. A modern calorimeter with such stringent provisions is shown in Fig. 9.2.

The specific heat C_{P} is calculated from the formula

$$C_{\mathrm{P}} = (M/m)(Vit/(T_{\mathrm{f}} - T_{\mathrm{i}})), \tag{9.4}$$

where M is the molecular weight, m the weight of the specimen, V the heater voltage, i the current, t the time and T_{i}, T_{f} the initial and final temperatures. The

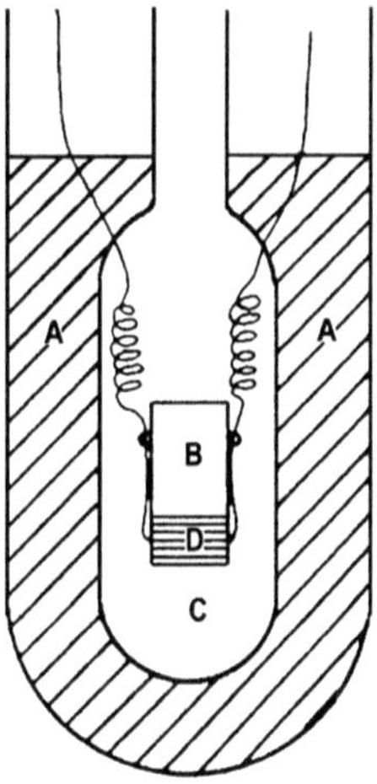

Fig. 9.1 Nernst vacuum calorimeter

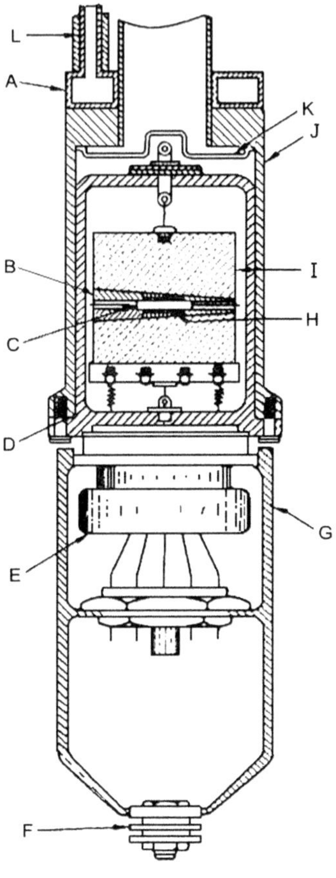

Fig. 9.2 Low temperature calorimeter: A vapour pressure thermometer bulb, B beryllium-copper tapered plug, C carbon resistance thermometer, D gold 'O' ring, E wire-glass vacuum seal, F helium bath heater, G plastic holder, H manganin resistance heater, I sample, J vacuum-tight outer brass container, K radiation shield, L vacuum-jacketed thermometer tube

measured value of C_P is associated with the mean temperature $T = (T_i + T_f)/2$. The values of these quantities in a typical experiment [9.1] are:

$$V = 0.12149\,\text{V}$$
$$i = 0.06908\,\text{mA}$$
$$t = 18.673\,\text{s}$$
$$T_f - T_i = 0.0267\,\text{K}$$
$$T = (T_i + T_f)/2 = 1.898\,\text{K}$$

These figures are quoted to emphasise that specific heat determination involves precision measurements.

9.2.3 Experimental Results and Trends

Let us first consider the experimental results at room temperature. The values of the gram-atomic heats of several elemental solids are given in Table 9.1. It can be seen that they are all in the range 5.5–6.5 cal/g-atom-deg with an average value of 5.97 cal/g-atom-deg. In fact, it was stated by Dulong and Petit [9.3] that the gram-atomic specific heat (product of specific heat per gram and the gram atomic weight) is a constant independent of temperature. This is known as the Dulong–Petit law. Be, B, C (graphite) and C (diamond) with gram-atomic heats 3.93, 2.64, 2.06 and 1.46, respectively, are deviations from the Dulong–Petit law.

A more serious deviation from the Dulong–Petit law is that the specific heat shows a strong variation with temperature, particularly at low temperature. As an example, the temperature variation of specific heat of silver and sodium chloride is shown in Fig. 9.3. The following features are noted:

1. The specific heat curve becomes nearly flat at high temperature; in this region the specific heat $\approx$ 6 cal/g-atom-deg for monatomic and twice this value for diatomic solids.
2. A steep fall is noticed at low temperatures.
3. At very low temperatures, it is empirically found that the specific heat varies as T^3.
4. As $T \to 0$, the specific heat tends to zero.

Any theory of specific heats should explain these features.

Table 9.1 Specific heats C_P (cal/g-atom-deg) of some elemental solids [9.2]

Solid	C_P	Solid	C_P	Solid	C_P	Solid	C_P
Li	5.65	Ca	6.29	Cu	5.85	Ag	6.09
Na	6.74	Ti	5.98	Zn	6.07	Cd	6.21
Mg	5.92	Fe	5.98	Ge	5.47	W	5.84
Al	5.82	Co	5.95	As	5.90	Au	6.06

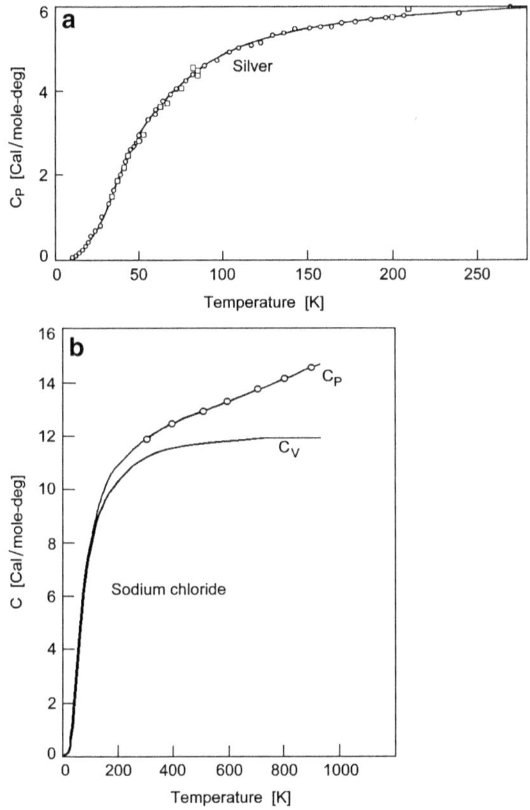

Fig. 9.3 Specific heat as a function of temperature: (**a**) silver, (**b**) NaCl

9.2.4 Theories

9.2.4.1 Classical Theory

Eighty years after Dulong and Petit stated their law, Richarz [9.4] attempted its explanation. Treating an atom as a harmonic oscillator, its energy ε may be written as

$$\varepsilon = (p^2/2m) + (m\omega^2 x^2/2), \tag{9.5}$$

where p is the momentum, m the mass, ω the angular frequency and x the displacement. Using classical statistical mechanics, the average energy $< \varepsilon >$ is

$$< \varepsilon >= \int_0^\infty \varepsilon \mathrm{e}^{-\varepsilon/k_B T} \mathrm{d}\varepsilon \bigg/ \int_0^\infty \mathrm{e}^{-\varepsilon/k_B T} \mathrm{d}\varepsilon = k_B T. \tag{9.6}$$

Considering that an atom is equivalent to three oscillators, the total energy U per mole is

$$U = 3N_{\mathrm{A}}k_{\mathrm{B}}\,T, \tag{9.7}$$

where N_{A} is the Avogadro number. It follows that the specific heat C_{V} is

$$\begin{aligned} C_{\mathrm{V}} = \left(\frac{\partial U}{\partial T}\right)_V &= 3N_{\mathrm{A}}k_{\mathrm{B}} = 3R \\ &= 5.96\,\mathrm{cal/mol\text{-}deg} \qquad (9.8) \\ &= (5.96 \times 4.186)\ \mathrm{J/mol\text{-}deg}. \end{aligned}$$

This value is close to the experimental values of the molar specific heat of most elemental solids (Table 9.1). For diatomic solids like NaCl, the value will be $6R$.

This model explains the Dulong–Petit law. But it does not explain the low values of the specific heats of Be, B, graphite and diamond nor does it explain the observed temperature variation.

9.2.4.2 Einstein's Theory

Einstein [9.5] (Box 9.1) was the first to propose a theory to account for the temperature variation of specific heats and also to explain the deviations from the Dulong–Petit law in the case of some elements.

Einstein assumed that all atoms in a solid vibrate with the same frequency ν. The vibrations are governed by Planck's quantum hypothesis. Each atom is equivalent to three mutually perpendicular Planck oscillators; a Planck oscillator has discrete quantities of energy $nh\nu$ where h is Planck's constant and n an integer. Thus, Einstein extended Planck's quantisation of the radiation energy to the energy of atomic oscillators. Quantised radiation energy is called a photon; the quantised atomic vibration energy is called a phonon.

According to classical statistics, the probability that an oscillator has energy $nh\nu$ is proportional to $\exp(-nh\nu/_{k_{\mathrm{B}}}T)$. Since n can take different integral values, the mean energy of $< \varepsilon >$ of an oscillator is

Box 9.1

Albert Einstein, a German national, is considered the greatest scientist after Newton. He proposed the law of photoelectricity for which he was awarded the Nobel Prize for physics in 1921. He is remembered for the theory of relativity. In solid state physics his contribution is the Einstein model for the specific heat.

A. Einstein (1879–1955)

$$< \varepsilon >= \left[\sum_{n=0}^{\infty} nh\nu e^{-nh\nu/k_B T} \Big/ \sum_{n=0}^{\infty} e^{-nh\nu/k_B T} \right]. \tag{9.9}$$

Substituting $e^{-h\nu/k_B T} = x$, (9.9) becomes

$$< \varepsilon >= h\nu x \left(\sum_{n=0}^{\infty} nx^{n-1} \Big/ \sum_{n=0}^{\infty} x^n \right) = h\nu x \frac{d}{dx} \left(\log \sum_{n=0}^{\infty} x^n \right). \tag{9.10}$$

But $\sum\limits_{n=0}^{\infty} x^n = \frac{1}{1-x}$ for $0 < x < 1$. Hence, (9.10) becomes

$$< \varepsilon >= \frac{h\nu}{e^{h\nu/k_B T} - 1}. \tag{9.11}$$

Further, the average number of quanta $< n >$ associated with a given frequency ν is given by

$$< n >= \frac{< \varepsilon >}{h\nu} = \frac{1}{e^{h\nu/k_B T} - 1}. \tag{9.12}$$

It can be seen that at high temperatures $< n >$ approximates to $k_B T/h\nu$

The total energy U of the N_A three-dimensional oscillators is

$$U = 3N_A < \varepsilon >, \tag{9.13}$$

and the specific heat is

$$C_V = \left(\frac{\partial U}{\partial T} \right)_V = 3N_A \frac{k_B (h\nu/k_B T)^2 e^{h\nu/k_B T}}{(e^{h\nu/k_B T} - 1)^2}. \tag{9.14}$$

The frequency ν eventually came to be called the Einstein characteristic frequency ν_E. The parameter $(h\nu_E/k_B)$ which has dimensions of temperature is called the Einstein characteristic temperature θ_E. Equation (9.14) can now be written as

$$\begin{aligned} C_V &= 3N_A \left[k_B \left(\frac{h\nu_E}{k_B T} \right)^2 \frac{e^{h\nu_E/k_B T}}{(e^{h\nu_E/k_B T} - 1)^2} \right] \\ &= 3N_A E(\nu_E, T). \end{aligned} \tag{9.15a}$$

Substituting θ_E for $h\nu_E/k_B$ in (9.15a), we get

$$\begin{aligned} C_V &= 3N_A \left[k_B (\theta_E/T)^2 \frac{e^{(\theta_E/T)}}{(e^{\theta_E/T} - 1)^2} \right] \\ &= 3N_A E(\theta_E, T). \end{aligned} \tag{9.15b}$$

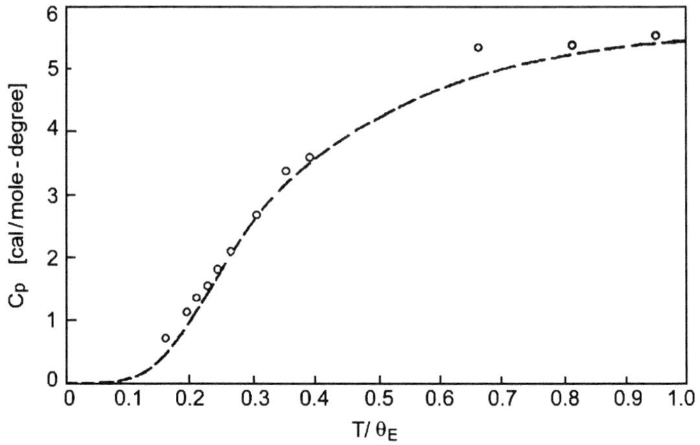

Fig. 9.4 Comparison of experimental values of the specific heat of diamond (*circles*) and values calculated from the Einstein model, using $\theta_E = 1{,}320$ K (*dashed curve*)

The function $E(\nu_E, T)$ is known as the Einstein specific heat function. It represents the contribution of a single oscillator to the specific heat of the crystal. The function $E(\theta_E, T)$ is useful in estimating values of θ_E from experimental data on specific heats. In Fig. 9.4, the observed data on the specific heat of diamond is compared with the values calculated from (9.15b) with a value of $\theta_E = 1{,}320$ K. The agreement is generally fair but at very low temperatures the calculated values tend to be lower than the experimental values.

Let us see to what extent the Einstein model accounts for the observed features enumerated in Sect. 9.3.

1. From (9.15), it can be seen that $C_V \to 0$ as $T \to 0$. This is also consistent with the Nernst heat theorem (otherwise known as the third law of thermodynamics).
2. At very low temperatures ($T << \theta_E$), (9.15) approximates to

$$C_V = 3N_A k_B (\theta_E/T)^2 e^{-\theta_E/T}.$$

3. At very high temperatures ($T >> \theta_E$), $C_V \to 3R$ (the Dulong–Petit value). It may be noted that $E(\theta_E, T) \to k_B$ when $(T/\theta_E) \to 1$ or when $T \approx \theta_E$. In most solids, $\theta_E \sim 200$–400 K and so C_V takes the Dulong–Petit value near about room temperature. But in diamond, Be and boron, θ_E is very large and $T/\theta_E = 1$ at very high temperatures. Consequently, C_V is much less than the Dulong–Petit value at room temperature.

Einstein has to be credited with having, at least qualitatively, explained the temperature variation of specific heats. However, at very low temperatures, there is a distinct difference between experimental values and values calculated from Einstein's formula (9.15). Further, Einstein's assumption of a single common frequency for the atomic oscillators is unrealistic.

Nernst and Lindemann [9.6] showed that experimental data for a number of solids at very low temperatures give a better fit to the following empirical formula:

$$C_V = \frac{3R}{2}\left\{\frac{(h\nu/k_BT)^2}{[e^{(h\nu/k_BT)-1}]^2} + \frac{(h\nu/2k_BT)}{[e^{(h\nu/2k_BT)-1}]^2}\right\}. \qquad (9.16)$$

This formula assumes two frequencies ν and $\nu/2$ as against Einstein's single frequency. This suggested that the atomic oscillators have more than one frequency. Einstein [9.7] himself concluded that the assumption of a single frequency was responsible for the failure of his formula at low temperatures and suggested that a whole band of frequencies be employed in a theoretical model.

9.2.4.3 Debye's Theory

Debye [9.8] (Box 9.2) assumed the solid to be an elastic continuous medium. Such a solid can sustain standing sound waves between its boundaries. Debye assumed the medium to be dispersionless, i.e. the velocity of the waves is independent of the frequency. These sound waves, called phonons, are described by a propagation velocity c_0, a wavelength λ or a propagation vector $|\mathbf{k}|$ $(= 2\pi/\lambda)$ and an angular frequency $\omega = 2\pi\nu = c_0 k$ where ν is the frequency.

Instead of a single frequency assumed in the Einstein model, the Debye model assumes a whole range of frequencies. In calculating the number of modes of vibrations by the solid, it is convenient to introduce a distribution function $g(\nu)$ which is also called the vibration spectrum or the spectral function or the number density. The number of modes of vibration in the frequency range ν and $\nu + d\nu$ is given by $g(\nu)d\nu$. The derivation of the distribution function $g(\nu)$ has been discussed by De Launay [9.9], Brown [9.10], Dekker [9.11] and Kittel [9.12].

The wave equation in three dimensions is

$$\frac{\partial^2 u}{\partial x^2} + \frac{\partial^2 u}{\partial y^2} + \frac{\partial^2 u}{\partial z^2} = \frac{1}{c_0{}^2}\frac{\partial^2 u}{\partial t^2}, \qquad (9.17)$$

Box 9.2

Petrus Josephus Wilhelmus Debye was a German Scientist. His studies include dipole moment and charge distribution, specific heat of solids and X-ray diffraction. Debye was awarded the Nobel Prize for chemistry in 1936.

P. J. W. Debye (1884–1966)

where u is the displacement and c_0 the velocity. We shall assume the elastic medium to be a cube of side L with its faces fixed. We shall assume the following standing wave solutions:

$$u(x, y, z, t) = A \sin(n_x \pi x/L) \sin(n_y \pi y/L) \sin(n_z \pi z/L) \cos 2\pi \nu t, \tag{9.18}$$

where n_x, n_y, n_z are positive integers ≥ 1. Substituting (9.18) in (9.17), we get

$$(\pi^2/L^2)({n_x}^2 + {n_y}^2 + {n_z}^2) = (4\pi^2/{c_0}^2)\nu^2, \tag{9.19}$$

or

$$({n_x}^2 + {n_y}^2 + {n_z}^2) = (4L^2/{c_0}^2)\nu^2 = R^2. \tag{9.20}$$

This is an equation to a sphere of radius R. If we construct a network with n_x, n_y, n_z as the coordinate points, each set of these integers corresponds with one of the modes of vibration. The number of frequencies $g(\nu)\mathrm{d}\nu$ is then the number of coordinate points in a shell with radii R and $R + \mathrm{d}R$. Thus,

$$g(\nu)\mathrm{d}\nu = \frac{1}{8} \times 4\pi R^2 \mathrm{d}R. \tag{9.21}$$

Since the coordinates n_x, n_y, n_z have only positive values, we consider only an octant; hence the factor (1/8). Substituting for R and $\mathrm{d}R$ from (9.20), we get

$$g(\nu)\mathrm{d}\nu = \left(\frac{4\pi V}{{c_0}^3}\right) \nu^2 \mathrm{d}\nu, \tag{9.22}$$

where $V = L^3$ is the volume of the solid. For elastic waves, each frequency is associated with one longitudinal and two transverse waves. These have velocities c_l and c_t, respectively. Thus, finally,

$$g(\nu)\mathrm{d}\nu = a\nu^2 \mathrm{d}\nu, \tag{9.23}$$

with

$$a = 4\pi V \left(\frac{1}{{c_l}^3} + \frac{2}{{c_t}^3}\right). \tag{9.24}$$

It is to be noted that the Debye distribution function $g(\nu)$ is parabolic in ν.

At this stage, Debye introduced in his model the discrete nature of the solid by limiting the total number of vibrational modes to $3N$, N being the total number of atoms. Thus, the vibration spectrum is 'cut off' at some maximum frequency ν_D such that

$$\int_0^{\nu_\mathrm{D}} g(\nu)\mathrm{d}\nu = 3N. \tag{9.25}$$

Substituting from (9.23), (9.24), integrating and rearranging, we get

$$\nu_{\rm D}{}^3 = \frac{9N}{4\pi V}\left(\frac{1}{c_l{}^3} + \frac{2}{c_t{}^3}\right)^{-1}. \tag{9.26}$$

The cut-off frequency ν_D is called the maximum frequency or, more often, the Debye frequency. Its value is about 10^{13} per sec; the associated wavelength is about twice the interatomic distance. In Debye's model, ν_D is the same for the longitudinal and transverse modes. Further, the velocities c_l and c_t are independent of the wavelength. Debye introduced the quantity θ_D defined by

$$\theta_{\rm D} = (h\nu_{\rm D}/k_{\rm B}). \tag{9.27}$$

The parameter θ_D has dimensions of temperature; it is known as the Debye characteristic temperature.

The specific heat is obtained by integrating over the entire vibration spectrum the product of the contribution of each frequency to the specific heat and the distribution function, i.e.

$$C_{\rm V} = \int_0^{\nu_{\rm D}} E(\nu,T)g(\nu)d\nu. \tag{9.28}$$

Substituting from (9.15) and (9.23), we get

$$C_{\rm V} = (3Nk_{\rm B})\cdot 3\left(\frac{T}{\theta_{\rm D}}\right)^3 \int_0^{\xi_m} \frac{\xi^4 e^{\xi}}{(e^{\xi}-1)^2}d\xi, \tag{9.29}$$

where $\xi = h\nu/k_B T$ and $\xi_m = h\nu_D k_B T$. Identifying N as the Avogadro number N_A, (9.29) may be written as

$$C_{\rm V} = (3R)\cdot D(\theta_{\rm D}/T), \tag{9.30}$$

where

$$D(\theta_{\rm D}/T) = 3\left(\frac{T}{\theta_{\rm D}}\right)^3 \int_0^{\xi_m} \frac{\xi^4 e^{\xi}}{(e^{\xi}-1)^2}d\xi. \tag{9.31}$$

The function $D(\theta_D/T)$ is called the Debye specific heat function; its values are available in tabular form for several values of θ_D/T [9.13, 9.14]. The experimental values of the specific heat of aluminium and those calculated from the Debye expression (9.29) are shown in Fig. 9.5. The agreement is good down to very low temperatures.

Debye estimated the values of the Debye temperature of several solids by fitting experimental data to (9.29). From (9.26) it follows that θ_D can be calculated from sound velocities, i.e. from elastic properties. Debye showed that

$$\left(\frac{1}{c_l^3} + \frac{2}{c_t^3}\right) = \rho^{3/2}\psi^{3/2}\left\{2\left[\frac{2(1+\sigma)}{3(1-\sigma)}\right]^{3/2} + \left[\frac{(1+\sigma)}{3(1-\sigma)}\right]^{3/2}\right\}, \tag{9.32}$$

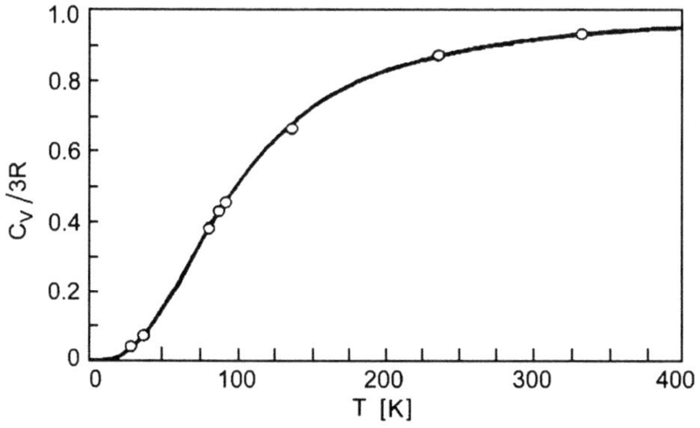

Fig. 9.5 Specific heat of aluminium; *circles* are experimental values, continuous curve calculated from (9.29) with $\theta_D = 396\,K$

Table 9.2 Debye temperatures θ_D of some solids [9.2]

Solid	θ_D [K]		Solid	θ_D [K]	
	Specific heat	Elastic properties		Specific heat	Elastic properties
Ag	228	227	Ga	317	–
Al	423	428	Ge	378	375
As	236	–	Mg	396	387
Au	165	162	Ni	427	476
B	1,315	–	Pb	102	105
Be	1,160	1,462	Pt	234	229
$C_{graphite}$	1,550	–	Sb	150	187
$C_{diamond}$	2,240	2,240	Si	647	649
Cd	252	212	Ti	426	373
Co	452	446	W	388	384
Cu	342	345	Zn	316	324
Fe	457	477			

where ρ is the density, ψ the compressibility and σ the Poisson's ratio. Alers [9.15] has reviewed more refined methods of calculating θ_D from single-crystal elastic constants. The values of θ_D calculated from specific heats and elastic constants are given in Table 9.2; the agreement is fair. In view of the importance of θ_D, several methods have been developed for estimating θ_D from other physical properties like thermal expansion and melting points. These have been reviewed by Blackman [9.16] and Gopal [9.14].

Let us examine the predictions from (9.31) in different temperature ranges. At high temperatures ($T >> \theta_D$), $D(\theta_D/T) \rightarrow 1$ and C_V approaches the Dulong–Petit value. It should be noted that C_V is controlled by the value of (θ_D/T) and not T itself. Since the value of θ_D differs from crystal to crystal, the Dulong–Petit value is attained at different temperatures. For diamond the Debye temperature is 1,860 K and so its C_V reaches the Dulong–Petit value at $T \approx 2{,}000\,K$. It is thus understandable that its specific heat is much less than $3R$ at room temperature.

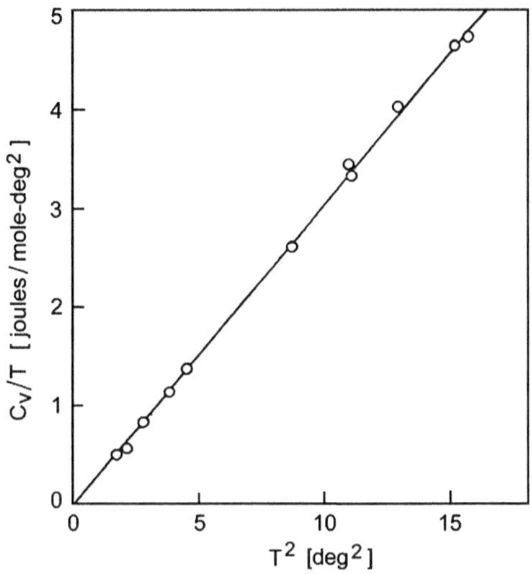

Fig. 9.6 Plot of (C_V/T) against T^2 at very low temperatures for KCl

At very low temperatures ($T << \theta_D/50$), (9.29) approximates to

$$C_V \approx (12\pi^4/5)R(T/\theta_D)^3. \tag{9.33}$$

This is known as the Debye-T^3 law. It may be recalled that this feature was empirically noted much before the advent of Debye's theory. A linear plot of C_V/T vs. T^2 for KCl is shown in Fig. 9.6 as an illustration of the Debye-T^3 law. It may be noted that the temperature variation of C_V predicted by (9.33) is close to the observed variation and is lesser than the exponential variation predicted from Einstein's equation (9.15). Finally, as $T \to 0$, $C_V \to 0$ which is consistent with the third law of thermodynamics.

9.2.4.4 Electronic Specific Heat

At very low temperatures, the specific heat is expected to follow the T^3 law. However, in metals, it varies according to the following law:

$$C_V \approx AT + BT^3. \tag{9.34}$$

In fact, at very low temperatures, the linear term predominates. It can be separated from the total specific heat by plotting C_V/T against T^2; it will be a linear plot with intercept equal to A. Figure 9.7 shows such a plot for silver; the plot has an intercept $A = 0.61 \times 10^{-3}$J/mol-deg^2. This linear term arises from the contribution of the electron gas and is, therefore, called the electronic specific heat.

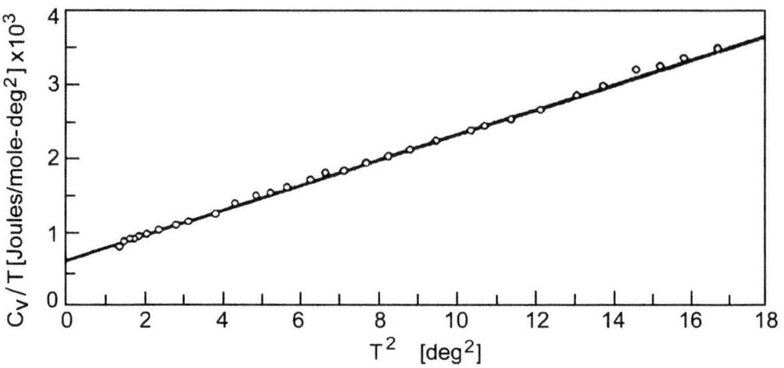

Fig. 9.7 Plot of (C_V/T) against T^2 for silver at very low temperatures

9.2.4.5 Limitations of Debye's Theory

Debye's theory shows a much better overall agreement with experiment than Einstein's theory. Conceptually also the assumption of a range of frequencies is better than Einstein's assumption of a single frequency. Yet, it has its own drawbacks.

Firstly, at the conceptual level, though it takes into account the discrete nature of the solid by limiting the number of frequencies, it essentially assumes an elastic continuum and ignores crystal structure and interatomic forces. Further, it assumes that the solid medium is dispersionless, i.e. the propagation velocity of the waves is independent of frequency.

Secondly, Debye assumed that the Debye temperature is a constant independent of temperature. By fitting (9.29) to the experimental value of C_V at a given temperature, we can get a value of θ_D appropriate to that temperature. By repeating this procedure at different temperatures, a θ_D vs. T plot can be constructed. Ideally, such a plot would be a straight line parallel to the temperature axis. But actual θ_D–T plots (Fig. 9.8) obtained from experimental data reveal that θ_D is not independent of T. Further, the shape of the θ_D–T curve is different for different crystals. Blackman [9.16] pointed out that the observed temperature variation of θ_D is dependent on the actual vibration spectrum of a crystal which is much more complicated than the simple parabolic vibration spectrum which results from Debye's assumption of a continuum. A more elaborate treatment will be discussed in the next chapter.

9.3 Thermal Expansion

9.3.1 General

Thermal expansion is the dilatation of a solid due to change in temperature. It is a property with both technical and scientific significance. Bimetallic strips made of

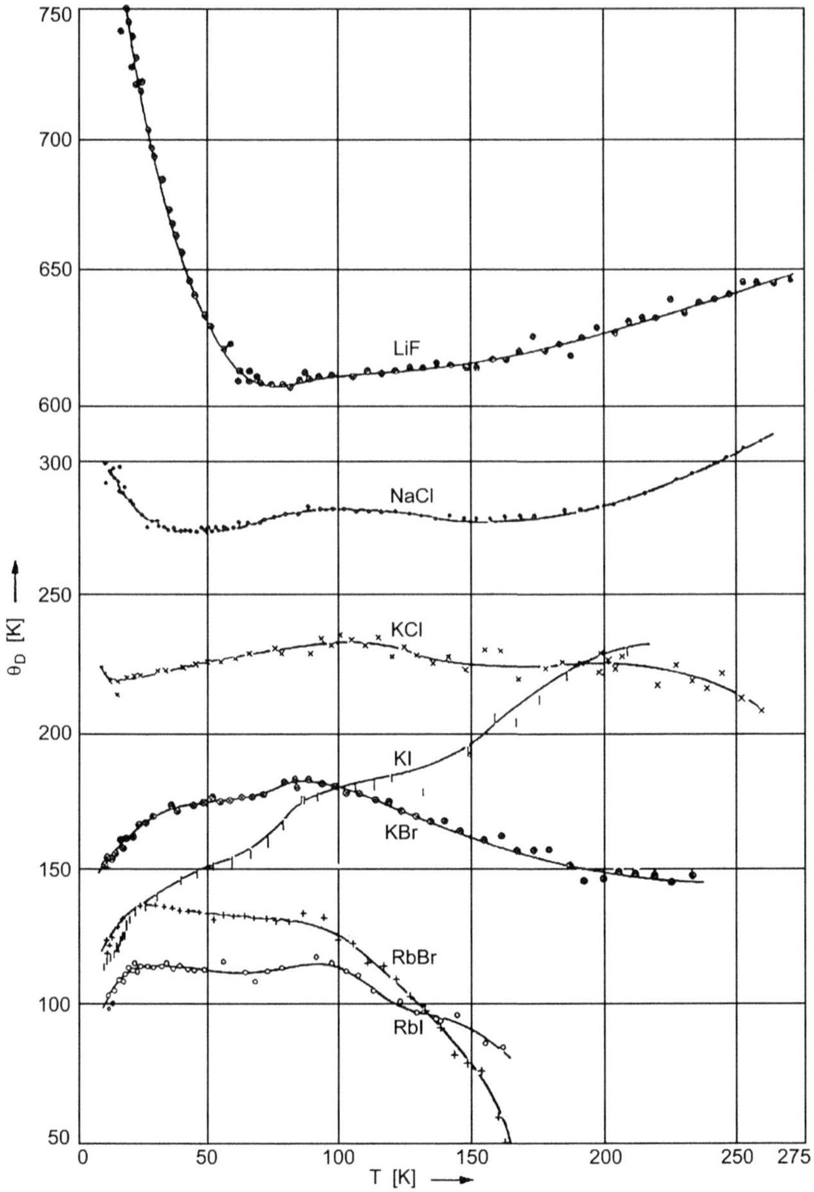

Fig. 9.8 $\theta_D - T$ curves for alkali halides (from [9.16])

metals with different expansions are used in temperature controllers and metals with very low (near-zero) expansion are used in length and time standards. Its scientific importance lies in its direct relation with the anharmonicity of the crystal potential.

In macroscopic terms, the linear coefficient of thermal expansion α is defined as

$$\alpha = (1/L_0)(\Delta L/\Delta T), \tag{9.35}$$

where L_0 is the length at a reference temperature and ΔL the change in length due to a temperature change ΔT. It can also be defined in microscopic terms as

$$\begin{aligned}\alpha &= (1/r_0)(\Delta r/\Delta T)\\ &= (1/a_0)(\Delta a/\Delta T),\end{aligned} \tag{9.36}$$

where r_0 and a_0 are the interatomic distance and lattice constant, respectively, at a reference temperature and Δr, Δa, the corresponding changes. The volume coefficient of thermal expansion β is

$$\beta = (1/V_0)(\Delta V/\Delta T), \tag{9.37}$$

where V_0 is the volume at a reference temperature and ΔV the change in it. β is simply three times α.

A property which relates a second-rank tensor with a scalar is itself a second-rank tensor. Thermal expansion connects the strain caused by change of temperature which is a second-rank tensor to temperature which is a scalar quantity; hence, thermal expansion is a second-rank tensor. The non-vanishing independent components of the thermal expansion tensor for different crystal systems are given in Table 9.3.

9.3.2 *Experimental Methods*

There are several methods of measurement of thermal expansion. We shall consider three of them which are used for accurate results.

9.3.2.1 Optical Method

This is perhaps the earliest method; it was used by Fizeau in late nineteenth century to measure the thermal expansion of a number of mineral crystals. In this method [9.17], interference fringes are formed by reflection of light from two surfaces; the fringe width depends on the separation of the surfaces which, in turn, is controlled by the length of the specimen. When the specimen is heated, its length and hence the fringe width changes. While this is the basic principle, a number of improvements have been made like the use of laser light and use of photo devices to record the fringe width (or shift of fringes).

One such set-up [9.18] is shown in Fig. 9.9. The source of light is a helium discharge tube. The interferometer consists of two fused quartz discs A and B. Interference fringes are produced by reflection of light from the bottom surface of A and the top surface of B which are optically flat. The top surface of A is inclined at a small angle to its bottom surface and the bottom surface of B is frosted to ensure

Table 9.3 The independent components of the thermal expansion tensor for the seven crystal systems

	System	No. of constants	Form of tensor
1.	Triclinic	6	$\begin{bmatrix} \alpha_{11} & \alpha_{21} & \alpha_{31} \\ \alpha_{21} & \alpha_{22} & \alpha_{32} \\ \alpha_{31} & \alpha_{32} & \alpha_{33} \end{bmatrix}$
2.	Monoclinic	4	$\begin{bmatrix} \alpha_{11} & 0 & \alpha_{31} \\ 0 & \alpha_{22} & 0 \\ \alpha_{31} & 0 & \alpha_{33} \end{bmatrix}$
3.	Orthorhombic	3	$\begin{bmatrix} \alpha_{11} = \alpha_1 & 0 & 0 \\ 0 & \alpha_{22} = \alpha_2 & 0 \\ 0 & 0 & \alpha_{33} = \alpha_3 \end{bmatrix}$
4.	Tetragonal	2	
5.	Trigonal	2	$\begin{bmatrix} \alpha_{11} = \alpha_1 & 0 & 0 \\ 0 & \alpha_{22} = \alpha_1 & 0 \\ 0 & 0 & \alpha_{33} = \alpha_3 \end{bmatrix}$
6.	Hexagonal	2	
7.	Cubic	1	$\begin{bmatrix} \alpha_{11} = \alpha & 0 & 0 \\ 0 & \alpha_{22} = \alpha & 0 \\ 0 & 0 & \alpha_{33} = \alpha \end{bmatrix}$

that fringes are not formed due to reflection from these surfaces. The specimen is a single crystal carefully cut so that it has three projections of equal length; it rests on B and supports A. Thus, the separation between A and B is defined by the length of the projections (called feet or legs of the sample). The optics is controlled by the mirror and the lenses L_1, L_2 and L_3. The filter chooses the intense line of wavelength 5,876 Å. Finally, the fringes are recorded by the photomultiplier tube. The sample can be heated or cooled by a suitable arrangement. If M is the fringe shift due to a change ΔT in temperature, the change in length ΔL is

$$\Delta L = M\lambda/2, \tag{9.38}$$

and the coefficient of expansion is

$$\alpha = (1/L_0)(\Delta L/\Delta T) = \frac{(M\lambda/2)}{L_0} \cdot \frac{1}{\Delta T}. \tag{9.39}$$

9.3.2.2 Capacitance Dilatometer

Another very accurate method is the capacitance dilatometer. The capacitance C of a parallel plate condenser is

$$C = \varepsilon A/d, \tag{9.40}$$

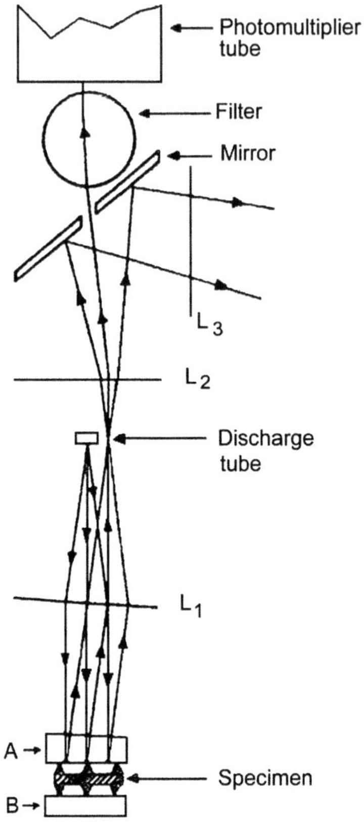

Fig. 9.9 Optical system of the interferometric dilation recording apparatus

where ε is the static dielectric constant of the medium between the plates, A the area of the plates and d the separation between them. When such a condenser is used in this experiment, the medium between the plates is air or vacuum for which $\varepsilon = 1$. One plate is fixed and the other is moveable. This latter plate is kept in communication with the sample. If ΔL is the change in the length L_0 of the sample due to a change of temperature ΔT, then Δd, the change in d is also ΔL. The resulting change ΔC in the capacitance is

$$\Delta C = -(\varepsilon A/d^2)\Delta d. \tag{9.41}$$

The coefficient of expansion is given by

$$\alpha = (1/L_0)(\Delta L/\Delta T) = -\left(\frac{d}{L_0 C}\right)\frac{\Delta C}{\Delta T}. \tag{9.42}$$

A three terminal set-up used in a capacitance dilatometer is shown in Fig. 9.10. The capacitance is measured between electrodes 1 and 2. Electrode 1 is the sample.

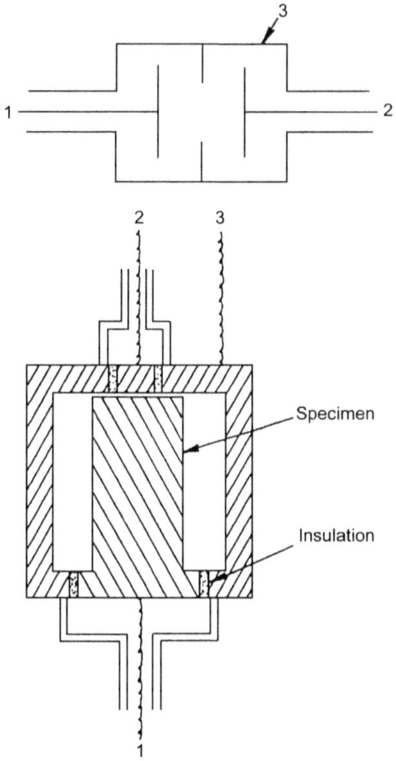

Fig. 9.10 Schematic diagram for three-terminal parallel plate capacitor technique

Electrodes 2 and 3 are generally made of copper. Electrode 3 completely surrounds the other two electrodes; it acts as an earth shield and eliminates the effect of lead wires. The separation d between the top surface of the specimen and the lower surface of 2 is adjusted such that the value of C is ~10pF which can be measured with an accuracy of 10^{-7}pF. The measurement of C and ΔC is made with a standard capacitance bridge like Scherring Bridge. For accurate results, corrections have to be applied for (1) border effects, (2) non-parallelism of the two surfaces and (3) the expansion of the plates.

9.3.2.3 X-ray Diffraction Method

X-rays incident on a crystal are diffracted by its lattice planes. The wavelength λ of the X-rays, the Miller indices h, k, l of the lattice planes and the angle of diffraction θ are related by the Bragg equation

$$a_0 = \lambda(h^2 + k^2 + l^2)^{1/2}/2\sin\theta, \qquad (9.43)$$

where a_0 is the lattice constant of the crystal. When the crystal temperature changes by ΔT, a_0 changes by Δa and so the Bragg angle changes by $\Delta\theta$. By differentiating (9.43), it can be shown that

$$\Delta a/a_0 = -\cot\theta.\Delta\theta, \tag{9.44}$$

and

$$\alpha = -\cot\theta(\Delta\theta/\Delta T). \tag{9.45}$$

This is the principle of the X-ray method of determining thermal expansion. A simple arrangement for recording the X-ray pattern is shown in Fig. 9.11. It consists of a sample holder and a flat-film holder, mounted on a bench. X-rays pass through the collimator and are diffracted back on the flat film. This arrangement is called the flat film back-reflection camera; the back reflection region ($\theta > 45°$) is preferred as errors are least at high Bragg angles. The shift of the lines as the sample is heated to different temperatures is shown in Fig. 9.12 for NaCl.

For better accuracy, cameras are available which allow recording of several reflections at high angles. Some of these cameras are discussed by Peiser et al. [9.19] and, more recently, by Sirdeshmukh et al. [9.20].

By applying corrections for systematic errors, the lattice constants a_0 can be determined accurately; α is then determined from (9.36).

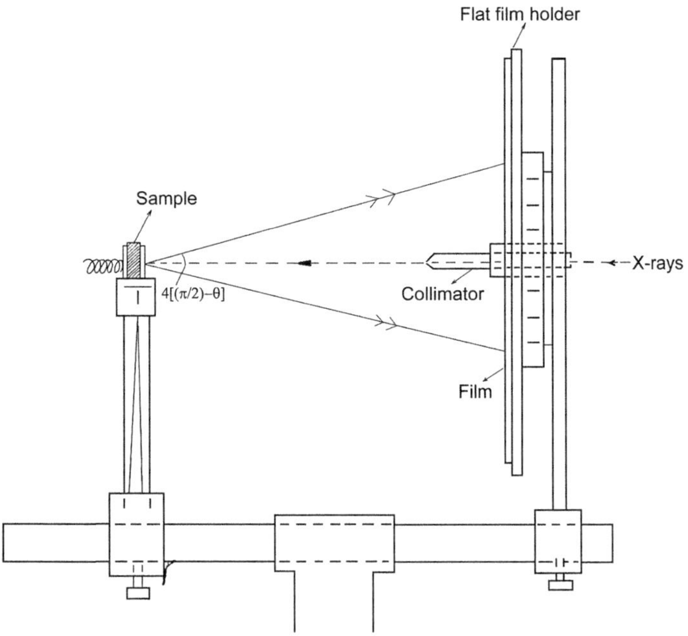

Fig. 9.11 Geometry of the flat film back-reflection camera

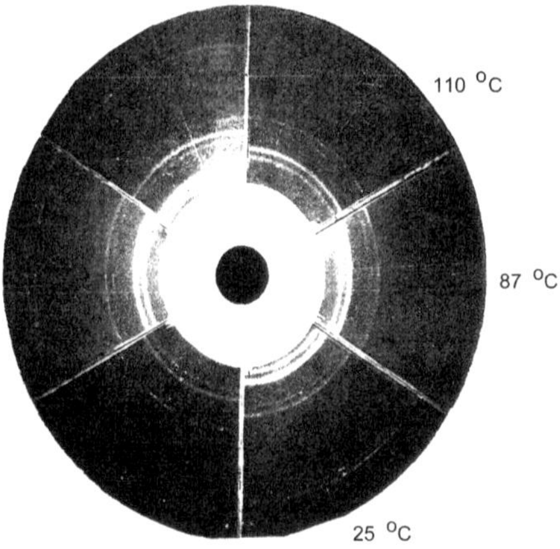

Fig. 9.12 Back-reflection photographs of NaCl at various temperatures

9.3.2.4 Comparison of Methods

We shall first consider sample requirements. In the optical method a single crystal is required which, further, should be machined to provide a three-legged sample. For the capacitance method also a single crystal is required. In both these methods, samples with different orientations are required if the crystal is anisotropic. In contrast, the X-ray method requires milligram quantity of the sample in powder form. Further, and more importantly, the thermal expansion in different directions can be determined in a single experiment as the X-ray film records reflections which represent different directions in the crystal.

With regard to the relative accuracy $\Delta\alpha$ of expansion measurement, it is 10^{-6}, 10^{-7} and 10^{-10} (per degree), respectively, in the X-ray, optical and capacitance methods. For a typical crystal, the expansion coefficient α is $\sim 10^{-6}$ per degree from room temperature down to ~ 30 K but at very low temperatures (~ 3 K), it is $\sim 10^{-9}$ per degree. Hence at low temperatures, the capacitance dilatometer is the most appropriate technique.

9.3.3 Experimental Results and Trends

Thermal expansion is a second-rank tensor; as shown in Table 9.3, the number of independent coefficients is six for triclinic, four for monoclinic, three for orthorhombic, two for trigonal, hexagonal and tetragonal and one for cubic crystals. The values of the thermal expansion coefficients for a few orthorhombic, tetragonal and cubic crystals are given in Table 9.4.

The temperature variation of thermal expansion coefficient of copper is shown in Fig. 9.13. Most solids show a similar behaviour. We may note the following features from this typical curve.

Table 9.4 Values of the coefficients of thermal expansion (α_i) of some crystals at room temperature [9.21]

Crystal	α_1	α_2	α_3
	[10^{-6} per degree]		
Orthorhombic			
Barite	13.6	23.9	14.0
Rochelle salt	58.3	35.5	42.1
Tetragonal	$\alpha_1 = \alpha_2$		
Rutile	7.14		9.19
Potassium dihydrogen phosphate	21.6		34.3
Cubic	$\alpha_1 = \alpha_2 = \alpha_3$		
Diamond	0.87		
LiF	33.2		
NaCl	39.2		

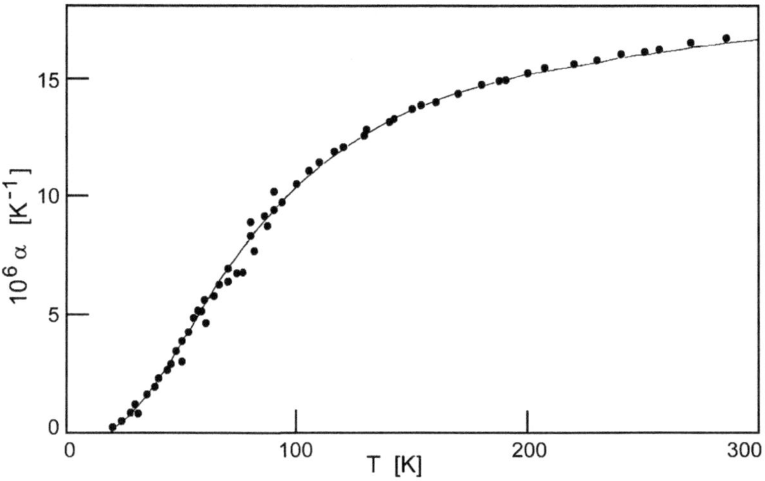

Fig. 9.13 The linear thermal expansion coefficient α of copper below room temperature

1. The curve shows a very slow increase in α at high temperatures.
2. There is a steep decrease in α at low temperatures.
3. $\alpha \to 0$ as $T \to 0$.

Any theory of thermal expansion will have to explain these features.

9.3.4 *Theories*

9.3.4.1 Anharmonicity and Thermal Expansion

Let us assume that the interatomic potential ϕ in the crystal is harmonic; then, the ϕ vs. interatomic distance curve is a symmetric parabola (Fig. 9.14). With increasing

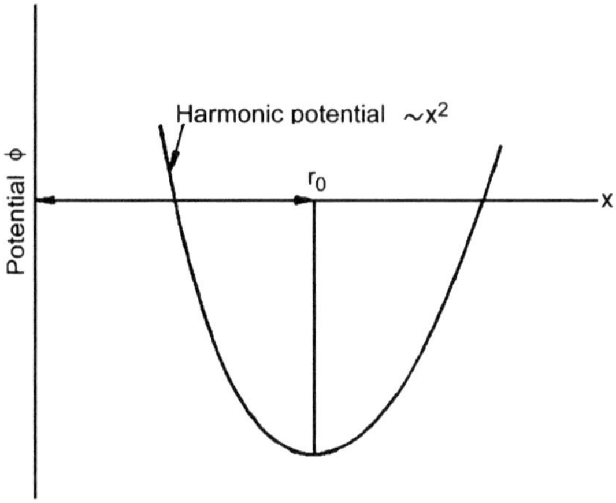

Fig. 9.14 Mean position in a harmonic potential

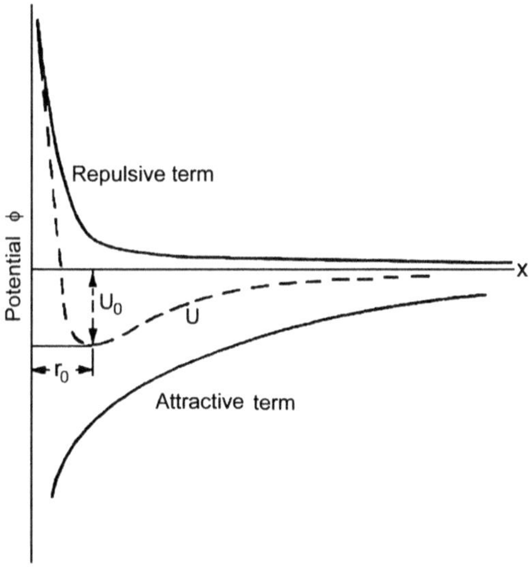

Fig. 9.15 The contributions to the potential of an ionic crystal. The lattice constant r_0 is determined by the potential energy minimum

temperature, the energy of an atomic oscillator increases. As a result its amplitude of vibration also increases. Yet, because of the symmetric nature of the curve, the mean position of the atom r_0 remains unchanged and, hence, there is no expansion. On the other hand, because of the attractive and repulsive forces in the crystal, the potential becomes asymmetric (Fig. 9.15). The amplitude of vibration of an oscillator in such a potential is unequal on the two sides of its original mean position. As a result, as temperature increases, the mean position keeps changing (Fig. 9.16). This results in a non-zero mean displacement $< x >$ and, hence, in thermal expansion.

Let the true potential ϕ (Fig. 9.15) be represented by

$$\begin{array}{ccc} \phi = bx^2 & -gx^3 & -fx^4 \\ \uparrow & \uparrow & \uparrow \\ \text{Harmonic term} & \text{Anharmonic} & \text{terms} \end{array} \tag{9.46}$$

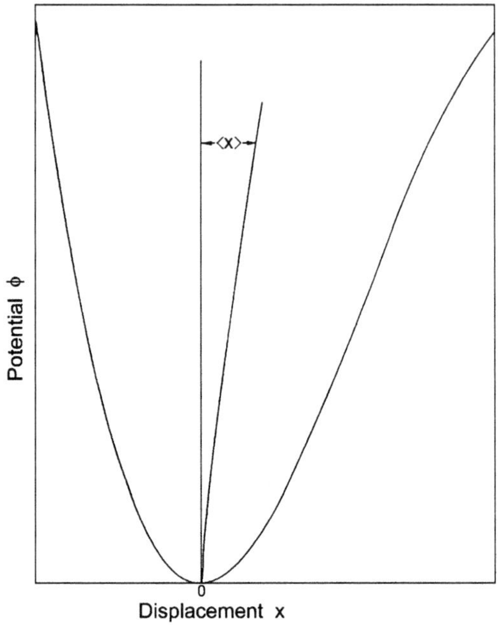

Fig. 9.16 Mean position in anharmonic potential

Here, x is the displacement and b, g and f are constant coefficients. Using Maxwell–Boltzmann statistics, it can be shown [9.22] that the mean value $< x >$ of the displacement is

$$< x >= \frac{\int_0^{\infty} x\mathrm{e}^{-(bx^2-gx^3-fx^4)/k_{\mathrm{B}}T}\mathrm{d}x}{\int_0^{\infty} \mathrm{e}^{-(bx^2-gx^3-fx^4)/k_{\mathrm{B}}T}\mathrm{d}x}. \tag{9.47}$$

Since the anharmonic terms are small, (9.47) approximates to

$$< x >= (3/4)(g/b^2)k_{\mathrm{B}}T, \tag{9.48}$$

and

$$\alpha = \frac{1}{r_0}\frac{\mathrm{d} < x >}{\mathrm{d}T} = \left(\frac{1}{r_0}\right)\left(\frac{3}{4}\right)\left(\frac{g}{b^2}\right)k_{\mathrm{B}}. \tag{9.49}$$

Here r_0 is the equilibrium interatomic distance. The important point to be noted is that thermal expansion is directly related to the anharmonic coefficient g; if $g = 0$ (no anharmonicity), α becomes zero.

Equation (9.49) predicts that α is a constant independent of temperature. This is a consequence of using $k_B T$ as the energy which pertains to high temperature. If instead, the quantum expression for energy is used, (9.48), (9.49) get modified to

$$< x >= (3g/4b^2)(h\nu)/[\exp(h\nu/k_{\mathrm{B}}T) - 1], \tag{9.50}$$

and

$$\alpha = (1/r_0)(3g/4b^2)(h\nu/k_B T)^2 \exp -(h\nu/k_B T). \tag{9.51}$$

Equation (9.51) predicts a steep fall in α at low temperature in agreement with experiment (Fig. 9.13). Further (9.51) approximates to $\alpha \to 0$ as $T \to 0$; this is consistent with experimental observation and also with the third law of thermodynamics. However, this model is only of qualitative value; it assumes a single frequency and the parameters b and g cannot be evaluated in a straightforward manner.

9.3.4.2 Gruneisen's Theory

A theory of thermal expansion was proposed by Gruneisen [9.23]. Though it has undergone some refinements in due course, yet it remains the only full-fledged theory. Full derivation of Gruneisen's theory is given by Roberts [9.24], Wallace [9.25] and in an English translation of the original paper [9.26]. We shall give here a brief outline of Gruneisen's theory.

There are several equations of state of solids. The one known as the Mie-Gruneisen equation of state is

$$PV + G(V) = \gamma E_{th}. \tag{9.52}$$

Here $G(V)$ is a function of volume and E_{th} the thermal energy. The constant γ is defined as

$$\gamma = -d \ \log \nu / d \ \log V, \tag{9.53}$$

where ν is a vibration frequency. γ has come to be known as the Gruneisen constant or Gruneisen parameter. Gruneisen showed that phenomenologically

$$\gamma = 3\alpha V/\psi C_V. \tag{9.54}$$

At $P = 0$, (9.52) simplifies to

$$G(V) = \gamma E_{th}. \tag{9.55}$$

Introducing $\delta = V - V_0$ (where V is the volume at temperature T and V_0 its value at a standard temperature), $G(V)$ may be expanded in a power series in δ.

$$G(V) = G(V_0) + \delta G'(V_0) + \frac{\delta^2}{2!} G_0''(V_0) + \ldots \tag{9.56}$$

It can be shown that

$$\begin{aligned} G(V_0) &= 0 \\ G'(V_0) &= 1/\psi \\ G_0'' &= -C/\psi V_0. \end{aligned} \tag{9.57}$$

Here, C is a constant which is related to the interatomic potential. Substituting (9.56) and (9.57) in (9.55), we get

$$G(V) \approx (\delta/\psi)[1 - (\delta/V_0)] = \gamma E_{\mathrm{th}},$$

or

$$\delta = \left[\frac{\gamma\psi\, E_{\mathrm{th}}}{1 - C(\gamma\psi/V_0)E_{\mathrm{th}}}\right]. \tag{9.58}$$

Differentiating (9.58) with respect to T and introducing

$$q = V_0/\gamma\psi, \tag{9.59}$$

we have

$$\begin{aligned} 3\alpha &= (1/V_0)(\mathrm{d}V/\mathrm{d}T) \\ &= (1/V_0)(\mathrm{d}\delta/\mathrm{d}T) \\ &= qC_{\mathrm{V}}/[q - CE_{\mathrm{th}}]^2. \end{aligned} \tag{9.60}$$

This is Gruneisen's equation for thermal expansion. The quantities q and C are obtained by fitting (9.60) to experimental values of α at some select low temperatures. Another method to obtain these quantities is to use the following relations:

$$\begin{aligned} \gamma &= 3\alpha V/\psi C_{\mathrm{V}} \\ C &= \gamma + (2/3) \\ q &= (V_0/\gamma\psi) \\ E_{\mathrm{th}} &= \int_0^T C_{\mathrm{V}}\mathrm{d}T. \end{aligned} \tag{9.61}$$

C_{V} is taken either from experimental data or calculated from θ_{D} using tables of C_{V} and T/θ_{D}.

In Fig. 9.17, the thermal expansion values for Al from experiment are compared with those calculated from (9.60) using values $q = 8.4 \times 10^4$ cal/mol, $C = 2.3$ and $\gamma = 2.06$. The agreement is satisfactory.

It can be shown that (9.61) predicts (1) a slow variation of α at $T >> \theta_{\mathrm{D}}$, (2) a steep decrease in α with temperature at low temperatures ($T < \theta_{\mathrm{D}}$) and (3) finally, $\alpha \to 0$ as $T \to 0$.

9.3.4.3 The Gruneisen Parameter

The Gruneisen parameter is an important aspect of Gruneisen's theory. Gruneisen's theory is a quasi-harmonic theory in which the vibrations are assumed to be

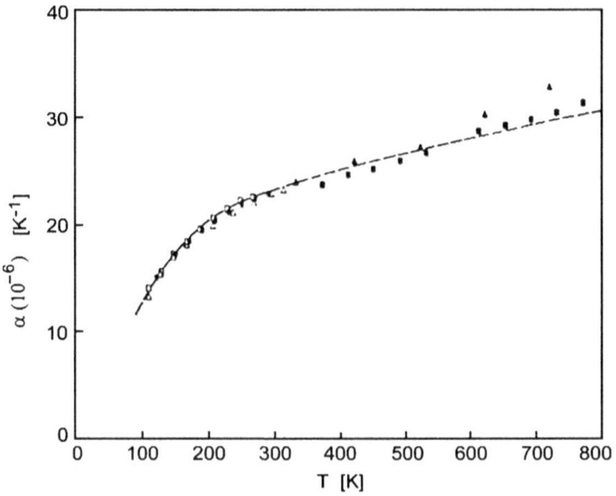

Fig. 9.17 Temperature dependence of the thermal expansion coefficient of Al. The theoretical line was calculated with (9.60)

Table 9.5 Values of the Gruneisen parameter γ [9.23]

Crystal	γ	Crystal	γ
Na	1.25	NaCl	1.33
K	1.34	KCl	1.60
Al	2.17	CaF_2	1.70
Cu	1.96	FeS_2	1.47

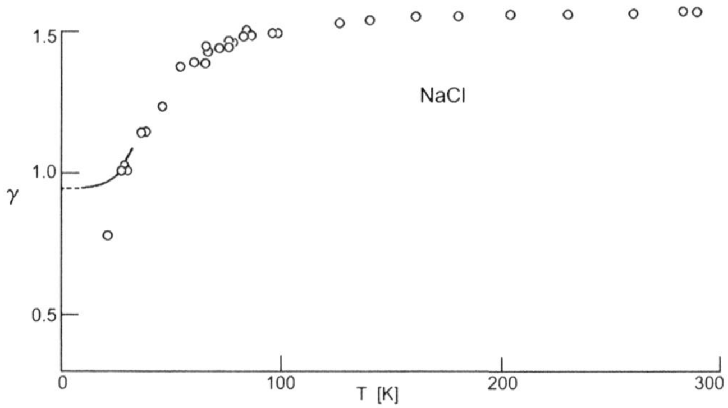

Fig. 9.18 Variation of Gruneisen parameter γ with temperature for sodium chloride

harmonic but the frequencies are volume dependent. The microscopic and phenomenological definitions of γ are given in (9.53) and (9.54). The values of γ for some crystals are given in Table 9.5.

Gruneisen assumed γ to be the same for all frequencies and to be independent of temperature. However, it shows a strong temperature dependence (Figs. 9.18 and 9.19). Further, the temperature dependence varies from crystal to crystal. This aspect will be discussed in the next chapter.

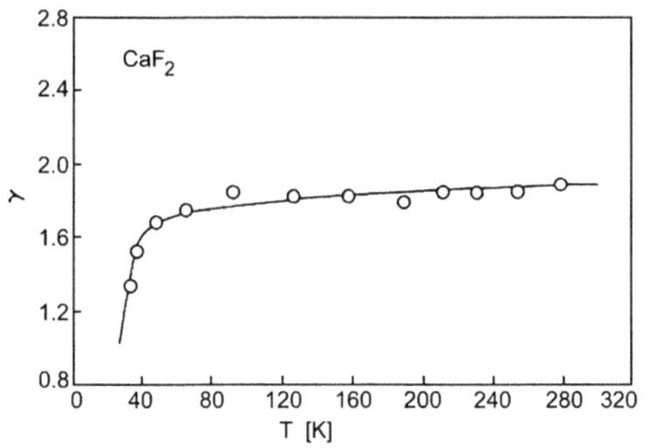

Fig. 9.19 Temperature variation of the Gruneisen parameter γ of calcium fluoride

9.4 Thermal Conductivity

9.4.1 General

Like thermal expansion, thermal conductivity is also directly related to the anharmonicity of lattice vibrations. In the absence of anharmonicity, thermal expansion becomes zero whereas thermal conductivity becomes infinite.

The coefficient of thermal conductivity K is defined by

$$Q = -K(\Delta T/\Delta x), \tag{9.62}$$

where Q is the heat current density (thermal energy flow across unit area in unit time) in the direction of the thermal gradient $(\Delta T/\Delta x)$. The units of K are cal/deg-cm-s or W/deg-cm.

The heat flow Q is a vector and so is the thermal gradient $(\Delta T/\Delta x)$. Thus, the coefficient of thermal conductivity relates two vectors. Hence, it is a symmetric second-rank tensor. The thermal conductivity tensors for different crystal systems will be exactly like the thermal expansion tensors (Table 9.3).

9.4.2 Experimental Methods

Most of the methods of determination of thermal conductivity are based on the defining equation (9.62). Heat is provided to a sample at one end and the resulting temperature gradient is measured.

An experimental set-up based on what is known as the absolute method is shown in Fig. 9.20. The specimen is a bar of rectangular or circular cross-section. A heat source is located at one end and a heat sink at the lower temperature end. Heat flows axially along the bar. The measurements to be made are the heat flow into and out

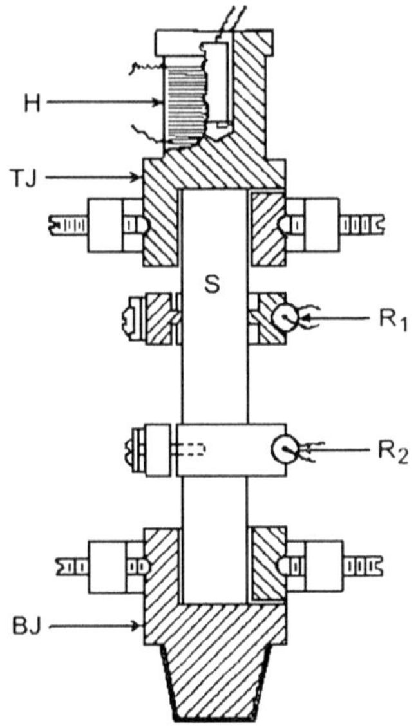

Fig. 9.20 Thermal conductivity set-up: H heater, TJ top jaw, BJ bottom jaw, R_1, R_2 thermometers, S sample

of the bar, the cross-sectional area, the temperature at, at least, two points along the bar and the distance between the points of temperature measurement. To reduce heat losses at high temperatures, guard heaters are used. At low temperatures, radial losses are negligible and insulation is not necessary. The temperatures T_1 and T_2 along the bar have to be measured accurately. The thermometers R_1 and R_2 are either carbon resistance thermometers or gas thermometers. In Slack's [9.27] experiment on KCl, the sample length was 2–3 cm, the area of cross-section $\sim$0.4 cm^2 and the distance between R_1 and R_2 was $\sim$1.6 cm. The difference in temperature at two points $(T_1 - T_2)$ was $\sim$0.01 K. The coefficient of thermal conductivity was calculated from

$$K(T) = \Phi b / A \Delta T, \tag{9.63}$$

where $\Phi = Vi$ (V is the voltage of the heater and i the current), b the distance between R_1 and R_2, A the area of cross-section and $\Delta T = T_1 - T_2$. The measured value of K is associated with the mean temperature $T = (T_1 + T_2)/2$.

In the comparator method (also called the divided bar method), a rod of reference material of known thermal conductivity is placed in series with the sample, both having the same cross-section. The temperature gradients in the reference as well as the sample are measured. If K_R and K_S are the coefficients of thermal conductivity of the reference material and sample and $(\Delta T/\Delta x)_R$ and $(\Delta T/\Delta x)_S$ the temperature gradients, we have

$$K_S = K_R (\Delta T/\Delta x)_R / (\Delta T/\Delta x)_S. \tag{9.64}$$

In deriving (9.64) it is assumed that the heat flow is the same in the reference and the sample.

Table 9.6 Values of the coefficients of thermal conductivity (K_i) of some crystals at room temperature [9.21]

Crystal	K_1	K_2	K_3
	$[10^{-3}$ cal/deg-cm-s]		
Orthorhombic			
Barite	3.93	3.76	3.53
Rochelle salt	1.10	1.46	1.34
Tetragonal	$K_1 = K_2$		
Rutile	21		30
Potassium dihydrogen phosphate	4.2		3.1
Cubic	$K_1 = K_2 = K_3$		
Diamond	300		
LiF	34		
NaCl	15		

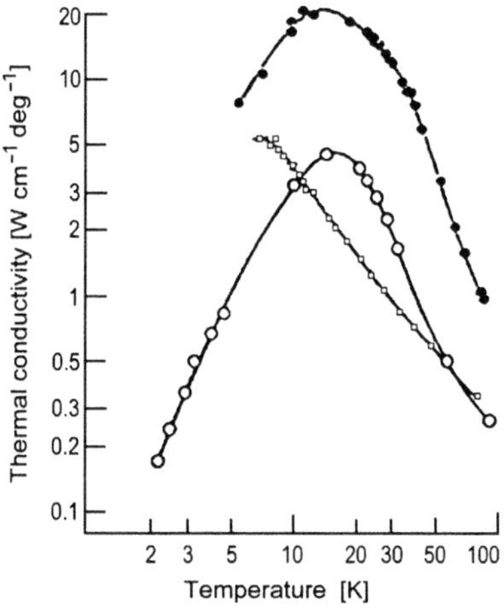

Fig. 9.21 The thermal conductivities of rutile (*open circle*), lithium fluoride (*filled circle*) and KCl (*open square*)

9.4.3 Experimental Results and Trends

Since the coefficient of thermal conductivity is a second-rank tensor, the number of independent coefficients varies from one through six from the cubic to triclinic crystal systems. Experimental values of the coefficients of thermal conductivity for some crystals at room temperature are given in Table 9.6.

The temperature variation of thermal conductivity of some insulators is shown in Fig. 9.21 and typical qualitative trends are shown in Fig. 9.22. As can be seen, the thermal conductivity varies as T^{-1} at high temperatures ($T >> \theta_D$), exponentially in the temperature range $\theta_D/20 < T < \theta_D/10$ and as T^3 at very low temperatures ($T < \theta_D/20$). In between, the conductivity reaches a maximum. Finally, $K \rightarrow 0$ as $T \rightarrow 0$.

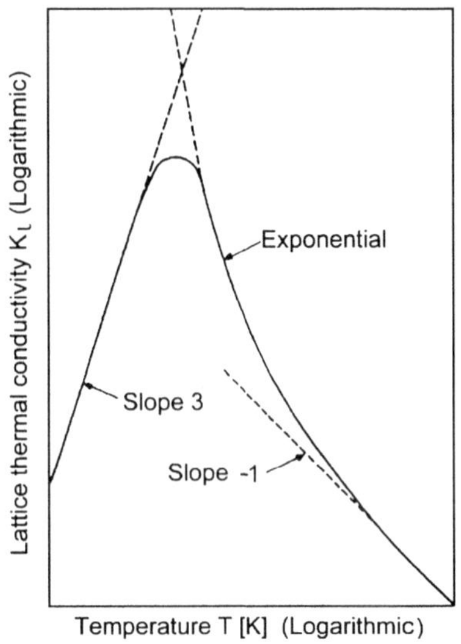

Fig. 9.22 Typical variation of thermal conductivity with temperature

9.4.4 *Theory*

Important contributions to the theory of thermal conductivity have been made by Debye [9.28], Peierls [9.29, 9.30], Casimir [9.31] and Klemens [9.32]. The theory of thermal conductivity is quite involved and complicated and a detailed discussion is not possible. Instead, we shall consider only the concepts and results.

9.4.4.1 The Debye Formula

In gases, heat transport is through collisions of molecules. The kinetic theory gives

$$K = (1/3)C\upsilon l, \tag{9.65}$$

where C is the specific heat, υ the average molecular velocity and l the inter-collision length. For solids, Debye (1914), derived the formula

$$K = (1/3)C_V\upsilon_s\Lambda, \tag{9.66}$$

which is similar to (9.65). Here C_V is the molar specific heat of the solid and υ_s the sound velocity (velocity of the elastic waves). In solids, heat transport is through collisions and consequent scattering of phonons. This scattering is possible because of the coupling between the waves caused by their anharmonic nature. As a measure of the coupling, Debye introduced the phonon mean free path Λ. The phonon mean free path may be considered as the average distance travelled by the phonons between successive collisions. It is also defined as the distance travelled by

Table 9.7 Thermal conductivity K (W/cm-deg) and phonon mean path Λ (cm) in some solids at 273 K [9.34]

Solid	K	Λ	Solid	K	Λ
Si	1.3	4.3×10^{-6}	CaF_2	0.11	7.2×10^{-7}
Ge	0.7	3.3×10^{-6}	NaCl	0.064	6.7×10^{-7}
Quartz	0.14	9.7×10^{-7}	LiF	0.10	3.3×10^{-7}

the elastic wave in which its intensity is attenuated by a factor e [9.11, 9.28, 9.33]. The phonon mean free path may also be expressed as $\Lambda = \upsilon_s \tau$ where τ is called the relaxation time for phonon collision. Then (9.66) becomes

$$K = (1/3) C_V \upsilon_s^2 \tau. \tag{9.67}$$

Values of the phonon mean free path for some solids are given in Table 9.7.

9.4.4.2 Normal and Umklapp Processes

Let us suppose that two phonons with wave vectors $\mathbf{k}_1$ and $\mathbf{k}_2$ collide (interact) such that a third phonon with wave vector $\mathbf{k}_3$ results (Fig. 9.23a). In this process, the energy and momentum are conserved so that

$$\hbar\omega_1 + \hbar\omega_2 = \hbar\omega_3, \tag{9.68}$$

and

$$\hbar\mathbf{k}_1 + \hbar\mathbf{k}_2 = \hbar\mathbf{k}_3. \tag{9.69}$$

Such a process is called a normal process (or N-process). Peierls [9.29] made the important suggestion that due to anharmonicity, two phonons may interact such that, together with the resulting phonon, they satisfy the relation

$$\hbar\mathbf{k}_1 + \hbar\mathbf{k}_2 = \hbar\mathbf{k}_3 + \mathbf{G} \tag{9.70}$$

where $\mathbf{G}$ is a reciprocal lattice vector. The relationship between the four vectors is shown in Fig. 9.23b. Such a process is called an Umklapp process (or U-process); Umklapp in German means flip-over.

In an N-process, the direction of energy flow is not altered; hence, there is no thermal resistance and the thermal conductivity is infinite. On the other hand, in a U process, the direction of $\mathbf{k}_3$ is broadly the reverse of those of $\mathbf{k}_1$ and $\mathbf{k}_2$; while $\mathbf{k}_1$ and $\mathbf{k}_2$ convey energy to the right, $\mathbf{k}_3$ carries it to the left. Such an interaction results in thermal resistance and a finite thermal conductivity. Another way of expressing the difference between the two processes is to state that in the N-process the momentum is conserved whereas in the U-process it is not. The difference in energy flow in the two processes is shown in Fig. 9.24. It is the U-process that determines the value of the phonon mean path at low temperatures.

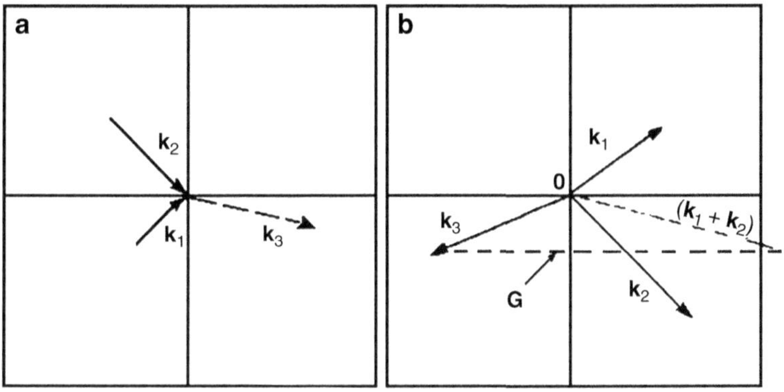

Fig. 9.23 Phonon interaction: (**a**) normal and (**b**) Umklapp processes

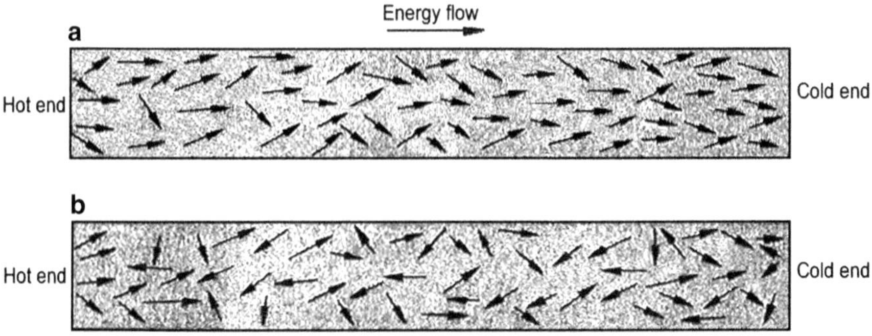

Fig. 9.24 Energy flow in (**a**) normal and (**b**) Umklapp processes. As we move from left to right, the net flux of phonons remains unchanged in (**a**) but is reduced in (**b**)

9.4.4.3 Temperature Variation of Thermal Conductivity

We shall now discuss the temperature variation of thermal conductivity K on the basis of (9.67). Since the sound velocity υ_s has a slow smooth dependence on temperature, the temperature dependence of K is determined mainly by C_V and Λ.

At high temperatures ($T >> \theta_D$), C_V is nearly constant. The phonon mean free path decreases with the number of times a given phonon can interact with other phonons. At high temperature, the number of excited phonons available for interaction is proportional to T (9.12). Hence, K varies as T^{-1}as observed experimentally (Fig. 9.22).

At low temperatures the value of Λ is determined mainly by the U-processes. Peierls [9.29] showed that in the range $\theta_D/20 < T < \theta_D/10$, the thermal conductivity is given by

$$K = C'(T/\theta_D)\exp(\theta_D/2T), \qquad (9.71)$$

where C' is a constant. Thus, K essentially increases exponentially as T decreases; the effect of the pre-exponential T-term is masked by the exponential term. This explains the observed trend at low temperatures (Fig. 9.22).

Equation (9.71) predicts $K \to \infty$ as $T \to 0$. But at very low temperatures $\sim \theta_D/20$, the effects of scattering of phonons by the boundaries of the sample and by defects become dominant making Λ nearly independent of temperature. Thus, at these low temperatures, K depends mainly on C_V which varies as T^3; hence K also varies as T^3 (Fig. 9.22). Further, since $C_V \to 0$ as $T \to 0$, so also $K \to 0$ as $T \to 0$.

The temperature variation has opposite trends above and below $\theta_D/20$. Hence, in between, a maximum in K is observed. Thus, all the observed features in Fig. 9.22 are explained.

9.5 Review of Concepts

Before closing this chapter, we shall briefly review some of the new concepts. Though these concepts were introduced mainly in the theories of thermal properties, in course of time, their influence has extended to solid state physics as a whole.

9.5.1 Phonons

Planck's idea of quantised electromagnetic radiation energy particles (photons) was extended by Einstein to the energy of atomic oscillators in solids. In Debye's theory, these oscillations are identified as elastic waves. These are called phonons. A phonon is described by a frequency ν, energy $h\nu$ and h/λ or $\mathbf{k}$ ($\mathbf{k}$ being the propagation vector). Two other quantities associated with the phonon are the mean energy $<\varepsilon>$ of an oscillator and $<n>$, the mean number of phonons per oscillator. These are given by

$$<\varepsilon>= \frac{h\nu}{e^{h\nu/k_BT}-1}, \tag{9.11}$$

and

$$<n>= \frac{<\varepsilon>}{h\nu} = \frac{1}{e^{h\nu/k_BT}-1}. \tag{9.12}$$

At high temperatures $<n>$ becomes proportional to T.

Debye pointed out that phonons have frequencies in the acoustic as well as optical range. At low temperatures, only acoustic phonons are excited; as the temperature increases, optical phonons are also excited. Phonons are created by increasing the temperature and are annihilated by decreasing it. When this happens, energy flows into or out of the crystal by heat conduction so that the total energy is conserved. Phonons behave like particles; they interact with other phonons, electrons, impurities and defects in crystals. Phonons are governed by Bose-Einstein statistics.

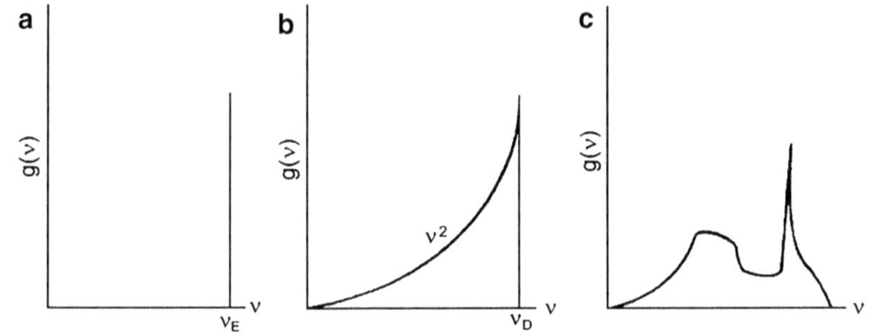

Fig. 9.25 Vibration spectra in different models: (**a**) Einstein, (**b**) Debye and (**c**) actual lattice

9.5.1.1 Vibration Spectrum

Debye introduced the idea of a vibration spectrum for a solid. The vibration spectrum is represented by a distribution function $g(\nu)$. Einstein assumed that all atoms in a solid vibrate with the same frequency. Thus, his 'vibration spectrum' consists of a single frequency. On the other hand, Debye's vibration spectrum is a parabola limited at a maximum frequency. It will be seen later that the actual vibration spectrum of a solid is much more complicated (Fig. 9.25) and differs from crystal to crystal.

9.5.1.2 Debye Temperature

Debye cut off the parabolic spectrum at a maximum frequency ν_D; this is expressed as a temperature by writing $\theta_D = (h\nu_D/k_B)$. Though the Debye temperature originally appeared in the Debye theory of specific heat, it was subsequently found to affect several other physical properties.

Debye assumed θ_D to be a constant independent of temperature but experiments show that it has a strong temperature dependence. This is considered as a drawback of the Debye model. However, the concept of the Debye model was not abandoned. Instead, more elaborate models were developed to account for the observed temperature variation of θ_D.

9.5.1.3 Anharmonicity of Vibrations

If the interatomic potential was a quadratic function of atomic displacements, the atomic vibrations would be harmonic. But terms higher than the quadratic term are present in the potential which result in the anharmonicity of vibrations.

Anharmonicity has a drastic effect on some properties. The thermal expansion of a harmonic crystal would be zero; anharmonicity makes it non-zero. In a harmonic crystal, elastic waves (phonons) would travel without interaction (scattering); this would make the thermal conductivity infinite. But due to anharmonicity, the elastic waves interact in such a way (Umklapp process) that thermal conductivity remains finite. Anharmonicity makes the vibration frequencies volume dependent. This is responsible for the temperature dependence of Debye temperature and also for non-zero Gruneisen constant. Other effects of anharmonicity are inequality of adiabatic and isothermal elastic constants, the pressure variation and temperature variation of elastic constants.

Problems

1. The specific heat C_P of copper at 300 K is 5.87 cal/g-atom-deg. Calculate $C_V(\psi = 0.776 \times 10^{-12}\,\text{cm}^2/\text{dyn}, \beta = 49.21 \times 10^{-6}$ per degree and $V = 7.6\,\text{cm}^3/\text{g-atom})$.
2. In deriving the expression for C_V from Debye's theory, we start with the assumption that the energy of a quantised oscillator is $nh\nu$. Rigorously speaking, the energy of an oscillator is $(n + 1/2)h\nu$. Starting with this value for the energy, rederive C_V. What is the effect?
3. Calculate θ_D for metal for which $C_V = 0.0015$ cal/mol-deg at $T = 4$ K. ($R = 1.987$ cal/mol-deg).
4. Evaluate the Gruneisen constant γ at room temperature for copper using the data given in problem 1.
5. The linear expansion for calcite is $\alpha_1 = \alpha_2 = -5.6 \times 10^{-6}$ per degree and $\alpha_3 = 25 \times 10^{-6}$ per degree. Calculate the thermal expansion in a direction lying in the x_3x_2 plane and making 64.7° with the x_3 axis.

References

9.1. L. Mackinnon, *Experimental Physics at Low Temperatures* (Wayne State University Press, Detroit, 1966).
9.2. K.A. Gschneidner, Jr., Solid State Phys. **16**, 275 (1964).
9.3. P.L. Dulong, A.T. Petit, Ann. Chim. Phys. **2**, 395 (1819).
9.4. F. Richarz, Ann. Phys. **48**, 708 (1899).
9.5. A. Einstein, Ann. Phys. **22**, 180 (1907).
9.6. W. Nernst, F.A. Lindemann, Z. Electrochem. **17**, 817 (1911).
9.7. A. Einstein, Ann. Phys. **31**, 679 (1911).
9.8. P. Debye, Ann. Phys. **39**, 789 (1912).
9.9. J. de Launay, Solid State Phys. **2**, 219 (1956).
9.10. F.C. Brown, *The Physics of Solids* (W.A. Benjamin Inc., New York, 1967).
9.11. A.J. Dekker, *Solid State Physics* (Macmillan Press, New York, 1981).
9.12. C. Kittel, *Introduction to Solid State Physics* (John Wiley, New York, 1996).
9.13. AIP *Handbook* (McGraw-Hill, New York, 1963).
9.14. E.S.R. Gopal, *Specific Heats at Low Temperatures* (Plenum Press, New York, 1966).
9.15. G.A. Alers, *Physical Acoustics*, **IIIB** (Academic Press, New York, 1965).

9.16. M. Blackman, *Handbook of Physics* **Vii/i** (Springer Verlag, Berlin, 1955).
9.17. H. Fizeau, Ann. Chim. Phys. **8**, 146 (1864).
9.18. F.D. Enck, J.G. Dommel, J. Appl. Phys. **36**, 839 (1965).
9.19. H.S. Peiser, H.P. Rooksby, A.J.C. Wilson, *X-ray Diffraction by Polycrystalline Materials* (Institute of Physics, London, 1955).
9.20. D.B. Sirdeshmukh, L. Sirdeshmukh, K.G. Subhadra, *Micro- and Macro-Properties of Solids* (Springer, Heidelberg, 2006).
9.21. R.S. Krishnan, *Progress in Crystal Physics* (S. Viswanathan, Madras, 1958).
9.22. R.A. Levy, *Principles of Solid State Physics* (Academic Press, New York, 1968).
9.23. E. Gruneisen, *Handbuch der Phys.* Vol. **10** (Springer, Berlin, 1926).
9.24. J.K. Roberts, *Heat and Thermodynamics* (Blackie and Sons, Glasgow, 1960).
9.25. D.C. Wallace, *Thermodynamics of Crystals* (Wiley, New York, 1972).
9.26. E. Gruneisen, English Translation of [9.23], (NASA, Washington, 1959).
9.27. G.A. Slack, Phys. Rev. **105**, 832 (1957).
9.28. P. Debye, *Vortrage liber die Kinetic Theory de Materie und der Electrizitat* (Tuebner, Berlin, 1914).
9.29. R. Peierls, Ann. Phys. **3**, 1055 (1929).
9.30. R. Peierls, *Quantum Theory of Solids* (Oxford University Press, New York, 1955).
9.31. H.B. Casimir, Physica, **5**, 495 (1938).
9.32. P.G. Klemens, Proc. R. Soc. Lond. **A208**, 108 (1951).
9.33. R. Berman, Proc. R. Soc. (Lond.), **A208,** 90 (1951).
9.34. J.S. Blakemore, *Solid State Physics* (W.B. Saunders Co., London, 1969).

Chapter 10
Lattice Vibrations

10.1 Introduction

While discussing the thermal properties of solids, it was pointed out that Debye's theory treats all solids alike and ignores the differences in structure and interatomic forces. At about the same time as Debye, another theory of lattice vibrations was proposed by Born and Von Karman [10.1] that does take into account the structure and forces in solids. This theory was further developed by Born [10.2, 10.3] (Box 10.1) and Blackman [10.4, 10.5].

We start with the Born–Von Karman treatment of the vibrations of the linear chain lattice. This is discussed in detail as it provides an insight into the physics of lattice vibrations and also introduces us to new concepts. Two- and three-dimensional lattices are also discussed. The most important parameter in the evaluation of the vibration spectrum is the force constant. Theoretical and experimental methods of determining the force constants are discussed. The experimental determination of the vibration spectrum by the technique of neutron inelastic scattering is also described.

Although, historically, theories of lattice vibrations were associated with the specific heat of solids, in due course it was realised that lattice vibrations influence several physical properties of solids as indicated in Fig. 10.1.

Box 10.1

Max Born, a German physicist was a pioneer in the development of quantum mechanics. He is also known for his contribution in the field of solid state physics and optics. Born laid the foundation for the theory of ionic solids and lattice dynamics. He was awarded the Nobel Prize for physics in 1954.

Max Born (1882–1970)

D.B. Sirdeshmukh et al., *Atomistic Properties of Solids*, Springer Series in Materials Science 147, DOI 10.1007/978-3-642-19971-4_10,

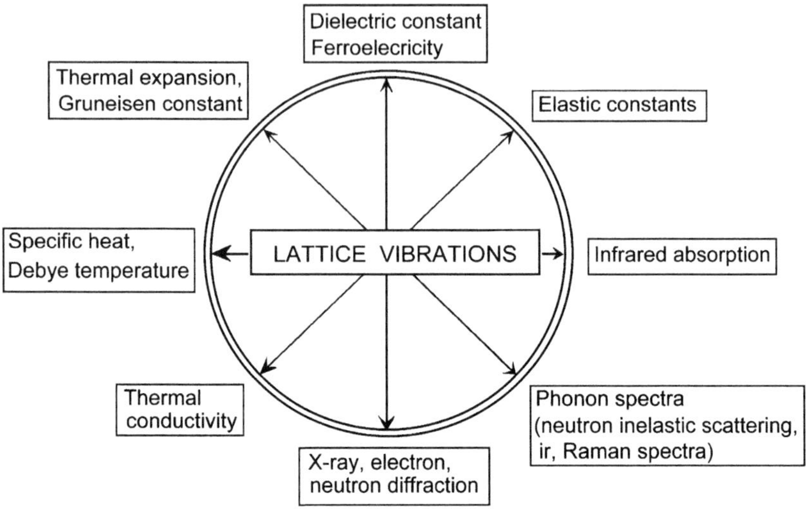

Fig. 10.1 Interrelations between lattice vibrations and physical properties

The problem of vibrations of monatomic and diatomic linear chain lattices is discussed in most text books on solid state physics. Vibrations of two- and three-dimensional lattices are discussed by Born and Huang [10.3], Blackman [10.6], De Launay [10.7], Donovan and Angress [10.8] and Ghatak and Kothari [10.9].

10.2 Vibrations of Linear Lattices

We shall consider the vibrations of a variety of linear lattices. We shall see that even in such an oversimplified model, the vibration spectrum depends on the "structure" and interatomic forces.

10.2.1 Simple Monatomic Chain

Let us consider a simple, infinite, monatomic chain (Fig. 10.2). The length of the chain is along the x-direction. The atoms have mass M and an identical spacing a. There are restoring forces only between neighbouring atoms. These are Hooke's law forces described by a force constant α. As shown in the figure, three consecutive atoms are indexed as $n-1, n$ and $n+1$ respectively. We shall consider longitudinal vibrations. We shall denote the displacements of these atoms in the x-direction as u_{n-1}, u_n and u_{n+1}. Then, the equation of motion for the nth atom is

$$\begin{aligned} M\,\ddot{u}_n &= -\alpha\,[u_n - u_{n-1}] - \alpha\,[u_n - u_{n+1}] \\ &= \alpha\,[u_{n+1} + u_{n-1} - 2u_n]. \end{aligned} \tag{10.1}$$

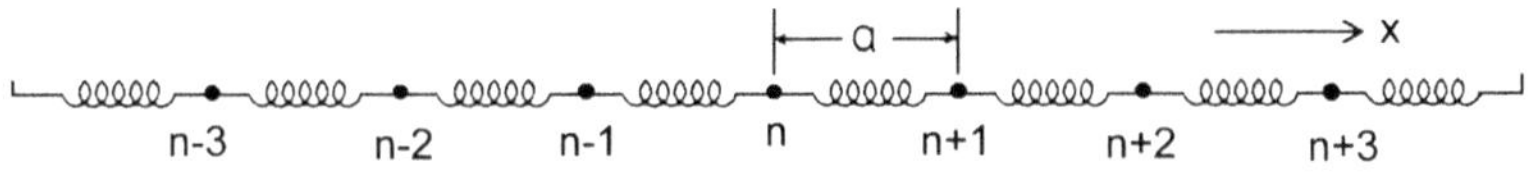

Fig. 10.2 An infinite simple monatomic linear lattice

There will be one such equation for each atom in the chain.

We shall assume the following running wave solution:

$$u_n = A\,\mathrm{e}^{\mathrm{i}(\omega t - kx_n)}, \tag{10.2}$$

where A is the amplitude, ω the angular frequency (equal to 2π times the frequency ν) and $k(= 2\pi/\lambda)$ is the magnitude of the propagation vector $\mathbf{k}$. Since there is no matter between the particles, x_n takes values na. Thus, (10.2) becomes

$$u_n = A\,e^{\mathrm{i}(\omega t - nka)}. \tag{10.3}$$

Substituting in (10.1), we get

$$-M\omega^2 = \alpha\,[\mathrm{e}^{ika} + \mathrm{e}^{-ika} - 2]$$

or

$$\omega_{\mathrm{L}} = \omega_{0\mathrm{L}} \left|\sin\frac{ka}{2}\right|, \tag{10.4}$$

with

$$\omega_{0\mathrm{L}} = (4\alpha/M)^{1/2}. \tag{10.5}$$

In (10.4), we consider only the absolute value of $\sin(ka/2)$ since ω is always positive irrespective of whether the running wave travels in the positive or negative direction.

We have so far considered longitudinal vibrations. Transverse vibrations can also be treated in the same way [10.9]. The corresponding dispersion relation will be

$$\omega_{\mathrm{T}} = \omega_{0\mathrm{T}} \left|\sin\frac{ka}{2}\right|, \tag{10.6}$$

where

$$\omega_{0\mathrm{T}} = (4\beta/M)^{1/2}. \tag{10.7}$$

Here, β is the force constant for transverse displacements and $\omega_{0\mathrm{T}}$ the corresponding maximum frequency at $k = \pi/a$. In general, a linear monatomic lattice will have one longitudinal branch and two degenerate transverse branches related to two mutually perpendicular vibrations.

10.2.1.1 Dispersion Relations

We shall note some important features of (10.4–10.7). In Debye's theory, the frequency is a linear function of k, i.e. the medium was dispersionless. Equations

Fig. 10.3 Dispersion relation for a one-dimensional chain

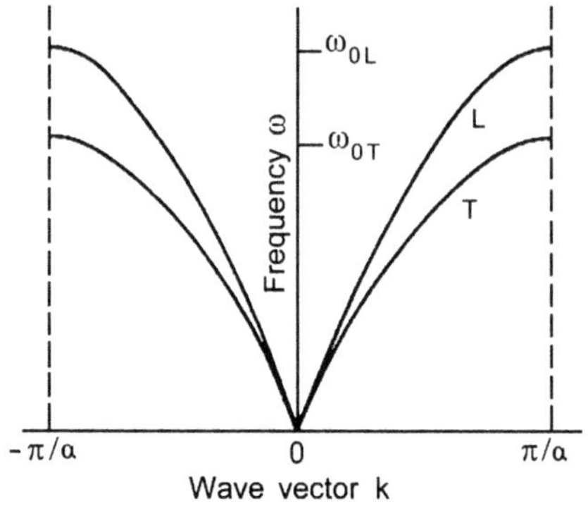

(10.4, 10.6) are nonlinear relations between ω and k. Thus, there is dispersion. Relations like (10.4, 10.6) connecting the frequency with k are called *dispersion relations*.

It can also be seen that the maximum value of ω is ω_{0L} and it occurs at $k = \pm\pi/a$. The same is true about ω_{0T}. The values $-\pi/a$ and $+\pi/a$ of k define the limits of what is called the first *Brillouin zone*. The values $k = \pm\pi/a$ to $\pm 2\pi/a$ define the second Brillouin zone and so on. If we replace any k value by $k + (2\pi m/a)$ where $m = \pm 1, \pm 2, \ldots$, the value of ω remains the same. Thus, it is enough to consider the values of ω only in the first Brillouin zone. The dispersion relations for this lattice are shown in Fig. 10.3.

10.2.1.2 Finite Lattice

Let us consider a finite lattice (Fig. 10.4). It is similar to the lattice in Fig. 10.2 with two differences. Firstly, the end atoms are fixed. Secondly, we have adopted a more convenient indexing scheme. Let the total number of atoms be $(N+1)$ and the length of the chain be L. We index the atoms as 0, 1, 2, ..., N. Since the lattice spacing is "a", it follows that $L = Na$. The displacement of any atom is the resultant of the displacements due to two running waves in opposite directions. Hence, for the nth atom,

$$u_n = A_1\, e^{i\omega t}\, e^{inka} + A_2\, e^{i\omega t}\, e^{-inka} \tag{10.8}$$

Now, $u_n = 0$ for $n = 0$; this is possible if $A_1 = -A_2$. Then, (10.8) becomes

$$u_n = 2iA_1\, e^{i\omega t} \sin nka. \tag{10.9}$$

This represents a standing wave and k now takes only positive values from 0 to π/a; thus the k values are confined to half the first Brillouin zone.

Again, $u_n = 0$ for $n = N$, i.e. for $x = Na$. Now (10.9) becomes

$$\sin Nka = 0. \tag{10.10}$$

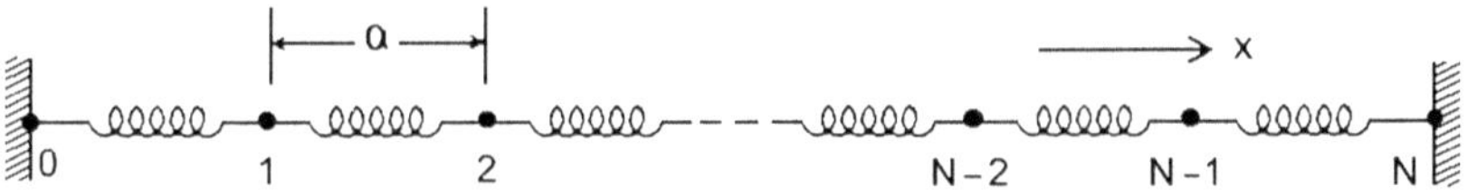

Fig. 10.4 A one-dimensional chain consisting of $N + 1$ identical particles with the two end particles fixed

This is possible if

$$k = J(\pi/Na) = J(\pi/L), \tag{10.11}$$

where $J = 1, 2, \ldots, (N - 1)$; the values $J = 0$ and N are precluded since these values correspond to all atoms at rest. The dispersion relation (10.4) now becomes

$$\omega_{\mathrm{L}} = \omega_{0\mathrm{L}} \left|\sin J(\pi/2N)\right|. \tag{10.12}$$

Thus, while k took continuous values in the range $-\pi/a < k < \pi/a$ in the case of the infinite lattice, it now takes discrete values in the range $0 < k < \pi/a$. In other words, the vibration spectrum now consists of closely spaced lines (frequencies). Also, it may be noted that, while the total number of atoms is $(N + 1)$, the total number of modes from (10.11) is $(N - 1)$, i.e. the number of moveable atoms.

10.2.1.3 Frequency Distribution Function

In Chap. 9, we defined the frequency distribution function $g(\nu)$ as the number of modes or frequencies in unit range of frequencies, i.e. the number of frequencies in the range ν to $\nu + d\nu$ is $g(\nu)\mathrm{d}\,\nu$. If we define $f(k)$ as the density of allowed k-values in the k-space,

$$g(\nu)\mathrm{d}\,\nu = f(k)\mathrm{d}\,k. \tag{10.13}$$

Since the values are equally spaced,

$$f(k) = \frac{\text{Total no. of modes}}{\text{Range in k-space}} = \frac{(N-1)}{(\pi/a)}. \tag{10.14}$$

From (10.13) and (10.14), we get

$$g(\nu) = \frac{(N-1)}{(\pi/a)}(\mathrm{d}\,k/\mathrm{d}\,\nu). \tag{10.15}$$

Substituting $(\mathrm{d}\,k/\mathrm{d}\,\nu)$ from (10.4) in (10.15), we get

$$g(\nu) = \left(\frac{2}{\pi}\right)\frac{(N-1)}{(\nu_{0L}^2 - \nu^2)^{1/2}}, \tag{10.16}$$

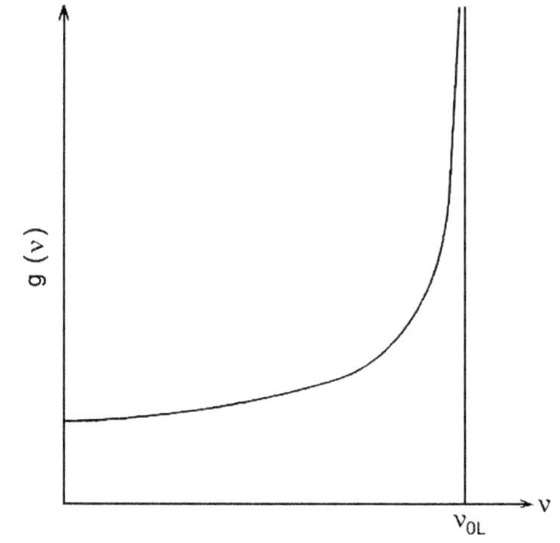

Fig. 10.5 The frequency distribution function for a one-dimensional chain

where $\nu_{\mathrm{0L}} = \omega_{\mathrm{0L}}/2\pi$. A plot of $g(\nu)$ versus ν according to (10.16) is shown in Fig. 10.5. It may be noted that $g(\nu) \to \infty$ as $\nu \to \nu_{\mathrm{0L}}$, i.e. there is a singularity at $\nu = \nu_{\mathrm{0L}}$. There will be a similar curve for the transverse vibrations.

10.2.1.4 Specific Heat

The specific heat is calculated from the following general equation

$$C_V = \int_0^{\nu_{\max}} E(\nu)g(\nu)\mathrm{d}\,\nu, \tag{10.17}$$

where $E(\nu)$ is the Einstein function given by

$$E(\nu) = k_{\mathrm{B}}\xi^2\mathrm{e}^{\xi}\,/(\mathrm{e}^{\xi}-1)^2\,. \tag{10.18}$$

Here $\xi = h\nu/k_{\mathrm{B}}T$ and $\nu_{\max}$ is the frequency at the Brillouin zone boundary.

We have noted that for this lattice the $g(\nu)$ curve has a singularity at $\nu = \nu_{\max}$. This creates a difficulty in the evaluation of the integral. This difficulty is overcome by arbitrarily assigning a finite value to $g(\nu)$ at $\nu_{\max}$. This finite value is chosen by normalising the total number of modes $\int_0^{\nu_{\max}} g(\nu)\mathrm{d}\,\nu$ with the Debye exprcssion. It can be shown [4.10] that

$$C_V = \frac{k_{\mathrm{B}}T}{a\theta_L}\int_0^{\theta_L/T}\frac{\xi^2\mathrm{e}^{\xi}}{(\mathrm{e}^{\xi}-1)^2}\mathrm{d}\,\xi, \tag{10.19}$$

where θ_L is a parameter analogous to the Debye temperature; its value depends on the arbitrarily chosen value of $g(\nu)$. The interesting point to note is that at very low temperatures, the integral in (10.19) becomes approximately equal to $(\pi^2/3)$ and, hence, C_V shows a T dependence (unlike the T^3 dependence in Debye's theory).

This treatment of the specific heat of the linear chain lattice is not entirely unrealistic. Some crystals, e.g. Se and Te do have a chain-like structure and their experimental C_V values do show a T dependence at low temperatures.

10.2.2 Linear Monatomic Lattice with a Basis

In the simple lattice considered in Sect. 10.2.1, there was one atom per unit cell. The coordinates of these atoms were $x = na$, a being the lattice constant. We shall now consider a linear chain lattice with a basis (Fig. 10.6a). All atoms are of the same mass M and the length of the chain is in the x direction. There are two atoms per unit cell. The coordinates of one set of atoms are given by $x = na$ whereas the coordinates of the other set are $x = na + b$; b is the magnitude of the basis vector **b**. If we consider an atom of the first type, its neighbours are at distances b and $(a - b)$.

We shall index an atom of the first type as 0 and its neighbours as 1 and 2 (Fig. 10.6b). For convenience, we shall index an atom of the second type as $0'$ and its neighbours as $1'$ and $2'$. For clarity, the coordinates and displacements of the atoms are given in Table 10.1.

Assuming only central forces and nearest-neighbour interactions, we shall denote the force constant between atoms 0 and 1 (and also $0'$ and $2'$) by α_1 and that between 0 and 2 (and also $0'$ and $1'$) by α_2. The equations of motion for atoms 0 and $0'$ are, respectively,

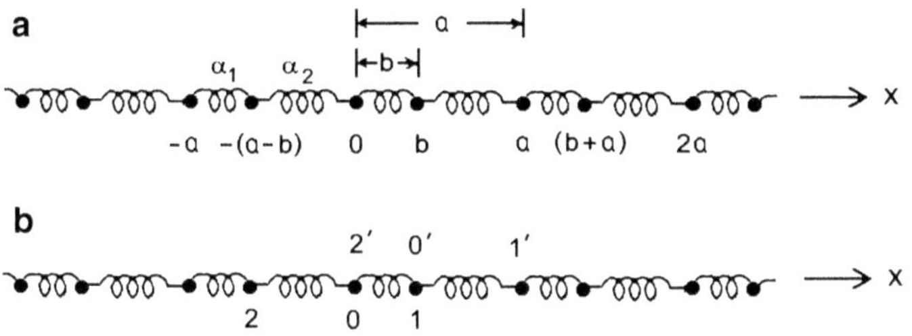

Fig. 10.6 (**a**) A one-dimensional lattice with a basis; (**b**) the scheme of indexing the pertinent atoms in the one-dimensional lattice with a basis

Table 10.1 Coordinates of the indexed atoms

Index	Coordinate
0	0
1	b
2	$-(a-b)$
$0'$	b
$1'$	a
$2'$	0

$$\begin{aligned} M\,\ddot{u}_0 &= -\alpha_1(u_0 - u_1) - \alpha_2(u_0 - u_2), \\ M\,\ddot{u}_{0'} &= -\alpha_1(u_{0'} - u_{2'}) - \alpha_2(u_{0'} - u_{1'}). \end{aligned} \tag{10.20}$$

We shall assume two running wave solutions (10.2) but with amplitudes A and B for the waves passing through the two types of atoms. These solutions give

$$\begin{aligned} u_0 &= A\,\mathrm{e}^{\mathrm{i}\omega t} \\ u_1 &= B\,\mathrm{e}^{\mathrm{i}(\omega t - kb)} \\ u_2 &= B\,\mathrm{e}^{\mathrm{i}[\omega t + k(a-b)]} \\ u_{0'} &= B\,\mathrm{e}^{\mathrm{i}(\omega t - kb)} = u_1 \\ u_{1'} &= A\,\mathrm{e}^{\mathrm{i}(\omega t - ka)} \\ u_{2'} &= A\,\mathrm{e}^{-\mathrm{i}\omega t} = u_0. \end{aligned} \tag{10.21}$$

Substituting these solutions in (10.20), we get

$$\begin{aligned} (\alpha_1 + \alpha_2 - M\omega^2)A - [\alpha_1\,\mathrm{e}^{-\mathrm{i}kb} + \alpha_2\,\mathrm{e}^{\mathrm{i}k(a-b)}]B &= 0 \\ -[\alpha_1\,\mathrm{e}^{\mathrm{i}kb} + \alpha_2\,\mathrm{e}^{-\mathrm{i}k(a-b)}]A + (\alpha_1 + \alpha_2 - M\omega^2)B &= 0. \end{aligned} \tag{10.22}$$

These equations will have non-trivial solutions only if the determinant of the coefficients of A and B vanishes, i.e.

$$\begin{vmatrix} (\alpha_1 + \alpha_2 - M\omega^2) & -[\alpha_1\mathrm{e}^{-\mathrm{i}kb} + \alpha_2\mathrm{e}^{\mathrm{i}k(a-b)}] \\ -[\alpha_1\mathrm{e}^{\mathrm{i}kb} + \alpha_2\mathrm{e}^{-\mathrm{i}k(a-b)}] & (\alpha_1 + \alpha_2 - M\omega^2) \end{vmatrix} = 0. \tag{10.23}$$

Such an equation is called a “secular equation”. Equation (10.23) expands into a quadratic equation in ω^2. The roots of this quadratic equation are:

$$M\omega^2 = \alpha_1 + \alpha_2 \pm \sqrt{(\alpha_1 + \alpha_2)^2 - 4\alpha_1\alpha_2\sin^2(ka/2)}. \tag{10.24}$$

Equation (10.24) constitutes the dispersion relations for this lattice. These two roots yield two sets of frequencies shown in Fig. 10.7. The upper branch obtained with the positive radical is called the “optic” branch and the lower one obtained with the negative radical is called the “acoustic” branch.

At very small values of k, i.e. for long waves, (10.24) becomes

$$(M\omega^2)_{\text{optic}} = 2(\alpha_1 + \alpha_2) - \frac{2\alpha_1\alpha_2}{(\alpha_1 + \alpha_2)}\left(\frac{ka}{2}\right)^2$$

and

$$(M\omega^2)_{\text{acoustic}} = \frac{2\alpha_1\alpha_2}{(\alpha_1 + \alpha_2)}\left(\frac{ka}{2}\right)^2. \tag{10.25}$$

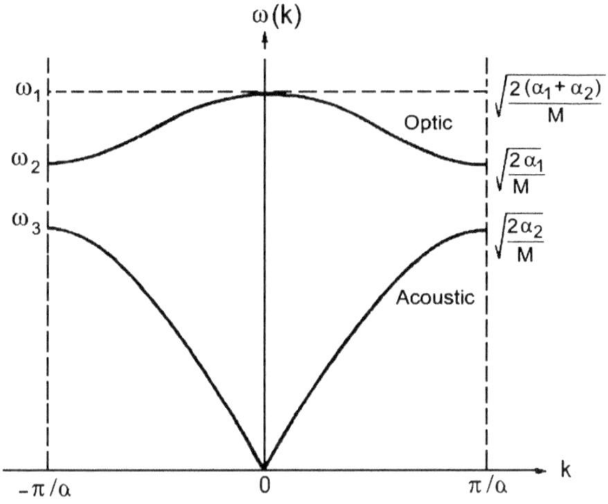

Fig. 10.7 Dispersion relation for a lattice with a basis

For $k = 0$, the limiting frequencies are

$$(\nu_{\text{optic}})_{k=0} = \frac{1}{2\pi}\sqrt{\frac{2(\alpha_1 + \alpha_2)}{M}} = \nu_1$$

and

$$(\nu_{\text{acoustic}})_{k=0} = 0. \tag{10.26}$$

The limiting frequencies at the end of the Brillouin zone are obtained by substituting $k = \pi/a$ in (10.24). These are

$$(\nu_{\text{optic}})_{k=\pi/a} = \frac{1}{2\pi}\sqrt{\frac{2\alpha_1}{M}} = \nu_2$$

and

$$(\nu_{\text{acoustic}})_{k=\pi/a} = \frac{1}{2\pi}\sqrt{\frac{2\alpha_2}{M}} = \nu_3. \tag{10.27}$$

These limiting frequencies are shown in Fig. 10.7. The lattice does not sustain waves with frequencies in the range $(1/2\pi)\sqrt{2\alpha_2/M}$ and $(1/2\pi)\sqrt{2\alpha_1/M}$; this range is called the forbidden gap.

The frequency distribution function is evaluated according to the general principle discussed in Sect. 10.2.1 viz.

$$g(\nu) = f(k)(\mathrm{d}k/\mathrm{d}\nu).$$

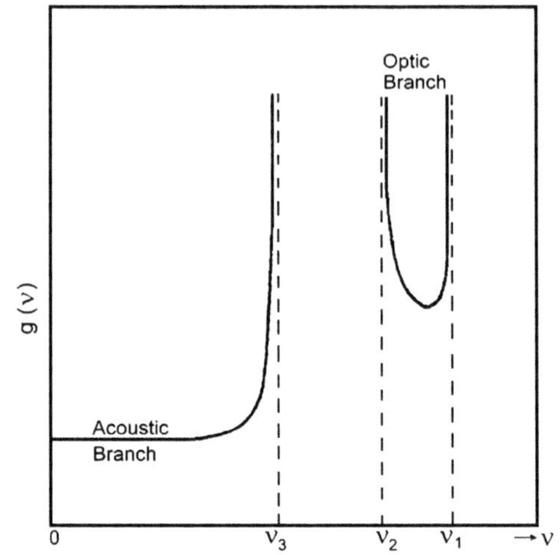

Fig. 10.8 The two branches of the distribution function corresponding to the two branches shown in Fig. 10.7 for the one-dimensional lattice with a basis

In the present case, we write

$$g(\nu) = f(k)\left[\left(\frac{\mathrm{d}k}{\mathrm{d}\nu}\right)_{\text{acoustic}} + \left(\frac{\mathrm{d}k}{\mathrm{d}\nu}\right)_{\text{optic}}\right]. \tag{10.28}$$

The actual evaluation of (10.28) from (10.24) is complicated and the result is not in closed form; we have merely reproduced the result in Fig. 10.8. We have considered only the longitudinal vibrations; there are transverse vibrations also. Thus there will be longitudinal optic and acoustic branches (LO, LA) and also transverse branches (TO, TA), the latter being doubly degenerate.

10.2.3 A Linear Diatomic Lattice

We shall now consider a linear diatomic lattice (Fig. 10.9). This is also a lattice with a basis with two atoms per unit cell but the two atoms have different masses M and $m(M > m)$. The atoms with mass m are located midway between the atoms with mass M. The lattice constant is a and the interatomic distance is $d(=a/2)$. The atoms with mass M have coordinates $x = 2nd$ and those with mass m have coordinates $x = (2n+1)d$. We shall consider longitudinal vibrations. Only central forces and nearest-neighbour interactions are assumed. The force constant between two neighbouring atoms is α. Denoting the displacements of the two types of atoms by u_{2n} and u_{2n+1}, the equations of motion are:

$$\begin{aligned} M\,\ddot{u}_{2n} &= \alpha[u_{2n+1} + u_{2n-1} - 2u_{2n}], \\ m\,\ddot{u}_{2n+1} &= \alpha[u_{2n} + u_{2n+2} - 2u_{2n+1}]. \end{aligned} \tag{10.29}$$

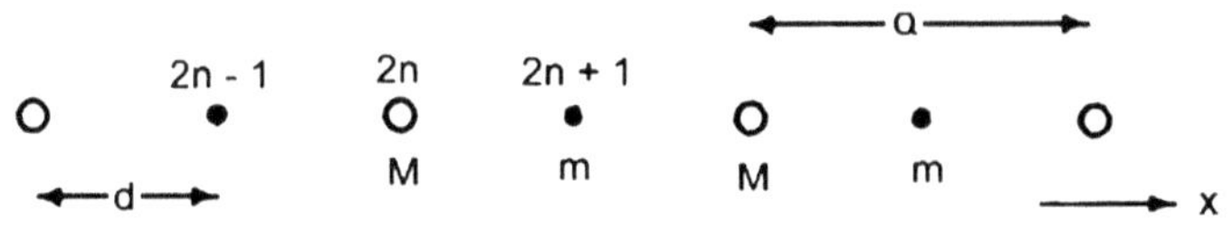

Fig. 10.9 A linear diatomic lattice

We assume the following running wave solutions:

$$u_{2n} = A\,\mathrm{e}^{\mathrm{i}[\omega t - 2nkd]},$$
$$u_{2n+1} = B\,\mathrm{e}^{\mathrm{i}[\omega t - (2n+1)kd]}. \tag{10.30}$$

where A and B are the amplitudes of the two waves. Substituting (10.30) in (10.29) and simplifying, we get

$$[-M\omega^2 + 2\alpha]A - [\mathrm{e}^{\mathrm{i}kd} + \mathrm{e}^{-\mathrm{i}kd}]\alpha B = 0,$$
$$-[\mathrm{e}^{\mathrm{i}kd} + \mathrm{e}^{-\mathrm{i}kd}]\alpha A + [-m\omega^2 + 2\alpha]B = 0. \tag{10.31}$$

These equations will have solutions only if the determinant of the coefficients of the amplitudes A, B vanishes, i.e.

$$\begin{vmatrix} (-M\omega^2 + 2\alpha) & -(\mathrm{e}^{\mathrm{i}kd} + \mathrm{e}^{-\mathrm{i}kd})\alpha \\ -(\mathrm{e}^{\mathrm{i}kd} + \mathrm{e}^{-\mathrm{i}kd})\alpha & (-m\omega^2 + 2\alpha) \end{vmatrix} = 0. \tag{10.32}$$

Expanding the determinant and substituting $2\cos kd$ for $(\mathrm{e}^{\mathrm{i}kd} + \mathrm{e}^{-\mathrm{i}kd})$, we get

$$[-M\omega^2 + 2\alpha][-m\omega^2 + 2\alpha] + 4\alpha^2\cos^2 kd = 0. \tag{10.33}$$

The two roots of (10.33) are

$$\omega_{\pm}^2 = \alpha\left[\left(\frac{1}{m} + \frac{1}{M}\right) \pm \sqrt{\left(\frac{1}{m} + \frac{1}{M}\right)^2 - \frac{4}{mM}\sin^2 kd}\,\right]. \tag{10.34}$$

Equation (10.34) relates the frequencies to the propagation vector and hence is the dispersion relation for this lattice. As can be seen, ω_+ and ω_- represent two branches of frequencies called the optic and acoustic branches. These two branches are shown in Fig. 10.10. We shall now discuss several aspects of (10.34).

1. In (10.34), if k is replaced by

$$k_j = k + j\frac{\pi}{d} \qquad (j = \pm 1, \pm 2, \ldots) \tag{10.35}$$

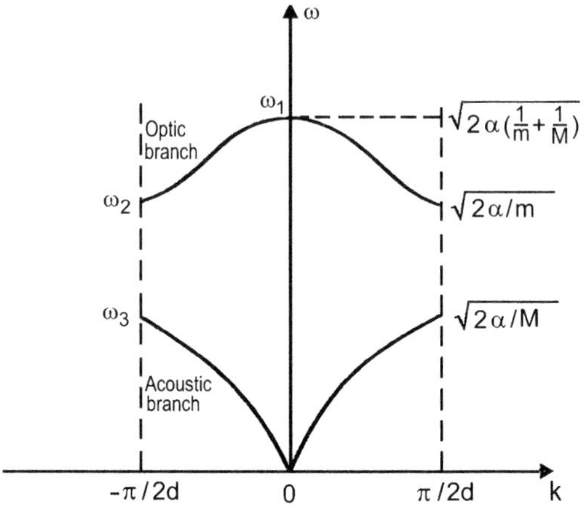

Fig. 10.10 The dispersion relation for a diatomic chain with $M > m$

the dispersion relation is unaffected. Thus the total range of k is π/d or, in other words, the boundaries of the first Brillouin zone are defined by the interatomic distance d and not the lattice constant a. Also, the width of the first Brillouin zone is half of what it would be if all atoms are identical.

2. For small values of k, (10.34) approximates to

$$\omega_+ = \sqrt{2\alpha\left(\frac{1}{m}+\frac{1}{M}\right)\left[1-\frac{mM}{(m+M)^2}k^2d^2\right]}$$

and

$$\omega_- = \sqrt{\frac{2\alpha d^2}{(m+M)}}k. \qquad (10.36)$$

Thus, at small k (i.e. long waves) the optic branch frequencies decrease parabolically with k; on the other hand, the acoustic branch frequencies increase linearly (Fig. 10.10). Further, the limiting frequencies at $k = 0$ are $\omega_1 = \sqrt{2\alpha\left(\frac{1}{m}+\frac{1}{M}\right)}$ for the optic branch and 0 for the acoustic branch.

3. At the Brillouin zone boundary, i.e. at $k = \pi/2d$, the limiting frequencies are $\omega_2 = \sqrt{2\alpha/m}$ for the optic branch and $\omega_3 = \sqrt{2\alpha/M}$ for the acoustic branch. The lattice does not sustain frequencies in the range ω_2 to ω_3; this is called the forbidden gap. The width of the forbidden gap depends on the difference between masses m and M; the gap closes for $m = M$.
4. Let us consider the relative amplitudes of the acoustic and optic vibrations. From (10.31), we get

$$\frac{A}{B} = \frac{2\alpha - m\omega^2}{[\mathrm{e}^{ikd}+\mathrm{e}^{-ikd}]\alpha} = \frac{2\alpha - m\omega^2}{2\alpha\cos kd}. \qquad (10.37)$$

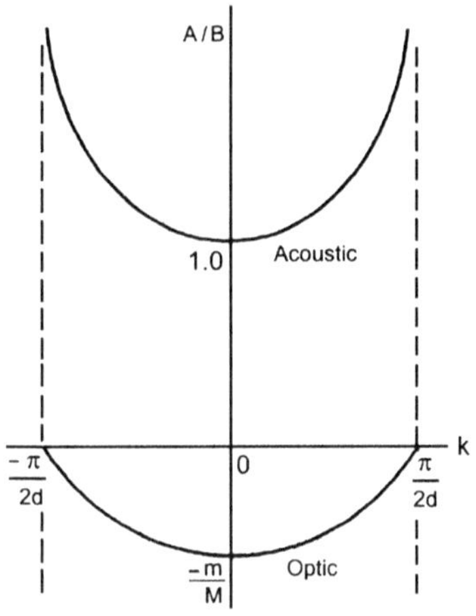

Fig. 10.11 Ratio of the amplitudes of vibration A/B as a function of the wave vector for acoustical and optical modes

Substituting for ω^2 from (10.34), we get

$$\frac{A}{B} = \left(\frac{1}{2M\cos kd}\right)\left[(M-m) \mp \sqrt{M^2 + m^2 + 2Mm\cos 2kd}\right]. \qquad (10.38)$$

The variation of (A/B) with k is shown in Fig. 10.11. From (10.38) and Fig. 10.11, we may plot the relative displacements for some specific modes (Fig. 10.12); for easy appreciation, the displacements are shown in the transverse direction. In the ω_2 and ω_3 modes only one set of atoms (m or M) vibrates and the other is at rest. The ω_1 mode is interesting; in this mode the two types of atoms are vibrating "against" each other, i.e. they are out of phase by π. These represent the internal vibrations of the chain; they play an important role in infrared absorption.

5. The distribution function $g(\nu)$ is calculated from the general formula $g(\nu) = f(k)(\mathrm{d}k/\mathrm{d}\nu)$, where $f(k)$ is the density of propagation vectors. As in the case of the monatomic chain with a basis, here also, $g(\nu)$ cannot be obtained in a closed form; it is shown in Fig. 10.13. It has singularities at ν_1, ν_2 and ν_3.
   One may notice a qualitative similarity between the dispersion relations (Figs. 10.7 and 10.10) and the distribution functions (Figs. 10.8 and 10.13) for the monatomic chain with a basis and the diatomic chain. It needs to be noted that the values of the frequencies are very different. The difference between the frequencies ν_1 and ν_2 is caused in the former by the difference in the force constants α_1 and α_2 and in the latter by the difference in the masses m and M.
6. We have considered longitudinal vibrations. Transverse vibrations can be treated in a similar way and would lead to similar results but with a different force constant. Thus the lattice will have two longitudinal branches (LO and LA) and four (effectively two doubly degenerate) transverse (TO, TA) branches.
7. The specific heat C_V is calculated from the dispersion relation using the general expression (10.17). Since there are two branches in the present case, the expression becomes

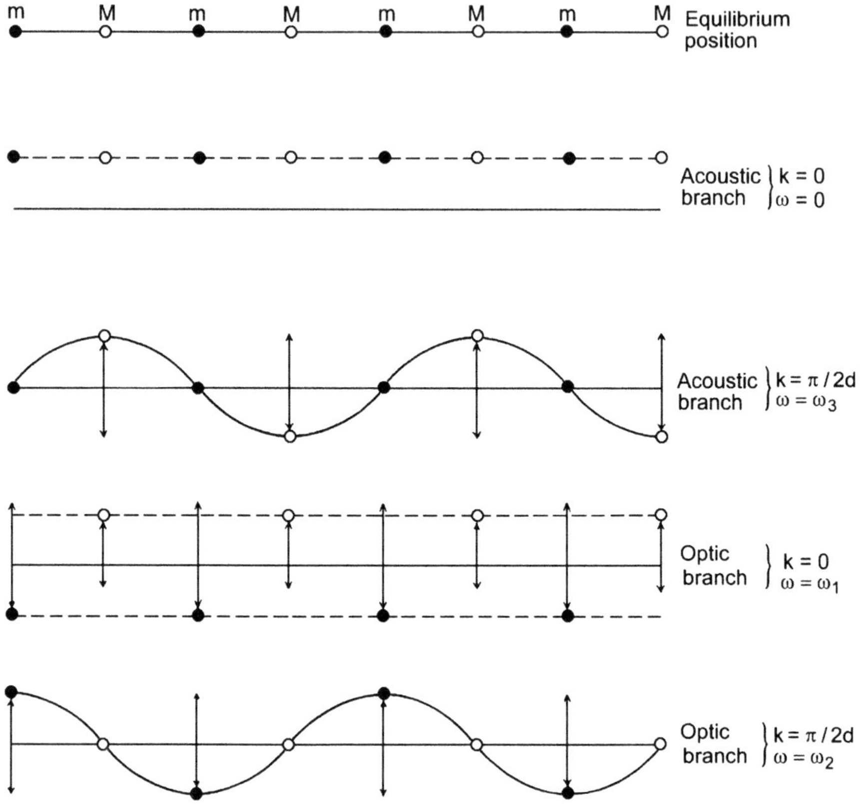

Fig. 10.12 Transverse displacements of the chain for the acoustical and optical branches corresponding to $k = 0$ and $k = \pi/2d$

$$C_V = \int E(\nu)\left[\left(f(k)\frac{\mathrm{d}k}{\mathrm{d}\nu}\right)_{\text{acoustic}} + \left(f(k)\frac{\mathrm{d}k}{\mathrm{d}\nu}\right)_{\text{optic}}\right]\mathrm{d}\nu, \tag{10.39}$$

where $E(\nu)$ is the Einstein function. As seen in Fig. 10.13, the distribution function has singularities at ν_1, ν_2 and ν_3. The procedure of overcoming this difficulty was discussed in Sect. 10.2.1. Following the same procedure, the specific heat is obtained [10.2] as

$$C_V = 2L\,k_\mathrm{B}(T/\theta_\mathrm{D})\int_0^{\theta_\mathrm{D}/T}\frac{\xi^2 \mathrm{e}^{\xi}}{(\mathrm{e}^{\xi}-1)^2}\mathrm{d}\,\xi, \tag{10.40}$$

where L is the number of unit cells in the length of the chain and θ_D is a parameter analogous to the Debye temperature (Chap. 9). Again, at very low temperatures, the integral equals a constant and C_V shows a T dependence.

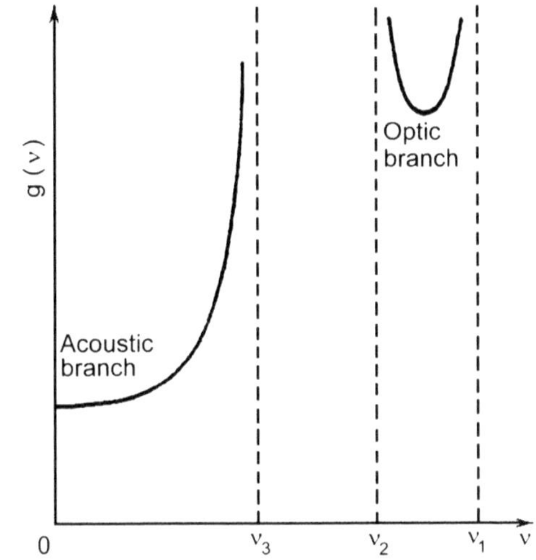

Fig. 10.13 The frequency distribution function for a diatomic lattice

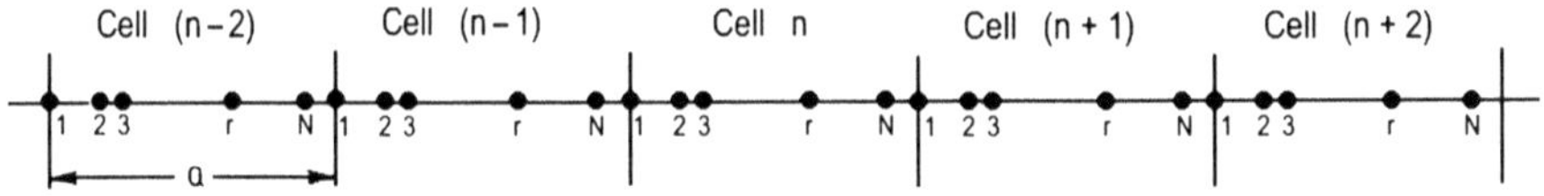

Fig. 10.14 A row of polyatomic molecules

10.2.4 Polyatomic Linear Chain

Let us consider the lattice shown in Fig. 10.14. Each unit cell contains N atoms. The unit cell is shown by vertical lines which mark the position of the first atom in the cell. The lattice constant is "a". We shall denote a cell by the index n and the atom in the cell by p. Thus the index n, r indicates the rth atom in the nth unit cell. We shall assume that an atom in a cell interacts with all atoms in another cell. Following Brillouin [4.11], we shall denote the Hooke's law force between atoms (n, r) and $(n + p, s)$ by $f_{n,r,n+p,s}$ and the corresponding force constant by α_{prs}. If the longitudinal displacements are represented by $u_{n,r}$, we have

$$f_{n,r;n+p,s} = \alpha_{prs}(u_{n+p,s} - u_{n,r}). \tag{10.41}$$

It may be noted that the force constant is independent of the cell index n. The total force $f_{n,r}$ acting on atom r in the nth cell is given by

$$f_{n,r} = \sum_p \sum_s \alpha_{prs}(u_{n+p,s} - u_{n,r}). \tag{10.42}$$

Let us assume a running wave solution

$$u_{n,r} = A_r \mathrm{e}^{\mathrm{i}(\omega t - kx)}. \tag{10.43}$$

The distance (x) of the origin of the cell from the origin of the lattice is given by $x = na$. Thus (10.43) becomes

$$u_{n,r} = A_r \mathrm{e}^{\mathrm{i}(\omega t - nka)}. \tag{10.44}$$

Substituting (10.44) in (10.42), we get

$$f_{n,r} = \mathrm{e}^{\mathrm{i}(\omega t - nka)} \sum_{p,s} \alpha_{prs}(A_s \mathrm{e}^{-\mathrm{i}kp} - A_r). \tag{10.45}$$

But

$$f_{n,r} = M_r \ddot{u}_{n,r} = -\omega^2 M_r A_r \mathrm{e}^{\mathrm{i}(\omega t - nka)}, \tag{10.46}$$

where M_r is the mass of the rth atom in a cell. From (10.45) and (10.46) we get

$$-\omega^2 M_r A_r = \sum_s D_{r,s}(k) A_s. \tag{10.47}$$

The function $D_{r,s}(k)$ is defined by

$$D_{r,s}(k) = \sum_p \alpha_{prs} \mathrm{e}^{-\mathrm{i}kp}, \quad r \neq s, \quad \text{and}$$

$$D_{r,r}(k) = -\sum_{ps} \alpha_{prs} + \sum_p \alpha_{prr} \mathrm{e}^{-\mathrm{i}kp}. \tag{10.48}$$

The sum over s is to be taken over all atoms in a cell and the sum over p over all cells. Equation (10.47) relates the frequency ω with the propagation vector k and is, hence, the dispersion relation for this lattice.

For the acoustic branch, at $k = 0$, we have $A_r = A_s$. Substitution in (10.47) gives $\omega = 0$. For other values of k, (10.47) may be written as

$$\sum_s \left\{D_{r,s}(k) + \omega^2 M_r \delta_{r,s}\right\} A_s = 0, \tag{10.49}$$

where $\delta_{r,s}$ is Kronecker delta. Equation (10.49) is a set of N equations in A_s. It has solutions if the determinant of the coefficients of A_s vanishes, i.e.

$$\left|D_{r,s}(k) + \omega^2 M_r \delta_{r,s}\right| = 0. \tag{10.50}$$

The expansion of (10.50) is a N *th* degree equation in ω^2. There will be N values for ω^2 for each value of k. In other words, the ν versus k curve will have N branches.

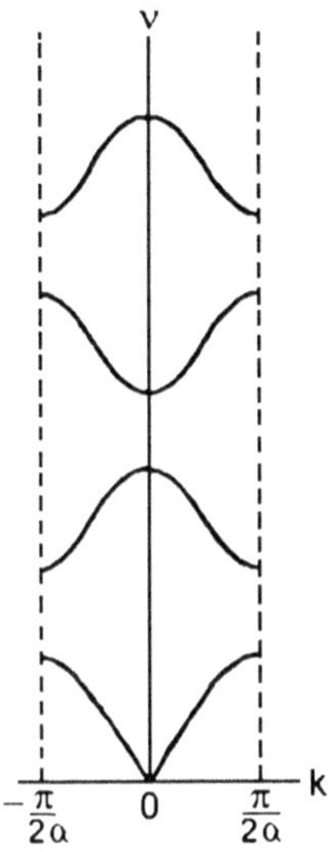

Fig. 10.15 Dispersion relation for a polyatomic lattice with four atoms per cell

One of these branches will be the acoustic branch and the remaining $(N - 1)$ branches will be optic branches. The dispersion (ν versus k) curve for $N = 4$ is shown in Fig. 10.15.

10.2.5 Linear Lattice with an Impurity (Local Modes)

To complete our study of linear lattices, let us consider the case of linear monatomic lattice with an impurity; for simplicity we shall assume the impurity atom to be an isotope of the main constituent of the chain.

We have seen that the pure monatomic chain lattice will have modes with a range of frequencies. The maximum frequency $\omega_0 = 4\alpha/M$, where α is the force constant and M the mass of an atom.

If the lattice contains impurity atoms, there will be additional modes of vibration. Some of them will be within the permitted range $(0 - \omega_0)$; these are called band modes. More interestingly, new modes appear with $\omega > \omega_0$ in the region that is prohibited for the pure lattice; such modes are called gap modes.

The approach to the problem would be to set up, and then solve the equation of motion for the linear lattice of the host atoms of mass M and the impurity atom (isotope) of mass $M - \Delta M$; the ratio $\varepsilon = \Delta M/M$ is a variable parameter. Simple though the problem may appear, it involves a high degree of mathematical complexity far beyond our scope. Hence we shall consider the results qualitatively [10.8].

The band modes for some chosen values of ε are shown in Fig. 10.16. It is seen that the modes are wave-like but with some distortion. More interesting is

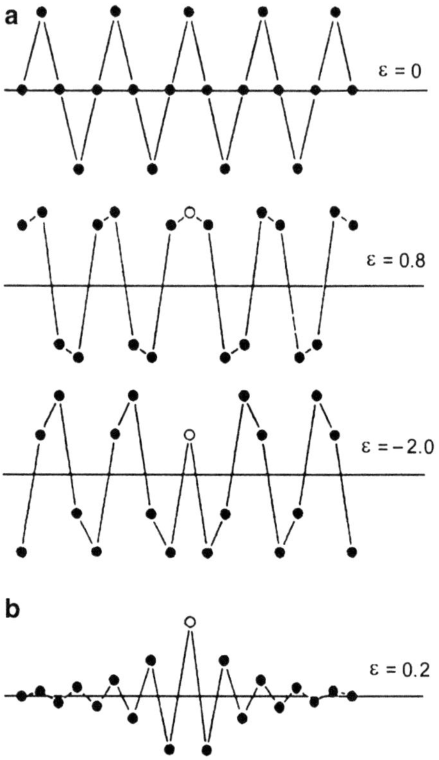

Fig. 10.16 Atomic displacements in (**a**) band modes and (**b**) a localised mode. The impurity atom is represented by the *open circle*

the gap mode with $\varepsilon = 0.2$ also shown in the same figure. It can be seen that the amplitude of the displacement falls off rapidly away from the impurity atom; in fact for $\varepsilon \to 1$, only the impurity atom vibrates. The disturbance associated with this mode does not propagate as a normal mode. In view of this spatial concentration of the displacement in the vicinity of the impurity atom, the gap mode is also called the localised mode. The frequency ω_l of this mode is given by

$$\omega_l^2 = \frac{\omega_0^2}{1-\varepsilon^2}, \tag{10.51}$$

which approximates, for $\Delta M \ll M$, to

$$\omega_l^2 = \frac{2\alpha}{M - \Delta M}.$$

The treatment may be extended to three-dimensional lattices containing impurities. Obviously, the theoretical treatment will be even more complicated. Again, focusing on the results, it may be stated that the presence of impurities leads to a continuum upto ω_0 together with sharp peaks at $\omega_l > \omega_0$ associated with local modes. Figure 10.17 [10.12] shows absorption spectrum of Si doped with equal

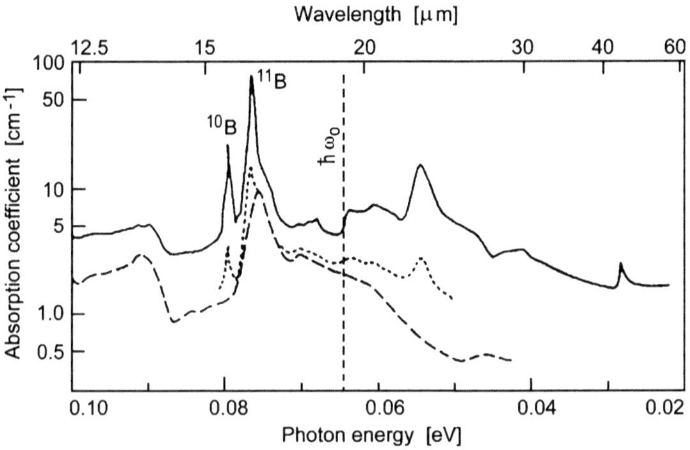

Fig. 10.17 Absorption spectrum of silicon doped with equal concentrations of boron and phosphorus. *Full line*: $D = 5 \times 10^{19}\mathrm{cm}^{-3}$; *dotted line*: $D = 5 \times 10^{18}\mathrm{cm}^{-3}$; *broken line*: pure silicon; D is the number of impurity atoms per unit volume

concentrations of B and P (this is necessary to maintain charge neutrality). A continuum is seen at $\omega < \omega_0$ and local bands due to the two isotopes of B at $\omega > \omega_0$. Besides, an in-band mode for P is seen at 0.055 eV.

10.3 Vibrations of a Two-dimensional Lattice

We shall consider the lattice dynamics of a simple square lattice (Fig. 10.18). Its lattice constant is a and there is one atom per unit cell. The atoms have mass m. The reference atom at the origin is indexed 0. This reference atom has four nearest neighbours (numbered 1–4) at a distance a and four next-nearest neighbours (numbered 5–8) at a distance $\sqrt{2}a$. Following De Launay [10.7], we shall denote the coordinates of atoms by (x_n, y_n), the direction cosines by (λ_n, μ_n) and the displacements by (u_n, v_n). Assuming central forces, the force constants between nearest neighbours and next-nearest neighbours are denoted by α_1 and α_2. For clarity, the relevant data for atoms 1–8 are given in Table 10.2.

We shall denote the x- and y-components of the force on the 0th atom due to the nth atom by X_n and Y_n. Using the data in Table 10.2, the expressions for X_n are:

$$\begin{aligned} X_1 &= -\alpha_1(u_0 - u_1) \\ X_2 &= -\alpha_1(u_0 - u_2) \\ X_3 &= X_4 = 0 \end{aligned} \tag{10.52}$$

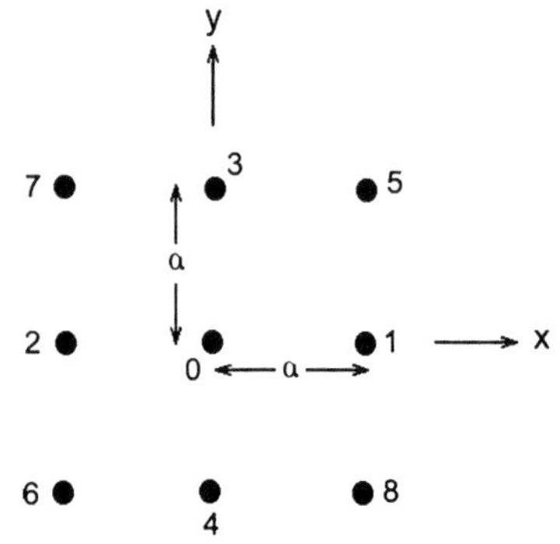

Fig. 10.18 The simple square lattice

Table 10.2 Data describing the atoms in the square lattice

Atom number	Force constant	x_n	y_n	λ_n	μ_n
1	α_1	a	0	1	0
2	α_1	$-a$	0	-1	0
3	α_1	0	a	0	1
4	α_1	0	$-a$	0	-1
5	α_2	a	a	$1/\sqrt{2}$	$1/\sqrt{2}$
6	α_2	$-a$	$-a$	$-1/\sqrt{2}$	$-1/\sqrt{2}$
7	α_2	$-a$	a	$-1/\sqrt{2}$	$1/\sqrt{2}$
8	α_2	a	$-a$	$1/\sqrt{2}$	$1/\sqrt{2}$

$$X_5 = -\frac{\alpha_2}{2}(u_0 - u_5 + v_0 - v_5)$$
$$X_6 = -\frac{\alpha_2}{2}(u_0 - u_6 + v_0 - v_6)$$
$$X_7 = -\frac{\alpha_2}{2}(u_0 - u_7 + v_0 - v_7)$$
$$X_8 = -\frac{\alpha_2}{2}(u_0 - u_8 + v_0 - v_8).$$

The equation of motion for the reference atom for displacement in the x-direction is:

$$M\,\ddot{u}_0 = \alpha_1(u_1 + u_2 - 2u_0) - \frac{\alpha_2}{2}\left[4u_0 - \sum_{n=5}^{8} u_n - v_5 - v_6 + v_7 + v_8\right]. \quad (10.53)$$

Similarly, by writing the y-components of the force $Y_1, \ldots, Y_8$, we get the equation of motion for the y-direction as:

$$M\ddot{v}_0 = \alpha_1(v_1 + v_2 - 2v_0) - \frac{\alpha_2}{2}\left[4v_0 - \sum_{n=5}^{8} v_n - u_5 - u_6 + u_7 + u_8\right]. \quad (10.54)$$

We shall adopt the following two-dimensional wave solutions

$$
\begin{aligned}
u_n &= A_x e^{i[\omega t - \mathbf{k}\cdot\mathbf{r_n}]} \\
v_n &= A_y e^{i[\omega t - \mathbf{k}\cdot\mathbf{r_n}]},
\end{aligned}
\tag{10.55}
$$

where

$$
\mathbf{k}\cdot\mathbf{r_n} = k_x x_n + k_y y_n. \tag{10.56}
$$

Substituting (10.55) in (10.53), (10.54) and introducing

$$
\begin{aligned}
C_1 &= \cos\, k_x a \\
C_2 &= \cos\, k_y a \\
S_1 &= \sin\, k_x a \\
S_2 &= \sin\, k_y a\,,
\end{aligned}
\tag{10.57}
$$

we get

$$
\begin{aligned}
&[2\alpha_1(1-C_1) + 2\alpha_2(1-C_1C_2) - \omega^2 M]A_x + 2\alpha_2 S_1 S_2 A_y = 0 \\
&\text{and} \\
&2\alpha_2 S_1 S_2 A_x + [2\alpha_1(1-C_2) + 2\alpha_2(1-C_1C_2) - \omega^2 M]A_y = 0.
\end{aligned}
\tag{10.58}
$$

These equations will have non-trivial solutions only if the determinant of the coefficients of A_x and A_y vanishes. Thus, we get the secular equation

$$
\begin{vmatrix}
[2\alpha_1(1-C_1) + 2\alpha_2(1-C_1C_2) - \omega^2 M] & 2\alpha_2 S_1 S_2 \\
2\alpha_2 S_1 S_2 & [2\alpha_1(1-C_2) + 2\alpha_2(1-C_1C_2) - \omega^2 M]
\end{vmatrix} = 0.
\tag{10.59}
$$

On expansion of the determinant, we get a quadratic equation in ω^2 giving two roots for ω as a function of k_x, k_y. Thus we get two branches of frequencies. We shall call them branches I and II.

k_x and k_y can take values within the first Brillouin zone, i.e. $\pm\pi/a$. Frequencies in these two branches can be calculated by substituting chosen values of k_x and k_y in (10.59); note that C_1, C_2, S_1, S_2 are functions of k_x, k_y. These frequencies are plotted in the k_x, k_y space and curves are drawn through select values of frequencies. Such isofrequency curves are shown in Fig. 10.19. It is enough to draw the frequency values in 1/4 of the square since the curves are symmetric about the k_x, k_y axes.

To get the frequency distribution function $g(\nu)$, the area between the curves for frequencies ν and $\nu + d\nu$ is calculated; this is equated to $g(\nu)d\nu$. The g(ν) versus ν curves for branches I and II and also for their sum are shown in Fig. 10.20.

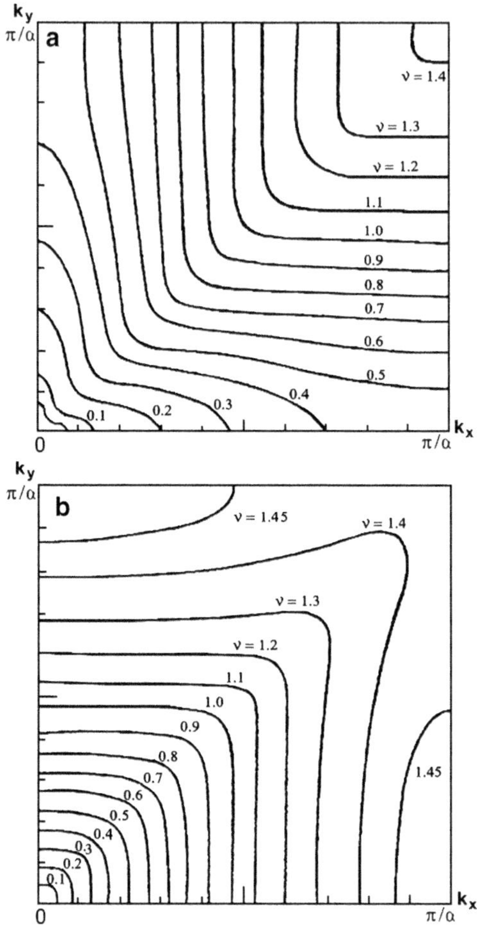

Fig. 10.19 Constant frequency curves of (**a**) branch I and (**b**) branch II; frequency is in arbitrary units

From the distribution function curves, C_V can be calculated from the general formula

$$C_V = \int E(\nu)g(\nu)\mathrm{d}\,\nu, \tag{10.60}$$

where $E(\nu)$ is the Einstein function. By equating the calculated value of C_V with the Debye expression, the value of θ_D can be calculated at different temperatures.

Such $\theta_\mathrm{D} - T$ curves are shown in Fig. 10.21; it is found that these curves depend on the value of α_1/α_2. Further, at low temperatures, C_V for this lattice is found to vary as T^2. Some crystals, e.g. Bi, As and graphite have a quasi-two-dimensional structure. It is indeed observed that the experimental specific heat data for these crystals show a T^2 dependence.

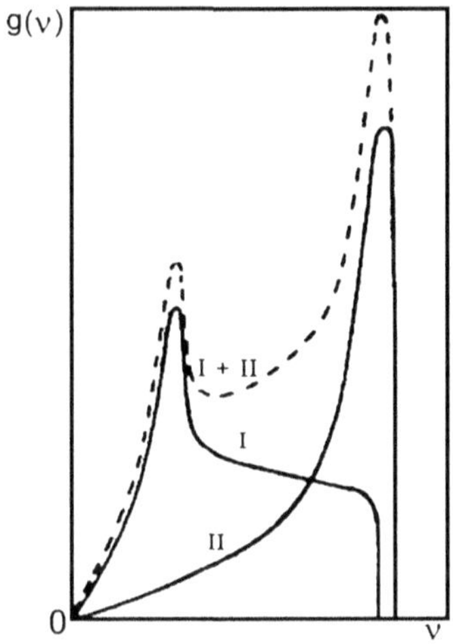

Fig. 10.20 The distribution function g(ν). The contributions from branches I and II are shown separately. The *broken curve* labelled I + II is the sum of the *solid curves*

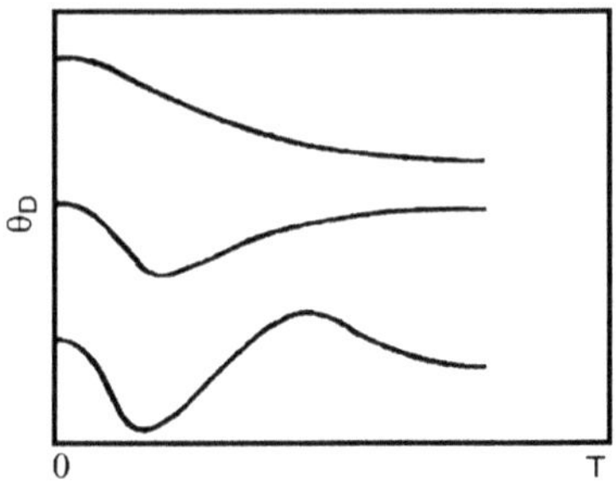

Fig. 10.21 Possible $\theta_D - T$ curves for the square lattice. The lowest curve corresponds to the ratio $\alpha_1/\alpha_2 = 10$

10.4 Vibrations of Three-dimensional Lattices

The lattices that we have considered so far are hypothetical like the one- and two-dimensional lattices. Real lattices are three-dimensional. We shall now consider the simplest three-dimensional lattice viz. the simple cubic lattice. The atomic positions are shown in Fig. 10.22. The lattice constant is a and there is one atom per unit cell. The mass of each atom is M. The displacements of an atom in the x, y, z directions are denoted by u_n, v_n and w_n, n being given appropriate value. Following Blackman [10.6] the atom at the origin is indexed l, m, n and the neighbouring atoms by indices differing by ± 1. The reference atom l, m, n has six nearest neighbours at distance a and 12 next-nearest neighbours at a distance $\sqrt{2}a$. The nearest-neighbour interaction is described by a force constant α_1 and the next-nearest neighbour interaction by α_2. If the x-component of the total force on atom l, m, n is denoted by $X_{l,m,n}$, the equation of motion for displacements in the x-direction will be

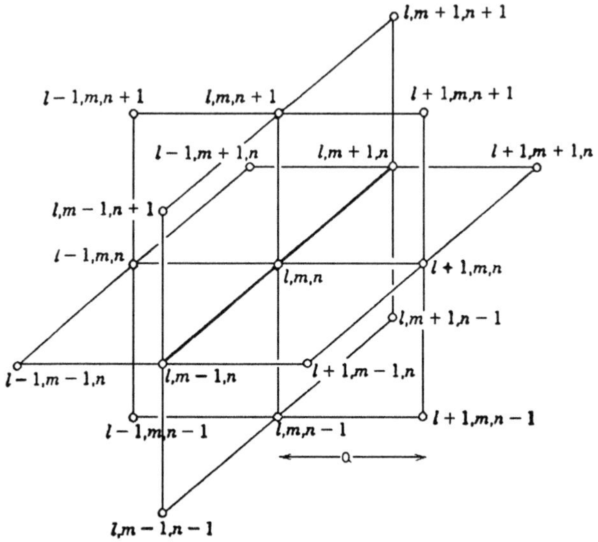

Fig. 10.22 The simple cubic lattice. With reference to the atom l, m, n, atom $l + 1, m, n$ is a typical nearest neighbour and the atom $l + 1, m, n + 1$ is a typical next-nearest neighbour

$$\begin{aligned} X_{lmn} &= M\,\ddot{u}_{lmn} \\ &= \alpha_1[u_{l+1,m,n} + u_{l-1,m,n} - 2u_{l,m,n}] \\ &\quad + \alpha_2[u_{l+1,m,n+1} + u_{l+1,m+1,n} + u_{l+1,m,n-1} + u_{l+1,m-1,n} - 4u_{l,m,n} \\ &\quad + u_{l-1,m,n+1} + u_{l-1,m+1,n} + u_{l-1,m,n-1} + u_{l-1,m-1,n} - 4u_{l,m,n} \\ &\quad + v_{l+1,m+1,n} + v_{l-1,m-1,n} - v_{l+1,m-1,n} - v_{l-1,m+1,n} \\ &\quad + w_{l+1,m,n+1} + w_{l-1,m,n-1} - w_{l+1,m,n-1} - w_{l-1,m,n+1}] \end{aligned} \tag{10.61}$$

with two similar equations for $M\ddot{v}_{lmn}$ and $M\ddot{w}_{lmn}$. We assume solutions of the form

$$u_{lmn} = A_x\,\mathrm{e}^{\mathrm{i}(\omega t - k_x x_l - k_y y_m - k_z z_n)} \tag{10.62}$$

with similar equations for v_{lmn} and w_{lmn}. Substitution of (10.62) in (10.61) leads to three simultaneous equations in A_x, A_y and A_z. These equations will be solvable if the determinant of the coefficients of A_x, A_y, A_z vanishes. This leads to the secular equation

$$\begin{vmatrix} -M\omega^2 + A_{11} & B_{12} & B_{13} \\ B_{21} & -M\omega^2 + A_{22} & B_{23} \\ B_{31} & B_{32} & -M\omega^2 + A_{33} \end{vmatrix} = 0, \tag{10.63}$$

where the A_{ij} and B_{ij} values are functions of α_1, α_2, k_x, k_y and k_z. The expansion of the determinant results in a cubic equation in ω^2. The solutions of this equation

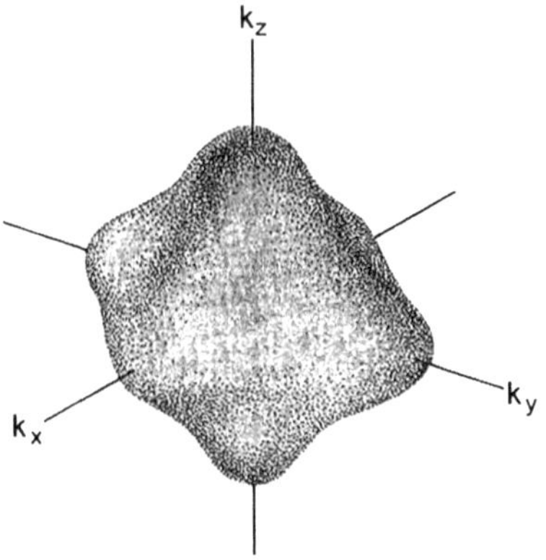

Fig. 10.23 A typical constant frequency surface in the k-space of a three-dimensional lattice

gives three roots in terms of M, α_1, α_2, k_x, k_y and k_z. These are the three acoustic branches I, II, III of the vibration spectrum for this lattice.

The next step is to calculate frequencies for different combinations of k_x, k_y, k_z. To give the reader an idea of the work involved, it may be mentioned that Blackman [109.5] who was the first to work out the vibration spectrum for a three-dimensional lattice (the simple cubic lattice), chose 30,000 points in the k-space, i.e. about 90,000 frequencies were calculated. To obtain the distribution function, we have to count the number of allowed frequencies between two close frequencies like ν and $\nu + \mathrm{d}\,\nu$. For this, all the k-values for a given frequency, say ν, are joined to form a surface. In earlier work, such a surface was constructed in plaster of Paris (Fig. 10.23). A similar surface was constructed for $\nu + \mathrm{d}\,\nu$. The difference in weight of these two solids was taken as a measure of $g(\nu)\mathrm{d}\,\nu$ for that frequency range (De Launay, 1956); obviously, this procedure is no longer necessary with the advent of computers which facilitate the direct evaluation of $g(\nu)\mathrm{d}\,\nu$. The plot of $g(\nu)$ for the three branches for the simple cubic lattice for an assumed value of $\alpha_2/\alpha_1 = 0.05$ is shown in Fig. 10.24; the resultant of the three branches is also shown as curve IV.

Once the distribution function is known, the specific heat is calculated from the general formula $C_V = \int E(\nu)g(\nu)\mathrm{d}\,\nu$ where $E(\nu)$ is the Einstein function; the limits of integration are the frequencies at the boundaries of the first Brillouin zone. Equating the calculated specific heat at a given temperature with the Debye expression, the effective Debye temperature θ_D at that temperature is calculated. By repeating this procedure at different temperatures, a $\theta_\mathrm{D} - T$ curve can be established. The $\theta_\mathrm{D} - T$ curve for the simple cubic lattice for this model with $\alpha_2/\alpha_1 = 0.05$ is shown in Fig. 10.25; the shape of the curve is quite sensitive to the value of α_2/α_1.

10.5 Summary of Vibration Spectra

To go back to the beginning of this chapter, the main criticism of Debye's theory is that it assumes the same parabolic vibration spectrum for all solids.

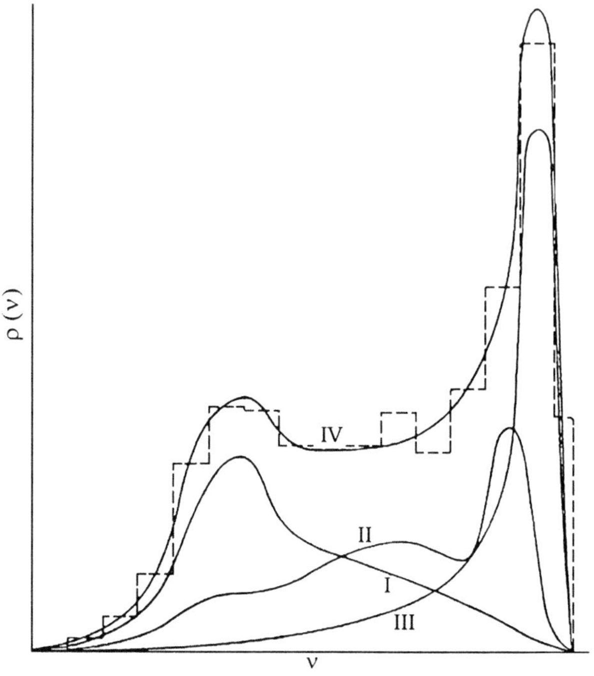

Fig. 10.24 The vibration spectrum of a simple cubic lattice ($\alpha_2/\alpha_1 = 0.05$), curve IV. The three frequency branches are shown separately as I, II and III

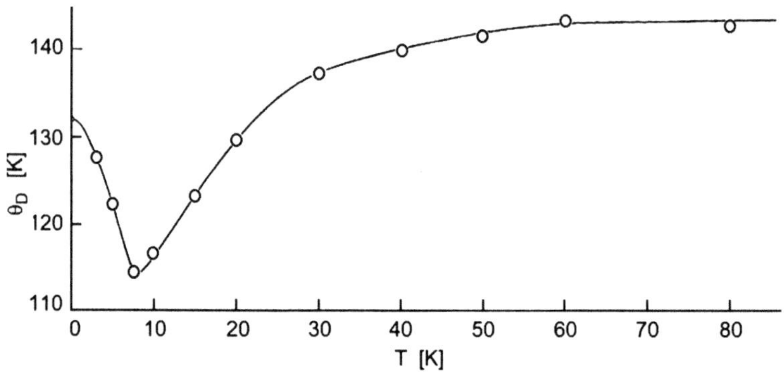

Fig. 10.25 The Debye θ_D value as a function of the temperature for a simple cubic lattice ($\alpha_2/\alpha_1 = 0.05$)

On the other hand, the theory of lattice vibrations developed by Born and Blackman yields vibration spectra which differ from crystal to crystal. Thus, even among the linear chain lattices, the vibration spectrum is very different for the simple monatomic lattice (Fig. 10.3), the diatomic lattice (Fig. 10.10) and the polyatomic lattice (Fig. 10.15). Further, the dispersion curve, distribution function and C_V values for the monatomic linear lattice with a basis depend on the value

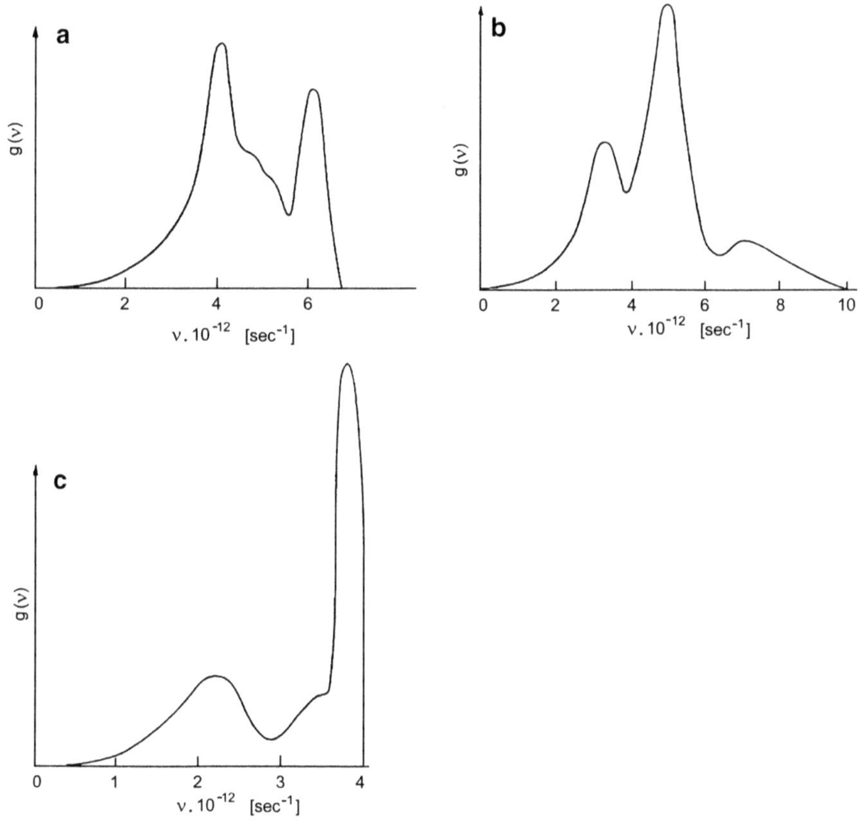

Fig. 10.26 Vibration spectra of some three-dimensional lattices: (**a**) tungsten, (**b**) NaCl and (**c**) diamond

of the ratio of the force constants and those for the diatomic linear lattice upon the values of the masses. Among the three-dimensional lattices also the vibration spectrum is different for each crystal (Fig. 10.26).

However, Blackman [10.6] pointed out that it is not possible to associate a typical vibration spectrum with a crystal family; two crystals having different structures may have similar vibration spectra and two crystals with same structure may have dissimilar spectra. This is not surprising since the vibration spectrum of a crystal is the result of combination of factors like the structure, the interatomic forces, the masses and the force constants.

10.6 Experimental Determination of Phonon Dispersion Curve

Initially, the phonon dispersion curves calculated from different theoretical models were tested by comparing the calculated $C_V - T$ curves or the $\theta_D - T$ curves with corresponding curves obtained from experimental data. But this is not a reliable method as the C_V values represent the cumulative effect of the vibration spectrum as a whole. Attempts to determine the phonon spectra by using diffuse scattering of X-rays were made in the late 1940s and early 1950s. In the late 1950s, the first measurement of phonon dispersion curves was attempted using neutron inelastic scattering.

10.6.1 X-ray Temperature Diffuse Scattering

The scattering of X-rays by crystals which results in sharp Bragg reflections is accompanied by diffuse scattering. The intensity of this diffuse scattering increases with increasing temperature; hence it is called temperature diffuse scattering (TDS). Obviously, it is related to the creation of phonons in the solid. This effect is discussed in detail by James [10.13], Blackman [10.6] and Warren [10.14].

An example of "thermal spots" caused by thermal diffuse scattering [10.13] is shown in Fig. 10.27. A progressive wave in a crystal associated with a phonon has the effect of disturbing the regular lattice arrangement. The effect of this disturbance is equivalent to the presence of an extra scattering point in each cell in the reciprocal lattice. This point is at distance r from the origin of the cell and is in the direction of the progressive wave. The intensity associated with this scattering is proportional to $(\bar{\varepsilon}/m\omega^2)(\xi \cdot \mathbf{H})^2$, where $\bar{\varepsilon}$ is the mean energy associated with the vibration, m the mass of the atom, ω the angular frequency, ξ the amplitude of scattering and $\mathbf{H}$ the reciprocal vector drawn to the point representing the disturbance. However, since there are three frequencies, one longitudinal and two transverse for every $\mathbf{H}$, the total intensity I is

$$I = \text{constant} \sum_{j=1}^{3} \frac{\bar{\varepsilon}_j}{m\omega_j^2}(\xi_j \cdot \mathbf{H}). \tag{10.64}$$

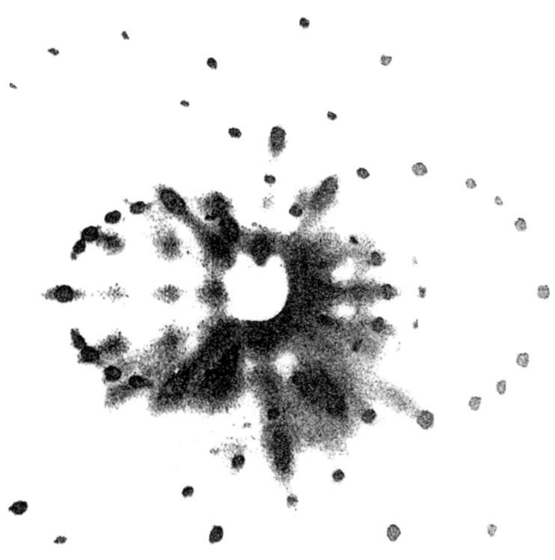

Fig. 10.27 X-ray Laue photograph of KCl; the *dark sharp spots* are the Laue spots and the *less dark diffuse spots* are the TDS spots

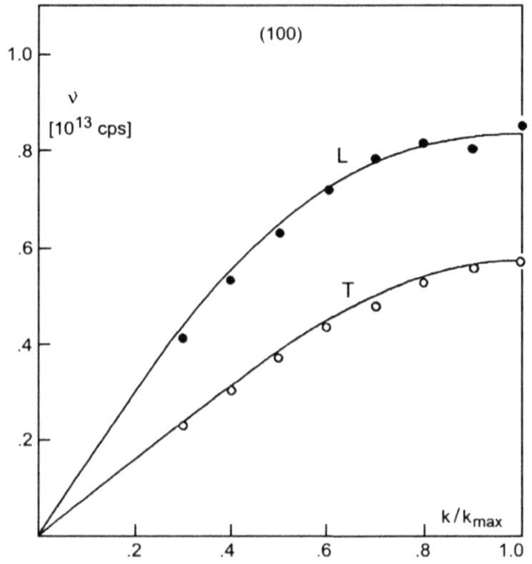

Fig. 10.28 Phonon dispersion curve for aluminium (100 direction) from X-ray TDS measurements

The different vibrations, now represented by points in the reciprocal cell fill the space (Brillouin zone). Thus, in principle, it is possible to relate the frequencies of the phonons to the intensities of the TDS spots.

This method was used for the first time by Olmer [10.15] to determine the phonon dispersion curve for Al. Curien [10.16] and Jacobsen [10.17] used the method to study Fe and Cu respectively. Measurements on Al have also been made by Walker [10.18]; the phonon frequencies are shown in Fig. 10.28. Accurate measurement of the TDS intensity is difficult and considerable work is involved in processing the results. The use of this method has been limited.

10.6.2 Neutron Inelastic Scattering

The energy of a phonon is of the order of 0.1 eV. Neutrons have comparable energy ($\sim 10^{-2}$ eV). It is therefore possible to observe any change in neutron energy due to their inelastic scattering by phonons. A detailed theoretical treatment by Placzek and Hove [10.19] indicated that the phonon spectrum could be determined from a study of the inelastic scattering of neutrons.

The method is based on the conservation equations:

$$\mathbf{k}_0 - \mathbf{k}' = \mathbf{Q} = 2\pi\,\boldsymbol{\tau} + \mathbf{k},$$
$$E_0 - E' = \pm h\nu_j(k). \qquad (10.65)$$

Here $\mathbf{k}_0$, $\mathbf{k}'$ and E_0, E' are the wave vectors and energies of the incident and scattered neutron, $\mathbf{Q}$ the momentum transfer vector, $\boldsymbol{\tau}$ a reciprocal lattice vector

Box 10.2

Bertram Neville Brockhouse (1918–2003) was a Canadian physicist. He developed the experimental technique of neutron inelastic scattering which helps to verify lattice dynamical models. Brockhouse was awarded the Nobel Prize for physics in 1994 for this work.

B. N. Brockhouse

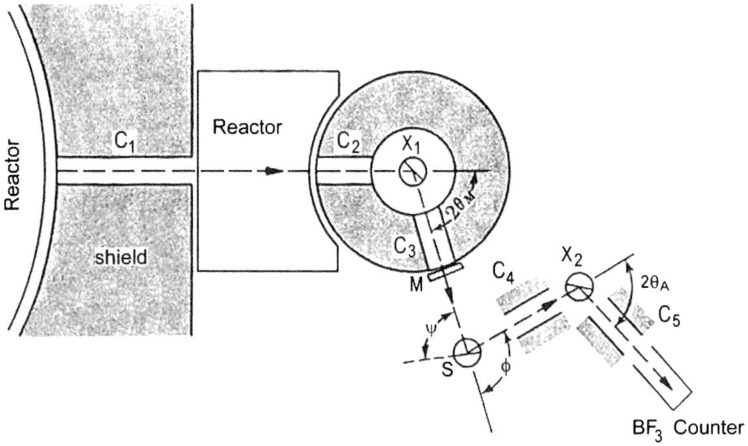

Fig. 10.29 Triple axis crystal spectrometer (schematic) showing monochromating crystal X_1, analysing crystal X_2 and specimen S

and **k** the propagation vector of a phonon of frequency ν_j. The energy distribution of the scattered neutron consists of peaks at which these equations are satisfied.

The experimental technique based on this principle was innovated by Brockhouse (Box 10.2) in the 1950s. Brockhouse's triple-axis spectrometer is shown in Fig. 10.29. C_1–C_5 are collimators. The neutron beam from a reactor passes through the heavily shielded collimators C_1 and C_2. The beam is monochromatised by diffraction at an angle θ_M by a suitable crystal X_1 (first axis). This monoenergetic beam is incident on the sample S mounted on a rotatable table at an angle ψ (second axis). The beam scattered by the sample at an angle ϕ falls on a crystal X_2 which diffracts it at an angle θ_A (third axis) and is received by a BF_3 counter. The monochromator angle θ_M yields values of ω_0 and k_0; the angles ϕ, ψ and θ_A help in determining ω and k'; $\omega_0 - \omega = 2\pi\nu$; and $\mathbf{k_0} - \mathbf{k}' = \mathbf{k}$. Thus by measurements at different settings, the ν versus k curve (dispersion relation) can be constructed.

The first phonon dispersion measurement was made on an aluminium crystal by Brockhouse and Stewart [10.20]. Subsequently, Brockhouse and his group determined the phonon dispersion curves of a variety of crystals like alkali metals,

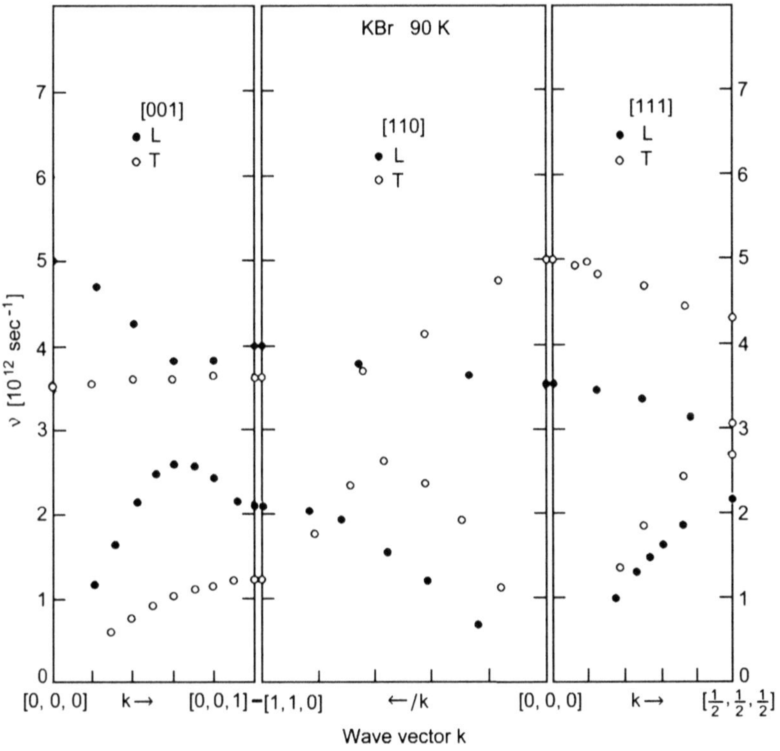

Fig. 10.30 The frequency spectrum of KBr as determined experimentally by neutron inelastic scattering

semiconductors and alkali halides. The phonon dispersion curve for KBr determined by Woods et al. [10.21] is shown in Fig. 10.30. The phonon dispersion relations (curves) for a large number of insulators have been brought together by Bilz and Kress [10.22]. Phonon dispersion curves determined from neutron inelastic scattering have become the standard test for lattice dynamical models.

10.7 Some Related Topics

So far we have presented the Born–Blackman formalism of lattice vibrations. We considered the simplest of lattices. This approach helped us to bring out the physics of the subject. We would hasten to clarify that the detailed vibration spectrum of a real crystal is far from simple. The complexity of the structure of each crystal, the number of atoms in the unit cell, the interatomic forces involved and the number of force constants to be handled add to the problem. As Blackman [10.23] commented, the further development of the methods of lattice dynamics (beyond the Born–Blackman formalism) is "concerned with mathematical rather than the physical

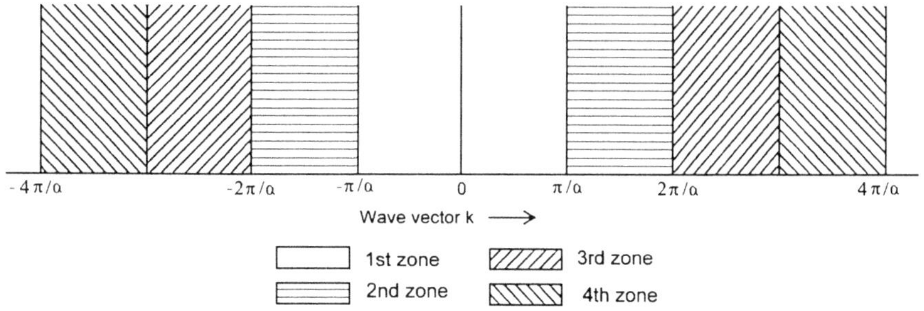

Fig. 10.31 Brillouin zones for a one-dimensional lattice (vertical space has no significance)

side". While we are constrained to limit ourselves to the physics of lattice vibrations, we shall, for completeness, touch briefly, upon some related aspects.

10.7.1 Brillouin Zones

While dealing with the linear lattice, we introduced the concept of the Brillouin zone. At that stage, we considered the vibrational modes pertaining to k-values in the range $-\pi/a < k < \pi/a$; we called these limits the boundaries of the first Brillouin zone. The first and higher Brillouin zones of a one-dimensional lattice are shown in Fig. 10.31. The concept of the Brillouin zone can be extended to two- and three-dimensional lattices. The first few Brillouin zones of a square lattice are shown in Fig. 10.32 and the first Brillouin zones of some three-dimensional lattices in Fig. 10.33.

The significance of the limiting value of k at the boundaries of the Brillouin zones in calculating the frequency values is that the frequencies simply repeat themselves at higher k-values because of the periodicity. Further, in two- and three-dimensional lattices, even within the first Brillouin zone, because of its symmetry, the calculations need to be done only for a part, say a quadrant, of the Brillouin zone; the results get replicated in the other parts, thereby reducing the calculations.

10.7.2 Born's Cyclic (or Periodic) Boundary Conditions

In Sect. 10.2.1, we considered the effect of finiteness of the lattice on the number of modes of vibrations of the linear chain lattice. We treated the end atoms as fixed with $u_0 = u_N = 0$, the total number of atoms being $(N + 1)$. As a result, the total number of modes turned out to be $(N - 1)$, i.e. the number of moveable atoms.

Another approach to this problem was suggested by Born; this is known as Born's cyclic (or periodic) boundary conditions. It is assumed that since N is generally

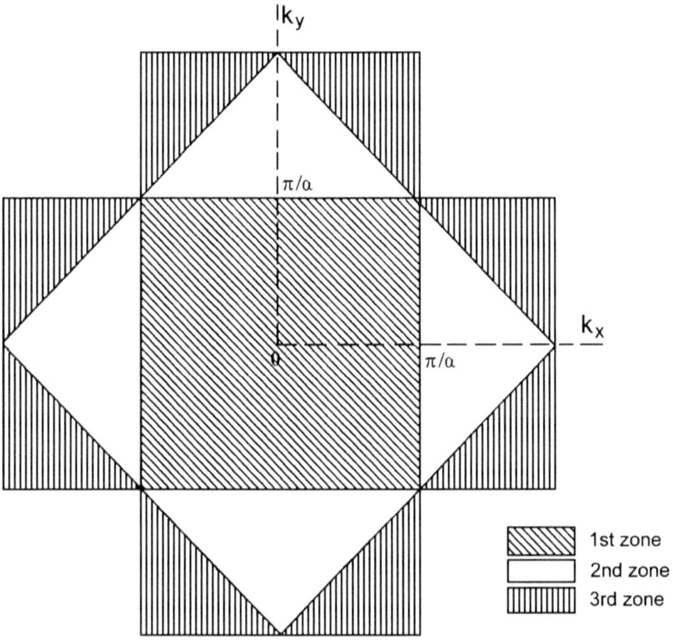

Fig. 10.32 The first three Brillouin zones for a simple square lattice

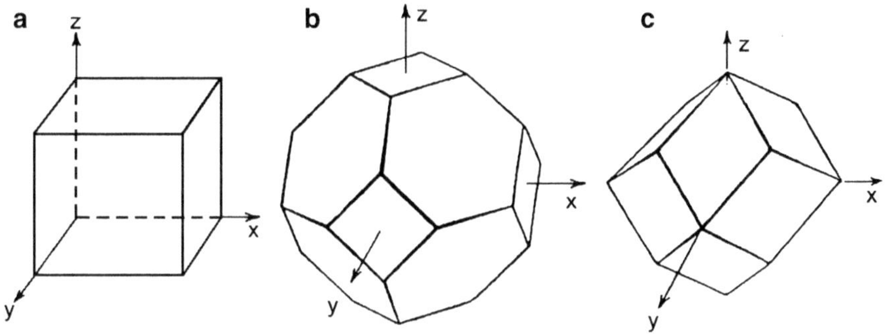

Fig. 10.33 The first Brillouin zone in some three-dimensional lattices: (**a**) simple cubic, (**b**) fcc, (**c**) bcc

large, the atoms in the chain may be imagined to be placed on the circumference of a circle such that the Nth atom coincides with the 0th atom (Fig. 10.34). Now the pth particle is the same as the $(p + N)$th particle. Hence,

$$u_p = u_{p+N}. \tag{10.66}$$

Using a running wave solution, we get

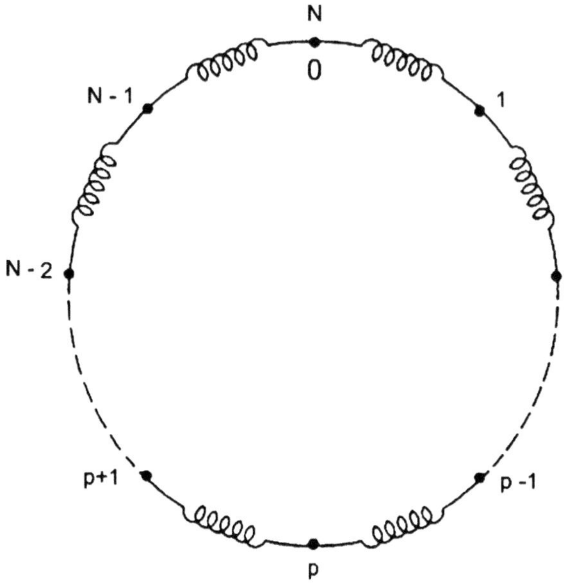

Fig. 10.34 A chain of identical particles placed on the circumference of a circle to satisfy Born–Von Karman's cyclic boundary condition

$$Ae^{i(\omega t - pka)} = Ae^{i[\omega t - (p+N)ka]}. \tag{10.67}$$

This leads to

$$e^{-iNka} = 1, \tag{10.68}$$

which is possible if

$$k = \left(\frac{2\pi}{Na}\right) J, \tag{10.69}$$

where J is an integer.

Assuming N as even and limiting the values of k to $-\pi/a < k < \pi/a$, the values of J giving independent solutions are

$$J = 0, \pm 1, \pm 2, \ldots, \pm\left(\tfrac{N}{2} - 1\right), \tfrac{N}{2}. \tag{10.70}$$

In view of the cyclic nature of the lattice, the solution for $-(N/2)$ is the same as that for $+(N/2)$; hence only the positive sign of $N/2$ is included in (10.70). Thus, the total number of modes is N, i.e. the total number of atoms in the chain. This result is the same as that obtained by the fixed-end method.

Though the fixed-end method and the cyclic boundary condition method lead to the same result in the case of the linear lattice, the latter is more general and is preferred in the treatment of three-dimensional lattices.

10.7.3 Normal Modes

The general solution for the displacement of atom with index n is the sum of various possible solutions. Thus,

$$u_n = \sum_k A_k e^{i(\omega_k t - kna)}. \tag{10.71}$$

The motion described by (10.71) is not simple harmonic but is multiply periodic. Calculations would be very much simplified if these oscillations could be decoupled. This is possible if we replace the coordinates u_n by normal coordinates Q_k defined by

$$u_n = \frac{1}{\sqrt{MN}} Q_k e^{ikna}, \tag{10.72}$$

where M is the mass of an atom and N the total number of atoms in the finite lattice. It may be shown [10.9] that the kinetic energy T and potential energy V can be expressed as

$$T = \frac{1}{2}\sum_k \dot{Q}_k \dot{Q}_k^*,$$

$$V = \frac{1}{2}\sum_k Q_k Q_k^* \omega^2(k), \tag{10.73}$$

where Q_k^* is the complex conjugate of Q_k. Writing the Hamiltonian $H = T + V$ and using Hamilton's equations of motion, we get

$$\ddot{Q}_k + \omega^2(k) Q_k = 0. \tag{10.74}$$

Thus, there is one simple harmonic oscillation for each k (each mode). Further, it can be shown that

$$Q_k = \sqrt{\frac{M}{N}} \sum_n u_n e^{-ikna}. \tag{10.75}$$

10.7.4 Quantisation of Normal Modes

If we treat the normal modes as quantised oscillators, we get the following important results:

$$E_n = (n + \tfrac{1}{2}) h\nu_k, \tag{10.76}$$

and

$$\bar{n} = \frac{1}{e^{h\nu/k_B T}} - 1. \tag{10.77}$$

Here E_n is the energy of the oscillator. The term $1/2(h\nu_k)$ in (10.76) is related to zero-point energy. $\bar{n}$ is the number of phonons excited at temperature T. It is called the phonon occupancy number. In the language of statistical mechanics, it is the probable number of states with energy $h\nu$ at temperature T occupied by particles which obey Bose–Einstein statistics. Thus, phonons are "Bosons".

10.7.5 Force Constants

In the very early work in lattice dynamics, vibration spectra were calculated by assuming arbitrary but reasonable values of force constants; these were varied until the θ_D–T curve derived from the theoretically calculated C_V data agreed with the curve derived from experimental C_V data.

We shall discuss a better approach to choose force constants. The force constant between two atoms arises from the interaction energy $U(r)$ between them which is given by

$$U(r) = U\left(|x_{n+m} - x_n|\right), \tag{10.78}$$

where x_{n+m} and x_n are the coordinates of the two atoms. The total potential energy V of the lattice is

$$V = \sum_n \sum_m U(x_{n+m} - x_n). \tag{10.79}$$

Now

$$x_{n+m} - x_n = ma + u_{n+m} - u_n, \tag{10.80}$$

where the u values are displacements and a the lattice constant. Substituting (10.80) in (10.79) and expanding V as a power series,

$$V = V_0 + \sum_n \sum_m (u_{n+m} - u_n)\, U'(ma) + \sum_n \sum_m (u_{n+m} - u_n)^2\, U''(ma) + \cdots, \tag{10.81}$$

where $U'(ma)$ and $U''(ma)$ are $\mathrm{d}U(x)/\mathrm{d}x$ and $\mathrm{d}^2U/\mathrm{d}x^2$, respectively, evaluated at $x = ma$. In (10.81), V_0 is a constant and, for a harmonic crystal, $U' = 0$ for equilibrium. If we define F_n as the force on atom n, we have

$$F_n = \frac{\partial V}{\partial u_n} = U''(ma)[u_{n+m} - u_n]. \tag{10.82}$$

But if α is the force constant, we also have

$$F_n = \alpha(u_{n+m} - u_n). \tag{10.83}$$

We thus have

$$\alpha = U''(ma) = U''(x_0). \tag{10.84}$$

Hence, in principle, a force constant is related to the second derivative of the interaction energy.

For different crystals, different types of interaction energies have been used. In the rigid ion model, the interaction consists of long-range Coulomb interaction between the ions and the short-range repulsion interaction. Kellerman [10.24] used such a model to derive the force constants and, finally, to work out the vibration spectrum of NaCl. Dick and Overhauser [10.25] proposed the "shell" model in which each ion consists of a spherical ionic shell coupled to its ionic core. In this model, the ionic charge has a value $z'e$ which is different from the formal ionic charge ze. Woods et al. [10.21] and several later workers used this model to evaluate the force constants and vibration spectra of alkali halides.

Another approach to evaluate the force constants is to relate them to the elastic constants. In metals, if α_1 and α_2 are the force constants between nearest and next-nearest neighbours, it has been shown that

$$\frac{\alpha_1}{a} = C_{44} \quad \text{and} \quad \frac{4\alpha_2}{a} = C_{11} - C_{12} - C_{44} \quad \text{for fcc metals,}$$

and

$$\frac{2\alpha_1}{3} = C_{44} \quad \text{and} \quad \frac{2\alpha_2}{a} = C_{11} - C_{12} \quad \text{for bcc metals.} \tag{10.85}$$

If interactions with more distant neighbours are included and/or non-central interactions are included, a larger number of force constants are required and additional input data like some prominent phonon frequencies are needed.

10.8 Effect of Lattice Vibrations on Physical Properties

The study of lattice vibrations was prompted by problems with regard to the specific heat of solids. However, we shall see that lattice vibrations influence several physical properties of solids besides specific heat.

10.8.1 Thermal Expansion

In the preceding chapter, it was pointed out that Gruneisen [10.26] assumed the Gruneisen parameter γ to be independent of temperature and to be the same for all vibrational modes. Experimental observations have shown that the mean γ calculated from thermal expansion data is temperature-dependent with limiting values γ_0 and γ_∞ at very low and very high temperatures (Table 10.3).

Table 10.3 Low temperature (γ_0) and high temperature (γ_∞) values of the Gruneisen parameter from [10.27]

Crystal	γ_0	γ_∞
NaCl	0.90	1.57
NaI	1.04	1.71
KCl	0.32	1.45
KBr	0.29	1.49

Instead of assuming that γ is the same for all vibrational modes, we shall assign a value γ_i to each frequency ν_i such that

$$\gamma_i = -\frac{\mathrm{d}\log \nu_i}{\mathrm{d}\log V}. \tag{10.86}$$

γ_i is called the mode Gruneisen parameter. The mean Gruneisen parameter at temperature T is now defined as

$$\gamma_T = \left(\sum_{i=1}^{3p} \gamma_i E_i\right) \Bigg/ \left(\sum_{i=1}^{3p} E_i\right) \tag{10.87}$$

Here p is the number of atoms in the unit cell and E_i is the Einstein specific heat function for frequency ν_i.

At high temperatures $E_i \to 1$ and (10.87) becomes

$$\gamma_\infty = \frac{1}{3p}\sum_{i=1}^{3p} \gamma_i. \tag{10.88}$$

On the other hand, at very low temperatures, the optic branch modes are not excited; only the acoustic modes are excited. Hence,

$$\gamma_0 = \left(\sum_{i=1}^{3} \gamma_i E_i\right) \Bigg/ \left(\sum_{i=1}^{3} E_i\right) \tag{10.89}$$

In fact, some mode gammas are negative. In certain crystals, these negative gammas contribute substantially to γ_0 making it negative. Thus γ_0 is negative at low temperatures in RbI and some crystals with ZnS structure.

From the phenomenological definition $\gamma = 3\alpha V/\psi C_V$ we may write

$$3\alpha = \frac{\psi}{V}\gamma C_V = \frac{\psi}{V}\sum_{i=1}^{3p} \gamma_i C_i, \tag{10.90}$$

where γ_i is the mode gamma and C_i the contribution of the ith frequency to the specific heat.

10.8.2 Thermal Conductivity

We have seen in Chap. 9 that the coefficient of thermal conductivity K is given by

$$K = \frac{1}{3} C_V \upsilon \Lambda, \qquad (10.91)$$

where C_V is the specific heat, υ the sound velocity and Λ the phonon mean path. Since each solid has its own vibration spectrum, each mode of vibration contributes to each term and (10.91) gets modified to

$$K = \frac{1}{3} \sum_j \sum_i C_{ji} \upsilon_{ji} \Lambda_{ji}, \qquad (10.92)$$

where the index j denotes the branch and i the particular wave vector of the mode.

10.8.3 Dielectric Properties

Lyddane et al. [10.28] showed that in diatomic ionic crystals the ratio of the low and high frequency dielectric constants $(\varepsilon_0/\varepsilon_\infty)$ is related to the ratio of the squares of the LO, TO mode frequencies. The Lyddane–Sachs–Taylor (LST) relation for diatomic crystals is

$$(\varepsilon_0/\varepsilon_\infty) = (\omega_{\mathrm{LO}}/\omega_{\mathrm{TO}})^2. \qquad (10.93)$$

The general form of the LST relation is

$$(\varepsilon_0/\varepsilon_\infty) = \prod_i (\omega_{\mathrm{LO}}/\omega_{\mathrm{TO}})_i^2. \qquad (10.94)$$

The product Π_I is over all optic modes.

Equation (10.94) has an important role in ferroelectric crystals in which some vibrational modes decrease with temperature; in some cases, they even assume a near-zero value. Such vibrational modes are called *soft modes*. When this happens, it can be seen from (10.94) that the static dielectric constant becomes enormously large.

Born and Mayer [10.29] showed that in diatomic ionic crystals like the alkali halides, the electronic contribution to the dielectric constant $(\varepsilon_0 - \varepsilon_\infty)$ is related to the TO frequency ω_{TO} by the relation

$$\varepsilon_0 - \varepsilon_\infty = (2\pi e^2 / r^3 {\omega_{\mathrm{TO}}}^2 \mu), \qquad (10.95)$$

where r is the interionic distance, e the electron charge and μ the reduced mass. Szigeti [10.30] refined this relation and showed that

$$\varepsilon_0 - \varepsilon_\infty = (2\pi e^{*2}/9r^3\omega_{TO}{}^2\mu)(\varepsilon_\infty + 2)^2. \tag{10.96}$$

Note that e in (10.95) is replaced by e^* which is known as the Szigeti effective ionic charge. Values of e^* for some crystals are given in Chapter 11; generally, $e^* < 1$.

10.8.4 Elastic Properties

The vibration spectrum can be worked out from the force constants. The force constants are related to the elastic constants. Examples of relations between force constants and elastic constants are given in Sect. 10.7.5. It was pointed out in the same section that the force constants can be calculated theoretically from the knowledge of interatomic interactions. Thus, if force constants are known, they can be used to estimate elastic constants. Kellerman [10.24] determined force constants for NaCl from the rigid ion model and used them to estimate the elastic constants.

Brout [10.31] derived the following relation between the phonon frequencies and the compressibility of crystals with NaCl structure:

$$\sum_i 4\pi^2 \nu_i^2(k) = (18r/\mu\psi). \tag{10.97}$$

Here ν_i is the frequency for a given wave vector and the summation is over all TO and LO modes with the same wave vector.

10.8.5 Infrared Absorption

Let us again consider the linear chain of atoms with alternate atoms having masses M and m (Sect. 10.2.3). Further, let the M and m atoms carry charges $-e$ and $+e$ respectively. Obviously, this is the one-dimensional analogue of the NaCl lattice. Let such a lattice be exposed to an electromagnetic wave

$$E = E_0 \mathrm{e}^{\mathrm{i}(\omega t - kx)}. \tag{10.98}$$

The ions will now experience this field besides the restoring forces. Equations (10.29) are now modified to

$$M\,\ddot{u}_{2n} = \alpha[u_{2n+1} + u_{2n-1} - 2u_{2n}] - \mathrm{e}E_0\mathrm{e}^{\mathrm{i}(\omega t - 2nkd)},$$

and

$$m\,\ddot{u}_{2n+1} = \alpha[u_{2n} + u_{2n+2} - 2u_{2n+1}] + \mathrm{e}E_0\mathrm{e}^{\mathrm{i}[\omega t - (2n+1)kd]}. \tag{10.99}$$

As was done earlier, we assume running wave solutions:

$$u_{2n} = A\,\mathrm{e}^{\mathrm{i}[\omega t - 2nkd]},$$
$$u_{2n+1} = B\,\mathrm{e}^{\mathrm{i}[\omega t - (2n+1)kd]}. \tag{10.100}$$

Substituting (10.100) in (10.99), we get

$$-\omega^2 MA = \alpha B\,[\mathrm{e}^{\mathrm{i}kd} + \mathrm{e}^{-\mathrm{i}kd}] - 2\alpha A - eE_0,$$
$$-\omega^2 mB = \alpha A\,[\mathrm{e}^{\mathrm{i}kd} + \mathrm{e}^{-\mathrm{i}kd}] - 2\alpha B + eE_0. \tag{10.101}$$

Now let us consider long waves, i.e. waves with k very small. With this approximation, (10.101) becomes

$$-\omega^2 MA = 2\alpha(B - A) - eE_0,$$
$$-\omega^2 mB = 2\alpha(A - B) + eE_0. \tag{10.102}$$

On solving these equations, we get for the amplitudes

$$A = -\frac{(e/M)E_0}{\omega_0^2 - \omega^2},$$
$$B = \frac{(e/m)E_0}{\omega_0^2 - \omega^2} \tag{10.103}$$

with

$$\omega_0^2 = 2\alpha\left(\frac{1}{M} + \frac{1}{m}\right). \tag{10.104}$$

These are resonance equations. When $\omega \to \omega_0$, $A(B) \to \infty$, i.e. there is maximum absorption of the electromagnetic wave in the vicinity of ω_0. This model, however simple, predicts that crystals like NaCl will have an absorption maximum at ω_0.

Note that only transverse modes will respond to electromagnetic waves. Thus, ω_0 is to be identified with the TO mode at $k \approx 0$. The transmission curve for NaCl crystal recorded by Barnes [10.32] is shown in Fig. 10.35. The absorption wavelengths for some alkali halides are given in Table 10.4.

10.8.6 X-ray Diffraction Intensities

The intensity of a Bragg reflection decreases with increasing temperature according to the relation

$$I_T = I_0 \mathrm{e}^{-2B(\sin\theta/\lambda)^2}. \tag{10.105}$$

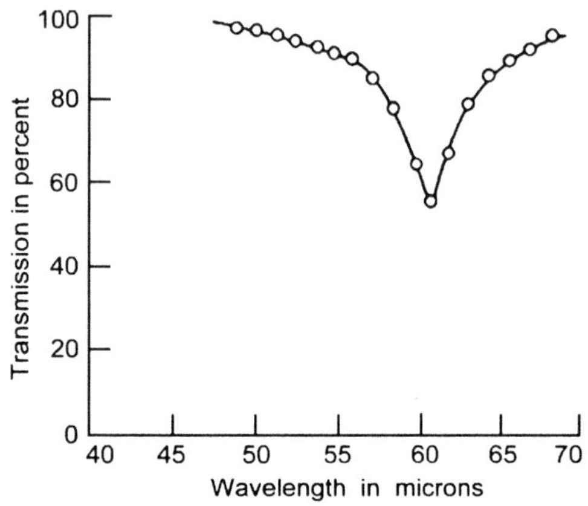

Fig. 10.35 Infrared transmission curve for NaCl crystal

Table 10.4 Infrared absorption parameters (angular frequency ω_0, frequency ν_w and wavelength λ_0) for some diatomic ionic crystals (from [10.3])

Crystal	$\omega_0[10^{13}/s]$	$\nu_0[10^{12}/s]$	$\lambda_0[\mu m]$
LiF	5.78	9.20	32.6
NaCl	3.09	4.92	61.1
KCl	2.67	4.25	70.7
CsCl	1.85	2.95	102
MgO	10.9	17.3	17.3
ZnS	5.7	9.1	33

Here I_T and I_0 are the intensities at temperatures T and $T = 0$; θ the Bragg angle and λ the wavelength. B is called the Debye–Waller factor; it can be determined from measured intensities.

The Debye–Waller factor represents the effect of thermal vibrations on the diffraction intensities. Debye [10.33] and Waller [10.34] showed that

$$B_l = (8\pi^2\hbar/3m_l N)\sum_{k,j} \frac{|e_\alpha(l\,|k,j)\,|\,[n(k,j)+\frac{1}{2}]}{\omega(k,j)}. \qquad (10.106)$$

Here, $e_\alpha(l\,|k,j)$ is the eigenvector for the lth atom, wave vector k and jth mode along the direction α, m_l is the mass of the lth atom, N the number of wave vector points in the first Brillouin zone, $\omega(k,j)$ the phonon frequency for the jth mode with wave vector k and $n(k,j)$ the phonon occupation number

$$n(k,j) = [e^{\hbar\omega(k,j)/k_B T} - 1]^{-1}. \qquad (10.107)$$

The derivation of (10.107) [10.13, 10.14] is long and cumbersome and beyond our scope. It was originally derived for diffraction of X-rays but is applicable to diffraction of electrons and neutrons as well. What matters for us is that B is directly

Table 10.5 Theoretical and experimental values of Debye–Waller factors; B_1 and B_2 relate to the first and second ions [10.35]

Crystal		B_1[Å^2]	B_2[Å^2]
NaCl	Theo.	1.58	1.31
	Exp.	1.81	1.49
LiF	Theo.	0.91	0.73
	Exp.	1.01	0.68
MgO	Theo.	0.30	0.33
	Exp.	0.31	0.34

related to the vibration spectrum of the crystal. Since the vibration spectrum can be derived from an assumed model, the value of B calculated from (10.106) will depend on the model. If different models are used, comparison of the model-based B values with the experimental value can throw light on the validity of the model. A few examples are given in Table 10.5. This method of testing a model is not very sensitive since, as in the case of the specific heat, the vibration spectrum appears in (10.106) in a cumulative manner.

Problems

1. How would you describe the vibrational behaviour of a simple monatomic chain lattice and a diatomic chain lattice in the terminology of electric filters?
2. Derive an expression for the amplitude ratio B/A for the linear diatomic lattice from (10.31). What is the value of B/A at $k = 0$ and $k = 2\pi/a$?
3. Derive an expression for the vibrational energy of a one-dimensional lattice of identical atoms at temperature T. Hence obtain an expression for the specific heat.
4. The phase velocity $\upsilon_p = \omega/k$ and the group velocity $\upsilon_g = \mathrm{d}\omega/\mathrm{d}k$. Find υ_p and υ_g for a one-dimensional lattice of identical atoms.
5. What is the number of LA, TA, LO and TO modes of vibrations in a three-dimensional crystal containing pN atoms (p atoms in the primitive cell and N basis atoms)?

References

10.1. M. Born, Th. Von Karman, Phys. Z. **13**, 297 (1912)
10.2. M. Born, *Atomtheorie des festen Zustandes* (Tebner, Leipzig, 1923)
10.3. M. Born, K. Huang, *Dynamical Theory of Crystal Lattices* (Clarendon Press, Oxford, 1954)
10.4. M. Blackman, Proc. Roy. Soc. Lond. **A148**, 384 (1935)
10.5. M. Blackman, Proc. Roy. Soc. Lond. **A159**, 416 (1937)
10.6. M. Blackman, Handbuch der Physik **VII/I** (Springer-Verlag, Heidelberg, 1955)

10.7. J. De Launay, Solid State Phys. **2**, 219 (1956)
10.8. B. Donovan, J.F. Angress, *Lattice Vibrations* (Chapman and Hall, London, 1971)
10.9. A.K. Ghatak, L.S. Kothari, *An Introduction to Lattice Dynamics* (Addison-Wesley, London, 1972)
10.10. J.S. Blakemore, *Solid State Phys.* (W.B Saunders Co., London, 1969)
10.11. L. Brillouin, *Wave Propagation in Periodic Structures* (McGraw Hill, New York, 1953)
10.12. J.F. Angress, A.R. Goodwin, S.D. Smith, Proc. Roy. Soc. Lond. **A287**, 64 (1965)
10.13. R.W. James, *Optical Principles of the Diffraction of X-rays* (G. Bell, London, !967)
10.14. W.E. Warren, *X-ray Diffraction* (Dover Publications, New York, 1969)
10.15. P. Olmer, Bull. Soc. Franc. Miner. **71**, 144 (1948)
10.16. H. Curien, Bull. Soc. Franc. Miner. **75**, 345 (1952)
10.17. E.H. Jacobsen, Phys. Rev. **97**, 654 (1955)
10.18. C.B. Walker, Phys. Rev. **103**, 547 (1956)
10.19. G. Placzek, L. Van Hove, Phys. Rev. **93**, 1207 (1954)
10.20. B.N. Brockhouse, A.T. Stuart, Phys. Rev. **100**, 756 (1955)
10.21. A.D.B. Woods, B.N. Brockhouse, R.A. Cowley, W. Cochran, Phys. Rev. **131**, 1025 (1963)
10.22. H. Bilz, W. Kress, *Phonon Dispersion Relations in Insulators* (Springer-Verlag, New York, 1979)
10.23. M. Blackman, Rep. Progr. Phys. **8**, 11 (1941)
10.24. E.W. Kellermann, Phil. Trans. Roy. Soc. Lond. **A238**, 513 (1940)
10.25. B.G. Dick, A.W. Overhauser, Phys. Rev. **112**, 90 (1958)
10.26. E. Gruneisen, Handbuch der Physik **X/1** (Springer-Verlag, Heidelberg, 1926)
10.27. R.S. Krishnan, R. Srinivasan, S. Devanarayanan. *Thermal Expansion* (Thomson, Faridabad, 1979)
10.28. R.H. Lyddane, R.G. Sachs, E. Teller, Phys. Rev. **59**, 673 (1941)
10.29. M. Born, G. Mayer, Handbuch der Physik **XXIV/2** (Springer-Verlag, Heidelberg, 1933)
10.30. B. Szigeti, Trans. Faraday Soc. **45**, 155 (1949)
10.31. R. Brout, Phys. Rev. **113**, 43 (1959)
10.32. R.B. Barnes, Zeit. Phys. **75**, 723 (1932)
10.33. P. Debye, Ann. Phys. **43**, 49 (1914)
10.34. I. Waller, Zeit. Phys. **17**, 398 (1923)
10.35. D.B. Sirdeshmukh, L. Sirdeshmukh, K.G. Subhadra, *Micro- and Macro-Properties of Solids* (Springer, Berlin, 2006)

Chapter 11
Dielectrics

11.1 Introduction

Solids can be divided into two groups on the basis of their response to an applied electric field: those which allow free flow of electric current and others which offer high resistance. The former are called conductors and the latter insulators or dielectrics. In this chapter, we are interested in the latter.

In conductors, the applied field sets the charge carriers into motion. In the dielectrics, on the other hand, the field causes displacement of charges and realignment of molecules which result in induced dipole moment; the dipole moment per unit volume is called the polarization. The parameter that describes these field-induced effects is the dielectric constant.

We begin with the explanation of the concepts of polarizability, polarization and dielectric constant. In 1837, Faraday showed that a dielectric substance in a capacitor can appreciably increase the ability of the capacitor to store charge. We use the parallel plate capacitor as a model to explain several basic aspects of dielectric behaviour. The behaviour of a dielectric is discussed first in a static field and subsequently in an alternating field. The frequency dependence of polarizability and dielectric constant is discussed and the phenomenon of spectral absorption is pointed out. A brief account of experimental methods for the measurement of the dielectric constant is given.

Fairly detailed treatments of the topic are given by Dekker [11.1], Kittel [11.2], Brown [11.3] and Levy [11.4]. Kubo and Nagamiya [11.5] comment that while quantum mechanics is important for understanding electric conduction and magnetism, dielectric phenomena which involve electrostatic forces are quite comprehensible in the classical theory. For this reason they used semiclassical methods in their treatment of dielectrics; we have followed the same approach. For convenience, CGS units are used. The conversion factors to convert various quantities to SI units are given in Appendix A.

D.B. Sirdeshmukh et al., *Atomistic Properties of Solids*, Springer Series in Materials Science 147, DOI 10.1007/978-3-642-19971-4_11,

11.2 Dielectric in a Static Field

11.2.1 Dielectric Polarization

11.2.1.1 Macroscopic Description of Polarization

The dielectric constant may be defined by considering the example of a parallel plate condenser. Let us consider two parallel plates each of surface area A separated by a distance d. If d is small compared to the dimensions of the plate, the electric field is treated as uniform and perpendicular to the surface. The field strength E_{vac} is given by

$$E_{\text{vac}} = 4\pi q, \tag{11.1}$$

where q is the surface charge density. The potential difference ϕ_{vac}between the plates is

$$\phi_{\text{vac}} = E_{\text{vac}} d. \tag{11.2}$$

The capacitance of the condenser C_{vac} is given by

$$C_{\text{vac}} = Aq/\phi_{\text{vac}}. \tag{11.3}$$

When a dielectric is introduced between the plates, the potential decreases and the capacitance increases. If ϕ and C are the potential and capacitance, respectively, in the presence of the dielectric, the dielectric constant ε is defined as

$$\varepsilon = \phi_{\text{vac}}/\phi = C/C_{\text{vac}}. \tag{11.4}$$

If the field strength in the presence of the dielectric is E,

$$E_{\text{vac}} = \varepsilon E = D. \tag{11.5}$$

D is called the electric displacement or the flux density given by $4\pi q$. A decrease in the field strength on introduction of the dielectric is interpreted as a reduction in the surface charge density from q to q' (11.1) where

$$q' = E/4\pi. \tag{11.6}$$

Thus we have

$$q - q' = (E_{\text{vac}} - E)/4\pi. \tag{11.7}$$

Denoting the reduction in surface charge density (q - q') by P and using (11.5) we have

$$4\pi P = (\varepsilon E - E). \tag{11.8}$$

The electric displacement **D** is given by

$$\mathbf{D} = \mathbf{E} + 4\pi \mathbf{P}. \tag{11.9}$$

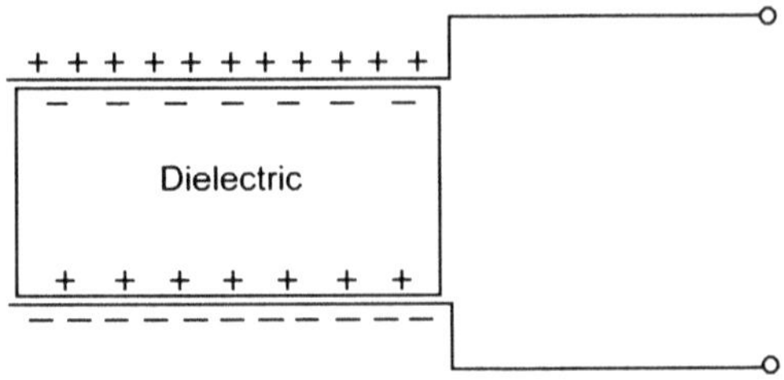

Fig. 11.1 Illustration of charges induced at the surface of a dielectric

P, the reduction in the surface charge density, is called the electric polarization. **E**, **D** and **P** are vectors. The polarization of the dielectric is the result of opposite charges induced on the surface of the dielectric at the plate–dielectric interface (Fig. 11.1).

From a macroscopic point of view, the polarization is a result of an apparent charge appearing on the surface of the dielectric. We shall see later that the polarization can also be interpreted as a phenomenon taking place inside the dielectric. For small fields, the amount of polarization induced in the solid is proportional to the field and may be expressed as

$$P \propto E \quad \text{or} \quad \mathbf{P} = \chi \mathbf{E}, \tag{11.10}$$

where χ is the electric susceptibility. Substituting for P from (11.8) in (11.10) we get

$$\varepsilon = 1 + 4\pi\chi. \tag{11.11}$$

It has been pointed out in Chap. 7 that a physical property of a crystal which linearly relates the components of two vector quantities is a symmetric second-rank tensor. From (11.10) it is seen that the vectors **P** and **E** are linearly related. Hence, the quantity χ which connects them is a second-rank tensor. Consequently the dielectric constant ε, being related to χ (11.11) is also a second-rank tensor. Like other second-rank tensor properties, the dielectric constant has six principal independent constants in triclinic crystals. Due to symmetry, this number is reduced to four in monoclinic crystals, three in orthorhombic crystals, two each in trigonal, hexagonal and tetragonal crystals and only one in cubic crystals. The form of the tensor is given in Table 11.1.

11.2.1.2 Microscopic Description of Polarization

In the macroscopic description discussed earlier, P stands for the amount of reduction of surface charge density on account of the presence of the dielectric between the plates. In the microscopic description, P is expressed in terms of the properties of the atoms and molecules of the dielectric. The basic property involved is the electric dipole moment. If the material is considered as a collection of point charges e_i located at the end of a set of vectors r_i, the dipole moment $\boldsymbol{\mu}$ is defined as

$$\boldsymbol{\mu} = \sum_i e_i \mathbf{r}_i. \tag{11.12}$$

Table 11.1 The dielectric constant tensor for different crystal systems

System	Number of independent coefficients	Form of tensor showing independent coefficients
Cubic	1	$\begin{bmatrix} \varepsilon & 0 & 0 \\ 0 & \varepsilon & 0 \\ 0 & 0 & \varepsilon \end{bmatrix}$
Tetragonal Hexagonal Trigonal	2	$\begin{bmatrix} \varepsilon_1 & 0 & 0 \\ 0 & \varepsilon_1 & 0 \\ 0 & 0 & \varepsilon_3 \end{bmatrix}$
Orthorhombic	3	$\begin{bmatrix} \varepsilon_1 & 0 & 0 \\ 0 & \varepsilon_2 & 0 \\ 0 & 0 & \varepsilon_3 \end{bmatrix}$
Monoclinic	4	$\begin{bmatrix} \varepsilon_{11} & 0 & \varepsilon_{31} \\ 0 & \varepsilon_2 & 0 \\ \varepsilon_{31} & 0 & \varepsilon_{33} \end{bmatrix}$
Triclinic	6	$\begin{bmatrix} \varepsilon_{11} & \varepsilon_{12} & \varepsilon_{31} \\ \varepsilon_{12} & \varepsilon_{22} & \varepsilon_{23} \\ \varepsilon_{31} & \varepsilon_{23} & \varepsilon_{33} \end{bmatrix}$

For a neutral system $\sum_i e_i = 0$. In such a situation $\boldsymbol{\mu}$ is independent of the origin. Hence, the dipole moment corresponding to two charges $+e$ and $-e$ separated by a distance x is given by ex. If $\boldsymbol{\mu} \neq 0$ in the absence of a field, the substance is a polar substance with a permanent dipole moment. On the other hand, if $\boldsymbol{\mu} = 0$ the substance is non-polar. In the presence of an applied field, non-polar dielectrics can only have an induced dipole moment which disappears when the field is withdrawn.

The induced dipole moment is given by

$$\mu_{\text{ind}} = ex, \tag{11.13}$$

where x is the displacement induced by the field. The induced dipole moment $\boldsymbol{\mu}_{\text{ind}}$ is proportional to the applied field $\mathbf{E}$. Hence

$$\boldsymbol{\mu}_{\text{ind}} = \alpha\mathbf{E}. \tag{11.14}$$

α is called the polarizability. The polarization is defined as the electric dipole moment per unit volume of the dielectric and is given by

$$\mathbf{P} = \alpha\mathbf{E}/V = N\alpha\mathbf{E}, \tag{11.15}$$

where N is the number of atoms or molecules per unit volume. In a general case, the polarization is the sum of three contributions:

$$\mathbf{P} = N(\alpha_{\text{e}} + \alpha_{\text{a}} + \alpha_{\text{d}})\mathbf{E}, \tag{11.16}$$

where α_e, α_a and α_d are the electronic, atomic and dipolar polarizabilities. We now try to understand the origin of these three contributions and their magnitudes.

Electronic Polarizability

In a free atom, the centre of gravity of the electron charge distribution coincides with the nucleus and the atom will not possess a dipole moment. When the atom is subjected to a uniform external field **E**, the force exerted on the positive nucleus and that on the electrons will be oppositely directed resulting in a displacement of the centres of positive and negative charge. The restoring force between the electrons and the nucleus tends to preserve a vanishing dipole moment. In the equilibrium situation, the atom acquires an induced dipole moment $\boldsymbol{\mu}_{ind}$ given by

$$\boldsymbol{\mu}_{ind} = \alpha_e \mathbf{E}, \tag{11.17}$$

where α_e is called the electronic polarizability of the atom.

To find the magnitude of α_e, let us assume that in the absence of an electric field the atom consists of a point nucleus of charge $+ze$ surrounded by a homogeneous spherical electron cloud of charge $-ze$ and with radius r as shown in Fig. 11.2a. An electric field E will exert a force zeE on the charges and separate the centre of gravity of the negative and positive charges by a distance d as shown in Fig. 11.2b. From electrostatics, the restoring force F_{res} is

$$F_{res} = \frac{(ze)[ze(d^3/r^3)]}{d^2}. \tag{11.18}$$

Equating the force (zeE) due to applied field to the restoring force we have

$$zeE = (ze)^2 d/r^3. \tag{11.19}$$

Since zed is the induced dipole moment μ_{ind}, we have from (11.17) and (11.19)

$$\alpha_e E = r^3 E. \tag{11.20}$$

Hence, α_e is equal to r^3 for this simple model. We note from (11.20) that the electronic polarizability has the dimensions of volume. Atoms with many electrons will have higher polarizability than atoms with fewer electrons. Electrons in the outer shells will contribute more to α_e as they are not strongly bound to the nucleus in comparison to electrons in inner shells. Hence, negative ions will have larger polarizability compared to the corresponding neutral atoms as may be seen from Table 11.2. The reverse is true for positive ions.

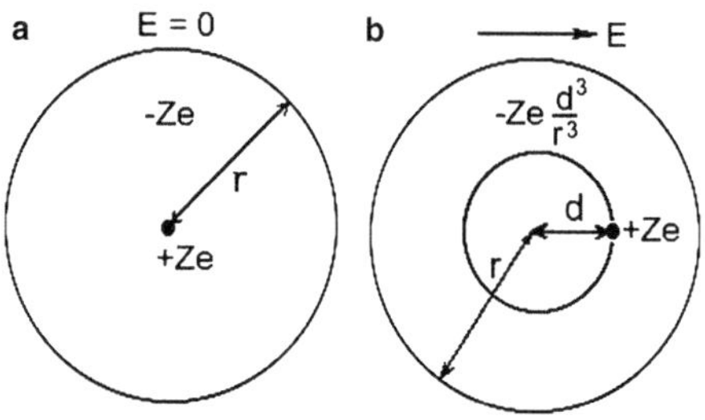

Fig. 11.2 Illustration of electronic polarizability: (**a**) centre of electron cloud and the nuclear charge at the centre in the absence of a field; (**b**) shift of the nuclear charge with respect to the centre of electron cloud of charge $-zed^3/r^3$ in the presence of a field

Table 11.2 Some electronic polarizabilities (α_e in 10^{-24}cm^3)

Atom	α_e	Ion	α_e	Ion	α_e
He	0.201	Li^+	0.02		
Ne	0.39	Na^+	0.22	F^-	0.85
Ar	1.62	K^+	0.97	CL^-	3.00
Kr	2.46	Rb^+	1.50	BR^-	4.13
Xe	3.99	Cs^+	2.42	I^-	6.16

Atomic or Ionic Polarizability

Under the action of an external field, the atoms or ions may be displaced from their equilibrium positions. For example, the positive ion lattice is displaced with respect to the negative ion lattice when an ionic crystal is placed in an external field. Consequently, the induced dipole moment will contribute an atomic or ionic polarizability component (α_a) to the total polarizability. Its magnitude depends on the nature of the material. The contribution α_a to the total polarizability of an ionic crystal will be discussed in Sect. 11.2.3.

Dipolar or Orientational Polarizability

α_d is the dipolar contribution to the polarizability for substances composed of molecules with permanent electric dipole moment. These are dipolar solids in which the dipoles have the ability to align themselves on the application of an electric field. α_d is given by

$$\alpha_d = \mu^2/3k_B T, \tag{11.21}$$

where μ is the permanent dipole moment of the molecule (see Appendix B).

In deriving (11.21) it is assumed that free rotation of the molecules is possible. In some dipolar solids, the dipoles may either be completely hindered or partially hindered. The dielectric constant will become temperature-dependent. The variation of dielectric constant with temperature is shown in Fig. 11.3 for two types of dipolar

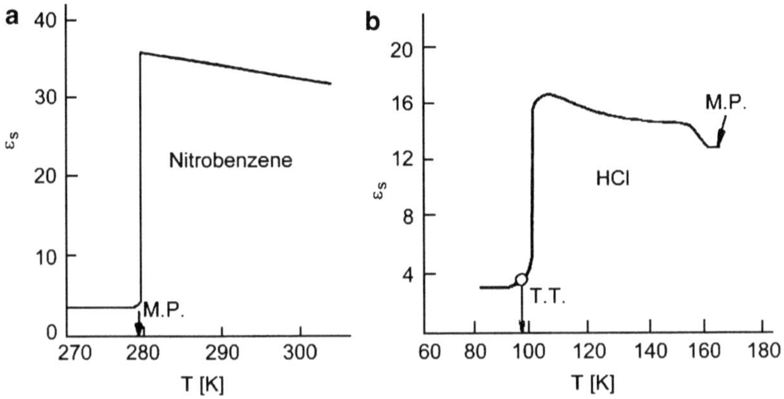

Fig. 11.3 Variation of dielectric constant with temperature for (**a**) nitrobenzene and (**b**) HCl

solids. For nitrobenzene ($C_6H_5NO_2$), the dielectric constant shows a large increase at the melting point (M.P.). This indicates that the dipoles are completely hindered below the melting point and the contribution P_d is zero. The slow decrease beyond the melting point is due to thermal motion. In the case of HCl, the dipoles exhibit partially hindered rotation before reaching the melting point. Hence, the dipolar contribution P_d is present well below the melting point. The temperature at which the large increase is observed is called the transition temperature (T.T.). For a solid with dipoles which are able to rotate only partially, the dipolar polarizability is modified from $\mu^2/3k_BT$ to μ^2/k_BT.

Another category is that of crystals in which the dipolar contribution may arise due to the orientation of crystal defects such as impurity–vacancy complexes. The polarization and the dielectric constant in such a solid will be dependent on the frequency of the applied field and temperature.

11.2.2 Local Electric Field (Lorentz Internal Field)

The field that is effective in polarizing an atom at the lattice site differs from the externally applied field. This is due to the fields produced by the dipoles surrounding the given atom. The field acting at the position of the atom is called the local field (E_{loc}) or internal field (E_{int}). To evaluate the local field, the method suggested by Lorentz [11.6] is discussed here.

We take the example of a parallel plate condenser filled with a dielectric. A small spherical region is selected from the dielectric with the atom, at which the field is to be evaluated, located at the centre (Fig. 11.4). The radius of the sphere is such as to allow the region outside the sphere to be treated as a continuum of dielectric constant ε. However, inside the sphere the actual structure of the medium is taken into account. The local field consists of the following contributions:

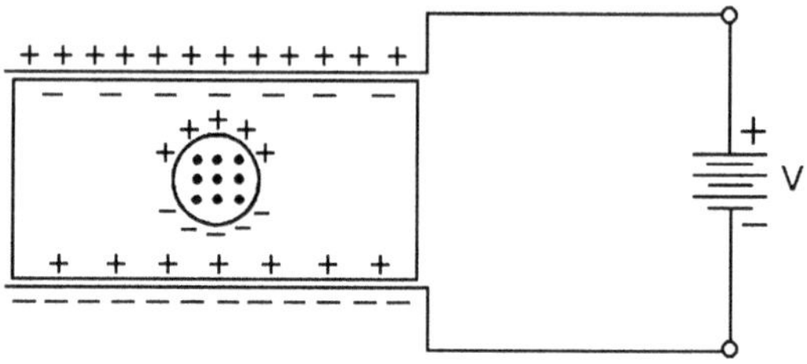

Fig. 11.4 Various charges contributing to the local field

E_1: The contribution from the surface charge density on the plates. This is given by $4\pi_q = D = E + 4\pi P$.

E_2: The contribution from the induced charges at the plate–dielectric interface. This contribution is $-4\pi P$ as opposite charges are induced on the surface of the dielectric.

E_3: Contributions from the charges induced at the spherical surface $= (4/3)\pi P$ (see Appendix C).

E_4: The contributions from the atomic dipoles of all atoms inside the spherical region (see Appendix D). For a crystal with cubic symmetry this reduces to zero.

The internal field E_{loc} is the sum of the contributions E_1, E_2, E_3 and E_4. As E_4 is zero, we have for E_{loc}

$$E_{\text{loc}} = (E + 4\pi P) - 4\pi P + (4/3)\pi P = E + (4/3)\pi P. \tag{11.22}$$

Substituting εE for $(E + 4\pi P)$ from (11.8) we get

$$E_{\text{loc}} = \frac{\varepsilon + 2}{3} E. \tag{11.23}$$

It may be seen from (11.23) that the local field or the internal field is always larger than the applied field.

11.2.3 Polarizability and Dielectric Constant

We shall now proceed to obtain a relation between the dielectric constant and polarizability. When a dielectric solid is subjected to an electric field, it is the local field that polarizes the atoms or ions. We define the polarizability of an atom at the ith site in the crystal as

$$\alpha_i = \frac{\mu_i}{E_{\text{loc}}}, \tag{11.24}$$

where i denotes the particular atom or ion. If there are N_i atoms of type i per unit volume the polarization P is given by

$$P = E_{\text{loc}} \sum_i N_i \alpha_i. \tag{11.25}$$

From (11.10) and (11.11) we have$(\varepsilon - 1)/4\pi = P/E$. Substituting for P and E from (11.22) and (11.25), we get

$$\frac{\varepsilon - 1}{4\pi} = \frac{E_{\text{loc}} \sum_i N_i \alpha_i}{E_{\text{loc}} - (4\pi/3) E_{\text{loc}} \sum_i N_i \alpha_i}, \tag{11.26}$$

or

$$\frac{\varepsilon - 1}{4\pi} = \frac{\sum_i N_i \alpha_i}{1 - (4\pi/3) \sum_i N_i \alpha_i}. \tag{11.27}$$

Rearranging the terms in (11.27) we get

$$\frac{\varepsilon - 1}{\varepsilon + 2} = (4\pi/3) \sum_i N_i \alpha_i. \tag{11.28}$$

If $\alpha_i = \alpha$ is the same for all atoms and N is the number of atoms per unit volume, we get

$$\frac{\varepsilon - 1}{\varepsilon + 2} = (4\pi/3) N\alpha. \tag{11.29}$$

The number of atoms per unit volume $N = N_{\text{A}}\rho/M$, where N_{A} is the Avogadro number, ρ the density and M the molecular weight. Replacing N in (11.29) we have

$$\frac{M}{\rho}\left(\frac{\varepsilon - 1}{\varepsilon + 2}\right) = (4\pi/3) N_{\text{A}}\alpha. \tag{11.30}$$

Equation (11.30) is the Clausius–Mosotti relation. The importance of (11.30) is that it shows the connection between the macroscopic parameter ε and the microscopic parameter α of the dielectric. It may be seen that the polarizability α can be calculated from (11.30) using known and measurable quantities.

In the optical region $\varepsilon = n^2$, where n is the refractive index. Substituting in (11.30) we get the relation

$$\frac{M}{\rho}\left(\frac{n^2 - 1}{n^2 + 2}\right) = (4\pi/3) N_{\text{A}}\alpha_{\text{e}}. \tag{11.31}$$

This relation is known as the Lorentz–Lorenz formula.

Dielectrics can be divided into three groups on the basis of the polarizability of the host lattice:

1. For some substances, the total polarization is entirely due to electronic displacements. They are necessarily elemental solids, e.g. diamond. For these materials α in (11.31) is replaced by α_{e}. α_{e} can be evaluated from (11.31) using the measured values of the high frequency dielectric constant ε_∞ which is equal to the square

of the refractive index. The electronic polarizabilities of some atoms and ions are given in Table 11.2.

2. For substances containing more than one type of atom but not permanent dipoles, both electronic and ionic polarizabilities contribute to the total polarizability α. These are ionic crystals. The dielectric constant ε_s measured at low frequencies or in a static field will have both the contributions α_e and α_a. In this case (11.30) may be expressed as

$$\frac{M}{\rho}\left(\frac{\varepsilon_s - 1}{\varepsilon_s + 2}\right) = (4\pi/3)N_A(\alpha_e + \alpha_a) \tag{11.32}$$

α_a is evaluated using (11.32) in conjunction with (11.31).

We now take the example of an alkali halide crystal. From the measured value of ε_s, $(\alpha_e + \alpha_a)$ is evaluated from (11.32). $(\alpha_e + \alpha_a)$ is the sum of the electronic and ionic polarizabilities of both positive and negative ions. We call this α_{total}. The electronic polarizability α_e is obtained by substituting the measured value of the optical dielectric constant ε_∞ for n^2 in (11.31). The electronic polarizability α_e thus obtained is the sum of the electronic polarizabilities of both ions. The difference of α_{total} and α_e gives the ionic polarizability α_a. α_a is the sum of the polarizabilities of both positive and negative ions. The ionic polarizabilities of the individual ions are obtained by the additivity rule according to which the total polarizability may be expressed as the sum of the individual polarizabilities of the ions. The ionic polarizabilities of alkali and halogen ions are given in Table 11.3.

A comparison of the ionic polarizabilities in Table 11.3 and the electronic polarizabilities in Table 11.2 shows that while the electronic polarizabilities show a large variation as the number of electrons increases, the corresponding variation in ionic polarizabilities is relatively less.

3. For solids containing permanent dipoles or dipoles caused by crystal defects or impurities, the polarizability will have three contributions and α in (11.30) is expressed as $\alpha_e + \alpha_a + \alpha_d$.

To summarize, the electronic polarizability component is present in all types of materials. In polar materials the electronic as well as atomic polarizability will be present. Dipolar materials will have all the three contributions. The frequency dependence of various contributions for a dipolar material is shown in Fig. 11.5.

Table 11.3 Ionic polarizabilities (in 10^{-24}cm^3) of some alkali and halogen ions [11.7]

Ion	$\alpha_{ion}(\alpha_a)$	Ion	$\alpha_{ion}(\alpha_a)$
Li^+	0.91	F^-	1.20
Na^+	1.35	Cl^-	2.00
K^+	2.32	Br^-	2.32
Rb^+	3.06	I^-	2.55
Cs^+	3.27		

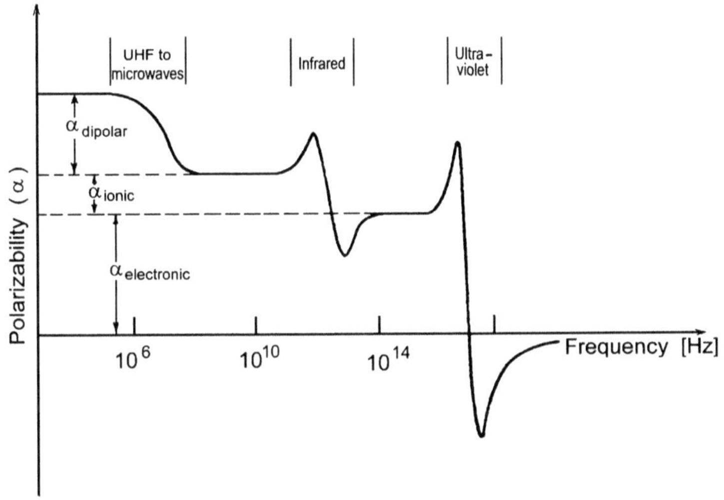

Fig. 11.5 Frequency dependence of the several contributions to the polarizability (Schematic)

11.3 Dielectric in an Alternating Field

11.3.1 Complex Dielectric Function

In Sect. 11.2, the dielectric constant is treated as a real quantity. This is because, under the influence of a static field, the polarization P sets in instantaneously on application of the field. On the other hand if the dielectric is subjected to an alternating electric field, the displacement D and the polarization P cannot follow the field. This is due to the inertial effects and losses associated with the polarizable entities of the dielectric. This introduces a phase lag between the applied field and the displacement.

Let an electric field $E^* = E_0 e^{i\omega t}$ act on the dielectric. Here, E_0 is the amplitude and ω the angular frequency of the field. The resulting displacement D^* will be of the same form as the field but with a phase lag δ and is given by $D^* = D_0 e^{i(\omega t - \delta)}$; here D_0 is the amplitude. Denoting the real part of E^* and D^* by E and D respectively, we have

$$E = E_0 \cos \omega t \tag{11.33}$$

and

$$D = D_0 \cos(\omega t - \delta). \tag{11.34}$$

To take into account the frequency dependence and phase lag of the displacement, we introduce a frequency-dependent dielectric constant ε^* given by

$$\varepsilon^* = \varepsilon'(\omega) - i\varepsilon''(\omega) \tag{11.35}$$

where $\varepsilon'(\omega)$ and $\varepsilon''(\omega)$ are the real and imaginary parts respectively. Expanding (11.34) we have

$$D = D_0 \cos\delta \cos\omega t + D_0 \sin\delta \sin\omega t. \tag{11.36}$$

Defining new coefficients $D_1 = D_0 \cos\delta$ and $D_2 = D_0 \sin\delta$, we have

$$D = D_1 \cos\omega t + D_2 \sin\omega t. \tag{11.37}$$

Since D_0 is proportional to E_0, we consider the coefficients D_1 and D_2 also to be proportional to E_0. Using (11.5) we may therefore write

$$\varepsilon'(\omega) = D_1/E_0 \quad \text{and} \quad \varepsilon''(\omega) = D_2/E_0. \tag{11.38}$$

Writing D_1 and D_2 in terms of the phase lag δ, we get

$$\varepsilon'(\omega) = (D_0/E_0)\cos\delta \tag{11.39}$$

$$\varepsilon''(\omega) = (D_0/E_0)\sin\delta, \tag{11.40}$$

and the ratio

$$\varepsilon''(\omega)/\varepsilon'(\omega) = \tan\delta. \tag{11.41}$$

Since both ε' and ε'' are frequency-dependent, the phase angle δ is also frequency-dependent. It can be shown that ε'', the imaginary part of ε^* is a measure of the average power loss in the dielectric. To obtain the power loss we define a complex current density I as

$$I = \sigma^* E \tag{11.42}$$

and

$$\sigma^* = \sigma' + \mathrm{i}\sigma'', \tag{11.43}$$

where σ^* is the complex conductivity and σ' and σ'' its real and imaginary parts.

Using (11.37), we get

$$I = \mathrm{d}q/\mathrm{d}t = (1/4\pi)\mathrm{d}D/\mathrm{d}t = (\omega/4\pi)(-D_1 \sin\omega t + D_2 \cos\omega t) \tag{11.44}$$

Writing $D = \varepsilon^* E$ in (11.44) and using (11.35) we have

$$I = (1/4\pi)d[(\varepsilon' - i\varepsilon'')E]/\mathrm{d}t. \tag{11.45}$$

Again, writing $E = E_0 e^{\mathrm{i}\omega t}$ in (11.45) we get

$$I = \left(\frac{\omega\varepsilon''}{4\pi} + \mathrm{i}\frac{\omega\varepsilon'}{4\pi}\right) E_0\, \mathrm{e}^{\mathrm{i}\omega t}. \tag{11.46}$$

Comparing (11.46) with (11.42) and using (11.43) we find that

$$\sigma' = \omega\varepsilon''/4\pi \quad \text{and} \quad \sigma'' = \omega\varepsilon'/4\pi. \tag{11.47}$$

σ' the component in phase with the applied field is due to free charge carriers which contribute to conduction. The out-of-phase component does not contribute to conduction.

The average power loss W is obtained using the relation

$$W = (\omega/2\pi)\int_0^{2\pi/\omega} IE\,\mathrm{d}t. \tag{11.48}$$

Substituting from (11.44) for I and using (11.33) in the above integral, we find that the term with D_1 vanishes and we are left with

$$W = (\omega/8\pi)D_2E_0 = (\omega/8\pi)E_0^2\varepsilon''. \tag{11.49}$$

Since $\varepsilon'' = (D_0/E_0)\sin\delta$,

$$W = (\omega D_0E_0/8\pi)\sin\delta \approx (\omega D_0E_0/8\pi)\tan\delta \tag{11.50}$$

for small values of δ.

Hence, $\tan\delta$ is called the loss factor and δ the loss angle.

11.3.2 Dielectric Constant and Loss

We have discussed the various types of polarization induced in a dielectric on the application of a static field in Sect. 11.2. We now consider the polarization and the resulting dielectric constant when an alternating field is applied. Let us consider a dielectric in which all the three contributions, P_e, P_a and P_d are present. The total polarization (P_s) in a static field is

$$P_s = P_e + P_a + P_d. \tag{11.51}$$

When a static field is applied, the values of P_e and P_a are attained instantaneously while a certain time is required for P_d to reach its saturation value. If we denote the saturation value of P_d as P_{ds}, then the instantaneous value $P_d(t)$ is given by

$$P_d(t) = P_{ds}(1 - e^{-t/\tau}). \tag{11.52}$$

Here, τ is the relaxation time defined as the time required for P_d to reach $1/e$ of the equilibrium value. Then

$$\mathrm{d}P_d/\mathrm{d}t = (1/\tau)[P_{ds} - P_d(t)]. \tag{11.53}$$

If an alternating field $E = E_0e^{i\omega t}$ is applied, P_{ds} will also be time-dependent and (11.53) becomes

$$\mathrm{d}P_d/\mathrm{d}t = (1/\tau)[P_{ds}(t) - P_d(t)]. \tag{11.54}$$

Let us denote the instantaneous dielectric constant due to the electronic and ionic polarization $(P_e + P_a)$ as ε_{ea}. Using (11.8) we have

$$P_s = (\varepsilon_s - 1)E/4\pi, \tag{11.55}$$

where ε_s is the static dielectric constant and

$$P_e + P_a = (\varepsilon_{ea} - 1)E/4\pi. \tag{11.56}$$

Hence,

$$P_{ds} = P_s - (P_e + P_a) = (\varepsilon_s - \varepsilon_{ea})E/4\pi. \tag{11.57}$$

Substituting for P_{ds} in (11.54) we get

$$\mathrm{d}P_d/\mathrm{d}t = (1/\tau)\left[\frac{\varepsilon_s - \varepsilon_{ea}}{4\pi}E_0 e^{i\omega t} - P_d(t)\right]. \tag{11.58}$$

Equation (11.58) is a standard differential equation with solution

$$P_d(t) = C e^{-t/\tau} + \frac{1}{4\pi}\frac{\varepsilon_s - \varepsilon_{ea}}{1 + i\omega\tau}E_0 e^{i\omega t}. \tag{11.59}$$

The first term on the right-hand side represents a transient which can be neglected. The total polarization is also a function of time and is given by

$$P(t) = P_e + P_a + P_d(t). \tag{11.60}$$

The displacement is now given by

$$D(t) = \varepsilon^* E(t) = E(t) + 4\pi P(t). \tag{11.61}$$

Substituting for $(P_e + P_a)$ from (11.56) and for P_d from (11.59)

$$\varepsilon^* E(t) = E(t) + 4\pi\left[\frac{\varepsilon_{ea} - 1}{4\pi} + \frac{\varepsilon_s - \varepsilon_{ea}}{4\pi(1 + i\omega\tau)}\right]E(t). \tag{11.62}$$

On simplification,

$$\varepsilon^* = \varepsilon_{ea} + \frac{\varepsilon_s - \varepsilon_{ea}}{(1 + i\omega\tau)}. \tag{11.63}$$

Since $\varepsilon^* = \varepsilon' - i\varepsilon''$, separating the real and imaginary terms in (11.63) we get

$$\varepsilon'(\omega) = \varepsilon_{ea} + \frac{\varepsilon_s - \varepsilon_{ea}}{(1 + \omega^2\tau^2)} \tag{11.64}$$

and

$$\varepsilon''(\omega) = \frac{(\varepsilon_s - \varepsilon_{ea})\omega\tau}{(1 + \omega^2\tau^2)}. \tag{11.65}$$

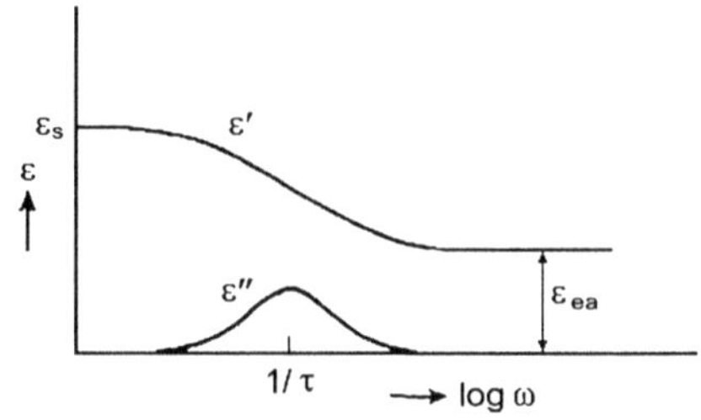

Fig. 11.6 Debye plots for ε' and ε'' as a function of frequency for a dielectric with a single relaxation time

Equations (11.64) and (11.65) are known as the Debye equations. Plots of the real part $\varepsilon'(\omega)$ and imaginary part $\varepsilon''(\omega)$ of the dielectric constant are shown as a function of the angular frequency ω in Fig. 11.6. It can be seen that at low frequencies ($\omega \ll 1/\tau$) the real part of the dielectric constant (ε') is equal to the static dielectric constant (ε_s). For frequencies higher than $(1/\tau)$, the dipoles cannot follow the field and ε' approaches ε_{ea}. ε'', representing the dielectric loss, shows a maximum at $\omega = (1/\tau)$.

The loss or tan δ consists of two contributions, one due to conduction and the other due to relaxation effects. From (11.47) the conductivity σ' is given by $\omega\varepsilon''/4\pi$. If ε'' is proportional to inverse of frequency, σ' will be frequency independent. Substituting for ε'' from (11.41), the loss is expressed as

$$\tan\delta = \frac{4\pi\sigma'}{\omega\varepsilon'}. \tag{11.66}$$

The presence of dipolar impurities or dipoles created by defects shows relaxation effects. Using (11.64) and (11.65), the loss due to relaxation is obtained as

$$\tan\delta = \varepsilon''(\omega)/\varepsilon'(\omega) = \frac{(\varepsilon_s - \varepsilon_{ea})\omega\tau}{\varepsilon_s + \varepsilon_{ea}\omega^2\tau^2}. \tag{11.67}$$

The frequency dependence of the loss due to relaxation effects differs from that of conduction loss. Unlike the conduction loss, which shows a linear plot of log(tan δ) versus log ω , the plot of relaxation loss versus frequency shows a peak at a certain frequency.

The total contribution to dielectric loss from both the effects is

$$\tan\delta = \frac{4\pi\sigma'}{\omega\varepsilon'} + \frac{(\varepsilon_s - \varepsilon_{ea})\omega\tau}{\varepsilon_s + \varepsilon_{ea}\omega^2\tau^2}. \tag{11.68}$$

11.3.3 Dielectric Dispersion

The variation of dielectric constant with frequency of the applied field for a typical ionic crystal is shown in Fig. 11.7. The low frequency (or the static) dielectric constant ε_s is found to be a constant up to a certain frequency. In this region both

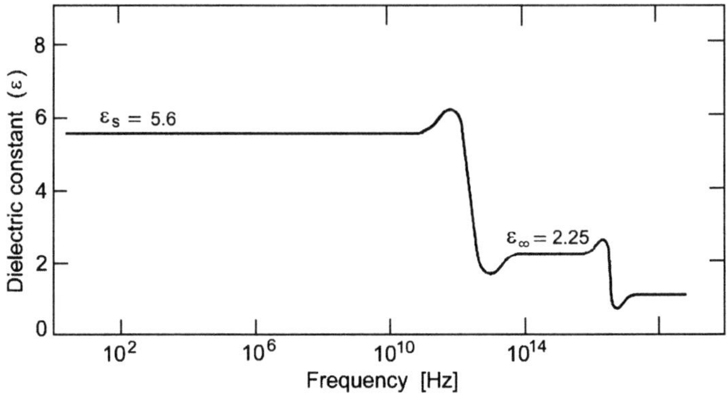

Fig. 11.7 Variation of dielectric constant with frequency for a typical ionic crystal (NaCl)

Table 11.4 Values of the static dielectric constant ε_s and the optical dielectric constant $\varepsilon_\infty = n^2$ for some crystals

	ε_s	$\varepsilon_\infty = n^2$		ε_s	$\varepsilon_\infty = n^2$
LiF	9.27	1.92	RbBr	5.0	2.33
NaF	5.3	1.75	RbI	5.0	2.63
NaCl	5.62	2.25	AgCl	12.3	2.33
NaBr	5.99	2.62	AgBr	13.1	2.69
NaI	6.60	2.91	CsCl	7.20	2.60
KCl	4.68	2.13	CsBr	6.51	2.87
KBr	4.78	2.33	TlCl	31.9	5.19
KI	4.94	2.69	TlBr	29.8	5.41
RbCl	5.0	2.19			

α_e and α_a contribute to the polarizability and hence to the dielectric constant. On increasing the frequency, a region is observed in which the dielectric constant varies rapidly with frequency. This dispersion is associated with ionic polarizability and is observed in the infrared range of the spectrum. On further increase of frequency, the ions cannot follow the field due to inertial effects and the ionic (atomic) contribution vanishes leaving only the electronic contribution. The dielectric constant in this region has a lower value and is referred to as the high frequency dielectric constant ε_∞ or optical dielectric constant ε_{opt}; $\varepsilon_{opt} = n^2$ where n is the refractive index. The region where this difference occurs is in the far infrared region (frequencies $10^{12}-10^{13}$Hz). As the frequency is increased further, a second dispersion associated with electronic polarizability is observed in the ultraviolet range of the spectrum. The static and high frequency dielectric constants for some selected materials are given in Table 11.4. The classical theory of dispersion is discussed in the following sections. It may be mentioned that the dispersion phenomenon is caused by electronic and ionic polarizations and not by dipolar polarization.

11.4 Polarization and Spectral Absorption

11.4.1 Electronic Polarizability and Optical Absorption

The equation of motion of an electrically bound electron subjected to an electric field $E = E_0\,\mathrm{e}^{\mathrm{i}\omega t}$ is given by

$$m\frac{\mathrm{d}^2x}{\mathrm{d}t^2} + \beta x = -eE_0\,\mathrm{e}^{\mathrm{i}\omega t}, \tag{11.69}$$

where m is the mass and e the charge of the electron and β the force constant. In the present case, β is given by $(ze)^2/r^3$. Assuming a solution of the type $x = x_0\mathrm{e}^{\mathrm{i}\omega t}$, we obtain

$$x_0 = -\frac{eE_0}{m(\omega_0^2 - \omega^2)}, \tag{11.70}$$

where ω_0 $[= (\beta/m)^{1/2}]$ is the characteristic frequency. For an electron, ω_0 is of the order of 10^{15}Hz. From (11.15), the polarization is

$$P = N\alpha_e E \quad \text{and also} \quad P = -Nex. \tag{11.71}$$

Then, using (11.70), we get

$$\alpha_e = -ex/E = \frac{e^2}{m(\omega_0^2 - \omega^2)}. \tag{11.72}$$

From (11.72) it follows that at frequencies $\omega \ll \omega_0$, $\alpha_\mathrm{e} \approx e^2/m\omega_0^2$. Taking $\alpha_\mathrm{e} \approx 10^{-24}\mathrm{cm}^3$ and substituting the values of e and m for an electron, we get $\omega_0 \approx 10^{15}$ Hz which corresponds to the visible–ultraviolet region. From (11.10) and (11.11), the dielectric constant is $\varepsilon = 1 + 4\pi(P/E)$. Further, from (11.25) $\varepsilon = 1 + E_\mathrm{loc}\sum N_i\alpha_i$. Assuming $E_\mathrm{loc} = E$ and all atoms have the same polarizability α_e, we get

$$\varepsilon_\mathrm{opt} = 1 + 4\pi\sum_i N_i\alpha_i = 1 + 4\pi N\alpha_\mathrm{e}, \tag{11.73}$$

where N is the number of atoms per unit volume. Hence

$$\varepsilon_\mathrm{opt} = 1 + 4\pi\frac{e^2}{m(\omega_0^2 - \omega^2)}. \tag{11.74}$$

Equation (11.74) shows that $\varepsilon_\mathrm{opt} \to \infty$ as $\omega \to \omega_0$. Hence this model does not correspond to the real situation. We consider a modified model by including a damping term $\gamma\mathrm{d}x/\mathrm{d}t$ in (11.69).

$$m\frac{\mathrm{d}^2x}{\mathrm{d}t^2} + m\gamma\frac{\mathrm{d}x}{\mathrm{d}t} + \beta x = -eE_0\mathrm{e}^{\mathrm{i}\omega t}. \tag{11.75}$$

The amplitude of the displacement $\bar{x}_0$ is given by

$$\bar{x}_0 = -\frac{eE_0}{m(\omega_0^2 - \omega^2 + i\gamma\omega)}. \tag{11.76}$$

$\bar{x}_0$ and so also the polarizability and dielectric constant are now complex quantities. We now have

$$\bar{\alpha}_e = \frac{e^2}{m(\omega_0^2 - \omega^2 + i\gamma\omega)} \tag{11.77}$$

and

$$\bar{\varepsilon}_{\text{opt}} = 1 + \frac{4\pi Ne^2}{m(\omega_0^2 - \omega^2 + i\gamma\omega)}. \tag{11.78}$$

Separating (11.78) into real and imaginary parts we get

$$\varepsilon'_{\text{opt}} = 1 + \frac{4\pi Ne^2}{m}\frac{\omega_0^2 - \omega^2}{(\omega_0^2 - \omega^2)^2 + \gamma^2\omega^2} \tag{11.79}$$

and

$$\varepsilon''_{\text{opt}} = \frac{4\pi Ne^2}{m}\frac{\gamma\omega}{(\omega_0^2 - \omega^2)^2 + \gamma^2\omega^2}. \tag{11.80}$$

The variation of dielectric constant as a function of frequency for a simple harmonic oscillator model is shown in Fig. 11.8. The real and imaginary parts of the dielectric constant for a damped harmonic oscillator (11.79) and (11.80) are shown in Fig. 11.9.

In the above treatment, the field acting on the atom was assumed to be the applied field E. This is true if the dielectric sample is a sphere for which $E_{\text{loc}} = E$. But in a general case $E_{\text{loc}} = E + (4\pi P/3)$. If this local field is used in place of E the same result will be obtained with the difference that in (11.78) ω_0^2 is replaced by $(\omega_0^2 - 4\pi N e^2/3m)$. With this change the absorption frequency will be displaced.

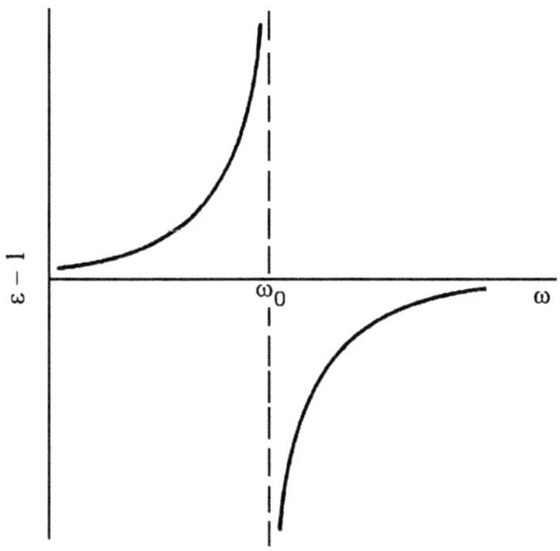

Fig. 11.8 The optical dielectric constant as a function of frequency for a simple harmonic oscillator model

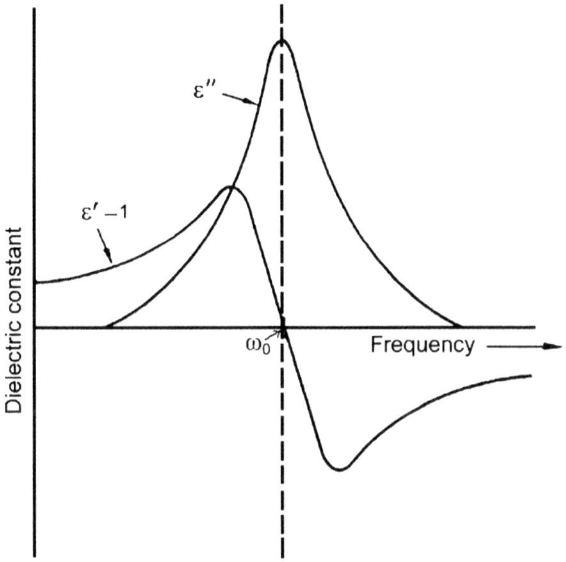

Fig. 11.9 The real and imaginary parts of the optical dielectric constant for a damped harmonic oscillator model in the vicinity of ω_0

11.4.2 Ionic Polarization and Infrared Absorption

We now consider the mechanism of ionic polarization based on the theory of lattice vibrations. Let us consider a one-dimensional diatomic lattice with two atoms per unit cell. On application of a field $E = E_0\, e^{i\omega t}$, let the displacements of ions 1 and 2 be x_1 and x_2. The solution of the equation of motion for the two ions is given in the chapter on lattice vibrations. We shall use the results needed for our discussion. We have

$$x_1 = -\frac{e}{M_1}\frac{E_0}{(\omega_0^2 - \omega^2)}$$

and

$$x_2 = \frac{e}{M_2}\frac{E_0}{(\omega_0^2 - \omega^2)} \tag{11.81}$$

where M_1 and M_2 are the masses of atoms 1 and 2 respectively. The characteristic frequency ω_0 is given by

$$\omega_0^2 = 2\beta\left(\frac{1}{M_1} + \frac{1}{M_2}\right) = \frac{2\beta}{M_R}, \tag{11.82}$$

where β is the force constant and M_R is the reduced mass. We consider the displacement in the frequency range where the ions can follow the field, i.e. $\omega \ll \omega_0$. The maximum displacement x_0 is given by

$$x_0 = x_2 - x_1 = \frac{eE_0}{\omega_0^2}\left(\frac{1}{M_R}\right). \tag{11.83}$$

For the three-dimensional case, an extrapolation of the one-dimensional theory is made as follows. For an ionic crystal, the volume per ion pair is $2a^3$ where a is the distance between the nearest neighbours. The contribution to the polarization per unit volume (P_a) due to ionic displacement is given by

$$P_a = \frac{e(x_2 - x_1)}{2a^3}, \tag{11.84}$$

or

$$P_a = \frac{e^2 E_0}{2\omega_0^2 a^3 M_R}. \tag{11.85}$$

The contribution to dielectric constant due to the ionic polarization P_a is given by the difference between the static dielectric constant ε_s and the high frequency dielectric constant ε_∞. Using (11.11),

$$\varepsilon_s - \varepsilon_\infty = \Delta\varepsilon = \frac{4\pi P_a}{E_0}. \tag{11.86}$$

From (11.83) and (11.84) we have

$$\varepsilon_s - \varepsilon_\infty = \frac{2\pi e^2}{\omega_0^2 a^3} \frac{1}{M_R}. \tag{11.87}$$

In (11.87), the difference between the static and high frequency dielectric constant is expressed in terms of observable quantities. The characteristic frequency ω_0 is the long wavelength transverse optical frequency ω_t which is obtained as the maximum absorption frequency (Restrahlen) in the infrared region.

Equation (11.87) obtained by Born and Mayer [11.8] was later modified by Szigeti [11.9] by taking into account the following factors:

1. The local field correction.
2. In an ideal crystal consisting of deformable ions, each ion carries a formal charge given by ze. But in the real situation in addition to ionic displacement, ionic distortion also takes place. Hence the net charge carried by an individual ion differs from its formal charge. This net charge is known as the effective ionic charge e^*. Hence the charge e in (11.87) is replaced by e^*. Szigeti's formula for $\varepsilon_s - \varepsilon_\infty$ is given by

$$\begin{aligned}\varepsilon_s - \varepsilon_\infty &= \frac{2\pi e^{*2}}{\omega_0^2 a^3}\left(\frac{\varepsilon_\infty + 2}{3}\right)^2\left(\frac{1}{M_1} + \frac{1}{M_2}\right)\\ &= \frac{4\pi e^{*2}}{9\upsilon\omega_0^2}(\varepsilon_\infty + 2)^2\frac{1}{M_R}\end{aligned} \tag{11.88}$$

where υ is the volume per ion pair. Values of e^*/e for some crystals are given in Table 11.5.

Table 11.5 Values of the transverse optical frequency ω_t and effective charge e^*/e for some crystals

	$\omega_t[s^{-1}]$	e^*/e
LiF	5.8×10^{13}	0.87
NaF	4.4	0.93
NaCl	3.08	0.74
NaBr	2.55	0.69
NaI	2.20	0.71
KCl	2.71	0.80
KBr	2.18	0.76
KI	1.91	0.69
RbCl	2.24	0.84
RbBr	1.69	0.82
RbI	1.42	0.89
AgCl	1.94	0.78
AgBr	1.51	0.73
CsCl	1.87	0.85
CsBr	1.39	0.78
TlCl	1.19	0.80
TlBr	0.90	0.82

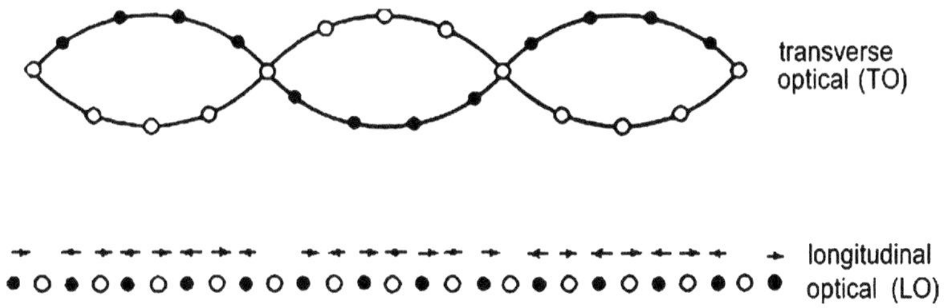

Fig. 11.10 The transverse and longitudinal mode for a diatomic linear lattice

11.4.3 Polarization Waves in Ionic Crystals

We consider a diatomic lattice like an alkali halide crystal. When a light wave is incident on the crystal, the resulting polarization is associated with both transverse and longitudinal optical modes (Fig. 11.10). The light wave interacts with the transverse optical mode and displaces the +ve and −ve ions. In addition, the longitudinal mode also induces polarization due to the overlap of displaced ions.

If the positive and negative ions are displaced in opposite directions by x_+ and x_- respectively, the ionic polarization per unit volume is given by

$$P_a = ex/2a^3, \tag{11.89}$$

where the displacement $x = (x_+ + x_-)$.

The equation of motion for the transverse and longitudinal cases may be written separately. For the transverse mode, the restoring force $F = -M_R\omega_t^2 x$ and the equation of motion is

$$\ddot{x} + \omega_t^2 x = 0, \tag{11.90}$$

where ω_t is the transverse optical mode frequency. For the longitudinal case, the restoring force F_l is given by

$$F_l = -M_R\omega_t^2 x + Ee. \tag{11.91}$$

The first term is due to the displacement of ions and the second term is due to the electric field E associated with the long wavelength longitudinal vibration. The equation of motion for this case is given by

$$\ddot{x} + \omega_t^2 x - \frac{eE}{M_R} = 0. \tag{11.92}$$

The total polarization for an ionic crystal is given by $P = P_e + P_a$ where P_e given by (11.8) is

$$P_e = (\varepsilon_\infty - 1)E/4\pi. \tag{11.93}$$

Then,

$$P = \frac{(\varepsilon_\infty - 1)E}{4\pi} + \frac{ex}{2a^3}. \tag{11.94}$$

From (11.9), $D = E + 4\pi P$. For the longitudinal component, the displacement $D = 0$ and hence, $E = -4\pi P$. The field is then given by

$$E = -4\pi\left[\frac{(\varepsilon_\infty - 1)E}{4\pi} + \frac{ex}{2a^3}\right]. \tag{11.95}$$

On simplification, we get

$$E = -4\pi\frac{ex}{2a^3\varepsilon_\infty}. \tag{11.96}$$

Substituting for E in (11.92), the equation of motion takes the form

$$\ddot{x} + \left(\omega_t^2 + \frac{4\pi e^2}{2a^3 M_R\varepsilon_\infty}\right)x = 0. \tag{11.97}$$

This equation has the same form as (11.90) if we write it as

$$\ddot{x} + \omega_l^2 x = 0, \tag{11.98}$$

where

$$\omega_l^2 = \left(\omega_t^2 + \frac{4\pi e^2}{2a^3 M_R\varepsilon_\infty}\right). \tag{11.99}$$

Identifying ω_0 in Born's equation (11.87) with ω_t, we get,

$$\frac{4\pi e^2}{2a^3 M_R} = (\varepsilon_s - \varepsilon_\infty)\omega_t^2. \tag{11.100}$$

Substituting in (11.97) we get,

$$\omega_l^2 = \omega_t^2 + \left(\frac{\varepsilon_s - \varepsilon_\infty}{\varepsilon_\infty}\right)\omega_t^2 \tag{11.101}$$

or

$$\frac{\omega_l^2}{\omega_t^2} = \frac{\varepsilon_s}{\varepsilon_\infty}. \tag{11.102}$$

Equation (11.102) is the Lyddane–Sachs–Teller relation or LST relation [11.10] and relates the static and high frequency dielectric constants $(\varepsilon_s, \varepsilon_\infty)$ to the transverse optical and the longitudinal optical frequencies (ω_t, ω_l). As discussed earlier for ionic crystals $\varepsilon_s > \varepsilon_\infty$ indicating a difference between ω_l and ω_t whereas for a covalent crystal $\varepsilon_s = \varepsilon_\infty$ and the two frequencies ω_l and ω_t are identical; this has been experimentally verified. Equation (11.102) holds good for cubic diatomic crystals; these crystals have an LO mode and a doubly-degenerate TO mode. For a more general case, (11.102) is modified to

$$\frac{\varepsilon_s}{\varepsilon_\infty} = \prod_i \frac{(\omega_l^2)_i}{(\omega_t^2)_i}. \tag{11.103}$$

The product Π is over all the optic modes. Some of the applications of the LST relation are:

1. Generally ω_l is difficult to determine experimentally. In many cases only ω_t is available. In such cases, ω_l can be estimated from the LST relation.
2. The LST relation has application in identifying ferroelectric behaviour of a crystal.

11.5 Experimental

For the measurement of dielectric constant and loss, the particular method chosen depends on the frequency range and the nature of the substance. The method generally involves the measurement of the change in the capacitance of a condenser when a sample is introduced. A.C. capacitance bridge is commonly used for the measurements in the frequency range 10 Hz to 10 MHz. In the higher range of frequency 10–100 MHz, resonance circuit methods are used. For frequencies beyond 100 MHz, microwave techniques are employed.

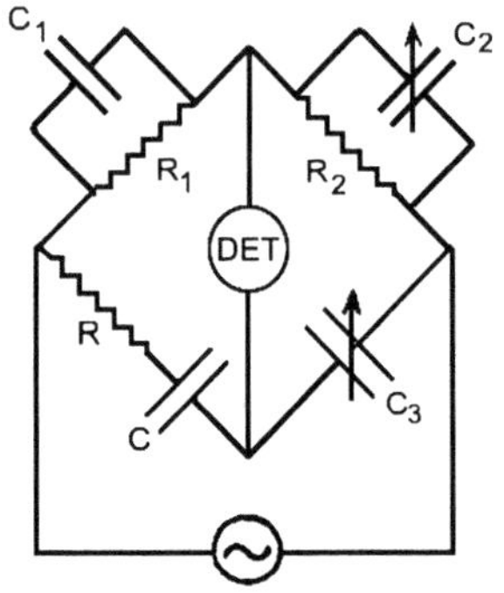

Fig. 11.11 Schering bridge

11.5.1 *Measurement in Audio Frequency and Radio Frequency Range*

11.5.1.1 Capacitance Bridge

The Schering bridge shown in Fig. 11.11 is widely used. C represents the condenser with the dielectric and R is the resistance due to the dielectric loss. The resistances R_1 and R_2 are generally kept equal. C_2 and C_3 are variable capacitors. Initially, a balance is obtained with empty condenser to get C_0. The bridge is again balanced with the condenser with dielectric to obtain C and R in terms of the other known components. ε' is given by C/C_0 and the loss $\varepsilon'' = \varepsilon' \tan\delta$ is obtained as $\tan\delta = \omega C_2 R_2$. The stray capacitance C_1 across R_1 is neglected as an approximation. Using transformer ratio arms in place of the resistors R_1 and R_2, improvement in the accuracy is achieved.

11.5.1.2 Resonance Circuits

A resonance circuit is shown in Fig. 11.12. The capacitor C with the dielectric is made a part of the circuit which is loosely coupled to an oscillator. The other two capacitors are the tuning capacitor C_T and the trimmer capacitor C_{Tr}. Tuning is done once with the dielectric in C and again without the sample. The difference is the sample capacitance C_S. C_0 is calculated from the geometry of the cell. The dielectric constant $\varepsilon = C_S/C_0$. When the dielectric is in the cell, the width of the resonance curve is determined by varying the trimmer capacitor. The dielectric loss factor $\tan\delta$ is obtained from the quality factor Q (the ratio of the energy stored to the energy dissipated). Q is related to the bandwidth $2\delta\omega$ and the resonance frequency as

$$\tan\delta = 1/Q = 2\delta\omega/\omega_0 = \omega_0 C_{\text{total}} R_T, \tag{11.104}$$

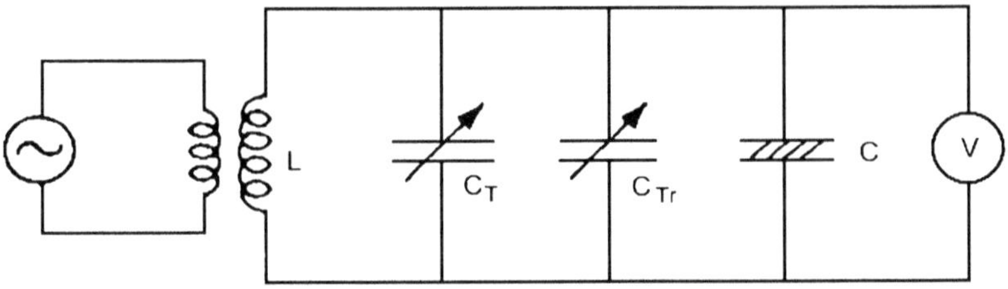

Fig. 11.12 Circuit for resonance method

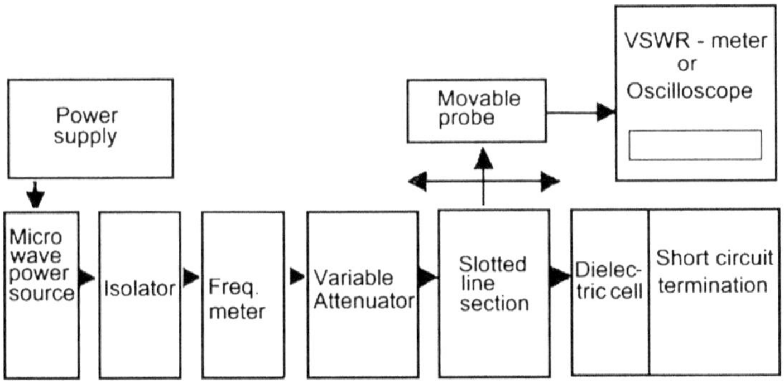

Fig. 11.13 Schematic diagram for experimental set up for standing wave measurements

where

$$C_{\text{total}} = (C_{\text{T}} + C_{\text{Tr}} + C) \quad \text{and} \quad R_{\text{T}} = (R_{\text{C}} + R_{\text{L}}). \tag{11.105}$$

Here, R_{C} is the resistance' due to dielectric loss and R_L the resistance of the coil.

11.5.2 Measurements in the Microwave Range

For measurements beyond 100 MHz, the transmission line technique and the cavity perturbation technique (also called the resonance technique) are used.

1. The transmission line technique is suitable when the material is available in sufficient quantity. The method is based on the fact that the wavelength of a standing wave set-up in a coaxial transmission line changes on introduction of a dielectric material.

   The experimental set-up is shown in Fig. 11.13. It consists of a microwave bench with various parts required for the measurement of wavelength. The sample holder is a waveguide filled with the material. At one end of the waveguide, the circuit is terminated using a short circuit terminator. As a result the power transmitted is reflected back to form a standing wave pattern.

The technique involves the measurement of wavelength by noting the minima of the standing wave with and without the sample in the waveguide. For detecting the minima either a voltage standing wave ratio meter (VSWR) or an oscilloscope is used.

If λ_0 is the free space wavelength of the microwave, λ_C the cut off wavelength of the waveguide and λ_d the wavelength inside the medium, the dielectric constant ε as given by Chandra [11.11] is

$$\varepsilon = (\lambda_0/\lambda_C)^2 + (\lambda_0/\lambda_d)^2 \tag{11.106}$$

2. The cavity perturbation technique, also called the resonance technique is used when a small quantity of the material is available. The resonant cavities are designed in the TM (transverse magnetic) or TE (transverse electric) mode of electromagnetic field. It is based on the shift in the resonance frequency and the change in the width of absorption characteristics on the insertion of the dielectric sample. The shift in frequency is directly related to the dielectric constant and the absorption to the quality factor and thereby the dielectric loss.

   The experimental set-up consists of the standard microwave bench to which the resonant cavity is attached. The size of the cavity is designed for a particular resonant frequency. The sample is kept at the centre of the cavity where the electric field is maximum. The frequency is measured using a frequency counter.

   If f_0 and Q_0 are the resonant frequency and quality factor without the sample and f_S and Q_S the corresponding parameters with the sample, the dielectric constant ε' and the loss ε'' as given by Von Hippel [11.12] are

$$\varepsilon' = 1 - 0.539\frac{V_0}{V_S}\left(\frac{\Delta f}{f_0}\right), \tag{11.107}$$

$$\varepsilon'' = 0.269\frac{V_0}{V_S}\left(\frac{1}{Q_S} - \frac{1}{Q_0}\right). \tag{11.108}$$

   where V_0 is the volume of the cavity, V_S the volume of the sample and Δf is the shift in the frequency (Δf is always negative).

11.5.3 Numerical Data on Static Dielectric Constant

Valucs of the dielectric constants of some crystals are given in Table 11.6. These values give an idea of the typical magnitudes of dielectric constants of ionic crystals and also illustrate the anisotropy and tensor nature of the dielectric constant discussed in Sect. 11.2.

Table 11.6 Static dielectric constants of some isotropic and anisotropic crystals

Crystal system	Crystal	ε_1	ε_2	ε_3
Orthorhombic	$NaNO_2$	6.8	6.4	7.8
	$BaSO_4$	7.6	12.5	8.5
	Mg_2SiO_4	6.9	7.4	6.7
	$RbHSO_4$	7.0	8.0	10.0
		$\longleftarrow \varepsilon_1 = \varepsilon_2 \longrightarrow$		
Tetragonal	MgF_2	5.5		4.8
	TiO_2	86		170
Trigonal	Al_2O_3	9.4		11.6
	SiO_2	4.5		4.6
Hexagonal	CdS	9.4		10.4
	ZnO	9.3		8.2
		$\longleftarrow \varepsilon_1 = \varepsilon_2 = \varepsilon_3 \longrightarrow$		
Cubic	NaCl	5.9		
	CaF_2	6.8		
	BaO	34.0		
	$Sr(NO_3)_2$	5.3		

Problems

1. A parallel plate capacitor of area A separated by a distance d holds a rod of a material of dielectric constant 5 up to (a) half and (b) a quarter of its length. Find the ratio of the capacitance in the two cases.
2. Explain why the polarizability of an ionic crystal is greater than that of a covalent crystal. Evaluate the ionic polarizability for a KCl crystal ($\varepsilon_s = 4.68$ and $\varepsilon_\infty = 2.13$). Verify your result using the infrared absorption frequency for KCl ($\omega_0 = 2.7 \times 10^{13}$ Hz).
3. A parallel plate capacitor has circular electrodes of diameter 4 cm separated by 4 mm. If the capacitor is filled with two dielectrics each of thickness 2 mm and dielectric constant ε_1 and ε_2, obtain an expression for the capacitance of the system.
4. Evaluate the Szigeti charge (e^*/e) for LiBr crystal given $\varepsilon_s = 12.1$, $\varepsilon_\infty = 3.2$, $\omega_0 = 3.2 \times 10^{13}$ Hz and $a = 2.75$ Å.
5. Show that the electronic polarizability for an atom has the dimensions of volume. Obtain the electronic polarizability for diamond given its refractive index is 2.4.

Appendix A: Conversion Factors from CGS Units to SI Units for Various Dielectric Properties

Quantity	CGS units	SI units
Polarizability	α	$4\pi\varepsilon_0\alpha$
Polarization	$\mathbf{P}$	$\mathbf{P}/4\pi\varepsilon_0$
Dipole moment	$\boldsymbol{\mu}$	$\boldsymbol{\mu}/4\pi\varepsilon_0$
Susceptibility	$\chi = (\varepsilon - 1)/4\pi$	$\chi = \varepsilon - 1$
Electric displacement	$\mathbf{D} = \varepsilon\mathbf{E}$	$\mathbf{D} = 4\pi\varepsilon_0\varepsilon\mathbf{E}$
Local field	$\mathbf{E} + (4/3)\pi\mathbf{P}$	$\mathbf{E} + \mathbf{P}/3\varepsilon_0$
C–M equation	$\frac{\varepsilon-1}{\varepsilon+2} = (4\pi/3)\sum_i N_i\alpha_i$	$\frac{\varepsilon-1}{\varepsilon+2} = (1/3\varepsilon_0)\sum_i N_i\alpha_i$

Note: The dielectric constant $\varepsilon_{\text{CGS}} = \varepsilon_{\text{SI}}$, ε_0 is the permittivity of free space, also called vacuum dielectric constant. $\varepsilon_0 = 8.854 \times 10^{-12}\text{Fm}^{-1} = 8.854\ \text{pFm}^{-1}$.

Appendix B: Orientational or Dipolar Polarizability (α_{d})

Let us consider an assembly of permanent dipoles residing at the lattice sites of a solid. Let us assume a situation where free rotation of the dipoles is possible. Under the influence of an electric field E the dipoles will tend to align in the direction of the field. On the other hand, the influence of the applied field is counteracted by the thermal motion of the dipoles. At any given temperature the dipoles will assume all possible orientations. The average component of the dipole moment in the direction of the field is to be evaluated.

The potential energy U of a dipole oriented at an angle θ is given by

$$U = -\boldsymbol{\mu} \cdot \mathbf{E} = -\mu E \cos\theta. \tag{11.109}$$

The effective dipole moment in the direction of the field is $\mu\cos\theta$. If there are N dipoles per unit volume, the polarization (defined as the total dipole moment per unit volume) is

$$P = N\mu\langle\cos\theta\rangle \tag{11.110}$$

where $\langle\cos\theta\rangle$ is the average over a distribution in thermal equilibrium. According to the Boltzmann distribution law, the probability of finding a dipole in an element of solid angle $\text{d}\Omega$ between θ and $\theta + \text{d}\theta$ is proportional to $\text{e}^{-U/k_{\text{B}}T}\text{d}\Omega$ where $\text{d}\Omega = 2\pi\sin\theta\text{d}\theta$. Now

$$\langle\cos\theta\rangle = \int_0^{\pi} \text{e}^{-U/k_{\text{B}}T}\cos\theta\text{d}\Omega \bigg/ \int_0^{\pi} \text{e}^{-U/k_{\text{B}}T}\cos\theta\text{d}\Omega \tag{11.111}$$

Hence the average component of the dipole moment along the field direction is

$$\mu\langle\cos\theta\rangle = \int_0^{\pi} \mu\cos\theta\sin\theta\text{e}^{\mu E\cos\theta/k_{\text{B}}T}\text{d}\theta \bigg/ \int_0^{\pi} \sin\theta\text{e}^{\mu E\cos\theta/k_{\text{B}}T}\text{d}\theta. \tag{11.112}$$

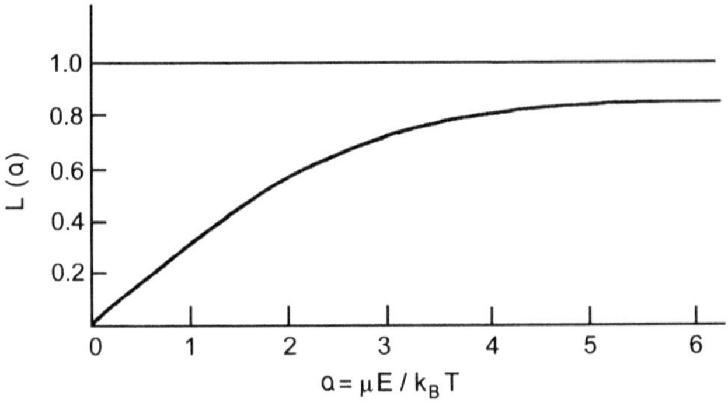

Fig. 11.14 The Langevin function $L(a)$

To evaluate the integrals let us denote $\mu E/k_{\mathrm{B}}T = a$ and $a\cos\theta = x$. Substituting these in (11.112) we have

$$\langle \cos\theta \rangle = \frac{1}{a}\int_{-a}^{+a} x\mathrm{e}^{x}\mathrm{d}x \bigg/ \int_{-a}^{+a} \mathrm{e}^{x}\mathrm{d}x, \tag{11.113}$$

or

$$\langle \cos\theta \rangle = \frac{\mathrm{e}^{a}+\mathrm{e}^{-a}}{\mathrm{e}^{a}-\mathrm{e}^{-a}} - \frac{1}{a} = L(a). \tag{11.114}$$

This equation was originally derived by Langevin [11.13] in connection with paramagnetism; hence $L(a)$ is called the Langevin function. Debye [11.14] postulated the presence of permanent electric dipoles in molecules and extended Langevin's treatment to dipolar solids. The function $L(a)$ is shown in Fig. 11.14.

For large values of a, $L(a) \rightarrow 1$. In such a case, the dipoles are completely aligned in the direction of the field. If the field strength is not too high and at temperatures not too low, expanding the exponentials in (11.114) we have

$$L(a) = \frac{2a}{3[2+(a^{2}/3)]}. \tag{11.115}$$

For $a \ll 1$, $L(a) \rightarrow a/3$. Hence

$$\mu\langle \cos\theta \rangle = (\mu^{2}/3k_{\mathrm{B}}T)E. \tag{11.116}$$

The polarization P is

$$P = (N\mu^{2}/3k_{\mathrm{B}}T)E. \tag{11.117}$$

Finally, the orientational or dipolar polarizability α_{d} is

$$\alpha_{\mathrm{d}} = \mu^{2}/3k_{\mathrm{B}}T \tag{11.118}$$

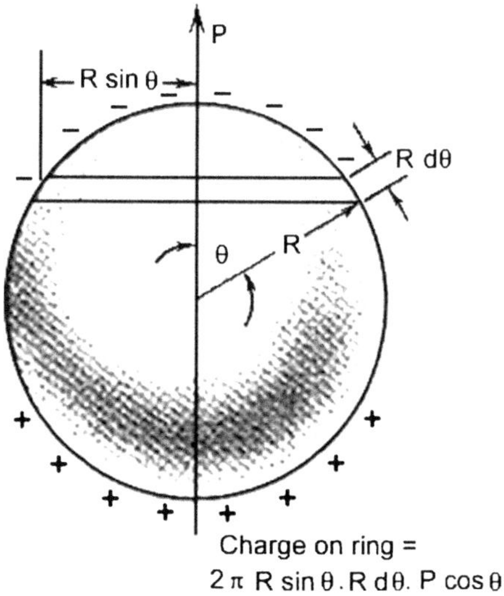

Fig. 11.15 Calculation of the field on the surface of a spherical cavity

Appendix C: Evaluation of E_3 — Field Due To Charges Induced at a Spherical Cavity in a Dielectric

We refer to Fig. 11.4. The atom is at the centre of the spherical cavity of radius R. The problem is to evaluate the field due to the induced charges in the direction of the applied field.

Consider a ring of area $2\pi R^2 \sin\theta \mathrm{d}\theta$ on the inner surface of the sphere (Fig. 11.15). The surface charge density on the spherical surface along the direction towards the centre making an angle θ with the applied field E is $-P\cos\theta$. The charge Q_2 on the ring is,

$$Q_2 = -(2\pi R^2 \sin\theta \mathrm{d}\theta) P \cos\theta \tag{11.119}$$

If we assume unit charge Q_1 to be situated at the centre and Q_2 to be the charge on the ring, the field F at the centre is $Q_1 Q_2 / R^2$. Hence, on substitution for Q_2, we have the field F along the direction of the radius as

$$F = -(2\pi R^2 \sin\theta \mathrm{d}\theta) P \cos\theta / R^2 \tag{11.120}$$

The field along the direction of the applied field **E** is obtained as

$$-F\cos\theta = P\cos^2\theta (2\pi R^2 \sin\theta \mathrm{d}\theta)/R^2 \tag{11.121}$$

Hence, the contribution (E_3) from the charges on the spherical surface is

$$E_3 = 2\pi \int_0^{\pi} P\cos^2\theta \sin\theta \mathrm{d}\theta \tag{11.122}$$
$$= -2\pi P \int_{+1}^{-1} x^2 \mathrm{d}x$$

where $\cos\theta = x$. On simplification, (11.122) yields $4\pi P/3$ for the contribution of the induced surface charges to the local field at the centre.

Appendix D: Evaluation of E_4

We refer to Fig. 11.4 (Sect. 11.2.2). E_4 is the contribution from the atomic dipoles of all the atoms inside the spherical region. This contribution depends on the dipole moment of each atom and the crystal structure. We consider a crystal with cubic symmetry and make the following assumptions:

1. The applied field acts along the x-axis.
2. The induced dipoles are all parallel to the field.

Let the atoms have coordinates x_k, y_k, z_k and their dipole moment components be $\mu_{kx}, \mu_{ky}, \mu_{kz}$. From electrostatics, the field due to an electric dipole is given by

$$\mathbf{E}_{\text{dipole}} = \frac{3(\boldsymbol{\mu}\cdot\mathbf{r})\mathbf{r} - r^2\boldsymbol{\mu}}{r^5}. \tag{11.123}$$

The contribution E_4 along the direction of the field E is given by

$$E_4 = \sum_k \left(\mu_{kx}\frac{3x_k^2 - r_k^2}{r^5} + \mu_{ky}\frac{3x_k y_k}{r^5} + \mu_{kz}\frac{3x_k z_k}{r^5} \right). \tag{11.124}$$

For a cubic lattice of like atoms $\mu_{ky} = \mu_{kz} = 0$ and μ_{kx} will be same for all atoms. Hence

$$E_4 = \sum_k \left(\mu_{kx}\frac{3x_k^2 - r_k^2}{r^5} \right). \tag{11.125}$$

Further, for the atoms in the spherical cavity

$$\sum x_k^2 = \sum y_k^2 = \sum z_k^2 \quad \text{and} \quad \sum r_k^2 = 3\sum x_k^2. \tag{11.126}$$

Substituting in (11.125), we get,

$$E_4 = 0. \tag{11.127}$$

The relation $E_4 = 0$ is valid for cubic lattices of like atoms such as bcc, fcc lattices and crystals of NaCl type. It does not hold good for all cubic lattices. A well-known example is barium titanate. In this crystal since oxygen is surrounded by titanium ions, E_4 does not vanish. In a general case, the environment of each atom is different and has its own internal field. In such a crystal, the internal field for atoms of type 1, 2, 3,... may be expressed as $E_{loc(1)} = E + \gamma_1 P$, $E_{loc(2)} = E + \gamma_2 P, \ldots$, where γ values are the internal field constants. Only if $E_4 = 0, \gamma = 4/3$ and the local field is $E_{loc} = E + (4\pi P/3)$.

References

11.1. A.J. Dekker, *Solid State Physics* (Prentice Hall, New York, 1957)
11.2. C. Kittel, *Introduction to Solid State Physics* (John Wiley and Sons, New York, 1996)
11.3. F.C. Brown, *The Physics of Solids* (Benjamin Inc., New York, 1967)
11.4. R.H. Levy, *Principles of Solid State Physics* (Academic Press, New York, 1968)
11.5. R. Kubo, T. Nagamiya, *Solid State Physics* (McGraw-Hill, New York, 1968)
11.6. H.A. Lorentz, *Theory of Electrons* (Teubner, Leipzig, 1909)
11.7. D.B. Sirdeshmukh, L. Sirdeshmukh, K.G. Subhadra, *Alkali Halides – A Handbook of Physical Properties* (Springer-Verlag, Berlin, 2001)
11.8. M. Born, G. Mayer, Handbuch der Phys. **24/2**, 623 (1933)
11.9. B. Szigeti, Trans. Faraday Soc. **45**, 135 (1949)
11.10. R.H. Lyddane, R.G. Sachs, E. Teller, Phys. Rev. **59**, 673 (1941)
11.11. S. Chandra, Can. J. Phys. **47**, 969 (1969)
11.12. A.R. Von Hippel, *Dielectric Materials and Measurements* (John Wiley and Sons, New York, 1954)
11.13. P. Langevin, J. Physique **4**, 678 (1905)
11.14. P. Debye, Phys. Z. **13**, 97 (1912)

Chapter 12
Piezo-, Pyro- and Ferroelectricity

12.1 Introduction

In the preceding chapter, we discussed aspects of dielectric behaviour which are common for all insulators. Some electric properties are displayed only by restricted groups of insulators. These are piezo-, pyro- and ferroelectricity. The common feature in these crystals is that they lack a centre of symmetry.

Symmetry aspects of these properties are discussed by Nye [12.1] and Sirotin and Shaskoslaya [12.2]. While general treatments are given by Levy [12.3], Dekker [12.4] and Kittel [12.5], specialized treatments are given by Cady [12.6], Von Hippel [12.7], Fatuzzo and Merz [12.8], Zheludev [12.9] and Lines and Glass [12.10].

For convenience, we shall discuss piezo- and pyroelectricity in part A and ferroelectricity in part B; antiferroelectricity is briefly discussed for completeness.

Part A: Piezo- and Pyroelectricity

12.2 Piezoelectricity

12.2.1 Definitions

In some crystals, the application of stress results in electric polarization. This effect, called piezoelectricity, was discovered by Pierre and Jacques Curie in 1880. They observed the effect in quartz and tourmaline. If the applied stress is σ and the resulting polarization is P, it is found that $P \propto \sigma$ or

$$P = d\sigma, \tag{12.1}$$

D.B. Sirdeshmukh et al., *Atomistic Properties of Solids*, Springer Series in Materials Science 147, DOI 10.1007/978-3-642-19971-4_12,

where d is a constant called the direct piezoelectric stress modulus (or coefficient). This effect occurs only in crystals lacking a centre of symmetry; thus, it is limited to 21 point groups.

We have seen in Chap. 7 that in general, the stress on a crystal is described by a symmetric second-rank tensor σ_{jk}. The polarization, on the other hand, is a vector **P** with three components P_i. Each component of **P** is linearly related to all the stress components. Thus, the component P_1 is given by

$$\begin{aligned} P_1 &= d_{111}\sigma_{11} + d_{112}\sigma_{12} + d_{113}\sigma_{13} \\ &\quad + d_{121}\sigma_{21} + d_{122}\sigma_{22} + d_{123}\sigma_{23} \\ &\quad + d_{131}\sigma_{31} + d_{132}\sigma_{32} + d_{133}\sigma_{33}, \end{aligned} \tag{12.2}$$

with similar equations for P_2 and P_3. Equation (12.2) may be shortened to

$$\begin{aligned} P_1 &= d_{1jk}\sigma_{jk} \quad (j, k = 1, 2, 3), \\ P_2 &= d_{2jk}\sigma_{jk}, \\ P_3 &= d_{3jk}\sigma_{jk}. \end{aligned} \tag{12.3}$$

Equation (12.3) may be further generalized as

$$P_i = d_{ijk}\sigma_{jk} \quad (i, j, k = 1, 2, 3). \tag{12.4}$$

The d_{ijk} values are called the piezoelectric stress coefficients. It is obvious that there are 27 coefficients. We shall see later that the actual number is smaller.

12.2.2 The Piezoelectric Tensor and Matrix

Let us consider the physical nature of the piezoelectric coefficients. The components of the polarization P_i and the stress components σ_{jk} are defined with respect to the coordinate system $Ox_1x_2x_3$. Let us rotate the coordinate system to a new system $Ox_1'x_2'x_3'$. Let the polarization components, stress components and piezoelectric coefficients in the new coordinate system be P_i', σ_{jk}' and d_{ijk}' respectively. In the new coordinate system, (12.4) may be written as

$$P_i' = d_{ijk}'\sigma_{jk}'. \tag{12.5}$$

Following the notation in Chap. 7, P_i' can be written as

$$P_i' = \sum_{l} \alpha_{il} P_l = \sum_{lmn} \alpha_{il} d_{lmn}\sigma_{mn} = \sum_{lmnjk} \alpha_{il} d_{lmn}\alpha_{jm}\alpha_{kn}\sigma_{jk}'. \tag{12.6}$$

Table 12.1 Components of the piezoelectric tensor d_{ijk}

1st layer $i = 1$			2nd layer $i = 2$			3rd layer $i = 3$		
d_{111}	d_{112}	d_{113}	d_{211}	d_{212}	d_{213}	d_{311}	d_{312}	d_{313}
(d_{121})	d_{122}	d_{123}	(d_{221})	d_{222}	d_{223}	(d_{321})	d_{322}	d_{323}
(d_{131})	(d_{132})	d_{133}	(d_{231})	(d_{232})	d_{233}	(d_{331})	(d_{332})	d_{333}

Equation (12.6) may be written as

$$P_i' = \sum_{jk} d_{ijk}' \sigma_{jk}', \tag{12.7}$$

where

$$d_{ijk}' = \sum_{lmn} \alpha_{il}\alpha_{jm}\alpha_{kn} d_{lmn}. \tag{12.8}$$

From Chap. 7, we know that this is the transformation law for a third-rank tensor. Thus the piezoelectric tensor constitutes a third-rank tensor. It has 27 components which can be represented as three arrays (or layers) in the shape of a cube; these are shown in Table 12.1. The stress tensor is symmetric ($\sigma_{ij} = \sigma_{ji}$); as a consequence, $d_{ijk} = d_{ikj}$. The d_{ikj} are shown in the table in parentheses. As a result, the 27 d_{ijk} coefficients are reduced to 18 independent coefficients.

It is convenient to reduce the two index stress components to a single-index notation as follows:

$$\sigma_{11} = \sigma_1;\ \sigma_{22} = \sigma_2;\ \sigma_{33} = \sigma_3;\ \sigma_{23,32} = \sigma_4/2;\ \sigma_{13}, \sigma_{31} = \sigma_5/2,\ \sigma_{12}, \sigma_{21} = \sigma_6/2.$$

With this notation, we get

$$\begin{aligned} P_1 &= d_{11}\sigma_1 + d_{12}\sigma_2 + d_{13}\sigma_3 + d_{14}\sigma_4 + d_{15}\sigma_5 + d_{16}\sigma_6 \\ &= d_{1j}\sigma_j \quad (j = 1, 2, \ldots, 6) \end{aligned} \tag{12.9}$$

and

$$P_i = d_{ij}\sigma_j\ , \quad i = 1, 2, 3;\ j = 1, 2, \ldots, 6. \tag{12.10}$$

These new coefficients d_{ij} can be represented by the array given below:

$$\begin{pmatrix} d_{11} & d_{12} & d_{13} & d_{14} & d_{15} & d_{16} \\ d_{21} & d_{22} & d_{23} & d_{24} & d_{25} & d_{26} \\ d_{31} & d_{32} & d_{33} & d_{34} & d_{35} & d_{36} \end{pmatrix}.$$

It may be noted that whereas the d_{ijk} coefficients form a tensor of third rank, the d_{ij} coefficients form a (6×3) matrix.

Equation (12.4) represents the direct piezoelectric effect. There also exists an inverse piezoelectric effect which is the deformation or strain ε_{jk} produced in a crystal to which an electric field component E_i is applied. The two are related by the equation

$$\varepsilon_{jk} = d_{ijk} E_i. \quad i = 1, 2, 3;\ j = 1, 2, \ldots, 6. \tag{12.11}$$

The coefficients d_{ijk} are the same as in (12.4).

12.2.3 Effect of Crystal Symmetry

To start with, let us consider the effect of the presence of a centre of symmetry on the piezoelectric property of a crystal. Let a stress be applied to the crystal which, as we have seen, produces electric polarization. Let us now subject the whole system to the process of inversion about the centre of symmetry. Since we have assumed that the crystal has a centre of symmetry, inversion leaves the crystal unaffected. The stress also remains unaffected since it is a symmetric tensor. The polarization components, on the other hand, become negative because of inversion. Thus, on inversion, the crystal and the applied stress remain unchanged and, yet, the polarization reverses sign. This anomaly can be avoided if the polarization is zero, i.e. the d_{ijk} values are zero. Thus, we have arrived at the important result that the piezoelectric effect can occur only in crystals in which the centre of symmetry is absent. Out of the 32 point groups, the centre of symmetry is absent in 21 point groups. Piezoelectricity can, in principle, occur only in these 21 crystal classes.

Let us now consider the effect of symmetry on the number of piezoelectric coefficients in a given crystal class. As an example we shall consider monoclinic class 2. This crystal class has a diad axis which we shall assume to be parallel to the Ox_3 axis. A twofold rotation about the Ox_3 axis transforms the coordinate indices as follows: $1 \to -1$, $2 \to -2$, $3 \to 3$. The symbol $\to$ reads "transforms to". The indices in the coefficients d_{ijk} will change in the same manner. Thus, $d_{133} \to -d_{133}$. Since the diad axis is a symmetry element of the crystal, the coefficient should remain unaffected according to Neumann's principle. Conformity with Neumann's principle can be ensured by equating d_{133} to zero. We will arrive at the same result if we consider other coefficients with two 3's in the indices; only the coefficients with one or three 3's will have non-zero values. Thus the non-zero tensor components for class 2 are

$$\underset{\text{1st layer}}{\begin{bmatrix} 0 & 0 & d_{113} \\ 0 & 0 & d_{123} \\ d_{131} & d_{132} & 0 \end{bmatrix}} \quad \underset{\text{2nd layer}}{\begin{bmatrix} 0 & 0 & d_{213} \\ 0 & 0 & d_{223} \\ d_{231} & d_{232} & 0 \end{bmatrix}} \quad \underset{\text{3rd layer}}{\begin{bmatrix} d_{311} & d_{312} & 0 \\ d_{321} & d_{322} & 0 \\ 0 & 0 & d_{333} \end{bmatrix}}$$

In the two-index notation, the piezoelectric matrix becomes

$$\begin{pmatrix} 0 & 0 & 0 & d_{14} & d_{15} & 0 \\ 0 & 0 & 0 & d_{24} & d_{25} & 0 \\ d_{31} & d_{32} & d_{33} & 0 & 0 & d_{36} \end{pmatrix}.$$

The number of independent d_{ij} coefficients is 8. Some crystals in class 2 have the diad axis oriented along the Ox_2 axis. It can be easily shown that in such a case, coefficients like d_{122} will be reduced to zero and only coefficients having one or three 2's will remain leading to a total of eight independent non-zero d_{ij} coefficients.

Following the above procedure, the independent non-zero coefficients for the other 20 crystal classes can be worked out. It will be found that the number of coefficients varies from 1 to 18. Only in cubic class 432, all the coefficients vanish though it is a non-centrosymmetric class. The scheme of piezoelectric matrices for all the 21 non-centrosymmetric crystal classes is given in Table 12.2 [12.1].

12.2.4 Experimental Measurement

There are several methods for the measurement of the piezoelectric coefficients. Most of them involve the measurement of the frequency of a piezoelectric crystal included in a resonant circuit. The frequency depends on the experimental conditions. For example, the frequency is different for an electroplated crystal (f_p) and for a bare crystal (f_b). Mason [12.11] has shown that for a 45° Z-cut KDP crystal, the d_{36} coefficient is given by

$$1 - \frac{d_{36}^2}{4}\left(\frac{4\pi}{k_3 S_{11}}\right) = \left(\frac{f_p}{f_b}\right)^2, \tag{12.12}$$

where k_3 is the dielectric constant in the x_3 direction and S_{11} the elastic compliance. Similarly for an X-cut KDP crystal, d_{14} can be obtained from

$$d_{14}^2 = 4\left(\frac{k_1 S_{22}}{4\pi}\right)\left[1 - \left(\frac{f_p}{f_b}\right)^2\right], \tag{12.13}$$

where k_1 and S_{22} are the appropriate dielectric constant and elastic compliance.

Again, to allow room for the crystal to vibrate, there will be a gap between the crystal surface and the capacitance plate; the frequency depends on the width of the gap. If the frequency is f_0 for a small gap and f_∞ for a large gap, Cady [12.6] has shown that for an X-cut bar of Rochelle salt,

$$d_{12}^2 = \left(\frac{k}{4\pi}\right)\left(\frac{1}{4l^2\rho}\right)\left(\frac{1}{f_0{}^2} - \frac{1}{f_\infty{}^2}\right), \tag{12.14}$$

where k is the dielectric constant in the particular direction, l the length of the bar and ρ the density of the crystal.

Table 12.2 Piezoelectric matrices (d_{ij}) for different crystal classes

KEY TO NOTATION

·	Zero modulus
•	Non-zero modulus
•—•	Equal moduli
•—o	Moduli numerically equal, but opposite in sign
	A modulus equal to minus 2 times the heavy dot modulus to which it is joined

TRICLINIC

Class 1

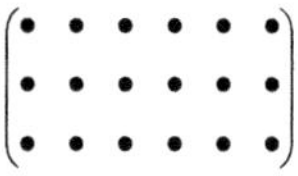

(18)

MONOCLINIC

$2 \parallel x_2$ (standard orientation)

Class 2

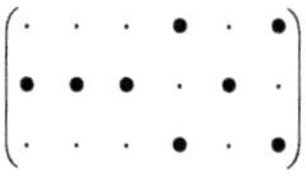

(8)

$2 \parallel x_3$

Class 2

(8)

$m \perp x_2$ (standard orientation)

Class m

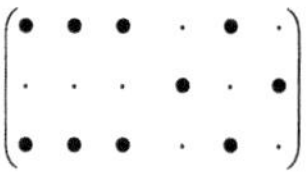

(10)

$m \perp x_3$

Class m

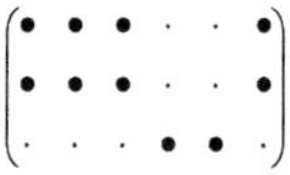

(10)

ORTHORHOMBIC

Class 222

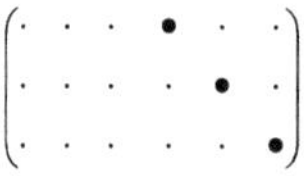

(3)

Class $mm2$

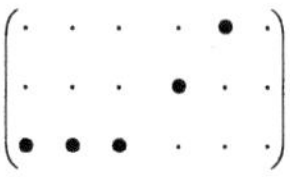

(5)

TETRAGONAL

Class 4

(4)

Class $\bar{4}$

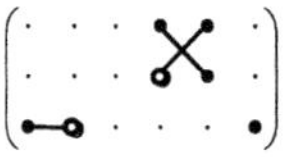

(4)

Class 422

(1)

Class $4mm$

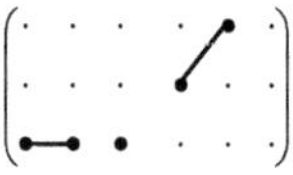

(3)

$2 \parallel x_1$ Class $\bar{4}2m$

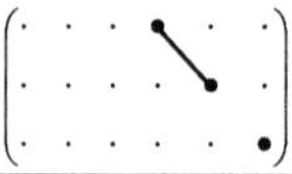

(2)

(*continued*)

Table 12.2 (continued)

			TRIGONAL				
	Class 3		(6)		Class 32	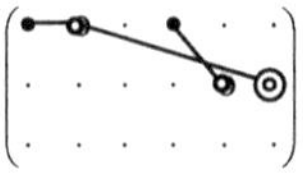	(2)
$m \perp x_1$ (standard orientation)	Class 3*m*	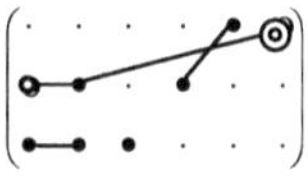	(4)	$m \perp x_2$	Class 3*m*	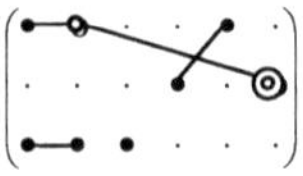	(4)
			HEXAGONAL				
	Class 6	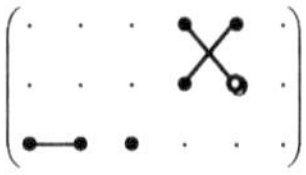	(4)		Class 6*mm*		(3)
	Class 622		(1)		Class $\bar{6}$	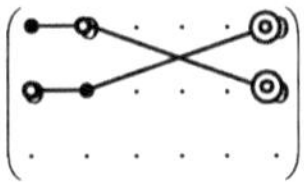	(2)
$m \perp x_1$ (standard orientation)	Class $\bar{6}m2$	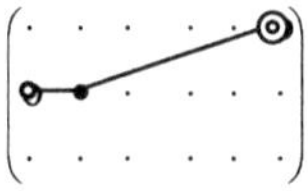	(1)	$m \perp x_2$	Class $\bar{6}m2$	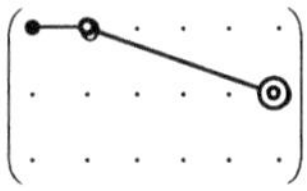	(1)
			CUBIC				
	Class 432	 All moduli vanish	(0)		Classes $\bar{4}3m$ and 23	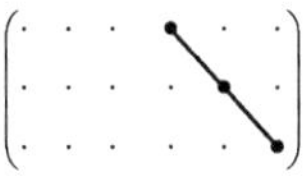	(1)

Sophisticated circuitry permits accurate determination of the frequencies. In view of the technical importance of piezoelectric crystals, instruments are now available for quick and easy measurements in which a particular coefficient (d_{ij}) can be read off once the crystal is inserted into the holder.

12.2.5 Experimental Results

In view of the technical importance of piezoelectric crystals, experimental data are available for a large number of crystals. In order to have an idea of the relative

Table 12.3 Piezoelectric constants of some crystals [12.9]

Crystal	d_{ij} $[10^{-8}$ cgs esu$]$							
	d_{14}	d_{16}	d_{21}	d_{22}	d_{23}	d_{25}	d_{34}	d_{36}
Potassium tartrate	−25.0	6.5	−2.2	8.5	−10.4	−22.5	29.4	−66.0
Lithium sulphate monohydrate	14.0	−12.5	11.6	−45.0	− 5.5	16.5	−26.4	10.0
	d_{15}	d_{24}	d_{31}	d_{32}	d_{33}			
Resorcinol	53.9	55.3	−12.4	−12.8	16.8			
	d_{15}	d_{22}	d_{31}	d_{33}				
Tourmaline	10.9	1.0	1.03	5.5				
	d_{15}	d_{31}	d_{33}					
$BaTiO_3$ (R.T.)	1,176	−104	250					
	d_{14}	d_{25}	d_{36}					
Rochelle salt (R.T.)	1,150	−160	35					
	d_{14}	d_{36}						
KDP (R.T.)	4.2	69.6						
ADP (R.T.)	5.2	−145.0						
	d_{11}	d_{14}						
Quartz (α)[a]	−6.76	2.56						
	d_{14}							
$NaClO_3$	6.1							
ZnS (sphalerite)	−9.70							

[a]Some authors refer to it as β-quartz

magnitudes of different coefficients of a crystal and of different crystals, values of piezoelectric coefficients of some crystals are given in Table 12.3. The units are often given in literature in cgs esu; these can be converted to mks units (Coul/N) with a multiplying factor of 3×10^4. It may be noted that there is a difference in the orientation scheme (left-handed or right-handed) followed by different authors resulting in a difference in sign, e.g. d_{14} for $NaClO_3$ is given as -4.8×10^{-8} by Cady [12.6] and as 6.1×10^{-8} by Zheludev [12.9], both in cgs esu.

Many piezoelectric crystals are water-soluble or become hydrated easily; for technical applications, water-insoluble and otherwise stable materials are preferable. Also, for several applications, materials with large values of piezoelectric coefficients are desirable. In Table 12.3, only barium titanate and Rochelle salt have large d values. A breakthrough in this direction was achieved when Jaffe et al. [12.12] prepared ceramic solid solutions of lead titanate and lead zirconate with added impurities and observed that the solid solution $Pb(Zr_{.45}Ti_{.55})O_3$ with 30% $PbO{:}SnO_2$ has an average d_{31} value of 222×10^{-8} cgs esu. This compound has the perovskite structure and is known by the abbreviated name PZT. This discovery was the starting point of studies of many ceramic piezoelectric materials. The National Physical Laboratory, New Delhi fabricates several PZT materials; one composition in particular, called NPL-ZT-5R, is found to have an average value of $1{,}275 \times 10^{-8}$ cgs esu for d_{33}. Polycrystalline ceramic materials are called piezoelectric "textures".

12.2.6 Applications

Piezoelectricity is one of the properties which has found numerous applications. Piezoelectric crystals like quartz, Rochelle salt and PZT have found use in stabilized oscillatory circuits, as frequency and time standards (quartz clock) and in ultrasonic generators. They are also used in highly selective filters and in electromechanical transducers used to detect mechanical vibrations.

12.3 Pyroelectricity

Pyroelectricity is the property which enables a crystal to develop electric polarization when its temperature changes. If a change in temperature ΔT results in polarization ΔP_i, then

$$\Delta P_i = p_i \, \Delta T. \tag{12.15}$$

The p_i values are called pyroelectric coefficients. The p_i values relate a vector ΔP_i to a scalar ΔT; hence the p_i values constitute a first-rank tensor or a vector. Pyroelectricity is one of the very few vector physical properties of crystals. Obviously, there are three pyroelectric coefficients.

Let us assume that a crystal having a centre of symmetry is pyroelectric. Let its temperature be changed by ΔT and let it develop a polarization ΔP_i. If the crystal is now subjected to inversion, the coordinates x_1, x_2, x_3 of a point will change to $-x_1$, $-x_2$, $-x_3$. ΔP_i will then become negative and so will the p_i values. But since the crystal has a centre of symmetry, the p_i values should not be affected by inversion. This can be ensured if all the p_i values are equal to zero, i.e. the crystal is not pyroelectric. Thus the pyroelectric effect can exist only in crystals lacking a centre of symmetry. This was the condition for existence of piezoelectricity. We must hasten to add that the absence of centre of symmetry is not a sufficient condition for pyroelectricity. Being a vector property, pyroelectricity tends to appear in a unique direction in a crystal (like a two-, three-, four-, sixfold axis). These two conditions (absence of centre of symmetry and presence of a unique axis) are satisfied by only 10 of the 21 non-centrosymmetric crystal classes. Thus while all pyroelectric crystals are piezoelectric, the converse is not true. The components of the pyroelectric vector **p** in the ten crystal classes [12.1] are:

Triclinic
Class 1: No symmetry restriction on the direction of **p**: (p_1, p_2, p_3).
Monoclinic; x_2 parallel to the diad axis, rotation or inversion, (y).
Class 2: **p** parallel to the diad axis: $(0, p, 0)$.
Class m: **p** has any direction in the symmetry plane: $(p_1, 0, p_3)$
Orthorhombic; x_1, x_2, x_3 parallel to crystallographic x, y, z respectively.
Class $mm2$: **p** parallel to the diad axis: $(0, 0, p)$.

Tetragonal, trigonal, hexagonal; x_3 parallel to z.
Classes 4, $4mm$, 3, $3m$, 6, $6mm$: **p** parallel to the 4, 3 or 6 axis: $(0, 0, p)$.

The pyroelectric coefficients can be experimentally determined by heating the crystal uniformly and measuring the resulting polarization (electric charge). The experiment has to be done carefully as the charged crystal gets easily neutralized by attachment of atmospheric charges to the crystal surface.

Tourmaline which was the first crystal in which pyroelectricity was observed has a pyroelectric coefficient of value -1.28 cgs esu $(°\mathrm{C})^{-1}$. Triglycine sulphate, barium titanate, Rochelle salt, lithium niobate, lithium sulphate decahydrate and resorcinol are other examples of pyroelectric crystals.

Part B: Ferroelectricity

12.4 General

In 1920, Joseph Valasek observed that the crystalline compound Rochelle salt has unusual dielectric properties like a high dielectric constant, a spontaneous polarization which can be reversed by an electric field and hysteresis. Valasek pointed out the qualitative similarity between this dielectric behaviour and the behaviour of ferromagnetic materials. Such dielectric materials are called "*ferroelectrics*". Subsequently, ferroelectric behaviour has been observed in potassium dihydrogen phosphate, barium titanate, triglycine sulphate and a host of other materials. In this part, we shall discuss experimental and theoretical aspects of ferroelectricity.

12.5 Characteristic Properties of Ferroelectrics

Ferroelectric materials have the following characteristic properties.

12.5.1 Spontaneous Polarization

A ferroelectric exhibits polarization in the absence of an external field. The direction of the spontaneous polarization may be reversed by the application of electric field. The relation between the polarization and the externally applied field is a hysteresis loop. This is analogous to the behaviour of a ferromagnetic material. The observation of hysteresis is explained on the basis of "*domains*". The domains are regions within which the dipoles are aligned in the same direction but the direction of polarization is different in different domains.

12.5.2 Phase Transition

Ferroelectric behaviour is exhibited below a certain temperature above which the ferroelectricity disappears. The temperature at which the change occurs is called the transition temperature T_C. Above T_C the crystal is said to be in the paraelectric phase. The transition from the paraelectric to ferroelectric state is accompanied by a change of crystal structure and sharp changes in the dielectric constant. The structure in the ferroelectric phase is always of a lower symmetry compared to the structure in the paraelectric phase. For instance, barium titanate changes from cubic structure in the paraelectric phase to tetragonal structure in the ferroelectric phase.

12.5.3 Temperature Variation of Dielectric Constant

Above the transition temperature, the dielectric constant obeys Curie–Weiss law

$$\varepsilon = \frac{A}{T - \theta}, \tag{12.16}$$

where A is a constant and θ is the Curie–Weiss temperature.

The dielectric susceptibility χ is given by

$$\chi = \frac{\varepsilon - 1}{4\pi} \approx \frac{C}{T - \theta}, \tag{12.17}$$

where C is the Curie constant.

It can be seen from (12.16) that the dielectric constant assumes large values (a few hundreds or even thousands) in the vicinity of θ. Further, (12.16) predicts that a $(1/\varepsilon)$ versus T plot would be linear with an intercept θ on the T-axis. Such a plot for a representative ferroelectric is shown in Fig. 12.1. It may be mentioned that θ is equal to T_C or in some cases, a few degrees less.

12.5.4 Symmetry Considerations

We have seen in Sect. 12.3 that the lack of a centre of symmetry is a condition for a crystal to be a piezoelectric. Out of 32 crystal classes (point groups), 21 are non-centrosymmetric and are piezoelectric. Further, if the point group symmetry includes a unique polar axis, the crystal is pyroelectric. There are only ten classes satisfying the two conditions. Thus if the symmetry elements of a crystal are known it is possible to predict whether the crystal is piezoelectric or pyroelectric.

Following Weinreich [12.13], these symmetry requirements are shown in Fig. 12.2. The first diagram (a) has a centre of symmetry at point C. A crystal

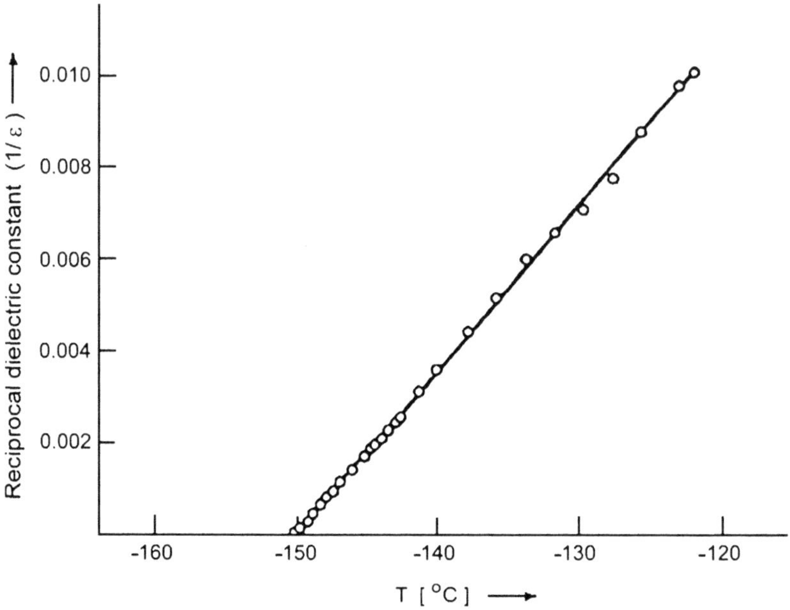

Fig. 12.1 Reciprocal dielectric constant as a function of temperature for KH_2PO_4

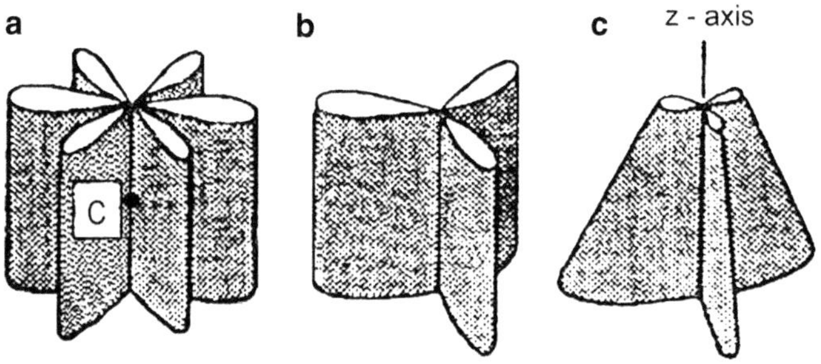

Fig. 12.2 Symmetry requirements for occurrence of ferroelectricity (**a**, **b**, **c** explained in text)

with this symmetry cannot be piezoelectric or ferroelectric. Diagram (b) has no centre of symmetry but has three equivalent piezoelectric directions. A crystal with these symmetry elements can be piezoelectric but not ferroelectric. Finally, diagram (c) has no centre of symmetry and has a single unique axis (z-axis). A crystal with these symmetry elements qualifies to be a ferroelectric. In addition, occurrence of ferroelectricity depends on the nature of charge distribution and the changes on application of electric field.

12.6 Classification of Ferroelectrics

There are different approaches for the classification of ferroelectrics. Some of them are:

12.6.1 Classification Based on Crystal Symmetry

(a) Ferroelectrics with a single axis along which spontaneous polarization can occur, e.g. Rochelle salt: sodium potassium tartrate tetrahydrate ($NaKC_4H_4O_6 \cdot 4H_2O$) and potassium di hydrogen phosphate (KH_2PO_4).
(b) Ferroelectrics with several equivalent axes along which spontaneous polarization can occur, e.g. barium titanate ($BaTiO_3$).

12.6.2 Classification Based on the Piezoelectric Property

(a) Ferroelectrics which are piezoelectric in the unpolarized state, e.g. Rochelle salt and related tartrates and KH_2PO_4-type crystals.
(b) Ferroelectrics which are not piezoelectric in the unpolarized state, e.g. $BaTiO_3$.

12.6.3 Classification Based on Theoretical Concepts

(a) The order–disorder class which includes KH_2PO_4-type crystals in which each proton corresponding to a hydrogen bond makes a tunnelling motion between two equilibrium positions.
(b) The displacive class of ferroelectrics which include $BaTiO_3$-type crystals. In this class of crystals each atom vibrates about an equilibrium configuration.

12.7 Behaviour of Some Representative Ferroelectrics

The crystal structures of KH_2PO_4, Rochelle salt and $BaTiO_3$ are shown in Fig. 12.3.

12.7.1 Potassium Dihydrogen Phosphate KH_2PO_4

The crystal undergoes a single-phase transition from a paraelectric to ferroelectric phase at 121 K. The crystal structure changes from tetragonal structure ($D_{2d} - \overline{4}2m$) in the paraelectric phase to orthorhombic structure (of lower symmetry) in the

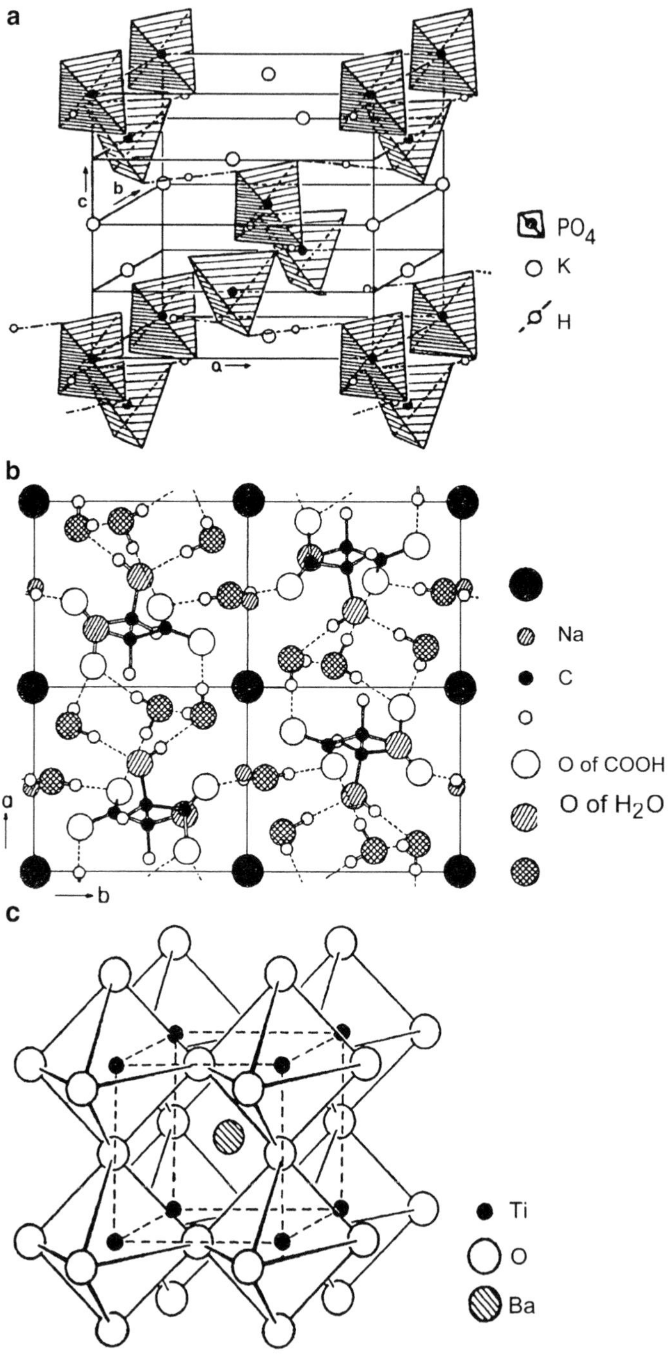

Fig. 12.3 Crystal structures of (**a**) KH_2PO_4, (**b**) Rochelle salt and (**c**) $BaTiO_3$

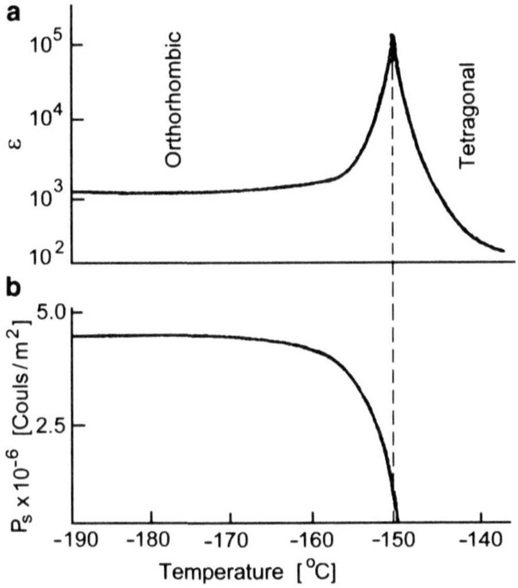

Fig. 12.4 Temperature variation of (**a**) dielectric constant and (**b**) spontaneous polarization for KH_2PO_4

Table 12.4 Transition temperatures and spontaneous polarization in some KDP-type crystals

Crystal	Transition temperature	Spontaneous polarization	
	T_C [K]	$P_s[\mu C/cm^2]$	at temp. [K]
KH_2PO_4	121	4.75	96
KD_2PO_4	213	4.83	180
KH_2AsO_4	97	5.00	78
KD_2AsO_4	162	–	–
RbH_2PO_4	146	5.60	90
RbD_2PO_4	218	–	–

ferroelectric phase. It has only one ferroelectric axis, i.e. the c-axis. The temperature variation of dielectric constant and spontaneous polarization are shown in Fig. 12.4. From the structure, it is seen that the PO_4 group forms a tetrahedron with the phosphorus atom at the centre and oxygen atoms at the corners. The phosphate groups are bound together by hydrogen bonds. Different possible arrangements of hydrogen atoms result in different orientations of $(H_2PO_4)^-$ dipoles. When hydrogen is replaced by deuterium, the transition temperature changes from 121 to 213 K. This change in the transition temperature shows the involvement of hydrogen bonds in polarization of the crystal. The transition temperature and spontaneous polarization for some KH_2PO_4-type crystals are listed in Table 12.4.

12.7.2 Rochelle Salt

Rochelle salt in the unpolarized state belongs to the orthorhombic crystal class $D_{2d} - 222$. The orthorhombic a axis is the ferroelectric axis. ε_a shows peaks at

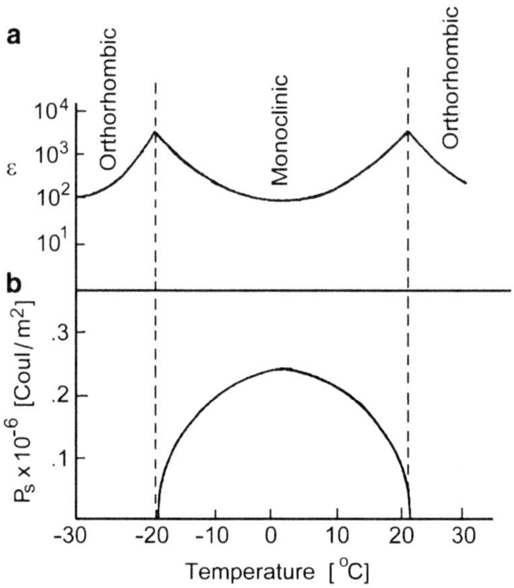

Fig. 12.5 Temperature variation of (**a**) dielectric constant and (**b**) spontaneous polarization for Rochelle salt

two temperatures 255 and 297 K. The crystal is spontaneously polarized only in the region between the two transition temperatures. The crystal symmetry is lowered to monoclinic in the ferroelectric phase. The variation of dielectric constant and spontaneous polarization of Rochelle salt with temperature are shown in Fig. 12.5. When hydrogen is replaced by deuterium, the temperature range over which the material is ferroelectric and the magnitude of spontaneous polarization are found to change.

12.7.3 Barium Titanate ($BaTiO_3$)

$BaTiO_3$ is the representative crystal of the perovskite group. This has a simple structure compared to the two groups mentioned earlier. The crystal has a cubic symmetry ($O_h - m3m$) in the unpolarized state because of which the crystal has many sets of equivalent axes along which polarization can take place. Perovskites exhibit three transition temperatures. For $BaTiO_3$, the highest transition temperature corresponds to a transition from paraelectric to ferroelectric state at 408 K. The transition is accompanied by a change to tetragonal structure with the direction of polarization along the polar axis [001]. The crystal structure of a unit cell in the unpolarized state is shown in Fig. 12.6a. The cubic structure has Ba^{2+} ions at the corners, O^{2-} ions at the face-centres and a Ti^{4+} ion at the body-centre. Below the transition temperature Ba^{2+} and Ti^{4+} ions are displaced with respect to the position of O^{2-} ion. Oxygen ions are also displaced to a small extent in the opposite direction. The deformed structure is the tetragonal structure. Figure 12.7 shows the change in polar axis as the temperature is lowered. The transitions at the temperatures 278 and 193 K correspond to the successive changes

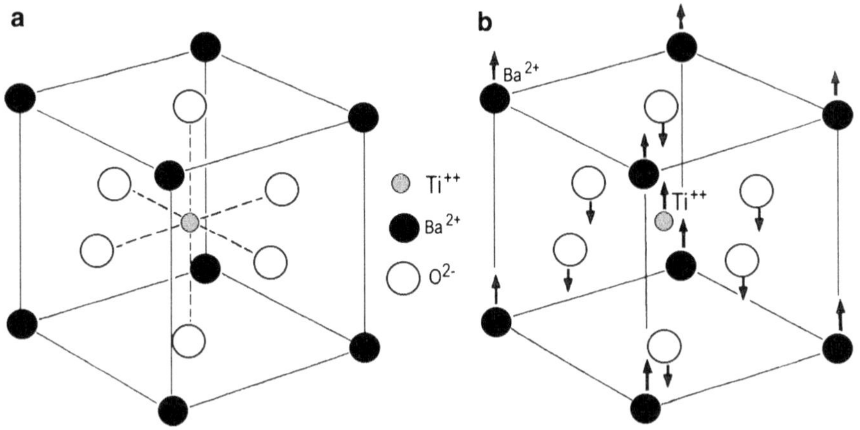

Fig. 12.6 (**a**) The crystal structure of $BaTiO_3$ and (**b**) displacement of ions below the transition temperature for $BaTiO_3$

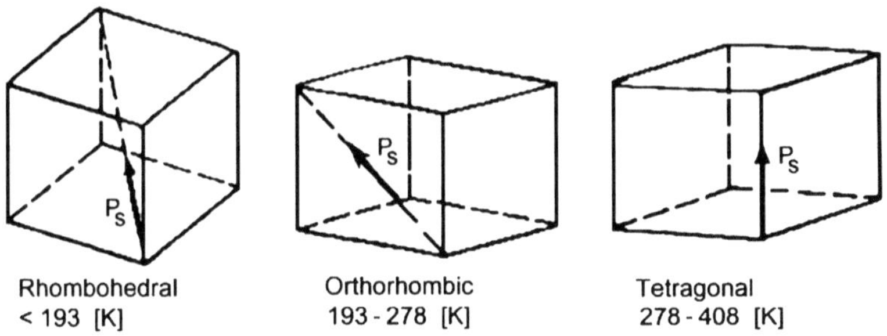

Fig. 12.7 Polar axis and the change in the structure as a function of temperature for $BaTiO_3$

in polar axis to [011] and [111]. This is accompanied by a change in the crystal structure to orthorhombic and rhombohedral respectively. The temperature variation of dielectric constant and spontaneous polarization are shown in Fig. 12.8.

$BaTiO_3$ exhibits thermal hysteresis, i.e. the actual transition temperature is found to occur at a slightly higher value if reached from the low temperature side instead of from high temperature side. This can be seen in Fig. 12.8a. Transition temperatures and spontaneous polarization for a few crystals of $BaTiO_3$ type are listed in Table 12.5.

12.8 Theoretical Aspects

12.8.1 Dipole Theory

For a crystal to be spontaneously polarized, the dipole moment of different unit cells should be oriented along a common direction. The dipole moment may be due to permanent dipoles or arising out of electronic and ionic displacements.

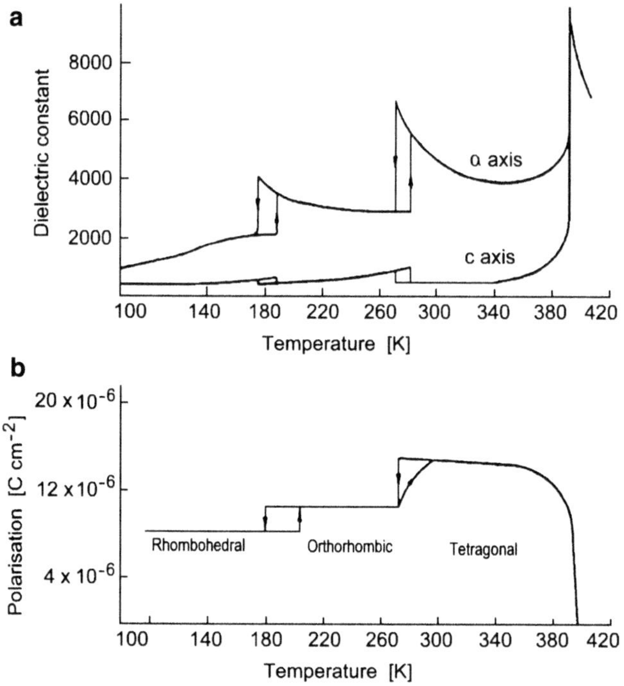

Fig. 12.8 Temperature variation of (**a**) dielectric constant and (**b**) spontaneous polarization for $BaTiO_3$

Table 12.5 Transition temperatures and spontaneous polarization in some perovskites

Crystal	Transition temperature $T_C[K]$	Spontaneous polarization $P_s[\mu C/cm^2]$	at temp. [K]
$BaTiO_3$	408	26.0	296
$KNbO_3$	708	30.0	523
$PbTiO_3$	763	>50	296
$LiTaO_3$	938	50	221
$LiNbO_3$	1,480	71	296

In the dipole theory, the presence of permanent dipoles is assumed. It is assumed that the internal field E_i required to orient a dipole along the direction of the applied field is given by

$$E_i = E + \gamma P, \tag{12.18}$$

where E is the applied field, γ the internal field constant and P the polarization.

Initially, at a certain temperature the dipoles are assumed to be oriented at random. As the temperature is lowered the internal field E_i tends to orient the dipoles along the direction of the applied field leading to paraelectric–ferroelectric transition. The behaviour of the dipoles and the phase transition are explained in a way similar to Langevin–Weiss theory of ferromagnetism.

The polarization P is obtained by using the Langevin function (Chap. 11, Appendix B) as

$$P = N\left(\frac{\mu^2}{3k_BT}\right)E_i, \tag{12.19}$$

where N is the number of dipoles per unit volume with dipole moment μ. This is true for $\mu E_i << k_BT$, i.e. at high temperatures. Using (12.18) and (12.19) the susceptibility χ is obtained as

$$\chi = \frac{P}{E} = \left(\frac{N\mu^2/3k_BT}{1-\gamma\,(N\mu^2/3k_BT)}\right)E_i. \tag{12.20}$$

Equation (12.20) may be written as

$$\chi = \frac{\theta/\alpha}{T-\theta} = \frac{C}{T-\theta}, \tag{12.21}$$

which has the form of Curie–Weiss law (12.17) with Curie temperature $\theta = \gamma(N\mu^2/3k_B)$ and Curie constant $C = \theta/\gamma$.

The dipole theory could account for ferroelectric behaviour in Rochelle salt with permanent dipole moment due to H_2O molecules and also for KH_2PO_4 in which the structure involves intermolecular hydrogen bonds O–H···O leading to formation of $H_2PO_4^-$ dipoles.

The theory was found to be only qualitatively successful.

12.8.2 Theory Based on Ionic Displacements

We examined the Clausius–Mosotti equation (see Chap. 11) and showed that ferroelectric behaviour may be obtained on the basis of electronic and ionic displacements only without the assumption of permanent dipoles.

The Clausius–Mosotti relation in macroscopic form may be written as

$$\frac{M}{\rho}\frac{\varepsilon-1}{\varepsilon+2} = \frac{4\pi}{3}N_A\alpha, \tag{12.22}$$

where M is the molecular weight, ρ is the density and N_A the Avogadro number. Equation (12.22) can be written as

$$\frac{\varepsilon-1}{\varepsilon+2} = \frac{4\pi}{3}\frac{\alpha}{V}, \tag{12.23}$$

where α is the total polarizability of a macroscopic, small sphere of the substance of volume V.

Assuming α to be independent of temperature and differentiating (12.23) with respect to T, we get

$$\frac{3}{(\varepsilon+2)(\varepsilon-1)}\frac{\mathrm{d}\varepsilon}{\mathrm{d}T} = -\frac{1}{V}\frac{\mathrm{d}V}{\mathrm{d}T} = -3\lambda, \tag{12.24}$$

where λ is the linear coefficient of expansion of the solid. For ferroelectrics $\varepsilon >> 1$. Hence, we obtain from (12.24)

$$\int \frac{\mathrm{d}\varepsilon}{\varepsilon^2} = -\lambda\,\mathrm{d}T \tag{12.25}$$

or

$$\varepsilon = \frac{1/\lambda}{T-\theta}, \tag{12.26}$$

where θ is a constant with dimensions of temperature. Note that θ arises out of a constant of integration. Equation (12.26) has the form of Curie–Weiss equation with $1/\lambda$ as the Curie constant and θ the Curie–Weiss temperature.

We now examine whether Clausius–Mosotti relation leads to other characteristics of ferroelectrics. Let us consider Clausius–Mosotti equation in microscopic form:

$$\frac{\varepsilon-1}{\varepsilon+2} = \frac{4\pi}{3}\sum N_i\alpha_i, \tag{12.27}$$

where α_i is the sum of electronic and ionic polarizabilities of an ion of type i and N_i is the number of ions of type i per unit volume. We write (12.27) in the form

$$\varepsilon = \frac{1+(8\pi/3)\sum N_i\alpha_i}{1-(4\pi/3)\sum N_i\alpha_i}. \tag{12.28}$$

From (12.28) we find that, if $(4\pi/3)\ \Sigma N_i\alpha_i = 1$, the dielectric constant becomes infinite and it results in spontaneous polarization. The condition $\varepsilon \to \infty$ is termed "polarization catastrophe". Thus the theory based on ionic displacements accounts for the Curie–Weiss law behaviour and spontaneous polarization.

The question still remains as to why a large number of crystals of perovskite structure exhibit ferroelectricity. For this we consider the local field assumption in deriving Clausius–Mosotti relation (see Chap. 11). When all the atoms in a crystal are in cubic environment, the local field is given by $(E + 4\pi P/3)$. In $BaTiO_3$, the titanium and barium ions are in cubic environment. But, only two titanium ions are present on either side of the oxygen ions. Hence oxygen ions are in non-cubic environment. The local field is not $4\pi P/3$ at all the sites. The field has to be evaluated at the individual sites. Evaluation of the internal field using a detailed procedure given by Slater [12.14] shows that the interaction between titanium and oxygen ions is very strong and the local field at these ions is nearly eight times $4\pi P/3$. Hence the condition for polarization catastrophe is satisfied even for small values of $\Sigma\ N_i\alpha_i$.

12.8.3 Thermodynamic Theory

Landau [12.15] proposed a phenomenological theory of phase transitions. The theory is based on thermodynamical aspects. According to his theory symmetry-lowering phase transition is associated with an "order parameter". He postulated that the free energy of a system depends on the order parameter. This theory was applied to phase transition in $BaTiO_3$ by Ginzberg [12.16] and later by Devonshire [12.17]. The theory developed by Devonshire is now discussed.

The theory is independent of any particular atomic model. It involves the study of the behaviour of a substance at the transition temperature. Let us consider a solid which is ferroelectric for temperatures $T < T_C$. The order parameter is the polarization P. We assume that the spontaneous polarization occurs along a single axis when a field is applied along that axis. This is applicable in the case of Rochelle salt, KH_2PO_4 and for the cubic-to-tetragonal transition in $BaTiO_3$.

Let G_0 represent the free energy of the unpolarized crystal. The free energy G of the polarized crystal is expressed as a function of polarization P as

$$G = G_0 + \tfrac{1}{2}\alpha^* P^2 + \tfrac{1}{4}\gamma P^4 + \tfrac{1}{6}\delta P^6 + \cdots . \tag{12.29}$$

The coefficients α^*, γ, δ are functions of temperature; the numerical coefficients are included for mathematical convenience; the symbol α^* is introduced to distinguish it from α which is generally used to represent polarizability. Only even powers of P are included to ensure that the free energy is the same for positive and negative polarizations along the polar axis.

The change in the internal energy $\mathrm{d}U$ of a system subjected to electric field E in which polarization takes place is given by

$$\mathrm{d}U = T\mathrm{d}S + E\mathrm{d}P, \tag{12.30}$$

where S is the entropy. The free energy G is given by

$$G = U - TS. \tag{12.31}$$

On differentiating (12.31) and using (12.30), we have

$$\mathrm{d}G = -S\mathrm{d}T + E\mathrm{d}P. \tag{12.32}$$

The applied field is given by

$$E = (\partial G/\partial P)_T = \alpha^* P + \gamma P^3 + \delta P^5. \tag{12.33}$$

Above the transition temperature, P is small and so the higher-order terms of P in (12.33) may be neglected. Hence

$$\left(\frac{\mathrm{d}E}{\mathrm{d}P}\right) = \alpha^*. \tag{12.34}$$

Under thermal equilibrium, G must be minimum. Hence $(\partial G/\partial P)_T = 0$ and the spontaneous polarization P_S satisfies the equation

$$\alpha^* P_S + \gamma P_S^3 + \delta P_S^5 = 0. \tag{12.35}$$

$P_S = 0$ corresponds to the only minimum of free energy at $E = 0$; hence spontaneous polarization cannot occur.

We now proceed to examine the condition for the spontaneous polarization to occur. If the first differential $(\partial G/\partial P)_T = 0$ and the second differential $(\partial^2 G/\partial P^2)_T = \alpha^*$ is positive, G is minimum and the corresponding state is a stable state for $P_S = 0$. On the other hand, if the second differential α^* is negative, G is maximum for $P_S = 0$ and it is not a stable state. However, there will exist a stationary state for a non-zero value of P_S.

According to Devonshire if α^* changes continuously from a positive to a negative value the stable state of the system will change from one with $P = 0$ to a state with P non-zero. Thus a ferroelectric transition is possible. The coefficient α^* is assumed to be a linear function of temperature having the form

$$\alpha^* = \beta(T - T_C). \tag{12.36}$$

Substituting for α^*, we have from (12.33)

$$E = (\partial G/\partial P)_T = \beta(T - T_C)P + \gamma P^3 + \delta P^5. \tag{12.37}$$

The equilibrium of the crystal depends on the change of free energy with temperature. This in turn depends on the sign of the coefficients in (12.37). Here, β is positive, δ is known to be positive in all known ferroelectrics. γ can be either positive or negative; we now consider these two cases.

(a) Let us consider γ as positive. β and δ are positive. Figure 12.9a shows the variation of free energy with P. The curves correspond to different temperatures as α^* varies from positive to negative values. It can be observed that the minima of the free energy correspond to $P = P_S$. P_S varies continuously with temperature as shown in Fig. 12.9b. Neglecting δ in (12.37), the spontaneous polarization P_S is given by

$$\beta(T - T_C)P_S + \gamma {P_S}^3 = 0. \tag{12.38}$$

The solutions are either $P_S = 0$ or ${P_S}^2 = -(\beta/\gamma)(T - T_C)$. For $T > T_C$, the only root is $P_S = 0$. For $T < T_C$, the minimum of the free energy is at

$$|P_S| = (\beta/\gamma)^{1/2}(T_C - T)^{1/2}. \tag{12.39}$$

Further, for this kind of transition no latent heat is observed and the specific heat shows a discontinuity. These are the characteristics of a second-order phase transition. KH_2PO_4 and Rochelle salt show second-order phase transition.

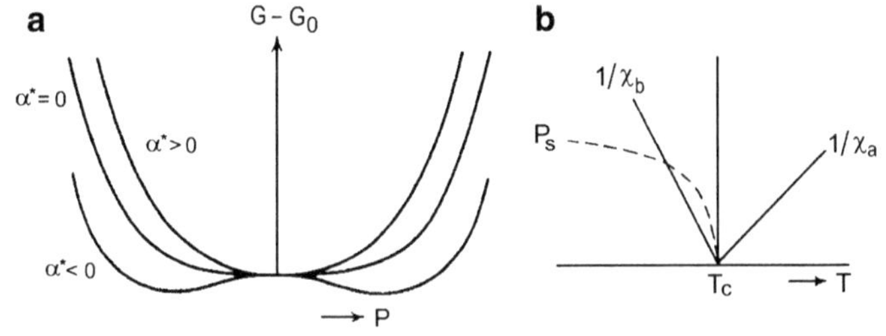

Fig. 12.9 (**a**) Variation of the free energy as a function of polarization for γ positive and (**b**) spontaneous polarization and reciprocal susceptibility as a function of temperature for a second-order transition

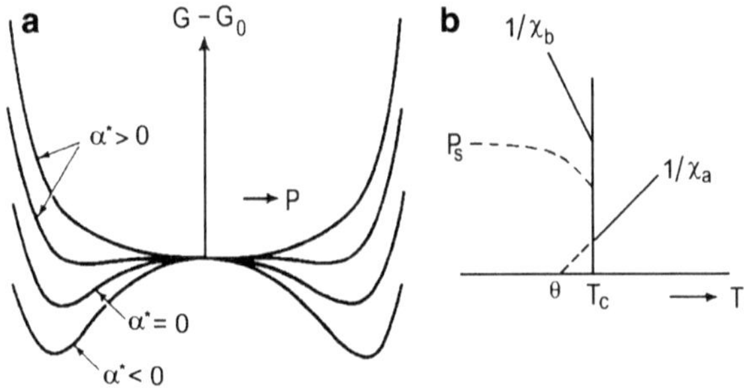

Fig. 12.10 (**a**) Variation of the free energy as a function of polarization for γ negative and (**b**) spontaneous polarization and reciprocal susceptibility as a function of temperature for a first-order transition

(b) Let us now consider γ as negative while β and δ are positive. Variations of free energy when α^* changes from positive to negative values are shown in Fig. 12.10a. The curves show that the minimum of the free energy for $P_S = 0$ and minima for a polarized state of non-zero P_S coexist. A transition will result when the two minima are equal. The polarization will jump at the transition temperature showing a discontinuity as shown in Fig. 12.10b. A discontinuity in entropy accompanied by latent heat is observed at the transition temperature. These are the characteristics of first-order phase transition. The transition in $BaTiO_3$ is an example of a first-order phase transition.

Let us consider the variation of susceptibility with temperature for the two cases discussed above.

Second-Order Phase Transition

If χ_a is the susceptibility above the transition temperature, we have from (12.34)

$$\frac{1}{\chi_a} = \frac{dE}{dP} = \alpha^*. \tag{12.40}$$

The susceptibility above the transition temperature is given by Curie–Weiss law $\chi_a = C/(T - \theta)$. At the transition $\alpha^* = 0$ and $\theta = T_C$. Hence

$$\frac{1}{\chi_a} = \frac{T - T_C}{C}. \tag{12.41}$$

In the ferroelectric region, P has higher values and so the higher-order terms in (12.33) cannot be neglected. Considering the first two terms of (12.33)

$$E = \alpha^* P + \gamma P^3. \tag{12.42}$$

When spontaneous polarization occurs, $E = 0$ and $P = P_S$, hence,

$$P_S^2 = -\alpha^*/\gamma. \tag{12.43}$$

If χ_b is the susceptibility below the transition temperature

$$\frac{1}{\chi_b} = \frac{dE}{dP} = \alpha^* + 3\gamma P_S^2. \tag{12.44}$$

Substituting for ${P_S}^2$ from (12.43) and assuming the temperature dependence of α^* is still given by (12.41) below T_C

$$\frac{1}{\chi_b} = -2\alpha^* = \frac{2(T_C - T)}{C}. \tag{12.45}$$

First-Order Phase Transition

We have seen earlier that the minimum for $P_S = 0$ is compatible with the existence of spontaneous polarization $P_S \neq 0$ for a first-order phase transition. At the critical temperature T_C, $G(T_C) = G_0(T)$ and we have from (12.29)

$$\tfrac{1}{2}\alpha^* + \tfrac{1}{4}\gamma P_S^2(T_C) + \tfrac{1}{6}\delta P_S^4(T_C) = 0. \tag{12.46}$$

In the absence of external field $E = 0$, $P = P_S$ and $(\delta G/\delta P)_T = 0$. Then

$$\alpha^* + \gamma P_S^2 + \delta P_S^4 = 0. \tag{12.47}$$

From (12.46) and (12.47), we get the relations:

$$P_S^2(T_C) = -\frac{3}{4}\frac{\gamma}{\delta}, \qquad P_S^4(T_C) = \frac{3\alpha^*}{\delta}, \qquad \alpha^* = \frac{3}{16}\frac{\gamma^2}{\delta}. \tag{12.48}$$

χ_a, the susceptibility above the transition temperature, is obtained by taking only the first term in (12.33), similar to that in second-order phase transition, and

$$\frac{1}{\chi_a} = \alpha^* = \frac{T-\theta}{C}. \tag{12.49}$$

For the region below the transition temperature,

$$E = \alpha^* P_S + \gamma P_S^3 + \delta P_S^5. \tag{12.50}$$

Hence,

$$\frac{dE}{dP} = \frac{1}{\chi_b} = \alpha^* + 3\gamma P_S^2 + 5\delta P_S^4. \tag{12.51}$$

Substituting for various terms in (12.51) from (12.48), and assuming Curie–Weiss law to hold, we get

$$\frac{1}{\chi_b} = 4\alpha^* = 4\frac{T-\theta}{C}. \tag{12.52}$$

At the transition temperature,

$$\frac{1}{\chi_a} = \frac{T_C-\theta}{C} \quad \text{and} \quad \frac{1}{\chi_b} = 4\frac{T_C-\theta}{C}. \tag{12.53}$$

For the first-order phase transition θ is slightly smaller than T_C. Plots of $1/\chi_a$ and $1/\chi_b$ for the first- and second-order transitions close to the transition temperature are shown in Figs. 12.9b and 12.10b.

12.8.4 Lattice Dynamical Theory

A major step in the microscopic understanding of ferroelectricity is the lattice dynamical theory proposed independently by Cochran [12.18] and Anderson [12.19]. The theory accounts for the ferroelectric behaviour in perovskite structure. As the detailed treatment is quite complicated only the essence of the theory is given here.

We consider the normal vibrational modes associated with a crystal (Chap. 10). The stability of a crystal at any temperature depends on the short range and Coulomb forces. For a ferroelectric crystal, at a certain temperature for one particular mode, short range and Coulomb forces almost cancel each other and the crystal becomes unstable. Such a mode is called a "ferroelectric mode" or "soft mode". Since the net restoring force is very small, the frequency of the soft mode is lower than those of other modes. The phenomenon of decrease of frequency of the ferroelectric mode with temperature is known as "softening". The soft mode is identified as the low frequency transverse optical phonon (ω_{TO}^s). As the temperature of a ferroelectric crystal is changed, the crystal becomes unstable at the transition temperature and drives itself to change over to a structure with lower symmetry.

We shall now consider the Lyddane, Sachs and Teller (LST) relation [12.20]

$$\frac{\varepsilon_s}{\varepsilon_\infty} = \frac{\omega_{LO}^2}{\omega_{TO}^2}, \tag{12.54}$$

where ε_s and ε_∞ are the static and high frequency dielectric constant respectively. ω_{LO} and ω_{TO} are the frequencies of the longitudinal and transverse optical modes respectively. Cochran [12.21] derived a generalized LST relation appropriate for ferroelectrics:

$$\frac{\varepsilon_s}{\varepsilon_\infty} = \frac{\prod_i (\omega_{LO})^2}{\prod_i (\omega_{TO})^2}, \tag{12.55}$$

where the product is over all the optic mode frequencies.

From (12.55) we find that if one of the frequencies is a soft mode, $\omega_{TO}^s \to 0$ as $T \to T_C$ and $\varepsilon_s \to \infty$. We can write (12.55) as

$$(\omega_{TO}^s)^2 \varepsilon_s = \varepsilon_\infty \left[\frac{\prod_i (\omega_{LO})^2}{\prod_i' (\omega_{TO})^2} \right] = A, \tag{12.56}$$

where the product $\prod'$ is taken over all TO frequencies except ω_{TO}^s and A is a constant. Hence

$$(\omega_{TO}^s)^2(T) = \frac{A}{\varepsilon_s(T)}. \tag{12.57}$$

Since, from the Curie-Weiss law

$$\varepsilon_s(T) = \frac{C}{T - T_C}, \tag{12.58}$$

we get

$$(\omega_{TO}^s)^2(T) = B(T - T_C). \tag{12.59}$$

Here, A/C is replaced by another constant B. Equations (12.57–12.59) indicate that both ε_s^{-1} and $(\omega_{TO}^s)^2$ vary linearly with temperature.

Historically, the study of lattice vibration modes in relation to phase transitions goes back to 1940 when Raman and Nedungadi [12.22] observed the softening of an optic phonon in the structural phase transition in quartz. Later Cochran's lattice dynamical studies led to a similar phenomenon in ferroelectric phase transitions. Cowley [12.23] discovered a soft mode in neutron scattering experiments on $SrTiO_3$ and many other perovskites including $KTaO_3$, $PbTiO_3$ and $BaTiO_3$. The soft mode in $BaTiO_3$ was studied by Luspin et al. [12.24] from i.r. reflectivity measurements. The results shown in Fig. 12.11 confirm that the temperature variation follows (12.59). The inverse dielectric constant (Fig. 12.11, inset) shows a similar variation as expected from (12.58).

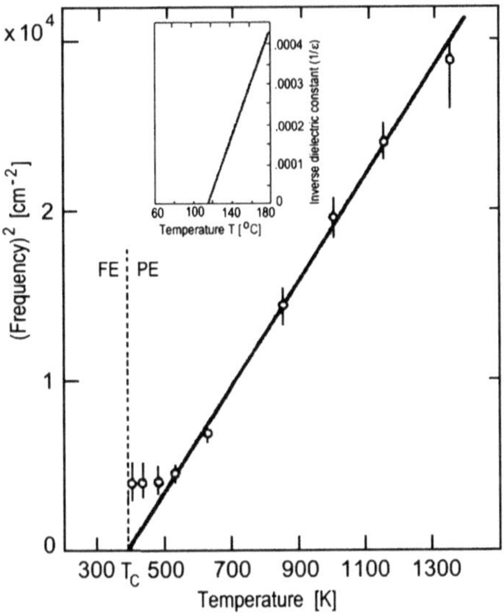

Fig. 12.11 Square of optic mode frequency and reciprocal dielectric constant (*inset*) as a function of temperature for $BaTiO_3$

12.9 Ferroelectric Domains

12.9.1 Description of Domain Structure

A ferroelectric crystal is composed of a number of domains which contain a large number of dipoles aligned in the same direction within a domain. Consequently, each domain has a macroscopic spontaneous polarization. A ferroelectric crystal splits into domains because such a splitting lowers the free energy by reducing the electrostatic energy of the spontaneous polarization charges. The domains are separated by domain walls which are the loci of the points where the dipole orientation suddenly changes. Some ferroelectrics have only two possible orientations of dipoles. For them the domain walls separate antiparallel domains and are called 180° walls. Some have more than one type of wall. For instance $BaTiO_3$-like ferroelectrics have more than one orientation in the tetragonal phase. The six possible directions result in three pairs of antiparallel directions along the cube edges. This gives rise to two different types of walls. They are the walls separating the antiparallel dipoles (called the 180° walls) and those separating the dipoles which are at right angles to each other (called the 90° walls). Figure 12.12a shows the atomic displacements on either side of a boundary between domains polarized in opposite directions. The 180° walls are illustrated in Fig. 12.12b. The 180° and 90° walls for a $BaTiO_3$ crystal are shown in Fig. 12.13. Domain structure of a KH_2PO_4 crystal cut perpendicular to the c-axis is shown in Fig. 12.14. For Rochelle salt, a group of domains at the crystal surface are shown in Fig. 12.15.

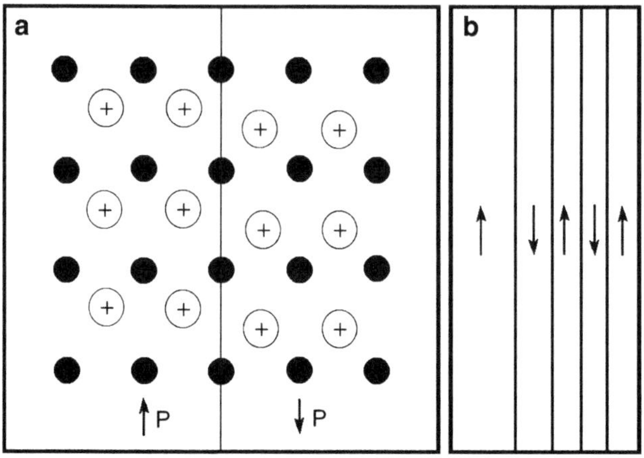

Fig. 12.12 (**a**) Schematic drawing of atomic displacements for adjacent domains polarized in opposite directions and (**b**) domain structure with 180° wall

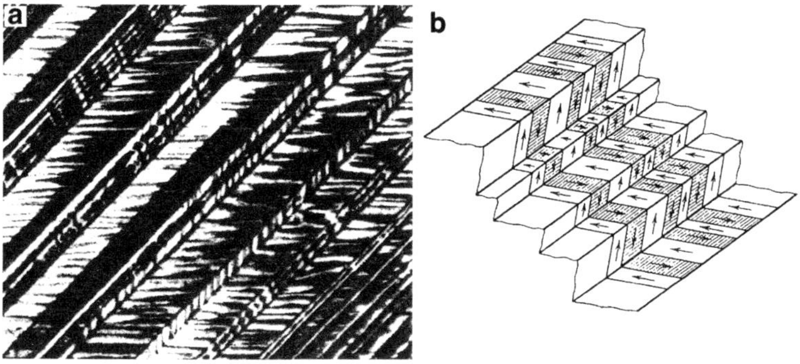

Fig. 12.13 (**a**) Photograph and (**b**) line diagram of 180° and 90° domain walls on the surface of a $BaTiO_3$ crystal

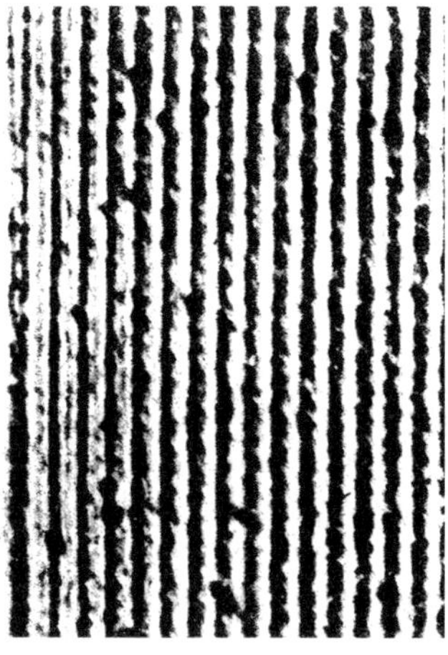

Fig. 12.14 Domain structure of a KH_2PO_4 crystal surface

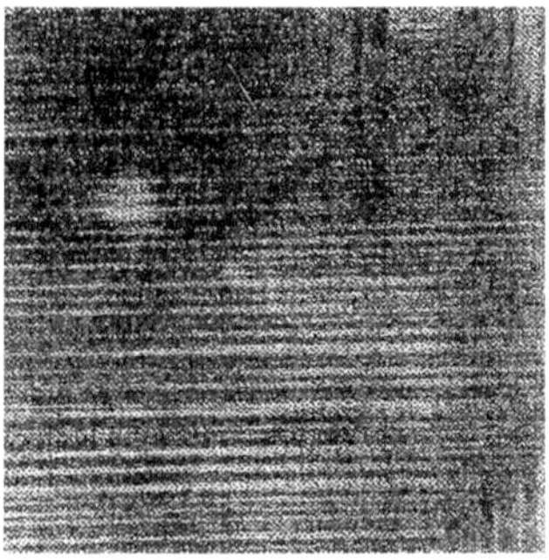

Fig. 12.15 Domain structure of a Rochelle salt crystal surface

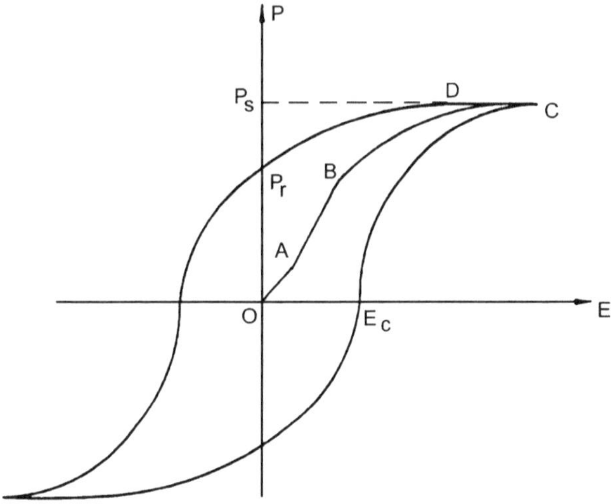

Fig. 12.16 Hysteresis loop

12.9.2 Domains and Hysteresis

The occurrence of the hysteresis loop pattern for a ferroelectric can be explained on the basis of domains.

Consider the hysteresis loop shown in Fig. 12.16. Initially, the polarization is zero when the dipole moments of the individual domains cancel one another. At very low fields, in the region OA the polarization shows a linear variation. On an increase in the applied field, the domains with dipole moment components along the direction of the applied field grow and the polarization increases along AB. In the region BD, the number of domains aligned in the direction of the field merge into a single domain with saturated polarization. On further increase of the field, the polarization shows a slow linear increase with field. The extrapolation of the linear part CD to zero field gives the spontaneous polarization P_s. When the field is reduced to zero the point P_r on the P axis is the remnant polarization. P_r is less than P_s. When the applied field is reversed, the field E_C required to reduce the polarization to zero is called the coercive field.

12.9.3 Display of Hysteresis Loop

Hysteresis loop may be observed on an oscilloscope using the Sawyer–Tower circuit (Fig. 12.17). Resistance R acts as voltage divider. C_s is a capacitor with the sample. C_0 is a standard linear capacitor. If V_s and V_0 represent the voltage acting on the capacitors C_s and C_0 respectively, we have the relation

$$C_s V_s = C_0 V_0. \tag{12.60}$$

The polarization P is given by the surface charge density $\sigma = Q/A$ where Q is the charge and A the surface area. Hence

$$P = C_0 V_0 / A. \tag{12.61}$$

From (12.61) it is seen that the polarization is proportional to V_0. When the crystal is in the polarized state, the resulting oscillogram has an ordinate proportional to the charge and abscissa proportional to the applied voltage. Figure 12.18 shows oscillograms obtained at different temperatures for Rochelle salt. As the temperature

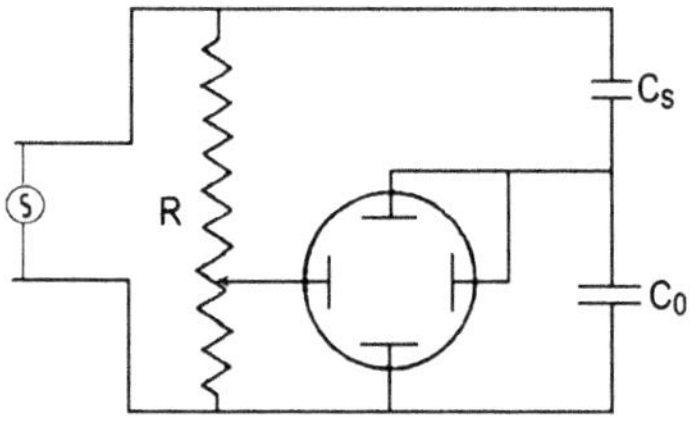

Fig. 12.17 Sawyer–Tower circuit for the observation of hysteresis loop

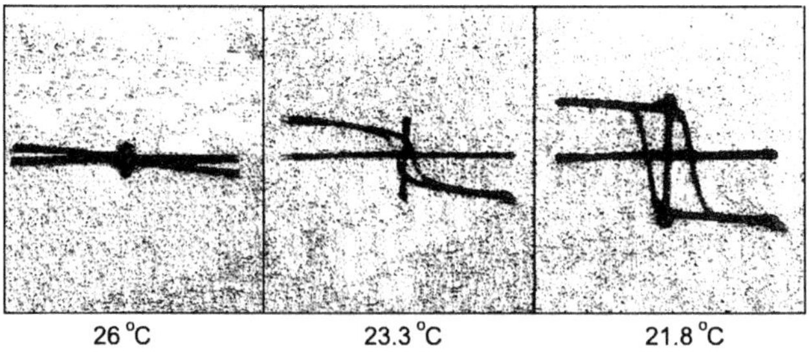

Fig. 12.18 Hysteresis loop at different temperatures for a Rochelle salt crystal

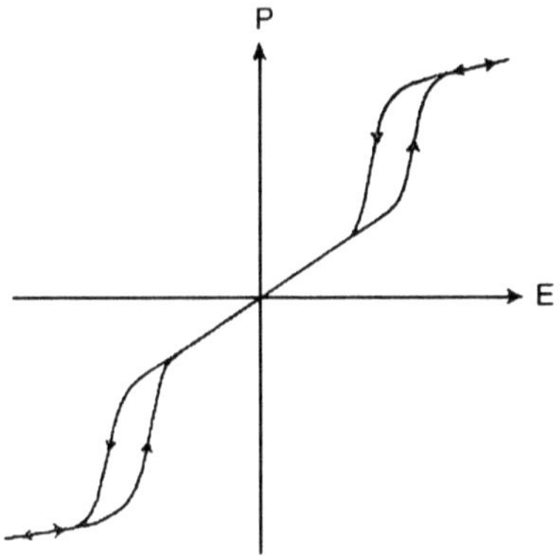

Fig. 12.19 A typical double loop pattern

is decreased from 26°C the voltage and charge increases and a hysteresis loop is observed at the transition temperature of 24°C.

Several modified versions of ferroelectric loop tracers using computer-based automatic analysers are discussed by Singh et al. [12.25].

A transition in a ferroelectric crystal may be induced at a temperature slightly above the transition temperature by application of a strong electric field. The transition so induced shows a "double loop" pattern as shown in Fig. 12.19. The double loop pattern is observed for crystals showing a first-order phase transition. The appearance of a double loop may be taken as evidence for a first-order phase transition.

12.9.4 Methods for Observation of Domain Structure

12.9.4.1 Charged Powder Method

Charged powder particles are attracted to bound charges by electrostatic interaction. For instance, red lead oxide gets deposited on the positive end of a dipole and sulphur on the negative end. Since they have different colours, a mixture of the two powders is used for observing the domain structure. The domain structures are made visible by direct deposition of the powders on the surface of a ferroelectric crystal or by using colloidal suspensions in insulating organic liquids.

12.9.4.2 Etching Technique

This method depends on the fact that the etching rates of opposite ends of domains are different. The etchants are to be selected after several trials for a particular crystal. For example for $BaTiO_3$, HCl is used as etchant while for triglycine sulphate domains are made visible by etching with water–ethanol solution. This method was earlier used for studying static behaviour of domains. The limitation of this method is that it cannot be used to study the dynamic characteristics such as the motion of domain walls.

12.9.4.3 Optical Methods

Domain structure can be observed using a polarizing microscope. The method depends on the fact that the optic axes of the adjacent domains differ. This method can be used for barium titanate since the crystal is transparent and the refractive index differs slightly in different directions. For instance at room temperature, the difference in refractive index along the c and a axes $n_c - n_a = -0.055$. The optical method can be used to study the temperature-induced changes also.

The antiparallel domains become visible if the sample is placed between crossed polaroids and observed using white light. In $BaTiO_3$, the reversed domains appear darker in shade when compared to the surrounding region. This method is used to observe the motion of domain walls continuously.

12.9.4.4 Electron Microscopy

Optical techniques can be used to observe domains of the order of micron dimensions. By transmission electron microscopic technique, domains of dimensions of the order of a few nanometres may be observed. Domain structures of $BaTiO_3$ and Rochelle salt have been observed by this technique.

12.9.4.5 X-ray Topography

Anomalous dispersion of X-rays causes a difference between the X-ray intensity reflected from the positive and negative ends of domains. By using X-rays of suitable wavelength of the constituent elements large difference in the intensity can be achieved. This method has been used to observe domains of $BaTiO_3$ and $LiNbO_3$ crystals.

12.10 Applications

Ferroelectric materials are useful because of two significant properties: the high dielectric constant and the reversible spontaneous polarization. Apart from this, the piezoelectric and pyroelectric properties have made them useful in modern technology as devices.

12.10.1 Capacitors

A ferroelectric is an insulator with a high dielectric constant. Small-sized capacitors with large capacitance are produced using ferroelectric materials. Thin film capacitors have led to fabrication of integrated circuits.

12.10.2 High Capacity Computer Memories

The property of bistable polarization has made ferroelectrics the prime materials for binary memories. To record the information, the polarization is reversed or reoriented by application of an electric field greater than the coercive field. To read the information, the stored information is retrieved by using two methods viz., by electrical means and by the use of electro-optic effect which involves light-induced polarization charges.

Ferroelectric random access memories (Fe RAM) are non-volatile ferroelectric-based memories. The information is stored using the bistable polarization states of capacitors such as PZT ($PbZrTiO_3$). These are used in smart card chips, cellular phones and audio/video memory storage devices.

12.10.3 Thin Film Technology

Ferroelectric ceramics and thin films have some advantages. Unlike single crystals, it is possible to prepare ceramic materials with wide range of compositions to suit required characteristics. For example, ferroelectric thin films with their high dielectric, piezoelectric and pyroelectric properties are used as Fe RAMS, as electro-optic devices, ultrasonic-based imaging and surface acoustic wave (SAW) devices used in telecommunication. They are also used as infrared detectors and more recently in cameras to produce flashes.

12.11 Antiferroelectricity

Antiferroelectrics are analogous to the antiferromagnetics among magnetic materials. In an antiferroelectric crystal, one ionic chain is displaced along a particular line resulting in a dipole moment but the adjacent ionic chain is displaced in the opposite direction to the first. Thus, the adjacent unit cells have opposite orientation of polarization. Hence the crystal as a whole is not spontaneously polarized.

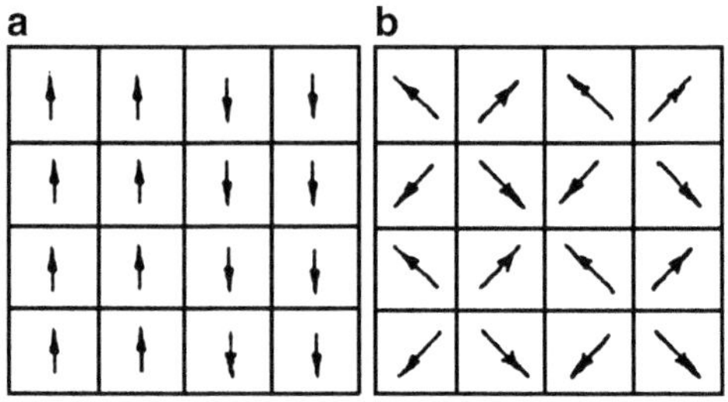

Fig. 12.20 Orientation of unit-cell dipoles in (**a**) ferroelectrics, (**b**) antiferroelectrics

The orientation of unit cell dipoles is illustrated in Fig. 12.20. In ferroelectrics, the direction of spontaneous polarization P_s is same within a domain but differs from one domain to another. In antiferroelectrics, even neighbouring unit cells have different orientations. Hence $P_s = 0$.

Antiferroelectrics show a characteristic phase transition with temperature with dielectric anomaly similar to that observed in ferroelectrics. The anomaly is small if the dielectric constant in the paraelectric phase is small. At high temperatures they exhibit a paraelectric behaviour with an increase in dielectric constant with a decrease in temperature. The change in dielectric constant follows Curie–Weiss law (12.16). Below a certain temperature (the antiferroelectric temperature T_a), a decrease in the dielectric constant is observed. The transition involves an anomaly in specific heat and also in latent heat if the transition is of first order.

The phase transition in antiferroelectrics is accompanied by structural changes as in the case of ferroelectrics. The structure of the crystal changes to that of lower symmetry in the antiferroelectric phase similar to that of changes in ferroelectrics. A schematic representation of a ferroelectric and an antiferroelectric phase transition is shown in Fig. 12.21.

In many cases a ferroelectric and an antiferroelectric are isomorphous, e.g. KDP and ADP. A phase transition between antiferroelectric and ferroelectric state is observed on application of high electric field at a critical field E_{crit}. This suggests that the free energies of ferroelectric state and antiferroelectric state are not very different from each other. This is particularly so in the case of crystals with perovskite structure. The critical switching field E_{crit} increases linearly with temperature.

In Table 12.6, some antiferroelectrics along with their transition temperatures are listed. We shall discuss two prominent antiferroelectrics $PbZrO_3$ and WO_3 in some detail.

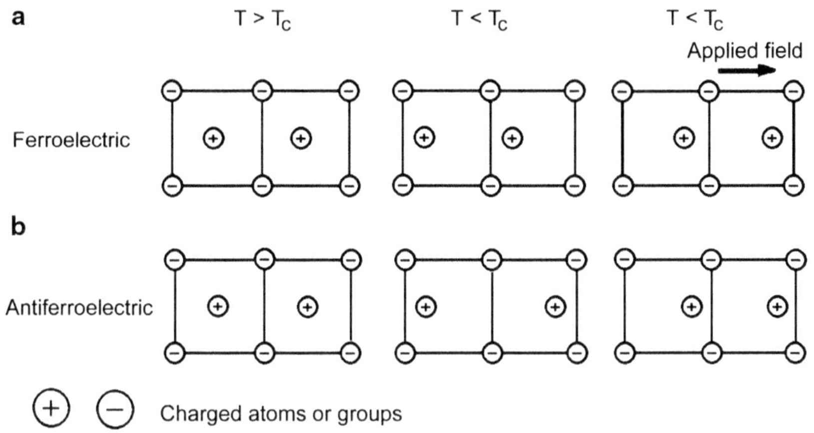

Fig. 12.21 Schematic representation of structural phase transition from a centrosymmetric to (**a**) ferroelectric, (**b**) antiferroelectric unit cell

Table 12.6 Transition temperatures of some antiferroelectric crystals

Crystal	Transition temperature T_a[K]
$PbZrO_3$	506
$NaNbO_3$	793
$NH_4H_2PO_4$	148
$ND_4D_2PO_4$	242
$NH_4H_2AsO_4$	216
$ND_4D_2AsO_4$	304
WO_3	1,010

12.11.1 Lead Zirconate ($PbZrO_3$)

Amongst the antiferroelectrics with perovskite structure, $PbZrO_3$ is most widely investigated. It undergoes a phase transition from ferroelectric to antiferroelectric at a temperature of 506 K. Above the transition temperature T_a it has a cubic structure which changes to orthorhombic structure in the low temperature phase. The dielectric constant in the paraelectric phase follows a Curie–Weiss law with $C = 1.6 \times 10^5$ K and Curie–Weiss temperature θ at 475 K. The dielectric constant is $\approx$ 5,000 at the peak. The transition is of the first order as in the case of $BaTiO_3$. Figure 12.22 shows the variation of dielectric constant and reciprocal dielectric constant with temperature.

The polarization in the ferroelectric phase is described by a double loop hysteresis pattern (Fig. 12.23) as expected for a first-order phase transition. The variation of critical field E_{crit} with temperature is shown in Fig. 12.24.

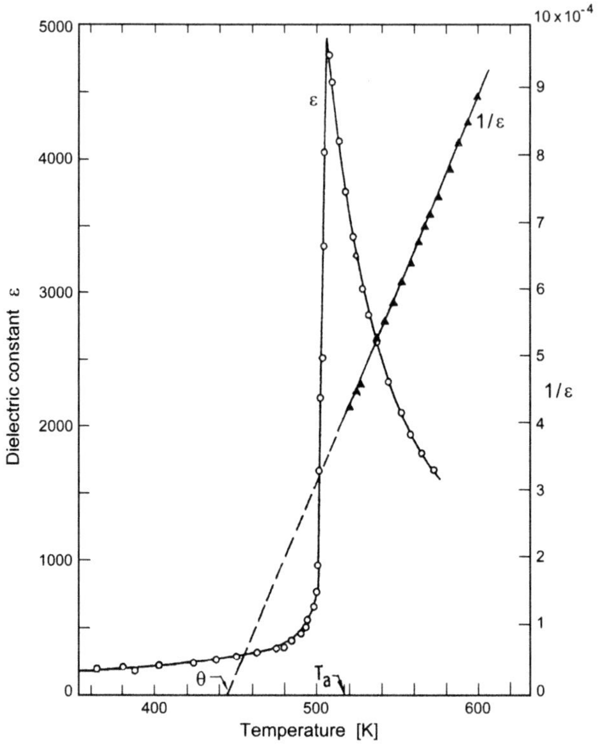

Fig. 12.22 Variation of dielectric constant and reciprocal dielectric constant with temperature for $PbZrO_3$

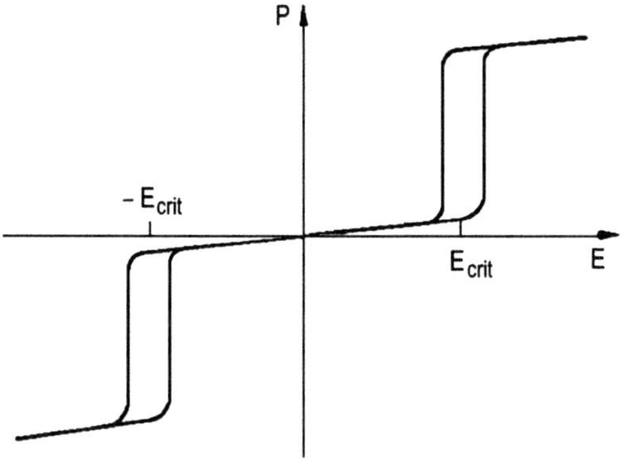

Fig. 12.23 Double hysteresis loop pattern for $PbZrO_3$

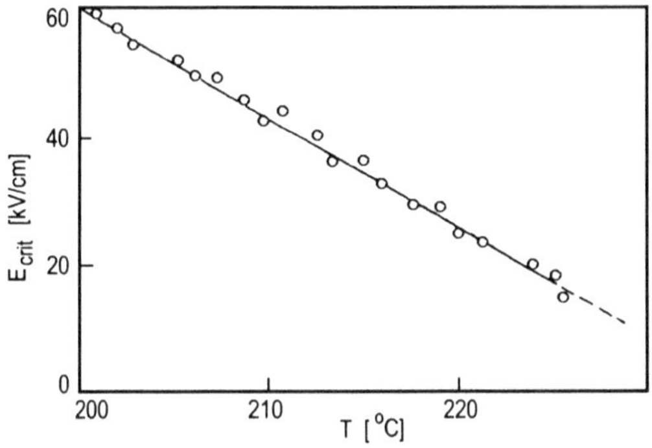

Fig. 12.24 Variation of critical field E_{cr} with temperature for $PbZrO_3$

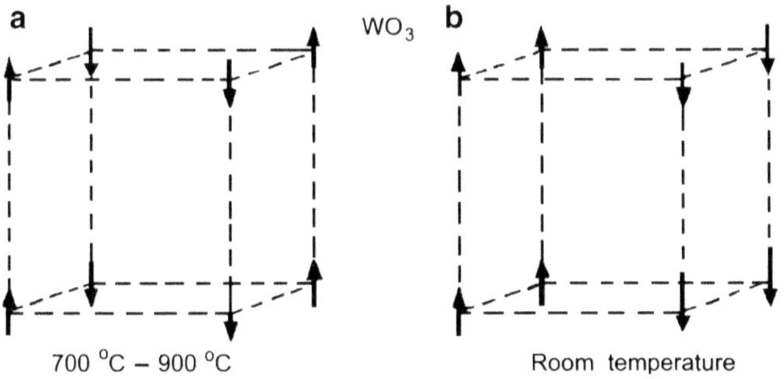

Fig. 12.25 Dipole arrangement in WO_3: (**a**) tetragonal phase, (**b**) monoclinic phase

12.11.2 Tungsten Oxide (WO_3)

WO_3 exhibits antiferroelectric as well as ferroelectric phases. The structure of WO_3 is closely related to perovskite structure. The transition temperature, phase and the corresponding structure are:

Temperature	700–900°C	Room temperature	At $-50°C$
Phase	Antiferroelectric	Antiferroelectric	Ferroelectric
Symmetry	Tetragonal	Monoclinic	Higher than monoclinic

The two antiferroelectric phases are shown schematically in Fig. 12.25. It is observed that the polarization alternates in adjacent lattice lines as shown in Fig. 12.25a. In Fig. 12.25b, the polarization alternates in adjacent lattice planes.

Problems

1. The pyroelectric coefficient p of tourmaline along the triad axis is $p = 4 \times 10^{-6}$ Coulomb $\mathrm{m}^{-2}(°\mathrm{C})^{-1}$. Estimate the electric field required to produce the same polarization (dielectric susceptibility of tourmaline along the triad axis is 6.1).
2. Derive the piezoelectric matrix for point group $\bar{4}2m$.
3. Which of the following crystal classes are

   (a) neither piezoelectric nor pyroelectric
   (b) piezoelectric but not pyroelectric
   (c) piezoelectric and also pyroelectric.

   (1) $mm2$, (2) 222, (3) $3m$, (4) $m3$, (5) $\bar{4}$, (6) $m3m$.
4. The dielectric constant of a ferroelectric crystal at some temperatures in its paraelectric state is given below.

t[°C]	175	159	147	37
ε	2,600	3,300	4,700	6,700

   Estimate the Curie temperature.
5. Assuming that the polarization of $BaTiO_3$ crystal is caused only by the displacement of the Ti^{4+} ions, calculate saturation polarization at room temperature (lattice constant = 4Å, displacement of Ti^{4+} = 0.2 Å).

References

12.1. J.F. Nye, *Physical Properties of Crystals* (Oxford University Press, London, 1957)
12.2. Y.I. Sirotin, M.P. Shaskoslaya, *Fundamentals of Crystal Physics* (Mir Publishers, Moscow, 1982)
12.3. R.A. Levy, *Principles of Solid State Physics* (Academic Press, New York, 1968)
12.4. A.J. Dekker, *Solid State Physics* (Macmillan Press, New York, 1981)
12.5. C. Kittel, *Introduction to Solid State Physics* (Wiley, New York, 1995)
12.6. W.G. Cady, *Piezoelectricity* (McGraw-Hill, New York, 1946)
12.7. A.R. Von Hippel, *Dielectric Materials and Applications* (Wiley, New York, 1954)
12.8. E. Fatuzzo, W.J. Merz, *Ferroelectricity* (North-Holland Pub. Co., Amsterdam, 1967)
12.9. I.S. Zheludev, *Physics of Crystalline Dielectrics*, vols. 1, 2 (Plenum Press, New York, 1971)
12.10. M.E. Lines, A.M. Glass, *Principles and Applications of Ferroelectrics and Related Materials* (Clarendon Press, Oxford, 1979)
12.11. W.P. Mason, Phys. Rev. **69**, 173 (1946)
12.12. B. Jaffe, R.S. Roth, S. Marzullo, J. Res. Natl. Bur. Standards **55**, 239 (1955)
12.13. G. Weinreich, *Solids: Elementary Theory for Advanced Students* (Wiley, New York, 1965)
12.14. J.C. Slater, Phys. Rev. **78**, 748 (1950)
12.15. L.D. Landau, Zh eksp teor Fiz. **7**, 627 (1937)
12.16. V.L. Ginzberg, Zh eksp teor Fiz. **19**, 36 (1949)
12.17. H.F. Devonshire, Adv. Phys. **3**, 85 (1954)
12.18. W. Cochran, Adv. Phys. **9**, 387 (1960)

12.19. P.W. Anderson, *Fizike Dielectrikov* (Akad Nauk, Moscow, 1960)
12.20. R.H. Lyddane, R.G. Sachs, E. Teller, Phys. Rev. **59**, 673 (1941)
12.21. W. Cochran, Adv. Phys. **18**, 157 (1969)
12.22. C.V. Raman, T.M.K. Neddungdi, Nature **145**, 147 (1940)
12.23. R.A. Cowley, Phys. Rev. **134**, A981 (1964)
12.24. Y. Luspin, J.L. Servoin, F. Gervais, J. Phys. C Solid State Phys. **13**, 3761 (1980)
12.25. K. Singh, S.S. Limaye, R.U. Tiwari, S. Nath, S.S. Bhoga, Ferroelectrics **189**, 9 (1996)

Chapter 13
Optical Properties of Insulators

13.1 Introduction

An electromagnetic wave travels in vacuum with velocity c and in a medium with velocity $v(v < c)$. The ratio c/v is called the refractive index n.

In crystals, the refractive index is direction-dependent. It is isotropic in cubic crystals and anisotropic in crystals of all other symmetries. This anisotropy gives rise to the property of birefringence. Crystals show other interesting properties like photoelasticity (piezo-optic effect), electro-optic effect and optical activity. Lasers provide radiation with intense power. This facilitates observation of nonlinear optical effects, the most important among them being harmonic generation.

The tensor nature of optical properties is treated in detail by Nye [13.1], Bhagavantam [13.2] and Sirotin and Shaskoslaya [13.3]. Lasers and nonlinear optical effects are discussed by Bloembergen [13.4], Lines and Glass [13.5], Thyagarajan and Ghatak [13.6] and Dmitriev et al. [13.7].

13.2 Propagation of Light Through Crystals

The refractive index n is isotropic only in cubic crystals like the alkali halides. In tetragonal crystals like rutile (TiO_2), the refractive index has values n_0 along the c-axis and n_e along the a- and b-axes. Let us consider light travelling along the b-axis. If the light is polarized in the bc plane, it experiences a refractive index n_0. On the other hand, if the light is polarized in the ba plane, it experiences a refractive index n_e. This phenomenon is called double refraction or birefringence. This is true of any other direction except the c-direction in which there is no birefringence. The c-direction is called the optic axis. In tetragonal crystals as well as hexagonal and trigonal crystals (like ZnO and calcite, respectively), there is only one optic axis and it is along the c-axis. Such crystals are called uniaxial. In crystals of lower symmetry (orthorhombic, monoclinic and triclinic, e.g. borax, olivine and topaz),

D.B. Sirdeshmukh et al., *Atomistic Properties of Solids*, Springer Series in Materials Science 147, DOI 10.1007/978-3-642-19971-4_13,

Table 13.1 Refractive indices of some crystals

Crystal	Refractive index n		
Isotropic crystals	n_0		
CaO	1.82		
CaF_2	1.42		
Uniaxial crystals	n_0		n_e
Calcite	1.658		1.486
$LiNbO_3$	2.272		2.187
α-quartz	1.544		1.553
TiO_2 (rutile)	2.616		2.903
Biaxial crystals	n_1	n_2	n_3
Borax	1.447	1.469	1.472
Olivine	1.640	1.660	1.680
$CaTiO_3$	2.300	2.340	2.380
Topaz	1.618	1.620	1.627

the refractive index has three values n_1, n_2 and n_3. It can be shown that in these crystal systems, there are two directions of zero birefringence, i.e. there are two optic axes; such crystals are called biaxial. The refractive indices of some crystals are given in Table 13.1.

13.2.1 The Optical Indicatrix

In Chap. 7, we saw that a symmetric second-rank tensor can be represented by an ellipsoidal surface called the representation quadric whose equation is

$$S_{11}x_1^2 + S_{22}x_2^2 + S_{33}x_3^2 + 2S_{31}x_3x_1 + 2S_{23}x_2x_3 + 2S_{12}x_1x_2 = 1. \tag{13.1}$$

Here S_{ij} values are the components of a tensor. The magnitude of the radius vector in any direction is related inversely to the square root of the magnitude of the property in that direction. Thus if (13.1) represents the dielectric constant, then $S_{11} = 1/\varepsilon_1$, $S_{22} = 1/\varepsilon_2$ and $S_{33} = 1/\varepsilon_3$, where ε_1, ε_2 and ε_3 are the principal dielectric constants. According to the electromagnetic theory, at high frequencies (optical frequencies) $\varepsilon = n^2$. Hence, it should be possible to represent n^2 also by an equation like (13.1). By convention, the coefficients are now denoted by B and the surface is called the optical indicatrix. The general equation for the optical indicatrix is:

$$B_{11}x_1^2 + B_{22}x_2^2 + B_{33}x_3^2 + 2B_{31}x_3x_1 + 2B_{23}x_2x_3 + 2B_{12}x_1x_2 = 1. \tag{13.2}$$

In the single-index notation (13.2) becomes

$$B_1x_1^2 + B_2x_2^2 + B_3x_3^2 + 2B_4x_2x_3 + 2B_5x_3x_1 + 2B_6x_1x_2 = 1. \tag{13.3}$$

If the axes of the property ellipsoid are oriented to coincide with the coordinate axes, (13.3) reduces to

$$B_1 x_1^2 + B_2 x_2^2 + B_3 x_3^2 = 1 \tag{13.4}$$

for biaxial crystals. In such a case, the property is said to be referred to principal axes. For isotropic crystals, $B_1 = B_2 = B_3 = B^0$ and (13.4) reduces to

$$B^0 \left(x_1^2 + x_2^2 + x_3^2\right) = 1, \tag{13.5}$$

where $B^0 = 1/n_0^2$. For uniaxial crystals, $B_1 = B_2 \neq B_3$ and (13.4) reduces to

$$B_1 \left(x_1^2 + x_2^2\right) + B_3 \left(x_3^2\right) = 1, \tag{13.6}$$

where $B_1 = 1/n_0^2$ and $B_3 = 1/n_e^2$. It should be noted that the refractive index n is not a tensor in as much as it is not a coefficient relating two vector quantities [13.1] and yet the coefficients B in (13.2) and (13.3) may be treated as the components of a tensor because of their relation with the dielectric constant which is a tensor [13.2].

13.2.2 Propagation of Polarized Light Through Crystals

We shall consider the transmission of polarized light through crystals. Here, we shall see that the optical indicatrix plays an important role. To start with, let us consider a cubic crystal. Because the refractive index is isotropic in a cubic crystal, the optical indicatrix is a sphere according to (13.5). A ray of light travels with a velocity and in a direction determined by the refractive index. The passage of a ray in a cubic crystal is similar in any direction in view of the spherical shape of the indicatrix. The passage of light in a cubic crystal is, thus, rather uneventful.

On the other hand, the indicatrix for uniaxial crystals is an ellipsoid of revolution about the principal symmetry axis. Its shape, which follows from (13.5), is shown in Fig. 13.1 for $n_e > n_0$ (positive uniaxial) and $n_e < n_0$ (negative uniaxial). A beam of plane-polarized light incident on the crystal in any direction, in general, splits into two plane-polarized beams having mutually orthogonal planes of polarization travelling with two refractive indices n_e and n_0. Let us consider the situation shown in Fig. 13.2.

The indicatrix being an ellipsoid of revolution, the central section normal to the optic axis is a circle whereas any other central section is an ellipse. The elliptical section intersects with the circular section with one axis (the minor axis in the figure) passing through the two points of intersection. If plane-polarized light passes along the optic axis, both the resolved components lie in the circle and have equal refractive indices. On the other hand, if the wave normal is in any other direction, as shown in the figure, one component lies in the circle and has refractive index n_0 but the other component which is along the axis of the ellipse has refractive index less than n_e. As the direction of the wave normal changes, one component of the beam

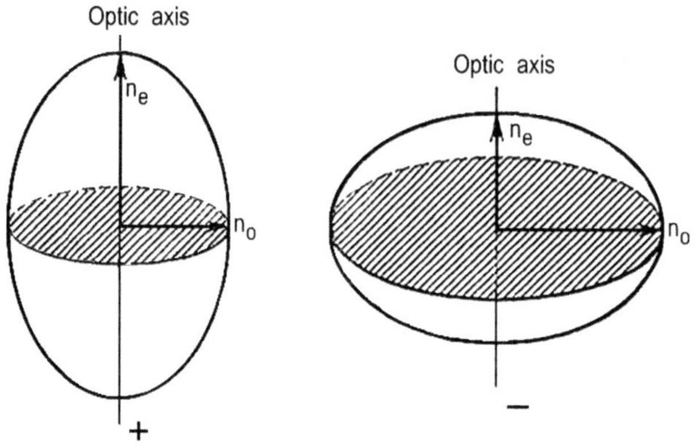

Fig. 13.1 Optical indicatrices of positive and negative uniaxial crystals

Fig. 13.2 Central section perpendicular to optic axis (*circular*) and another oriented with respect to the optic axis (*elliptical*) in a positive uniaxial crystal

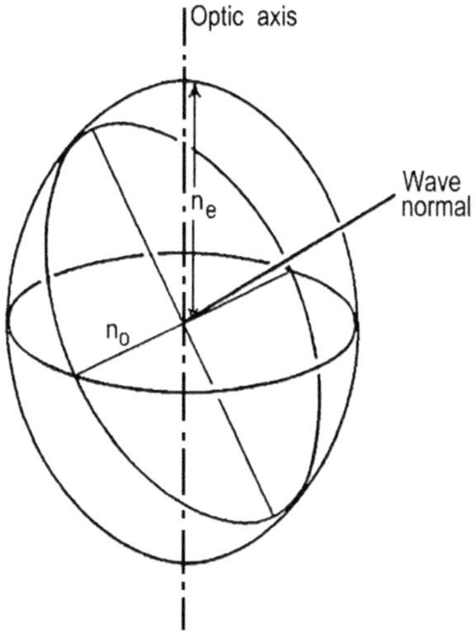

always travels with refractive index n_0 whereas the other travels with refractive index varying from n_0 to n_e. This is called double refraction or birefringence which exists in all directions except along the optic axis. The ray which travels with refractive index n_0 is called the ordinary ray and the other is called the extraordinary ray. The difference in refractive indices $\Delta n = n_e - n_0$ is called the birefringence. The birefringence is small, generally $\sim 1\%$, though in some cases like calcite it is $\sim 10\%$.

An interesting effect of double refraction in calcite observed by Bartholinus in 1699 was quoted by Magie [13.8]. This is shown in Fig. 13.3. When a calcite crystal DEFGHIJK is placed on a pencil dot A on a piece of paper, two images B and C are seen when viewed from above. The image C is caused by the ordinary ray and B by the extraordinary ray. If the calcite prism is now rotated about the vertical, the image C remains stationary whereas the image B moves around C in a circle.

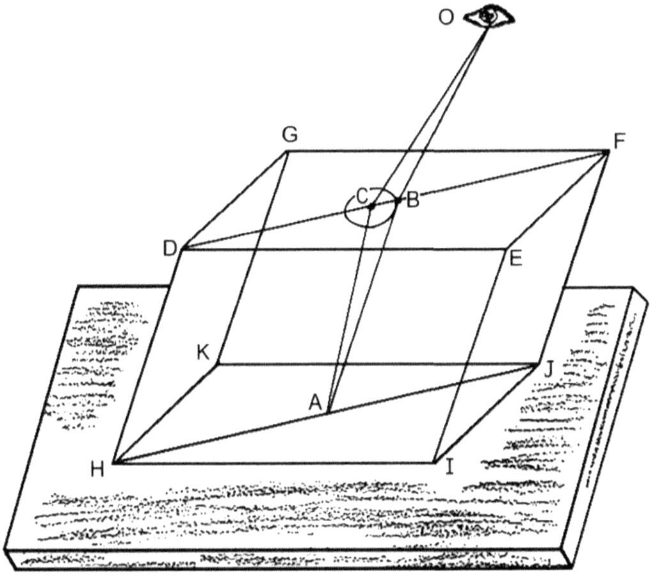

Fig. 13.3 Double images in calcite

Birefringence has been used to design optical devices. A commonly used optical component is a quarter wave plate or a half wave plate. A birefringent crystal like calcite or quartz is cut such that its optic axis is in the face of the plate. A plane-polarized ray incident on the face such that its plane of polarization makes an angle with the optic axis gets resolved into components parallel and perpendicular to the optic axis. These components travel through the plate with refractive indices n_e and n_0. When they emerge from the plate, they would have developed a phase difference $2\pi t(n_e - n_0)/\lambda$, where t is the thickness of the plate and λ the wavelength. The two rays now combine to form a plane-polarized wave. If this phase difference equals π the plane of polarization of the resulting wave is at right angles to the plane of polarization of the incident wave. By proper choice of t and the birefringent material, wave plates that can provide the desired rotation of the plane of polarization can be designed. These are called octadic wave plate ($\lambda/8$), quarter wave plate ($\lambda/4$), half wave plate ($\lambda/2$) and full wave plate (λ).

Another popular device is a Nicol prism which is used as a polarizer (or analyser). As shown in Fig. 13.4, it is made of two parts of a calcite prism cut and joined with a thin layer of Canada balsam between them. The ordinary ray O which has refractive index 1.658 undergoes total internal reflection at the balsam layer which has refractive index 1.55. This reflected ray leaves out of the calcite prism. The extraordinary ray E with refractive index 1.486 does not undergo reflection and travels across the balsam layer and exits out of the second part of the prism as a single plane-polarized beam.

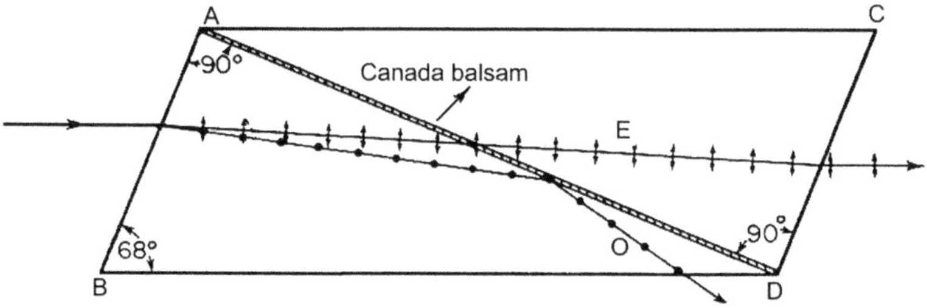

Fig. 13.4 Principle of a Nicol prism

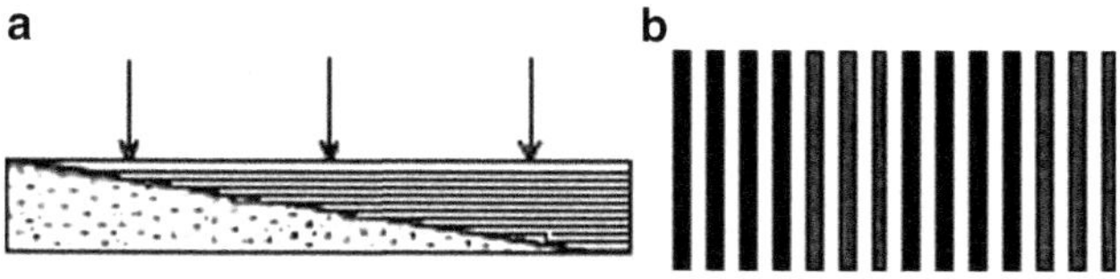

Fig. 13.5 Principle of Babinet compensator: (**a**) quartz wedges and (**b**) fringe pattern

Finally, we shall consider the Babinet compensator which is an instrument useful in the measurement of birefringence. It consists of two wedge-shaped quartz prisms with a small angle. The two prisms are cut with their edges parallel and perpendicular to the optic axis (Fig. 13.5a). One wedge is moveable. When plane-polarized light is incident normally on the wedge pair, with the vibration direction making an angle with the optic axis of the wedge, it splits into two waves, the ordinary (O) and extraordinary (E) waves. These waves travel through the wedge in the same direction but with an optical path difference due to the difference in the refractive indices. This path difference varies from point to point along the length of the wedge. As a result, there is an interference fringe pattern in the field of view when viewed through an eyepiece (Fig. 13.5b). The path difference in the rays interfering at two dark fringes is λ. If there is relative displacement of the wedges, the fringe system moves. Thus, the compensator can be calibrated, the fringe width β being a measure of the path difference.

The indicatrix for the low symmetry crystals is a triaxial ellipsoid (Fig. 13.6). It has two central circular sections and hence two optic axes (hence, also, the term biaxial). It can be seen that in these crystals an incident polarized ray gets split into two plane-polarized rays travelling with different refractive indices but, now, both refractive indices are "extraordinary"; their magnitudes vary with direction. One optic axis makes angle V with the major principal axis of the optical indicatrix and the other optic axis makes angle $2V$, i.e. the angle between the two optic axes is V; this is given by

$$\tan V = \left[\left(n_{\mathrm{p}}^{-2} - n_{\mathrm{m}}^{-2}\right) / \left(n_{\mathrm{m}}^{-2} - n_{\mathrm{g}}^{-2}\right)\right]^{1/2}. \tag{13.7}$$

Here, n_{g} is the largest among the three refractive indices n_1, n_2, n_3 and n_{p} is the least; n_{m} is the mean of the three.

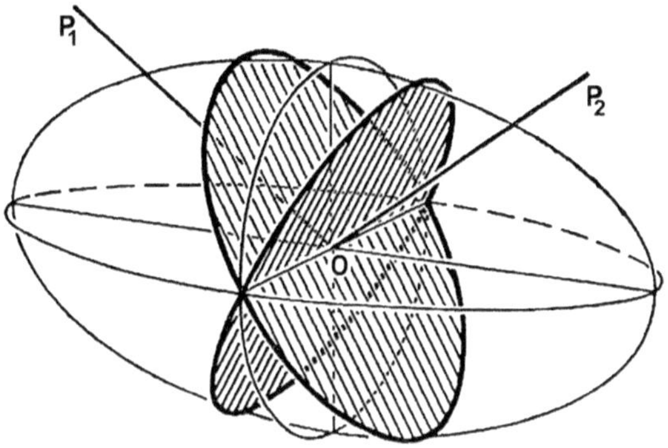

Fig. 13.6 Optical indicatrix of a biaxial crystal. Note the two circular principal sections; OP_1 and OP_2 are the two optic axes

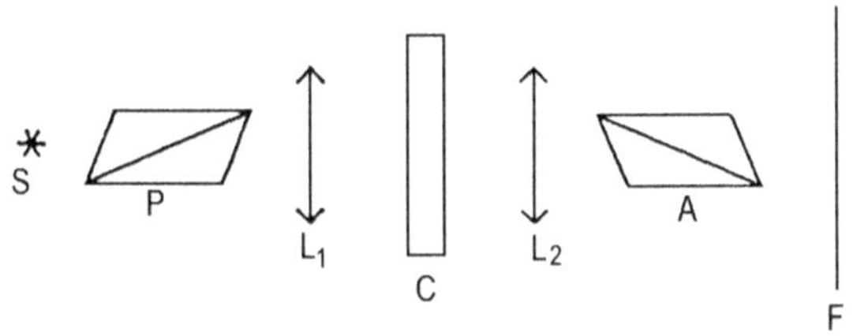

Fig. 13.7 Set-up to observe a crystal in convergent polarized light

13.2.3 *Observation of Optical Anisotropy*

Let us consider how anisotropic crystals look in convergent light. The set-up shown in Fig. 13.7 consists of a source of light S, a polarizer P, lenses L_1, L_2 and an analyser A. C is the crystal plate and F the screen (or viewing plane). The source of light is not a point source but one that gives a broad beam of light; this can be achieved, for instance, by placing a point source behind a plate of frosted glass. The light polarized at P and rendered convergent by the lens L_1 is incident on the crystal C. In the crystal, it splits into two polarized beams with mutually orthogonal planes of polarization. The lens L_2 focuses the beams onto the screen.

What is seen on the screen is the result of the interference between the polarized beams with a path difference caused by unequal refractive indices. The observed pattern is called a conoscopic picture. The conoscopic picture depends on the symmetry of the crystal, the orientation of the optical indicatrix with respect to the overall direction of the beam, the magnitude of birefringence and the thickness of the crystal plate. A typical conoscopic picture consists of a set of dark bands called isogyres and a set of bright (or coloured) bands called isochromes. If the crystal is rotated, the isogyres change form but not the isochromes.

The conoscopic picture of a uniaxial crystal cut perpendicular to its optic axis is shown in Fig. 13.8a. As expected, it shows isogyres in the form of a dark cross superposed over isochromes in the form of dark and bright bands. The appearance is different (Fig. 13.8b) if the crystal is cut with the optic axis contained in the face. The conoscopic picture of a biaxial crystal is shown in Fig. 13.9. The isogyres are in the shape of hyperbolas. Rings of isochromes surround two points which are exit

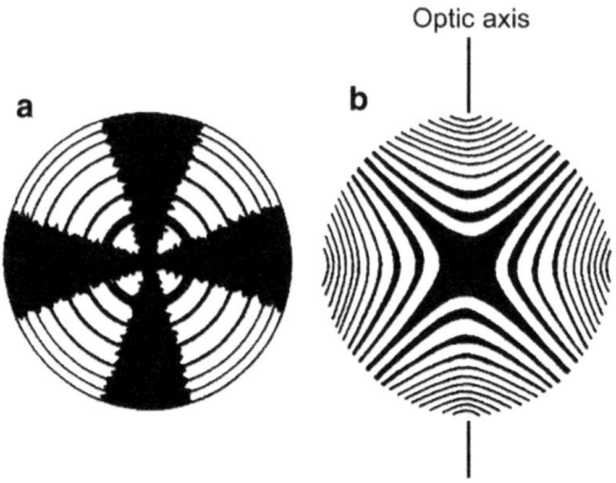

Fig. 13.8 Conoscopic picture of a uniaxial crystal (a) with optic axis perpendicular to crystal face and (b) with optic axis contained in the crystal face

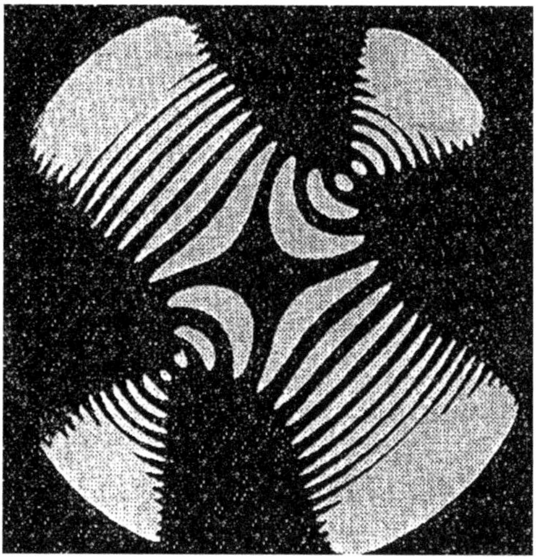

Fig. 13.9 Conoscopic picture of a biaxial crystal observed at right angles to the bisector of the angle between the two optic axes

points of the two optic axes. The distance between these points is a measure of the angle between the two optic axes. Once the set-up (Fig. 13.7) is ready, one can test whether a crystal is uniaxial or biaxial almost instantaneously from the appearance of the conoscopic pattern.

13.3 Photoelastic Effect

Let us consider a cube of an isotropic material with refractive index n_0. Two rays of light travelling along the y-direction with planes of polarization along the x- and z-directions will travel in the cube with the same refractive index n_0. If the cube is now stressed in the z-direction, the refractive index in all the three directions will change to new values n_x, n_y, n_z. The light with plane of polarization parallel to x-direction will now travel with refractive index n_x and that with polarization parallel to z-direction with refractive index n_z. Thus, the material has become birefringent under stress. If the material was already birefringent in the absence of stress, the magnitude of birefringence increases with the application of stress. The introduction of birefringence in a non-birefringent crystal (or the increase in birefringence in an originally birefringent crystal) due to applied stress is called the photoelastic (or piezo-optic) effect.

13.3.1 The Nature of the Photoelastic Effect

We have seen that the optical indicatrix of a crystal has the equation

$$B_{11}x_1^2 + B_{22}x_2^2 + B_{33}x_3^2 + 2B_{31}x_3x_1 + 2B_{23}x_2x_3 + 2B_{12}x_1x_2 = 1. \qquad (13.2)$$

If a stress is applied to the crystal, each refractive index changes and hence the B coefficients in (13.2) also change. Let us denote the new values by primes. Then the equation to the optical indicatrix of the crystal under stress is

$$B'_{11}x_1^2 + B'_{22}x_2^2 + B'_{33}x_3^2 + 2B'_{31}x_3x_1 + 2B'_{23}x_2x_3 + 2B'_{12}x_1x_2 = 1. \qquad (13.8)$$

Taking the difference between (13.2) and (13.8), we have

$$\Delta B_{11}x_1^2 + \Delta B_{22}x_2^2 + \Delta B_{33}x_3^2 + 2\Delta B_{31}x_3x_1 + 2\Delta B_{23}x_2x_3 + 2\Delta B_{12}x_1x_2 = 0, \qquad (13.9)$$

where $\Delta B_{ij} = B'_{ij} - B_{ij}$. We have noted earlier that though the refractive index is not a tensor, the B_{ij} coefficients in (13.2) can be treated as components of a second-rank symmetric tensor. For the same reason, the ΔB_{ij} terms can also be treated as tensor components. The ΔB_{ij} values are induced by the stress which also is a second-rank tensor. The components ΔB_{ij} are related to the components of the stress tensor $[\tau_{kl}]$ by the relation

$$[\Delta B_{ij}] = [\pi_{ijkl}][\tau_{kl}]. \qquad (13.10)$$

Similarly, the ΔB_{ij} values are related to the components of the strain tensor $[\varepsilon_{kl}]$ by the relation

$$[\Delta B_{ij}] = [p_{ijkl}][\varepsilon_{kl}]. \qquad (13.11)$$

The coefficients π_{ijkl} constitute a fourth-rank tensor. They are called piezo-optic coefficients or stress optical coefficients. The units for π_{ijkl} are $10^{-12}\text{m}^2/\text{N}$ (10^{-13} cm^2/dyn). Similarly, the coefficients p_{ijkl} are called elasto-optical coefficients or strain-optical coefficients; they are dimensionless. The two types of coefficients are mutually related through the relations

$$\pi_{ijkl} = p_{ijrs}s_{rskl}; \quad p_{ijrs} = \pi_{ijkl}c_{klrs}, \qquad (13.12)$$

where the c and s values are the elastic stiffness constants and elastic compliances, respectively (Chap. 8). As in the case of the elastic constants, we may use a contracted-index system and write (13.10) and (13.11) as

$$\begin{aligned} \Delta B_m &= \pi_{mn}\tau_n \quad (m, n = 1, 2, \ldots, 6) \\ &= p_{mn}\varepsilon_n \quad (m, n = 1, 2, \ldots, 6) \end{aligned} \tag{13.13}$$

The π and p values are now related as

$$\pi_{mn} = p_{mr}s_{rn}; \quad p_{mn} = \pi_{mr}c_{rn}. \tag{13.14}$$

It may be noted that the p_{mn} and π_{mn} values are matrices unlike the p_{ijkl} and π_{ijkl} values which are tensors.

13.3.2 Effect of Symmetry on Number of Constants

The photoelastic tensor is a fourth-rank tensor. It has $3^4 = 81$ components. In (13.9), $\Delta B_{ij} = \Delta B_{ji}$. Hence

$$\pi_{ijkl} = \pi_{jikl}. \tag{13.15}$$

Similarly, the stress tensor being a symmetric tensor, $\tau_{kl} = \tau_{lk}$. Hence

$$\pi_{ijkl} = \pi_{ijlk}. \tag{13.16}$$

These relations reduce the number of coefficients from 81 to 36. In crystal elasticity (Chap. 8), we saw that the number is further reduced from 36 to 21 because of the nature of the strain energy function. That argument is not applicable in photoelasticity. Hence, the number of independent photoelastic coefficients for a triclinic crystal is 36. The number is smaller for other crystal systems and can be worked out by the methods discussed in Chaps. 7 and 8. The results are given in Table 13.2 [13.1].

It may be mentioned that the number of independent constants for any physical property can be calculated by using group theory. Generally, the group theory results agree with the results obtained by conventional methods. But it was pointed out by Bhagavantam [13.9] that in photoelasticity there were differences in some crystal classes. These differences are given in Table 13.3. These results were eventually verified experimentally. The results in Table 13.2 for these classes pertain to the group theory calculations.

13.3.3 Stress-Induced Birefringence in a Cubic Crystal

We shall consider the photoelastic birefringence in a cubic crystal. For simplicity, we shall consider a crystal belonging to one of the classes for which there are only

Table 13.2 Forms of the (π_{ij}) and (p_{ij}) photoelastic matrices

KEY TO NOTATION

In both π and p matrices

·	Zero component
•	Non-zero component
•—•	Equal components
•—o	Components numerically equal, but opposite in sign

In the π matrices

⊙	A component equal to twice the heavy dot component to which it is joined
◎	A component equal to minus 2 times the heavy dot component to which it is joined
x	$(\pi_{11} - \pi_{12})$

In the p matrices

⊙	A component equal to the heavy dot component to which it is joined
◎	A component equal to minus the heavy dot component to which it is joined
x	$(\frac{1}{2})(p_{11} - p_{12})$

TRICLINIC

Both classes

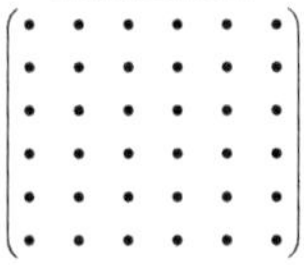

(36)

MONOCLINIC

All classes

Diad ∥ x_2 (standard orientation)

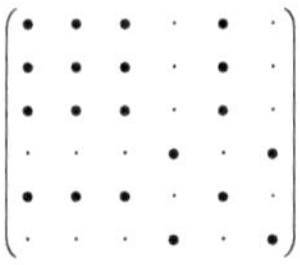

(20)

Diad ∥ x_3

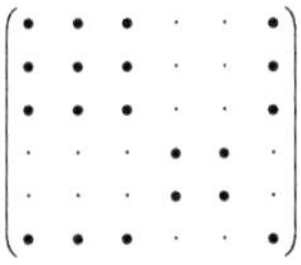

(20)

ORTHORHOMBIC

All classes

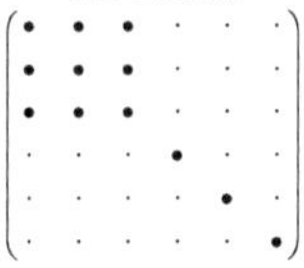

(12)

TETRAGONAL

Classes 4, $\bar{4}$, 4/*m*

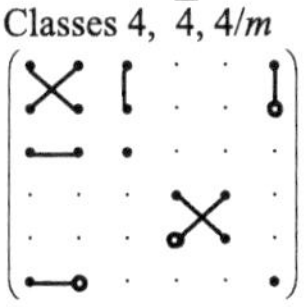

(10)

Classes 4*mm*, $\bar{4}$2*m*, 422, 4/*mmm*

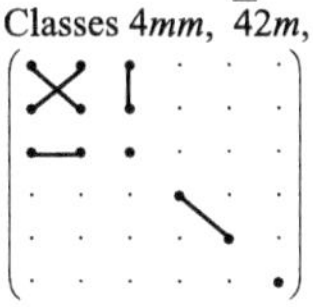

(7)

(*continued*)

Table 13.2 (continued)

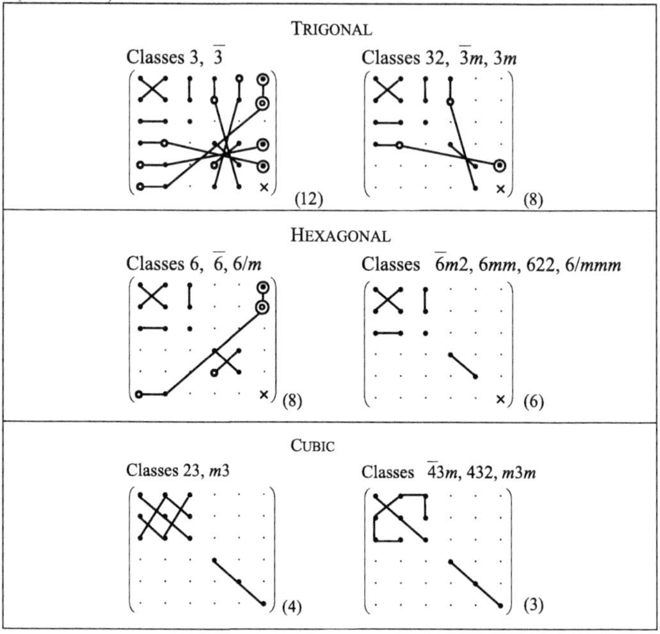

Table 13.3 Number of photoelastic constants of some crystal classes

Crystal class	Earlier work	Group theory	Extra constants
3, $\bar{3}$	11	12	π_{61}
4, $\bar{4}$, $4/m$	9	10	π_{45}
$\bar{6}$, 6, $6/m$	6	8	q_{61}, π_{45}
23, $m3$	3	4	$\pi_{12} \neq \pi_{13}$

three photoelastic constants. The crystal is oriented with its cubic axes along the coordinate axes Ox_1, Ox_2, Ox_3. Let plane-polarized light travel along the Ox_2 direction and let stress τ be applied in the Ox_1 direction. In the absence of stress, the indicatrix is a sphere with equation

$$B^0\left(x_1^2 + x_2^2 + x_3^2\right) = 1 \tag{13.5}$$

with $B^0 = 1/n_0^2$. On application of stress, the indicatrix takes the form

$$B_1x_1^2 + B_2x_2^2 + B_3x_3^2 + 2B_4x_2x_3 + 2B_5x_3x_1 + 2B_6x_1x_2 = 1. \tag{13.3}$$

The matrix equation for the photoelastic effect is

$$\Delta B_m = \pi_{mn}\tau_n. \tag{13.13}$$

Table 13.4 Birefringence of crystals under uniaxial stress for different directions of observation

Direction of uniaxial stress	Direction of observation	Birefringence ($n_{\|}-n_{\perp}$) Classes 23, $m3$	Classes $\bar{4}3m$, 432, $m3m$
[100]	[010]	$-\frac{1}{2}(n_0)^3\tau(\pi_{11}-\pi_{12})$	$-\frac{1}{2}(n_0)^3\tau(\pi_{11}-\pi_{12})$
	[001]	$-\frac{1}{2}(n_0)^3\tau(\pi_{11}-\pi_{12})$	$-\frac{1}{2}(n_0)^3\tau(\pi_{11}-\pi_{12})$
	[011]	$-\frac{1}{2}(n_0)^3\tau\{\pi_{11}-\frac{1}{2}(\pi_{12}+\pi_{13})\}$	$-\frac{1}{2}(n_0)^3\tau(\pi_{11}-\pi_{12})$
	[01$\bar{1}$]	$-\frac{1}{2}(n_0)^3\tau\{\pi_{11}-\frac{1}{2}(\pi_{12}+\pi_{13})\}$	$-\frac{1}{2}(n_0)^3\tau(\pi_{11}-\pi_{12})$
[111]	All directions $\perp$ to [111]	$-\frac{1}{2}(n_0)^3\tau\,\pi_{44}$	$-\frac{1}{2}(n_0)^3\tau\,\pi_{44}$

Its expanded form is

$$\begin{pmatrix}\Delta B_1\\ \Delta B_2\\ \Delta B_3\\ \Delta B_4\\ \Delta B_5\\ \Delta B_6\end{pmatrix}=\begin{pmatrix}B_1-B^0\\ B_2-B^0\\ B_3-B^0\\ B_4\\ B_5\\ B_6\end{pmatrix}=\begin{pmatrix}\pi_{11}&\pi_{12}&\pi_{12}&0&0&0\\ \pi_{12}&\pi_{11}&\pi_{12}&0&0&0\\ \pi_{12}&\pi_{12}&\pi_{11}&0&0&0\\ 0&0&0&\pi_{44}&0&0\\ 0&0&0&0&\pi_{44}&0\\ 0&0&0&0&0&\pi_{44}\end{pmatrix}\begin{pmatrix}\tau\\ 0\\ 0\\ 0\\ 0\\ 0\end{pmatrix}=\begin{pmatrix}\pi_{11}\tau\\ \pi_{12}\tau\\ \pi_{12}\tau\\ 0\\ 0\\ 0\end{pmatrix}. \tag{13.17}$$

In (13.17), $B_4 = B_5 = B_6$. Hence the axes of the indicatrix continue to be Ox_1, Ox_2, Ox_3. Since $B_1 = 1/n_1^2$, we have $\Delta B_1 = -(2/n_1^3)\Delta n_1$. Putting $n_1 \approx n_0$, we may write

$$\Delta n_1 = -(n_0^3/2)\Delta B_1 = -(n_0^3/2)\pi_{11}\tau$$

and (13.18)

$$\Delta n_3 = -(n_0^3/2)\Delta B_3 = -(n_0^3/2)\pi_{12}\tau.$$

For light travelling along Ox_2, the birefringence is

$$n_{11} - n_{\perp} = (\Delta n_1 - \Delta n_3) = -(n_0^3/2)(\pi_{11} - \pi_{12})\tau. \tag{13.19}$$

Here n_{11} and $n_{\perp}$ are the refractive indices parallel and perpendicular to the direction of stress. The stress birefringence for different directions of stress and observation for different classes of cubic crystals is given in Table 13.4.

13.3.4 Experimental Method and Results

A simple experimental arrangement for the measurement of photoelastic constants of cubic crystals is shown in Fig. 13.10. A beam of monochromatic light from source S passes through a slit and lens L_1 and is polarized by the polarizer P. It then passes

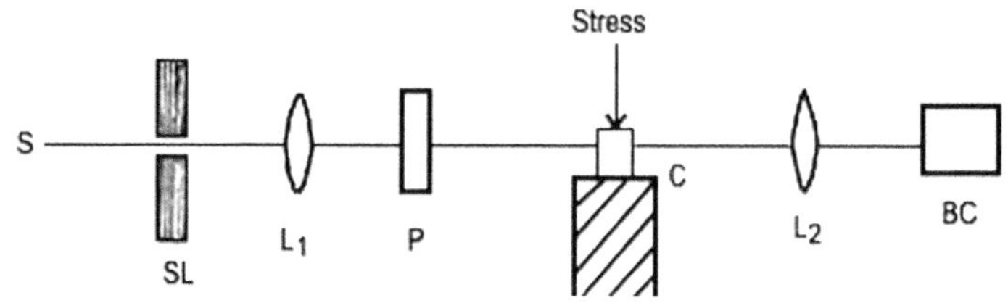

Fig. 13.10 Experimental set-up for determination of photoelastic constants

Table 13.5 Photoelasticity of some cubic crystals

Values of the piezo-optical coefficients for sodium D light (π_{ij} in 10^{-12} m^2/N)

Crystal	Class	π_{11}	π_{12}	π_{13}	π_{44}	$\pi_{11}-\pi_{12}$	$\pi_{11}-\pi_{13}$
Potassium alum	$m3$	3.7	9.1	8.5	−0.65	−5.43	−4.82
Barium nitrate	$m3$	18.11	40.0	35.2	−1.69	−23.84	−17.13
Lead nitrate	$m3$	70.21	89.34	82.05	−1.39	−19.13	−11.84
Sodium chloride	$m3m$	0.25	1.46		−0.85	−1.21	
Fluorite	$m3m$	−0.29	1.16		0.698	−1.45	
Diamond	$m3m$	−0.43	0.37		−0.27	−0.80	

Values of the elasto-optical coefficients (dimensionless) for sodium D light

Crystal	Class	p_{11}	p_{12}	p_{13}	p_{44}	$p_{11}-p_{12}$	$p_{11}-p_{13}$
Potassium alum	$m3$	0.27	0.35	0.34	−0.0056	−0.0792	−0.0704
Barium nitrate	$m3$	2.49	3.40	3.20	−0.0205	−0.992	−0.713
Lead nitrate	$m3$	8.50	8.78	8.67	−0.0191	−0.281	−0.174
Sodium chloride	$m3m$	0.137	0.178		−0.0108	−0.0408	
Fluorite	$m3m$	0.0558	0.228		0.0236	−0.1722	
Diamond	$m3m$	−0.125	0.325		−0.11	−0.45	

through the crystal C which is oriented such that the $\langle 001 \rangle$ axes are along the vertical and horizontal. Lenses L_1, L_2 first make the rays parallel and then focus them onto the Babinet compensator (BC). The view seen through the compensator is a set of interference fringes (Fig. 13.5). When stress τ is applied, the fringes shift. The birefringence is related to the fringe shift S. The working equation is

$$(S\lambda/\beta) = -\left(n_0^3/2\right) t(\pi_{11} - \pi_{12})\tau, \tag{13.20}$$

where λ is the wavelength, β the fringe-width and t the thickness of the crystal. The stress τ is calculated from the load applied and the geometry of the loading device. For determination of π_{44}, a different crystal orientation is used (Table 13.4).

It may be noted that these measurements yield only the difference $(\pi_{11} - \pi_{12})$. To get the individual values of π_{11} and π_{12} additional measurements have to be made using ultrasonic diffraction methods which give the ratio π_{11}/π_{12}. Experimental results for some crystals are given in Table 13.5.

13.4 Electro-optic Effect

The application of electric field has an effect on the refractive indices of a crystal. This, in turn, results in changes ΔB_{ij} in the coefficients that describe the optical indicatrix. The relation between ΔB_{ij} and the components E_k of the electric field is

$$\Delta B_{ij} = z_{ijk} E_k. \tag{13.21}$$

This is the electro-optic effect, also known as the Pockel's effect. The coefficients z_{ijk} are called electro-optic coefficients. Since z_{ijk} connects a second-rank tensor ΔB_{ij} with a first-rank tensor E_k, the z_{ijk} coefficients constitute a third-rank tensor similar to the piezoelectric tensor $[d_{ijk}]$ (Chaps. 7 and 12). One would expect $[z_{ijk}]$ to have $3^3 = 27$ components. But since $\Delta B_{ij} = \Delta B_{ji}$, $z_{ijk} = z_{jik}$ and the number of independent z_{ijk} values reduces to 18. As in the case of $[d_{ijk}]$, it can be shown that the components of $[z_{ijk}]$ reduce to zero if the crystal has a centre of symmetry.

Again, as in the case of $[d_{ijk}]$, we can employ the contracted notation and write (13.21) as

$$\Delta B_i = z_{ij} E_j, (i = 1, 2, \ldots, 6; j = 1, 2, 3). \tag{13.22}$$

It may be noted that (13.22) is a matrix equation. The z_{ij} matrices for various point groups are the same as the d_{ij} matrices (Chap. 12, Table 12.2) with two differences (a) the index i takes values 1, 2, …, 6 and the index j takes values 1, 2, 3. Thus the d_{ij} arrays apply to the z_{ij} arrays if i is the column number and j the row number, (b) the factor 2 does not appear in changing from three-index to two-index notation. Hence the double circle in the matrices simply means numerically equal but with reversed sign (same as the single open circles).

The typical magnitude of z_{ij} is 10^{-12}m/V. The measurement of the electro-optic effect is complicated by the existence of two types of electro-optic effects – primary (or true) and secondary (or false). The primary effect is observed when the crystal is clamped, i.e. it does not develop any strain. If the crystal is free, the applied electric field creates strain due to the reverse piezoelectric effect and this strain creates a photoelastic effect. Thus, what is observed is the combined electro-optic and photoelastic effect. The photoelastic effect can be reduced by employing a high frequency AC field as in this case the strains are small.

The electro-optic effect has many applications; we shall mention only one of them. Let us consider KDP which is uniaxial at room temperature. Let a laser beam travel down the optic axis and let the electric field also be applied in the same direction. When the electric field is switched on, the crystal becomes biaxial so that a phase difference develops between the ordinary and extraordinary rays; this phase difference increases as the beam travels down the crystal. The variation in E is converted into variation of the phase difference, i.e. the crystal acts like a modulator [13.10].

13.5 Optical Activity

When a beam of plane-polarized monochromatic light passes through a medium, its plane of polarization undergoes a rotation. This is called optical activity. Some crystals exhibit optical activity. When the light is travelling towards the observer, if the rotation is clockwise, the rotation is called right-handed; if it is anticlockwise, the rotation is called left-handed. Depending on whether the rotation in a crystal is right-handed or left-handed, the crystal is called optically positive or optically negative. The rotation ϕ is proportional to t, the path travelled. The optical rotatory power ρ is defined as

$$\rho = \phi/t. \tag{13.23}$$

By convention, t is taken in mm.

13.5.1 The Mechanism of Optical Activity

When a plane-polarized beam enters the optically active crystal, it gets resolved into two circularly polarized waves, one right-handed and the other left-handed. These two waves travel with different refractive indices n_r and n_l. As they travel through the crystal, they develop a phase difference $(2\pi/\lambda)(n_l - n_r)t$. When they emerge out of the crystal, the two circularly polarized waves combine to form a plane-polarized wave whose plane of polarization is at an angle

$$\phi = \frac{\pi}{\lambda}(n_l - n_r)t \tag{13.24}$$

with the original plane of polarization. The difference $(n_l - n_r)$ is very small ($\sim 10^{-4}$). But since t/λ is large, ϕ is measurable. Thus, for quartz $(n_l - n_r) \sim 7 \times 10^{-5}$ and for a 1-mm-thick quartz plate, ϕ is $\sim 21°$at $\lambda = 5{,}893$Å.

13.5.2 The Nature of Optical Activity

As a convention, the term $(n_l - n_r)$ is expressed as $G/\bar{n}$ where $\bar{n}$ is the mean of n_l and n_r. Equation (13.22) may now be written as

$$\rho = \frac{\phi}{t} = \frac{\pi}{\lambda}\frac{G}{\bar{n}}. \tag{13.25}$$

The parameter G is less than unity and is direction-dependent. The direction-dependence of G is given by

$$G = g_{11}l_1^2 + g_{22}l_2^2 + g_{33}l_3^2 + 2g_{23}l_2l_3 + 2g_{31}l_3l_1 + 2g_{12}l_1l_2, \tag{13.26}$$

where the l_i values are direction cosines of the direction of propagation with respect to the axes of the crystal. The coefficients g_{ij} form a second-rank axial tensor (or a pseudo-scalar). We shall assume its transformation law which is

$$g'_{ij} = \pm a_{ik} a_{ji} g_{kl}, \quad (13.27)$$

where g_{ij} is called the gyration tensor.

13.5.3 Effect of Symmetry

Let us consider the effect of symmetry on the g_{ij} coefficients. For this, we shall use the direct inspection method (Chaps. 7, 8 and 12) wherein the indices of a coefficient change in the same way as those of the axes on rotation. In addition, we should also note whether the hand of the axes changes, in which case the sign of the coefficient would change.

If the crystal has a centre of symmetry, then on inversion, $1 \to -1, 2 \to -2, 3 \to -3$. Also, the hand changes. Thus

$$g'_{ij} = -g_{ij}. \quad (13.28)$$

But since the crystal has a centre of symmetry, there should be no change and

$$g'_{ij} = g_{ij}. \quad (13.29)$$

We can reconcile (13.28) and (13.29) by equating g_{ij} to zero. Thus centrosymmetric crystals do not have optical activity. This is the most important aspect of optical activity. We have seen in Chap. 5 that this condition is useful in the determination of space groups.

Several other results can be derived by considering the symmetry elements of each point group. The $[g_{ij}]$ tensor for non-centrosymmetric point groups are given in Table 13.6. Though the point groups $4mm$, $\bar{4}3m$, $3m$, $6mm$, $\bar{6}$ and $\bar{6}m2$ are non-centrosymmetric, all components of $[g_{ij}]$ vanish; these point groups are not included in the table.

13.5.4 Experimental Method and Results

A simple set-up for the determination of optical activity in crystals is shown in Fig. 13.11. S is a source of monochromatic light, SL a slit and L_1, L_2 are lenses. P_1 and P_2 are polarizers. P_2 is mounted on a graduated circular scale. With P_1 fixed, P_2 is rotated till there is extinction. The crystal plate C is then inserted. This disturbs

Table 13.6 Forms of the gyration tensor $[g_{ij}]$

KEY TO NOTATION

$\cdot$	Zero component
$\bullet$	Non-zero component
•—•	Equal components
•—o	Components numerically equal, but opposite in sign

All tensors are symmetrical about the leading diagonal

TRICLINIC

Class 1

$$\begin{bmatrix} \bullet & \bullet & \bullet \\ & \bullet & \bullet \\ & & \bullet \end{bmatrix} \ (6)$$

MONOCLINIC

Class 2

$2 \parallel x_2$ {standard orientation)

$$\begin{bmatrix} \bullet & \cdot & \bullet \\ & \bullet & \cdot \\ & & \bullet \end{bmatrix} \ (4)$$

Class 2

$2 \parallel x_3$

$$\begin{bmatrix} \bullet & \bullet & \cdot \\ & \bullet & \cdot \\ & & \bullet \end{bmatrix} \ (4)$$

Class m

$m \perp x_2$ {standard orientation)

$$\begin{bmatrix} \cdot & \bullet & \cdot \\ & \cdot & \bullet \\ & & \cdot \end{bmatrix} \ (2)$$

Class m

$m \perp x_3$

$$\begin{bmatrix} \cdot & \cdot & \bullet \\ & \cdot & \bullet \\ & & \cdot \end{bmatrix} \ (2)$$

ORTHORHOMBIC

Class 222

$$\begin{bmatrix} \bullet & \cdot & \cdot \\ & \bullet & \cdot \\ & & \bullet \end{bmatrix} \ (3)$$

Class $mm2$

$$\begin{bmatrix} \cdot & \bullet & \cdot \\ & \cdot & \cdot \\ & & \cdot \end{bmatrix} \ (1)$$

TETRAGONAL

Classes 4, 422

(2)

Class, $\bar{4}$

(2)

Class $\bar{4}2m$

$2 \parallel x_1$

$$\begin{bmatrix} \cdot & \bullet & \cdot \\ & \cdot & \cdot \\ & & \cdot \end{bmatrix} \ (1)$$

TRIGONAL AND HEXAGONAL

Classes 3, 32, 6, 622

(2)

CUBIC

Classes 432, 23

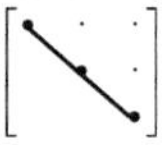

(1)

the extinction which is restored by rotating P_2. The angular rotation ϕ is noted. The specific optical rotation of some crystals is given in Table 13.7 [13.1].

There is limited experimental data on the individual g_{ij} values of optically active crystals. The g_{ij} values for α-quartz are: $g_{11} = g_{22} = \mp 5.82 \times 10^{-5}$; g_{33}

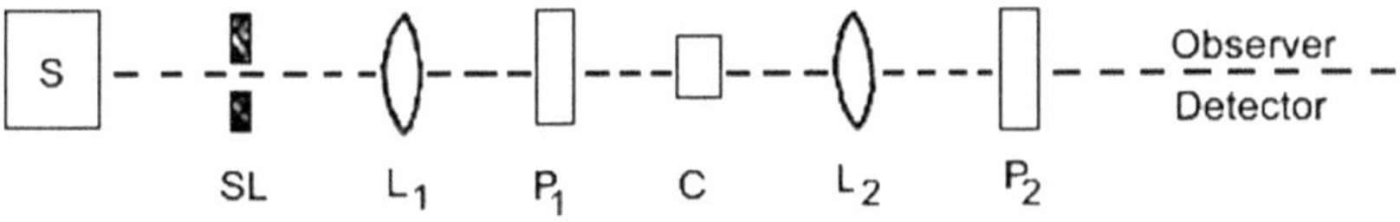

Fig. 13.11 Arrangement for determination of optical activity

Table 13.7 Specific optical rotation (ρ) for some crystals (for sodium D light)

Crystal	ρ [°/mm]
Sodium chlorate ($NaClO_3$)	3.5
Sodium bromate ($NaBrO_3$)	2.6
BGO ($Bi_{12}GeO_{20}$)	44
Quartz (SiO_2) along optic axis	21

$= \pm 12.96 \times 10^{-5}$ [13.11]. Here, the upper signs are for right-handed quartz and the lower ones for left-handed quartz.

13.6 Solid State Lasers

A laser is a source of light with the following features (1) high directionality, (2) high intensity, (3) high monochromaticity, (4) high coherence and (5) high degree of polarization. It is beyond our scope to go into the optics and physics of lasers; rather we are interested in the nonlinear properties of solids which became observable only with the high intensity radiation provided by lasers. We shall therefore discuss solid state lasers briefly as a prelude to nonlinear effects.

13.6.1 Models of Laser Action

We shall consider two common models of laser action. The three-level model is shown in Fig. 13.12. Here, the population of atoms in three levels E_1, E_2 and E_3 is as shown in the figure. By providing energy externally, some atoms from level E_1 are excited into level E_3 (Fig. 13.12a); this process is called "pumping". Some of these excited atoms decay into level E_2 by a non-radiative process. When this happens, the population of level E_2 increases with respect to that of level E_1 (Fig. 13.12b). This is called population inversion. This is a necessary condition for lasing action to take place. The energy released in transition from level E_2 to level E_1 is the laser energy.

Another model is the four-level model (Fig. 13.13). Here, atoms from E_1 are pumped into E_4. These excited atoms jump to the metastable level E_3 by a non-radiative transition. In this system, there is a level E_2 located between E_1 and E_3. Lasing action takes place between E_3 and E_2.

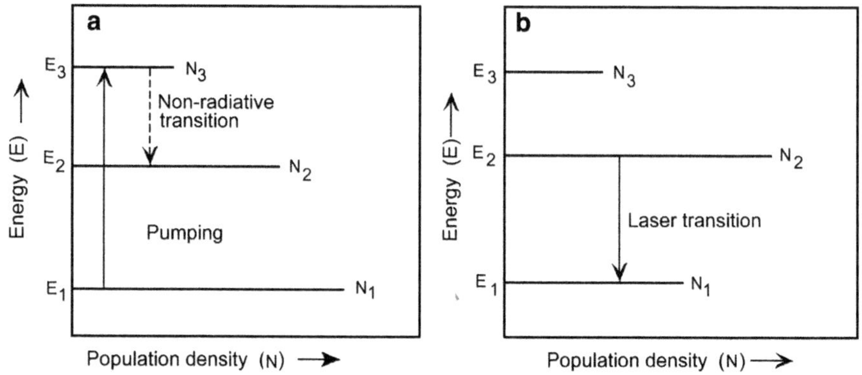

Fig. 13.12 Three-level model: (**a**) pumping and (**b**) laser action

Fig. 13.13 Four-level model

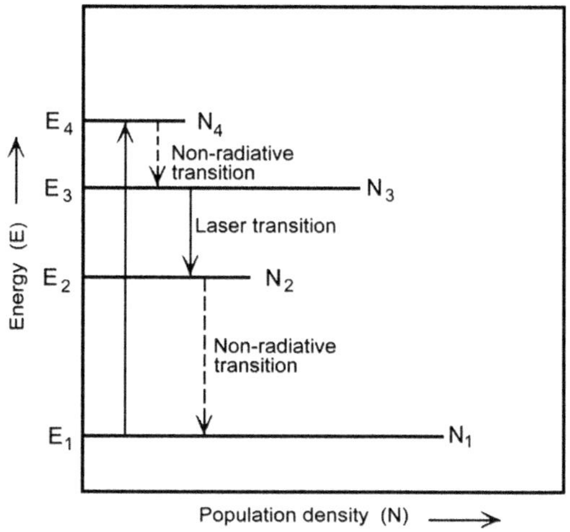

13.6.2 Some Solid State Lasers

Maiman [13.12] designed the first laser. He used a ruby single crystal which is Al_2O_3 containing 0.05% Cr^{3+} ions as impurity. The characteristic red colour of ruby is due to the Cr^{3+} ions which absorb blue and green. The design of the ruby laser is shown in Fig. 13.14. It shows a ruby rod surrounded by a Xenon flash lamp. For efficiency, the ruby rod and the flash lamp are enclosed in an elliptical reflector, the ruby rod and the lamp being at the foci of the ellipse.

The three-level energy diagram for Cr^{3+} is shown in Fig. 13.15. The E_3 level is actually made up of two levels E_3 and E_3'. The flash lamp excites the Cr^{3+} ions from E_1 to the levels E_3 and E_3'. The excited ions relax into the metastable E_2 level. Finally, the transition from E_2 to E_1 results in laser radiation of wavelength 6,943 Å. The ruby laser operates in pulses and reaches output levels of kW–mW power.

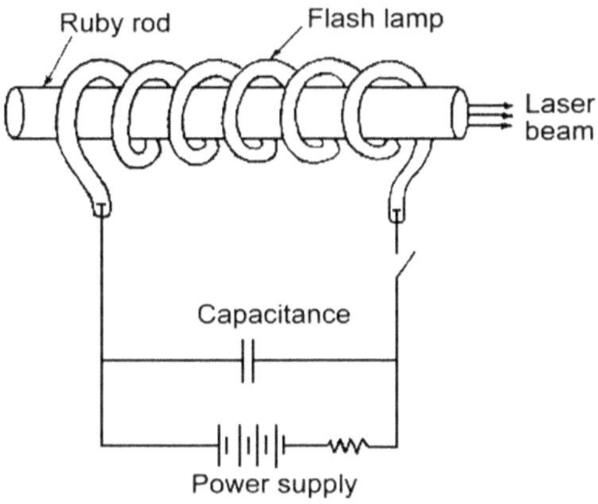

Fig. 13.14 Schematic of a ruby laser

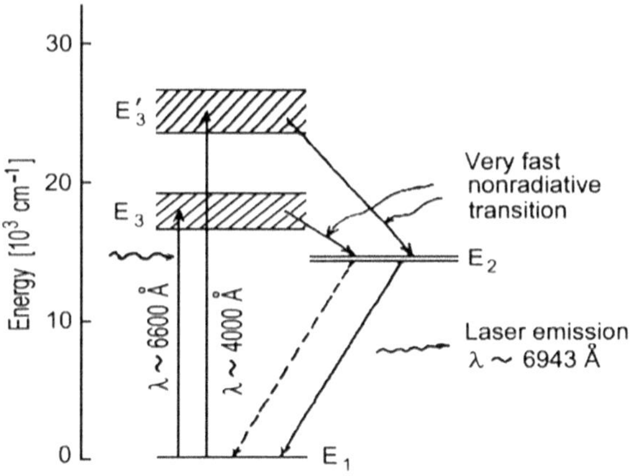

Fig. 13.15 Energy levels of Cr^{3+} ions in a ruby crystal

Another commonly used solid state laser is the YAG laser. This uses a yttrium aluminium garnet ($Y_3Al_5O_{12}$) crystal containing Nd^{3+} ions. Its working is based on the four-level model. The energy levels are shown in Fig. 13.16. It may be seen that the E_4 level consists of several closely spaced levels. The Nd ions are excited by a flash lamp from level E_1 to levels E_4. They then relax into level E_3. Population inversion takes place between E_3 and E_2. Lasing action between E_3 and E_2 results in infrared radiation of 1.06 μm with an intensity of $\sim 10^{16}$ W/cm^2. Nd ion lasers with other hosts like $CaWO_4$ crystal and glass are also in use.

13.7 Nonlinear Optical Properties of Solids

In Chap. 11, we have seen that the electric polarization P produced in a crystal is related to the applied field E by the relation

$$P_i = \varepsilon_{ij} E_j. \tag{13.30}$$

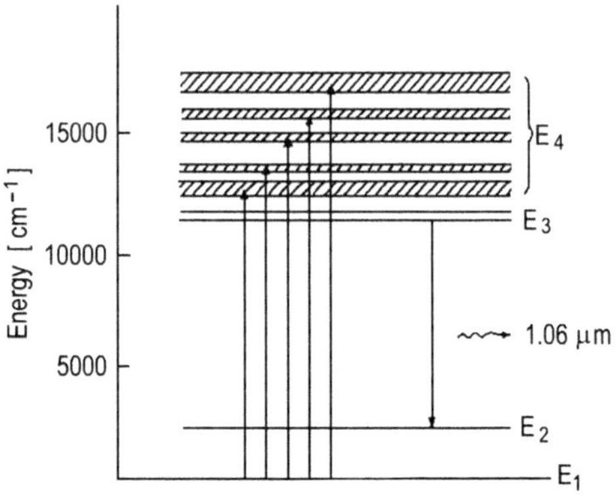

Fig. 13.16 Energy levels of Nd^{3+} ions in a YAG crystal

The coefficients ε_{ij} constitute the second-rank tensor $[\varepsilon_{ij}]$ of dielectric constants. We note here that at high frequencies, $\varepsilon = n^2$, where n is the refractive index. Equation (13.30) is valid at electric fields of ordinary intensities. At high fields, such as those made available by lasers, the relation between P and E becomes

$$P_i = \varepsilon_{ij} E_j + \chi_{ijk} E_j E_k + \chi_{ijkl} E_j E_k E_l + \cdots . \tag{13.31}$$

The coefficients χ are called nonlinear susceptibilities. We shall confine ourselves to the second term in (13.31). Then we have

$$P_i = \varepsilon_{ij} E_j + \chi_{ijk} E_j E_k . \tag{13.32}$$

Equation (13.32) represents the second-order nonlinear optical effect. As the coefficient connects the components of a vector P_i to the product $E_j E_k$ of two vectors, χ_{ijk} is a third-rank tensor (very much like the piezoelectric tensor). The χ_{ijk} tensor shares the following properties with the piezoelectric tensor:

1. The non-zero independent components of χ_{ijk} for various point groups are the same as those for the piezoelectric $[d_{ijk}]$ tensor (Chap. 12).
2. In centrosymmetric crystals each of the χ_{ijk} values vanishes, i.e. the nonlinear optical effect is observed only in non-centrosymmetric crystals.
3. As in piezoelectricity, the χ_{ijk} values can be represented in the two-index notation as χ_{ij} ($i = 1, 2, 3$; $j = 1, 2, \ldots, 6$).

13.7.1 Harmonic Generation

Let us consider (13.32). If the applied field is $E_0 \cos \omega t$, the second term may be written as

$$(E_0 \cos \omega t)^2 = E_0^2 (\tfrac{1}{2} + \tfrac{1}{2} \cos 2\omega t). \tag{13.33}$$

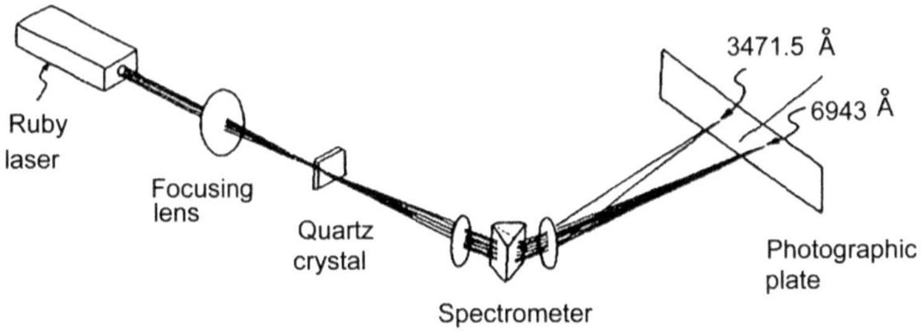

Fig. 13.17 Schematic of the set-up used by Franken for the observation of second harmonic generation in a quartz crystal

Thus the resulting oscillatory polarization contains frequencies ω (from the first term) and 2ω from the second term. This is called second harmonic generation. Supposing the applied field is made up of two waves of frequencies ω_1 and ω_2, the output will contain ω_1, ω_2, $2\omega_1$, $2\omega_2$, $(\omega_1 + \omega_2)$ and $(\omega_1 - \omega_2)$. The application of second harmonic generation (and also of the sum and difference frequencies) is very apparent; it helps to convert radiation of one frequency into that of another. We may mention here that if the higher order susceptibilities in (13.31) are of significant value, they generate higher order harmonics.

In general, crystals are weakly nonlinear, i.e. the χ_{ijk} values have a low value. The nonlinear effects can be observed (1) with intense fields, (2) fairly long crystals and (3) when there is a matching of the wave vectors of the participating waves (phase matching).

13.7.2 Observation of Harmonics

Second harmonic generation (SHG) was observed for the first time by Franken et al. [13.13]. Their set-up is shown in Fig. 13.17. Light of wavelength 6,943 Å from a ruby laser was focussed to a spot ~1 mm in diameter on a quartz crystal. The radiation from quartz was recorded with a quartz spectrograph on a red insensitive film; a line was observed in the ultraviolet at 3,471 Å. In later confirmatory experiments, a grating spectrograph was used where the first-order diffraction in the UV could be compared with the second-order diffraction of the red line in the same exposure.

As mentioned in the earlier section, several orders of harmonics can be observed depending on the values of the susceptibilities and the power of the incident field. Some relevant information is given in Table 13.8.

Table 13.8 Generation of harmonics of Nd: YAG laser radiation with $\lambda = 1.318\,\mu m$ [13.7]

Number of harmonic	λ [nm]	Crystal	l [mm]	Output	Energy conversion efficiency [%]
2	659.4	$LiNbO_3$	16	85 kW	10
3	439.6	KDP	30	3.4 kW	0.4
4	329.7	KDP	30	6 kW	0.6
2	659.4	$LiIO_3$	10	1 W	100
3	439.6	$LiIO_3$	8	1.4 mJ	1.2

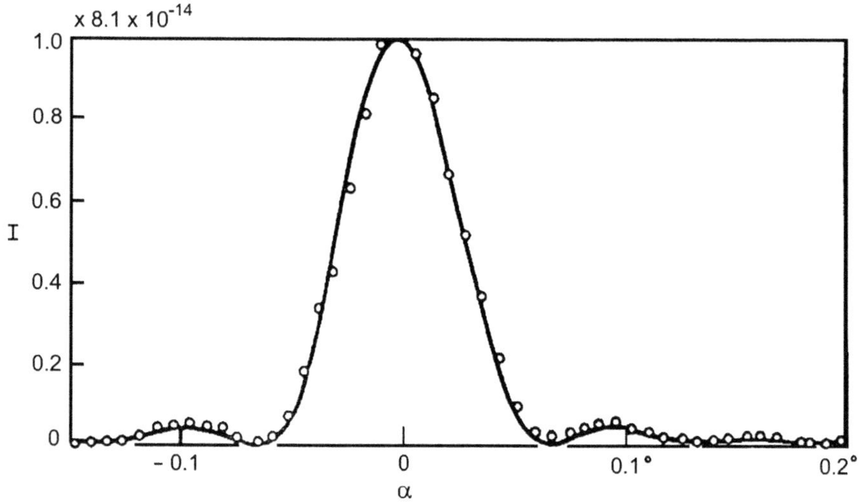

Fig. 13.18 Intensity profile of the second harmonic in KDP for determination of χ_{36}

13.7.3 Measurement of Susceptibility

The susceptibility χ_{ij} can be measured from the intensity profile of the second harmonic. The expression for the intensity (I) for the KDP crystal is given by

$$I = \chi_{36} \sin(\theta + \alpha) E_x E_y, \tag{13.34}$$

where θ is the angle which the direction of phase propagation makes with the z-axis of the crystal and α is a small angle which the direction of energy propagation makes with the direction of phase propagation. A plot of I against α for KDP is shown in Fig. 13.18. By fitting the curve to (13.34), a value of $\chi_{36} = 12 \times 10^{-9}$ esu is obtained for KDP. This value is taken as a standard in comparing χ_{ij} values of different crystals (Table 13.9).

Table 13.9 Values of the nonlinear susceptibility elements $\chi_{ij}(2\omega = \omega + \omega)$ for some crystals. These are relative values expressed in terms of χ_{36} for KDP ($= 12 \times 10^{-9}$ esu) as 1

Crystal	Susceptibility for SHG For $\lambda = 6{,}943$ Å	For $\lambda = 1\,\mu$m
KH_2PO_4	$\chi_{36} = 1.00$	
(tetragonal)	$\chi_{14} = 0.95 \pm 0.06$	
KD_2PO_4	$\chi_{36} = 0.75 \pm 0.02$	
	$\chi_{14} = 0.76 \pm 0.04$	
$NH_4H_2PO_4$	$\chi_{36} = 0.93 \pm 0.06$	
	$\chi_{14} = 0.89 \pm 0.04$	
Quartz		$\chi_{11} = 0.82 \pm 0.04$
		$\chi_{14} = 0.00 \pm 0.05$
$BaTiO_3$		$\chi_{15} = 35.0 \pm 3$
		$\chi_{31} = 37.0 \pm 3$
		$\chi_{33} = 14.0 \pm 1$

13.7.4 Efficiency of SHG Conversion

The efficiency of second harmonic generation η_{SHG} is (given by Bhat [13.14])

$$\eta_{\text{SHG}} = \frac{P_{2\omega}}{P_\omega} = 2\left(\frac{\mu}{\varepsilon_0}\right)^{3/2} \omega^2 \left(\frac{d^2 l^2}{n^3}\right) \left[\frac{\sin^2(\Delta k l/2)}{(\Delta k l/2)^2}\right] \left(\frac{P_\omega}{A}\right), \tag{13.35}$$

where P_ω, $P_{2\omega}$ are the powers of incident and SH beams, d is the effective nonlinear coefficient in the phase-matched direction, l the crystal length, n the refractive index, μ the optical absorption, ε_0 the vacuum dielectric constant, A the area of cross-section of the crystal and Δk the wave vector difference. The important parameters in (13.35) are the length l which occurs as l^2 and the figure of merit (d^2/n^3). The conversion efficiencies for some harmonics generated by some crystals are included in Table 13.8.

Problems

1. Derive the gyration tensor for class m.
2. Derive an expression for the birefringence in a crystal of class 432 subject to uniaxial stress.
3. Explain the changes which a beam of monochromatic light undergoes as it passes through each optical component in Fig. 13.10.
4. Discuss how a fourth harmonic is generated.
5. Discuss the use of optical properties in determination of crystal symmetry.

References

13.1. J.F. Nye, *Physical Properties of Crystals* (Oxford University Press, Clarendon, 1957)
13.2. S. Bhagavantam, *Crystal Symmetry and Physical Properties* (Academic Press, New York, 1966)
13.3. Y.I. Sirotin, M.P. Shaskoslaya, *Fundamentals of Crystal Physics* (Mir Publishers, Moscow, 1982)
13.4. N. Bloembergen, *Nonlinear Optics* (Benjamin-Cummings, Massachusetts, 1965)
13.5. M.E. Lines, A.M. Glass, *Principles and Applications of Ferroelectrics and Related Materials* (Oxford University Press, Clarendon, 1977)
13.6. K. Thyagarajan, A.K. Ghatak, *Lasers* (McMillan, New York, 1981)
13.7. V.G. Dmitriev, G.G. Gurzadyan, D.N. Nikogosyan, *Handbook of Nonlinear Optical Materials* (Springer, Berlin, 1991)
13.8. W.F. Magie, *A Source Book in Physics* (McGraw Hill, New York, 1935)
13.9. S. Bhagavantam, Proc. Ind. Acad. Sci. **A16**, 359 (1942)
13.10. V.K. Wadhawan, *Introduction to Ferroic Materials* (Gordon Breach Science Pub., Amsterdam, 2000)
13.11. G. Szivessy, C. Munster, Ann. Phys. **20**, 703 (1934)
13.12. T.H. Maiman, Nature **187**, 494 (1960)
13.13. P.A. Franken, A.E. Hill, C.W. Peters, G. Weinreich, Phys. Rev. Lett. **7**, 118 (1961)
13.14. H.L. Bhat, Bull. Mater. Sci. **17**, 1233 (1994)

Chapter 14
Defects in Crystals I (Point Defects)

An ideal crystal is defined as an infinite arrangement of atoms in three dimensions which satisfies certain symmetry conditions; such an array is called a crystal lattice. An essential feature of an ideal crystal lattice is that the environment of an atom in the crystal is identical to the environment of any other equivalent atom. But defects (or imperfections) do occur in crystals. An imperfection is any deviation from a regular lattice. These imperfections may be classified according to their dimensions. Thus we may mention:

1. Zero-dimensional or point defects: these include vacancies, interstitials, Schottky defects, Frenkel defects, impurities and colour centres.
2. One-dimensional or line defects: these include dislocations of different types.
3. Two-dimensional or planar defects: these include the surface of the crystal, stacking faults and grain boundaries.
4. Three-dimensional or volume defects: these include gross defects like voids and cracks in bulk crystals.

In this chapter, we shall limit ourselves to the study of three types of point defects. In part A, we shall consider vacancies and interstitials which are primary point defects; and in part B, we shall consider colour centres which are induced point defects.

Part A: Primary Point Defects

14.1 General

Ionic solids are essentially electrical insulators, yet they have a feeble conductivity. This conductivity is typically of the order of $10^{-9}\Omega^{-1}\text{cm}^{-1}$ at ordinary temperatures and $10^{-2}\Omega^{-1}\text{cm}^{-1}$ at temperatures close to the melting point. It is difficult to account for this conductivity if the crystal is perfect and if the conduction is assumed to be through exchange of neighbouring ions. To quote Lidiard [14.1],

D.B. Sirdeshmukh et al., *Atomistic Properties of Solids*, Springer Series in Materials Science 147, DOI 10.1007/978-3-642-19971-4_14,

"the drift of ions under the action of an electric field is inconceivable if the crystal is perfect......... Transport of charge could only take place if anion–cation exchanges were possible. But in NaCl and AgBr, the energy which would have to be supplied to effect an exchange of places between an anion and cation is so large that in one gram-molecule such an event would occur only in some fantastically long time such as 10^{30} years (!)".

The conductivity of ionic crystals could be explained only by introducing the concept of point defects in thermal equilibrium. Frenkel [14.2] and Schottky [14.3] introduced two types of imperfections (now known, respectively, after their names) which went a long way in accounting for the observed features of conductivity of ionic crystals.

In this part of the chapter, we describe the various types of point defects. The expression for the equilibrium concentration of defects is derived. The explanation of ionic conductivity in terms of the point defects is treated in detail. The effect of point defects on many properties other than conductivity is also discussed. Finally, direct evidence of the existence of vacancies and interstitials is cited.

The subject of point defects in crystals is basic to the understanding of solid state behaviour. Consequently, treatments of varying depths are found in standard text books like Levy [14.4], Dekker [14.5] and Kittel [14.6]. Somewhat more detailed treatments are given by Lidiard [14.1], Kubo and Nagamiya [14.7] and Franklin [14.8].

14.2 Types of Point Defects

We shall first consider point defects in elemental crystals. These are shown in Fig. 14.1. Figure 14.1a shows a vacancy; a vacancy can be imagined to have been formed by removing an atom from the interior of the crystal and placing it at a lattice point on the surface. Figure 14.1b shows an interstitial which is an atom present at a place not normally occupied by an atom. When an interstitial occurs in conjunction with a vacancy as shown in Fig. 14.1c, it is called a Frenkel defect. It may be noted that in all these cases, the atomic environment in the vicinity of the defect is different from the normal environment away from it.

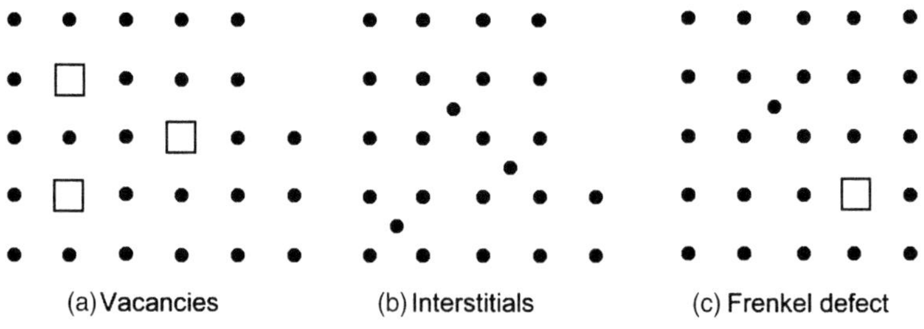

Fig. 14.1 Models of (**a**) vacancy, (**b**) interstitial and (**c**) Frenkel defect in elemental solids

We shall now consider the corresponding defects in ionic crystals. Figure 14.2a shows vacancies in an ionic crystal like NaCl. If a positive ion is taken out from its site and placed on the surface, then a positive charge will build up on the surface. To maintain charge neutrality, the creation of a positive ion vacancy must be accompanied by the creation of a negative ion vacancy. Such a pair of vacancies is called a Schottky defect. The defect shown in Fig. 14.2b is called a Frenkel defect. Here an ion leaves its site causing a vacancy and then settles in an interstitial position. In the figure, the vacancies necessary to provide interstitial ions are shown to be on the positive ion lattice. This is what generally happens in an ionic crystal with NaCl structure. Formation of Frenkel defects with negative ion vacancies is not ruled out [14.9]. However, occurrence of Frenkel defects simultaneously on the positive and negative ion lattices is not favoured [14.10].

Energy is required to form a defect. The formation energy depends on the type of defect and also on the host crystal. The prominent defect is that which has a lower formation energy.

The difference between the symmetry of the environment around a vacancy and an interstitial may be noted. A vacancy is formed at a normal lattice point. It can be seen from Fig. 14.2a that in NaCl lattice, the vacancy has an octahedral symmetry. On the other hand, the symmetry of the environment around an interstitial is tetrahedral. This cannot be visualised from Fig. 14.2b but is made clear in Fig. 14.3.

Aggregates (or combination) of defects also occur though not commonly. They will not be discussed here but will be considered whenever they show their effects.

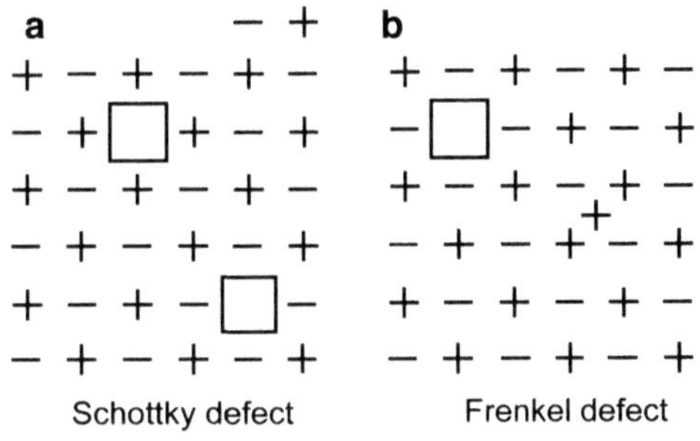

Fig. 14.2 Models of (**a**) Schottky and (**b**) Frenkel defects in ionic solids

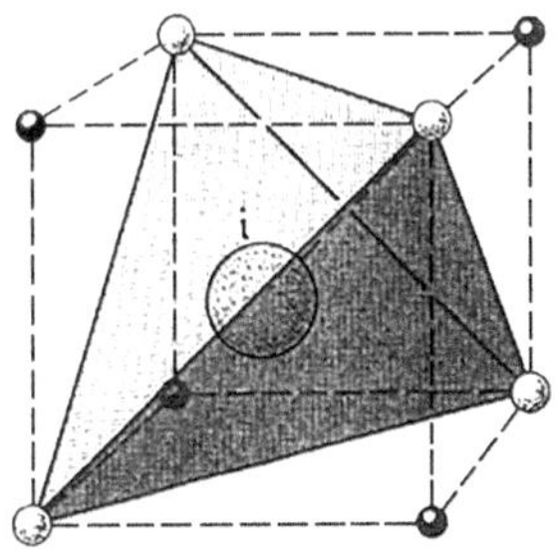

Fig. 14.3 Tetrahedral environment of an interstitial

14.3 Equilibrium Concentration of Defects

The equilibrium concentration of Frenkel defects and Schottky defects was originally derived by Frenkel [14.2] and Schottky [14.3]. The same derivations are given in different degrees of detail in the sources mentioned in Sect. 14.1.

If n is the number of defects and N the number of sites where they can be formed, (n/N) is called the equilibrium concentration. The derivation is based on the thermodynamic principle that for a system to be in equilibrium, its free energy F should be a minimum. The free energy is given by

$$F = U - TS, \tag{14.1}$$

where U is the internal energy, T the temperature and S the entropy.

14.3.1 *Qualitative Treatment*

For simplicity, we shall consider vacancies. Let ϕ_{V} be the energy of formation of a single vacancy. If there are n vacancies, the internal energy $U = n\phi_{\mathrm{V}}$; thus U increases linearly with (n/N). The entropy S has two contributions: thermal and configurational. The first, S_{th}, is the entropy due to the distribution of thermal energy and is given by

$$S_{\mathrm{th}} = 3Nk_{\mathrm{B}}\left[1 + \log\left(\frac{k_{\mathrm{B}}T}{h\nu}\right)\right], \tag{14.2}$$

where k_{B} is the Boltzmann constant and ν the Einstein characteristic frequency. The other contribution, S_{cf}, is determined by the number of ways in which atoms may be arranged over available sites. If a lattice consists of N_{A} number of atoms of type A and N_{B} number of atoms of type B and if all sites are equivalent,

$$S_{\mathrm{cf}} = k_{\mathrm{B}}\log\left[\frac{(N_{\mathrm{A}} + N_{\mathrm{B}})!}{N_{\mathrm{A}}!N_{\mathrm{B}}!}\right]. \tag{14.3}$$

If we identify N_{A} with the total number of sites N and N_{B} with the number of vacancies n, (14.3) becomes

$$S_{\mathrm{cf}} = k_{\mathrm{B}}\log\left[(N + n)!/N!n!\right], \tag{14.4}$$

and the total entropy S is

$$S = S_{\mathrm{th}} + S_{\mathrm{cf}}. \tag{14.5}$$

Let us now consider the variation of U, S and F with the vacancy concentration (n/N) at temperature $T > 0$. As mentioned, U increases linearly with (n/N) (Fig. 14.4). On the other hand, the creation of vacancies increases disorder in the

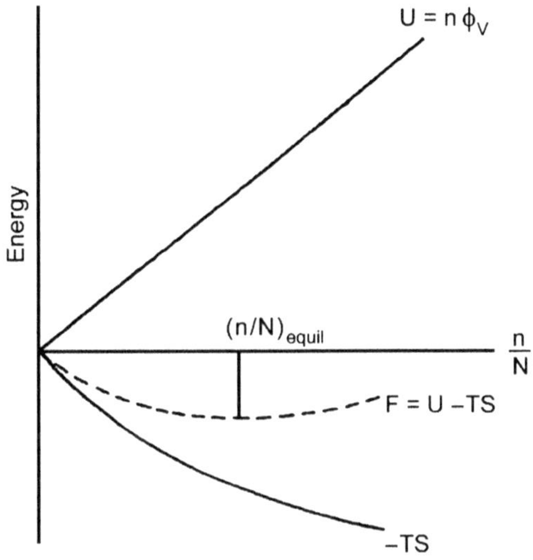

Fig. 14.4 Dependence of free energy on concentration of vacancies

crystal which leads to an increase in S_{cf} and a decrease in the term $(-TS_{cf})$ and hence in TS. This variation is also shown in the figure. Thus there are two opposing contributions to F, one due to U increasing with (n/N) and the other due to S_{cf} decreasing with (n/N). The curve for F has a minimum at a certain value of (n/N) which is the equilibrium concentration. Thus, for thermodynamic stability, a crystal has to necessarily have this concentration of vacancies.

14.3.2 Quantitative Treatment

14.3.2.1 Vacancies

Let us denote the free energy of a perfect crystal with N similar atoms at temperature T by $F(T)$. Let n vacancies be created each requiring an energy ϕ_V; this increases the energy of the crystal with vacancies by $n\,\phi_V$ in comparison with the crystal without any vacancies. Let the increase in thermal entropy per vacancy be ΔS_{th}. The expression for S_{cf} is given by (14.4). The full expression for the free energy of the crystal containing n vacancies is

$$F(n,T) = F_{perfect}(T) + n\phi_V - nTS_{th} - k_B T \log\left[(N+n)!/N!n!\right]. \qquad (14.6)$$

Applying the condition for thermal equilibrium $(\partial F/\partial n)_T = 0$ to (14.6). we get

$$\begin{aligned}(n/N) &= \exp(\Delta S_{th}/k_B)\exp(-\phi_V/k_B T)\\ &= A\exp(-\phi_V/k_B T). \qquad (14.7)\end{aligned}$$

In deriving (14.7), it is assumed that $n << N$ and $\log(x!) \approx x \log x$ for $x >> 1$. The term $A = \exp(\Delta S_{\text{th}}/k_{\text{B}})$ is called the pre-exponential factor. Using the Einstein model of atomic vibrations, it can be shown that

$$A = \left(\frac{\nu}{\nu'}\right)^{3x}, \tag{14.8}$$

where x is the number of neighbours of a vacancy and ν and ν' are the Einstein frequencies of an atom away from and in the vicinity of a vacancy.

Equation (14.7) shows that there is a finite probability that (n/N) sites are vacant at temperature T and that this probability is given by a Boltzmann factor which includes the formation energy of the vacancy. If the pre-exponential factor is ignored, for a metal with $\phi_{\text{V}} = 1$ eV and $T \approx 1{,}000$ K, $(n/N) \approx \mathrm{e}^{-12} \approx 10^{-5}$, i.e. one site in every 1,00,000 sites is vacant.

14.3.2.2 Frenkel Defects

Let n be the number of Frenkel pairs (a vacancy and an interstitial), ΔS_{th} the increase in the thermal entropy per Frenkel pair and ϕ_{F} the energy of formation of a Frenkel pair. Then, the second term in (14.6) becomes $n\phi_{\text{F}}$ and the third term is the same. Regarding the configurational entropy, we have to consider two contributions: the distribution of vacancies and the distribution of interstitials. Let N and N' be the number of sites available for the occupation of vacancies and interstitials respectively. From (14.3) and (14.4) the configurational entropies for the vacancies and interstitials will be $k_{\text{B}}T \log[(N+n)!/N!n!]$ and $k_{\text{B}}T \log[(N'+n)!/N'!n!]$. According to the laws of probabilities, the total probability will be not the sum but the product of the two probabilities. Thus, the last term in (14.6) becomes $k_{\text{B}}T \log[(N+n)!/N!n!][(N'+n)!/N'!n!]$. Making these changes in (14.6) and again applying the equilibrium condition $(\partial F/\partial n)_T = 0$, we get

$$\begin{aligned} n\big/(NN')^{1/2} &= \exp(\Delta S_{\text{th}}/2k_{\text{B}})\exp(-\phi_{\text{F}}/2k_{\text{B}}T) \\ &= A' \exp(-\phi_{\text{F}}/2k_{\text{B}}T), \end{aligned} \tag{14.9}$$

where A' is the pre-exponential factor. Two differences may be noted between (14.7) and (14.9). Firstly, N in (14.7) is replaced by $(NN')^{1/2}$. Secondly, a factor 2 enters the denominators of the thermal entropy term and the formation energy term. This is because two processes (creation of vacancies and creation of interstitials) are involved in the creation of a Frenkel defect with consequent changes in the configurational entropy.

14.3.2.3 Schottky Pairs in Ionic Crystals

Let us consider an ionic crystal of composition A^+B^-. The process of creating a positive ion vacancy is the same as described earlier, i.e. a positive ion is removed from its site and is taken to a lattice site on the surface. Further, to maintain charge neutrality of the surface, an equal number of negative ion vacancies are created. The formation energies of the two types of vacancies may be unequal and yet the number of positive and negative ion vacancies will have to be the same. Hence, what matters is ϕ_S the energy of formation of a Schottky pair. Since two types of vacancies have to be arranged, the configurational entropy S_{cf} is now given by

$$S = k_B \log[(N+n)!/N!n!]^2. \tag{14.10}$$

The free energy is

$$F(n,T) = F_{perfect}(T) + n\phi_S - nT\Delta S_{th} - k_B T \log[(N+n)!/N!n!]^2. \tag{14.11}$$

Applying the equilibrium condition, we get

$$\begin{aligned} n/N &= \exp(\Delta S_{th}/2k_B)\exp(-\phi_S/2k_B T) \\ &= A'' \exp(-\phi_S/2k_B T) \end{aligned} \tag{14.12}$$

where A'' is the pre-exponential factor.

14.3.2.4 Frenkel Defects in Ionic Crystals

In an ionic crystal, a positive ion may leave its site causing a vacancy and move to an interstitial site; the process of creation of such a Frenkel defect may require an energy ϕ_F. Frenkels may also form on negative ions requiring an energy different from ϕ_F. However, there is no requirement that an equal number of these two types are formed since there is no charge build-up on the surface. The type of Frenkel defect formed depends on the type which needs lesser energy; in the silver halides, for example, Frenkel defects are formed on the positive ion lattice.

Apart from this, there is no difference in the Frenkel defect formation in an ionic crystal and an elemental crystal. The expression for the equilibrium concentration is similar viz.

$$n/(NN')^{1/2} = A''' \exp(-\phi_F/2k_B T) \tag{14.13}$$

where $A''' = \exp(\Delta S_{th}/2k_B)$ is the pre-exponential factor.

14.4 Electrical conductivity

In the introduction (Sect. 14.1), attention was drawn to the difficulty in explaining the electrical conductivity in a perfect ionic crystal. We shall now see how the same can be explained by invoking the presence of defects.

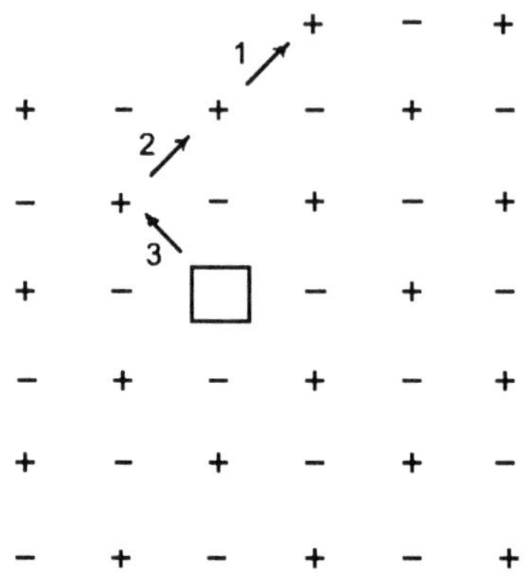

Fig. 14.5 Process of creation of a positive ion vacancy through jumps

14.4.1 Conductivity due to Schottky and Frenkel Defects

We shall assume that the conductivity in ionic crystals like alkali halides is due to the mobility of positive ion vacancies. The creation of a vacancy takes place in a series of jumps of the positive ion between neighbouring positive ion sites. As shown in Fig. 14.5, a positive ion on the surface jumps to form a new surface layer (jump 1) forming a vacancy in layer 1. An ion in the next atomic layer jumps into the vacancy in layer 1 (jump 2). This continues until a stable vacancy is formed (jump 3).

Let us consider a positive ion vacancy in a NaCl lattice (Fig. 14.6). This vacancy can jump to any of the 12 surrounding positive ion sites. If p is the probability of any jump, the probability of jump in the x-direction is $p/3$. The variation of potential along the x-direction in the absence of any external field is shown in Fig. 14.7a. The potential has minima at the planes A and B and a maximum in the C-plane midway between the A and B planes; the height of this maximum is the potential barrier ε_j. For a jump to take place, the ion has to overcome the potential barrier ε_j and the probability p is ν exp $(-\varepsilon_j/k_{\text{B}}T)$, ν being the Einstein frequency of the ion. Let us see what happens if an electric field E is applied in the x-direction (Fig. 14.7b). The effect of the field is to lift the minimum at A and depress the minimum at B by equal amounts. As a result, the potential experienced by the vacancy is modified by $-\frac{1}{2}aeE$ on its left and by $\frac{1}{2}aeE$ on the right, "a" being the interionic distance. This modifies the probabilities of jump. If we denote the probabilities of jump on the left and right by $p_{\leftarrow}$ and $p_{\rightarrow}$, these are given by

$$p_{\leftarrow} = \tfrac{1}{3}\nu \exp[-(\varepsilon_j - \tfrac{1}{2}aeE)/k_{\text{B}}T], \tag{14.14}$$

$$p_{\rightarrow} = \tfrac{1}{3}\nu \exp[-(\varepsilon_j + \tfrac{1}{2}aeE)/k_{\text{B}}T].$$

The current density I (net flux of charge passing across unit area per second) is

$$I = (1/2a^2)(n/N)(p_{\leftarrow} - p_{\rightarrow})e. \tag{14.15}$$

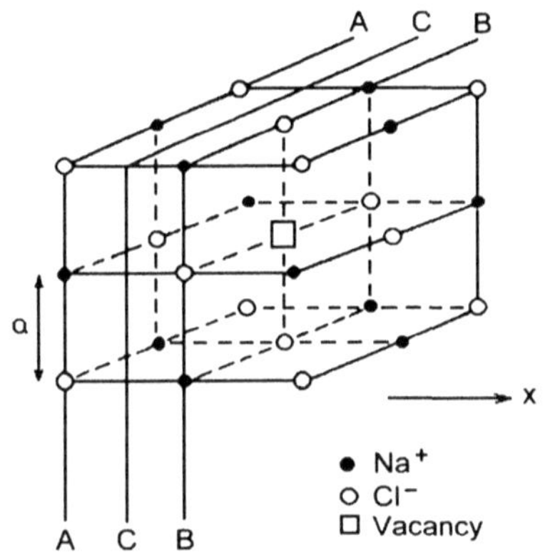

Fig. 14.6 Jump probability of a positive ion vacancy

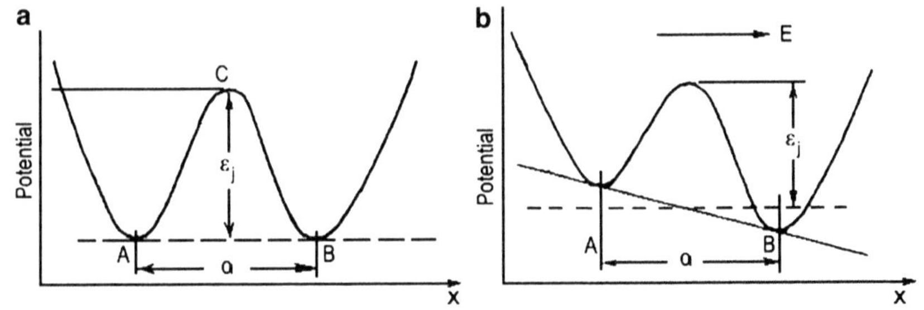

Fig. 14.7 Potential experienced by a vacancy in a crystal (**a**) in the absence and (**b**) in the presence of an electric field

Here $(1/2a^2)$ is the number of positive ion sites in a plane of unit area perpendicular to the x-axis and (n/N) is the probability that a site is vacant. Generally $aeE << k_BT$. With this assumption (14.14) and (14.15) lead to

$$\begin{aligned} I &= (n/N)\nu e^2(1/6ak_BT)E\exp[-(\varepsilon_j/k_BT)] \\ &= \sigma E \end{aligned} \tag{14.16}$$

where σ is the conductivity. Using the expression for (n/N) for Schottky pairs from (14.12), we get

$$I = A''(e^2\nu/6ak_BT)E\exp[-\{\varepsilon_j + (\phi_S/2)\}/k_BT]. \tag{14.17}$$

Thus the ionic conductivity of a crystal like NaCl can be explained in terms of the motion of the ions through the mechanism of the jump or migration of the vacancies. The conductivity is dependent on two processes: one related to the jump or migration of ions and the other to the thermal generation of Schottky pairs. It may be mentioned that a similar equation holds for the conductivity due to Frenkel defects in ionic crystals.

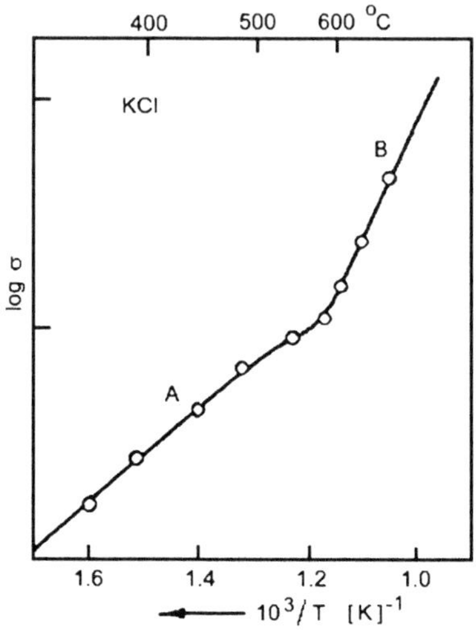

Fig. 14.8 Plot of log σ against T^{-1} for pure KCl

14.4.2 Comparison with Experiment

The temperature variation of conductivity is given by (14.17). The variation of the term T^{-1} in the pre-exponential factor is much slower than the exponential term. Therefore, it may appear that a plot of log σ against T^{-1} may yield a simple straight line. However, generally, such a plot consists of two straight regions (marked A and B in Fig. 14.8 which represents the conductivity data for KCl). The low temperature region A is called the extrinsic region. At low temperatures, the number of thermally generated vacancies is too small and the main contribution to conduction is through the jump or migration of ions through spurious vacancies; the origin of these vacancies could be in impurities or quenching of the crystal after its growth. Thus the slope of the plot A is simply (ε_j/k_B). As the temperature increases, the number of thermally generated Schottky pairs increases and they contribute predominantly to the conductivity in the high temperature region B. For this reason, this region is called the intrinsic region. The slope of the plot in this region is as given by (14.17), i.e. $[\varepsilon_j + (\phi_S/2)]/k_B$. From these two slopes, the values of the jump (or migration) energy ε_j and the formation energy ϕ_S can be evaluated. The slopes of the intrinsic (B) and extrinsic (A) regions in Fig. 14.8 lead to the values $\varepsilon_j = 1.04\,\text{eV}$ and $\phi_S = 2.26\,\text{eV}$ for KCl.

The analysis of conductivity curves for ionic crystals containing Frenkel defects like the silver halides is done in a similar way. It may be noted that conductivity data alone cannot differentiate between the two types of defects (Schottky and Frenkel). This point will be discussed again in a later section.

14.4.3 Conductivity of Doped Ionic Crystals

Let us now consider doped crystals, i.e. crystals which contain deliberately added impurities. We shall consider ionic crystals like alkali halides doped with aliovalent impurities. If, for example, KCl is doped with $SrCl_2$, the Sr^{2+} ion displaces two K^+ ions (to maintain charge neutrality). Sr^{2+} occupies one of the two K^+ sites and the other K^+ site remains vacant. Thus the number of positive ion vacancies, and, hence, the conductivity are enhanced, the enhancement being proportional to the concentration of the dopant. Specimens with different concentrations of the dopant will have different log σ versus T^{-1} plots in the A region but all of them will have the same slope (ε_j/k_B). At higher temperatures, however, the number of vacancies due to the dopant is much less than the number of thermally generated vacancies. Hence all the A plots merge into a single B plot with slope $[\varepsilon_j + (\phi_S/2)]/k_B$. Figure 14.9 shows log σ versus T^{-1} plots for several samples of KCl doped with $SrCl_2$.

14.4.4 Resistivity of Metals

Vacancies affect the resistivity of metals in much the same way as they affect the conductivity of ionic crystals. However, metals have a resistivity even in the absence

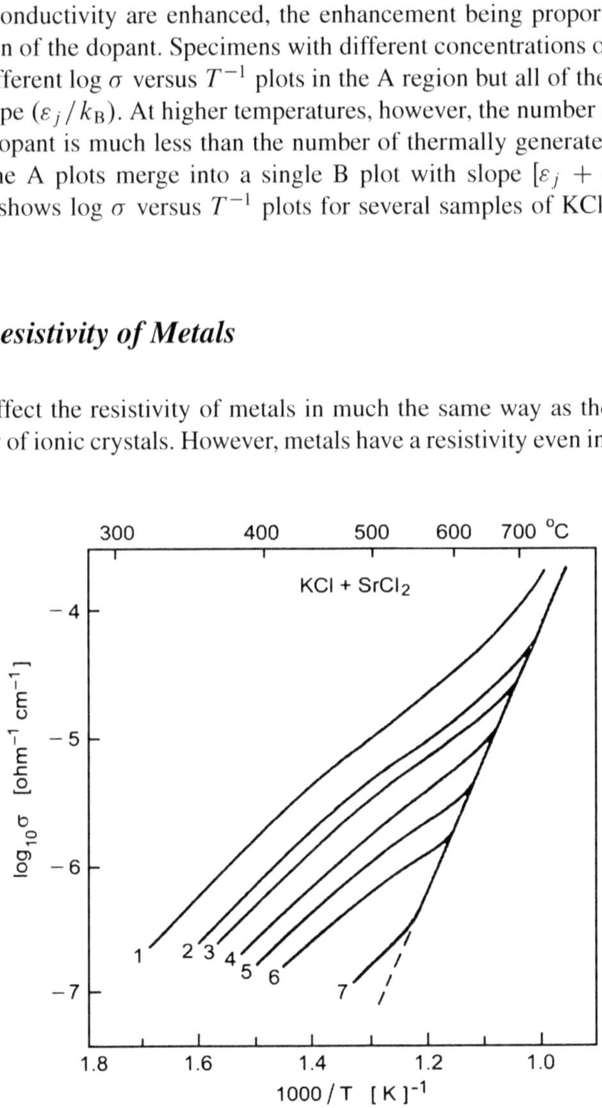

Fig. 14.9 Plot of log σ versus T^{-1} of KCl crystals containing various amounts of $SrCl_2$. In units of 10^{-5} the numbers refer to the following mole fractions: $M_1 = 19$; $M_2 = 8.7$; $M_3 = 6.1$; $M_4 = 3.5$; $M_5 = 1.9$; $M_6 = 1.2$; $M_7 = 0$

of vacancies. Thus the total resistivity $\rho(T)$ of a metal may be represented by

$$\rho(T) = \rho_L(T) + \rho_V(T) \tag{14.18}$$

where $\rho_L(T)$ is the lattice resistivity, i.e. the resistivity due to scattering of electrons by phonons and $\rho_V(T)$ the resistivity due to scattering of electrons by vacancies; the latter contribution is obviously proportional to the concentration of vacancies, i.e. $A \exp(-\phi_V/k_B T)$. There are different formulations of $\rho_L(T)$; the one based on the Einstein model predicts

$$\rho_L(T) = CT/\theta_E^2 \tag{14.19}$$

where C is a constant, T the temperature and θ_E the Einstein characteristic temperature. The temperature variation of the term $(\rho_L(T)/T)$ can be obtained by differentiating (14.19); this yields

$$\mathrm{d}[\log(\rho_L/T)]/\mathrm{d}T = 2\beta\gamma \tag{14.20}$$

where β is the volume coefficient of thermal expansion and γ the Gruneisen constant. Equation (14.18) may be rewritten as

$$\rho_V(T) = \rho(T)_{\text{meas.}} - \rho_L(T)_{\text{calc.}}. \tag{14.21}$$

A plot of log $[\rho(T)/T]_{\text{meas.}}$ and log $[\rho_L(T)/T]_{\text{calc.}}$ for gold is shown in Fig. 14.10. Since $\rho_V(T)$ is proportional to $\exp(-\phi_V/k_B T)$, a plot of $\log\{[\rho(T)]_{\text{meas.}} - [\rho_L(T)]_{\text{calc.}}\}$ against T^{-1} yields a linear plot with slope (ϕ_V/k_B).

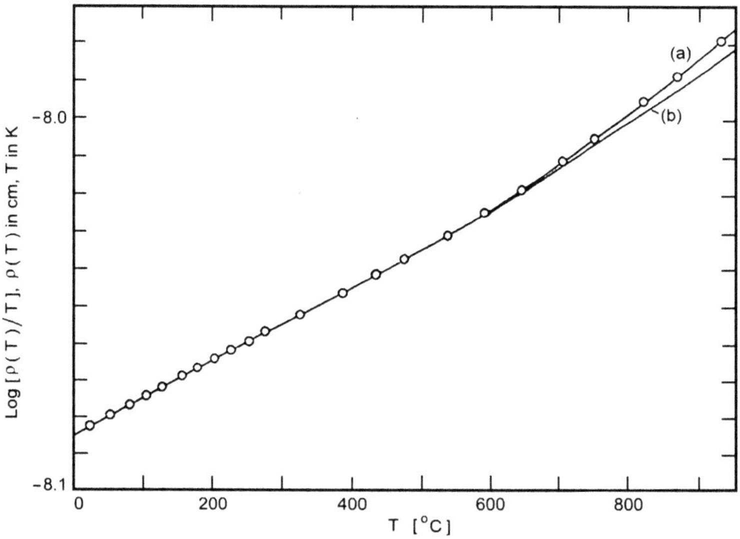

Fig. 14.10 Temperature dependence of log $[\rho(T)/T]$ for gold. Curve (**a**) is the experimental curve and curve (**b**) is an estimate of the ideal lattice contribution

It is not possible to measure the energy of migration ε_j from these measurements. ε_j is determined separately by measuring $\rho(T)$ as a function of time for different annealing temperatures [14.8].

14.5 Effect on Other Properties

Though the concept of point defects was introduced mainly to account for the electrical conductivity, it is found that they have a significant effect on a few other physical properties. We shall now discuss some of these properties.

14.5.1 Diffusion

The impurities or vacancies in a solid will be distributed uniformly when the solid is in equilibrium. But if there is a concentration gradient of the vacancies or impurities, there will be a flux or diffusion. Let us denote the number of vacancies or impurity atoms by n^* (the reason for using an asterisk will be clarified later). Then the flux J_{n^*} is proportional to the concentration gradient or grad n^*. Thus,

$$J_{n^*} = -D \text{ grad } n^*, \tag{14.22}$$

where J_{n^*} is the number of atoms crossing unit area in a second. The constant D is called the diffusion constant; its units are cm^2s^{-1}. The negative sign indicates that the migration or diffusion is away from the region of higher concentration. Equation (14.22) is the statement of Fick's law of diffusion.

Starting with Fick's law, we shall derive an expression for the distribution of the species n^* as a function of depth (x) in the solid. Applying the continuity condition $-(\partial n^*/\partial t) = \text{div } J_{n^*}$ to (14.22) and assuming the flow to be in the x direction, we get

$$\begin{aligned}(\partial n^*/\partial t) &= -\text{div } J_{n^*} \\ &= \text{div}(D \text{ grad } n^*) \\ &= D\,(\partial^2 n^*/\partial x^2).\end{aligned} \tag{14.23}$$

The solution of (14.23) is

$$n^*(x,t) = [n_0{}^* / (\pi D t)^{1/2}] \exp(-x^2/4Dt). \tag{14.24}$$

The pre-exponential factor is obtained with the assumption that the sample is a thin slab. In (14.24), $n^*(x,t)$ is the concentration of species n^* at depth x and time t and n_0^* its value at $x = 0$ and $t = 0$.

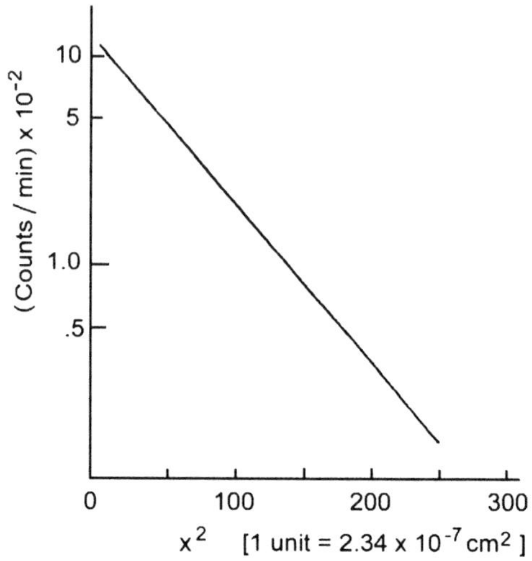

Fig. 14.11 Plot of counts against x^2 for distribution of radioactive Na in NaCl

Equation (14.24) can be verified by introducing radioactive atoms, say of Na, in a crystal of NaCl; hence the use of n^* in [14.22] to denote these species. The radioactive atoms are introduced by depositing a layer of a radioactive Na salt on one face of a crystal. The crystal is kept at a constant temperature for a time t during which the radioactive atoms diffuse through the crystal. The crystal is then cleaved or sliced into several thin sections and the content of the radioactive species in each section is counted by standard methods. A plot of the logarithm of the counts against x^2 is shown in Fig. 14.11; it is a linear plot in agreement with (14.24). From the plot, the value of the diffusion constant D can be obtained.

Let us now consider the temperature dependence of the diffusion constant. Experiments show that the relation between D and T is

$$D = D_0 \exp(-\varepsilon_D / k_B T) \tag{14.25}$$

where D_0 is a constant and ε_D the activation energy for the diffusion process. We shall now derive (14.25).

Let E be the field in the crystal and $\phi(r)$ the associated potential. Then

$$E = -\mathrm{grad}\, \phi(r). \tag{14.26}$$

If there are charged particles in the crystal carrying charge q, their distribution is given by the Maxwell-Boltzmann law

$$n(r) = C \exp[-q\phi(r)/k_B T], \tag{14.27}$$

and the gradient of $n(r)$ is

$$\mathrm{grad}\, n(r) = -(nq/k_B T)\mathrm{grad}\, \phi(r) = nqE/k_B T. \tag{14.28}$$

In terms of the diffusion coefficient D and the mobility μ of the particles, the equilibrium condition is that there is no flow. Then

$$nq\mu E - qD\,\mathrm{grad}\,n = 0. \tag{14.29}$$

Combining (14.28) and (14.29), we get

$$\mu/D = q/k_{\mathrm{B}}T. \tag{14.30}$$

This is known as the Nernst–Einstein equation. Since conductivity $\sigma = nq\mu$, (14.30) may be rearranged as

$$D = (k_{\mathrm{B}}T/nq^2)\sigma. \tag{14.31}$$

If the crystal is an ionic crystal like NaCl, the charged particles may be identified with the cations. Then, we may use the familiar quantities e for q and N for n, where N is the number of cations per unit volume and (14.31) becomes

$$D = (k_{\mathrm{B}}T/Ne^2)\sigma. \tag{14.32}$$

Thus the temperature variation of the diffusion constant is similar to that of conductivity. Substituting (14.17) in (14.32), we may write

$$D = (k_{\mathrm{B}}T/Ne^2)A''(e^2\nu/6ak_{\mathrm{B}}T)\exp-[(\varepsilon_j + \frac{\phi_{\mathrm{S}}}{2})/k_{\mathrm{B}}T]. \tag{14.33}$$

Alternate derivations of (14.33) are given by Franklin [14.8] and Dekker [14.5]. Comparing (14.33) with (14.25), it can be seen that the activation energy for diffusion ε_{D} is $[\varepsilon_j + (\phi_{\mathrm{S}}/2)]$. Thus a plot of log D against T^{-1} would be a straight line with slope $[\varepsilon_j + (\phi_{\mathrm{S}}/2)]/k_{\mathrm{B}}$. Such a plot for diffusion data on NaBr is shown in Fig. 14.12. At lower temperatures, we do observe a linear plot with a smaller slope but this plot cannot be explained just in terms of the migration of the ions (as in the case of conductivity) and other mechanisms are also involved. Hence, it may be emphasized that the log D versus T^{-1} plot yields the value of $[\varepsilon_j + (\phi_{\mathrm{S}}/2)]$; to obtain the value of ϕ_{S} from diffusion data, the value of ε_j from some other technique has to be used.

14.5.2 Volume

Eshelby [14.11] pointed out that point imperfections (vacancies, interstitials and substitutional impurities) act as "centres of dilatation", i.e. they affect the volume of the solid. Assuming an elastic continuum model, Eshelby showed that the change in volume ΔV is given by

$$\Delta V = 4\pi\,\Omega\,C \tag{14.34}$$

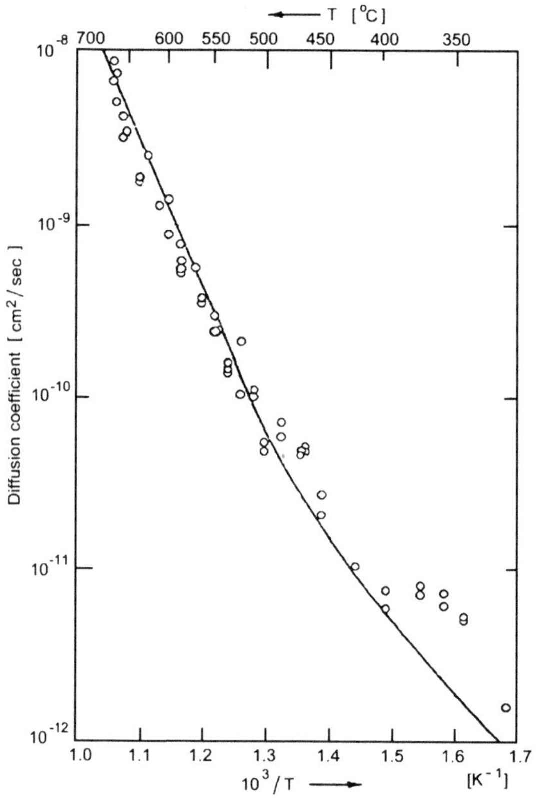

Fig. 14.12 Temperature dependence of diffusion coefficient of sodium ions in NaBr

where C is a constant which depends on the type of the defect and also the host lattice; the parameter Ω is given by

$$\Omega = 3(1 - \sigma)/(1 + \sigma). \tag{14.35}$$

Here, σ is the Poisson's ratio. Since $\sigma \approx (1/3), \Omega \approx 1.5$. An important result of Eshelby's theory is that ΔV is the same whether it is measured by following length changes $(\Delta L/L)$ in bulk samples or by following lattice constant changes $(\Delta a/a)$ using X-ray diffraction.

The volume change is measurable when the defects are present in sufficient numbers. One way of ensuring sufficient concentration is to dope an ionic crystal with aliovalent impurities, e.g. $SrCl_2$ in NaCl or $LaCl_3$ in $SrCl_2$. Such doping results in creation of positive ion vacancies. The measured density changes in such doped crystals are consistent with the mechanism of volume change as suggested by Eshelby. Another situation where the volume effect can be detected is high temperatures at which the concentration of thermally generated defects is large enough to show up as "anomalous thermal expansion".

14.5.3 Thermal Expansion

Thermal expansion measurements are capable of providing rich information regarding formation and concentration of point defects. There are two types of experimental methods: (1) X-ray diffraction measurements at different temperatures which give lattice constants as a function of temperature; these measurements yield the lattice expansion $(\Delta a/a)$ where "a" is the lattice constant and (2) length measurements on single crystal samples using either a capacitance dilatometer or an optical interferometer; these measurements yield the bulk expansion $(\Delta L/L)$ where L is the length of the sample. We shall consider two approaches to the study of point defects through thermal expansion measurements.

The first approach is based on the fact mentioned in the preceding section that point defects cause a volume change which shows up in both techniques of measurement. This volume change is over and above the thermal expansion due to thermal vibrations and is referred as "anomalous thermal expansion". It is obviously proportional to the number of point defects which increases exponentially with temperature. If we denote the anomalous expansion by $(\Delta\alpha_{\text{anom}})$, a plot of log $(\Delta\alpha_{\text{anom}})$ against T^{-1} will be a straight line with slope (ϕ_V/k_B) in the case of vacancies or $(\phi_S/2k_B)$ in the case of Schottky defects and $(\phi_F/2k_B)$ in the case of Frenkel defects. This anomalous expansion can be separated from the intrinsic expansion in either of two ways: (1) The temperature variation of the expansion $(\Delta a/a$ or $\Delta L/L)$ observed at moderately high temperatures may be extrapolated to high temperatures and then subtracted from the observed expansion. Figure 14.13 shows the thermal expansion data for NaCl measured on a single crystal by Enck and Dommel [14.12] using an optical interferometer and also that obtained by extrapolation of data at moderate temperatures. A plot of log $(\Delta\alpha_{\text{anom}})$ versus T^{-1} is shown in Fig. 14.14. (2) In the second method, the observed expansion at low and moderate temperatures is fitted to the Gruneisen equation and the parameters of the Gruneisen equation thus obtained are used to calculate the expansion at high temperatures. This is subtracted from the observed expansion. Figure 14.15 shows the observed thermal expansion data on AgCl obtained by Nicklow and Young [14.13] using the X-ray method and calculated data; Fig. 14.16 shows the linear plot of log $(\Delta\alpha_{\text{anom}})$ against T^{-1}.

The anomalous expansion effect is shown both by Schottky and Frenkel defects. The type of analysis discussed above merely indicates presence of defects and yields an activation energy without distinguishing between the two types of defects. For this purpose, we have to adopt the second approach. This is based on the fact that vacancies (and Schottkies) result in creation of additional unit cells on the surface and thus add to the bulk volume whereas Frenkel defects do not create any extra unit cells.

The difference in the change of volume obtained from bulk expansion measurements $(3\Delta L/L)$ and X-ray diffraction measurements $(3\Delta a/a)$ is simply equal to the concentration of defects at that temperature. Thus if the defects are vacancies,

Fig. 14.13 Anomalous expansion $(\Delta\alpha)_{\text{anom}}$ of NaCl

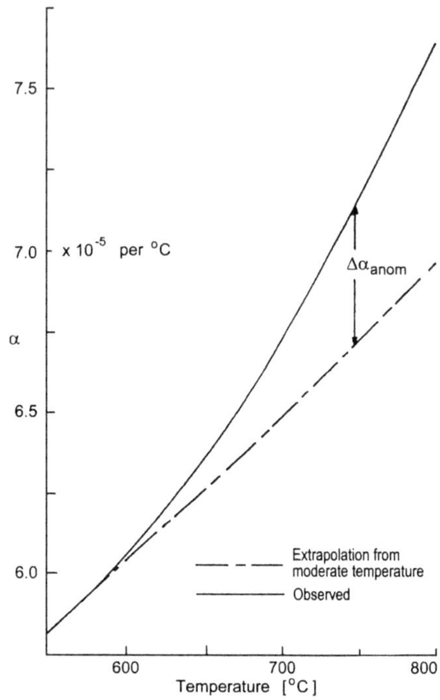

$$3[(\Delta L/L) - (\Delta a/a)] = n/N = A\exp(-\phi_V/k_B T). \tag{14.36}$$

A plot of log $3[(\Delta L/L) - (\Delta a/a)]$ versus T^{-1} will be a straight line with slope (ϕ_V/k_B). Simmons and Balluffi [14.4] made simultaneous measurements of $\Delta L/L$ and $\Delta a/a$ on Al. Their results are shown in Figs. 14.17 and 14.18. Clearly, vacancies are the predominant defect in aluminium.

In the case of ionic crystals like alkali halides where the defects are Schottkies, the exponential term in (14.36) will be $\exp[-(\phi_S/2k_B T)]$. A more important point to be noted is that if the defects were Frenkels, the two curves in Fig. 14.17 would not have separated at high temperatures.

14.5.4 Specific Heat

The specific heat C_p is the energy supplied to the crystal for exciting thermal vibrations and thereby, to increase its temperature. The creation of point defects also needs energy. Hence the effective specific heat is more than that in the absence of defects. If ΔC_p is this excess specific heat, then

$$\Delta C_p = (\partial \Delta H/\partial T)_p, \tag{14.37}$$

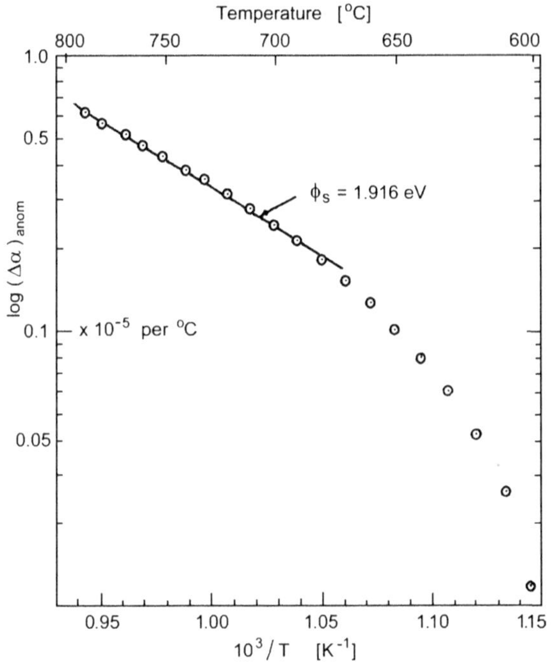

Fig. 14.14 Plot of log $(\Delta\alpha)_{\mathrm{anom}}$ against T^{-1} for NaCl

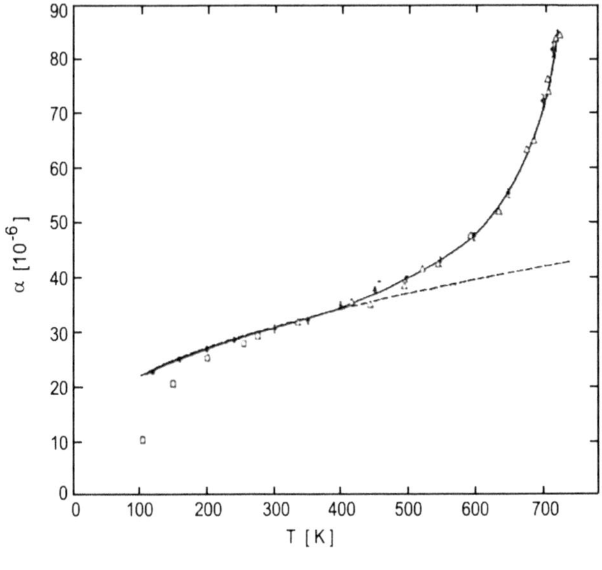

Fig. 14.15 Thermal expansion of AgCl: *continuous curve* – observed data; *dashed line* – calculated from Gruneisen theory

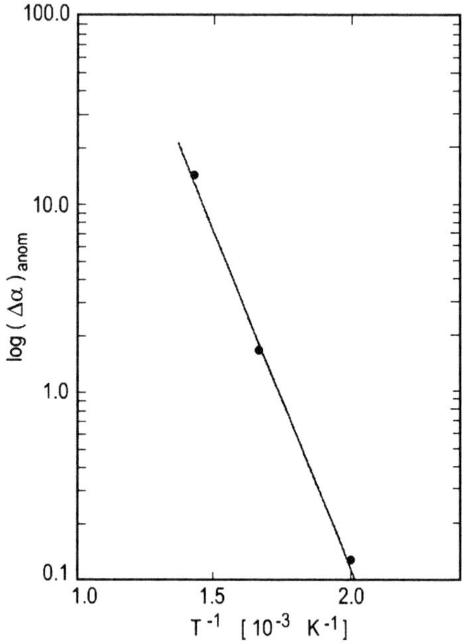

Fig. 14.16 Plot of log $(\Delta\alpha)_{\mathrm{anom}}$ against T^{-1} for AgCl

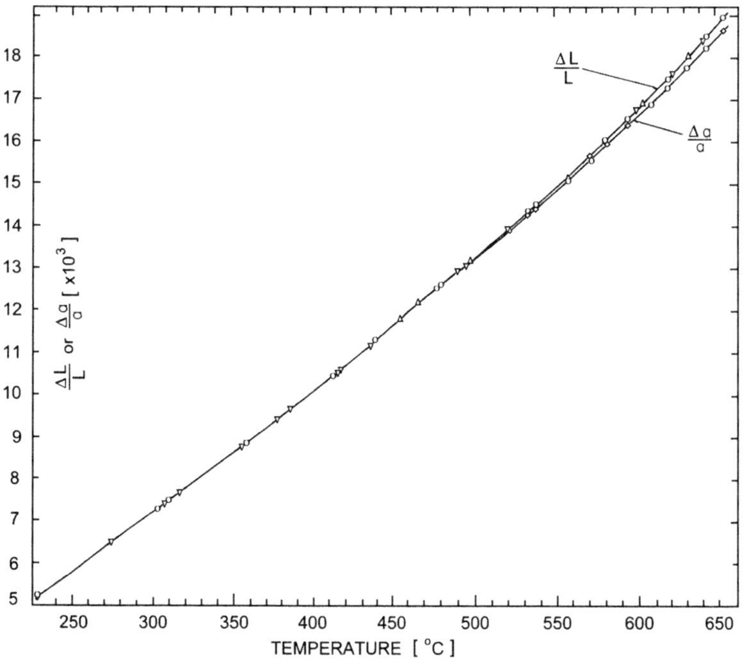

Fig. 14.17 Plot of $\Delta a/a$ and $\Delta L/L$ against temperature for aluminium

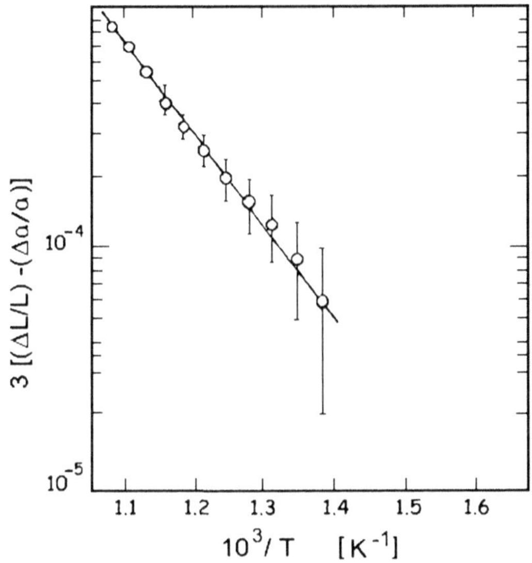

Fig. 14.18 Plot of log $3[(\Delta L/L)–(\Delta a/a)]$ against temperature for aluminium

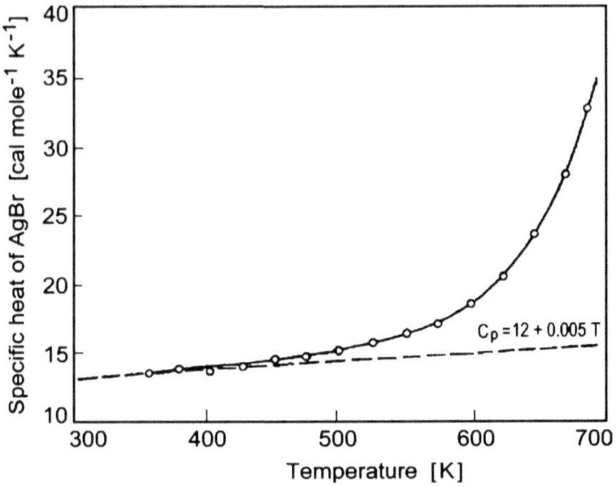

Fig. 14.19 Specific heat C_p of AgBr; continuous curve: observed; *dashed line* – extrapolation from moderate temperatures

where $\Delta H = n_i \phi_i$ where n_i and ϕ_i are the number and energy of formation of defects of type i. If the defects are Frenkel defects, using the appropriate expression for n_F (14.13), we get

$$\begin{aligned} \Delta C_p &= [(\phi_F)^2/2k_B T^2]n_F \\ &= A'''[(\phi_F)^2/2k_B T^2]\exp(-\phi_F/2k_B T). \end{aligned} \tag{14.38}$$

The specific heat data for AgBr is shown in Fig. 14.19; ΔC_p is the difference between the observed specific heat at high temperature and that extrapolated from the specific heat at moderate temperatures. From (14.38), it follows that a plot of

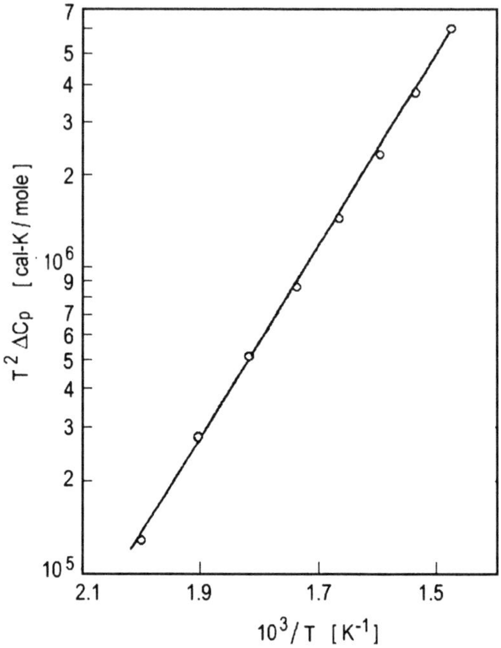

Fig. 14.20 Plot of log ($T^2\Delta C_p$) against T^{-1} for AgBr

log ($T^2\Delta C_p$) against T^{-1} should be a straight line with slope ($\phi_F/2k_B$); such a plot for AgBr is shown in Fig. 14.20.

14.6 Theoretical Calculation of Energy of Formation of Defects

To start with, let us consider the calculation of the energy of formation of a Schottky pair in an ionic crystal like NaCl. The first step in the creation of a Schottky pair is the removal of an ion, say a positive ion, from the interior of the crystal to infinity, keeping the charge distribution in the crystal undisturbed. From Born's theory, the energy ϕ_L involved in this process is given by

$$\phi_L = (Ae^2/r)(1 - \frac{1}{m}), \tag{14.39}$$

where A is the Madelung constant, r the interionic distance and m is the index in the Born repulsion term B/r^m; B is a constant. The second step is the shifting of the ion back from infinity to a new lattice site on the surface; this process involves energy equal to $\phi_L/2$. Thus ϕ_+ the energy to create a positive ion vacancy is $[\phi_L - (\phi_L/2)] = (\phi_L/2)$. The energy ϕ_- to create a negative ion vacancy is also the same. Thus ϕ_S, the energy to create a Schottky pair, is

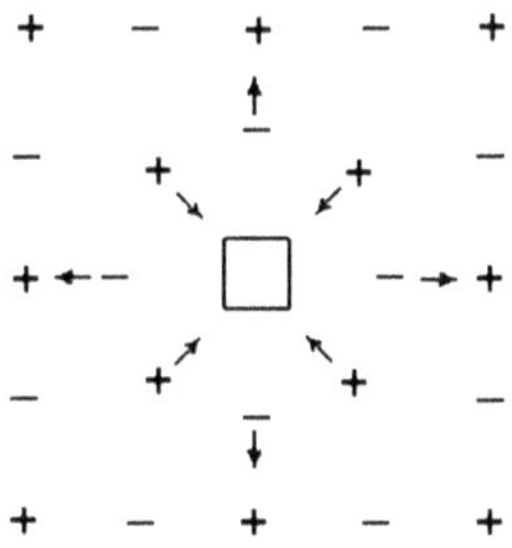

Fig. 14.21 Displacement of ions around a vacancy

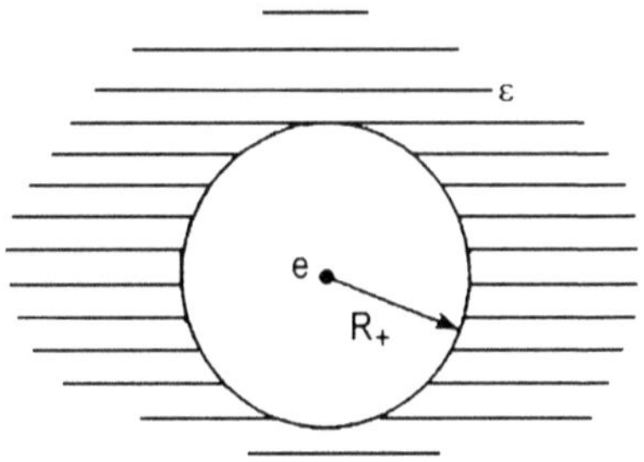

Fig. 14.22 Model to calculate polarization energy due to presence of a positive ion vacancy.

$$\phi_S = \phi_+ + \phi_- = (\phi_L/2) + (\phi_L/2) = \phi_L. \tag{14.40}$$

For NaCl, $\phi_L = 7.94$ eV; hence, ϕ_S calculated from (14.40) is 7.94 eV. This value is too large in comparison with the experimental value of ~2 eV.

The above oversimplified picture ignores an important mechanism that accompanies the creation of vacancies viz. the relaxation and consequent polarization of the medium around the vacancy. This may be understood from Fig. 14.21. The positive ion vacancy is equivalent to the presence of a negative ion in its place. Due to this, the positive ions around the vacancy are drawn towards it and the negative ions are drawn away from it. Consequently, dipoles are created and the material in the vicinity of the vacancy becomes polarized. This effect involves not just the nearest neighbours but several more distant neighbours; this polarization reduces the value of ϕ_S much below the value of ϕ_L given by (14.40). The calculation of this effect is important but complicated. It has been discussed in detail in [14.2, 14.3, 14.15–14.17]. We shall consider only the main steps in the derivation.

As shown in Fig. 14.22, the positive ion vacancy may be considered as a spherical hole in a medium of dielectric constant ε. The hole has a radius R_+ and a charge e (due to the missing ion). This charge polarizes the dielectric medium around it and creates a potential V at the location of the charge; this potential is given by

$$V = \frac{e}{R_+}\left(1 - \frac{1}{\varepsilon}\right). \tag{14.41}$$

The polarization energy P_+ is

$$P_+ = \frac{1}{2}eV = \frac{1}{2}\frac{e^2}{R_+}\left(1 - \frac{1}{\varepsilon}\right). \tag{14.42}$$

The formation energy ϕ_+ of the positive ion vacancy is now given by

$$\phi_+ = \frac{1}{2}\frac{Ae^2}{r}\left(1 - \frac{1}{m}\right) - \frac{1}{2}\frac{e^2}{R_+}\left(1 - \frac{1}{\varepsilon}\right). \tag{14.43}$$

The corresponding formation energy ϕ_- of a negative ion vacancy is

$$\phi_- = \frac{1}{2}\frac{Ae^2}{r}\left(1 - \frac{1}{m}\right) - \frac{1}{2}\frac{e^2}{R_-}\left(1 - \frac{1}{\varepsilon}\right) \tag{14.44}$$

where R_- is the effective radius of the negative ion hole. Thus, the formation energy ϕ_S of the Schottky pair is

$$\begin{aligned}\phi_s = \phi_+ + \phi_- &= \frac{1}{2}\frac{Ae^2}{r}\left(1 - \frac{1}{m}\right) - \frac{1}{2}e^2\left(1 - \frac{1}{\varepsilon}\right)\left(\frac{1}{R_+} + \frac{1}{R_-}\right)\\ &= \phi_L - P_+ - P_-.\end{aligned} \tag{14.45}$$

The terms P_+, P_- reduce ϕ_S from the value ϕ_L (14.40). The calculation of P_+ and P_- requires the calculation of R_+ and R_-. These calculations are beyond our scope. We shall note their magnitudes for NaCl: $\phi_L = 7.94\,\text{eV}$; $R_+ = 0.58r$; $R_- = 0.95\,r$; $P_+ = 3.32\,\text{eV}$, $P_- = 2.76\,\text{eV}$ and $(\phi_S)_{\text{calc.}} = 1.86\,\text{eV}$. This value is close to the experimental value of $\sim$2 eV.

Similar calculations have been made for the formation energy of Frenkel pairs in ionic crystals. It turns out that $\phi_S < \phi_F$ in alkali halides; hence the predominant defects are Schottky pairs. On the other hand, $\phi_S > \phi_F$ in silver halides; hence the predominant defects in these crystals are Frenkel pairs.

With regard to metals, the calculations differ from those from ionic crystals in two respects. Firstly, the relaxation is comparatively small in extent and, secondly, the relaxation is caused by repulsion between closed electron shells. The formation energies of vacancies in metals are of the same order as in ionic crystals but are smaller in magnitude.

14.7 Comparison of Formation Energy Values

Values of the energy of formation of defects obtained by different methods for several crystals of different types are given in Table 14.1. The experimental values are quoted from Franklin [14.8]. Theoretical values for metals are from Kubo and Nagamiya [14.7] and those for ionic crystals from Barr and Lidiard [14.18].

Table 14.1 Energy of formation of defects

Crystal	Defect	Jump or migration energy	Formation energy per defect [eV]				
			Experimental property				
			Conductivity or resistivity	Thermal expansion	Specific heat	Diffusion	Theory
Ar	Vacancy	–	–	–	0.052	–	–
Kr	Vacancy	–	–	0.077	0.077	–	–
Au	Vacancy	–	0.80	0.56	1.0	–	0.6
Cu	Vacancy	–	1.06	1.17	1.05	1.03	0.9
Al	Vacancy	–	0.79	0.74	–	–	–
Na	Vacancy	–	0.4	0.42	0.46	–	<0.3
NaCl	Schottky	0.80	2.12	2.02	–	–	1.86
KCl	Schottky	0.84	2.22	–	–	–	2.02
AgCl	Frenkel	0.055	1.44	1.4	–	–	1.96
AgBr	Frenkel	0.058	1.13	–	1.27	–	–

The data in Table 14.1 reveal a few features. The experimental values obtained from different properties are consistent among themselves. Further, the values obtained from experiment agree fairly with theoretically calculated values. The energy of formation of defects is in the range 1–2 eV for ionic crystals, 0.4–1 eV for metals and 0.05–0.08 eV for inert gas solids.

14.8 Direct Evidence of Point Defects

We have seen that the concept of point defects has been able to explain not only the observed features of ionic conductivity but also other phenomena like diffusion, anomalous thermal expansion and excess specific heat. But these are all indirect evidences of the existence of point defects and one may ask whether there is any direct evidence. Point defects are atomic scale features and their observation would require very high resolution. Such a technique was invented by Muller [14.19] and is known as the field ion microscope (FIM).

The principle of FIM is shown in Fig. 14.23. The sample is in the form of a needle with a hemispherical tip. The specimen (which should be a conductor) is held in an evacuated chamber at a distance of ~5 cm from a phosphor screen. A small trace of Ne gas is let into the chamber and a positive voltage is applied to the system. The Ne atoms are attracted to the specimen and if the applied voltage is sufficiently high, the Ne atoms lose electrons. The positively charged neon atoms are now repelled by the positively charged specimen. They move towards the phosphor screen where they produce an image. The angular distribution of the repelled Ne ions reflects the microscopic structure of the specimen surface. The magnification is the ratio of the radii of the collecting electrode (screen) and the specimen tip. A FIM micrograph of a region of a metallic surface including a vacancy is shown in Fig. 14.24. The image is in the form of dots; each dot represents an atom in the specimen surface. The dots

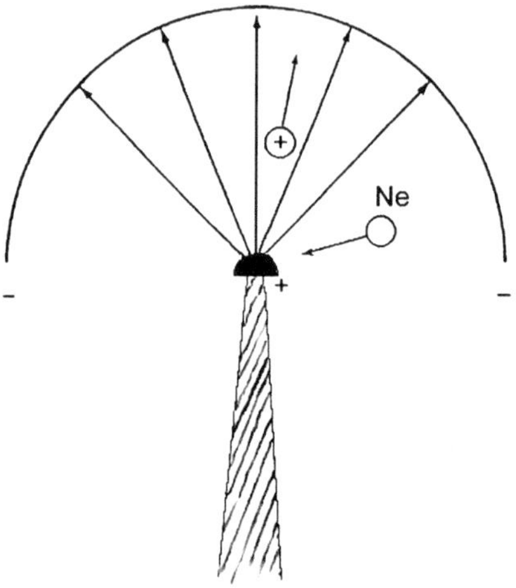

Fig. 14.23 Principle of field ion microscope

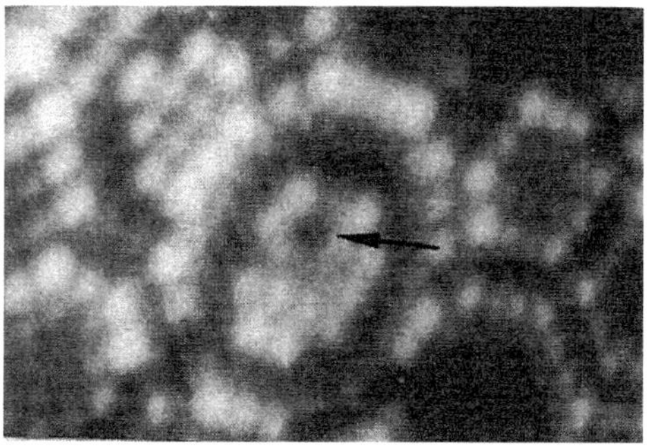

Fig. 14.24 FIM micrograph of a vacancy (shown by *arrow*) on a platinum surface

form an array of rings arising from the intersection of the atomic planes with the hemispherical surface of the specimen.

Part B: Induced Point Defects

14.9 Colour Centres

Pure alkali halides are transparent in the visible part of the spectrum but become coloured under some conditions. Goldstein [14.20] was the first to observe that alkali halide crystals darken when subjected to cathode ray bombardment. A systematic study of the colouration of alkali halides was taken up by Pohl [14.21]. When alkali halides which are otherwise transparent become coloured; the entities responsible for the optical absorption are called "colour centres". The subject of colour centres has now become a full-fledged branch of solid state physics. We will have to limit our treatment to the main features. Detailed treatments of the topic are given in review articles by Seitz [14.22, 14.23] and Compton and Rabin [14.24] and in books by Markham [14.25], Brown [14.9], Fowler [14.26] and Dekker [14.5].

In this part of the chapter, we shall start with the methods of colouring crystals. Various aspects of the F centres in alkali halides will be discussed like its model, the Ivey–Mollow relation, Smakula's equation and the magnetic properties of F centres. Several other colour centres will be briefly discussed.

14.10 Production of Colour Centres

Colour centres are produced mainly by three methods: by irradiation with high energy radiation, by additive colouration and by electrolysis.

14.10.1 *Irradiation*

Irradiation of a transparent crystal by X-rays or γ-rays results in colouration. Bombardment by electrons or neutrons also produces colour centres. Crystals coloured by irradiation bleach (discolour) easily and have to be protected from exposure to light.

14.10.2 *Additive Colouration*

In this method, excess metal ions are introduced into an alkali halide crystal. A typical experimental arrangement is shown in Fig. 14.25. An alkali halide crystal such as KCl is heated in an atmosphere of potassium vapour. Transparent KCl crystals are sealed in an evacuated quartz (or copper) tube together with potassium metal. The tube is held in a double furnace such that the KCl crystal and the potassium material are in the two zones of the furnace. The KCl crystals are heated to a temperature T_1 ($\sim$500°C) whereas the potassium vapour is maintained at temperature T_2($\sim$500–700°C). Temperature T_2 determines the vapour pressure of the potassium vapour and this, in turn, determines the concentration of the resulting colour centres. After a while, the tube is taken out and the coloured crystals are quenched to room temperature.

14.10.3 *Electrolysis*

In this method, electrons are directly injected into the crystal. Figure 14.26 shows an experimental arrangement. The crystal is held between a flat electrode and a pointed electrode. A high D.C. voltage of a few hundred volts is applied to the

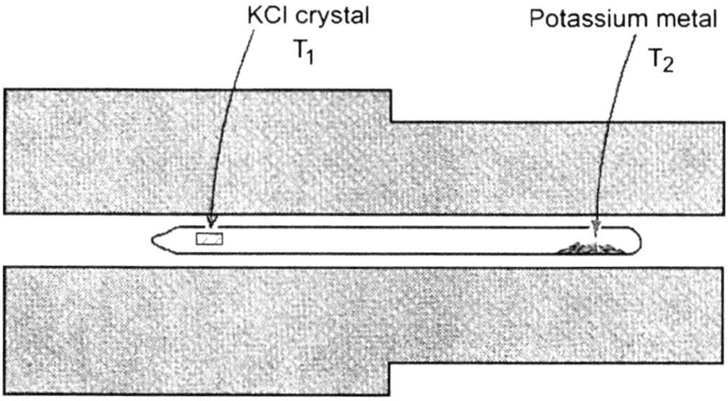

Fig. 14.25 Set-up for additive colouration

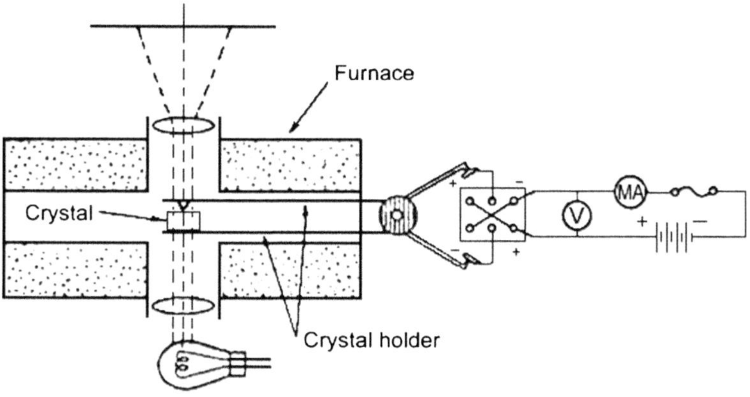

Fig. 14.26 Set-up for electrolytic colouration

pointed electrode and the crystal is heated in a furnace. At a certain temperature (400–600°C), a discharge of electrons starts from the pointed electrode. This is indicated by the appearance of colour at the point of contact of the electrode with the crystal. The colour spreads progressively into the crystal (Fig. 14.27). The exact temperature at which colouration commences depends on factors like the purity and history of the specimen. An interesting feature of this method is that the colouration can be completely withdrawn by reversing the voltage.

In all these methods, excess electrons are introduced into the crystal. Thus, high energy radiation knocks off electrons from the ions in the crystal. In additive colouration, the extra metal atoms in the vapour enter into the crystal as metal ions and electrons. In the electrolysis method, a stream of electrons is directly injected into the crystal.

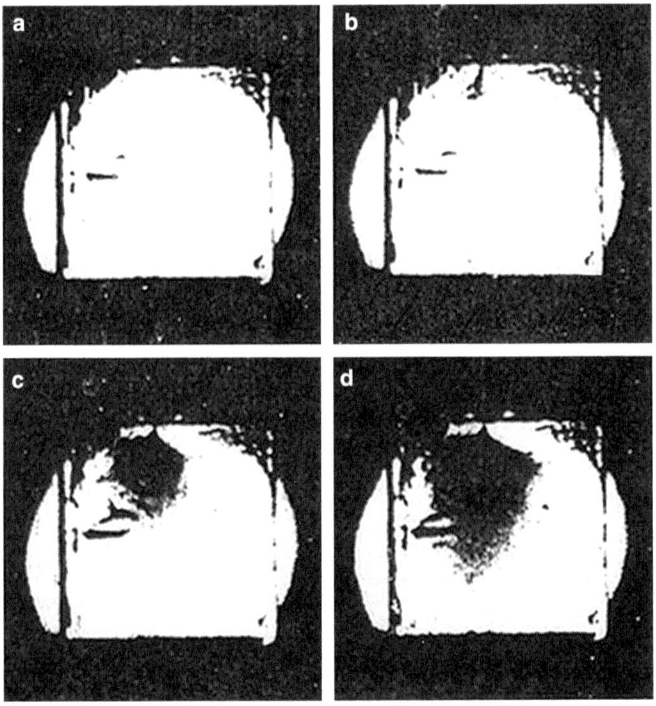

Fig. 14.27 Spread of colour during electrolysis

14.11 F Centres

14.11.1 General

Alkali halides coloured by any of the above methods display individual colours. Thus NaCl heated in Na vapour appears yellow, KCl heated in K vapour shows a magenta colour and LiF heated in Li vapour looks pink. KBr coloured by X-ray irradiation looks blue. These colours are due to absorption by defects now known as F centres. The German word for colour is Farbe and so the colour centres are called Farbezentren. As we shall see, there are several types of colour centres but the letter F has stuck to the most prominent band in the optical absorption spectrum of coloured alkali halides. As an example, the F band in RbCl is shown in Fig. 14.28. Its peak is at 624 nm and it has a half-width of 0.145 eV at 4.2 K. Its shape may be qualitatively described as bell-shaped; analytically, different workers have treated the shape as Lorentzian, Gaussian or Poissonian. The peak positions of the F centre in several alkali halides are given in Table 14.2. We shall now consider a few important aspects of F centres.

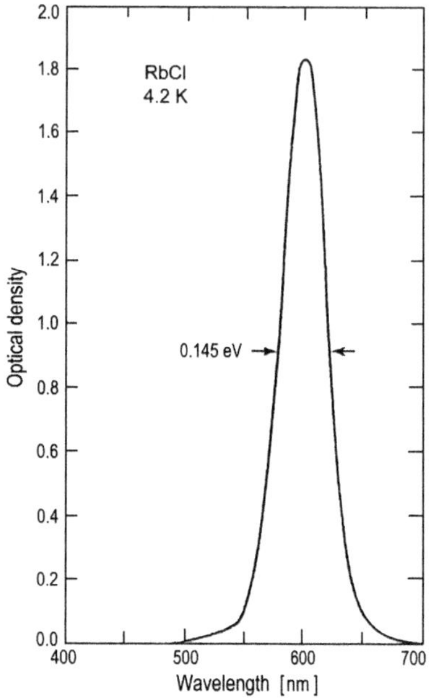

Fig. 14.28 F band in an X-ray irradiated RbCl crystal at 4.2 K

Table 14.2 Peak wavelengths (λ_F) and peak energies (ε_F) of F bands in alkali halides

Crystal	λ_F[nm]	ε_F[eV]	Crystal	λ_F[nm]	ε_F[eV]
LiF	250	4.96	RbCl	624	1.98
LiCl	385	3.22	RbBr	720	1.72
NaF	340	3.65	RbI	775	1.60
NaCl	465	2.67	CsCl	603	2.06
NaBr	540	2.30	CsBr	675	1.84
NaI	588	2.11	CsI	785	1.58
KF	455	2.72			
KCl	563	2.20			
KBr	630	1.97			
KI	685	1.81			

14.11.2 The F Centre Model

We shall consider the F centre model proposed by de Boer [14.27]. As an example, let us consider the lattice configuration in an additively coloured KCl crystal (Fig. 14.29). An atom from the potassium vapour is adsorbed on the surface and occupies lattice site A; it does so by splitting into a positive ion and an electron. To maintain charge neutrality of the surface, the negative ion at site B moves to the position next to A on the surface, leaving a negative ion vacancy at B. This vacancy is filled by the negative ion at C and, then, by the one at D. Finally, we have a stable negative ion vacancy at D. This negative ion vacancy behaves electrostatically like

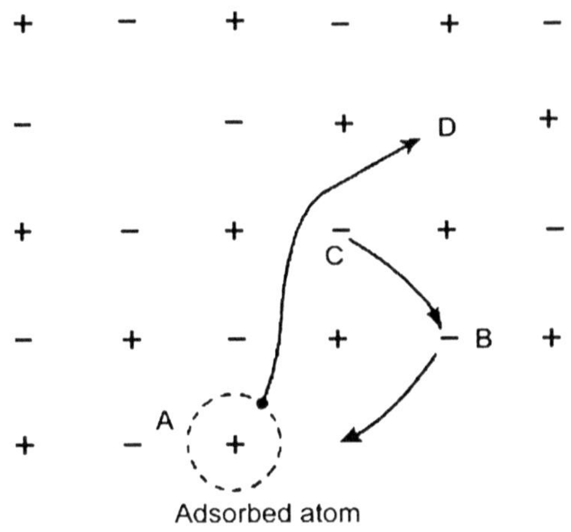

Fig. 14.29 The de Boer model of F centre

a positive charge. The electron released initially by the potassium atom diffuses into the lattice and gets trapped in the region around the negative ion vacancy. Such a negative ion vacancy with a trapped electron constitutes an F centre.

The main feature of the above model is the creation of free electrons and negative ion vacancies. There is considerable evidence in favour of the de Boer model. Firstly, according to this model, it is immaterial whether the excess metal ions introduced in the KCl crystal are K ions or Na ions. It is indeed observed that in either case, the same absorption band results. Secondly, the introduction of excess K ions in KCl results in creation of negative ion vacancies. It is known that vacancies cause increase in volume, i.e. decrease in density; such decrease in density has been experimentally verified. Additional evidence from magnetic properties will be discussed later.

It may be mentioned that a slightly modified model for the F centre has been proposed by Seitz (1946) but de Boer's simple model explains many of the observed properties of F centres.

The energy level diagram for the F centre absorption is shown in Fig. 14.30. The first excited state of the F centre is close but below the conduction band. F centre absorption arises from the excitation of the F centre electron from the ground state (1s) to the first excited state (2p).

14.11.3 Mollow–Ivey Relation

Figure 14.31 is a plot of the lattice constant "a" against the peak wavelength λ_F of the F centres in NaCl-type alkali halides; a gradation between the two parameters is clearly seen. Mollow [14.28] proposed the relation

$$\lambda_F a^{-2} = \text{constant}. \tag{14.46}$$

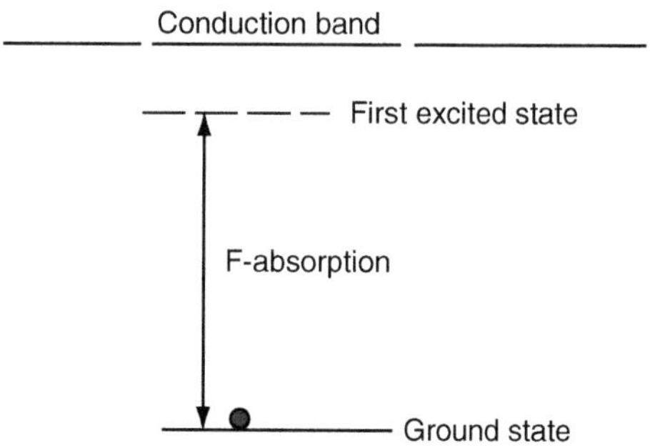

Fig. 14.30 Energy level diagram for F centre

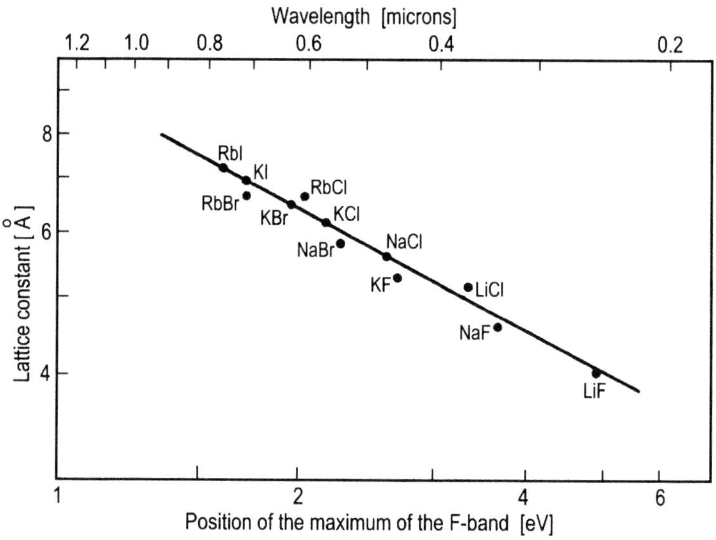

Fig. 14.31 Plot of peak wavelength λ_F against lattice constant

From a more careful analysis, Ivey [14.29] showed that

$$\lambda_F a^{-1.86} = \text{constant}. \quad (14.47)$$

In general, for each crystal family, there exists a relation of the type

$$\lambda_F a^{-m} = \text{constant} \quad (14.48)$$

where the index m has a different value for each family.

14.11.4 Smakula's Equation

The optical density of a coloured crystal depends on the concentration of colour centres. The concentration of colour centres and hence the optical density varies with various conditions like growth time, bleaching, temperature, etc. It is desirable to have a method to estimate the concentration of colour centres. Assuming (1) classical dispersion theory, (2) a damped harmonic oscillator model and (3) a Lorentzian shape for the F centre, Smakula [14.30] derived the following formula:

$$fN_{\mathrm{F}} = 1.29 \times 10^{17} \left[\frac{n'}{(n'^2 + 2)} \right] \alpha_m (\mathrm{cm}^{-1}) W_{1/2} (e\mathrm{V}). \tag{14.49}$$

Here N_{F} is the number of F centres in unit volume, n' the refractive index, α_m the absorption coefficient at peak position and $W_{1/2}$ the full width at half-maximum. The parameter f is called the oscillator strength. The derivation of the formula [14.9, 14.25] is quite involved and we shall conveniently accept Markham's view that "it is useful to regard Smakula's equation.....(as) an experimental law with an adjustable constant". The oscillator strength f is obtained from independent techniques. The values of f obtained from magnetic susceptibility for some alkali halides and by different methods for KCl are given in Table 14.3; the values are generally in the range 0.8–1.

14.11.5 *Magnetic Properties*

14.11.5.1 Magnetic Susceptibility

Alkali halides in their uncoloured state are diamagnetic with a negative, temperature-independent, susceptibility. Coloured alkali halides contain F centres. The unpaired spin of the trapped electron contributes a paramagnetic susceptibility which follows the Curie law. Thus the total susceptibility χ of a coloured crystal is

$$\begin{aligned} \chi &= \chi_{\text{diamagnetic}} + \chi_{\text{paramagnetic}} \\ &= \chi_{\text{diamagnetic}} + (C/T), \end{aligned} \tag{14.50}$$

where the Curie constant C is

$$C = N_{\mathrm{F}} g^2 \mu_{\mathrm{B}}^2 / 4k_{\mathrm{B}}. \tag{14.51}$$

Table 14.3 Values of f for some alkali halides

Crystal	Method			
	Ionic conductivity	Chemical	ESR	Magnetic susceptibility
KCl	0.9	0.9	0.85	0.66
KBr				0.85
KI				0.87
NaCl				0.87
LiF				0.82

Here N_F is the number of spins (i.e. F centres), g the gyromagnetic ratio for the electron and μ_B the Bohr magneton. Knowing the diamagnetic susceptibility and the constant C, and assuming the values of $g(= 2)$ and μ_B, the value of N_F can be estimated. Substituting this value of N_F in Smakula's equation (14.49), the value of the oscillator strength f can be estimated; the other quantities α_m and $W_{1/2}$ are obtained from the F-band. Some f-values determined in this way are given in Table 14.3.

14.11.5.2 Electron Spin Resonance

The ground state of the F centre gives rise to an electron spin resonance spectrum. The F centre ESR spectrum was first recorded by Hutchinson [14.341] on a neutron irradiated LiF single crystal. Using a frequency of 9,354 Mc/s, resonance was observed at a field strength of 3,345 Gauss. From the resonance condition $g = h\nu/\mu_B H$, a value of 2 was obtained for g which is the free electron value.

To obtain a good ESR spectrum, the sample should have an F centre density of 10^{18} centres/cm^3 or more. ESR spectra of a few alkali halides are shown in Fig. 14.32. It is seen that the ESR spectrum of KCl consists of a single broad line without any structure. Similar spectra are obtained for several other alkali halides (LiCl, NaCl, NaBr, NaI, KBr, KI, RbBr, RbI). On the other hand, in RbCl and NaF, the broad line has a structure with a varying number of components. So is the case with LiF and CsCl. Some important results from ESR studies are:

1. The g values are close (though not exactly equal) to the free electron value of 2.002.
2. The distribution of the excess (trapped) electron is largely on the six positive ions adjacent to the vacancy (Fig. 14.33).
3. There is a good correlation between the area under the ESR absorption curve and the number of spins (F centres) obtained from optical absorption.

These results confirm the de Boer model.

14.12 Other Colour Centres

The F band is the most prominent band in the optical absorption spectrum of a coloured crystal. But a number of other absorption bands are also observed in different conditions of the crystal like irradiation, purity and temperature.

If an additively coloured crystal, say KCl, is illuminated (bleached) with F light or is heated to about 100°C for a few hours, new bands appear on the long wavelength side of the F band (Fig. 14.34). The centres responsible for these bands are called R, M and N centres; these are aggregates of F centres. On the other hand, if the bleaching is carried out at a low temperature of about 125 K, a band appears close to the F band on its long wavelength side (Fig. 14.35). This is called the F′ band; it is attributed to two electrons trapped in a single vacancy.

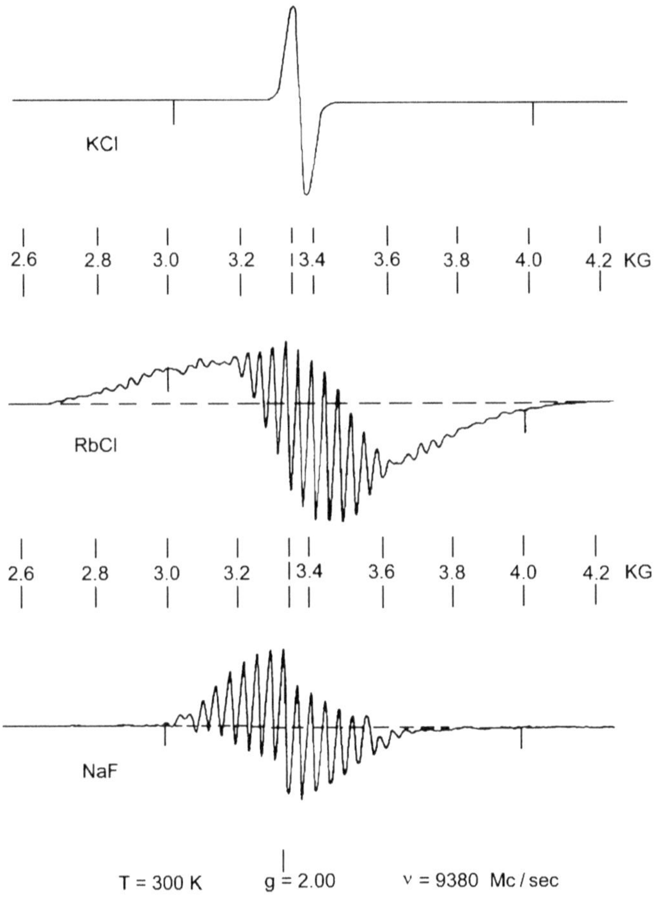

Fig. 14.32 ESR spectra of F centres in KCl, RbCl and NaF

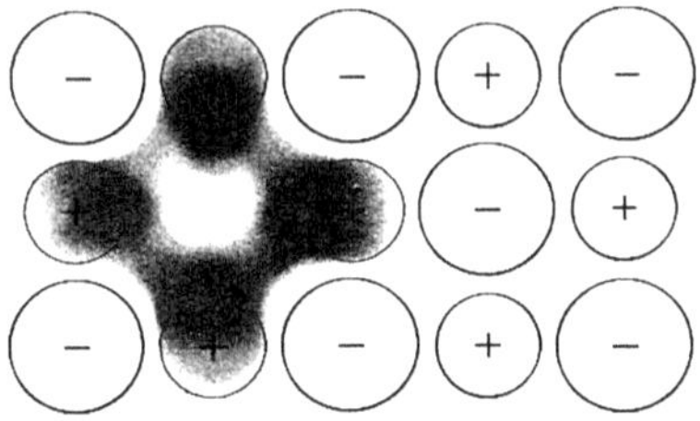

Fig. 14.33 Distribution of the trapped electron charge over the neighbours of the vacancy

The X-ray irradiation of alkali halides at liquid helium temperatures results in the appearance of the V_k bands. These appear at 3,650 and 7,500 Å in KCl and at 3,850, 7,500 and 9,000 Å in KBr. The V_k centre is believed to be a self-trapped hole. The hole is shared by two halogen ions which have left their sites and are oriented

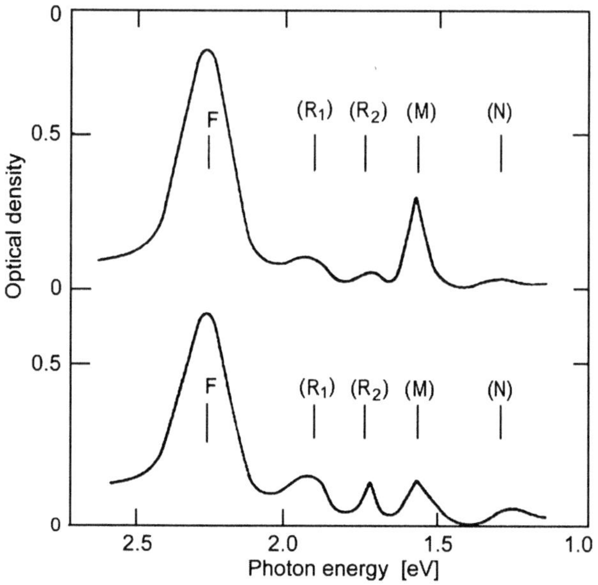

Fig. 14.34 The M, N, R_1 and R_2 bands formed in KCl by irradiating in the F band at room temperature. The two curves were taken following successive stages of irradiation

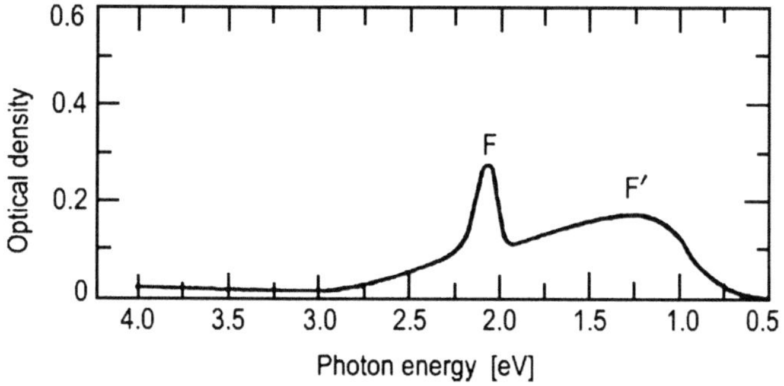

Fig. 14.35 F′ band in a coloured KBr crystal bleached with F light at 125 K

in the [110] direction. The centre can also be considered as a halogen-molecule-ion (Cl^{2-}). Another centre caused by irradiation at liquid helium temperature is the H centre which is a hole trapped at an interstitial halogen ion. Here the hole is associated not with two Cl ions, as in the case of the V_k centre, but with four Cl ions; again, it is aligned along the [110] direction.

Let us now consider some absorption bands caused by impurities in crystals. Hydrogen can be incorporated into alkali halides by heating the crystal in hydrogen gas or by adding an appropriate alkali hydride during the growth of the crystal, e.g. KH in KBr. Irradiation of these crystals produces a band called the U band

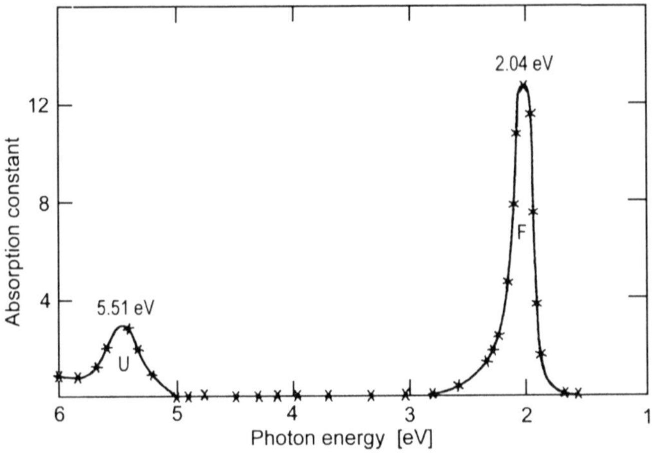

Fig. 14.36 U band in KBr

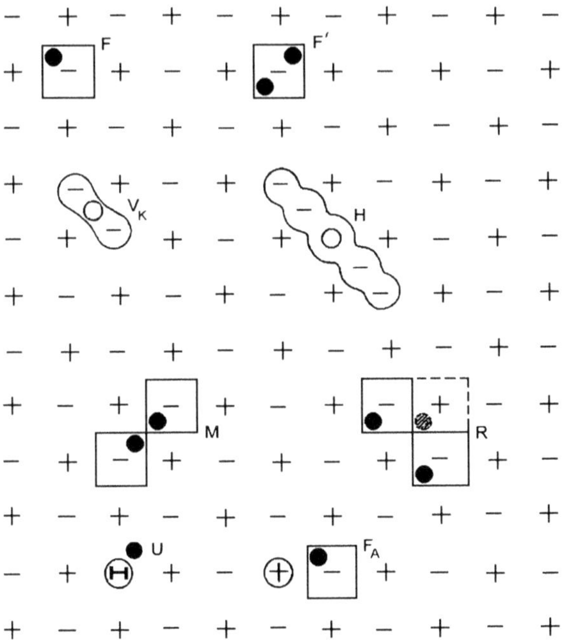

Fig. 14.37 Models for several important colour centres in a NaCl type lattice. Negative ion vacancy, • electron, O hole, ⊖ hydrogen atom, ⊕ monovalent impurity ion such as Na in KCl

in the ultraviolet (Fig. 14.36). Lastly, mention may be made of the formation of F_A centres which are F centres with one of the six neighbours replaced by a monovalent impurity, e.g. Li in KCl. Models of some of the colour centres are shown in Fig. 14.37.

Problems

1. Discuss how the various terms in the expression for the free energy of a solid vary with the concentration of vacancies. Hence show that a certain number of vacancies makes a crystal stable.
2. If the free energy of formation of a vacancy in a metal is 1 eV, find the ratio of the number of vacancies in the metal at 600 K and 300 K.
3. The expressions for equilibrium concentration of vacancies and Frenkel defects in a metal are of the form $A\exp(-\phi_V/k_BT)$ and $A'\exp(-\phi_F/2k_BT)$, where ϕ_V and ϕ_F are formation energies of the two types of defects. Explain the origin of the factor 2 in the second expression.
4. The lattice constant and F centre wavelength of NaCl are 5.64 Å and 4,750 Å respectively. If the lattice constant of KCl is 6.29 Å, estimate its F centre wavelength.
5. The parameters of the F band in NaCl are: peak wavelength 4,750 Å, absorption coefficient at peak maximum 21.6 cm^{-1}, refractive index 1.54 and half-width 0.3 eV, Estimate the F centre concentration assuming the oscillator strength to be 1.

References

14.1. A.B. Lidiard, Handbuch der Physik **20/II**, 246 (1957)
14.2. J. Frenkel, Z. Physik. **35**, 652 (1926)
14.3. W. Schottky, Z. Phys. Chem. **B29**, 335 (1935)
14.4. R.A. Levy, *Principles of Solid State Physics* (Academic Press, New York, 1968)
14.5. A.J. Dekker, *Solid State Physics* (Macmillan, New York, 1981)
14.6. C. Kittel, *Introduction to Solid State Physics* (Wiley and Sons, New York, 1996)
14.7. R. Kubo, T. Nagamiya, *Solid State Physics* (McGraw-Hill Book Co. Inc., New York, 1968)
14.8. A.D. Franklin, *Point Defects in Solids*, Vol. 1 (Plenum Press, New York, 1972)
14.9. F.C. Brown, *The Physics of Solids* (Benjamin Inc., New York, 1967)
14.10. N.W. Ashcroft, N.D. Mermin, *Solid State Physics* (Sanders College, Philadelphia, 1987)
14.11. J.D. Eshelby, J. Appl. Phys. **25**, 255 (1954)
14.12. F.D. Enck, J.G. Dommel, J. Appl. Phys. **36**, 389 (1965)
14.13. R.M. Nicklow, R.A. Young, Phys. Rev. **129**, 1936 (1963)
14.14. R.O. Simmons, R.W. Balluffi, Phys. Rev. **117**, 52 (1960)
14.15. W. Jost, J. Chem. Phys. **1**, 466 (1933)
14.16. N.F. Mott, M.J. Littleton, Trans. Faraday Soc. **34**, 485 (1935)
14.17. P. Brauer, Z. Naturforsch. **7a**, 372 (1952)
14.18. L.W. Barr, A.B. Lidiard, *Physical Chemistry – An Advanced Treatise*, Vol. 10 (Academic Press, New York, 1970)
14.19. E.W. Müller, Z. Physik. **131**, 136 (1951)
14.20. E. Goldstein, Sitz. Berliner Acad. Wiss. **16**, 937 (1894)
14.21. R.W. Pohl, Proc. Phys. Soc. (London) **49**, 3 (1937)
14.22. F. Seitz, Rev. Mod. Phys. **18**, 384 (1946)
14.23. F. Seitz, Rev. Mod. Phys. **26**, 7 (1954)
14.24. W.D. Compton, H. Rabin, Solid State Phys. **16**, 121 (1964)
14.25. J.J. Markham, *F Centres in Alkali Halides* (Academic Press, New York, 1966)

14.26. W.B. Fowler, *Physics of Colour Centres* (Academic Press, New York, 1968)
14.27. J.H. de Boer, Rec. Trav. Chim. **56**, 301 (1937)
14.28. E. Mollow, Z. Physik. **85**, 56 (1933)
14.29. H.F. Ivey, Phys. Rev., **72**, 341 (1947)
14.30. A. Smakula, Z. Physik. **59**, 603 (1930)
14.31. C.A. Hutchinson, Phys. Rev. **75**, 1769 (1949)

Chapter 15
Defects in Crystals II: Dislocations

15.1 Introduction

It was realised quite early that materials, particularly metals, can undergo permanent deformation (plastic deformation) at low stresses. A study of this intriguing phenomenon led to the discovery of dislocations. It was found that apart from providing an explanation for plastic deformation, dislocations play a role in various other aspects of crystal behaviour like crystal growth, hardness, fracture, thermal properties, etc. The subject of dislocations is thus of considerable importance.

In this chapter, we start with the situation before the discovery of dislocations. The concept of dislocations is discussed and the various types of dislocations are described. Several related aspects like dislocation multiplication, climb, grain boundary formation and whiskers are discussed. Dislocations are not just an idea but a reality; a number of methods of observing dislocations are discussed. Finally, the effect of dislocations on some physical properties of crystals is discussed.

A fairly detailed treatment of this topic is given by Cottrell [15.1], Read [15.2], Friedel [15.3], Weertman and Weertman [15.4] and Kovacs and Zsoldos [15.5].

15.2 Concept of Dislocations

15.2.1 Critical Resolved Shear Stress (CRSS)

Let a tensile force **F** be applied to a crystal in the form of a rod of length l and area of cross-section A; the tensile stress σ is $\mathbf{F}/A$. If this stress causes an increase $\mathrm{d}l$ in length, the strain is $\mathrm{d}l/l$. As the stress increases, the strain will also increase. A typical stress–strain curve is shown in Fig. 15.1. This curve consists of two regions. In region OA, the strain is linearly related to the stress and is reversible or elastic. Beyond the point A, the strain is nonlinear. A small increase in the stress

D.B. Sirdeshmukh et al., *Atomistic Properties of Solids*, Springer Series in Materials Science 147, DOI 10.1007/978-3-642-19971-4_15,

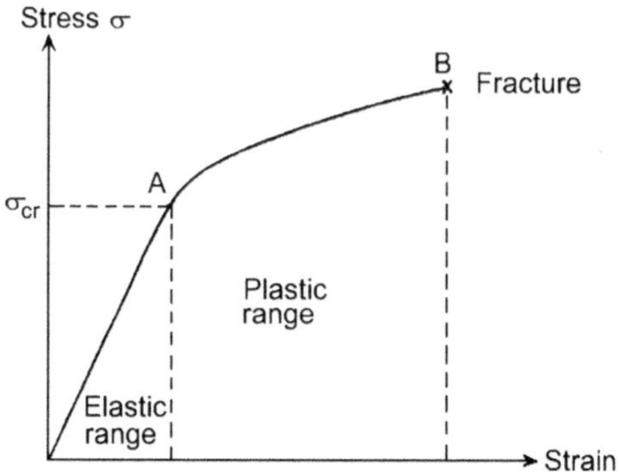

Fig. 15.1 A typical stress–strain curve

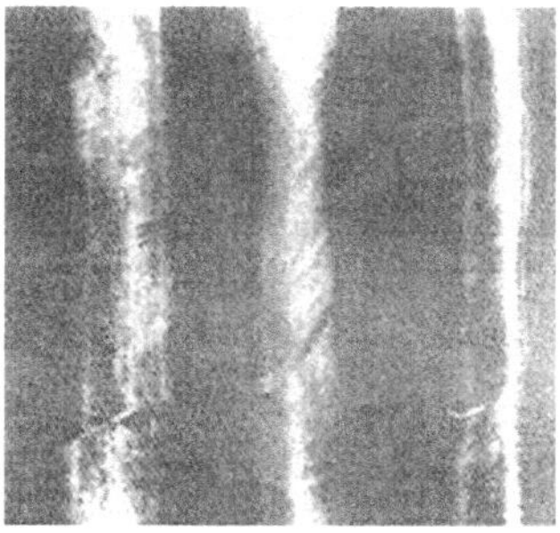

Fig. 15.2 Glide (slip) in a zinc crystal

causes a large increase in the strain and, in fact, at the point B the solid fractures. More importantly, the strain in the region AB is irreversible or plastic. The stress at the point A is called critical stress ($\boldsymbol{\sigma}_{\mathrm{cr}}$).

If the rod which has undergone plastic deformation is examined closely, it is found that the deformation is neither homogeneous nor isotropic. Some parts of the crystalline rod get displaced with respect to the adjoining parts (Fig. 15.2). The material in between consecutive displaced regions has normal undeformed lattice structure. The displacement is called glide or slip. The displacement is along specific planes (glide planes) and in specific directions (glide directions). For instance, in hcp crystals (Fig. 15.3), the glide plane is $\{0001\}$ and the glide direction is $\langle 2\bar{1}\bar{1}0\rangle$. The glide planes and glide directions in some crystals are given in Table 15.1.

The effective stress causing glide is different from $\boldsymbol{\sigma}_{\mathrm{cr}}$. The relative orientation of the tensile force **F**, the glide plane and the glide direction is shown in Fig. 15.4. The angle between **F** and the glide plane is α and that between **F** and the glide direction is β. The component of **F** in the glide direction is **F** cos β and the effective area of the glide plane is $A/\cos\alpha$, where A is the area of cross-section of the rod. Thus, the stress $\boldsymbol{\tau}$ acting along the glide direction is

$$\begin{aligned}\boldsymbol{\tau} &= \frac{\mathbf{F}\cos\beta}{A/\cos\alpha} = (\mathbf{F}/A)\cos\alpha\cos\beta \\ &= \boldsymbol{\sigma}\cos\alpha\cos\beta. \qquad (15.1)\end{aligned}$$

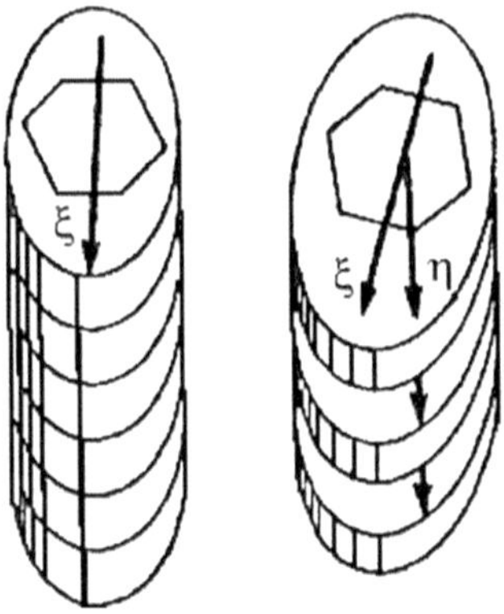

Fig. 15.3 Glide planes (ξ) and glide directions (η) in a hexagonal system

Table 15.1 Glide systems in some crystals

Crystal	Lattice type	Glide direction	Glide plane
Al, Cu, Ag, Au, Ni, α-CuZn	A1-type fcc	<110>	{111}
NaCl, KCl, KI, KBr	B1-type fcc	<110>	{110}
PbTe	B1-type fcc	<100>	{110}
Si, Ge	A4-type fcc	<110>	{111}
		<111>	{110}
α - Fe	A2-type bcc	<111>	{112}
NH_4Cl, TlCl	B2-type cubic	<100>	{100}
Cd, Zn, Mg	A3-type hcp	$<2\bar{1}\bar{1}0>$	{0001}

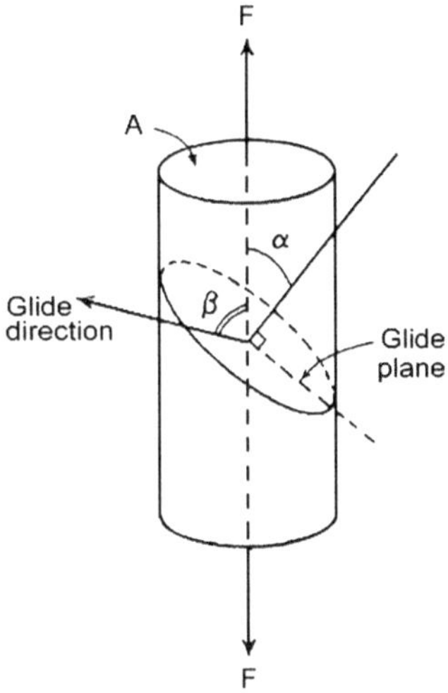

Fig. 15.4 Geometry of glide plane, glide direction and tensile force F

We have seen that glide takes place when $\boldsymbol{\sigma}$ reaches a critical value $\boldsymbol{\sigma}_{cr}$. Thus, the effective critical stress $\boldsymbol{\tau}_{cr}$ is

$$\boldsymbol{\tau}_{cr} = \boldsymbol{\sigma}_{cr} \cos\alpha \cos\beta. \quad (15.2)$$

Table 15.2 Shear modulus G and theoretical and experimental values of the critical resolved shear stress τ_{cr}

Metal	$G[10^{11}\text{dyn/cm}^2]$	$\tau_{cr}[10^{11}\text{dyn/cm}^2]$	
		Experimental	Theoretical (15.9)
Al	2.5	4×10^{-4}	8×10^{-2}
Ag	2.8	6×10^{-4}	9×10^{-2}
Cu	4.5	10×10^{-4}	15×10^{-2}
Zn	3.5	15×10^{-4}	12×10^{-2}
Mg	1.7	8×10^{-4}	6×10^{-2}

τ_{cr} is called the critical resolved shear stress (CRSS). The experimental values of τ_{cr} for some materials are given in Table 15.2.

Let us now estimate the CRSS assuming the crystal to be an elastic continuum. Figure 15.5 shows the model assumed by Frenkel [15.6]. Here the upper atomic layer glides over the layer below it by an amount x under the effect of a shearing stress $\tau(x)$. $\tau(x)$ will be zero when the atoms are at their normal positions (x=0) and also when the relative displacement of the two layers is $a/2$ (a being the interatomic distance). On the two sides of these zero-stress sites, the force acting on an atom differs in sign. In effect, the stress $\tau(x)$ is periodic; Frenkel assumed it to be sinusoidal. Thus,

$$\tau(x) = k \sin(2\pi x/a), \tag{15.3}$$

where k is the amplitude constant. At small values of x, $\tau(x)$ takes the Hooke's law form

$$\tau(x) \approx k(2\pi x/a). \tag{15.4}$$

When x is the displacement, the shear strain is (x/d) where d is the interlayer distance. If G is the shear modulus, then, by definition

$$\tau(x) = G(x/d). \tag{15.5}$$

From (15.4) and (15.5), we get

$$k = (G/2\pi)(a/d). \tag{15.6}$$

We may now rewrite (15.3) as

$$\tau(x) = (G/2\pi)(a/d)\sin(2\pi x/a). \tag{15.7}$$

The CRSS is the maximum value of $\tau(x)$, i.e.

$$\begin{aligned}\tau_{cr} &= (G/2\pi)(a/d) \\ &= (G/2\pi) \qquad \text{if}\, a = d.\end{aligned} \tag{15.8}$$

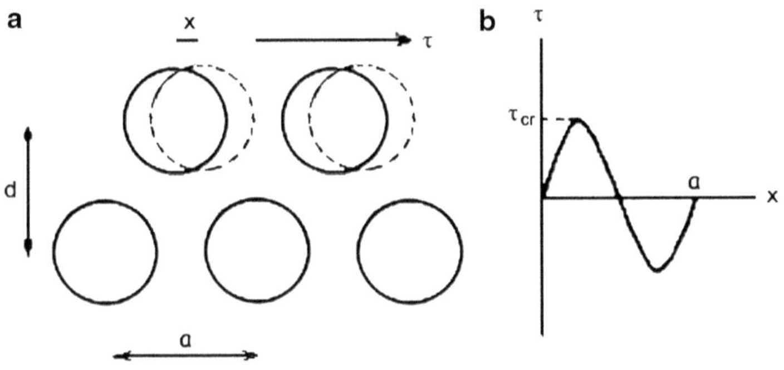

Fig. 15.5 (**a**) Displacement of an atomic layer due to shear stress, (**b**) periodic variation of shear stress

Thus, the CRSS is approximately $G/6$. As mentioned, Frenkel assumed a purely sinusoidal form of $\tau(x)$; a more realistic form gives

$$\tau_{cr} \approx G/30. \tag{15.9}$$

Values of τ_{cr} calculated from (15.9) using values of G from literature are given in Table 15.2. These values are much larger than the experimental τ_{cr} and the ratio (τ_{cr} exp / τ_{cr} calc.) is about 10^{-2}. In fact, for some materials [15.7], the ratio is as low as 10^{-8}. This is a serious discrepancy which could not be explained either in terms of errors in experimental values of G or τ_{cr} or in terms of the then existing theoretical concepts.

15.2.2 Concept of Dislocations

To explain the huge difference of several orders between the experimental and calculated values of CRSS, Taylor [15.8], Orowan [15.9] and Polanyi [15.10] independently proposed the concept of dislocations in crystals. Dislocations are line defects; these are lines around which the atomic arrangement is not the same as along other parallel lines in the lattice. The formation of such defects can be understood by imagining a cut-and-paste process.

Let us now imagine that a lattice is cut half way through its length (Fig. 15.6a) and an extra half lattice plane (ABCD) is inserted into the cut. Finally, the two parts are pasted together. This is a graphic representation of an edge dislocation. The dislocation line (AD) is normal to the plane of the paper. The boundary between the upper and lower parts is the glide plane. Note that the part above the glide plane has slipped by one interatomic distance in a direction normal to the dislocation line. The atomic arrangement around the dislocation line is different from that around a parallel line in a region away from the dislocation.

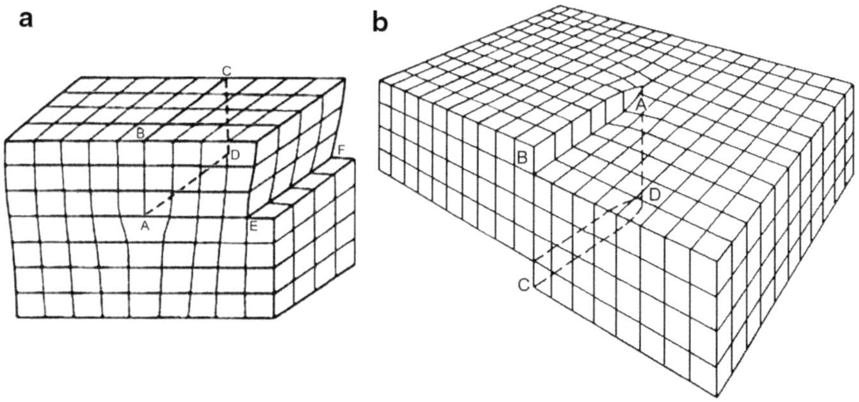

Fig. 15.6 (**a**) An edge dislocation in a simple cubic lattice, (**b**) a screw dislocation in a simple cubic lattice

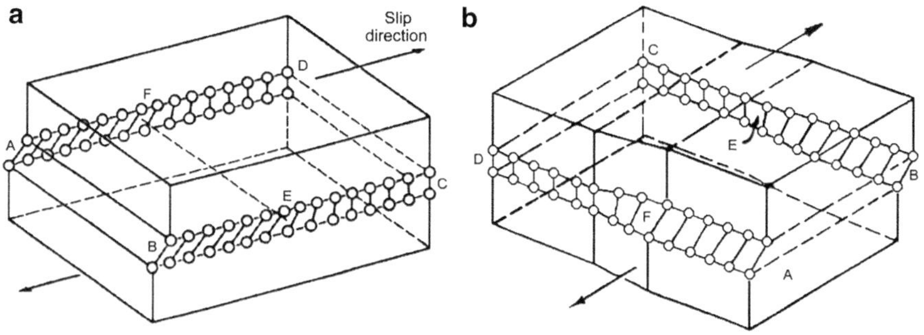

Fig. 15.7 Atomic configurations around (**a**) an edge dislocation and (**b**) a screw dislocation

Similarly, let us imagine that a cut ABCD is made in a lattice (Fig. 15.6b), the two parts are twisted or displaced relative to each other along the line of the cut and are then pasted together. This is the screw dislocation. The displacement of the material is along the dislocation line (AD); this is in contrast with what happens in an edge dislocation. The atomic arrangement about the dislocation line is different from that about a parallel line away from the dislocation. In the vicinity of the screw dislocation line, parallel lattice planes take the form of a spiral surface which winds through the crystal; when the dislocation intersects a surface, a step is produced.

Let us consider the atomic configurations in the vicinity of dislocations. Figure 15.7a shows an edge dislocation. ABCD is the glide plane. The atoms have been displaced by more than half a lattice constant in the slipped region ABEF and by less than half a lattice constant in the unslipped region FECD. EF is the dislocation line and AF the slip direction. Similarly, Fig. 15.7b shows a screw dislocation. Here again ABCD is the glide plane; the part ABEF has slipped parallel to the dislocation line EF. The presence of the screw dislocation transforms the atomic planes into the surface of a helix.

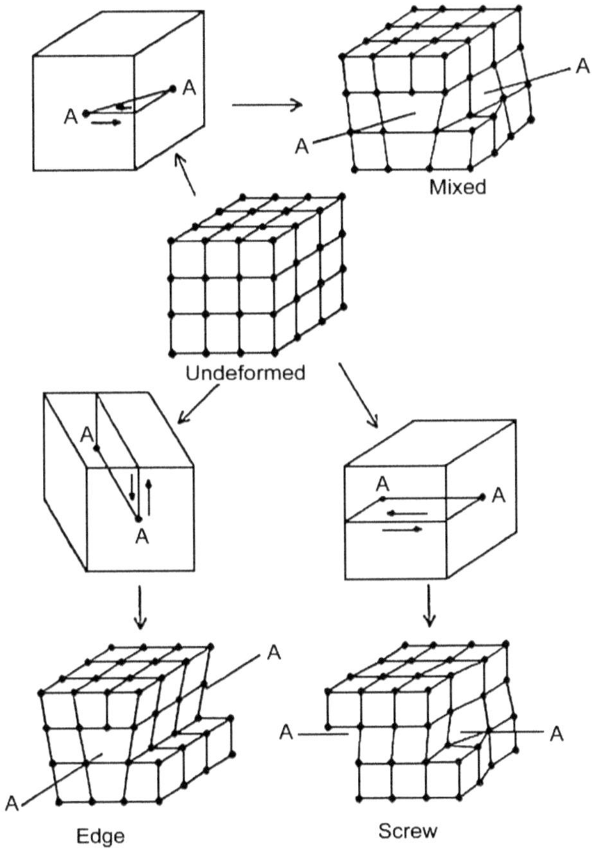

Fig. 15.8 Creation of an edge, a screw and a mixed dislocation

Yet another type of dislocation is the mixed dislocation which is a combination of an edge and a screw dislocation. The relationship between an undeformed lattice and lattices containing the three types of dislocations is shown in Fig. 15.8.

Finally, a closed dislocation loop could be created in a lattice by shifting atoms parallel to the plane of the loop (Fig. 15.9). As shown in the figure, the character of each segment of the dislocation loop changes continuously from pure edge to mixed to pure screw dislocation.

15.2.3 *The Burgers Vector*

Burgers [15.11] proposed a device by which a dislocation can be represented by a vector. This vector is called the Burgers vector and is generally represented by **b**; it will be seen that the Burgers vector is of the order of an interatomic distance.

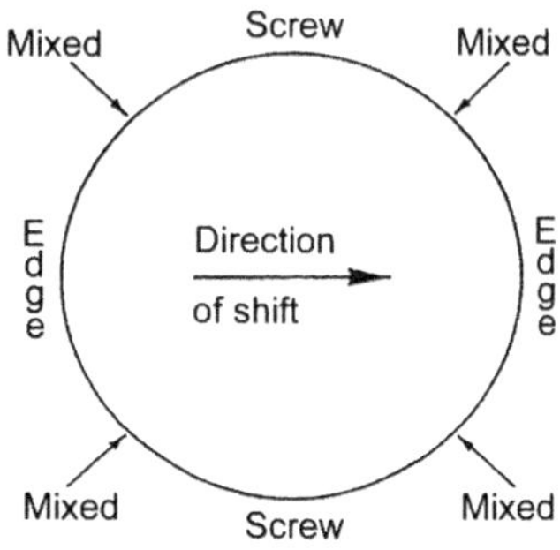

Fig. 15.9 A dislocation loop

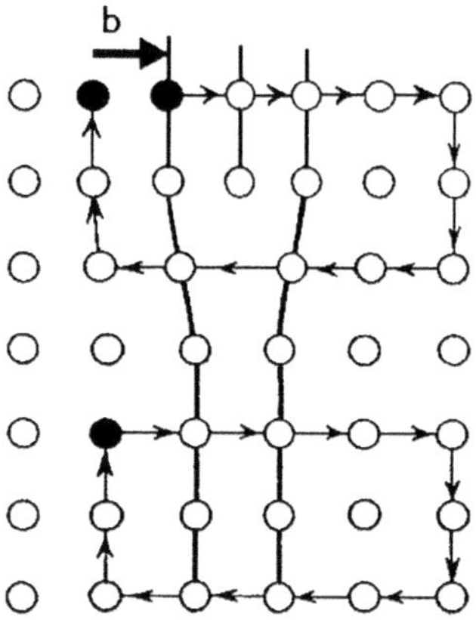

Fig. 15.10 Burgers circuit around an edge dislocation, the starting and end point atoms are shown *solid* in this and Fig. 15.11

Let us consider a section of a lattice which includes an edge dislocation. Let us draw a closed circuit made up of interatomic jumps. Such a circuit made up of 12 jumps is shown in the lower part of Fig. 15.10. This closed circuit is called a Burgers circuit. If we draw a similar circuit around the dislocation (upper part of the figure), the 12th jump does not close the circuit; there is a circuit failure. The circuit can be closed by drawing an extra jump. The vector required for this 13th jump, joining the initial point of the first jump and the end point of the 12th jump, is called the Burgers vector. It can be seen that the Burgers vector **b** of an edge dislocation is perpendicular to the dislocation line (which, in the present case, is normal to the plane of the paper). The edge dislocation can move in the direction of its Burgers vector.

Similarly, we shall consider a lattice which includes a screw dislocation. In Fig. 15.11, a ten-jump Burgers circuit is shown in a region away from the dislocation (left side of the figure). The corresponding circuit around the dislocation (right side of the figure) is open after the tenth jump and can be closed only by the 11th jump which represents the Burgers vector. In the case of a screw dislocation, the Burgers vector is along the dislocation line. A screw dislocation moves in a direction perpendicular to the Burgers vector. We can intuitively infer that in the case of a mixed dislocation, the Burgers vector will be at some angle to the dislocation line.

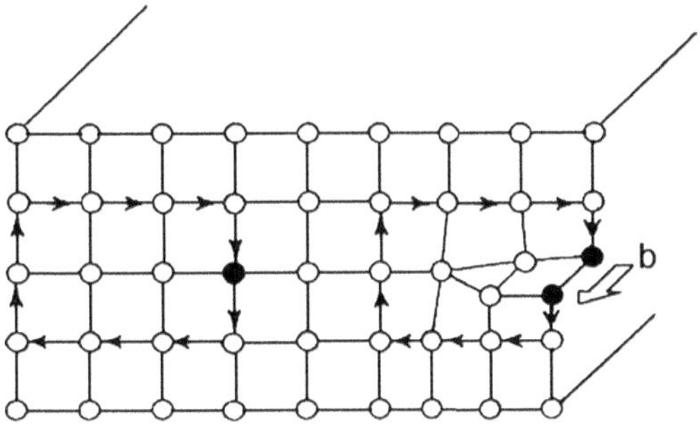

Fig. 15.11 Burgers circuit around a screw dislocation

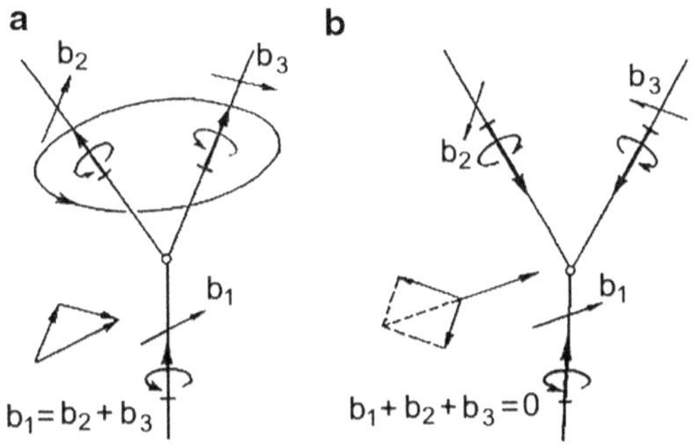

Fig. 15.12 Conservation of the Burgers vector on branching

Let us now consider the sign convention for dislocations. An edge dislocation is represented by the symbol ⊥; the vertical line denotes the extra half plane and the horizontal line the glide plane. The edge dislocation is positive if the extra half plane is above the glide plane (⊥) and negative if the extra half plane is below the glide plane (⊤). To assign a sign to a screw dislocation, Weertman and Weertman [15.4] proposed an elaborate procedure. Srivastava and Srinivasan [15.12] suggested the following simpler procedure. In Fig. 15.11, a clockwise Burgers circuit resulted in a Burgers vector pointing upward from the plane of the paper. If this screw dislocation is taken as positive, a negative screw dislocation would have a Burgers vector pointing inward.

Dislocations do not terminate within a crystal; they terminate at the surface or at a grain boundary. Inside the crystal, the dislocations either form loops or branch into two or more dislocations. A dislocation with Burgers vector $\mathbf{b}_1$ branching into two dislocations $\mathbf{b}_2$ and $\mathbf{b}_3$ in two different ways is shown in Fig. 15.12. The relations between the three Burgers vectors are:

$$\begin{aligned} &\text{Case(a)} \qquad \mathbf{b}_1 = \mathbf{b}_2 + \mathbf{b}_3 \\ &\text{Case(b)} \qquad \mathbf{b}_1 + \mathbf{b}_2 + \mathbf{b}_3 = 0. \end{aligned} \tag{15.10}$$

These equations are a statement of the law of conservation of the Burgers vector.

The Burgers vector defines the magnitude and direction of glide; it indicates how much above the glide plane and in what direction the lattice has shifted relative to

Table 15.3 Stable Burgers vectors in some common crystals

Lattice type	Materials	Stable Burgers vector	
		b	|**b**|
Cubic	CsCl	<100>	a
Bcc	Fe, W, Nb	1/2<111>	$a\sqrt{(3/2)}$
		<110>	$a\sqrt{2}$
Fcc	Cu, Al, NaCl, sphalerite	1/2<110>	$a/\sqrt{2}$
Hcp	Mg, Wurtzite	1/3<$2\bar{1}\bar{1}0$>	a
		<0001>	c

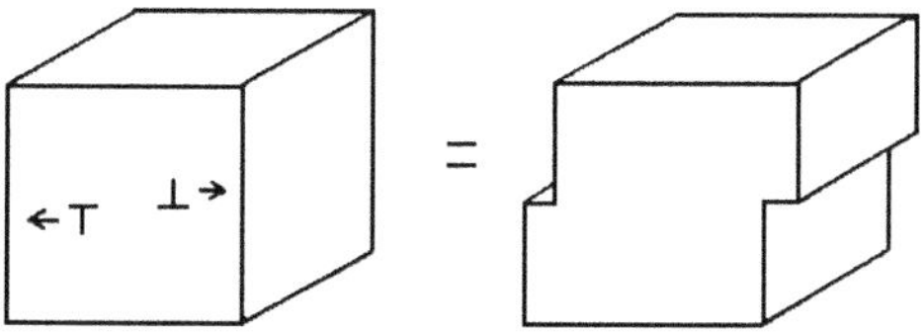

Fig. 15.13 Glide produced by motion of two edge dislocations

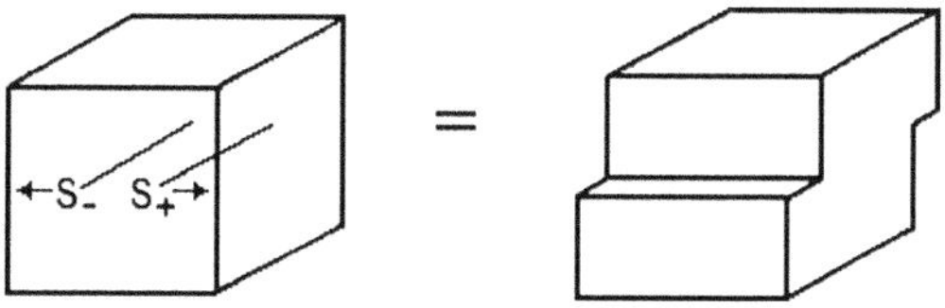

Fig. 15.14 Glide produced by motion of two screw dislocations

the lattice below the glide plane. The importance of the Burgers vector lies in the fact that if the Burgers vector and its orientation with respect to the glide plane are known, the dislocation is completely described. The possible stable Burgers vectors in some common lattices are given in Table 15.3.

15.2.4 Motion of Dislocations

Let us consider what happens when dislocations move across a glide plane. Figure 15.13 shows two edge dislocations of opposite sign. Under the effect of applied stress, these two dislocations move in opposite directions, being of opposite signs. In the case of edge dislocations, the shift of atoms is parallel to the direction of dislocation motion. Hence when the two dislocations move out, a step is produced on both sides of the crystal as shown in the figure.

Figure 15.14 shows two screw dislocations of opposite sign (or sense). These, again, move under the effect of an applied stress. It should be noted here that when a screw dislocation moves, the shift of atoms is perpendicular to the direction of dislocation motion. Thus, no steps are produced on the surfaces through which the dislocations pass out; instead, steps occur on surfaces where the dislocations terminate, as shown in the figure.

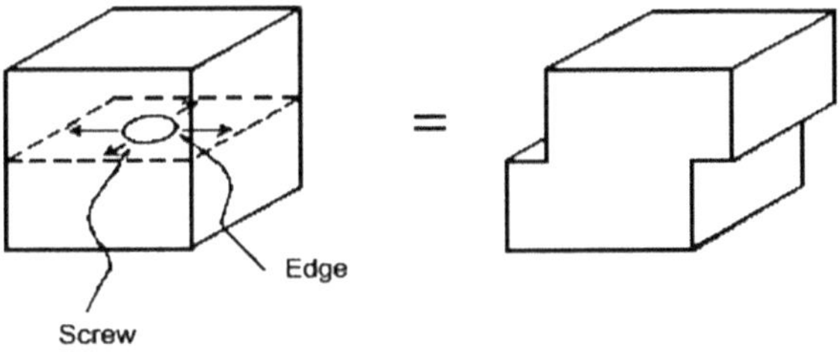

Fig. 15.15 Slip produced by expansion of a dislocation loop

Finally, let us consider the motion (or expansion) of a dislocation loop (Fig. 15.15). As can be seen, the loop is a combination of edge and screw dislocations. When the loop expands under the action of an applied stress, the effect is the same as if two edge and two screw dislocations have moved out. The directions must be noted carefully. The effect of motion of both types of dislocations is to produce steps on the side surfaces; there is no effect on the surfaces where the screw components terminate.

15.2.5 Dislocation as an Aid to Plastic Deformation

There are popular descriptions of the role of dislocations in plastic deformation. According to one description, the motion of a dislocation in a crystal is analogous to the passage of a ruck or wrinkle across a rug; the wrinkle moves with greater ease than the whole rug. According to another description, dislocation motion is analogous to the mode of locomotion of a caterpillar. The caterpillar forms a hump at its rear end by pulling in its last pair of legs; this hump is to be compared with the dislocation (Fig. 15.16). The caterpillar moves by pushing the hump forward. Finally, when the hump has reached the front end, the entire caterpillar has moved forward by the leg separation distance. The main point is that whereas in a perfect crystal, a whole plane of atoms has to glide over the next plane to cause a displacement by an interatomic distance, the same effect is achieved by the moving out of a single dislocation. The stress needed for the latter process should be much lower than that for the former. The concept of dislocations should, indeed, account for the discrepancy between the calculated and experimental values of critical shear stress. Actual estimate of the critical shear stress based on the theory of dislocations will be taken up in a later section.

15.3 Stress Fields Around Dislocations

The process of creation of dislocations described in Sect. 15.2.2 is bound to leave permanent stresses and strains in the region around a dislocation. We shall now see how these stresses can be estimated.

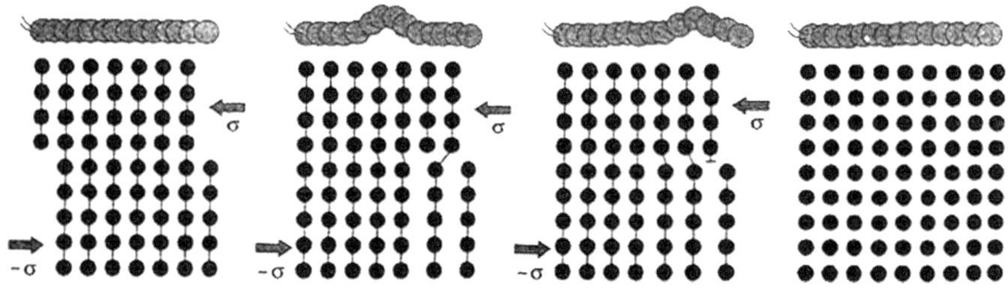

Fig. 15.16 Comparison of motions of a dislocation and a caterpillar

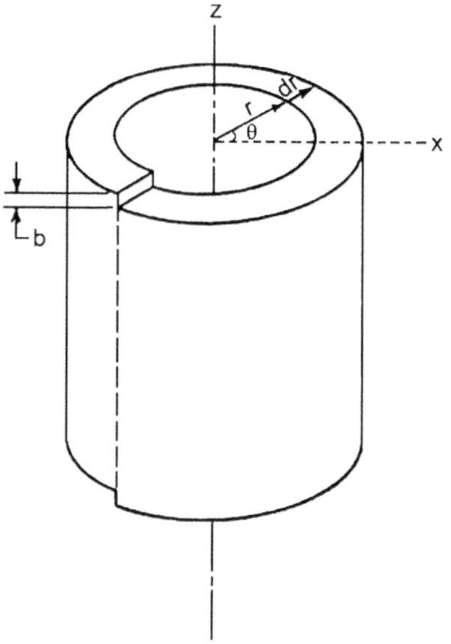

Fig. 15.17 Displacement of material in a cylindrical shell around a screw dislocation

15.3.1 Stress Field Around a Screw Dislocation

Let us consider a cylindrical shell bound by radii r and $r + \mathrm{d}r$ around a screw dislocation of length l (Fig. 15.17). The screw dislocation is along z. There is no strain along the r and θ directions. The strain is only in the z-direction. Along the line of the cut shown in the figure, the material has been displaced by b (the magnitude of the Burgers vector). This displacement is associated with the circumference of the shell. Hence, the shear strain is $(b/2\pi r)$. Assuming the crystal to be an elastic medium with shear modulus G, the shear stress σ is

$$\sigma = \text{strain} \times \text{shear modulus} = bG/2\pi r. \tag{15.11}$$

It can be seen that as $r \to 0$, $\sigma \to \infty$; this situation is to be avoided. Hence, (15.11) is to be used only for $r > r_0$ where r_0 is chosen arbitrarily; this region of radius r_0 is called the core of the dislocation. Let us now consider the strain energy E_{screw}. The strain energy $\mathrm{d}(E_{\text{screw}})$ for the shell of volume $\mathrm{d}V$ is given by

$$\begin{aligned} \mathrm{d}(E_{\text{screw}}) &= (1/2)\,\text{strain} \times \text{stress} \times \mathrm{d}V \\ &= (1/2)(b/2\pi r)(bG/2\pi r)(2\pi r l \mathrm{d}r) \\ &= (b^2G/4\pi)(\mathrm{d}r/r) \text{ for unit length of the dislocation.} \end{aligned} \quad (15.12)$$

The total strain energy is

$$E_{\text{screw}} = \int_{r_0}^{R} \mathrm{d}(E_{\text{screw}}) = \int_{r_0}^{R} (b^2G/4\pi)(\mathrm{d}r/r) = (Gb^2/4\pi)\log(R/r_0), \quad (15.13)$$

where R is some upper value for the radius of the shell limited by the dimensions of the crystal. The value of E_{screw} is not very sensitive to the value of the ratio (R/r_0) because of the logarithmic relation. Assuming $R = 1\,\text{cm}$, $b = 2.5 \times 10^{-8}\,\text{cm}$, $r_0 = 10^{-7}\,\text{cm}$ and $G = 4 \times 10^{11}\,\text{dyn/cm}^2$, we get $E_{\text{screw}} = 3.5 \times 10^{-4}$ erg or 5 eV for unit length of the screw dislocation.

15.3.2 Stress Field Around an Edge Dislocation

The derivation of the stress field around an edge dislocation is not so straightforward. A detailed discussion is given by Kovacs and Zsoldos [15.5]; we shall quote the results and discuss their significance.

Let us consider a cylindrical region in the crystal (Fig. 15.18); the z-axis of a Cartesian coordinate system is the cylinder axis. We shall now make a cut in the x–z plane from the outer surface up to the centre. Finally, we shall let the material above the x–z plane glide to the left by an amount b as shown by the dotted line.

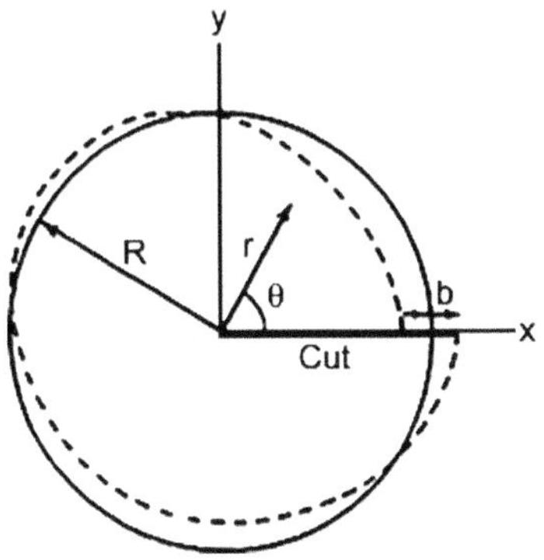

Fig. 15.18 An edge dislocation along the z-axis in a cylindrical piece of a solid

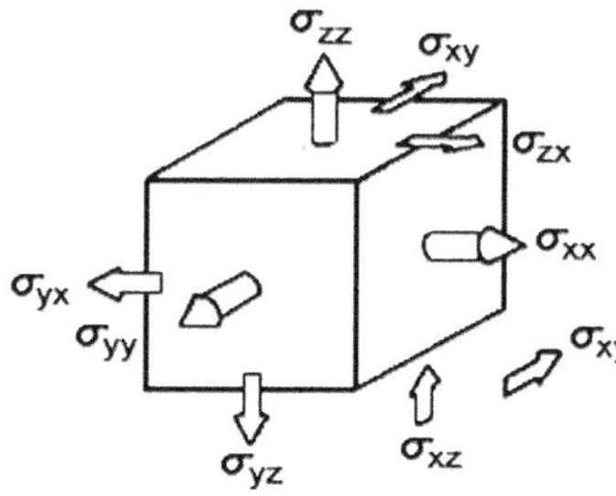

Fig. 15.19 Stress components acting on a crystal

We have, in effect, created an edge dislocation with Burgers vector **b** along the x-axis, with the x–z plane as the glide plane and the dislocation line along the z-axis.

In the Cartesian coordinate system, the stresses on a crystal are denoted by six stress components (Fig. 15.19; note that $\boldsymbol{\sigma}_{ij} = \boldsymbol{\sigma}_{ji}$). For an edge dislocation, these stress components are given by

$$\begin{aligned}
\sigma_{xx} &= [Gb/2\pi(1-\nu)][y(3x^2+y^2)/(x^2+y^2)^2],\\
\sigma_{yy} &= -[Gb/2\pi(1-\nu)][y(x^2-y^2)/(x^2+y^2)^2],\\
\sigma_{zz} &= [G\nu b/\pi(1-\nu)][y/(x^2+y^2)],\\
\sigma_{xy} &= -[Gb/2\pi(1-\nu)][x(x^2-y^2)/(x^2+y^2)^2],\\
\sigma_{xz} &= \sigma_{yz} = 0.
\end{aligned} \tag{15.14}$$

Here G is the shear modulus and ν the Poisson's ratio.

The stress components can also be expressed in terms of polar coordinates r, θ (Fig. 15.18). The radial and circumferential stresses are denoted by σ_{rr} and $\sigma_{\theta\theta}$ and the shear stress by $\sigma_{r\theta}$. These are given by

$$\sigma_{rr} = \sigma_{\theta\theta} = [Gb/2\pi(1-\nu)][\sin\theta/r] \quad \text{and} \tag{15.15}$$

$$\sigma_{r\theta} = [Gb/2\pi(1-\nu)][\cos\theta/r].$$

The spatial variations of the stress fields around edge dislocations with glide planes parallel to yz and xz plane, respectively, are shown in Fig. 15.20.

Following the same procedure as in Sect. 15.3.1, the strain energy E_{edge} of an edge dislocation is given by

$$E_{\text{edge}} = \int_{r_0}^{R} [Gb^2/2\pi(1-\nu)](\mathrm{d}r/r) = [Gb^2/2\pi(1-\nu)]\log(R/r_0), \tag{15.16}$$

where r_0 and R have the same significance as in (15.13). Using the same values for G, b, r_0 and R and with $\nu = 1/3$, we get $E_{\text{edge}} \sim 5.3 \times 10^{-4}$ erg or 7.5 eV for unit length of an edge dislocation line.

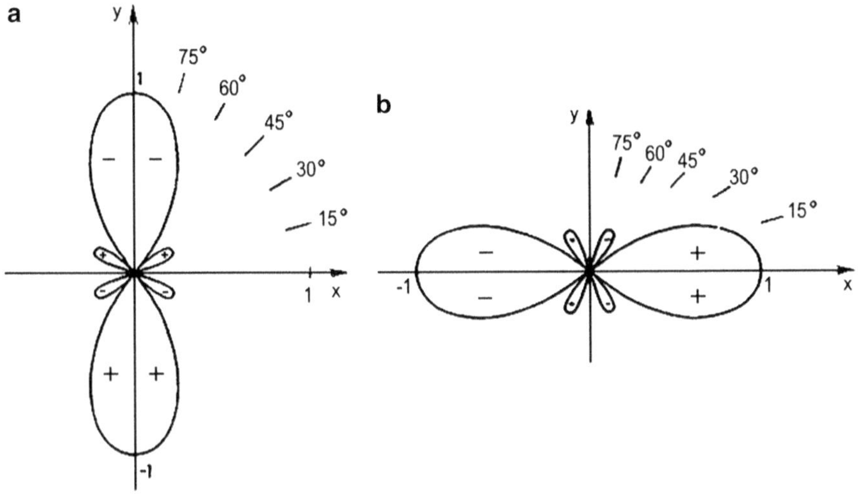

Fig. 15.20 Stress field of an edge dislocation (**a**) parallel to the y-axis, (**b**) parallel to x-axis

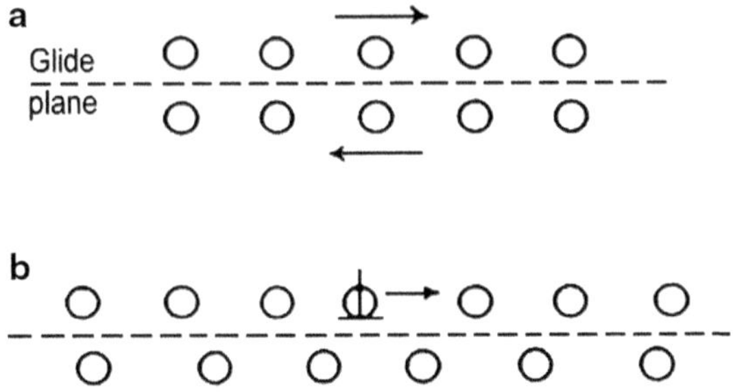

Fig. 15.21 Glide of atomic layers in a lattice (**a**) without a dislocation and (**b**) with a dislocation

15.4 Forces on Dislocations

Let us consider the situation depicted in Fig. 15.21. If a shear stress acts across a glide plane, it tends to move the atoms in the upper layer relative to those in the lower layer; of course such movement is resisted by the interatomic forces. On the other hand, if there is a dislocation line as shown in part (b) of the figure, the movement of the upper layer can be effected by the movement of the dislocation along the glide plane. We shall now consider in some detail the effect of external forces on dislocations.

15.4.1 Forces on Edge Dislocations and Dislocation Climb

Let us consider an external shear stress σ_{xy} acting on an edge dislocation in a slab of a crystal (Fig. 15.22). The dislocation has its Burgers vector in the x-direction.

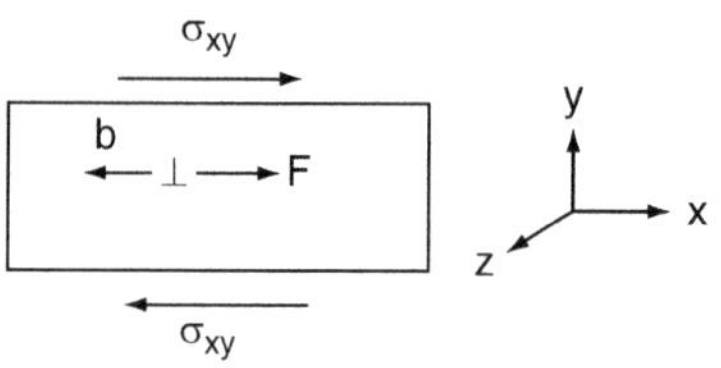

Fig. 15.22 A shear stress acting on an edge dislocation

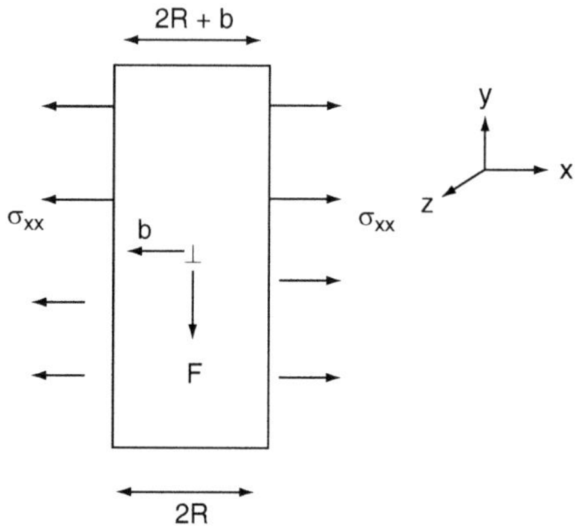

Fig. 15.23 Tensile stress acting on an edge dislocation

As the dislocation moves, it causes a relative displacement b between the material on either side of the glide plane. The external stress produces the force that pushes the dislocation along its journey. If the dislocation moves a distance L, the work done by the external stress will be $-\sigma_{xy}bL$. In the figure, **b** is in the negative x-direction; the minus sign indicates proper direction of motion of the dislocation. If we consider the work to have been done by an effective force F per unit length, then

$$FL = -\sigma_{xy}bL,$$

or,

$$F = -\sigma_{xy}b. \tag{15.17}$$

This force is parallel to the glide plane and is normal to the dislocation line.

An interesting phenomenon occurs when a tensile stress is applied to a crystal containing an edge dislocation and also vacancies. The tensile stress is applied in the x-direction as shown in Fig. 15.23. As shown in Fig. 15.24, if there is a vacancy near the dislocation, it tends to move to the bottom of the half plane and causes the dislocation to 'climb' upwards. Conversely, vacancies could be created at the bottom of the half plane and when they diffuse away, the dislocation 'climbs' down.

The climbing motion either inserts or removes a vertical plane of atoms and the width of the crystal changes by an amount b. Hence, the stress σ_{xx} does work or has work done on it. If the dislocation climbs a distance L, the work done by the stress is $\sigma_{xx}bL$. Again, if we attribute this work to an effective force F per unit length, then

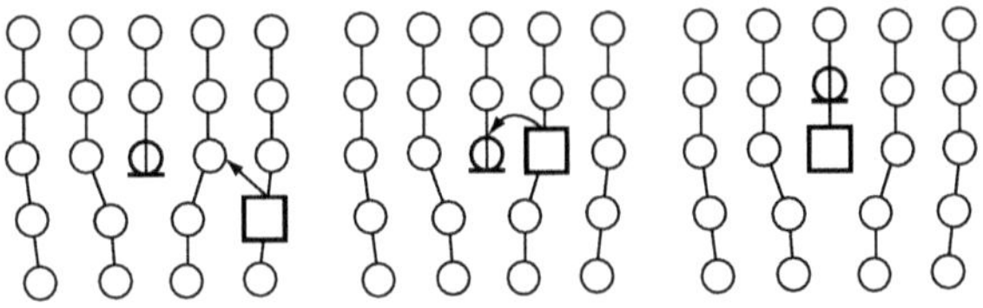

Fig. 15.24 The process of dislocation climb

$$FL = -\sigma_{xx} bL,$$

or,

$$F = -\sigma_{xx} b. \tag{15.18}$$

This force F acts in the y-direction.

15.4.2 Force on a Screw Dislocation

Let an external shear stress σ_{yz} act on a screw dislocation (Fig. 15.25) with the dislocation line along z. The work done in moving the dislocation through distance L is $-\sigma_{yz} bL$. If this is equated to the work done by an effective force F per unit length, then

$$FL = -\sigma_{yz} bL,$$

or,

$$F = -\sigma_{yz} b. \tag{15.19}$$

The force F is in the x-direction. The stress σ_{yz} is applied in the direction of the dislocation line and the screw dislocation moves on the glide plane in a direction normal to the dislocation line (although the stress is along the dislocation line).

15.4.3 Force on a Mixed Dislocation

We shall consider the force on a mixed dislocation. In a Cartesian coordinate system, let the dislocation be parallel to the z-axis and the glide plane normal to the y-axis.

The Burgers vector of such a mixed dislocation can be expressed as

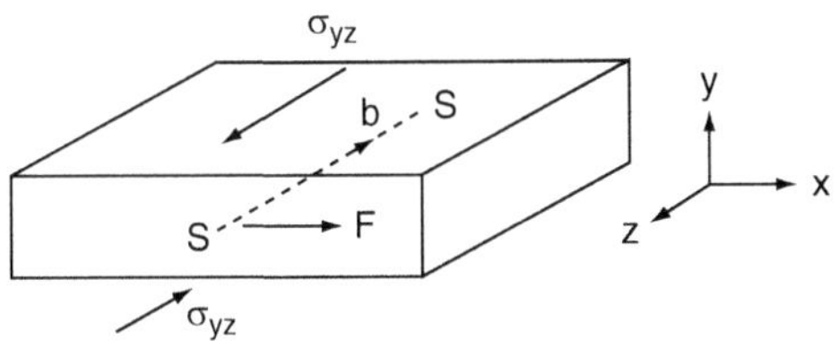

Fig. 15.25 Screw dislocation under shear stress

$$\mathbf{b} = b_x\mathbf{i} + b_z\mathbf{k}. \tag{15.20}$$

Here b_x and b_z are the edge and screw components of **b** and **i** and **k** are unit vectors along the x and y directions. An external stress acting on the crystal can be resolved into six components as shown in Fig. 15.19. Of these, only σ_{xz} and σ_{yz} can exert a force on the screw part of the dislocation and only σ_{xx} and σ_{xy} can exert a force on the edge part. The stresses σ_{yz} and σ_{xy} can cause the dislocation to move on the glide plane whereas σ_{xz} and σ_{xx} can cause it to move normal to the glide plane. σ_{zz} does no work and it exerts no force. Also, there is no displacement of material in the y-direction. Hence σ_{yy} does no work and does not contribute to the force on the dislocation. Thus, in analogy with (15.17–15.19), the force acting on the mixed dislocation may be written as

$$\mathbf{F} = -(\sigma_{xy}b_x + \sigma_{yz}b_z)\mathbf{i} + (\sigma_{xx}b_x + \sigma_{xz}b_z)\mathbf{j}. \tag{15.21}$$

Here, **j** is a unit vector parallel to y-direction. This force is normal to the dislocation line with one component parallel to the glide plane and the other normal to it.

15.4.4 The Peach–Koehler Equation

We shall now obtain a general equation for the force on a mixed dislocation. For this purpose, we shall define a unit vector **t** parallel to the dislocation line; the Burgers vector **b** may be in any chosen direction. The glide plane is obviously in the plane containing **t** and **b** (Fig. 15.26). The unit vector **n** normal to the glide plane is given by

$$\mathbf{n} = \frac{\mathbf{t}\times\mathbf{b}}{|\mathbf{t}\times\mathbf{b}|}. \tag{15.22}$$

Another unit vector **m** normal to both **t** and **n** may be defined by

$$\mathbf{m} = \mathbf{n}\times\mathbf{t}. \tag{15.23}$$

We can now make **m**, **n** and **t** coincide with the x, y, z axes (Fig. 15.27). The force **F** on unit length of the dislocation line is now given by

$$\mathbf{F} = (t_yG_z - t_zG_y)\mathbf{i} + (t_zG_x - t_xG_z)\mathbf{j} + (t_xG_y - t_yG_x)\mathbf{k}, \tag{15.24}$$

where

$$\begin{aligned} G_x &= \sigma_{xx}b_x + \sigma_{xy}b_y + \sigma_{xz}b_z \\ G_y &= \sigma_{yx}b_x + \sigma_{yy}b_y + \sigma_{yz}b_z \\ G_z &= \sigma_{zx}b_x + \sigma_{zy}b_y + \sigma_{zz}b_z. \end{aligned} \tag{15.25}$$

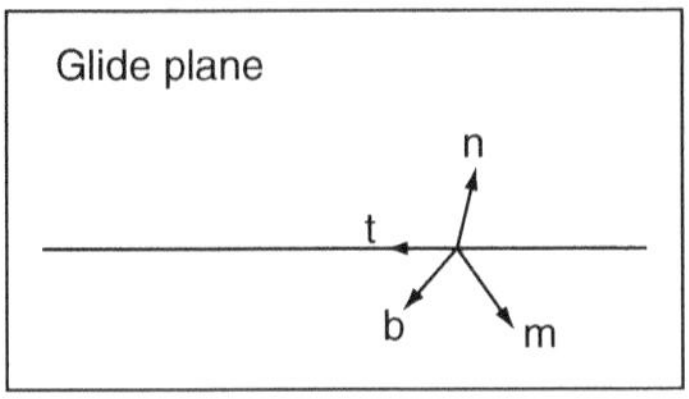

Fig. 15.26 Orientation of vectors **m**, **n** and **t**

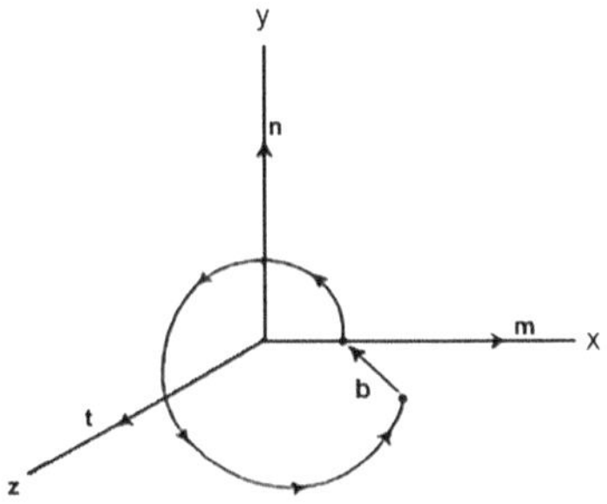

Fig. 15.27 The **m**, **n** and **t** vectors coinciding with the x, y, z system

Equations (15.24) and (15.25) define the force on a mixed dislocation in terms of the components of the vector **t**, the Burgers vector **b** and the stress components. If we treat (15.25) as representing the components of a vector **G**, (15.24) can be written as

$$\mathbf{F} = \begin{vmatrix} \mathbf{i} & \mathbf{j} & \mathbf{k} \\ t_x & t_y & t_z \\ G_x & G_y & G_z \end{vmatrix} = \mathbf{t} \times \mathbf{G}. \tag{15.26}$$

It may be noted that the force on a dislocation is always normal to the dislocation line.

Equation (15.26) is known as the Peach–Koehler equation. As in the original derivation [15.4], the symbol **G** is introduced to represent the 'vector' defined by (15.25); it may not be confused with the shear modulus. The Peach–Koehler equation has many applications in dislocation theory; thus, for example, it can be used to derive the forces between dislocations.

15.5 Interactions Between Dislocations

The energy required to create a dislocation in a crystal devoid of any dislocations is different from that required when dislocations are already present in the crystal. Because of the effect of dislocations, there will be an energy of interaction between two dislocations. The associated force will be given by the gradient of the interaction energy.

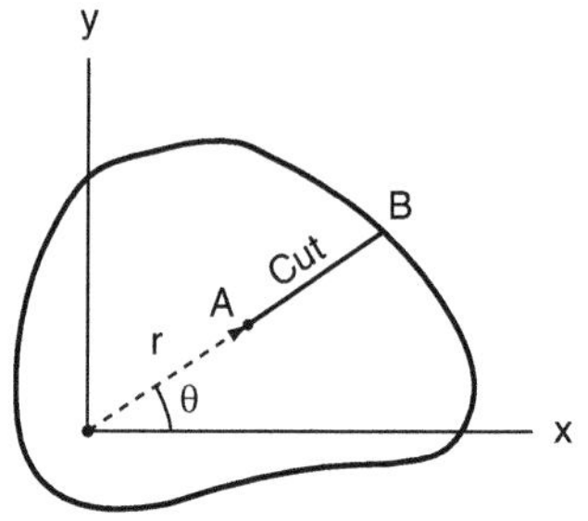

Fig. 15.28 Interaction of two screw dislocations running along the z-axis, one located at the origin and the other at A

15.5.1 Interaction Between Two Parallel Screw Dislocations

We shall consider the case of two parallel screw dislocations in a crystal. Let the first dislocation be along the z-axis (Fig. 15.28) and let another parallel dislocation be produced at the point A at a distance r from the origin. Let a cut be produced from point B to point A. Let the material on one side of the cut be displaced to the other side (i.e. in the z-direction) by a distance b. Let us calculate the work done in displacing the material in the presence of the stress field of the first dislocation which is at the origin. This interaction energy E_{i} is given by

$$E_{\mathrm{i}} = \int_{r}^{r_B} (Gb/2\pi r)b\mathrm{d}r. \tag{15.27}$$

where $(Gb/2\pi r)$ is the shear stress produced at A by the dislocation at the origin. The force F which is a function of r is given by

$$F(r) = -(\mathrm{d}E_{\mathrm{i}}/\mathrm{d}r) = (Gb^2/2\pi r). \tag{15.28}$$

This force varies as $(1/r)$; it is attractive for dislocations of opposite signs and repulsive for same signs.

15.5.2 Interaction Between Two Parallel Edge Dislocations

Let us consider two parallel edge dislocations both along the z-axis, one at the origin and the other at a point A which is a distance r from it (Fig. 15.29); their Burgers vector is along the x-axis. θ is the angle which r makes with the x-axis. In this case, the cut is along BA and there is a tangential force F_θ as well as a radial force F_{r}. These are given by

$$F_\theta = [Gb^2/2\pi(1-\nu)r]\sin 2\theta, \tag{15.29}$$
$$F_{\mathrm{r}} = [Gb^2/2\pi(1-\nu)r].$$

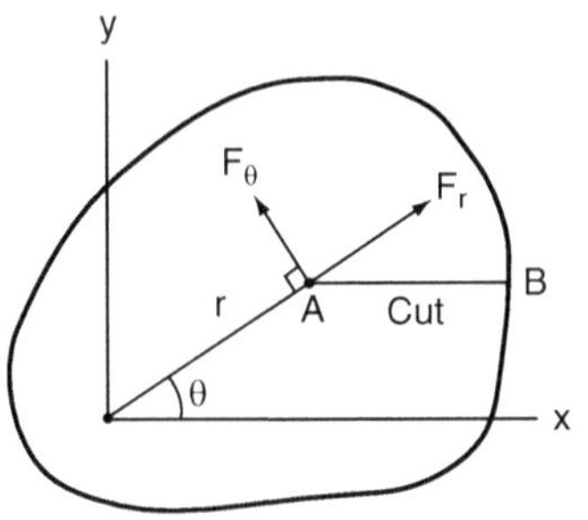

Fig. 15.29 Radial and tangential components of the force exerted by an edge dislocation at the origin on another at A

The motion of the dislocation is limited to the glide plane. Hence, we shall consider the force components along the x-direction. The net force F_x is

$$F_x = F_{\mathrm{r}} \cos \theta - F_\theta \sin \theta. \tag{15.30}$$

Substituting for F_{r} and F_θ, we get

$$F_x = [Gb^2 / 2\pi(1 - \nu)][x(x^2 - y^2) / r^4]. \tag{15.31}$$

It can be seen that F_x vanishes for $x = 0$ and $x = y$. Further, for $x > y$ and $\theta < 45°$, two parallel edge dislocations of the same sign repel each other in the direction of the glide plane whereas they attract each other for $x < y$ and $\theta > 45°$.

15.6 Some Dislocation Related Topics

15.6.1 *Peierls Stress*

In Sect. 15.2, we started with the calculation of the shear stress for an ideal crystal and found that it is several orders larger than the experimental value. To explain this discrepancy, the idea of dislocations was introduced. It was shown qualitatively that dislocations facilitate glide at lower stresses. We shall now examine the role of dislocations in initiating the process of glide.

Let us place two infinite half crystals side by side but with a relative displacement of $b/2$ (Fig. 15.30a); b is the magnitude of the Burgers vector. Now, let one side of the contacting plane be stretched and the other compressed. The two half crystals are now rejoined at some distance from the centre (Fig. 15.30b). If the displacement of an atom is denoted by $u(x)$, the upper and lower rows are displaced in opposite directions by an amount $u(x)$ and an edge dislocation has been created. The relative displacement of the two layers is $2u(x)$. Since the two layers were originally displaced by $b/2$, $|u(x)| \rightarrow b/4$ at large values of x. The position of the atom for $\alpha < 1/2$ and $\alpha = 1/2$ is shown in (c) and (d) of Fig. 15.30 where $x = \alpha b$.

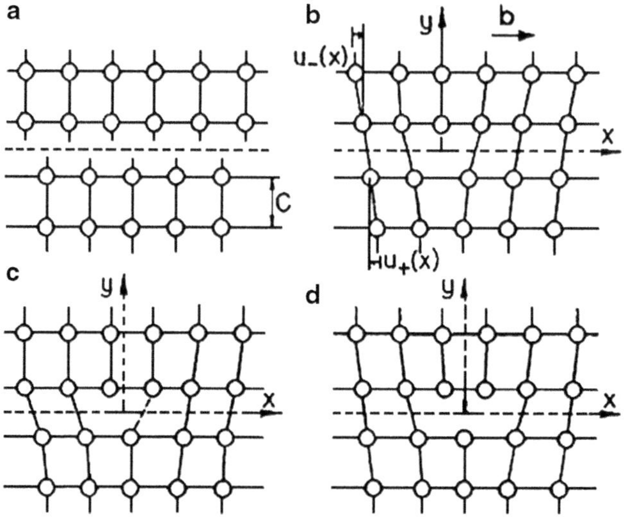

Fig. 15.30 Peierls model of dislocation (**a**) two atomic layers displaced by $b/2$ (**b**) displacement by $u(x)$ due to creation of an edge dislocation (**c**) displacement by $\alpha < 1/2$ and (**d**) displacement by $\alpha = 1/2$; $x = \alpha b$

The displacement $u(x)$ is in relation to the stress field σ_{xy} of the edge dislocation. It can be seen that σ_{xy} is a periodic function of $u(x)$ with period length $b/2$. As a first approximation, σ_{xy} may be written as

$$\sigma_{xy} = [Gb/2\pi c] \sin[4\pi u(x)/b], \tag{15.32}$$

where c is the distance between consecutive glide planes. This is the model of a dislocation proposed by Peierls [15.13].

The shear stress to move a dislocation can be expressed on the basis of the Peierls model. The derivation, which is rather, elaborate is given in books on dislocations, e.g. Kovacs and Zoldos (1973). Here we shall explain only the main steps. For a displacement du' of an atomic row of unit length, parallel to the glide plane and adjacent to the dislocation, the work done by the stress σ_{xy} is $b\sigma_{xy}\mathrm{d}u'$. The energy of the row at the end of full glide is

$$\int_0^u b\sigma_{xy}\mathrm{d}u' = [Gb^3/8\pi^2 c]\left[1 - \cos\frac{4\pi u(x)}{b}\right]. \tag{15.33}$$

This energy has now to be summed up for several rows on either side of the glide plane. The coordinates of the atomic planes with respect to the glide plane are

$$x = [\alpha + (m/2)]b \quad (m = 0, \pm 1, \pm 2, \ldots\ldots\ldots). \tag{15.34}$$

Table 15.4 Values of Peierls stress τ_P and the theoretical and experimental values of the critical resolved shear stress τ_{cr}; experimental and calculated values of τ_{cr} are reproduced from Table 15.2 for comparison

Metal	$\tau_{cr}[10^{11}$ dyn cm$^{-2}]$		τ_P (15.36) $[10^{11}$ dyn cm$^{-2}]$
	Experimental	Theoretical (15.9)	
Al	4×10^{-4}	8×10^{-2}	9×10^{-4}
Ag	6×10^{-4}	9×10^{-2}	10×10^{-4}
Cu	10×10^{-4}	15×10^{-2}	16×10^{-4}
Zn	15×10^{-4}	12×10^{-2}	12×10^{-4}
Mg	8×10^{-4}	6×10^{-2}	6×10^{-4}

The total energy U is obtained from

$$U = [Gb^3/8\pi^2 c] \sum_{m=-\infty}^{+\infty} \{1 - \cos 2[\arctan(\alpha + \frac{m}{2})](2b/s)\}, \qquad (15.35)$$

where $s = c/(1-\nu)$. The force F is given by the negative gradient of U. With some approximation, we get the critical shear stress from the Peierls model, for an edge dislocation τ_p as

$$\tau_p = F/b = [2G/(1-\nu)\exp[-2\pi c/b(1-\nu)]. \qquad (15.36)$$

Since the derivation was made by Nabarro using the Peierls model, (15.36) is known as the Peierls–Nabarro equation.

Values of τ_p calculated from (15.36) using $c = b$ and $\nu = 0.35$ are given in Table 15.4. These are much closer to the experimental values of τ_{cr} than those calculated for ideal crystals from (15.9). It may be mentioned that a similar equation can be derived for a screw dislocation [15.4].

15.6.2 Dislocation-Point Defect Interaction (Cottrell Pinning)

Forces can exist between dislocations and point defects (vacancies, interstitials, impurity atoms). Cottrell and Jawson [15.14] showed that the interaction between a point defect and the stress field of a dislocation can result in immobilisation of the dislocation. This effect is known as Cottrell pinning. We shall consider the effect briefly.

Let us consider an impurity atom located at a distance R from a dislocation in a crystal. Their relative positions are shown in Fig. 15.31. The point defect is located at the origin and the dislocation at $(-x, -y)$. The dislocation is parallel to the z-axis and its Burgers vector is in the negative x-direction; the glide plane is at $y = 0$. If an atom of radius r_0 leaves its site, it leaves behind a void of the same radius. An

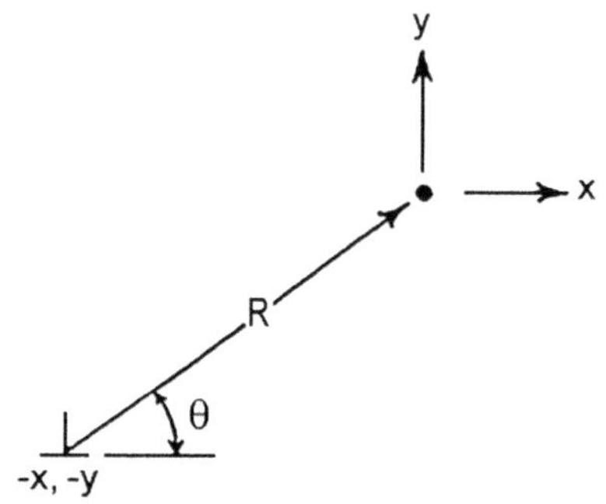

Fig. 15.31 Edge dislocation at $(-x, -y)$ and point defect at origin

impurity atom occupying this void may have radius $r^* = (1 + \varepsilon)r_0$ where ε is a dimensionless number. If ε is positive, $r^* > r_0$ and if it is negative, $r^* < r_0$.

Using elasticity theory, Cottrell showed that the forces acting on the dislocation because of the elastic displacement of the point defect are

$$F_x = 8G\varepsilon b r_0^3 xy/R^4 \tag{15.37}$$

and

$$F_y = -4G\varepsilon b r_0^3(x^2 - y^2)/R^4.$$

The energy of interaction E_{i} between the edge dislocation and the point defect is given by

$$E_{\mathrm{i}} = -4G\varepsilon b r_0^3 \sin\theta/R. \tag{15.38}$$

If it is remembered that a force is the negative of the gradient of the energy, (15.37) follows from (15.38).

If n_0 is the concentration of point defects in a lattice free of dislocations, the concentration n near a dislocation is given by

$$n = n_0 \exp(-E_{\mathrm{i}}/kT). \tag{15.39}$$

It can be seen that $n > n_0$ if E_{i} is negative. In such a case, an excess of point defects (impurities) will gather around a dislocation forming an impurity cloud (Fig. 15.32), which hinders the motion of the dislocation. This is called pinning of a dislocation or Cottrell pinning. Figure 15.33 is an electron micrograph showing the pinning of a dislocation in an MgO crystal by impurity particles.

15.6.3 Multiplication of Dislocations

Let us first consider the concept of line tension which is analogous to the surface tension in liquids. The line tension is the force acting tangentially on the end parts of a dislocation. Let us consider a line element dl of a curved dislocation (Fig. 15.34)

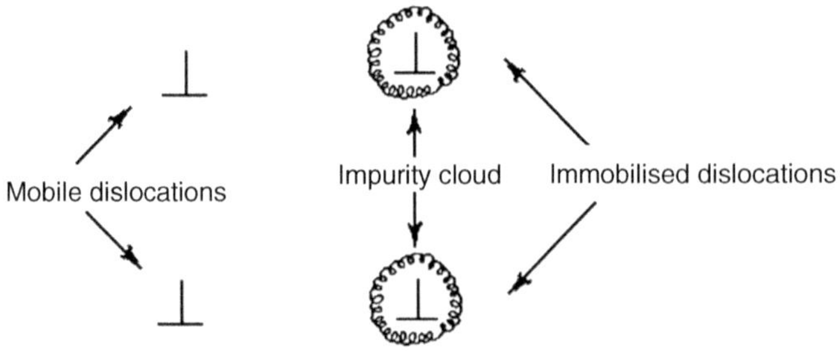

Fig. 15.32 Formation of impurity cloud around a dislocation

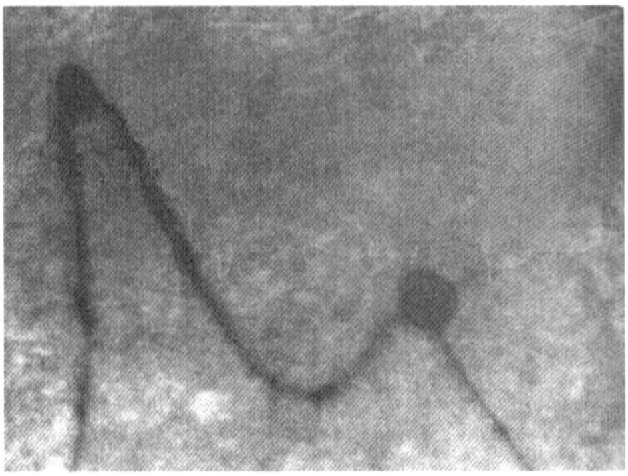

Fig. 15.33 Dislocations pinned by impurities in an MgO crystal

having a radius of curvature r_g and central angle $d\phi$. If T is the line tension, the force due to it in the direction of the Burgers vector b is $2T\sin(d\phi/2) \sim T d\phi \sim T dl/r_g$. On the other hand, if a shear stress τ acts in the glide plane in the direction of the Burgers vector, the resulting force is $\tau b dl$. Equilibrium is maintained when the two are equal, i.e.

$$\tau b dl = T dl/r_g,$$

or

$$\tau = T/br_g. \tag{15.40}$$

Using the theory of elasticity, the line tension can be worked out; the expression [15.5] is

$$T_{edge} = [Gb^2/8\pi(1-\nu)]\log(R/r_0), \tag{15.41a}$$

and

$$T_{screw} = [Gb^2/8\pi]\log(R/r_0), \tag{15.41b}$$

where G, ν, r_0 and R have the same significance as in Sect. 15.3.

Let us consider a dislocation segment AB (Fig. 15.35) pinned at points A and B. To start with, its position is given by 1 in the figure. When a stress is applied, the segment can move along the glide plane but with its end points pinned; in other

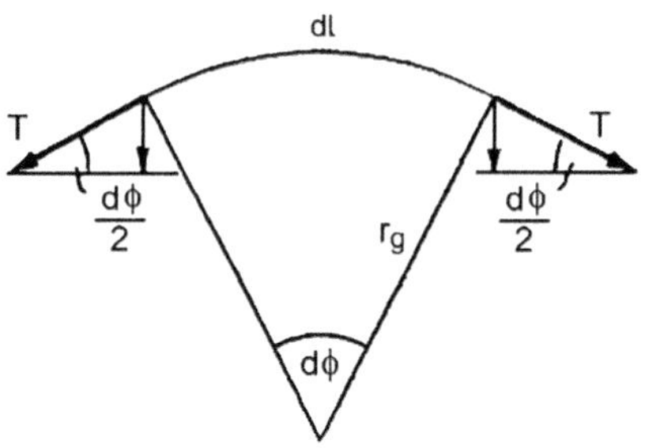

Fig. 15.34 Line tension in a curved dislocation

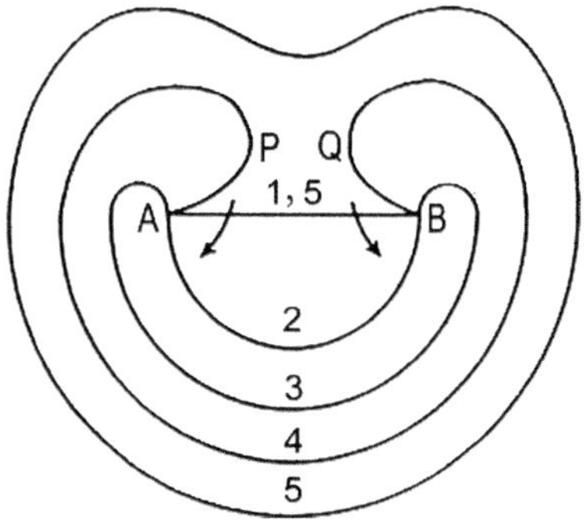

Fig. 15.35 Frank–Read mechanism of dislocation multiplication; note successive stages of the bending of the segment AB

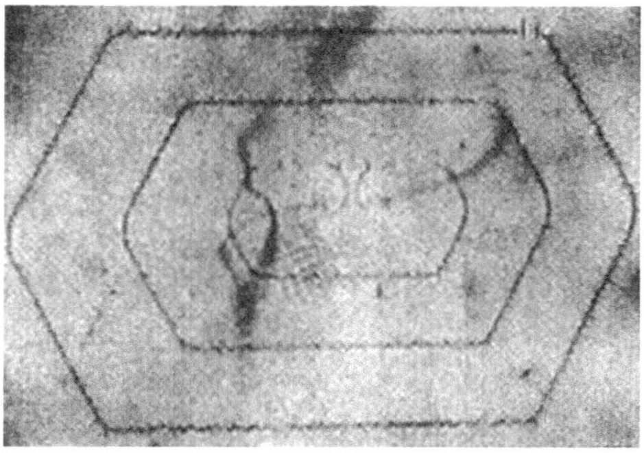

Fig. 15.36 A Frank–Read source in silicon decorated with Cu precipitates viewed with infrared light

words, it bows into successive positions 2, 3 and 4. Note that sections P and Q on the loop 4 are of opposite sign. Under continued effect of the stress, P and Q touch and annihilate each other. As a result, loop 4 splits into loop 5 and the segment 5 which is restored to its original position AB. This process continues leading to multiplication of dislocations. This mechanism was proposed by Frank and Read [15.15] and a source like AB is called a Frank–Read source. A Frank–Read source in a thin silicon sample decorated with copper particles observed with infrared illumination is shown in Fig. 15.36.

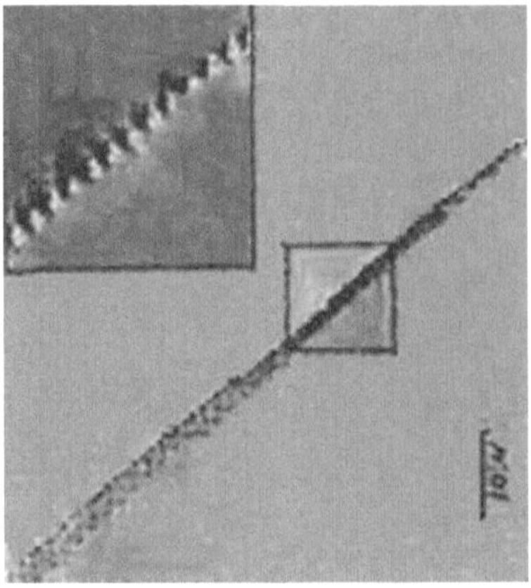

Fig. 15.37 A low angle grain boundary in Ge revealed by etch pits. The *marked region* is shown magnified in the inset

15.6.4 Low Angle Grain Boundary (LAGB)

Microblocks exist in crystals, each block making a small angle with the neighbouring block. This misorientation angle is typically 10–50 s of arc. The boundary between two such blocks is called a low angle grain boundary (LAGB). A LAGB in Ge revealed by etch pits is shown in Fig. 15.37. The formation of a LAGB can be imagined by the following process. Consider two normal crystal parts (Fig. 15.38a) with one face cut such that the new surface is not a crystallographic face; the angle of cut is small and same in both parts. These surfaces make an angle $\theta/2$ with the vertical lattice columns. It can be seen that at the points where a vertical column terminates into the new surface, an edge dislocation is created. In part (b) of the figure, these two crystals are joined. The edge dislocations in adjacent parts are oriented at the same angle as the blocks, i.e. θ. We have seen in Sect. 15.5.2 that two edge dislocations attract each other under some conditions. For stability, the edge dislocations assume positions and orientation shown in Fig. 15.38c.

If there are n dislocations in a length D of the crystal, then

$$nb = 2D\tan(\theta/2) \approx D\theta. \tag{15.42}$$

If d is the spacing between the dislocations, then

$$d = D/n = b/\theta. \tag{15.43}$$

The value of d can be estimated directly by measuring the distance between etch pits in a LAGB or by measuring the angle θ by X-ray diffraction experiments. In measurements on a Ge sample, the etch pit method gave $d = 4.7\,\mu\text{m}$ and the X-ray method gave $\theta = 17.5$ s of arc and $d = 5.3\,\mu\text{m}$. These results are consistent with the above model of the LAGB.

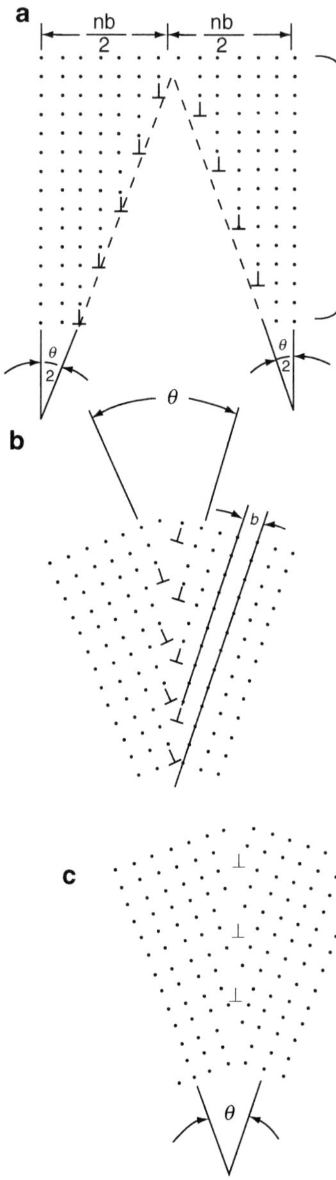

Fig. 15.38 Formation of a grain boundary

15.6.5 Obstacles to Dislocation Motion

A crystal contains a network of dislocation lines with an average spacing of about 10^4Å. When a stress is applied, the mobile dislocations move across the glide plane and initiate plastic deformation. However, there are factors which hinder dislocation motion. Such obstacles to the motion of dislocations are:

1. Difficulty in bending a dislocation line
2. Resistance to movement of a dislocation by irregularities like foreign atoms, precipitates, vacancies and other dislocations
3. Difficulty in transmitting glide across a grain boundary.

15.7 Observation of Dislocations

We have seen that the concept of dislocations was introduced in 1934. Several features of dislocations, such as dislocation multiplication, impurity pinning, LAGB, etc., were theoretically worked out in the 1930s and 1940s without the availability of any method for the observation of dislocations. Then in the mid-fifties (1952–1958) there was a sudden spurt of techniques to reveal dislocations. The first method to be reported (and also the simplest and most popular) is the etch pit method. This method is discussed in detail; a few other methods are briefly discussed.

15.7.1 The Etch Pit Method

The fact that dissolution figures are produced when a crystal surface is treated with a liquid was known long before the advent of the term 'dislocations'. Mineralogists routinely etched mineral crystals and used the shapes of the resulting 'etch pits' to draw information about the symmetry of the crystal. The etch pit pattern on fluorite (CaF_2) published in 1894 is shown in Fig. 15.39. The direct connection between etch pits and dislocations was established only when Horn [15.16] showed one-to-one correspondence between the centres of growth spirals (screw dislocations) and etch pits (Fig. 15.40).

Fig. 15.39 Etch figures on (111) face of CaF_2 produced with sulphuric acid

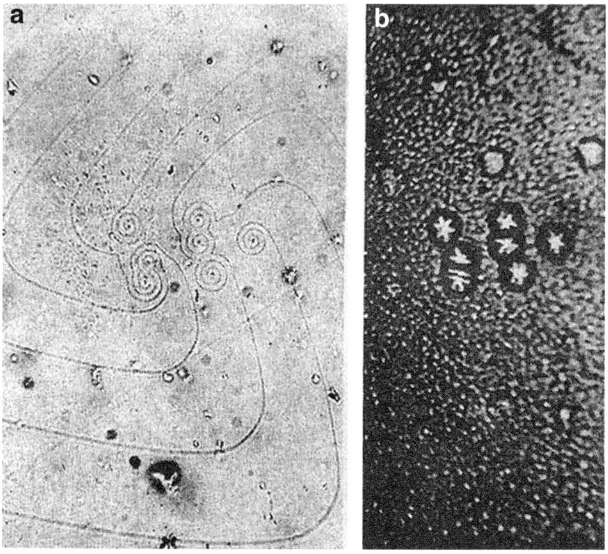

Fig. 15.40 Surface features on silicon carbide (**a**) growth spirals (**b**) etch pits

When a crystal is dipped in a liquid, raised, dried and then observed under a microscope, dissolution figures are seen; these are the 'etch pits'. The liquid may be water or some other, generally organic, liquid in which the crystal is soluble. Generally, a small quantity of an impurity is added to the liquid to improve the definition of the etch pits; the impurity is called a 'poison' and the liquid–poison combination is called an 'etchant'. The etching action is controlled by parameters like time of etching, temperature and the state of agitation. Etchants for some crystals are given in Table 15.5.

A liquid can produce etch pits only if the crystal has some solubility, however meagre, in the liquid. At a dislocation site, dissolution takes place with a velocity V_n along the depth by nucleation of steps and with velocity V_s along the surface by removal of steps across the surface (Fig. 15.41). For well-defined etch pits $(V_s/V_n) < 10$. This ratio can be varied by the addition of a poison. The changes in the appearance of etch pits in LiF with changes in the concentration of FeF_3 in water are shown in Fig. 15.42. It can be seen that the definition is best for a specific concentration of the poison (c in the figure).

Etch pits have a definite geometric form. The symmetry of etch pits reflects the symmetry of the crystal face. Several examples have been given in Chap. 7. The symmetry aspect of etch pits will therefore not be discussed here. We shall discuss the evidence that etch pits are indeed formed at dislocation sites.

1. Since a dislocation runs through a crystal, if the crystal is cleaved, the dislocation will intercept the new surfaces at identical points. Thus, the etch pits on two cleaved surfaces should show one-to-one correspondence. This can be seen in Fig. 15.43 which shows etch pits on cleaved faces of LiF.

Table 15.5 Dislocation etchants for some crystals

Crystal	Surface	Etchant	Etching time
Elements			
Bi	(111)	1% I_2 in methyl alcohol	15 s
C	(111)	Fused KNO_3 900°C	2 min
Sb	(111)	3HF: $5HNO_3$: $3CH_3COOH$: $0.3Br_2$	2 s
Te	$(10\bar{1}1)$	3HF : $5HNO_3$: $6CH_3COOH$	30 s
Semiconductors			
Ge	(111)	1HF : $1H_2O_2$: $4H_2O$	1 min
Si	(100)	1HF : $3HNO_3$: $10CH_3COOH$	1 min
GaAs	$(\bar{1}\bar{1}\bar{1})$	1HF : $1HNO_3$: $1H_2O$	10 min
GaSb	$(\bar{1}\bar{1}\bar{1})$	1HF: $2HNO_3$: $1CH_3COOH$	15 s
InAs	$(\bar{1}\bar{1}\bar{1})$	15HF: $75HNO_3$: $15CH_3COOH$: $0.06Br_2$	5 s
InSb	$(\bar{1}\bar{1}\bar{1})$	10HF: $25HNO_3$: $20CH_3COOH$: $1Br_2$	5 s
Alkali halides			
KBr	(100)	Glacial acetic acid	3 s
KCl	(100)	Ethyl alcohol+25% saturated with $BaBr_2$	5 s
KI	(100)	Isopropyl alcohol	25 s
LiF	(100)	$H_2O + FeF_3$	1 min
NaCl	(100)	Glacial acetic acid	0.1 s
Others			
$CaCO_3$	(010)	Formic acid	15 s
CaF_2	(111)	H_2SO_4	10 min
MgO	(100)	Conc. H_2SO_4(at 50°C)	1 in
PbS	(100)	1 HCl: 3 thiourea solution (100 g/L)	1–10 in
Quartz	$(10\bar{1}0)$	Saturated aqueous ammonium bifluoride	–
Yttrium iron garnet	(110)	Boiling concentrated HCl	1 h

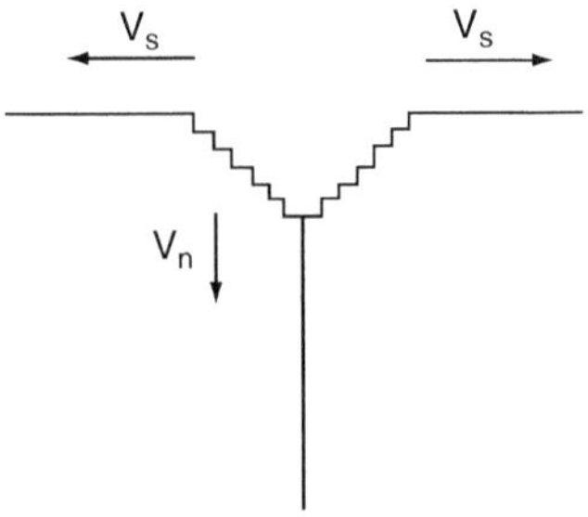

Fig. 15.41 Dissolution at a dislocation site

2. A crystal may be etched and the etch pits recorded. If the etched surface is dissolved (polished) till the etch pits disappear, re-etching the surface should reveal etch pits at the same positions. This can be seen in Fig. 15.44.
3. Similarly, if a crystal surface is repeatedly etched, the same etch pits should appear but with increased size without the appearance of any new pits. This is seen in Fig. 15.45.
4. In a thin crystal, a dislocation line will intersect the two opposite faces. Consequently, the etch pit pattern (distribution of etch pits) should be the same; this has been observed to be so.

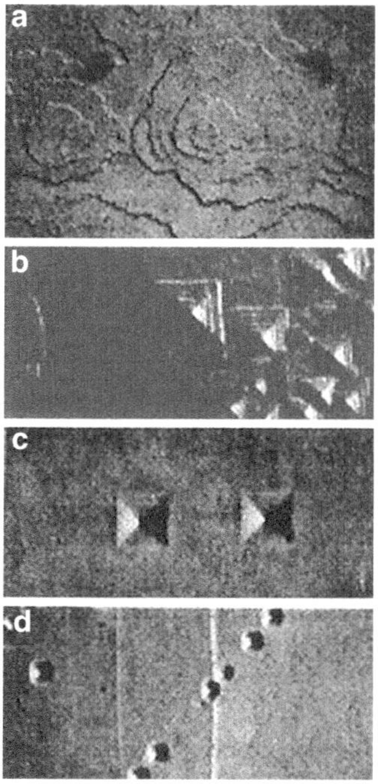

Fig. 15.42 Effect of poison (FeF_3) concentration in etchant (water) on the definition of etch pits on (100) face of LiF. The poison concentration increases from (**a**) to (**d**)

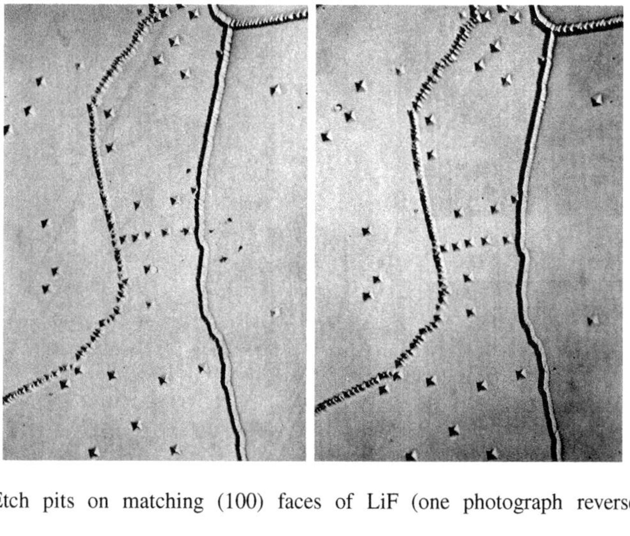

Fig. 15.43 Etch pits on matching (100) faces of LiF (one photograph reversed for easy comparison)

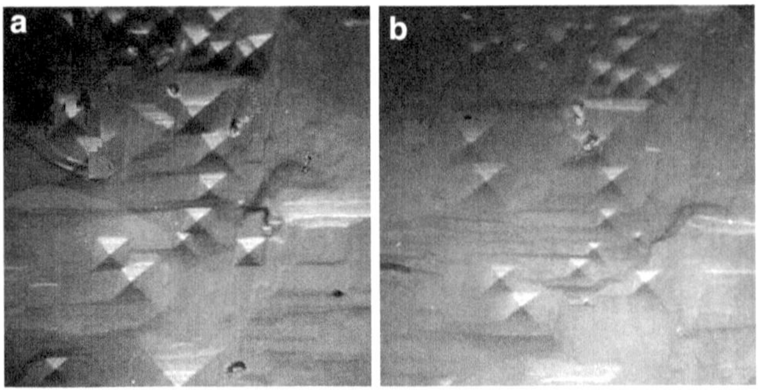

Fig. 15.44 Etch pits on (100) faces of $NaClO_3$ (**a**) before and (**b**) after polishing

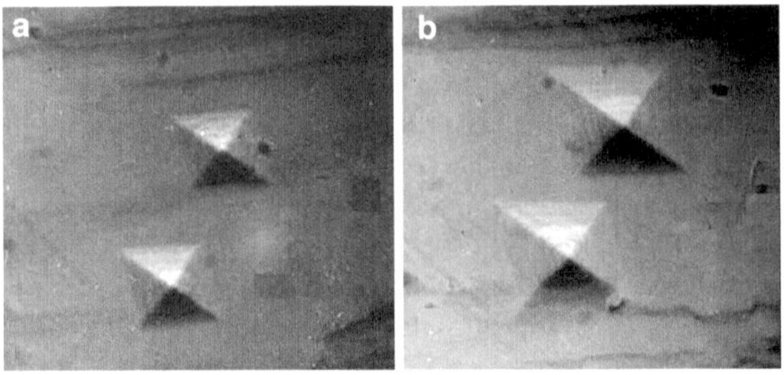

Fig. 15.45 Etch pits on (100) faces of $NaClO_3$ after repeated etching

5. According to theoretical considerations, when three tilt boundaries meet the linear dislocation densities ρ_i along the three boundaries must satisfy the equation

$$\sum_{i=1}^{3} \rho_i/(\cos\phi_i + \sin\phi_i) = 0, \qquad (15.44)$$

where ϕ_i is the angle which a boundary makes with a symmetrical boundary. If etch pits are formed at dislocations, then the density of dislocations in the boundaries should follow the same relation. This has been verified in NaCl and a few other crystals.

This evidence clearly established that etch pits are formed at dislocation sites.

15.7.2 Decoration Method

As in the case of the etch pit method of observing dislocations, decorated features in the inside of crystals were reported during 1905–1932. But the origin of these features was not understood. Hedges and Mitchell [15.17] were the first to use the

Fig. 15.46 Dislocations in KCl crystal revealed by decoration with silver atoms

decoration technique to reveal dislocations. They exposed AgBr crystals to light. The photolytically neutralised silver got deposited on dislocations which showed up in photographs. The routine method now is to heat a crystal with a chosen impurity so that it diffuses into the crystal and onto the dislocation network. A photograph showing dislocations in a KCl crystal decorated with silver atoms is shown in Fig. 15.46.

For some time, the use of the decoration technique was limited to transparent crystals. But Dash [15.18] used infrared light to observe dislocations in silicon. In fact, the beautiful photograph of a Frank–Read source shown in Fig. 15.36 was obtained in this way. The decoration method has been applied to several other alkali halides and fluorite.

15.7.3 Field Ion Microscope

The principle of the field ion microscope (FIM) has been discussed in the preceding chapter. It was pointed out that the FIM has a very high resolution (~2Å) and can, therefore, reveal atomic scale features like vacancies. It follows that the FIM would be able to reveal dislocations.

Drechsler et al. [15.19] were the first to use the FIM to observe dislocations on tips of tungsten, tantalum and nickel crystals. A FIM picture of an edge dislocation on a Mo tip is shown in Fig. 15.47.

15.7.4 Stress Birefringence Method

In Chap. 13, we have seen that when a crystal is stressed it develops birefringence. This causes the stressed region to look dark when the crystal is held between crossed Nicols. In Sect. 15.3 we have seen that the stress fields are associated with

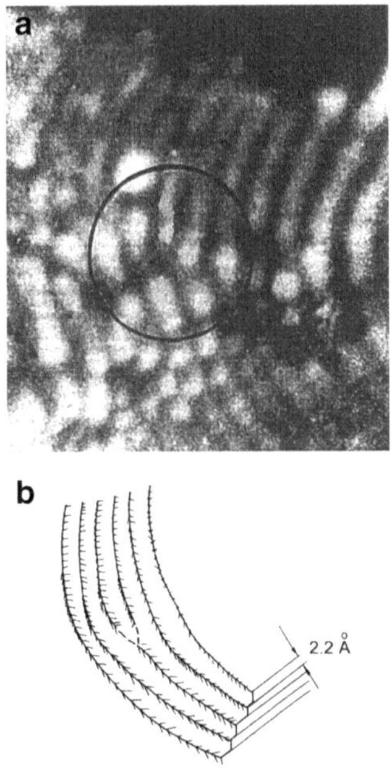

Fig. 15.47 (**a**) Picture taken with a field ion microscope of a Mo tip containing one dislocation, (**b**) line diagram showing details and step height

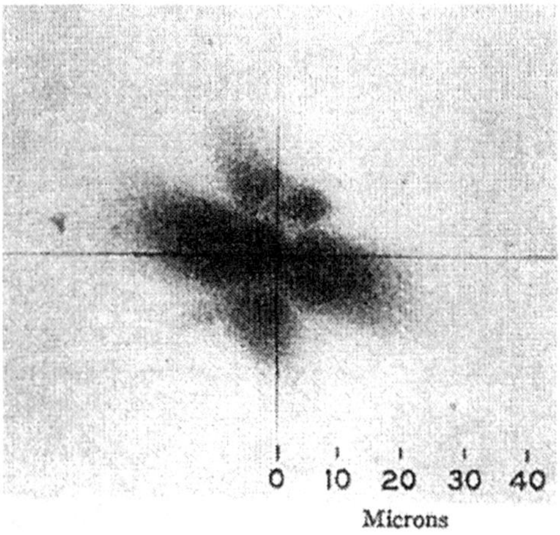

Fig. 15.48 Stress field around an edge dislocation in silicon crystal recorded with infrared radiation; the dislocation is inclined at 15° to the glide plane

dislocations; the variation of the stress field around edge dislocations was shown in Fig. 15.20. Thus, if a crystal containing dislocations is illuminated by polarized light and is viewed through a Nicol, the pattern would be similar to that in Fig. 15.20.

Bond and Andrus [15.20] observed the pattern shown in Fig. 15.48 when a silicon crystal illumined with near-infrared light was viewed through Nicol prisms. They also calculated the stress pattern for an edge dislocation whose glide plane is inclined at 15° with the polarizer axis (Fig. 15.49); the similarity between the patterns is obvious. The method has been used to observe dislocations and related phenomena in other crystals viz. Rochelle salt and corundum.

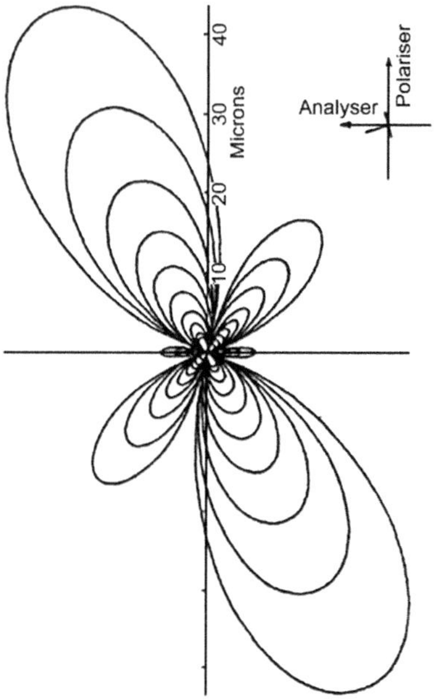

Fig. 15.49 Intensity of light calculated from stress field corresponding to Fig. 15.48 and photoelastic data

15.7.5 Transmission Electron Microscopy

An electron microscope uses a high velocity electron beam focussed by 'magnetic lenses' and detected by a fluorescent screen. The radiation used to 'see' objects is the de Broglie waves associated with the electrons having wavelength $\lambda = h/mv$ where v is the velocity and m the mass of the electron. By controlling the velocity, wavelengths as small as 0.4 Å are used (compared to $\lambda \sim 5{,}000$ Å in an optical microscope). This small wavelength bestows a high resolution on this technique; also, magnifications of 60,000× are possible.

Because of the high absorption of the electron beam, the technique was initially applied to thin metallic foils. In 1956, Hirsch et al. [15.21] and Bollman [15.22] observed dislocation structures in aluminium and other metals. Dislocation structure in a thin foil of stainless steel observed by Whelan et al. [15.23] is shown in Fig. 15.50. Subsequently, the technique was extended to observe dislocations in painstakingly prepared thin samples of non-metallic crystals like MgO, graphite, Bi_2Te_3, talc and muscovite. Parallel dislocation lines observed in Bi_2Te_3 are shown in Fig. 15.51. It may be mentioned that the pinning of a dislocation line shown in Fig. 15.33 was recorded by this technique.

Fig. 15.50 Transmission electron micrograph of dislocations in thin foil of stainless steel

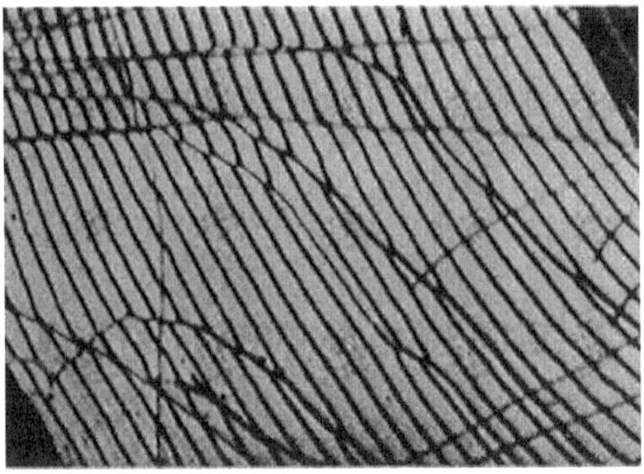

Fig. 15.51 Electron micrograph of parallel dislocations in Bi_2Te_3

15.7.6 X-ray Topography

It was pointed out in the chapter on X-ray diffraction that the intensity of X-rays diffracted by a crystal depends on its state of perfection, being more for an imperfect crystal than for a perfect one. X-rays diffracted from a region containing dislocations will be more intense than those from a nearby dislocation-free region. If the X-ray beam is made to scan a crystal, the diffraction image will consist of dark and less dark regions. This is the principle of X-ray topography.

In the method adopted by Lang [15.24], an X-ray microbeam is diffracted by a crystal (Fig. 15.52). It passes through a slit and is received on an X-ray film. There is an arrangement to move the crystal and the film together synchronously so that the incident X-ray beam scans the crystal and is diffracted by different parts of the crystal. The resulting X-ray image is a dislocation map of the crystal. The dislocations in a Si crystal recorded by Lang are shown in Fig. 15.53. By scanning the same crystal in different directions, it is possible to construct the three-dimensional distribution of dislocations in a crystal. Such a view of dislocations, slip bands and grain boundaries in a LiF crystal is shown in Fig. 15.54.

Arrangements different from Lang's have been proposed. Further, the use of synchrotron radiation has cut down exposure times by orders so that instantaneous

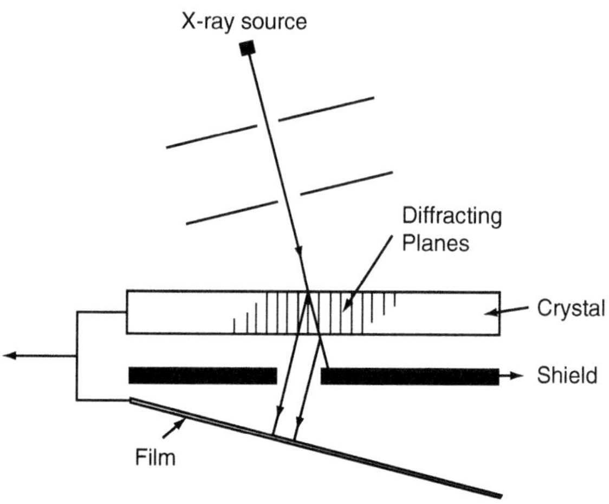

Fig. 15.52 Principle of Lang's method of X-ray topography

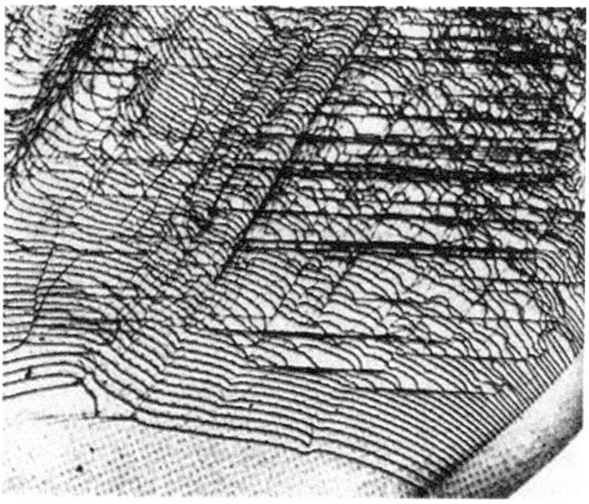

Fig. 15.53 X-ray topograph of dislocations in silicon

Fig. 15.54 Three sides of a strained LiF crystal with slip planes and grain boundaries constructed with X-ray topographs

records could be made. This has opened up the application of topographic method to study a crystal during its growth. The Lang camera is commercially available and has become a routine tool in characterising defects in crystals.

Table 15.6 Comparison of techniques of observation of dislocations

Technique	Maximum specimen thickness	Width of image of dislocation	Maximum detectable dislocation density	Remarks
Transmission electron microscope	500 Å	100 Å	3×10^{11} cm^{-2}	Requires very thin samples; offers high resolution and magnification
X-ray (Lang method)	0.1–1 mm	5 μm	2×10^{4} cm^{-2}	–
Decoration	10 μ (depth of focus)	0.5 μm	2×10^{7} cm^{-2}	Renders the sample impure; heating to high temperatures alters dislocation structure
Etch pit	No constraint	0.5 μm	4×10^{8} cm^{-2}	Simple and inexpensive; can be used for transparent as well as opaque samples

15.7.7 Comparison of Methods

We shall conclude this discussion with a comparison of the different techniques. The FIM can be used only for metallic samples made into the form of a tip. The stress-birefringence method can be employed only with crystals having a very low dislocation density as otherwise the stress fields overlap and create blurring of the pattern. Also, this method is not applicable to observation of screw dislocations along their axes. Features like the maximum sample thickness and maximum detectable dislocation density for the more commonly employed techniques are compared in Table 15.6.

15.8 Dislocation Densities

The density (ρ) of dislocations is defined as the number of dislocation lines intersecting unit area of the crystal surface; it is expressed as a number cm^{-2}. The creation of dislocations does not reduce the free energy of the crystal as the creation of vacancies does. Thus, the existence of dislocations is not a thermodynamic requirement and the existence of a totally dislocation-free crystal is not theoretically precluded. Yet, dislocations do exist in crystals with densities varying over a wide range up to 10^{12} cm^{-2}. We shall consider dislocation densities in crystals with different mechanical and growth histories.

15.8.1 Dislocation Densities in As-Grown Crystals

Dislocation densities in normal as-grown crystals are of the order 10^{4}–10^{8}cm^{-2}. Some factors responsible for the creation of dislocations are discussed below.

15.8.1.1 Thermal Stress

In materials which are good conductors of heat, temperature gradients which exist in the crystal at the time of growth create large thermal stresses. Dislocations are created under these conditions to release the stresses.

15.8.1.2 Collapse of Vacancies

If the crystal is grown from its melt, the concentration of vacancies at the melting point is about 1%. When the crystal is cooled to room temperature, the equilibrium concentration is negligible. Thus, about 10^{20} vacancies cm^{-3} must be removed to restore thermal equilibrium. There are no sinks in the crystal to accommodate so many vacancies. The vacancies collapse to form sheets whose edges are dislocations. 10^{20} vacancies cm^{-3} can cause the creation of half planes enough to create 10^6cm-cm^{-3} of edge dislocation line.

15.8.1.3 Impurities

Impurities do exist in crystal material. Though the overall concentration of an impurity may be small, there can be local clustering of impurity atoms. In such regions, the size difference between the impurities and the host atoms creates elastic stresses. These stresses are released by the creation of dislocations.

15.8.1.4 The Seed

The seed contains dislocations which radiate into the growing crystal. Some of these dislocations grow out of the surface of the crystal but those which are parallel to the growth axis of the crystal remain in the crystal.

15.8.2 Dislocation Densities in Nearly Perfect Crystals

From the discussion in the preceding section, it can be inferred that the dislocation density in an as-grown crystal can be minimised by reducing thermal gradients, ensuring purity, choosing a seed with low dislocation density and adopting a slow growth rate. Following such a strategy, success has been achieved in growing crystals (particularly Ge and Si) with dislocation densities well below $10^2\,cm^{-2}$. In fact, Dash [15.18] has grown dislocation free Si crystals which have been found to be 30 times stronger than ordinarily grown Si crystals.

Another approach to grow crystals with near-zero dislocation density is to reduce one dimension of the growing crystal to a value comparable with the average

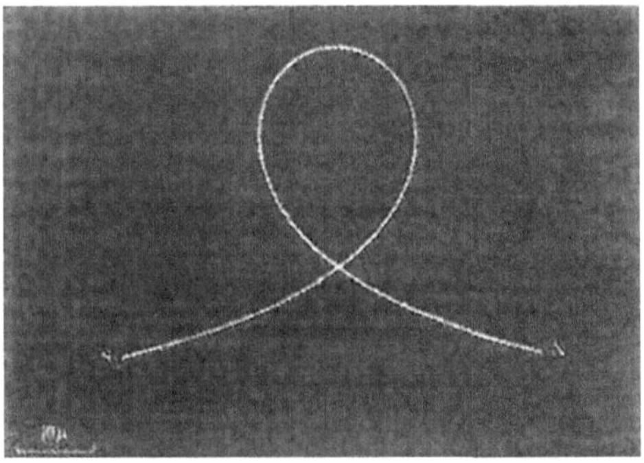

Fig. 15.55 A nickel whisker of diameter 1,000 Å

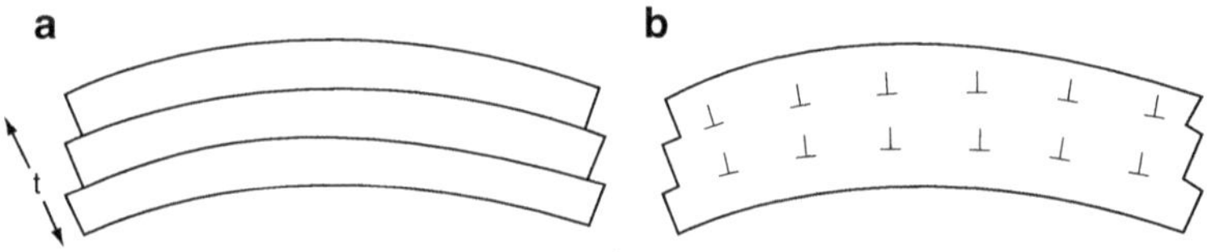

Fig. 15.56 (**a**) Plastically bent crystal and (**b**) corresponding dislocation model

spacing ($\sim 10^4$ Å) between dislocations in ordinary crystals. Very thin, wire-like crystals of dimensions 10^4 Å or less have been grown from Fe, Cu, Sn and Ni; these are called whiskers. A nickel whisker of diameter $\sim$1,000 Å is shown in Fig. 15.55. Whiskers are found to have strengths comparable with the theoretical strength for a dislocation-free crystal ($\sim G/30$). Most whiskers show the presence of single or a pair of screw dislocations along the growth axis.

15.8.3 Dislocation Densities in Deformed Crystals

Let us consider a crystal bent as in Fig. 15.56a. This bending is possible only due to slip (glide) facilitated by dislocations (Fig. 15.56b). If L is the length of the outer arc and t the thickness of the crystal bar, the length of the inner arc will be $L[1-(t/R)]$ where R is the radius of curvature. Hence, the total slippage is $L - L[1-(t/R)]$, i.e. Lt/R. But if the number of dislocations is n and b the magnitude of the Burgers vector, the total slippage is nb. Hence, the density of dislocations ρ is

$$\rho = n/Lt = (Lt/Rb)/Lt = 1/Rb. \tag{15.45}$$

If $R = 3\,\text{cm}$ and $b = 3 \times 10^{-8}\,\text{cm}$, then $\rho = 10^7\,\text{cm}^{-2}$. Thus, deformation creates a large dislocation density. Besides, the Frank–Read dislocation mechanism also contributes to the dislocation density in deformed crystals.

15.9 Effect of Dislocations on Physical Properties

We saw that there was a difference of several orders between the calculated and experimental values of the stress required to cause plastic deformation. The concept of dislocations was introduced to account for this difference. It was shown that the shear stress calculated from a model that assumes the presence of dislocations (the Peierls model) is much closer to the experimental value. The role of dislocations does not end there. It will be shown in the following sections that dislocations have an all round effect on several properties of crystals.

15.9.1 Hardness

Hardness is a mechanical property. It is generally measured by 'indentation', i.e. by pressing on the crystal surface a precisely cut diamond carrying a load. The shape of the commonly used indenter, known as the Vickers indenter, is shown in Fig. 15.57 and the impression created by it is shown in Fig. 15.58. In microhardness measurements, loads up to 200 g are used. The Vickers microhardness H_V is calculated from the formula

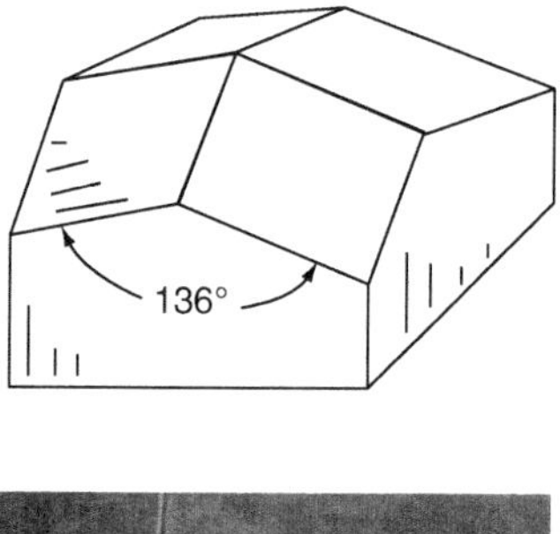

Fig. 15.57 Vickers diamond indenter

Fig. 15.58 Impression produced by a Vickers indenter

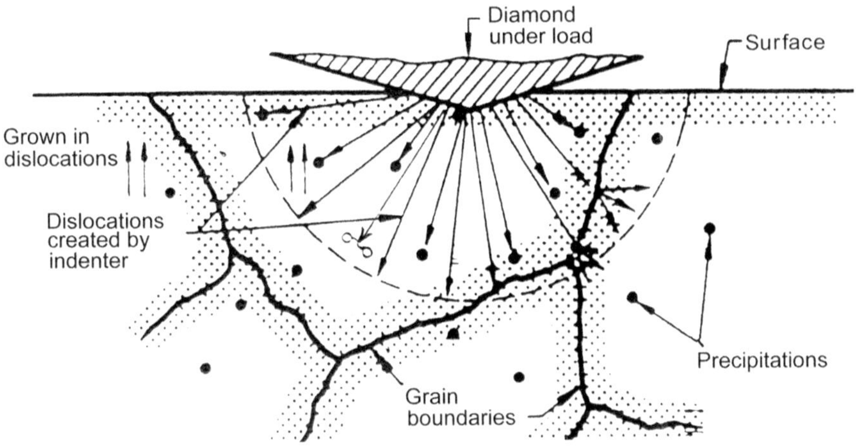

Fig. 15.59 Motion of dislocations created by an indenter

$$H_V = 1854P/d^2. \tag{15.46}$$

Here, P is the load in g and d the diagonal length of the impression in microns. The hardness calculated from the formula will be in units of kg mm^{-2}.

Figure 15.59 shows the creation of dislocations and their motion when the indenter presses on the crystal. The loaded indenter causes plastic deformation in the region of contact and creates dislocations. As the indentation continues, new dislocations are produced which push the dislocations produced earlier. These dislocations move into the crystal but their motion is hindered by several factors, some of which were mentioned in Sect. 15.6.5 and are shown in Fig. 15.59. Hardness is a measure of the resistance offered by the crystal to the motion of dislocations.

Let us consider a few examples of how the hindered motion of dislocations affects hardness. Since impurities cause pinning of dislocations, there is greater resistance to the motion of dislocations in an impure crystal than in a pure crystal. As an illustration, it may be pointed out that while the microhardness of pure NaCl is 26 kg/mm^2, that of a NaCl crystal containing Sr^{++} impurity is 35 kg/mm^2.

Since grain boundaries act as obstacles to dislocations, there should be an increase in hardness in the vicinity of a grain boundary. That such an increase actually takes place can be seen in Fig. 15.60. Also, if the dislocation density is high, a moving dislocation gets entangled in other dislocations resulting in enhanced hardness. Figure 15.61 shows Vickers indentations in different regions of an etched silver crystal. Diminution of the indentation (i.e. higher hardness) can be seen in the region where the number of etch pits is more.

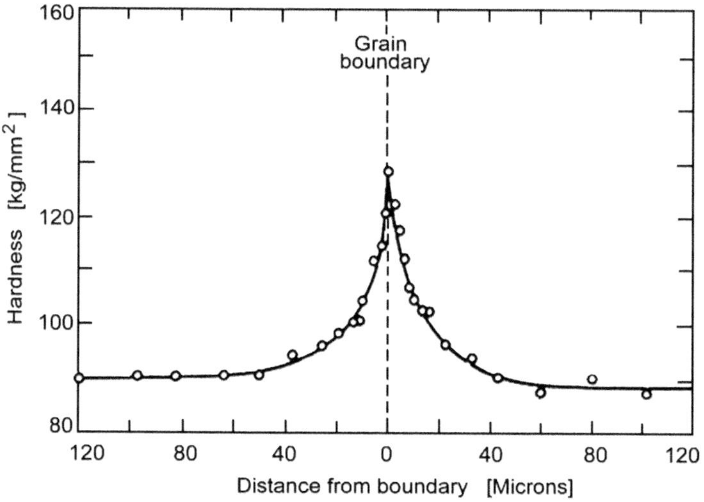

Fig. 15.60 Microindentation hardness profile in the vicinity of a grain boundary in iron crystal

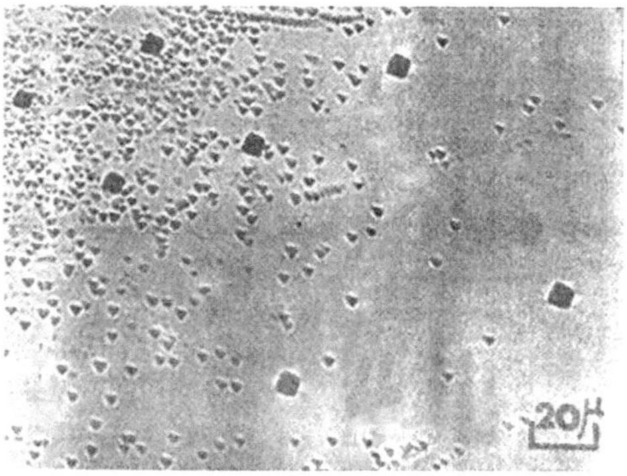

Fig. 15.61 Indentations in regions with different dislocation densities in a silver crystal

15.9.2 Thermal Properties

15.9.2.1 Thermal Expansion

A rigorous theoretical study of the effect of dislocations on thermal expansion of hcp metals was carried out by Nowick and Feder [15.25]. The difference between dilatometric and lattice expansion in the *a* and *c* directions is defined as

$$\Delta_a(T) = (\Delta L/L)_a - (\Delta a/a),$$
$$\Delta_c(T) = (\Delta L/L)_c - (\Delta c/c). \quad (15.47)$$

From a theoretical analysis, Nowick and Feder [15.25] arrived at the following results:

1. The ratio $(\Delta c/\Delta a)$ is dependent on dislocation climb
2. For diffusion-limited climb, $(\Delta c/\Delta a)$ is independent of temperature
3. For climb-rate limitation, $(\Delta c/\Delta a)$ is a function of temperature
4. $(\Delta c/\Delta a)$ is related to the distribution of dislocation and, hence, is sample dependent

Feder and Nowick [15.26] made accurate dilatometric and lattice expansion measurements on hcp metal Cd and found that $(\Delta c/\Delta a)$ is independent of temperature indicating diffusion-limited dislocation climb. They also found that $(\Delta c/\Delta a)$ had different values for different samples as predicted above.

15.9.2.2 Thermal Conductivity

Thermal conductivity is caused by the scattering of phonons. At low temperatures, the thermal conductivity decreases with temperature according to a T^3 law due to scattering of phonons by the lattice. The presence of isotopes contributes a T^{-1} term. Klemens [15.27] showed theoretically that dislocations participate in phonon scattering and this mechanism contributes a T^2 term to the temperature dependence of thermal conductivity. Sproull et al. [15.28] measured the thermal conductivity of deformed LiF crystals at low temperatures and found a T^2 dependence.

15.9.2.3 Melting

At some temperature, a solid melts into a liquid. At this point, the lattice structure collapses. There are several models of melting. According to the Lindemann model, melting takes place when the amplitude of the vibrating atoms reaches a certain critical value. Theoretically, this value is about 0.16 of the interatomic distance. But, experimentally, it is observed that melting takes place at a much lower amplitude, i.e. the experimental melting point is lower than the theoretical melting point.

It is now accepted that at high temperatures, dislocations are generated without affecting the free energy. This large number of dislocations causes the crystal structure to collapse at a much lower melting point. This qualitative explanation has been supported by a rigorous theory proposed by Kuhlman-Wilsdorf [15.29].

15.9.3 Electrical and Magnetic Properties

15.9.3.1 Electric Properties

It is observed that the electrical resistivity of metals and semiconductors are affected when they are deformed, i.e. when they contain dislocations. Thus, a copper sample with a dislocation density of $3 \times 10^{11}\ \text{cm}^{-2}$ undergoes a 2% change in electrical resistivity $(\Delta R/R)$. Electrical resistivity is caused by the scattering of the charge carriers by different scattering centres. Assuming that the edge dislocations participate in the scattering, Dexter [15.30] rigorously derived the expression

$$(\Delta R/R) = 2.46 \times 10^{-15} z^2 N, \tag{15.48}$$

where z is the effective atomic number (taken as unity) and N is the dislocation density. For a value of 2% for $(\Delta R/R)$, N turns out to be $8 \times 10^{12}\,\mathrm{cm}^{-2}$. In view of several assumptions in the theory, this estimate seems satisfactory.

15.9.3.2 Magnetic Properties

Magnetic properties of ferromagnetic materials are affected by the presence of plastic strain. The distortion of the lattice due to dislocations produces torques that tend to align the atomic magnets along certain directions. This results in a field dependence of the magnetisation J which now gets modified to

$$J \approx J_s - (a/H) - (b/H^2), \tag{15.49}$$

where J_s is the spontaneous magnetisation and H the field. By a rigorous theory, Brown [15.31] showed that the experimentally determined values of a and b can be accounted for with an assumed value of 2×10^{10}dislocations cm^{-2}.

15.9.4 *X-ray Diffraction*

15.9.4.1 X-ray Diffraction Line Width

If the crystal lattice is perfect, the width of an X-ray reflection should be ~5 s of arc. But, experimentally it is found that the line widths are several minutes of arc. This is because the crystal is made of 'mosaic blocks', a concept introduced by Darwin in 1914 much before the concept of dislocations. Each mosaic block is a perfect crystal but neighbouring blocks are misoriented. It is now agreed that the mosaic structure is only the three-dimensional network of dislocations.

Let us assume that L is the side of the crystal involved in the diffraction of the X-ray beam and let ρ be the dislocation density. Let λ be the edge of each mosaic block. If we associate one mosaic block with each dislocation line, $\lambda^2 = (1/\rho)$. The average angular misfit between the blocks is $\alpha = b/\lambda$ rad where b is the magnitude of the Burgers vector. As we cross the crystal, we cross L/λ blocks. The misfit can be positive or negative. Hence the probable misfit between the first and last block is $\alpha L/(\lambda)^{1/2}$. Identifying this with the observed width θ (in rad), we have

$$\theta = \alpha(L/\lambda)^{1/2},$$

or

$$\theta = bL^{1/2}\rho^{3/4}. \tag{15.50}$$

Assuming values of $L = 0.1\,\text{cm}$, $b = 3 \times 10^{-8}\,\text{cm}$ and $\rho = 10^{8}\,\text{cm}^{-2}$, we get $\theta = 10^{-2}$ rad and $\lambda = 10^{-4}\,\text{cm}$. These values agree with values obtained by other methods.

15.9.4.2 X-ray Diffraction Intensities

It has been pointed out in Chap. 4 that according to the dynamical theory, the intensity I of X-rays diffracted by a perfect crystal is given by

$$I \propto F, \tag{15.51}$$

where F is the structure factor. The corresponding expression for a mosaic crystal (crystal containing dislocations) given by the kinematical theory is

$$I \propto F^2. \tag{15.52}$$

It is clear that the intensity of an X-ray beam from an imperfect crystal is more than that from a perfect one. In fact, this difference is the basis of the X-ray topographic method of observing dislocations (Sect. 15.7.6).

15.9.5 Chemical Effects

Dislocations act as active centres for chemical reactions. Thus, it is observed that in Fe–Ni alloys heated in oxidising atmosphere, oxidation takes place preferentially at dislocation sites as revealed by formation of etch figures. Crystals containing water molecules, like alums, lose the water molecules on heating. This dehydration takes place at dislocation sites. Crystals like calcite which do not have a congruent melting point, decompose on heating. This decomposition is also initiated at dislocation sites. Sangwal [15.32] quotes several other examples of chemical reactions taking place at dislocation sites.

15.9.6 Crystal Growth

The role of screw dislocations in crystal growth was mentioned in the chapter on crystal growth. In view of its importance, we shall discuss this aspect again.

The screw dislocation emerging at a crystal surface provides steps continuously. These steps offer favourable sites for the atoms from the solution or vapour to attach

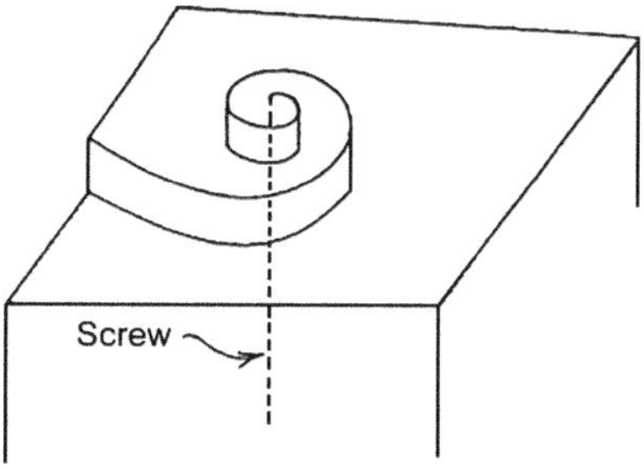

Fig. 15.62 Spiral ramp produced during crystal growth by an emergent screw dislocation

Fig. 15.63 Growth spiral on the surface of a paraffin crystal

themselves to the solid crystal. An individual atom experiences a stronger binding at a step than elsewhere on the surface since at a step, the crystal grows and the stepped surface advances giving the growing surface the appearance of a spiral ramp (Fig. 15.62). Such growth spirals on a paraffin crystal are shown in Fig. 15.63.

Problems

1. A tensile stress of 15 MPa applied on a silver single crystal is just sufficient to cause glide on the $\{1\bar{1}1\} < 0\bar{1}1 >$ system. Calculate the CRSS.
2. A crystal of size $1 \times 1 \times 1\,\text{cm}^3$ with $G = 4 \times 10^{11}\,\text{dyn/cm}^2$ and $\nu = 0.3$ contains dislocations with Burgers vector 3 Å, core radius 10 Å and crystal radius 1 cm. Calculate the strain energy per cm for (a) a screw dislocation and (b) an edge dislocation.
3. Estimate the Peierls critical stress for an edge dislocation gliding in a material with $\nu = 0.35$ and $G = 2.5 \times 10^{11}\,\text{dyn/cm}^2$.
4. Estimate the dislocation density in a crystal with Burgers vector 2.5 Å when it is bent in a curve of radius 4 cm.
5. Find the tilt angle of a grain boundary with Burgers vector 4 Å and a dislocation spacing of 1.3×10^{-4} cm.
6. Estimate the width of an X-ray reflection from a crystal of length 0.2 cm and Burgers vector 2.5 Å containing $10^8\,\text{cm}^{-2}$ dislocations.

References

15.1. A. Cottrell, *Dislocations and Plastic Flow in Crystals* (Oxford University Press, New York, 1953).
15.2. W.T. Read, Jr., *Dislocations in Crystals* (McGraw-Hill, New York, 1953).
15.3. J. Friedel, *Dislocations* (Pergamon Press, New York, 1964).
15.4. J. Weertman, J.A. Weertman, *Elementary Dislocation Theory* (Macmillan Co., New York, 1966).
15.5. I. Kovacs, L. Zsoldos, *Dislocations and Plastic Deformation* (Pergamon Press, New York, 1973).
15.6. J. Frenkel, Z. Phys. **37**, 572 (1926).
15.7. C. Kittel, *Introduction to Solid State Physics* (John Wiley, New York, 1996).
15.8. G.I. Taylor, Proc. R. Soc. (London) **A145**, 362 (1934).
15.9. E. Orowan, Z. Phys. **89**, 605 (1934).
15.10. M. Polanyi, Z. Phys. **89**, 660 (1934).
15.11. J.M. Burgers, Proc. Kon. Ned. Akad. Wet. **42**(293), 378 (1939).
15.12. C.M. Srivastava, C. Srinivasan, *Science of Engineering Materials* (Wiley Eastern, New Delhi, 1987).
15.13. R. Peierls, Proc. Phys. Soc. **52**, 34 (1940).
15.14. A.H. Cottrell, M.A. Jawson, Proc. R. Soc. (London) **A199**, 104 (1949).
15.15. F.C. Frank, W.T. Read, Phys. Rev. **79**, 722 (1950).
15.16. F.H. Horn, Philos. Mag. **43**, 1210 (1952).
15.17. J.M. Hedges, J.W. Mitchell, Philos. Mag. **44**, 223 (1953).
15.18. W.C. Dash, J. Appl. Phys. **27**, 1193 (1956).
15.19. M. Drechsler, G. Pankow, R. Vanselow, Z. Phys. Chem. **4**, 249 (1955).
15.20. W.L. Bond, J. Andrus, Phys. Rev. **101**, 1211 (1956).
15.21. P.B. Hirsch, R.W. Horne, M.J. Whelan, Philos. Mag. **1**, 677 (1956).
15.22. W. Bollman, Phys. Rev. **103**, 1588 (1956).
15.23. M.J. Whelan, P.B. Hirsch, R.W. Horne, W. Bollman, Proc. R. Soc. (London) **A240**, 524 (1957).
15.24. A.R. Lang, Acta Cryst. **12**, 249 (1959).
15.25. A.S. Nowick, R. Feder, Phys. Rev. **B5**, 1238 (1972).
15.26. R. Feder, A.S. Nowick, Phys. Rev. **B5**, 1244 (1972).
15.27. P.G. Klemen, Proc. R. Soc. (London) **A208**, 108 (1951).
15.28. R.L. Sproull, M. Moss, H. Weinstock, J. Appl. Phys. **30**, 334 (1959).
15.29. D. Kuhlman-Wilsdorf, Phys. Rev. **140**, A1599 (1965).
15.30. D.L. Dexter, Phys. Rev. **86**, 770 (1952).
15.31. W.F. Brown, Phys. Rev. **60**, 139 (1941).
15.32. K. Sangwal, *Etching of Crystals* (North-Holland, New York, 1987).

Chapter 16
Other Crystalline Forms

16.1 Introduction

Solids, liquids and gases are the three states of matter. Among solids, we have the crystalline solids and the amorphous solids. So far, we have considered the properties of crystalline solids. While Mohanty [16.1] calls liquid crystals the fourth state of matter, Dhar [16.2] suggests that even powders may be treated as a state of matter different from solids and fluids. Dhar [16.2] raised the broad issue of classifying matter in three states and pointed out the difficulties in making a clear-cut distinction. He suggested that any matter whose behaviour differs substantially from solids and fluids may be classified as a separate state of matter.

There are materials which are crystalline and yet their behaviour differs in some respect or the other from that of regular crystals. We shall avoid the use of the term "state" and instead we shall call them the "other crystalline forms". These materials are: (a) nanocrystals, (b) polycrystals, (c) thin films, (d) liquid crystals and (e) quasi crystals. The first three, while differing in properties from the parent crystal from which they are formed, retain the essential crystal structure. Liquid crystals show crystallinity which is different from that of the original crystal. Quasicrystals, on the other hand, flout the rules of crystallography.

The science of each of these materials has become an independent field. We provide only a brief introduction to these materials for the students of solid state physics.

16.2 Nanocrystals

16.2.1 General

"Nano" means 10^{-9} and 1 nm equals $10^{-3}\mu m$. Nanomaterials consist of small-sized grains having at least one dimension of nano size. The size, however, is not sharply

D.B. Sirdeshmukh et al., *Atomistic Properties of Solids*, Springer Series in Materials Science 147, DOI 10.1007/978-3-642-19971-4_16,

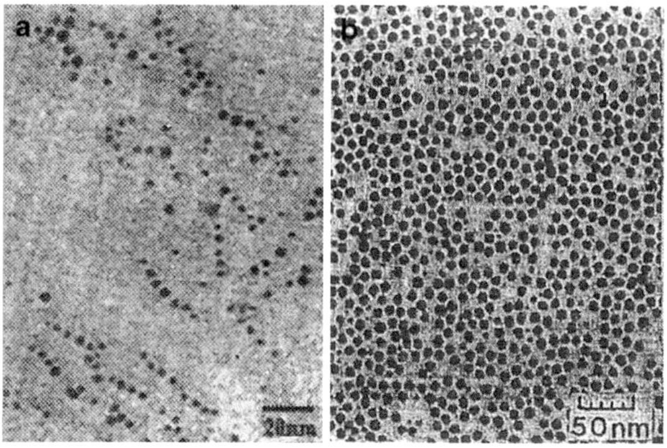

Fig. 16.1 Electron micrographs of some nanocrystals: (**a**) gold and (**b**) iron nitride

defined and, generally, particles with linear dimensions of upto 100 nm are termed nanoparticles.

Nanoparticles have been there in nature all the while, e.g. smoke particles, fine grains in minerals and nanoparticles in bacteria and micro-organisms. The DNA double helix has a diameter of ~2 nm. The first recorded laboratory synthesis of nanoparticles seems to have been carried out by Faraday [16.3] when he demonstrated colours of gold-sol. From the colour (ruby red), we can now estimate the particle size as sub-100 nm. However, a fresh impetus to scientists to look at possibilities in the nano-regime was provided by Feynman's [16.4] popular lecture entitled "There's plenty of room at the bottom".

Nanoscience has taken us by storm and today it has gone beyond the precincts of physics and chemistry into biology, medicine, engineering and information technology.

The optical microscope can observe particles as small as 1 μm, i.e. 10^3 nm. To observe nanoparticles, special instruments like scanning electron microscope, transmission electron microscope and scanning tunneling microscope are employed. Electron micrographs of some nanoparticles are shown in Fig. 16.1.

16.2.2 Synthesis of Nanocrystals

Nanocrystals can be synthesised by different methods. Some methods are discussed here.

16.2.2.1 Gas Evaporation Method

The inert gas evaporation method is particularly useful in the synthesis of single phase metals and oxides. The material is evaporated in a chamber (Fig. 16.2) in an

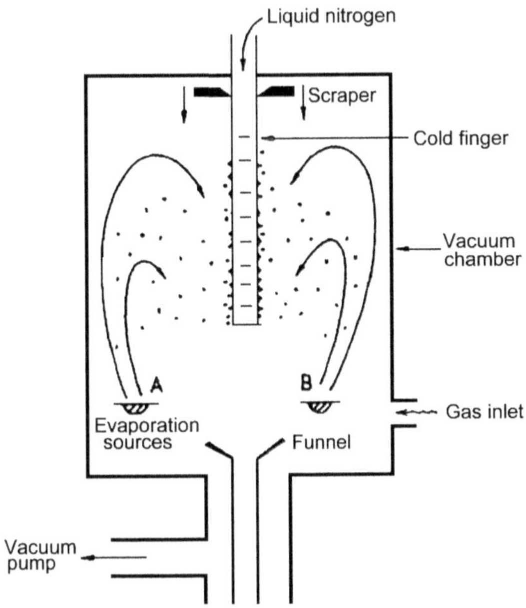

Fig. 16.2 Schematic of a gas evaporation unit

atmosphere of inert gas at a low pressure (<1 atm). The evaporated atoms collide with the inert gas atoms and form atom clusters. These clusters then condense on the outer surface of a cold finger. These clusters have to be prevented from coalescing. The condensed clusters are rapidly removed from the cold finger by a downward motion of a scraper and are collected in a receiver.

16.2.2.2 Ball-Milling Method

The starting powder is taken in a spherical steel container which can be rotated (Fig. 16.3). The container includes steel balls 1–2 cm in diameter. As the container rotates, the balls press against the powder and crush it into finer grains. It is possible to have an inert atmosphere inside the ball-mill. The method is generally used for production of nanoparticles in large quantities. There is a risk of the grains getting contaminated with impurities in the surfaces of the balls and/or the steel container.

16.2.2.3 Sputtering Method

In this method, an accelerated beam of inert gas atoms (Ar or He) is focussed on the material. In this process, atoms or clusters are ejected.

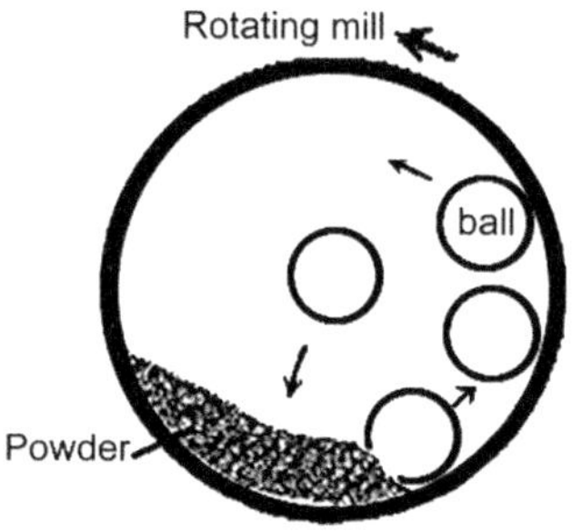

Fig. 16.3 Schematic of a ball-mill

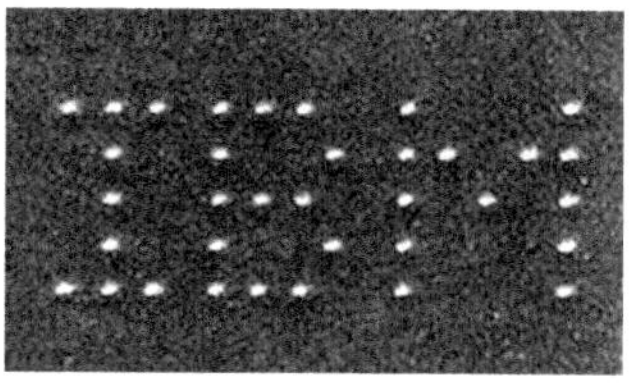

Fig. 16.4 IBM logo made of 35 Xenon atoms

16.2.2.4 Matrix Method

It is possible for clusters of nanoparticles to be formed by the entrapment of particles in interstices in host matrices of polymers, colloids, miscelles, glass and crystals. The resulting clusters have poorly defined surfaces and a broad size-distribution.

16.2.2.5 Self-Assembly Method

This is an example of a bottom-up method, i.e. creating assemblies or clusters of nanoparticles on an atom-by-atom basis. In a pioneering experiment, Eigler and Sweitzer [16.5] picked up Xenon atoms on a scanning transmission microscope tip and placed them at specific sites on a Ni (110) surface; they wrote the IBM logo (Fig. 16.4).

16.2.2.6 Precipitation from Solution

This is a reaction in aqueous or nonaqueous solutions. Once the solution is supersaturated with the solute, precipitation takes place around nuclei. The nuclei then grow by diffusion. For unagglomerated particles with a narrow size distribution, nuclei must form at the same time and growth should proceed without further nucleation.

16.2.2.7 Decomposition Method

The decomposition of iron pentacarbonyl [$Fe(CO)_5$] to yield iron particles is an example of this method. The decomposition is carried out in a boiling solvent like decalin. Thus, this is a case of thermal decomposition. It may be mentioned that decomposition can be carried out by ultrasound irradiation also.

16.2.2.8 Reduction Method

The most commonly employed strategy is chemical reduction. In many reactions, the metal atoms get released and, if the reaction is controlled, metal powder is formed; the grains are found to be 1–5 nm in size. An example of such reduction is

$$WCl_4 + 4NaBEt_3H \rightarrow W + 4NaCl + 4BEt_3 + 2H_2. \quad (16.1)$$

Several other examples are given by Gonsalves et al. [16.6].

Flow chart of synthesis of M50 steel

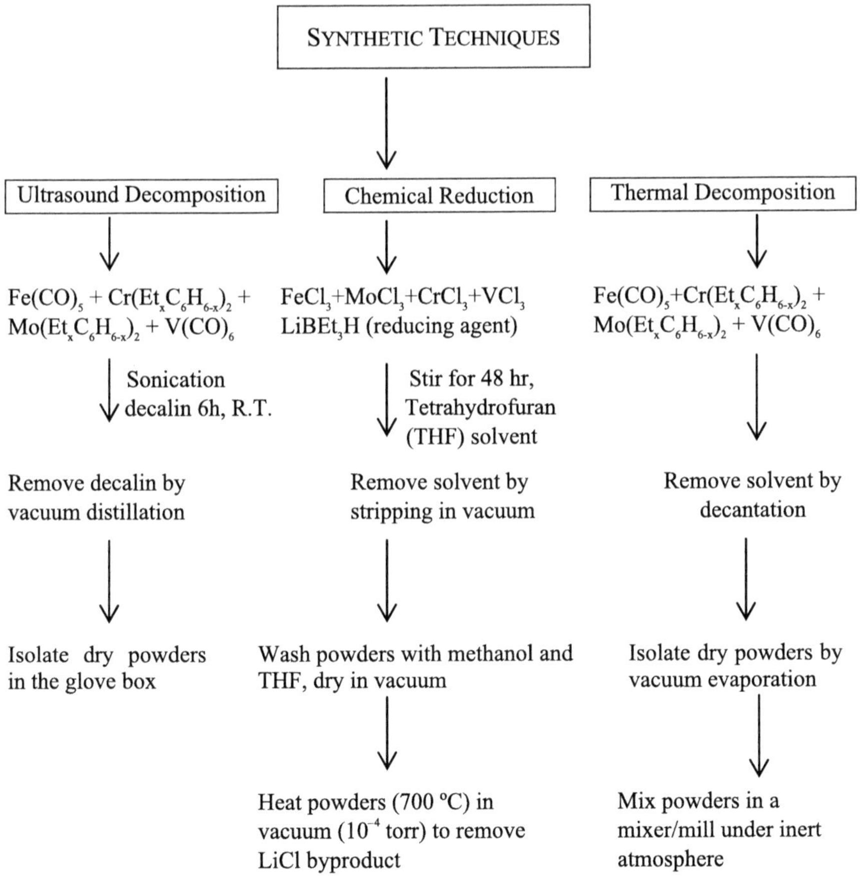

A given material can be obtained in the nanoform by choosing one or the other method. As an example the synthesis of nanopowders of M50 steel is shown in the flow chart. M50 steel is an important material with applications in aircraft industry. Its composition is 4% Cr, 4.5% Mo, 1.0% V and balance of Fe.

16.2.2.9 Hydrolysis

Hydrolysis is a reaction in which a water molecule breaks to release the desired end product. Some examples are given below:

$$2M^{3+}(H_2O)_6 \rightarrow M_2O_3 + 6H^+ + 9H_2O, \tag{16.2}$$

$$2[Ga(OH)_4] \rightarrow Ga_2O_3 + 3H_2O + 2(OH)^-. \tag{16.3}$$

It may be noted that the first reaction is basic whereas the second is acidic. The reaction

$$La_2O_3 + 6H^+ + 9H_2O \Leftrightarrow 2La^{3+}(H_2O)_6 \tag{16.4}$$

can proceed to the left only at 1,000 K.

16.2.2.10 Sol–Gel Method

Silica gel is formed in the following reaction:

$$\begin{array}{ll} Si(MeO)_4 + 4H_2O \rightarrow & Si(OH)_4 + 4MeOH \\ \text{(Tetramethoxysilane)} & \big\downarrow -2H_2O \\ & SiO_2 \text{ gel} \end{array} \tag{16.5}$$

The combination of a gel and nanoparticulate matter in a solution of colloidal form is called sol–gel. The solid particles are dispersed in nanosized pores in the gel. An example of nanoparticle formation in silica gel is

$$Si(OR)_4 + \text{Ru metal salt} \rightarrow Ru/SiO_2\,\text{gel} + \text{soluble products}. \tag{16.6}$$

Here R is isopropyl or isobutyl. The particles which are of 15-nm size are retrieved by calcination of the gel and subsequent thermal reduction under hydrogen. Some examples of metal nanoparticles synthesised by the sol–gel method are Cu, Ag, Co, Ni, Fe, Mo, Ge, Pt and Os. A number of other gels are also in use. The formation of nanocrystals of a large number of solids by the sol–gel method in different gels is discussed by Kwiatkowski and Lukehart [16.7].

16.2.2.11 Compaction of Nanoparticles

An important stage in nanoparticle synthesis is their final compaction into a disc. Compaction of nanoparticles offers difficulties as the size of the grains is affected by the compaction process. Methods of compaction of nanoparticles will be discussed in Sect. 16.3.

16.2.3 Properties of Nanocrystals

Most physical properties of nanocrystals show a size dependence. Thus, Nalwa [16.8] pointed out that 6 nm grains of Cu are five times harder than Cu in bulk. Similarly, the yield point of metals increases with reducing grain size.

Alivisatos [16.9] studied the variation of melting point of CdS with the grain size. This is shown in Fig. 16.5. He found that the melting point decreases substantially as the grain size decreases from bulk to 15 Å. Crystals with wurtzite structure transform to the rocksalt structure at high pressure. Alivisatos (Fig. 16.6) found that this transition pressure in CdSe increases from 3.6 GPa at a grain size of 21 Å to 4.9 GPa at 10 Å. Bhoraskar [16.10] observes that there is a systematic shift with particle size in the electroluminescence spectra of nanocrystals of ZnS/porous silicon junctions (Fig. 16.7). Murray [16.11] observed a systematic blue shift in the absorption spectrum of CdS nanocrystals (Fig. 16.8).

Third harmonic generation has been observed in CdS-doped polyvinyl alcohol. The nonlinear susceptibility responsible for this effect decreased from 3.3×10^{-10} esu for a particle size of 30 Å to 2.5×10^{-11} esu for a particle size of 15 Å.

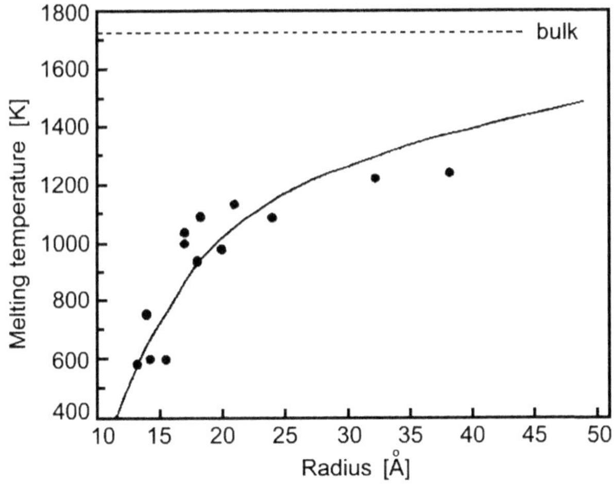

Fig. 16.5 Variation of melting point of CdS nanocrystals with size

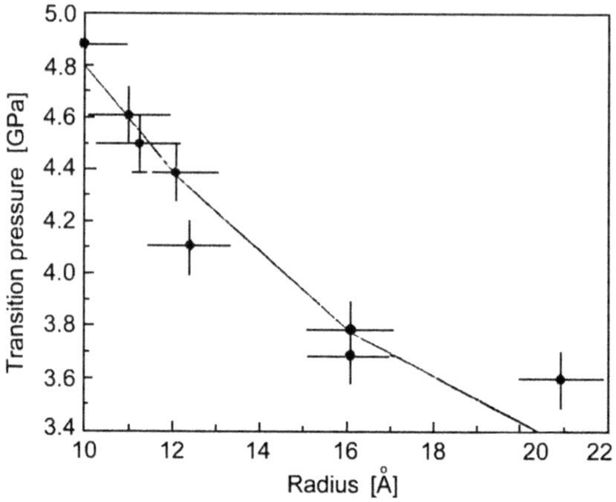

Fig. 16.6 Variation of transition pressure of CdSe nanocrystals with size

Fig. 16.7 Electroluminiscence spectra of ZnS for nanocrystals of different sizes: (i) 50 Å, (ii) 100 Å, (iii) 150 Å and (iv) 200 Å

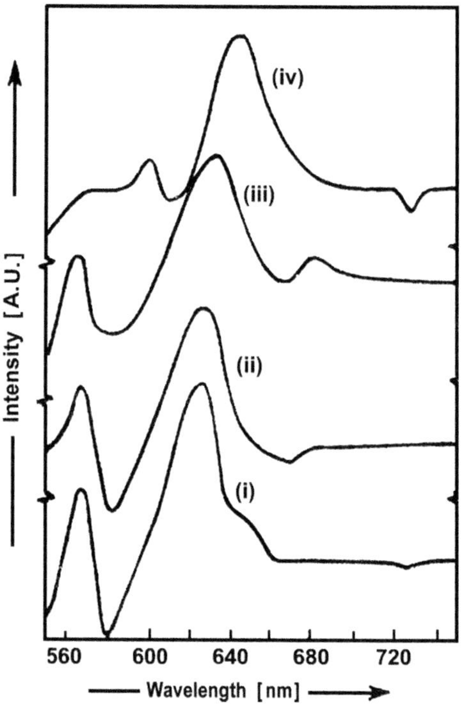

Thus, physical properties of nanocrystals show a strong dependence on the size of the grains. This size dependence makes it possible to have tailor-made nanomaterials.

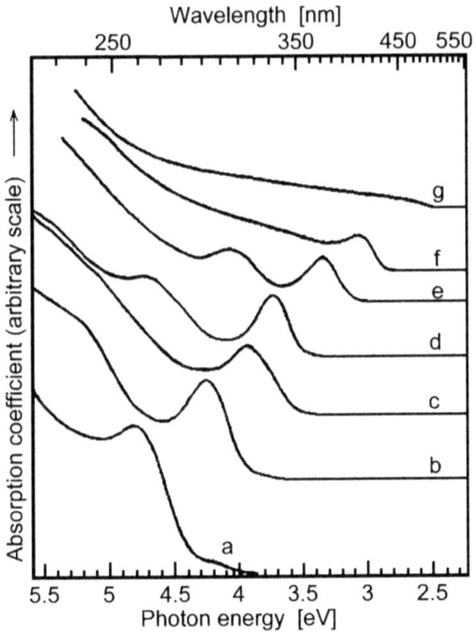

Fig. 16.8 Electronic spectra of CdS nanocrystals of different sizes: (**a**) 0.64 nm, (**b**) 0.72 nm, (**c**) 0.8 nm, (**d**) 0.93 nm, (**e**) 1.94 nm, (**f**) 2.8 nm and (**g**) 4.8 nm

16.2.4 Carbon Nanosystems

16.2.4.1 Fullerene

Fullerene is a molecule of carbon atoms. The idea of a large polyatomic carbon molecule with a closed structure originated in 1966. During the period 1966–1984, it remained a theoretical concept without any definite experimental evidence.

Experimental evidence of the existence of fullerene was obtained in 1985 by researchers at the Rice University led by Kroto (Box 16.1). In their experiments [16.12], a laser beam was directed on a graphite surface. This resulted in the release of carbon clusters. Mass-spectroscopic study revealed the presence of clusters of different masses but the most prominent among them had a mass of 720 units, i.e. C_{60}. This is called fullerene. The presence of the C_{60} molecule has been observed in soot particles and in cosmic dust. It is now possible to prepare fullerene in the laboratory by passing a large current between graphite electrodes in an inert atmosphere. Crystallised fullerene is obtained by the condensation of sublimated arc-produced graphite deposits. Millimetre-sized crystals of fullerene have been obtained by slow evaporation of saturated solution of fullerene in benzene. Crystallised fullerene is called fullerite. It has an fcc structure. Several other large carbon molecules like C_{30}, C_{70}, C_{80}, etc. have been observed. These molecules with the general formula C_n are called by the generic name "fullerenes".

Box 16.1

Sir Harold Walter Kroto (1939–), a British chemist, working at the Rice University discovered Fullerene in 1985 in collaboration with R.F. Curl Jr. and R.E. Smalley. Kroto, Curl and Smalley shared the Nobel Prize in chemistry in 1996.

H.W. Kroto

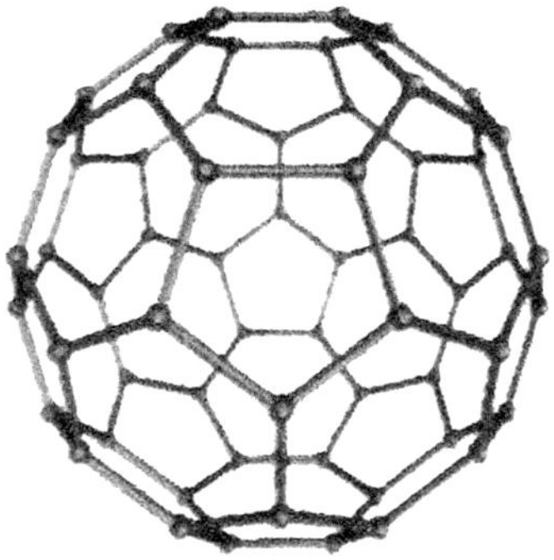

Fig. 16.9 Structure of fullerene

Investigations using several techniques, mostly spectroscopic, have shown that fullerene has a closed spheroidal soccer ball structure. It is a truncated icosahedron (Fig. 16.9) formed with 20 hexagons and 12 pentagons with a carbon atom at the vertices of each polygon and a bond along each polygon edge.

Fullerenes have many possible applications, e.g. in batteries and in medicine. Fullerenes, particularly those with intercalated alkali ions, show superconductivity at temperatures in the range 18–38 K.

16.2.4.2 Graphene

Graphene is the single-atom layer of graphite. As we know, this is a hexagonal honeycomb 2-D crystal lattice of covalently bonded carbon atoms. For quite some time such a single layer was considered thermodynamically unstable. But in 2004, researchers led by Geim (Box 16.2) at the Manchester University succeeded in isolating graphene [16.13].

The experimental method of obtaining graphene consists in sticking a scotch tape on to a graphite surface and pulling it off with a jerk. This results in a one-atom thick plane sheet which can be transferred to a substrate like a silicon wafer; the scotch tape is dissolved away. Some of the flakes produced in this way are not pure graphene but "few-layer graphene".

Box 16.2

A.K. Geim (1958–), a Russian by birth, working at the Manchester University was awarded the 2010 Nobel Prize in physics for the discovery of graphene along with K. Novoselov.

A.K. Geim

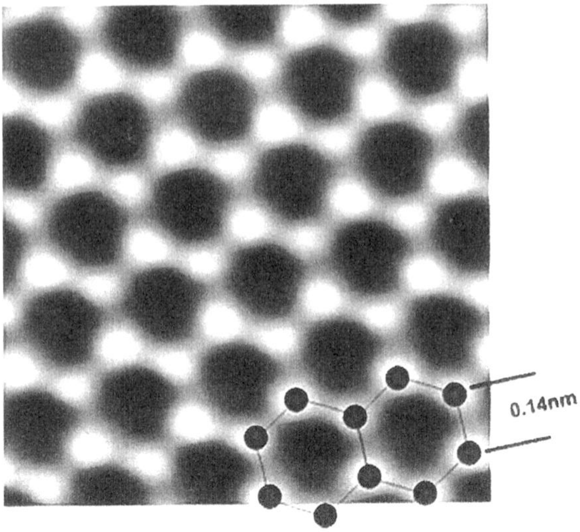

Fig. 16.10 Structure of graphene

The structure of graphene is shown in Fig. 16.10. As mentioned earlier, it consists of carbon atoms in hexagonal honeycomb arrangement. The C–C distance is ∼0.142 nm.

The thermal conductivity of graphene is $\sim$5,000 W m^{-1} K^{-1} compared to $\sim$1,000 W m^{-1} K^{-1} for the basal plane of graphite. It is one of the strongest materials. Its breaking strength is 200 times that of steel. Its Young's modulus is ∼0.5 TPa compared to ∼1.8 TPa for graphite.

16.2.4.3 Carbon Nanotubes

Carbon nanotubes (CNT) were discovered by Iijima [16.14] accidentally while examining surfaces of graphite electrodes in electric discharge. The soot at the end

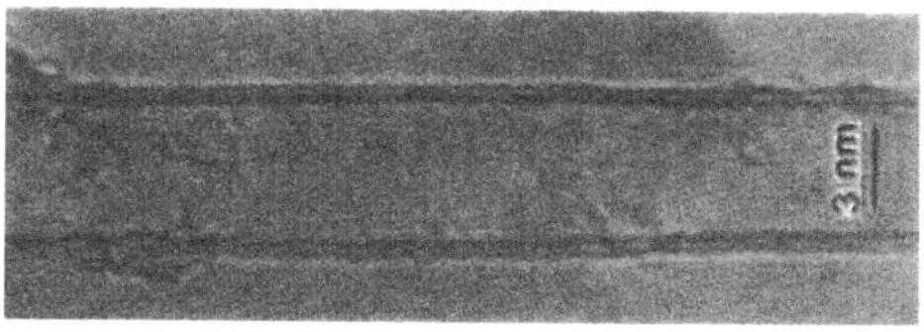

Fig. 16.11 Carbon nanotube

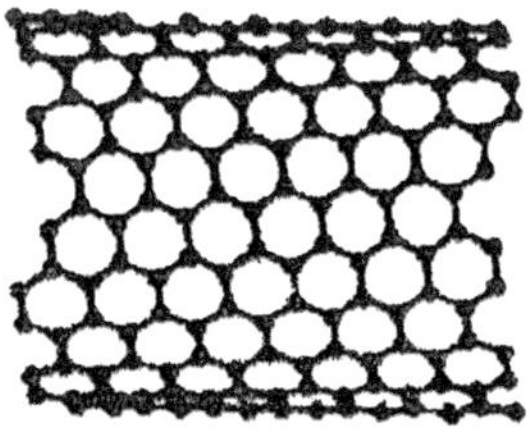

Fig. 16.12 Helical structure of carbon nanotube

of the electrodes consisted of carbon nanoparticles and CNTs in the form of needles. Typical working conditions for producing CNTs from electric discharge are: 20 V, 50–100 mA and a pressure of 500 Torr. CNTs can also be obtained by decomposition of hydrocarbon gases. The synthesis of CNTs is arduous. After a long process, one gets a minute quantity. The 2002 market price of CNT was $400 per gram.

The first transmission electron micrograph of a CNT obtained by Iijima [16.14] is shown in Fig. 16.11. CNTs can be single-walled or multi-walled. A structural diagram is shown in Fig. 16.12.

CNTs have a unique structure in which carbon hexagons are arranged on the surface of the tube in a honeycomb lattice. A single-walled CNT consists of single graphene (a single-atom graphitic layer) sheets with diameter 1–2 nm. Multi-walled tubes have thicker walls consisting of several graphene layers separated by 0.34 nm. The outer diameters of multi-walled tubes range between 2 and 25 nm and the inner diameter 1–8 nm. Electron diffraction patterns indicate that the top and bottom parts of each nanotube are sheared with respect to the tube axis. Thus, the hexagonal arrays have a helicity around the circumference. This can be seen in the figure.

Another important structural feature is that the tubes are closed at both ends. The arrangement of carbon atoms at the ends is pentagonal, hexagonal, heptagonal and a combination of 5–7 polygons.

It is known that the C–C covalent bond is the strongest bond. The CNT has a structure based on an arrangement of these bonds oriented along the tube axis. It is expected that such a structure would result in a strong material. The CNTs are indeed the strongest fibrous material that can be made with graphitic constituents. Experimental measurement of properties of CNTs is difficult but some theoretical estimates are available. Theoretically estimated value of the Young's modulus of single-walled CNT is 1–5 TPa. This is to be compared with a value of 1 TPa for graphite. An experiment by Dillon et al. [16.15] deserves mention for its ingenuity. Some CNTs got stuck in the holes of the specimen grid of a TEM. Treating the ends of such tubes projecting out of the holes as cantilevers, their vibration amplitude was estimated from the blurring of the images of the tube ends. From this, the Young's modulus could be estimated; it turned out to be 1.8 TPa. Atomic force microscope studies of the bending of CNTs attached to a substrate gave a similar value. Buckling of CNTs without breaking has been observed.

Hollowness makes the CNT material light. Density of single-walled CNT is about 0.8 $g\,cm^{-3}$ and of multi-walled CNT $\sim$1.88 $g\,cm^{-3}$ compared to 2.268 for graphite. Specific strength (strength/density) is two orders greater than that of steel. Due to this high strength and elasticity, CNTs offer themselves as nanoprobes, e.g. as tips of scanning probe microscopes.

Equally, if not more, exciting are the properties based on the electronic structure. This aspect is beyond our scope but we refer readers to books by Nalwa [16.8] and Rao et al. [16.16].

16.3 Polycrystals

16.3.1 General

A polycrystal (or a polycrystalline aggregate) is a solid sample made up of a large number of crystallites oriented at random. As shown in Fig. 16.13, within each crystallite, the atomic arrangement is the same as in a large single crystal of the substance but adjacent crystallites are differently oriented. One may be reminded of the "mosaic" structure discussed in Chap. 4. But there are differences. Firstly, the size of a crystallite is larger than that of a mosaic block by orders of magnitude. Also, the angular misorientation between mosaic blocks is very small; in a polycrystal, the crystallites assume all possible orientations. Another important feature of a polycrystal is that the crystallites are not loosely packed. Instead, they are tightly and homogeneously packed so that they form a continuous solid body in the form of a pellet or disc or brickette.

In spite of the tremendous advances in crystal growth, there are still situations, where we can learn about a crystal only through a study on its polycrystal. Such a situation arises, for instance, in the case of materials which are not amenable to crystal growth. As examples, we may mention the alkaline earth sulphides, cadmium oxide, boron nitride, the ferrites and the PZT materials. Most of the high temperature superconductors have been studied only as polycrystals. Most pharmaceuticals

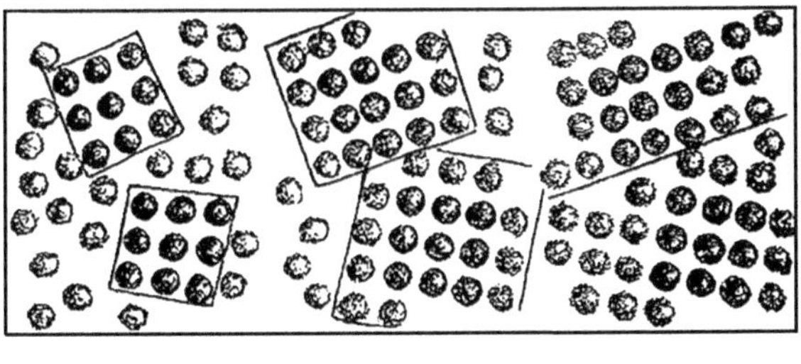

Fig. 16.13 Misoriented crystallites in a polycrystal

are crystalline materials converted for use as polycrystalline tablets. Another class of materials are the nanocrystals which, for applications, are compacted as polycrystalline discs.

16.3.2 Fabrication of Polycrystalline Aggregates

A simple set-up for the preparation of a polycrystalline aggregate consists of a cylindrical die (Fig. 16.14) placed in a hydraulic press (Fig. 16.15). The substance is taken in the form of a fine powder. The die is generally made of mild steel. The pressures applied are in the range 3,000–5,000 psi. Polycrystalline discs obtained with this arrangement are shown in Fig. 16.16.

The procedure described above may be termed room-temperature pressing. For hard and high melting-point materials, hot pressing is desirable. For this, three changes are made. Firstly, the die is surrounded by a heater to give high

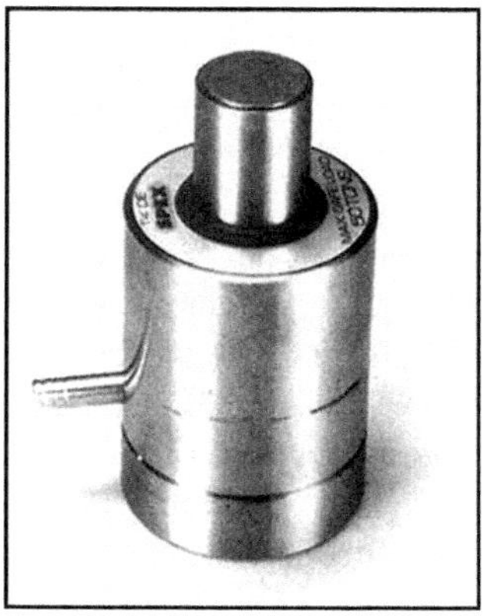

Fig. 16.14 A typical die

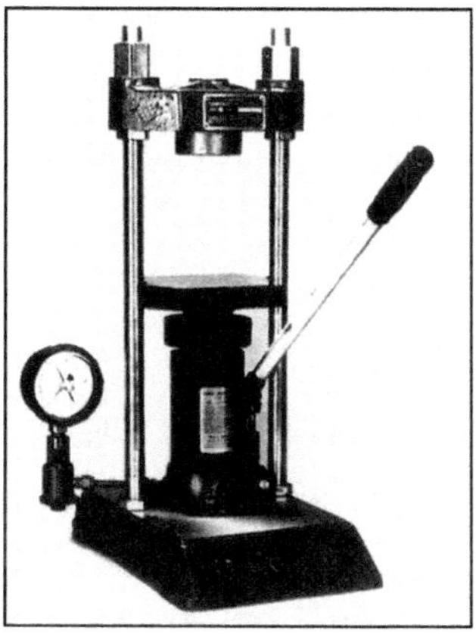

Fig. 16.15 Die in position in a hydraulic press

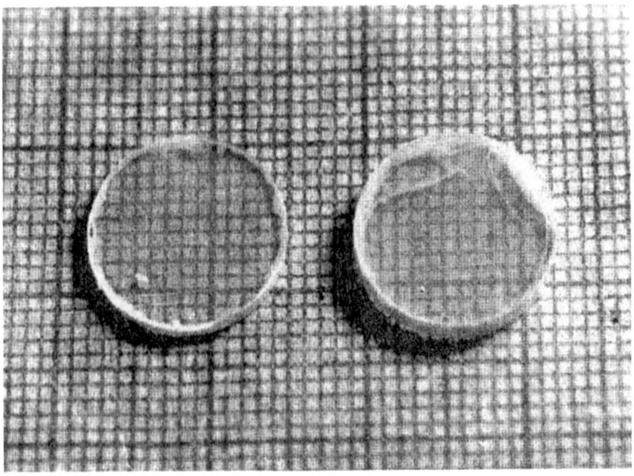

Fig. 16.16 Polycrystalline aggregates of KCl

temperatures. Secondly, the mild-steel die is replaced by graphite die. Thirdly, the press should be able to provide higher pressures in the range 30,000–50,000 psi.

In general, the density (ρ_m) of the polycrystal will be less than the density (ρ_0) of the single crystal of the parent material. ρ_m values are generally in the range (0.5–0.8) ρ_0. By using the hot-pressing technique, the Eastern Kodak Company succeeded in making nearly fully dense flats of materials like MgF_2, ZnS, CaF_2, ZnSe, MgO and CdTe. These materials are useful in infrared transmission and are called "Irtran" materials.

16.3.2.1 Compaction of Nanomaterials

The microstructure of two samples of MgO hot pressed at different temperatures is shown in Fig. 16.17. It was pointed out by Nielson and Leopald [16.17] that both the grain size and the pores between grains increase with the temperature of the pressing. This is a serious disadvantage in compaction of nanomaterials as it is essential to maintain nano grain size in compaction. For this purpose, low temperature pressing is adopted by enclosing the die in a cryogenic container [16.18]. Temperatures as low as liquid nitrogen temperature are employed. Such an arrangement is shown in Fig. 16.18.

Another problem in nanomaterial compaction is that the material is generally available in small quantity. In such a case, a suitable technique is the diamond-anvil (Fig. 16.19). The material is contained in the space between the anvils and is enclosed by a gold gasket with a hole. Pressure is applied by a plate below the cell, which is worked with a lever arrangement. Because of the small cross-section of the anvil ends, a high pressure is generated. The diamond-anvil cell, by its very design, is suitable to produce polycrystalline discs of small size (0.2 mm thick, 0.2 mm diameter).

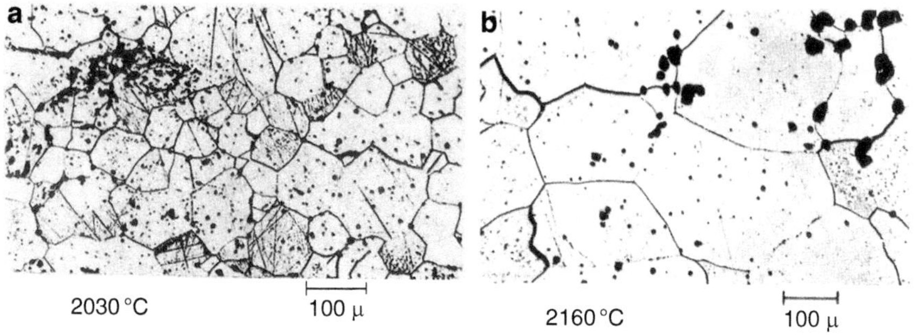

Fig. 16.17 Microstructure of polycrystalline discs of MgO pressed at (**a**) 2,030°C, (**b**) 2,160°C

Fig. 16.18 Set-up for cryogenic pressing: A moving piston, B die, C cryogenic container, D sample powder

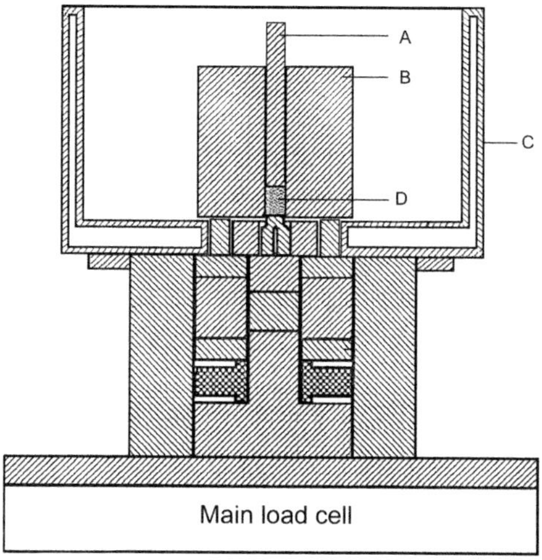

Fig. 16.19 Diamond anvil cell

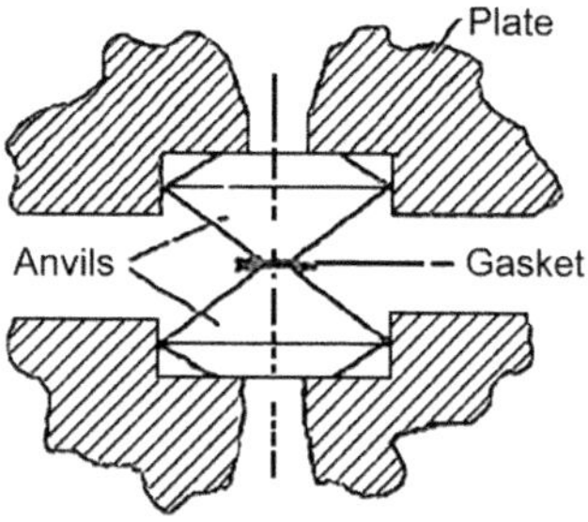

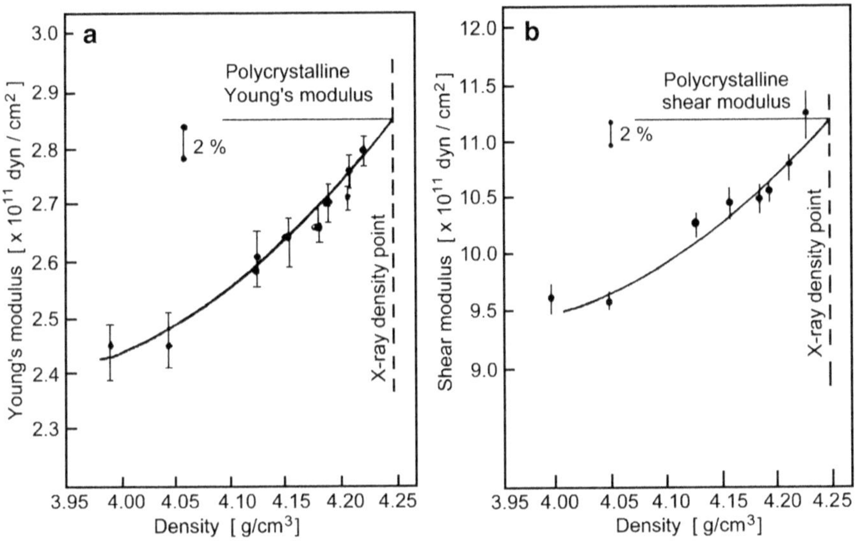

Fig. 16.20 Variation of (**a**) Young's modulus, (**b**) shear modulus of rutile (TiO_2) with density of aggregate

16.3.3 *Trends in Physical Properties*

As mentioned earlier, the measured density ρ_m of a polycrystalline aggregate differs from the true density ρ_0. The ratio $(\rho_m/\rho_0) = \delta$ is called the effective density or the packing faction. The term (1–δ) is called the porosity.

The physical properties of polycrystals show two prominent trends. Firstly, the properties show a dependence on the packing fraction, i.e. the measured property is not exactly the true property. Secondly, the property measured on a polycrystalline aggregate is always isotropic, irrespective of the anisotropic nature of the property of the parent single crystal. We shall now discuss these two features in detail.

16.3.3.1 Dependence of Properties on the Packing Fraction

As mentioned, the property P_m measured on the polycrystalline sample is not equal to the true property P_0 of the single crystal. P_m depends on the packing fraction δ of the sample. Most properties vary smoothly with δ. As an example, the variation with density of the Young's modulus and shear modulus of TiO_2 measured by Chung and Buessem [16.19] is shown in Fig. 16.20. The variation is smooth and slightly nonlinear. Extrapolation of the curves to $\rho_m = \rho_0$ gives the true value of the properties. Such an extrapolation was possible in this case because measurements were made on samples with values of ρ_m close to ρ_0, i.e. for values of δ close to

Table 16.1 Relation between measured property P_m, true property P_0 and packing fraction δ

Property (*P*)	Relation	Reference
Bulk modulus (*K*)	$\frac{1}{k_m} = \left(\frac{1}{k_0\delta}\right) + \frac{3(1-\delta)}{4G_0\delta}$	[16.20]
Shear modulus (*G*)	$\frac{G_0 - G_m}{G_0} = \frac{3(1-\delta)(3k_0 + 4G_0)}{(9K_0 + 8G_0)}$	[16.20]
Dielectric constant (ε)	$\varepsilon_0 = \left[1 + \frac{(\varepsilon_m^{1/3} - 1)}{\delta}\right]^3$	[16.21]
Electrical conductivity (σ)	$\sigma_0 = \sigma_m(3-\delta)/2\delta$	[16.22]
Hardness (*H*)	$H_0 = H_m/[1 - \exp(-c\delta)]$ where *c* is a constant	[16.23]
Hall coefficient (*R*)	$R_0 = R_m\,[4\delta/(3+\delta)]$	[16.24]
Thermal conductivity (κ)	$\kappa_0 = \kappa_m(3-\delta)/2$	[16.25]

unity. This is not always possible. Very often measurements are confined to δ values much less than unity. In such cases, an analytical approach is adopted.

The analytical approach is to establish a relation between the measured property P_m, the true property P_0 and the packing fraction δ. Let us denote the relation by

$$P_0 = f(P_m, \delta) \tag{16.7}$$

where f is a function of P_m and δ for each property. Such functions have been worked out for several properties; they are listed in Table 16.1. The advantage in using these relations is that one can predict the true property even if measurements are made on a single polycrystalline sample and that too with δ far less than 1. The predicted values are often accurate to within a few percent.

16.3.3.2 Isotropy of Physical Properties

Because of the random orientation of the crystallites in a polycrystalline aggregate, all measured properties become isotropic. The isotropic property of the polycrystal is related to the anisotropic properties of the single crystal. The isotropic property can be expressed in terms of the single crystal properties by making two different assumptions. The first assumption by Voigt [16.26] is that the strain throughout the polycrystal is continuous and the second by Reuss [16.27] is that the stress is continuous. These assumptions have been discussed in Chap. 8. They lead to two different expressions and different values (P_V and P_R) for a given property. These being extreme assumptions, the experimental values of the isotropic property measured directly on a polycrystal lie between these two bounds. A suitable average of the two values P_V, P_R leads to a value which is closer to the experimental value.

The application of thses ideas to elastic properties has been discussed in Chap.8. We shall only note that the expressions for the Voigt [16.26] and Reuss [16.27] isotropic values of the various elastic properties in terms of the single crystal elastic constants are quite complicated, particularly for the lower symmetry crystal systems. Chung [16.28] derived expressions for the isotropic values of pressure derivatives of elastic moduli from the pressure derivatives of single crystal elastic constants. Another property for which polycrystalline averages have been worked out is photoelasticity. Ranganath and Ramaseshan [16.29] derived the necessary expressions. The Voigt and Reuss values for elastic moduli show large quantitative differences. But in the case of the pressure derivatives of elastic moduli and also the photoelastic constants, the differences in the two averages are just not quantitative. It is observed that there is even a difference in the signs of the values in some cases like the rubidium halides.

Harshin [16.30] obtained the following expression for the thermal expansion coefficient α^* of a polycrystalline aggregate in terms of the principal expansion coefficients α_1, α_3 of tetragonal, trigonal and hexagonal crystals:

$$\frac{\alpha^* - \alpha_1}{\alpha_3 - \alpha_1} = \frac{S^* - S_1}{S_3 - S_1}. \tag{16.8}$$

Here S_1, S_3 are the elastic compliances in the two principal directions and S^* is the effective isotropic compliance for the aggregate.

16.4 Thin Films

16.4.1 General

A film is defined [16.31] as solid material contained between two parallel plane surfaces extended infinitely in two directions (x, y) and restricted along the third direction (z). The dimension along z, usually denoted by t or d, can range from $d \to 0$ to $d \sim 10\,\mu\text{m}$. Depending on the value of d, films are categorised as ultra thin ($d < 100$ Å), thin (100 Å $< d <$ 1,000 Å) and thick ($d >$ 1,000 Å to $d \sim 10\,\mu$m).

The use of solid state materials in the form of thin films is preferred in device technology because of several advantages, viz. they are compact, light, easy to fabricate and less expensive. On the other hand, thin films have some disadvantages, viz. they are highly sensitive to conditions of fabrication and to the presence of defects like impurities, holes and discontinuities. Further, the properties of thin films differ from those of the parent material in bulk; also, the properties are thickness dependent.

Thin film science and technology is a vast field. There are not only a number of books on different aspects of thin films but also book series on advances in the field. We shall briefly discuss the methods of fabrication of thin films, their characterisation and a few of their properties.

16.4.2 *Fabrication of Films*

16.4.2.1 Sputtering

This is the oldest method. In 1852, Grove [16.32] observed that the sputtering of a cathode in a discharge tube by high energy positive ions results in formation of thin films on objects and surfaces around the cathode. The cathode material is generally metallic but alloys, semiconductors and insulators can also be used. The impinging ions are of noble gases: He, Ar, Ne, Kr. The pressure in the discharge tube is of the order of 0.1 Torr. The mechanism of sputtering is the transfer of momentum of the high energy ions to the cathode atoms. The DC voltage applied is in the range 200–800 V. Often, a cathode shield is kept around the cathode at ground potential, quite close to the cathode. It has an opening through which the sputtered ions can escape and deposit themselves on a substrate.

Films formed by sputtering are susceptible to impurities in the cathode material as well as those in the sputtering gas. Thus, Ni films formed by sputtering of Ni cathode by Ar ions often show presence of nickel nitride rather than nickel itself [16.33]; it is known that commercial Ar gas contains N_2 as impurity. This effect vanishes when spectroscopically pure Ar gas is used.

16.4.2.2 Thermal Deposition in Vacuum

This is the most popular method. The metal is evaporated or sublimated by heating it in a chamber held at a vacuum of 10^{-5} Torr. The resulting vapours are allowed to deposit on a substrate. The metal is heated in a boat or a V-shaped element. The volume V of the vapour that strikes unit area of the substrate in unit time is

$$V = 3638(T_g/M)^{1/2}. \tag{16.9}$$

Here, T_g is the temperature of the gaseous species and M its molecular weight. The constant 3638 follows from the kinetic theory [16.31]. It is assumed that the vapour takes the shortest path from source to substrate. A higher substrate temperature improves the quality of the film. Examples of conditions under which thin films are obtained by this method are given in Table 16.2.

Table 16.2 Growth conditions of some films

Material	Melting point [°C]	Heating filament	Boat material
Ag	961	Ta, W	Ta, Mo
Au	1,063	W, Mo	Al_2O_3, graphite
CdS	1,750 at 100 atm	W, Pt	Mo, Al_2O_3
MgF_2	1,263	W, Mo	Graphite
Ta_2O_5	1,470	Ta, W	Ta, W
ZnS	1,830 at 150 atm	Mo, Ta	Mo, Ta

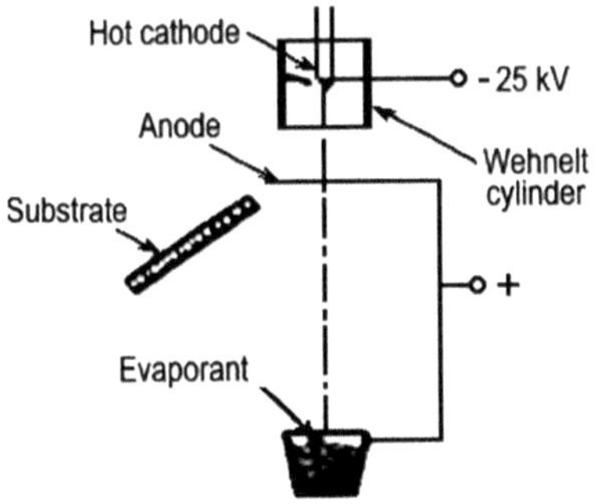

Fig. 16.21 Set-up for electron beam evaporation

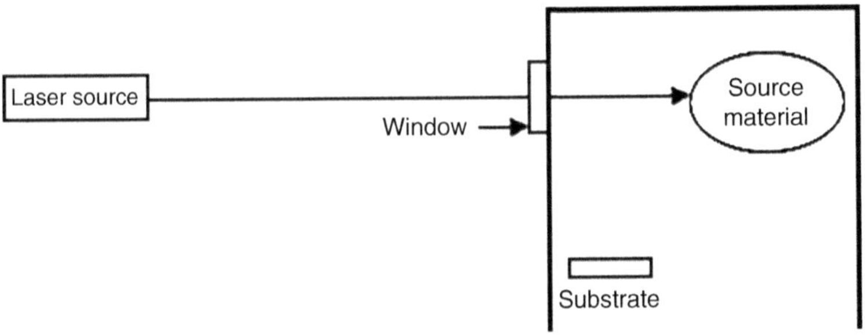

Fig. 16.22 Set-up for evaporation with a laser

Heating of the metal can also be carried out by using an electron beam. The electron beam from a hot cathode (Fig. 16.21) is directed on the metal charge and the evaporated atoms deposit on the substrate. Yet another method is to direct a laser beam into the chamber (Fig. 16.22).

16.4.2.3 Electrolytic Cathodic Deposition

A thin film can be obtained by electrolysis of a solution containing a salt of the metal. The metal ions in the electrolyte get deposited on the cathode surface. The rate of deposition is governed by Faraday's law:

$$W = Zit\alpha, \tag{16.10}$$

where W is the weight of the deposited metal, Z the electrochemical equivalent, i the current, t the time and α the cathode efficiency.

The value of α depends on factors like the temperature of the bath, the pH of the solution, the presence of other constituents in the bath like brighteners, complexing agents, etc. Assuming that the salt is divalent, the reaction taking place during electrolysis may be expressed as

$$M^{++} + 2e^{-} \rightarrow M \quad \text{at the cathode,} \tag{16.11}$$

where M is the divalent metal.

16.4.2.4 Electrolytic Anodic Deposition

By choosing a proper electrolyte and a metal as anode, oxide thin films can be formed at the anode. These films are hard and they adhere well. The use of this method is to provide protective oxide coating to aluminium (anodised aluminium) which is otherwise reactive. These films are stable enough to be taken off from the substrate. Some other metals which can be oxide coated by this method are: magnesium, titanium, tantalum, niobium, tungsten and zirconium.

16.4.2.5 Electrolytic Deposition Through Chemical Reaction

Electrolysis of a solution of silver nitrate and formaldehyde results in release of silver ions which deposit on substrates yielding silver films with mirror-like finish. Copper can also be deposited by a suitable reaction.

16.4.3 Characterisation of Thin Films

16.4.3.1 Structural Characterisation

The commonest techniques used for structural characterisation are X-ray diffraction and electron diffraction, the principles of which have been discussed in Chap. 4.

X-ray diffraction is useful in obtaining information about the lattice constant and the phase, i.e. whether the film material is the same as desired. Further, from the line width of the reflections, it is possible to estimate the grain size and the lattice strain.

However, the thinness of the film makes the intensity of X-ray reflections rather weak and electron diffraction is preferred. The main objective is to gain information regarding the crystallinity, i.e. whether the film is completely polycrystalline or fibre-like with preferred orientation or single crystalline. Figure 16.23 shows the electron diffraction pattern of a Ni film. It consists of continuous rings indicating that the grains in the film have a completely random orientation as in a powder sample. On the other hand, the pattern in Fig. 16.24 consists of distorted rings (or elongated spots) which is characteristic of axial strains. Finally, the electron diffraction pattern (Fig. 16.25) of another Ni film sample consists of spots which indicates that the film consists of oriented single crystals. Generally, this happens when the substrate is a single crystal.

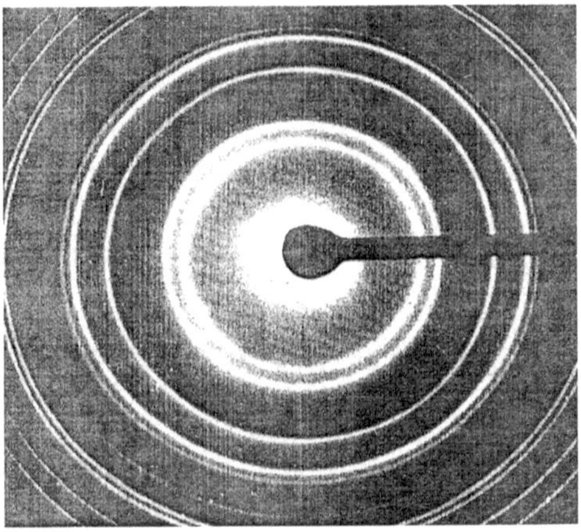

Fig. 16.23 Electron diffraction pattern of a polycrystalline Ni film

Fig. 16.24 Electron diffraction pattern of Ni film showing uniaxial strain

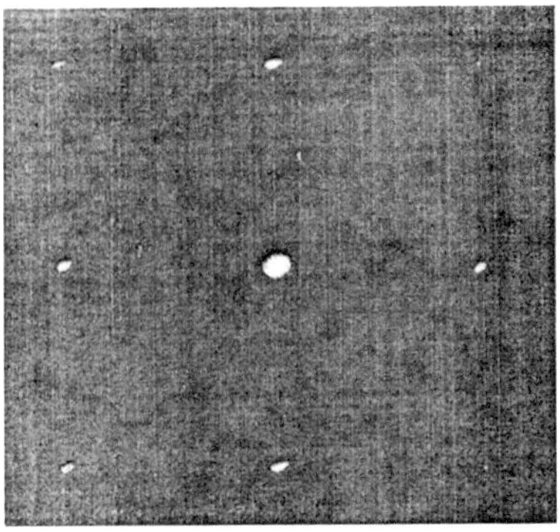

Fig. 16.25 Electron diffraction pattern of single crystal Ni film

16.4.3.2 Thickness Determination

Many properties of thin films show a dependence on the thickness. Hence a knowledge (and also control) of thickness is essential. There are several methods for the determination of thickness; some of them are discussed here.

From Mass Measurement

This method is based on the simple relation between the mass and thickness t;

$$t = w/A\rho, \tag{16.12}$$

where w is the mass of the deposited material, A the area of the film and ρ the density. The substrate is made one pan of a sensitive microbalance where all suspensions are made of quartz fibres [16.34].

Quartz Monitor

The oscillation frequency (f) of a quartz crystal is given by

$$f = c/t, \tag{16.13}$$

where c is called the frequency constant. The quartz crystal is made a part of an oscillator circuit. The film forms on the substrate as well on the quartz crystal. As the effective thickness of the quartz crystal changes because of film deposited on it, the frequency also changes. By measuring the change in frequency, the thickness can be estimated. In fact, the oscillator quartz crystal is installed inside the film deposition unit and functions as a thickness monitor.

From Absorption of Radiation

The law of absorption of radiation in passing through matter is

$$I = I_0 \mathrm{e}^{-\mu t}, \tag{16.14}$$

where I and I_0 are the intensities of the transmitted and incident beams, μ the absorption coefficient and t the thickness. If μ is already known, then t can be estimated from measurement of I and I_0. Depending on the material of the film, visible light, α rays, β rays, γ rays, X-rays or electrons can be employed.

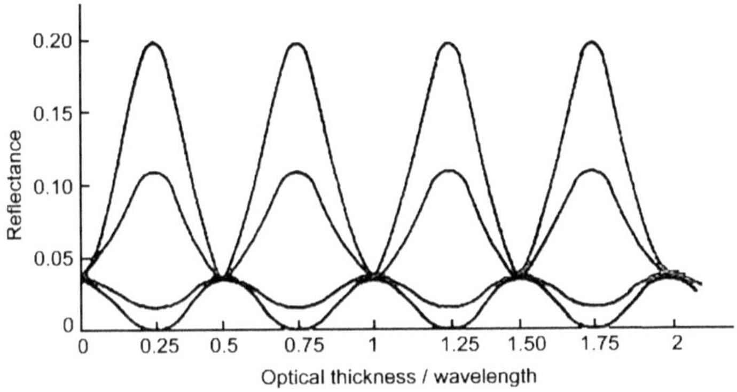

Fig. 16.26 Oscillatory variation of reflectance as a function of optical thickness/wavelength for films with different refractive indices

Optical Methods

In principle, reflectance, transmittance or interference of light incident on thin films can be used to determine the thickness. However, if the film is transparent, reflectance and transmittance cannot be used. But if a semitransparent film is formed on a transparent substrate, reflectance and transmittance can still be used. In this case, reflectance and transmittance results in an oscillatory variation with respect to thickness (Fig. 16.26) due to interference. If monochromatic light of wavelength λ is incident on the film at an angle θ, then

$$\begin{aligned} 2n_{\mathrm{f}}t \cos\theta &= m\lambda \text{ for bright fringes} \\ &= [(m-1)/2]\lambda \text{ for dark fringes,} \end{aligned} \tag{16.15}$$

where n_{f} is the refractive index of the film material. If the experiment is repeated with two wavelengths λ_1 and λ_2 at normal incidence, then for the mth order of λ_1 and $(m+1)$th order of λ_2 to occur at the same point,

$$2n_{\mathrm{f}}t = m\lambda_1 \quad \text{and} \quad 2n_{\mathrm{f}}t = (m+1)\lambda_2. \tag{16.16}$$

Then

$$t = \lambda_1\lambda_2/2n_{\mathrm{f}}(\lambda_1 - \lambda_2). \tag{16.17}$$

16.4.4 Properties of Thin Films

16.4.4.1 Mechanical Properties

The mechanical properties of thin films depend on the thickness, substrate temperature and annealing temperature. The tensile stress in iron films is shown in Fig. 16.27

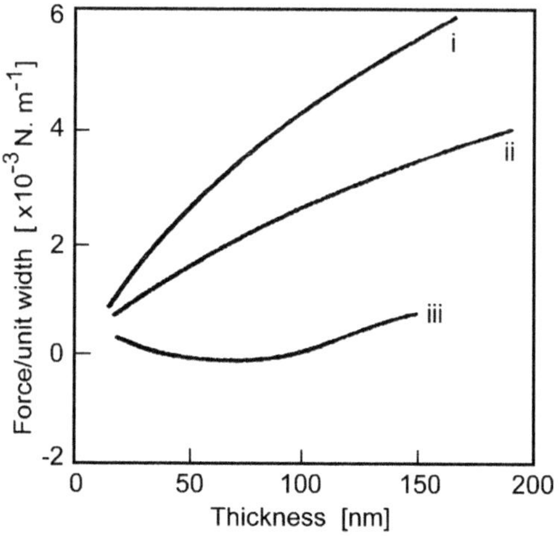

Fig. 16.27 Tensile stress in iron films as a function of thickness for different substrate temperatures: (i) 100°C, (ii) 165°C, (iii) 240°C

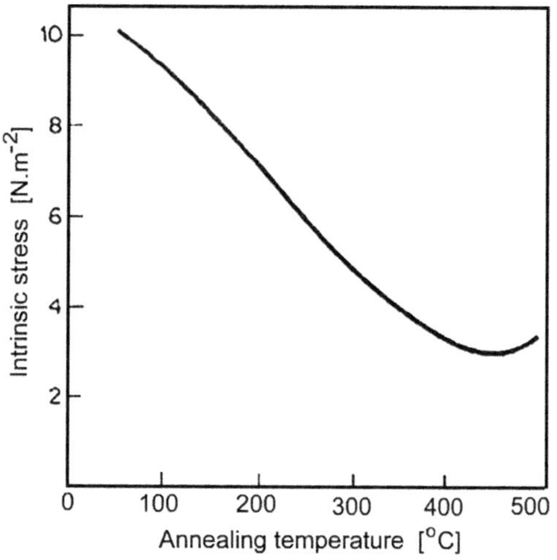

Fig. 16.28 Intrinsic stress of iron thin films as a function of annealing temperature

as a function of thickness and substrate temperature [16.33] and also as a function of annealing temperature (Fig. 16.28).

The strength of a solid depends on its resistance to the motion of dislocations. The restricted thickness of a film obstructs dislocation motion in that direction. Hence the mechanical strength of a film is, in general, greater than that of the same material in bulk.

16.4.4.2 Dielectric Properties

The dielectric constant is a property of a material in bulk. When the material is used in the form of a film, the dielectric constant shows dependence on thickness. The

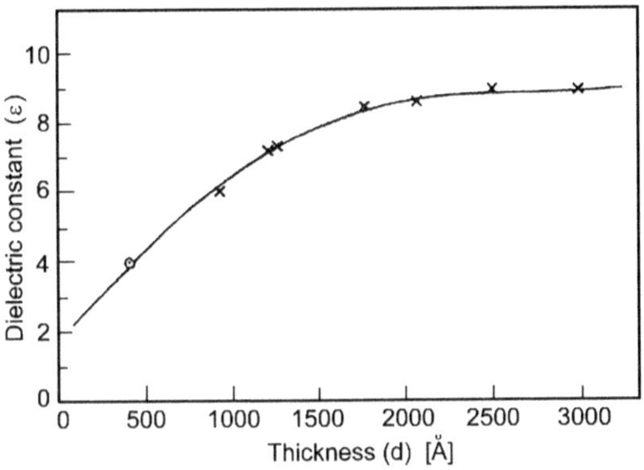

Fig. 16.29 Dielectric constant (ε) of ZnS films as a function of film thickness

variation of the dielectric constant of ZnS as a function of film thickness studied by Goswami and Goswami [16.35] is shown in Fig. 16.29.

The capacitance of a parallel plate condenser is directly proportional to the dielectric constant of the material in it and inversely proportional to the plate separation. A high ε material used as thin film can provide a high-capacitance condenser. Thin films of ZnS, MgF_2, TiO_2, Al_2O_3 and Ta_2O_3 are used for making capacitors.

The performance of thin film capacitors depends on the annealing temperature of the film. Another limiting factor is the breakdown voltage; beyond this voltage, conduction takes place between the plates.

16.4.4.3 Optical Properties

To start with, let us consider the optical behaviour of a single layer film. If the film is made of an absorbing nonmagnetic material, the conventional refractive index n is replaced by

$$n^* = n + ik, \tag{16.18}$$

where k is called the absorption coefficient. n and k together form the optical constants.

Let us consider a ray travelling through a medium of refractive index n_0 (Fig. 16.30), entering into a film of refractive index n_1, absorption coefficient k and emerging out into another medium with refractive index n_2. If the thickness of the film is d, the transmission coefficient T [16.36] is given by

$$T = \frac{16 n_0 n_2 (n_1^2 + k^2)^2}{[(n_1 + n_2)^2 + k^2][(n_0 + n_1)^2 + k^2]} \exp(-4\pi \nu k d), \tag{16.19}$$

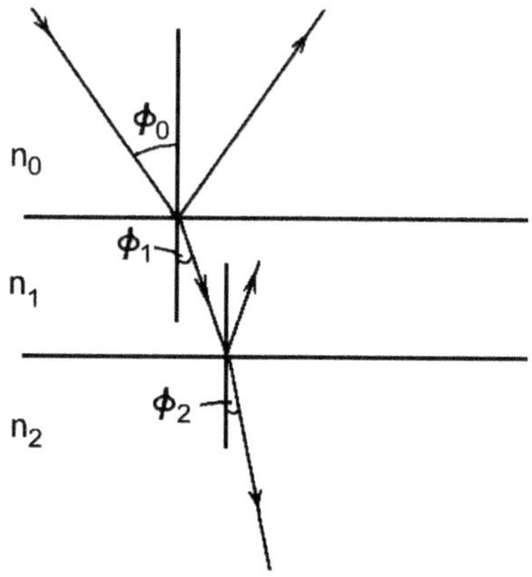

Fig. 16.30 Passage of a ray of light through a film

Table 16.3 Optical constants n_1, k of some thin film materials at 5,000 Å

Material	Thickness [Å]	n_1	k
Cr	900	3.23	1.81
Cu	800	0.82	2.64
CaF_2	5,000	1.25	–
SrF_2	1,500	1.2	–
In_2O_3	2,350	1.60	0.005
La_2O_3	1,420	1.84	0.03

where ν is the frequency of light of wavelength λ. It can be seen that if the transmission coefficient is measured on films with different thicknesses, then a plot of log T versus d will be a straight line with slope k. On the other hand, the reflection coefficient R is given by

$$R = \frac{(n_1 - n_0)^2 + k^2}{(n_1 + n_0)^2 + k^2}. \tag{16.20}$$

With k already determined, n_1 can be determined from R. Values of the optical constants for materials of some films are given in Table 16.3.

Let us now consider the behaviour of a multilayer film, which includes several films of equal thickness but alternate films having refractive indices n_H and n_L. For radiation of wavelength λ_0 which is four times the thickness of each single film, it can be shown [16.33] that the reflectance is high (Fig. 16.31) over a plateaux centred at ν_0 (corresponding to λ_0) and having width $\Delta\nu_0$ given by

$$\frac{\Delta\nu_0}{\nu_0} = \frac{4}{\pi} \arcsin \frac{(n_H - n_L)}{(n_H + n_L)}. \tag{16.21}$$

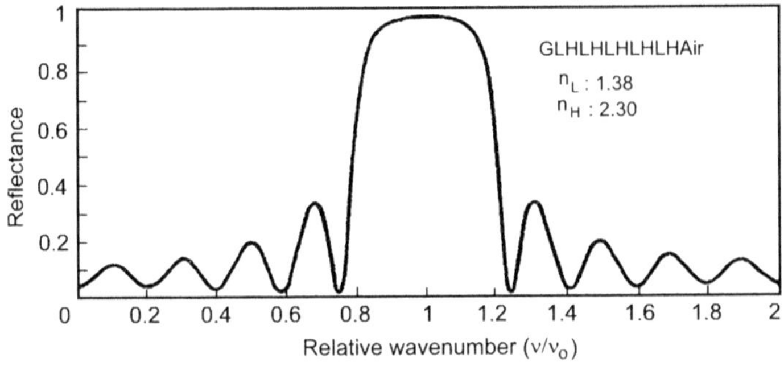

Fig. 16.31 Reflectance of a stack of films

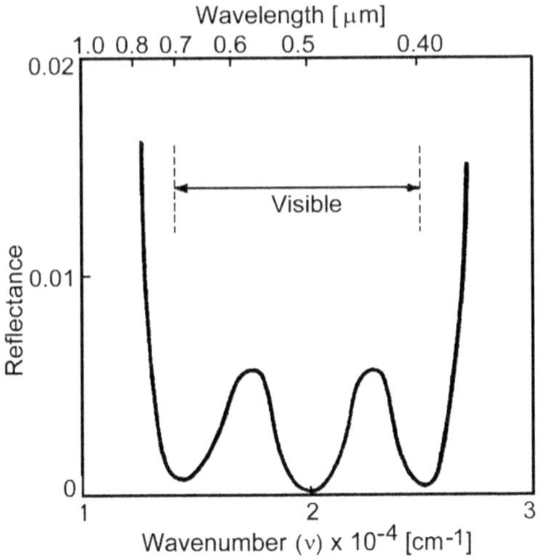

Fig. 16.32 Reflectance of a poorly reflecting system

The maximum value of R at λ_0 is given by

$$R_{2N} = \frac{(n_s f - n_0)}{(n_s f + n_0)}, \tag{16.22}$$

where n_s is the refractive index of the substrate and $f = (n_H/n_L)^{2N}$, $2N$ being the number of layers. For a multilayer system consisting of films of CeO_2 ($n_H = 2.42$) and MgF_2 ($n_L = 1.38$), $R_{2N} = 0.9996$ for $N = 15$. Thus, a multilayer stack provides very high reflectance over a narrow frequency band.

Similarly, a stack of dielectric films can be designed to give a number of zeros and a low average reflectance over a wavelength range. An example of the reflectance of a four-layer stack is shown in Fig. 16.32.

16.4.4.4 Other Properties

Films of materials with electrical, magnetic and semiconducting properties have applications like capacitors, resistors, thermistors, Hall probes, etc. but a discussion of these properties is beyond our scope.

16.5 Liquid Crystals

16.5.1 General

When a crystalline solid is heated, it generally melts at a well-defined temperature (the melting point) into a clear isotropic liquid. In 1888, Reinitzer [16.37] observed an unusual behaviour while studying an organic crystalline solid (cholesteryl benzoate); the substance seemed to have two melting points! At first, at 145°C the substance turned into a cloudy liquid and then at 178.5°C, the cloudy liquid turned into a clear isotropic liquid. Further studies by Lehmann [16.38] revealed that the intermediate "liquid" consisted of crystallites. This intermediate phase between the regular crystal and the clear liquid is called a liquid crystal. In liquid crystal terminology, it is called a mesophase. Several substances related to cholesteryl benzoate also showed similar behaviour. The characteristic properties of liquid crystals are: two melting points, crystallinity combined with ability to flow and optical properties like optical activity, birefringence and selective reflection of circularly polarized light. The molecular constituents of liquid crystals generally have an elongated shape. The transition from a crystalline solid to an isotropic liquid is shown in Fig. 16.33.

After these initial observations on cholesteryl benzoate and a number of other substances, there was a lull in further studies for nearly 80 years until, in 1969, Kelkar and Schuerle [16.39] synthesised N-p-methoxy benzyledene-p'-butylaniline

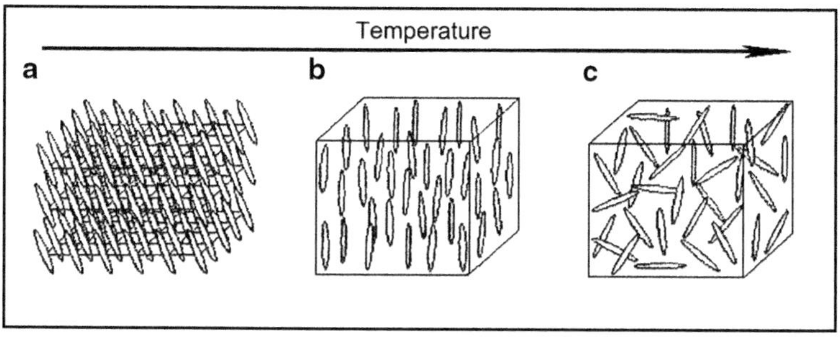

Fig. 16.33 Transition from (**a**) a crystal to (**b**) a liquid crystal and finally to (**c**) a liquid. In (**a**) there is a lattice and the molecules are oriented, in (**b**) there is no lattice but there is approximate orientation and in (**c**) there is no lattice and no orientation

(MBBA). This substance was found to be a liquid crystal at room temperature. This opened up the possibility of commercial applications of liquid crystals. Since then, liquid crystals have become a vast theatre of scientific and industrial activity. There is ample literature on the subject: by Chandrasekhar [16.40], Gennes and Prost [16.41], Collings and Hird [16.42] and the series on *Advances in Liquid Crystals* (Academic Press).

16.5.2 Broad Classification

There are now thousands of substances which show liquid crystal behaviour. These are broadly classified into three groups: thermotropic, lyotropic and metallotropic.

16.5.2.1 Thermotropic Liquid Crystals

These substances show liquid crystal behaviour within a certain temperature range. Para-azoxyanisole is an example of this type. They are made up of small, organic rod-like molecules.

16.5.2.2 Lyotropic Liquid Crystals

These liquid crystals are mixtures of two or more mesophases in different concentrations. A typical lyotropic molecule consists of two parts: a hydrophylic part connected to a hydrophobic part; together they form an amphiphilic system (Fig. 16.34a). In a lyotropic phase, these molecules are arranged in different patterns (Fig. 16.34b, c). The space between them is filled with nonmesophasic fluids like water or oil; this provides fluidity. Soap is an example. The ratio between the hydrophilic mesophase and the hydrophobic mesophase can vary, creating a number of different liquid crystals.

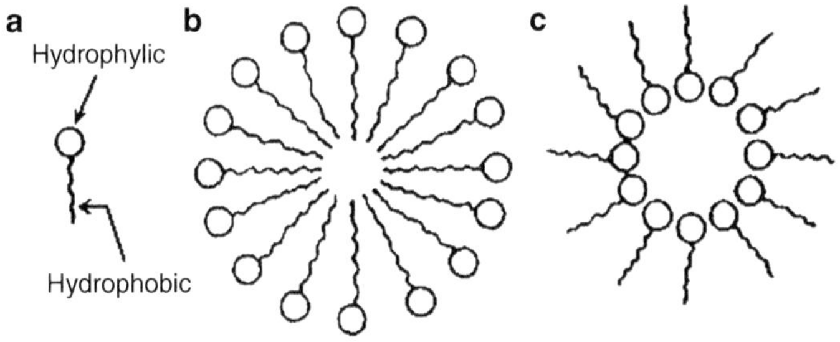

Fig. 16.34 Lyotropic liquid crystal (**a**) an amphiphilic molecule (**b**) and (**c**) lyotropic mesophases

16.5.2.3 Metallotropic Liquid Crystals

These are based on combinations of organic and inorganic substances. The inorganic component has a structure of linked polyhedra as in $ZnCl_2$. The behaviour of a metallotropic liquid crystal depends on the ratio of inorganic–organic components.

16.5.3 Types Among Thermotropic Liquid Crystals

The thermotropic group of liquid crystals is the most populous. There are several types within this group. Depending upon conditions, the phases transform from one type to another. We shall consider the various types of thermotropic liquid crystals.

16.5.3.1 Nematic Phase

It is the simplest of mesophases. It is composed of longish organic molecules. The long dimension of the molecules is approximately oriented along a preferred orientation called the director (Fig. 16.35). When viewed under a microscope, this phase shows thread-like features. These liquid crystals are optically uniaxial, i.e. they have two different refractive indices in two mutually perpendicular directions. So, they have an optic axis. Some nematics are biaxial, i.e. they have three different refractive indices; they have a secondary optic axis. This property renders them useful in optical applications.

As mentioned, the nematic liquid crystals have a preferred direction called the director and the molecules are roughly aligned along the director. The director is indicated by the unit vector **n**. An individual molecule makes an angle θ with the director (Fig. 16.36). The angular function $(3\cos^2\theta - 1)/2$ may be taken as a measure of the misorientation of a molecule with respect to the director. The value of this function will vary from molecule to molecule. The average value of this function for the entire assembly is taken as the order parameter S of the phase; S is defined as

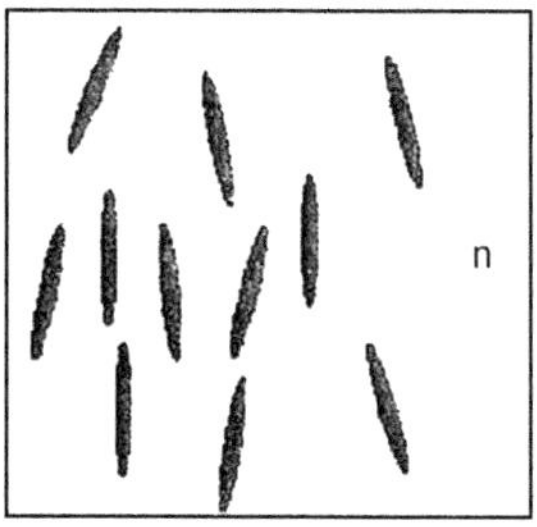

Fig. 16.35 Nematic phase

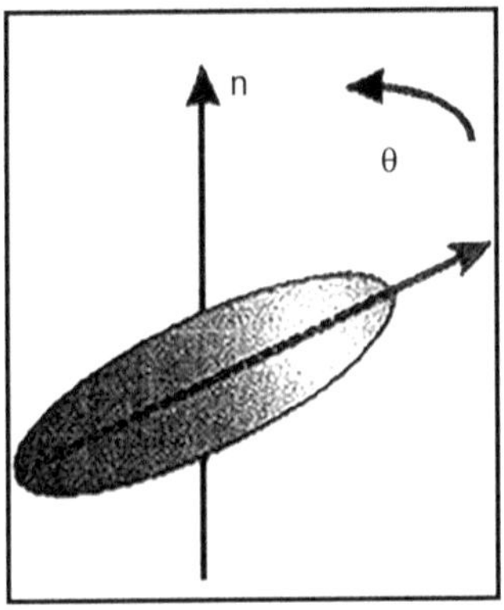

Fig. 16.36 Orientation of a molecule

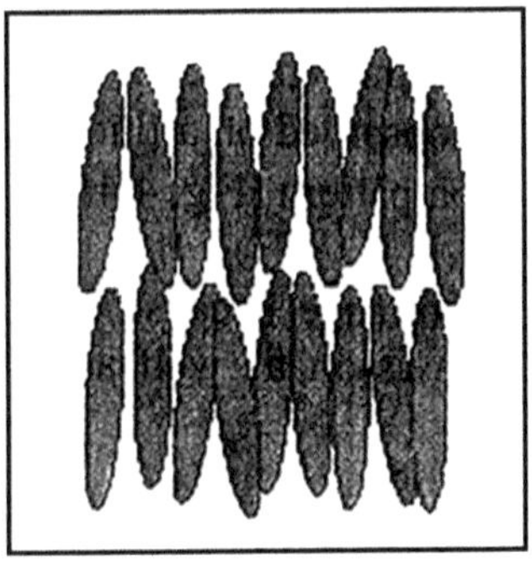

Fig. 16.37 A smectic phase

$$S = \langle (3\cos^2\theta - 1)/2 \rangle. \tag{16.23}$$

It can be seen that $S = 1$ for a regular crystal with perfect order and $S = 0$ for an isotropic liquid with total orientational disorder. For a given nematic phase, the order parameter is in the range $0.3 < S < 0.8$.

16.5.3.2 Smectic Phase

These phases form at temperatures lower than the nematic phases. Their molecules are also longish but they are arranged in layers (Fig. 16.37) which can slide on each other. Due to this, they display a soap-like behaviour. Among the smectics, again, there are two types: those in which the molecules are arranged along the layer normal (smectic A) and those in which the molecules are tilted with respect to the layer normal (smectic C).

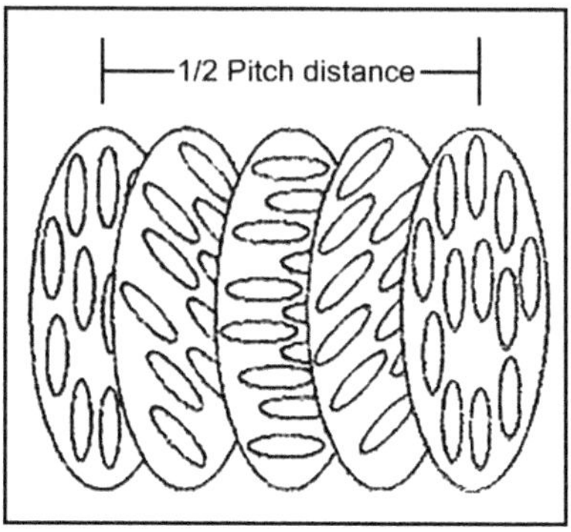

Fig. 16.38 A chiral phase

16.5.3.3 Chiral (Cholesteric) Phases

As the term indicates, these phases have chirality, i.e. a screw-like handedness. They were first observed in cholesterol derivatives; hence, the other name. The chiral phase (Fig. 16.38) has the characteristic structural feature that the molecular arrangement has an angular twist in a specific direction which repeats itself at regular intervals. This interval is called the pitch. This twist endows the phase with optical properties like optical activity. In some chiral liquid crystals, the pitch is of the order of the wavelength of visible light. This creates the interesting possibility of observing Bragg-like reflections of visible light.

16.5.3.4 Blue Phases

While studying some cholesteric substances between the cholesteric phase to the isotropic transition, Gray [16.43] observed that they assume green and blue colours. These are called blue phases. They have a cubic structure of defects with long lattice periods of several hundred nm. As in the chiral phases, in the blue phases also Bragg-type reflections of visible light can be observed. These phases have device applications, e.g. as light modulators but they have the disadvantage that they exist in very narrow temperature ranges of 0.5–2°C. Later, Coles and Pivnenko [16.44] discovered blue phases which are stable over a wider temperature range of 16–60°C.

16.5.3.5 Discotic Phases

Most of the thermotropic phases discussed earlier consist of elongated or rod-like molecules. It was observed by Chandrasekhar et al. [16.45] that some compounds consisting of disc-like molecules also form liquid crystals. These are called discotic or columnar mesophases. A schematic is shown in Fig. 16.39. Some examples are hexasubstituted esters of benzene and others of triphynelene. Typical transition temperatures of these phases are ~80°C.

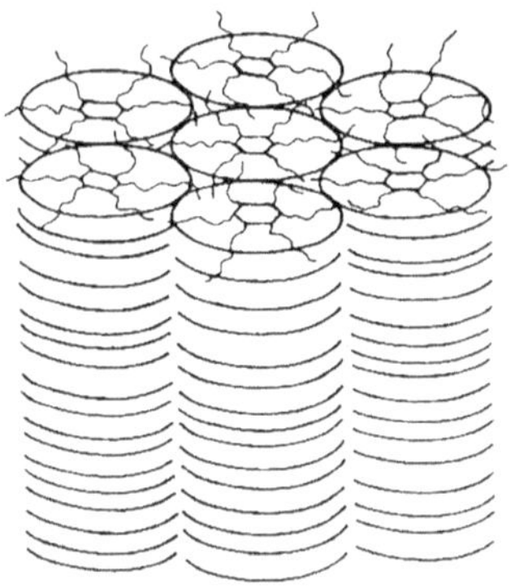

Fig. 16.39 Discotic phase; discs are spaced irregularly to form liquid-like columns

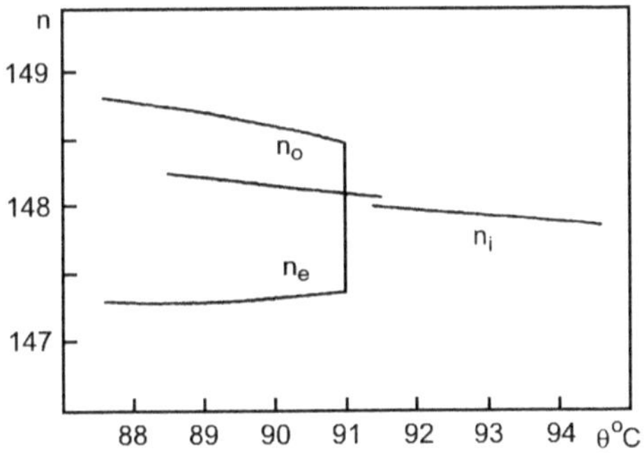

Fig. 16.40 Refractive indices (*right*) the isotropic liquid-like and (*left*) the anisotropic mesophase

16.5.4 Characterisation of Liquid Crystals

We shall discuss some of the techniques with which liquid crystals can be characterised.

16.5.4.1 Optical Characterisation

Liquid crystals have the optical properties of optical activity, birefringence and selective reflection of circularly polarized light. These can be studied by conventional methods with additional provisions of a heating or cooling stage and facilities for temperature control and accurate temperature measurement. The sudden appearance of optical anisotropy (different refractive indices) observed by Stagemeyer and Bergmann [16.46] is shown in Fig. 16.40.

Another important optical phenomenon that can be observed with a polarizing microscope is the appearance of textures in liquid crystals. A few typical textures observed by Destrade et al. [16.47] are shown in Fig. 16.41. These patterns occur because of the presence of defects in the liquid crystals. The texture patterns vary from mesophase to mesophase and, therefore, undergo changes whenever there is a phase transition.

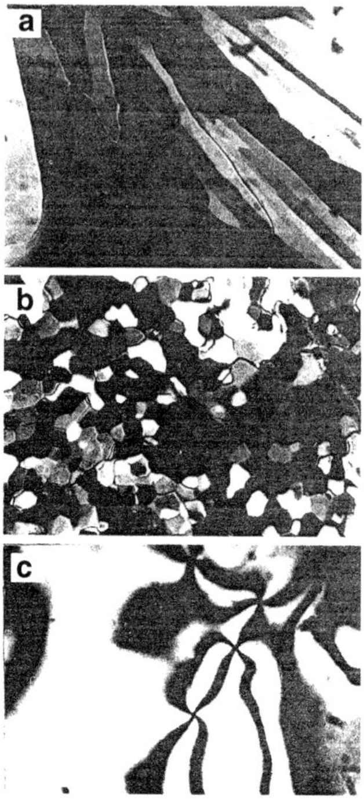

Fig. 16.41 Textures of different phases of hexa-*n*-hexyloxy benzoate of triphenylene (**a**) crystal at 175°C, (**b**) discotic phase at 188°C, (**c**) nematic phase at 202°C

16.5.4.2 Differential Scanning Calorimetry

Whenever a phase transition takes place, heat is required to drive it. This heat energy can be detected by the technique of differential scanning calorimetry. The method does not identify the new phase but records the transition temperature accurately. The appearance of new phases observed by Stagemeyer and Bergmann [16.46] using the DSC technique is shown in Fig. 16.42.

16.5.4.3 X-ray Diffraction

X-ray diffraction is used to give information regarding the structure of the parent crystal as well as of the crystalline components in the mesophase. The X-ray diffraction photographs of the mesophases [16.48] resemble powder photographs (Fig. 16.43) or fibre patterns (Fig. 16.44).

16.5.4.4 Other Techniques

Besides the above techniques, other techniques like magnetism, nuclear magnetic resonance, viscometry, light scattering, Mossbauer spectroscopy and neutron incoherent scattering are also employed in the characterisation of liquid crystals.

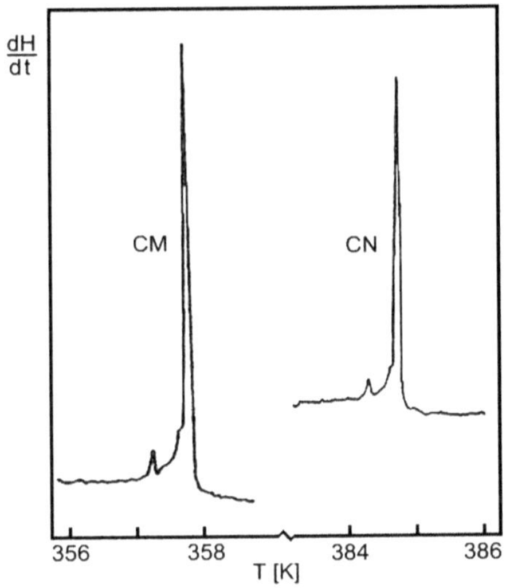

Fig. 16.42 DSC traces of mesophases cholesteryl myristinate (CM) and cholesteryl monoanoate (CN)

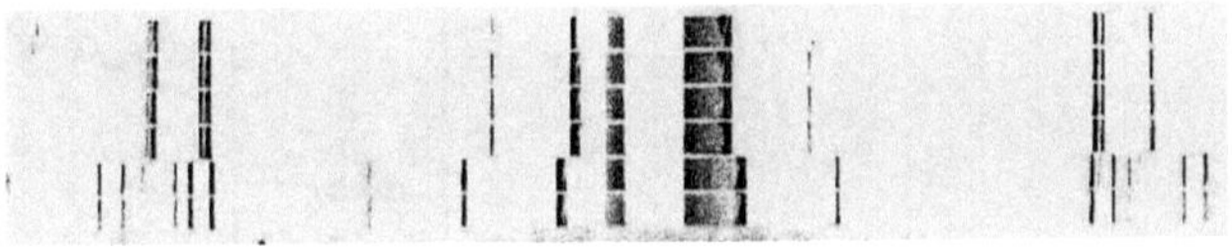

Fig. 16.43 Powder-like X-ray diffraction patterns of mesophases

Fig. 16.44 Fibre-like X-ray diffraction patterns of a mesophase

16.5.5 Some Properties of Liquid Crystals

Liquid crystals have a variety of properties and associated phenomena. We shall discuss a few of them.

16.5.5.1 Dielectric Properties

The dielectric behaviour of a material is described by three parameters, viz. the static dielectric constant ε, the dielectric loss (ε'') and the relaxation time (τ). We shall consider typical behaviour of these parameters in liquid crystals.

The static dielectric constant of a liquid crystal is anisotropic, the values ($\varepsilon_{||}$) along the director and ($\varepsilon_{\perp}$) perpendicular to it being different. Kresse [16.49] cites an example of the typical dielectric anisotropy that sets in at the isotropic–nematic transition (Fig. 16.45). The dielectric loss in liquid crystal 4-*n* decycloxyphenyl 4-*n* pentyloxybenzoate is shown in Fig. 16.46. Curves a–i pertain to different temperatures in the range 356–336 K. As can be seen, the curves are typical Debye curves. In this temperature range, the behaviour changes from nematic (N) to smectic A (SA) and smectic C (SC). The relaxation time calculated from

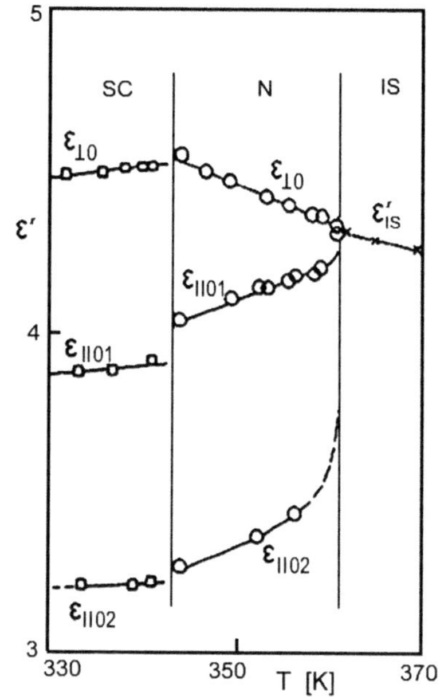

Fig. 16.45 Changes in dielectric constant from smectic (SC) to nematic (N) to isotropic (IS) phases

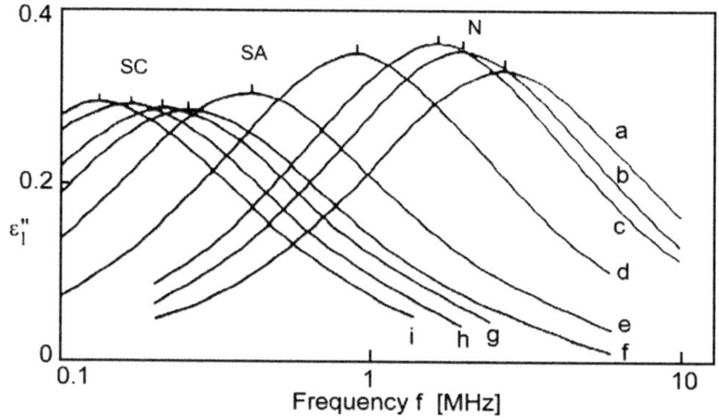

Fig. 16.46 Dielectric loss of smectic (SA, SC) and nematic (N) phases of 4-*n* decycloxyphenyl 4-*n* pentyloxybenzoate

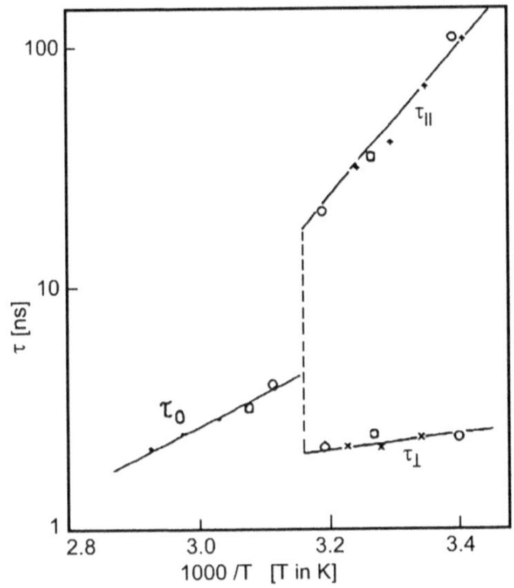

Fig. 16.47 Temperature variation of relaxation time (note the anisotropy at the transition)

these curves varies with temperature [16.50] and, further, becomes anisotropic in a mesophase (Fig. 16.47).

16.5.5.2 High Pressure Properties

In one of the earliest experiments on liquid crystals, Hulett [16.51] observed that the crystal–nematic and nematic–isotropic transition temperatures vary with pressure (Fig. 16.48). Further from studies on several systems, it is found that dT/dP has near-common values of 31.1 and 47.7 (kbar^{-1}) for the two transitions, respectively.

From measurement of change of volume with pressure, the compressibility β can be determined. Kim and Ogino [16.52] observed that β shows a peak at the nematic–isotropic transition (Fig. 16.49).

A very important finding by Chandrasekhar et al. [16.53] is that some substances which do not show liquid crystal behaviour at normal pressures, show liquid crystal behaviour at high pressures. The emergence of mesophases in *p*-methoxy benzoic acid which is normally nonmesophasic is shown in DTA curves (Fig. 16.50).

16.5.5.3 The Order Parameter

The order parameter S, defined in (16.23), can be experimentally determined by techniques like diamagnetism, Raman scattering, nuclear magnetic resonance and electron paramagnetic resonance. The order parameter is temperature dependent; the temperature variation is shown in Fig. 16.51. As temperature increases, the order parameter decreases slowly but jumps down to $S = 0$ at the nematic–isotropic

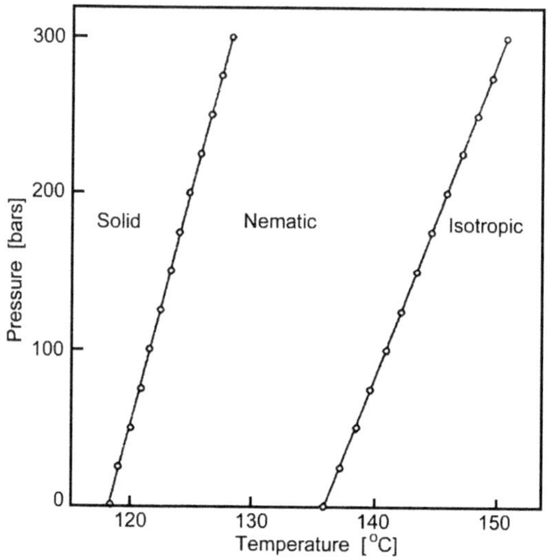

Fig. 16.48 Pressure variation of crystal–nematic and nematic–isotropic transition temperatures

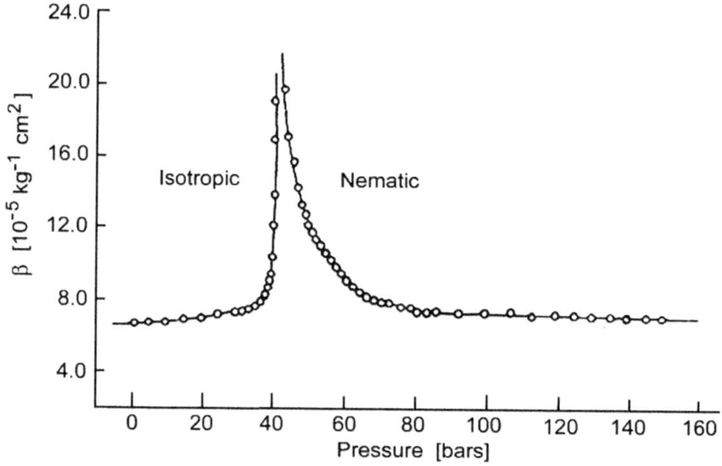

Fig. 16.49 Compressibility (β) as a function of pressure

transition temperature. On the other hand, NMR studies of the S parameter of p-azoxyanisole at the nematic–isotropic and solid–nematic transition temperatures by Deloche et al. [16.54] showed that the order parameter is virtually independent of pressure.

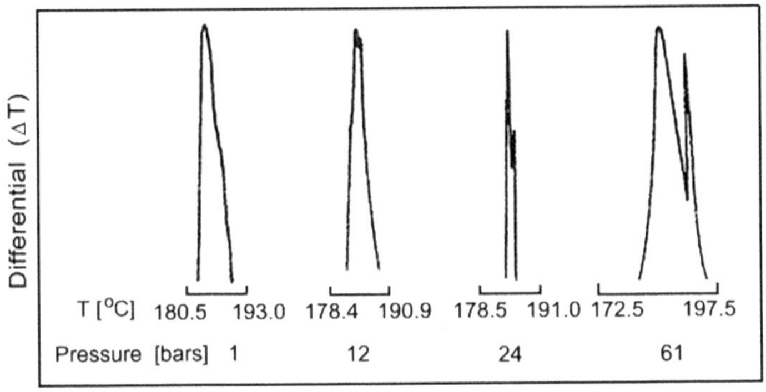

Fig. 16.50 DTA curves of a mesophase formed at high pressure

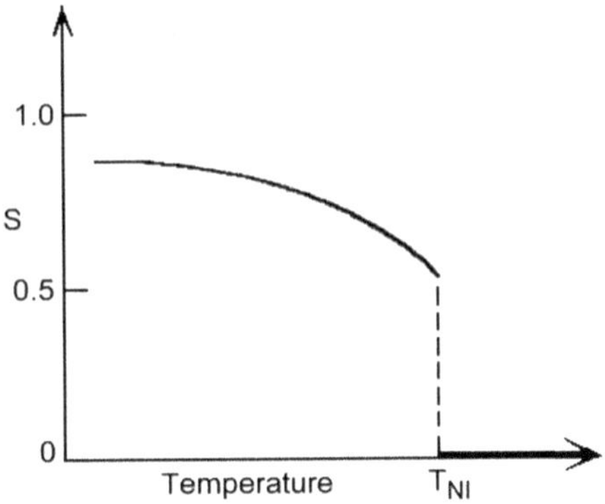

Fig. 16.51 Temperature variation of the order parameter S

16.5.5.4 Optical Properties

Several optical properties of liquid crystals have already been mentioned. Here, we shall discuss only the property which has led to the most important application of liquid crystals, viz. liquid crystal display.

We have mentioned that liquid crystals have optical activity. This is because, in the molecular arrangement in a mesophase there is a twist, i.e. the director is twisted. But, an external electric field has the effect of untwisting the molecular arrangement. In Fig. 16.52, let P_1 and P_2 be two crossed polarizers, i.e. light passing through P_1 cannot pass through P_2; the field of view when seen from P_2 is dark. Let us introduce a proper liquid crystal in the space between P_1 and P_2. Because of the twist in the director, any light passing through P_1 will undergo optical rotation so that it passes through P_2, and the field of vision is clear. But if an electric field is applied, the twist in the director is undone and the light is again stopped resulting in a dark field of vision. This is the principle of liquid crystal display.

The actual arrangement is as shown in Fig. 16.53. A and E are two crossed polarizers. B, C, D together form the liquid crystal display cell; B is the screen printed display grid with the figure 8 in seven segments. Each segment is connected

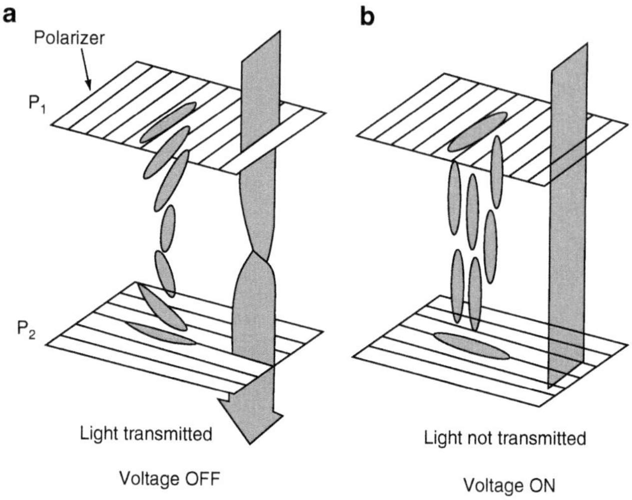

Fig. 16.52 Principle of liquid crystal display (*LCD*)

Fig. 16.53 Details of a LCD cell

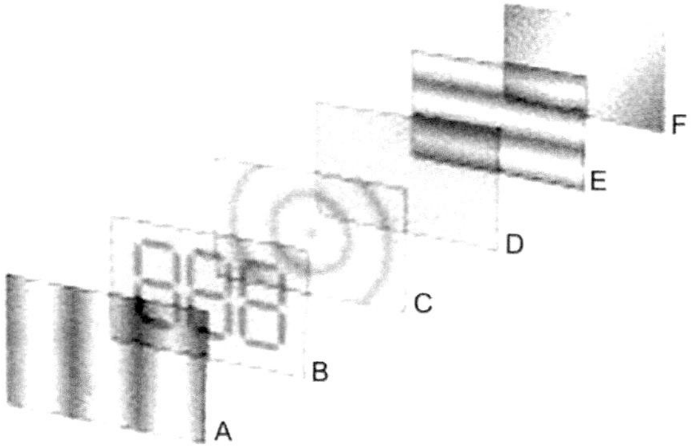

to the command unit with transparent tin indium oxide connections. C is a plate with a thin layer of the liquid crystal. D is a support plate and F is a reflector. Depending on the command, figures 0, 1, 2,. . . ,9 appear on the screen.

16.6 Quasicrystals

16.6.1 Discovery

A crystal can sustain rotational symmetry of order 2, 3, 4 and 6; rotational symmetry of other orders is not allowed. In this context, a startling discovery was made in 1984 when a fivefold axis was observed in a solid system.

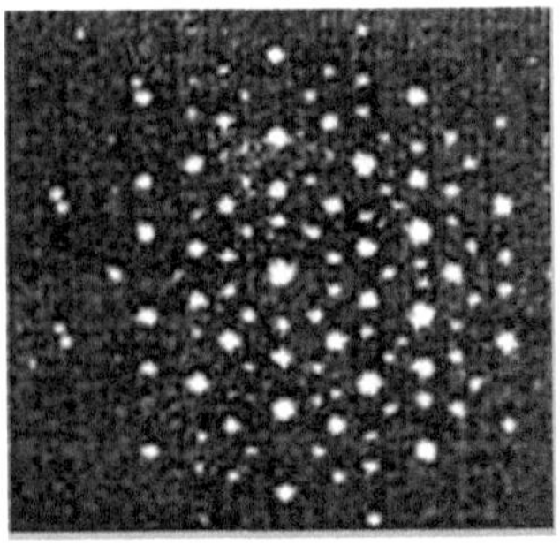

Fig. 16.54 The first electron diffraction pattern (Al–Mn alloy) showing fivefold symmetry

Shechtman et al. [16.55] grew samples of Al–Mn alloys by the melt-spinning method. This is a rapid cooling process (106 K s^{-1}). Generally, such rapid cooling results in glass formation. In the alloy of composition Al–14% Mn obtained by Shechtman et al., there were minute grains (~1 μm across) which, when subjected to selected area electron diffraction, gave a pattern with a fivefold symmetry (Fig. 16.54). By turning the grain on the goniometer, diffraction patterns with twofold and threefold symmetry were obtained. In all, the grain showed existence of 15 twofold axes, 10 threefold axes and 6 fivefold axes. These are the symmetry elements of the icosahedral point group $m\bar{3}\,\bar{5}$; it is not one of the allowed 32 point groups. Further, a powder X-ray diffraction pattern obtained by Shechtman et al. could not be indexed to any Bravais lattice.

When a solid yields a proper diffraction pattern, that is an evidence of the existence of long-range orientational order. Further, when the diffraction pattern conforms to one of the allowed Bravais lattices, that is evidence of the existence of translational order. When both conditions are satisfied, such a solid is called a proper crystal. What Shechtman et al. observed was a system with long range orientational order, but "no translational symmetry". Such a system is called a quasicrystal.

16.6.2 Further Work

The discovery of the Al–Mn icosahedral quasicrystal by Shechtman et al. was followed by further observations. Wang et al. [16.56] observed eightfold symmetry in V–Ni–Si alloys. In an Al–22% Mn alloy, Bendersky [16.57] observed tenfold symmetry in selected area diffraction patterns. This decagonal phase had point group symmetry $10/m$ (10/*mmm*) which is, again, disallowed. Ishimasa et al. [16.58] observed 12-fold dodecagonal symmetry in the Ni–Cr system. Dubost et al. [16.59] succeeded in growing single-crystalline grains of icosahedral Al–Cu–Si alloys. These were large samples of a few hundred μm size. With these samples, Denoyer et al. [16.60] obtained the first X-ray Laue photographs (Fig. 16.55) with fivefold symmetry. Bancel et al. [16.61] recorded a high-resolution powder pattern of the icosahedral phase of Al–Mn (Fig. 16.56). They observed that none of the reflections corresponded to the orthorhombic phase of Al–Mn.

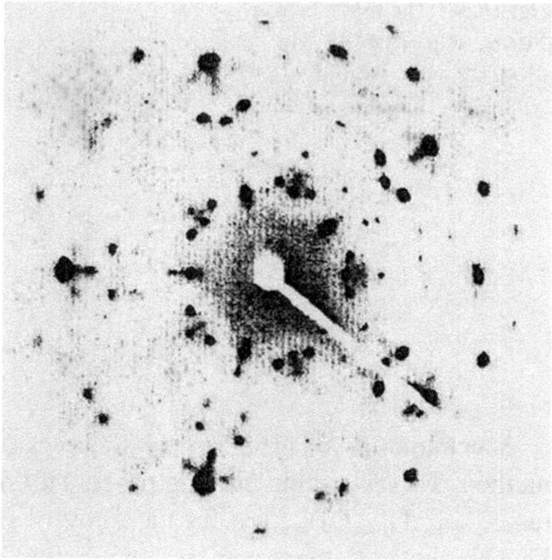

Fig. 16.55 X-ray photograph of Al_6Li_3Cu quasicrystal showing fivefold symmetry

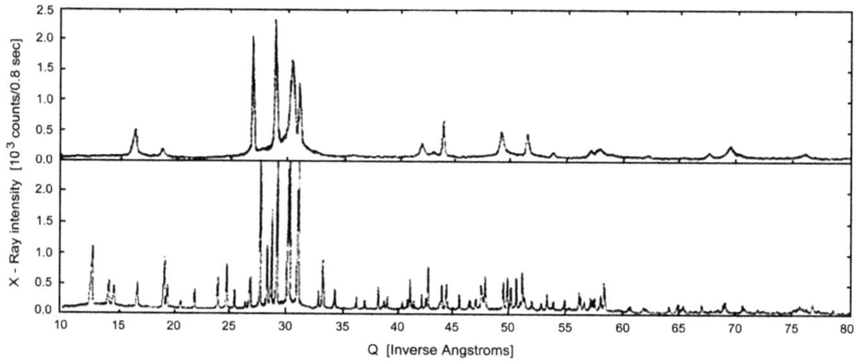

Fig. 16.56 X-ray powder pattern of Al–Mn powder; *top* Al–Mn quasicrystal; *bottom* Al–Mn annealed powder

Subsequently, quasicrystallinity has been observed in hundreds of systems. Some of them are:

Al–Cr	Cd–Yb	Al–Cu–Fe	Zn–Mg–Ho
Al–Fe	Al–Cu–V	Al–Cr–Si	Zn–Mg–Sc
Al–V	Al–V–Fe	V–Ni–Si	Pd–U–Si

16.6.3 Models

Several models have been proposed for quasicrystals. Some of them are highly mathematical and, hence, beyond our scope. We shall consider some of the simpler models.

16.6.3.1 Pauling's Model I

Pauling [16.62] proposed a cubic unit cell with 820 atoms. The cell is made up of 104-atom icosahedral complexes. Eight such polyhedra are arranged in the cubic β-W structure. This results in a giant unit cell with 820 atoms having space group $Pm3m$ and a lattice constant of ~23 Å.

16.6.3.2 Pauling's Model II

In another model, Pauling [16.63] proposed a unit cell with 1,012 atoms. In this model, the β-W structure is decorated with eight icosahedral clusters. There are two types of clusters, one with 104 atoms each and the other with 136 atoms each. These eight complexes share a total of 24 atoms. The structure has a unit cell of ~26 Å. Since the models have the β-W structure, their X-ray diffraction pattern can be indexed with three indices.

16.6.3.3 Quasicrystal Model

The quasicrystal model [16.61] has six basic vectors ε_1. These vectors are cyclic combinations of three numbers (1, τ, 0). Thus, $\varepsilon_1 = \mu(1, \tau, 0), \varepsilon_2 = \mu\ (1, -\tau,$ 0), etc. τ and μ are constants equal to 1.6180 and 0.5257, respectively. In the model, 12 vectors formed from the six basic vectors point towards the vertices of an icosahedron.

The above models are particularly useful in the interpretation of powder diffraction patterns of quasicrystals. Figure 16.57 shows a part of a powder diffraction pattern of quasicrystal Al–V–Fe [16.64]; its indexing according to the three models is given in Table 16.4.

16.6.4 Penrose Tiling

For a crystal lattice to be a proper lattice, it should have the space-filling property. This restricts the number of allowed lattices to 32. Penrose [16.65] suggested that

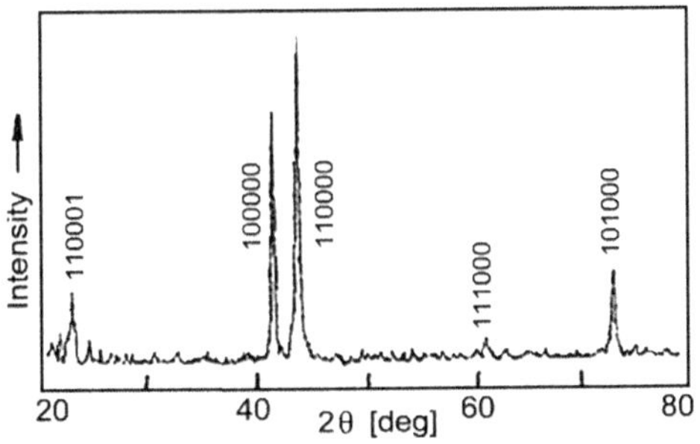

Fig. 16.57 X-ray powder pattern of Al–V–Fe quasicrystal indexed in the six index scheme

Table 16.4 Room temperature Cu Kα X-ray diffraction results for single-phase icosahedral $Al_{80}V_{12.5}Fe_{7.5}$

Measured			Quasicrystal model		Pauling model I		Pauling model II	
2θ [°]	d[Å]	I	Indices	a[Å]	Indices	a[Å]	Indices	a[Å]
22.67	3.923	27	110001	4.677	4 4 4	27.06	6 1 0	23.73
26.30	3.388	4	$1110\bar{1}0$		7 4 0		6 3 2	
35.14	2.554	8	$21\bar{1}001$		10 3 2		9 2 1	
40.99	2.202	76	100000		12 2 2		10 4 1	
43.19	2.095	100	110000		10 8 2		11 2 2	
50.87	1.795	8	210001		14 6 0		13 3 0	
61.10	1.517	6	111000		16 8 0		15 4 2	
62.97	1.467	4	111100		16 8 4		16 1 1	
73.14	1.294	25	101000		20 6 2		18 3 2	

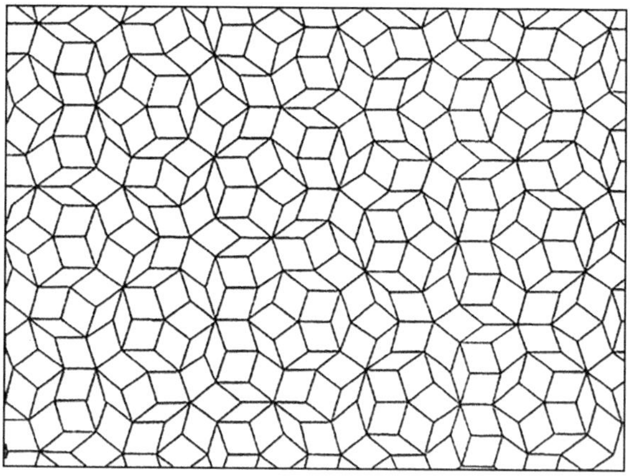

Fig. 16.58 Penrose tiling showing fivefold symmetry

a minor relaxation of this rule facilitates space filling with other unit cells. Thus, he showed that with two tiles (instead of one), 2-D space can be filled. Both these tiles are rhombuses. The pattern formed with them is shown in Fig. 16.58. It has an overall tenfold symmetry and many centres of local fivefold symmetry. Similarly, a pattern with sevenfold symmetry formed with three rhombuses is shown in Fig. 16.59. If the schemes are to be extended to 3-D space, rhombohedra will have to be used instead of rhombuses. The relevance of Penrose tiling to quasicrystals is obvious.

16.6.5 Some Properties

The quasicrystals grown so far have been of small size, often < μm. Dubost et al. [16.59] succeeded in growing grains of a few hundred μm in size by slow solidification. A quasicrystallite of Al–Li–Cu about 300 μm in size is shown in

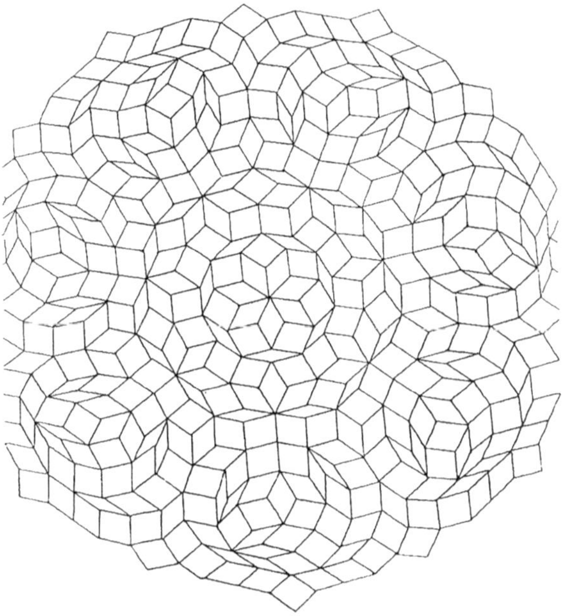

Fig. 16.59 Penrose tiling showing sevenfold symmetry

Fig. 16.60 A quasicrystallite of Al–Li–Cu showing fivefold symmetry

Fig. 16.60. Even this size is too small for meaningful measurement of properties. Besides the small size, quasicrystals are generally unstable; on heating they adopt the regular crystal structure of the parent alloy.

The alloys in quasicrystalline phase are very poor electrical conductors; in fact, they are near-insulators [16.66]. Besides diffraction studies, techniques of nuclear gamma ray resonance, extended X-ray absorption fine structure (EXAFS) and field ion microscopy have been used to study quasi crystals [16.67]. Some aspects of elasticity, dislocations, diffusion and magnetism are known through theoretical studies. Thus, to borrow Kittel's [16.66] words, quasicrystals have been more of "intellectual interest".

References

16.1. S. Mohanty, Resonance **8**, 52 (2003)
16.2. D. Dhar, Resonance **15**, 514 (2010)
16.3. M. Faraday, Trans. Roy. Soc. Lond. **147**, 145 (1857)
16.4. R. Feynman, Rept. Am. Phys. Soc. Meeting, Caltech, 1959
16.5. M. Eigler, E.K. Sweitzer, Nature **344**, 524 (1990)
16.6. K.E. Gonsalves, S.E. Rangarajan, J. Wang, *Nanostructured Material and Nano Technology* (Academic Press, New York, 2002)
16.7. K.C. Kwiatkowski, C.M. Lukehart, *Nanostructured Material and Nano Technology* (Academic Press, New York, 2002)
16.8. H.S. Nalwa, *Nanostructured Material and Nano Technology* (Academic Press, New York, 2002)
16.9. A.P. Alivisatos, J. Phys. Chem. **100**, 13226 (1996)
16.10. S. Bhoraskar, Phys. Educ. **19**, 19 (2002)
16.11. R.W. Murray, IBM J. Res. Dev. **45**, 47 (2001)
16.12. H.W. Kroto, J.R. Heath, S.C. O'Brien, R.F. Curl, R.K. Smalley, Nature **318**, 162 (1985)
16.13. K.S. Novoselov, A.K. Geim, S.V. Morozov, D. Jiang, Y. Zhang, S.V. Dubonos, I.V. Grigorieva, A.A. Firsov, Science **306**, 666 (2004)
16.14. S. Iijima, Nature **354**, 56 (1991)
16.15. A.C. Dillon, K.M. Jones, T. Bekkedahl, C.T.T. Kiang, D.S. Bethone, M.J. Heben, Nature **386**, 377 (1997)
16.16. C.N.R. Rao, A.K. Cheetham, A. Muller, *Chemistry of Nanomaterials* (John Wiley, New York, 2004)
16.17. T.N. Nielson, M.H. Leopald, J. Am. Ceram. Soc. **46**, 381 (1963)
16.18. W. Chen, A. Pachenik, S.J. Dapkunas, G.J. Piermarini, S.G. Malghan, J. Am. Ceram. Soc. **77**, 1005 (1994)
16.19. D.H. Chung, W.M. Buessem, J. Appl. Phys. **39**, 2777 (1968)
16.20. J.K. Mackenzie, Proc. Phys. Soc. **B63**, 2 (1950)
16.21. H. Looyenga, Physica **31**, 401 (1963)
16.22. B.H. Kerner, Proc. Phys. Soc. **B69**, 802 (1936)
16.23. H. Luenberger, Int. J. Pharm. **12**, 41 (1982)
16.24. H.J. Juretschke, R. Landauer, J.A. Swanson, J. Appl. Phys. **27**, 838 (1956)
16.25. A. Eucken, G. Kuhn, Z. Phys. Chem. **134**, 193 (1928)
16.26. W. Voigt, *Lehrbuch der Krist* (Taubuer, Leipzig, 1928)
16.27. A. Reuss, Z. Angew. Math. Mech. **9**, 49 (1929)
16.28. D.H. Chung, J. Appl. Phys. **38**, 5104 (1967)
16.29. G.S. Ranganath, S. Ramaseshan, Pramana, **1**, 78 (1973)
16.30. Z. Harshin, J. Mech. Phys. Solids **32**, 149 (1984)
16.31. A. Goswami, *Thin Film Fundamentals* (New Age International, New Delhi, 2007)
16.32. W.R. Grove, Phil. Trans. Roy. Soc. Lond. **142**, 87 (1852)
16.33. O.S. Heavens, *Thin Film Physics* (Methuen, London, 1970)
16.34. A.D. Houde, *Vacuum Microbalance Technique* (Plenum Press, New York, 1963)
16.35. A. Goswami, A.P. Goswami, Thin Solid Films **16**, 175 (1973)
16.36. O.S. Heavens, *Physics of Thin films* (Academic, New York, 1964)
16.37. F. Reinitzer, Monatsh. Chem. Wein, **9**, 421 (1888)
16.38. O. Lehmann, Z. Phys. Chem. **4**, 462 (1889)
16.39. H. Kelkar, B. Scheurle, Angew. Chem. **8**, 884 (1969)
16.40. S. Chandrasekhar, *Liquid Crystals* (Cambridge University Press, Cambridge, 1980)
16.41. P.G. de Gennes, J. Prost, *The Physics of Liquid Crystals* (Oxford University Press, Oxford, 1993)
16.42. J. Collings, M. Hird, *Introduction to Liquid Crystals* (Taylor and Francis, Bristol, 1997)
16.43. G.W. Gray, J. Chem. Soc. 3733 (1956)

16.44. H.J. Coles, M.N. Pivnenko, Nature **436**, 997 (2005)
16.45. S. Chandrasekhar, B.K. Sadashiva, K.A. Suresh, Pramana **9**, 471 (1977)
16.46. H. Stagemeyer, K. Bergmann,Springer Ser. Chem. Phys. **11**, 161 (1980)
16.47. C. Destrade, M.C. Bernand, H. Gasparoux, A.M. Levelut, N.H. Tinh, *Proc. Int. Liq. Cryst. Conf.*, Bangalore, 1979
16.48. D. Demus, S. Diele, S. Grande, H. Sackman, Adv. Liq. Cryst. **6**, 1 (1983)
16.49. H. Kresse, Adv. Liq. Cryst. **6**, 126 (1983)
16.50. D. Lippens, J.P. Parneix, A. Chapeton, J.Phys. (Fr.) **38**, 1465 (1977)
16.51. G.A. Hulett, Z. Phys. Chem. **28**, 629 (1899)
16.52. Y.B. Kim, K. Ogino, Phys. Lett. **A61**, 40 (1977)
16.53. S. Chandrasekhar, S. Ramaseshan, A.S. Reshamwala, B.K. Sadashiva, R. Shashidhar, *Proc. Int. Conf. Liq. Cryst.*, Bangalore, 1973
16.54. B. Deloche, B. Cabane, D. Jerome, Mol. Cryst. Liq. Cryst. **15**, 197 (1971)
16.55. D. Shechtman, T. Blech, D. Gratias, J.W. Cahn, Phys. Rev. Lett. **53**, 1951 (1984)
16.56. N. Wang, H. Chen, K. Kuo, Phys. Rev. Lett. **59**, 1010 (1987)
16.57. L. Bendersky, Phys. Rev. Lett. **55**, 1461 (1985)
16.58. T. Ishimasa, H.U. Nissen, Y. Fukano, Phys. Rev. Lett. **55**, 511 (1985)
16.59. B. Dubost, J.M. Lang, P. Tanaka, P. Sainfort, M. Audier, Nature **324**, 48 (1986)
16.60. F. Denoyer, G. Heger, M. Lambert, *Quasicrystals, Networks and Molecules of Five-fold Symmetry* (I. Hargittai, VCH Publ., New York, 1990)
16.61. P.A. Bancel, P.A. Hainey, P.W. Stephens, A.I. Goldman, P.M. Horn, Phys. Rev. Lett. **54**, 2422 (1985)
16.62. L. Pauling, Phys. Rev. Lett. **58**, 365 (1987)
16.63. L. Pauling, Proc. Natl Acad. Sci. USA **85**, 4587 (1988)
16.64. R.A. Dunlap, D.W. Lawther, V. Srinivas, *Quasicrystals, Networks and Molecules of Five-Fold Symmetry* (I. Hargittai, VCH Publ., New York, 1990)
16.65. R. Penrose, Bull. Inst. Math. Appl. **10**, 216 (1974)
16.66. C. Kittel, *Introduction to Solid State Physics* (John Wiley, New York, 1996)
16.67. S. Steinhardt, S. Ostland, *The Physics of Quasicrystals* (World Scientific, Singapore, 1987)

Index

D.B. Sirdeshmukh et al., *Atomistic Properties of Solids*, Springer Series in Materials Science 147, DOI 10.1007/978-3-642-19971-4,

GPSR Compliance

The European Union's (EU) General Product Safety Regulation (GPSR) is a set of rules that requires consumer products to be safe and our obligations to ensure this.

If you have any concerns about our products, you can contact us on ProductSafety@springernature.com

In case Publisher is established outside the EU, the EU authorized representative is:

Springer Nature Customer Service Center GmbH
Europaplatz 3
69115 Heidelberg, Germany

Batch number: 10614705

Printed by Printforce, the Netherlands